Handbook
of
Optofluidics

Handbook
of
Optofluidics

Edited by

Aaron R. Hawkins
Holger Schmidt

CRC Press
Taylor & Francis Group
Boca Raton   London   New York

CRC Press is an imprint of the
Taylor & Francis Group, an **informa** business

CRC Press
Taylor & Francis Group
6000 Broken Sound Parkway NW, Suite 300
Boca Raton, FL 33487-2742

First issued in paperback 2017

© 2010 by Taylor and Francis Group, LLC
CRC Press is an imprint of Taylor & Francis Group, an Informa business

No claim to original U.S. Government works

ISBN 13: 978-1-138-11360-2 (pbk)
ISBN 13: 978-1-4200-9354-4 (hbk)

Library of Congress Cataloging-in-Publication Data

Handbook of optofluidics / editors, Aaron R. Hawkins, Holger Schmidt.
    p. cm.
  "A CRC title."
  Includes bibliographical references and index.
  ISBN 978-1-4200-9354-4 (hardcover : alk. paper)
  1. Optofluidics. I. Hawkins, Aaron R. II. Schmidt, Holger. III. Title.

TJ853.4.O68H36 2010
621.36--dc22
                                         2009045579

Visit the Taylor & Francis Web site at
http://www.taylorandfrancis.com

and the CRC Press Web site at
http://www.crcpress.com

To Kellie and my three J's
        – Aaron R. Hawkins

To Jenny and Nina
        – Holger Schmidt

Contents

PART III Bioanalysis

Preface

Panta rhei—everything flows. This aphorism, commonly ascribed to the Greek philosopher Heraclitus (ca. 535–475 BC), originally referred to the ever-changing nature of our lives and the world around us. It can also be applied to the emerging field of optofluidics, which utilizes the flows of photons and fluids and is dynamically evolving from several established research areas.

"Optofluidics" is a rather recent addition to the scientific lexicon. It first appeared in 1985 in a keynote paper by Jones reviewing the use of fiber optics for sensors and systems, including the control of pneumatic valves [1]. The term did not gain traction, however, until a couple of decades later. Starting around 2002, the use of optofluidics in publications rose dramatically, as shown in Figure 1. This sudden surge can be attributed to the combination of integrated optics with miniaturization trends in the burgeoning areas of labs-on-chip and microfluidics. With the rise in scientific papers designated as optofluidic came dedicated journal issues [2] and a series of topical conferences and symposia [3].

Since optofluidics has only recently begun to take hold in the scientific community, its definition is still in flux. One motivation for putting together this handbook was to help define the scope of the field and its relation to other disciplines. At this time, the most comprehensive definition of optofluidics is *The combination of both integrated optical and fluidic components in the same miniaturized system*. This definition is actually quite broad. It encompasses the use of fluids to affect the function of integrated optical devices and the analysis of fluids by means of integrated optical elements. By this definition, the first attempt at building an optofluidic system dates back to the early 1970s when fluid-filled fibers were considered as viable candidates for implementing long-haul fiber communication systems [4]. Clearly, the notion of integrated optics is central to the field and provides a distinction from microfluidics and most labs-on-chip, which typically emphasize only the integration of the fluidic components. It also

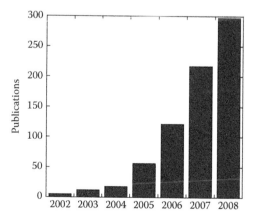

FIGURE 1 Rise in publications in optofluidics since 2002. (Data from Google Scholar, March 20, 2009.)

separates optofluidic approaches from biophotonics, which focuses exclusively on biological analysis and primarily employs conventional microscopy equipment. Optofluidics is a particularly diverse field because its devices and structures are rooted in both classical engineering and physics, but applications can differ fundamentally from established solid-state devices to include (analytical) chemistry, biology, or biomedicine, to name but a few.

With this *Handbook of Optofluidics*, we attempt to provide a snapshot of the state of the field, capture current exciting trends, and identify future challenges and opportunities. Moreover, because optofluidics represents a synergistic combination of integrated optics with other fields, another focus is to detail connections to these research areas and establish the scientific influences that shape the design and function of optofluidic systems.

In order to accomplish these goals, the book contains introductory reviews of the constituent fields followed by more specialized accounts of current research topics in optofluidics. All chapters were contributed by leading experts in their respective fields. The book is intended for researchers interested in optofluidics, but approaching it from different levels of education and backgrounds such as microfluidics, optical engineering, integrated optics, physics, analytical chemistry, and molecular/cellular biology. The level of the individual chapters is suitable for readers ranging from advanced undergraduate students to senior scientists in academia and industry. While not attempting to be a textbook, it is hoped that the handbook will serve as a starting point and reference for developing graduate courses in this rapidly expanding field.

Organization

Part I: Foundations of Optofluidics

Part I introduces the scientific foundations that contribute to optofluidics. These chapters are intended to provide condensed reviews of mature fields while emphasizing the aspects that are of particular relevance to optofluidics. They serve as reference materials and will help in the understanding of later, more specialized and detailed chapters. Chapter 1 by Mekala and Erickson introduces the physical concepts of micro- and nanofluidics with an emphasis on transport properties. Starting with a description of the physical laws governing fluid flow, they explore issues of scaling in micro- and nanosystems and the movement of liquid analytes with a direct bearing on many analytical applications. The chapter closes with an outlook on new trends in which light is actively used to move fluids and particles therein.

In Chapter 2, Hawkins et al. address building miniaturized optofluidic systems. Their overview of current fabrication methods displays the breadth of the field that combines complementary approaches from microfluidics and solid-state microfabrication. A thorough review of process steps starting from a bare wafer and ending with a packaged system is given and illustrated with representative examples from optofluidics.

Chapters 3 through 5 introduce relevant aspects of integrated optics and showcase the key role of this field in optofluidics. In Chapter 3, Janz provides a concise yet comprehensive review of passive integrated optics, covering materials and methods of conventional integrated optics that do not necessarily involve fluids. His chapter also provides a sense of the multitude of possible devices and functionalities (many of which are revisited using optofluidic approaches in later chapters). Chapter 4 by Benabid and Roberts complements Janz' discussion of planar optical waveguides by introducing the reader to the basic principles and characteristics of hollow-core photonic crystal fiber, a waveguide type that is well suited for many optofluidic applications. Chapter 5 on optoelectronics by Bernini and Zeni takes integrated optics one step further to active devices used for generating and collecting light. Due to the dominance of semiconductors in this area, the basic physics of semiconductors and lasers are introduced, followed by a thorough description of the use of these devices in spectroscopic applications and optofluidics. After a brief discussion of Bragg waveguides, the focus shifts to hollow-core fibers with two-dimensional photonic crystal claddings, their guiding mechanisms, and their practical limitations. A discussion of the ramifications of introducing liquids in hollow-core fibers provides a unique perspective on these structures that is essential for their use in optofluidic applications.

In Chapter 6, Zhang completes the coverage of optics background by reviewing spectroscopic methods that have traditionally been associated with microscopy and biophotonics. He provides an in-depth review of the vast field of spectroscopic techniques with a focus on Raman and fluorescence spectroscopy techniques that have found frequent use in optofluidic research.

Part I concludes with Chapter 7 by Chung and colleagues who provide a much-needed systems perspective. In this chapter, optofluidics is placed in context with the large and growing number of approaches to building labs-on-chip. They review standard lab-on-chip components and procedures with specific examples of optofluidic labs-on-chip, including the emerging area of scalable self-assembly.

Part II: Optical Elements and Devices

Part II explores the synthesis of fundamental concepts in optofluidics to create novel devices, specifically those with optical properties that are manipulated by fluids. A main theme that runs through this part is the dynamic reconfigurability made possible by flowing and reshaping fluids. Chapter 8 by Karnutsch and Eggleton begins with an overview of fluid-defined optical elements. This chapter is highlighted by a case study of planar 2D photonic crystal waveguides that can be filled with fluids and dynamically tuned at telecommunications wavelengths. In keeping with the theme of fluidic tuning, Chapter 9 by Mao et al. explores the use of fluids to create optical elements for imaging. These include lenses, mirrors, prisms, and sensing arrays that can be dynamically adjusted to alter focal lengths and reflecting angles.

Zamek et al. provide further examples of reconfigurable devices in Chapter 10, beginning with optical switches that can change the direction of light beams using dynamic fluid-steering elements. This chapter also introduces the idea of creating sensors that are based on the easily changeable nature of fluids. Exposure to external stimuli can alter a fluid's optical properties, which in turn vary an optofluidic device's output. Chapter 10 highlights such a sensor based on plasmonic properties (which are more fully explained in Part III) and Chapter 11 by Suter and Fan explores sensors based on tuning optical resonances. Ring resonators, in particular, are featured, including necessary background on their operation and design. Fluid-tuned ring resonators are demonstrated as sensitive elements in particle sensing and as light sources.

The final chapter in Part II, Chapter 12 by Kristensen and Mortensen, further investigates the use of fluidic elements as light sources. These include lasers that utilize fluids as gain media, especially in on-chip implementations. They show that the sensitive nature of laser resonances can also be exploited to convert optofluidic lasers into sensors that monitor their constituent lasing fluids.

Part III: Bioanalysis

In Part III, we review recent developments in particle detection and manipulation. Due to the high potential for future applications, a natural emphasis on biosensing and biomedical applications is present. The first two chapters provide some background and context for how optics have traditionally been used in particle sensing and manipulation. Chapter 13 by Cipriany and Craighead provides an in-depth review of single-molecule analysis, which is particularly suited for optofluidic implementation due to the small scales involved. Contemporary single-molecule detection techniques are introduced along with background information on commonly used materials and dyes. The treatment emphasizes the dominant fluorescence methods, including a discussion of emerging techniques with sub-diffraction "super"-resolution.

In Chapter 14, Chiou introduces the use of optical forces for particle manipulation. Starting with a review of optical and optoelectronic particle-trapping principles, the application of these ideas to waveguide-based trapping in optofluidic settings is discussed.

The two chapters that follow highlight the salient role of hollow-core optical waveguides in optofluidics. Building on the physical concepts introduced in Chapters 3 and 4, Barth and colleagues discuss microstructured cylindrical fibers with solid and hollow cores in Chapter 15. The chapter focuses on practical issues such as fiber fabrication and filling and losses, and reviews applications in biological and chemical sensing in optofluidics. In Chapter 16, Schmidt explores the role of liquid-core waveguides with a focus on planar integrated waveguides. He reviews guiding mechanisms and waveguide types, and highlights examples for the use of liquid-core waveguides as functional parts in optofluidic particle detection and manipulation.

Both Raman spectroscopy and plasmonics are gaining popularity in biosensing and related applications. Chapters 17 and 18 provide a comprehensive coverage of the optofluidic implementations of these nonfluorescence-based analyses methods. Raman detection is discussed in Chapter 17 by Benford et al. After reviewing conventional setups and experimental characteristics unique to Raman spectroscopy, they provide a detailed discussion of surface-enhanced Raman detection of proteins, peptides, and cardiovascular disease markers, illustrating the potential for medical applications of this technique in integrated settings.

Chapter 18 features an introduction by Sinton et al. to the very active area of plasmonics. They carefully introduce the physical foundations of plasmonics in electromagnetic theory, followed by a detailed look at state-of-the-art plasmonic structures for biosensing. The outlook to possible future plasmonic circuits establishes the connection to Chapter 19 by Chen and colleagues in which the authors discuss optofluidic approaches to flow cytometry and cell sorting. These two techniques are widespread, canonical methods for biological cell analysis. They introduce the principles and current status of this field and discuss challenges and possible approaches to miniaturized cell analysis. A concluding review of specific implementations based on liquid-core waveguides points the way toward future optofluidic cell analysis systems.

Throughout this book, the reader will find numerous thematic connections that are highlighted by the individual authors whenever possible. For example, Chapter 5 introduces optoelectronic elements, and their optofluidic implementations are described in Chapters 8 through 10, and 12. Likewise, optical waveguides are discussed in Chapters 3, 4, 8, 15, 16, and 19. The most relevant optical analysis techniques are also apparent throughout, including fluorescence (Chapters 6, 11, 13, 15, and 16) and Raman spectroscopy (Chapters 6, 15, and 17).

We thank all those who helped with the conception, editing, and assembly of this handbook: Luna Han and the staff at Taylor & Francis, Peter Minogue, Adam Woolley, and all of our colleagues who contributed chapters. We hope that this will serve as a concise reference for the growing optofluidics community and provide a stepping stone in the dynamic flow of optofluidics research.

References

1. Jones, B.E. 1985. Optical fibre sensors and systems for industry, *J. Phys. E* 17 18:770–782.
2. Erickson, D. (Ed.). 2008. Special issue on "Optofluidics," *Microfluid. Nanofluid.* 4:1–2.
3. *IEEE Summer Topical Meeting*, Quebec City, 2006; *IEEE Summer Topical Meeting*, Acapulco, MX, 2008; *Conference for Lasers and Electro-Optics* (*CLEO*), Baltimore, MD, 2009.
4. Payne, D.N. and Gambling W.A. 1972. New low-loss liquid-core fibre waveguide. *El. Lett.* 8:374–376.

Editors

Aaron R. Hawkins earned his bachelor's degree in applied physics from the California Institute of Technology, Pasadena, in 1994 and his PhD in electrical and computer engineering from the University of California, Santa Barbara, in 1998. He was a cofounder of Terabit Technology, Santa Barbara, California, and later worked as an engineer at CIENA (Linthicum, Maryland) and Intel. He is currently a professor at Brigham Young University, Provo, Utah, where he is the director of the Integrated Microfabrication Laboratory. His research interests include optofluidics, avalanche photodiodes, semiconductor devices, hollow optical waveguides, MEMS, and labs-on-a-chip. He is a member of the IEEE, the Optical Society of America, and the American Physical Society and serves as an associate editor for the *IEEE Journal of Quantum Electronics* and on program committees for Photonics West and CLEO. He has authored or coauthored over 200 technical papers. More information about him and his research group can be found at www.ee.byu.edu/faculty/ahawkins.

Holger Schmidt received his MS in physics from the University of Stuttgart, Germany, in 1994 and his MS and PhD in electrical and computer engineering from the University of California, Santa Barbara, in 1995 and 1999, respectively. After serving as a postdoctoral fellow at the Massachusetts Institute of Technology, Cambridge, he joined the University of California, Santa Cruz, in 2001, where he is currently a professor of electrical engineering and the director of the W.M. Keck Center for Nanoscale Optofluidics. He has authored or coauthored over 130 publications and several book chapters in various fields of optics. His research interests are integrated optofluidics for single-particle detection and analysis, integrated atomic spectroscopy and single-photon nonlinearities, and nano-magneto-optics. Dr. Schmidt is a member of the IEEE, the Optical Society of America, the Materials Research Society, and the American Physical Society. He was the recipient of a National Science Foundation CAREER Award in 2002 and a Keck Futures Nanotechnology Award in 2005.

Contributors

Hartmut Bartelt
Division for Optical Fibers and
Fiber Applications
Institute of Photonic
Technology
Jena, Germany

Michael Barth
Institute of Physics
Humboldt-Universität zu Berlin
Berlin, Germany

Fetah Benabid
Department of Physics
Centre for Photonics and
Photonic Materials
University of Bath
Bath, United Kingdom

Melodie Benford
Department of Biomedical
Engineering
Texas A&M University
College Station, Texas

Oliver Benson
Institute of Physics
Humboldt-Universität zu Berlin
Berlin, Germany

Romeo Bernini
Institute of Electromagnetic
Sensing of the Environment
National Research Council
Naples, Italy

Alexandre G. Brolo
Department of Chemistry
University of Victoria
Victoria, British Columbia,
Canada

Chun-Hao Chen
Department of Bioengineering
University of California, San
Diego
La Jolla, California

Eric Pei-Yu Chiou
Department of Mechanical and
Aerospace Engineering
University of California, Los
Angeles
Los Angeles, California

Sung Hwan Cho
Material Science and
Engineering Program
University of California, San
Diego
La Jolla, California

Sung Eun Choi
Department of Electrical
Engineering and Computer
Science
Seoul National University
Seoul, Republic of Korea

Su Eun Chung
Department of Electrical
Engineering and Computer
Science
Seoul National University
Seoul, Republic of Korea

Benjamin Cipriany
School of Applied and
Engineering Physics
Cornell University
Ithaca, New York

Gerard L. Coté
Department of Biomedical
Engineering
Texas A&M University
College Station, Texas

Harold Craighead
School of Applied and
Engineering Physics
Cornell University
Ithaca, New York

Benjamin J. Eggleton
Centre for Ultrahigh-
Bandwidth Devices for
Optical Systems
Institute of Photonics and
Optical Science
School of Physics
University of Sydney
Sydney, New South Wales,
Australia

David Erickson
Sibley School of Mechanical
    and Aerospace Engineering
Cornell University
Ithaca, New York

Yeshaiahu Fainman
Department of Electrical and
    Computer Engineering
University of California, San
    Diego
La Jolla, California

Xudong Fan
Biological Engineering
    Department
University of Missouri
Columbia, Missouri

Jessica Godin
Department of Electrical and
    Computer Engineering
University of California, San
    Diego
La Jolla, California

Reuven Gordon
Department of Electrical and
    Computer Engineering
University of Victoria
Victoria, British Columbia,
    Canada

Aaron R. Hawkins
Department of Electrical and
    Computer Engineering
Brigham Young University
Provo, Utah

Matthew R. Holmes
Department of Electrical and
    Computer Engineering
Brigham Young University
Provo, Utah

Tony Jun Huang
Department of Engineering
    Science and Mechanics
    and Department of
    Bioengineering
The Pennsylvania State
    University
University Park, Pennsylvania

Jisung Jang
Department of Electrical
    Engineering and Computer
    Science
Seoul National University
Seoul, Republic of Korea

Siegfried Janz
Institute for Microstructural
    Sciences
National Research Council
Ottawa, Ontario, Canada

Jun Kameoka
Department of Electrical and
    Computer Engineering
Texas A&M University
College Station, Texas

Christian Karnutsch
Department of Electrical
    Engineering and
    Information Technology
University of Applied Sciences
Karlsruhe, Germany

Mekala Krishnan
Sibley School of Mechanical
    and Aerospace Engineering
Cornell University
Ithaca, New York

Anders Kristensen
Department of Micro and
    Nanotechnology
Technical University of
    Denmark
Lyngby, Denmark

Sunghoon Kwon
Department of Electrical
    Engineering and Computer
    Science
Seoul National University
Seoul, Republic of Korea

Seung Ah Lee
Department of Electrical
    Engineering and Computer
    Science
Seoul National University
Seoul, Republic of Korea

Sung Hoon Lee
Department of Electrical
    Engineering and Computer
    Science
Seoul National University
Seoul, Republic of Korea

Uriel Levy
Department of Applied
    Physics
The Hebrew University of
    Jerusalem
Jerusalem, Israel

Yu-Hwa Lo
Department of Electrical and
    Computer Engineering
University of California, San
    Diego
La Jolla, California

Xiaole Mao
Department of Engineering
    Science and Mechanics
    and Department of
    Bioengineering
The Pennsylvania State
    University
University Park, Pennsylvania

N. Asger Mortensen
Department of Photonics
    Engineering
Technical University of
    Denmark
Lyngby, Denmark

Lin Pang
Department of Electrical and
  Computer Engineering
University of California, San
  Diego
La Jolla, California

Wook Park
Department of Electrical
  Engineering and Computer
  Science
Seoul National University
Seoul, Republic of Korea

Wen Qiao
Department of Electrical and
  Computer Engineering
University of California, San
  Diego
La Jolla, California

P. John Roberts
Department of Photonics
  Engineering
Technical University of
  Denmark
Lyngby, Denmark

Mikhail I. Rudenko
School of Engineering
University of California, Santa
  Cruz
Santa Cruz, California

Holger Schmidt
School of Engineering
University of California, Santa
  Cruz
Santa Cruz, California

Tao Shang
Department of Electrical and
  Computer Engineering
Brigham Young University
Provo, Utah

David Sinton
Department of Mechanical
  Engineering
University of Victoria
Victoria, British Columbia,
  Canada

Boris Slutsky
Department of Electrical and
  Computer Engineering
University of California, San
  Diego
La Jolla, California

Zackary S. Stratton
Department of Engineering
  Science and Mechanics
The Pennsylvania State
  University
University Park, Pennsylvania

Jonathan D. Suter
Biological Engineering
  Department
University of Missouri
Columbia, Missouri

Frank Tsai
Department of Electrical and
  Computer Engineering
University of California, San
  Diego
La Jolla, California

Miao Wang
Department of Electrical and
  Computer Engineering
Texas A&M University
College Station, Texas

Steve Zamek
Department of Electrical and
  Computer Engineering
University of California, San
  Diego
La Jolla, California

Luigi Zeni
Department of Information
  Engineering
Second University of Naples
Aversa, Italy

Jin Z. Zhang
Department of Chemistry and
  Biochemistry
University of California, Santa
  Cruz
Santa Cruz, California

Yue Zhao
Department of Electrical and
  Computer Engineering
Brigham Young University
Provo, Utah

I

Foundations of Optofluids

1

Introduction to Microfluidic and Optofluidic Transport

Mekala Krishnan
David Erickson

1.1 Introduction

Microfluidics is generally defined as the study of fluid flow and transport where the important characteristic dimension is between about 100 nm and 100 μm. Although the study of flows that meet this criteria dates back at least hundreds of years, and technologies that knowingly exploit the properties of such flows (e.g., lubrication) are similarly antiquated, the resurgence in research in this area has occurred largely due to the advancements in microfabrication, which led to the ability to construct two-dimensional microchannel networks. As a result of this resurgence, when we think about microfluidics now, we think of the well-known technology that evolved from this capability, namely, lab-on-a-chip.

In the 15 years that it has been developing, the lab-on-a-chip technology has been applied to a diverse range of applications from chemical and biological analytics to small-scale energy production. While analytical improvements associated with the scaling down of the size were originally thought to be the biggest advantage of these devices, further developments revealed other significant advantages including minimized consumption of reagents, increased automation, and reduced manufacturing costs (Kock et al. 2000). The latter two of these have benefited largely from the development of on-chip flow control elements, such as valves, pumps, and mixers (Unger et al. 2000; Stroock et al. 2002; Biddiss et al. 2004; Laser and Santiago 2004), and much simpler fabrication techniques and construction materials (McDonald et al. 2000; Becker and Locascio 2002; Ng et al. 2002; Gast and Fiehn 2003). Labs-on-a-chip are covered in more detail in Chapter 7 by Chung et al., and in numerous reviews available on modern microfluidic technology (Erickson and Li 2004; Dittrich et al. 2006; Whitesides 2006). Fundamentally governing the operation of these devices is, of course, microscale flow and transport. Put simply, flow at these scales greatly differs from that at large scales in that it is characterized by a strongly laminar behavior, a greater relevance of surface tension, an ability to exploit electrokinetic effects, and small volumes that can be manipulated. Laminar behavior makes these flows easier to control but introduces challenges such as difficulties in mixing different streams together to conduct a reaction. The greater importance of surface tension enables one to create discrete droplets, which can be independently manipulated on-chip and used as ultrasmall reaction vessels (Pollack et al. 2002; Cho et al. 2003). Electrokinetics, where the coupling of an externally applied electric field with surface charges at fluid–solid interfaces can be used to very precisely drive fluid motion (Li 2004), is too weak an effect to impact large-scale flows, but at the microscale provides a much more facile way of manipulating small volumes of fluids without the need for on-chip valves or pumps. The small volumes associated with microfluidics imply that smaller amounts of reagents are required to conduct a chemical reaction and that these reactions can occur much more quickly. Equally important for this book, these small volumes mean that optical forces, which are usually too small to have any significant effect on large-scale flow, can now play a significant role in driving and affecting the motion of fluids and particles contained therein.

In this chapter, we will focus on introducing some of the fundamentals and practical aspects of microfluidic flow and transport in a context relevant for optofluidics with the goal of enabling the reader to approach a proposed experiment in an informed way. We begin in Section 1.2 by introducing the fundamentals of microfluidic transport. We discuss commonly used transport methods at these scales, including pressure-driven flow and electrokinetics. The fluid dynamical equations governing microfluidic flow are also presented here. In Section 1.3, we introduce the reader to general optofluidic transport, or the use of optical forces to perform transport in microfluidic devices. In this section, we begin with a brief literature survey followed by a description of the fundamental equations that govern the coupling between optical and hydrodynamic forces. Methods for both manipulating particles within flows and microfluidic flows themselves are discussed.

1.2 Fundamentals of Microfluidic Transport

As discussed above, the main motivation for current microfluidic research came from the fields of biological and chemical analyses, microelectronics, and defense (Whitesides 2006). Broadly speaking, two kinds of systems are used to perform micro- and nanofluidic transport, namely, continuous micro-channel–based flows and discrete droplet–based manipulation (Pollack et al. 2002; Cho et al. 2003). Devices based on continuous microchannel flow, driven by either pressure or electrokinetics, are those most commonly encountered, and, therefore, this will be the focus of this chapter. The alternative paradigm of digital- or droplet-based microfluidics is usually carried out on open substrates and involves the manipulation of many discrete fluid droplets. The use of discrete droplets creates a "digitization" of matter that lends itself well to the automation of a number of bioanalytical applications, which involve pipetting and mixing large numbers of different samples. The most common method of actuating these droplet flows is through electrowetting (Pollack et al. 2002; Cho et al. 2003), though other approaches using thermal actuation have also been demonstrated (Darhuber et al. 2003). Although we do not cover this here explicitly, its effect does have an importance to optofluidics due to the success of adaptable electrowetting lenses (Kuiper and Hendriks 2004). For more details on different microscale transport mechanisms, see a recent comprehensive review by Stone et al. (2004). Lithographic techniques along with the use of soft polymers are the most common methods used to fabricate micro- and nanochannels. These channels usually have rectangular, rather than circular, cross sections because of the orthogonal nature of lithographic processing. Microfluidic channels range in size from a few microns across to a few hundred microns. Nanochannels are often very planar with large widths compared to their heights, though this is not always the case. Usually, a chip with nanochannels is fabricated with microscale paths to feed the flow to the smaller channels. For more details about the fabrication of micro- and nanoscale channels, readers are referred to McDonald et al. (2000), Quake and Scherer (2000), and Unger et al. (2000).

1.2.1 Brief Overview of Pressure and Electrokinetic Microscale Transport Techniques

As mentioned above, the majority of microfluidic devices exploit either pressure or electrokinetics as the primary transport mechanism. Pressure-driven flow is the most commonly used because it is the most robust technique, having very little dependence on fluid and surface properties, and requires very little external infrastructure. Additionally, very efficient and precise flow valving and pumping methods have been developed for pressure-driven flow at the microscale using techniques such as multilayer soft lithography (Unger et al. 2000). The main disadvantage of this method is that it does not scale well to channel sizes much smaller than 1 μm, since the flow velocity is proportional to the square of the channel size. The parabolic velocity profile characteristic of pressure-driven flow can result in dispersion of a transported chemical sample or tumbling of larger objects like cells (because the flow in the middle is faster than that near the wall). As we show below, electrokinetic flow exhibits a flat velocity profile (often referred to as plug flow) leading to minimal dispersion and flow vorticity. Because of this, it also tends to downscale much better than pressure, having an average velocity that is largely independent of channel height. While the actual manipulation of fluids on-chip using electrokinetic transport is conceptually very easy (requiring only the manipulation of external voltages), this technique is much less robust than pressure-driven transport, as appreciable flow velocities can only be obtained for low-ionic-concentration aqueous solutions and certain surface conditions.

1.2.2 Equations of Flow

Micro- and nanoscale flows are fundamentally characterized by having a Reynolds number, Re, which is much less than 1 ($Re = \rho L v_o / \eta$, where L is the characteristic length scale of the system, v_o is the characteristic velocity, and ρ and η are the fluid density and viscosity, respectively). For example, a channel containing water having a characteristic length of 10 μm and a flow velocity of 100 μm/s has a Reynolds number of the order of 0.001. Liquid flow at the length scales encountered in both microfluidic and nanofluidic devices can be well described by continuum mechanics (Israelachvili 1986), and the Navier–Stokes equations, used in traditional fluid mechanics, remain the governing equations for fluid transport. The equations of continuity and the Navier–Stokes equations (for momentum balance), respectively, for an incompressible Newtonian fluid with constant fluid properties are as follows:

$$\nabla \cdot \mathbf{v} = 0 \tag{1.1a}$$

$$\rho \left(\frac{\partial \mathbf{v}}{\partial t} + \mathbf{v} \cdot \nabla \mathbf{v} \right) = -\nabla p + \eta \nabla^2 \mathbf{v} \tag{1.1b}$$

where
  $\mathbf{v}$ is the fluid velocity
  p is the fluid pressure

Within the limit of a low Reynolds number, the Navier–Stokes equations reduce to the Stokes equations, with the time-dependent and convective transport terms on the left-hand side of the momentum, Equation 1.1b dropping out to yield

$$\nabla \cdot \mathbf{v} = 0 \tag{1.2a}$$

$$0 = -\nabla p + \eta \nabla^2 \mathbf{v} \tag{1.2b}$$

We note here that by dropping out the time-dependent term from the left-hand side of the equation results in microfluidic flows being characterized as time independent. Physically, this means that the transient period between when flow conditions are changed is very short, and, in many cases, it can be assumed that the flow response to changes in external conditions (e.g., input pressures) is instantaneous. For an electroosmotic flow, suitable modifications can be made to Equations 1.2, as discussed in Section 1.2.4.2.

1.2.3 Characteristics of Pressure-Driven Flow

Pressure-driven flow is the most robust microfluidic transport mechanism, and is therefore the most commonly implemented. Usually, off-chip pneumatics or syringe pumps are used to actuate the fluid in a microchannel directly and on-chip valves are used to manipulate flow locally. At present, the most common on-chip valve designs are those based on the use of multilayer soft lithography (Unger et al. 2000; Thorsen et al. 2002). These valves consist of a second layer of microchannels, which sit on top of the main microfluidic layer (all of which are fabricated in a soft polymer, such as poly(dimethylsiloxane), or PDMS). When pneumatically actuated, these valves inflate and press down on the main microchannel, collapsing it and stopping the flow.

Figure 1.1 shows a setup for pressure-driven flow with the chip placed on a microscope stage for observing within the chip. Two different mechanisms to drive this flow are shown in the figure, a syringe pump and a pressure manifold. The syringe pump works by pushing fluid at a given flow rate out of a syringe fitted on the pump. While, in principle, a device using such a syringe pump that produces a constant flow rate can be advantageous, it has the significant disadvantage that if a clogging problem occurs on-chip, the pressure will build up until eventual failure of the chip. Generally, therefore, it is preferential to use a pneumatic or constant-pressure technique, such as that from the pressure manifold shown in the figure. In this case, a constant-pressure air source is hooked up to the valve bank, and a tube runs from the outlet of the valve to a fluid reservoir (which is the scintillation vial in this figure). Air flows through the manifold into the sealed scintillation vial, pushing the fluid from the vial into the chip, as shown.

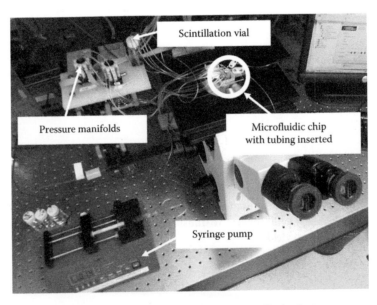

FIGURE 1.1 An experimental setup for pressure-driven flow in a microfluidic device.

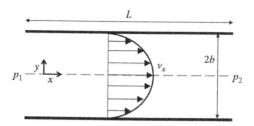

FIGURE 1.2   Schematic showing two-dimensional pressure-driven flow through a microchannel.

The flow velocity in a pressure-driven system can be calculated from Equations 1.2, using a set of generally applicable assumptions. Consider a channel such as that shown in Figure 1.2, with $x \gg y$, where pressure is applied along the x direction and the channel is sufficiently wide such that the sidewalls of the channel do not affect the flow profile. If we assign the channel a height $2b$ and the fluid does not slip at the walls, the solution to Equations 1.2 is that for the Poiseuille flow, as given by Equation 1.3:

$$v_x(y) = -\frac{1}{2\eta}\left(\frac{dp}{dx}\right)\left(b^2 - y^2\right) \tag{1.3}$$

where v_x is the fluid velocity along the x direction at position y. Here the term dp/dx represents an applied pressure gradient along the direction of the flow. For a simple channel geometry, such as that shown in Figure 1.2, $dp/dx = (p_2 - p_1)/L$. The flow velocity in pressure-driven flow exhibits a parabolic profile across the channel height such that the flow is faster in the center of the channel than near the edges, where it is zero (as enforced by the no-slip condition).

From Equation 1.3, the average velocity, v_{avg}, in the channel can be calculated as

$$v_{\text{avg}} = \frac{1}{2b}\int_{-b}^{b} -\frac{1}{2\eta}\left(\frac{dp}{dx}\right)\left(b^2 - y^2\right) dy = -\frac{1}{2\eta}\left(\frac{dp}{dx}\right)\left(\frac{2b^2}{3}\right) \tag{1.4a}$$

and the volumetric flow rate, Q, is given by

$$Q = \int_{-b}^{b} -\frac{w}{2\eta}\left(\frac{dp}{dx}\right)\left(b^2 - y^2\right) dy = -\frac{w}{2\eta}\left(\frac{dp}{dx}\right)\left(\frac{4b^3}{3}\right) \tag{1.4b}$$

where w is the channel width along the z direction. We note that both these equations are preceded by a negative sign since dp/dx is negative in the direction of flow. From these equations, it becomes clear that the average velocity scales with the square of the characteristic height, and the volume flow rate with its cube. Although the analysis presented above is for a simple, two-dimensional channel, we can deduce that as the channel cross section (height and width) becomes smaller, very large pressures need to be applied to the channel to achieve any significant flow. Thus, pressure-driven flow does not scale well with the decreasing channel dimensions, and this is one of the disadvantages of using this technique for nanofluidics. When the height of the channel is of the same order as its length, or the sidewalls of the channel are close enough to affect the velocity profile in the channel, the above equations do not apply, and one must search for more complete solutions to the Navier–Stokes equations (Happel and Brenner 1983).

1.2.4 Electrokinetics

Electrokinetics refers to a broad range of electrically induced transport phenomena, including electroosmosis, electrophoresis, dielectrophoresis, and streaming potential/current phenomena. It has emerged as an important technology in many different fields, including integrated micro- and nanofluidic systems (or labs-on-a-chip) (Erickson and Li 2004; Li 2004), small-scale energy production (Yang et al. 2003), and environmental decontamination (Acar et al. 1995). The main application of electrokinetic flow in micro- and nanofluidics is as a transport mechanism for lab-on-a-chip-type devices, and these flows have been used to transport, separate, and concentrate a range of species in solutions, from multicelled organisms to biomolecules (Hunter 1981; Lyklema 1991, 1995; Li 2004). In addition, there are many emerging application areas for electrokinetics, such as on-chip pressure sources based on electroosmosis (Zeng et al. 2001; Reichmuth et al. 2003; Wang et al. 2006), using the high shear rates in electroosmotic flow in combination with electrophoretic and joule heating effects, to discriminate single-nucleotide polymorphisms in microfluidic architectures (Erickson et al. 2005) and electrokinetic energy conversion (Yang et al. 2003; Daiguji et al. 2004). In Figure 1.3, we show a standard electrokinetic flow setup. The microfluidic chip is located in the middle of the setup, with electrical leads inserted into it to apply the voltage that drives the flow. This particular chip has an H-shaped channel into which four leads may be placed. V_1 through V_4 in the image indicates the different voltages that can be applied. Manipulation of these external voltages allows one to direct the flow on-chip, as will be described below. For further details on the practical implementation of electrokinetic transport on a microfluidic chip, it is suggested that readers consult Erickson and Krishnan (2009).

The two most significant disadvantages of electrokinetic flow compared to pressure-driven flow are the relatively high potential required to achieve significant flow velocities and the relatively stringent surface/solution conditions needed for this flow. For instance, an applied potential of 1000 V over a 10 cm channel is often required to reach the relatively meager transport speed of 100 μm/s. The use of such high potentials causes significant ohmic or joule heating (Erickson et al. 2003), which could lead to variations in the local viscosity and uneven thermal conditions for chemical reactions. Additionally, electrokinetic transport can generally not be used for organic solvents, limiting the applications of this transport mechanism.

In this section, we discuss the three elements of electrokinetics that are of primary importance to channel-based microfluidics, namely, electroosmosis, electrophoresis, and dielectrophoresis. Electroosmosis refers to the actual transport of the liquid in the channel, whereas electrophoresis and dielectrophoresis (in this context) act on any particles that may be present in the fluid. In this text, we use the word "particle" as a broad term to refer to any material being transported in a

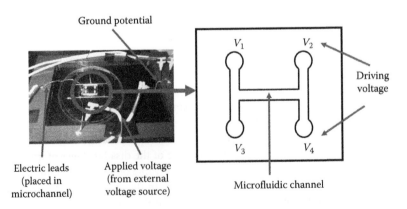

FIGURE 1.3   An experimental setup for electrokinetic flow.

solution that can be uniquely identified (e.g., organic/inorganic particles, cells, and biomolecules). The net particle velocity, v_p, is usually affected by all the above electrokinetic phenomena, and hence is given by

$$v_p = v_{flow} + v_{ep} + v_{dep} \qquad (1.5)$$

where
 v_{flow} is the velocity of the flow at the location of the particle
 v_{ep} is the electrophoretic velocity of the particle
 v_{dep} is the dielectrophoretic velocity of the particle

v_{flow} represents the speed at which the particle is convected with the bulk fluid motion, and can occur due to an applied external pressure (which is calculated as discussed in the previous section) and/or the electroosmotic flow velocity (which is described below). Figure 1.4 shows a schematic of a particle being moved by electrokinetic flow (by applying an electric field, E, across an anode and a cathode) in a microchannel.

1.2.4.1 Electrical Double Layer

The basis for electroosmotic motion is the electrical double layer (EDL), which is a very thin region near a two-phase interface having a net nonzero charge density. Figure 1.5 is a schematic showing an EDL at the wall of a microchannel. For our purposes here, the two-phase interface is usually a solid–liquid interface, either between the fluid and the solid microchannel walls or between the fluid and the solid particles dispersed in solution. The EDL usually forms due to the absorption or desorption of charged species from the channel surface and the resulting rearrangement of the free ions in the bulk solution, so as to maintain overall electroneutrality (Lyklema 1991, 1995). The EDL decays into the bulk solution from a region with a nonzero net charge density (at the channel walls) to a region of zero net charge density (in the bulk solution). The characteristic depth of the EDL is given by the inverse of the Debye–Hückel parameter ($1/\kappa$). The depth of the EDL depends on the ionic strength of the bulk phase solution, and can vary from a few nanometers to over a micrometer depending on the strength of the solution. This small size of the EDL is the reason why we do not observe electrokinetic effects at larger scales. Most microfluidic flows are characterized by double layers that are thin compared to the channel dimensions, with very little overlap between the double layers of the two channel walls. For nanofluidic flows though, a double-layer overlap tends to occur more frequently, changing the characteristics of the flow. A detailed discussion of the EDL is beyond the scope of this chapter, and the reader is referred to other texts (Lyklema 1991, 1995).

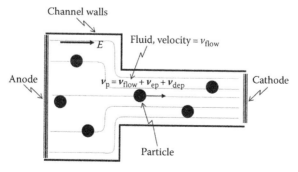

FIGURE 1.4 Schematic showing electrokinetic flow in a channel. Fluid and particle velocities are shown in the figure.

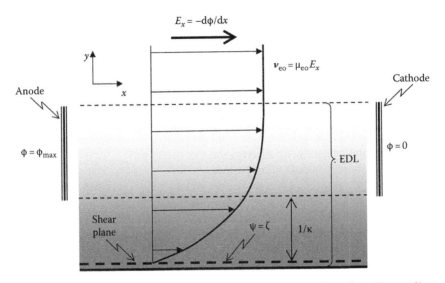

$E_x = -d\phi/dx$

y

x

$v_{eo} = \mu_{eo}E_x$

Cathode

Anode

$\phi = \phi_{max}$

$\phi = 0$

EDL

Shear plane

$\psi = \zeta$

$1/\kappa$

FIGURE 1.5 Schematic showing an EDL at the wall of a microchannel, and the resulting flow profile.

1.2.4.2 Electroosmosis

Electroosmotic flow occurs on applying an electric field across a channel, parallel to the channel surface and perpendicular to the decay direction of the EDL. The excess ions in the EDL experience a force because of the applied electric field, giving rise to a net body force on the fluid and pulling it along the surface walls. The resulting velocity profile consists of a region of high shear, where fluid velocity goes from zero at the channel walls, to the bulk channel velocity at the edge of the EDL, as shown in Figure 1.5. The flow equations describing electroosmotic flow can be modified from the Stokes equations (Equations 1.2) by adding an additional coulombic force term because of the ions in the EDL, giving

$$\nabla \cdot v = 0 \qquad\qquad (1.6a)$$

$$\eta \nabla^2 v = \nabla p + \rho_e \, \nabla \phi \qquad\qquad (1.6b)$$

where
ρ_e is the charge density in the EDL
$\nabla\phi$ is the gradient of the electrical potential

A solution to these equations needs a model for this last coulombic force term, which is in general difficult to compute accurately because it occurs on a much different length scale from the bulk flow. One popular method to simplify these equations significantly is the Helmholtz–Smoluchowski approximation (Hunter 1981), which can be applied in the limit of a thin EDL. In this case, one can approximate the electroosmotic velocity, v_{eo}, at the edge of the double layer as

$$v_{eo} = -\mu_{eo} \, \nabla \phi = \mu_{eo} E = -\frac{\varepsilon\varepsilon_o \zeta}{\eta} E \qquad\qquad (1.7)$$

where

μ_{eo} is the electroosmotic mobility
E is the applied electric potential
ζ is the zeta potential
ε_o is the permittivity of a vacuum ($\varepsilon_o = 8.854 \times 10^{-12}$ C/V m)
ε is the dielectric constant of the particle
η is the fluid viscosity

Since the double layer in this case is thin compared to the channel dimensions, the coulombic force term can be eliminated from Equation 1.6b, and instead the electroosmotic velocity calculated from Equation 1.7 can be applied as a slip boundary condition at the channel walls. As mentioned above, it is interesting to note that in the absence of any applied pressure, the solution to the flow equations for purely electrokinetic flow yields a plug flow, where the flow everywhere in the channel is given by Equation 1.7. The typical value of the electroosmotic mobility of a glass–water interface is approximately 4.0×10^{-8} m²/V s. For more details on the nature of the ζ potential, see Hunter (1981).

From Equation 1.7, it is also apparent that the electroosmotic velocity is independent of the channel cross section, and thus the electroosmotic flow tends to be preferable for nanoscale transport. Figure 1.6 shows a comparison of the flow velocity in a channel, between pressure-driven flow and electrokinetic flow. The flow velocity for the former is seen to decrease with the channel height, while in the latter case, the flow velocity is essentially constant with the channel height. In general this is true; however, in many nanofluidic environments, the EDL (which has a thickness between 10 and 100 nm) is no longer thin relative to the channel height, and the Helmholtz–Smoluchowski approximation behind Equation 1.7 is invalidated. For such cases, models exist describing the net charge density in the EDL based on electrostatics, and an accurate calculation of the coulombic force term in the above equations can be carried out if necessary (Lyklema 1991, 1995). In the case of double-layer overlap in particular, the above plug flow solution for electroosmotic flow is not valid and the flow profile is more complex. For more details on accurately calculating solutions for thick double layers, or for double-layer overlap, the reader is referred to Lyklema (1991, 1995).

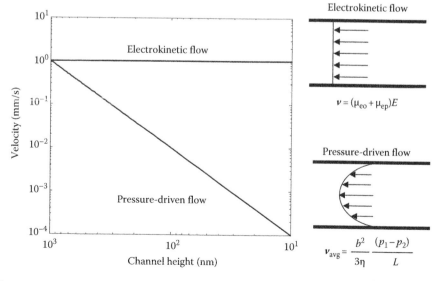

FIGURE 1.6 Comparison of electrokinetic and pressure-driven flow.

1.2.4.3 Electrophoresis

Electrophoresis occurs due to much the same basic phenomena as electroosmosis and acts on particles present in the flow. It is caused because of a Coulombic force exerted on a charged particle in an external electrical field. The electrophoretic velocity of a particle is usually represented as the electrophoretic mobility, μ_{ep}, multiplied by the gradient of the electric potential, as below:

$$v_{ep} = -\mu_{ep}\nabla\phi = \mu_{ep}E \tag{1.8}$$

where
$\nabla\phi$ is the gradient of the electric potential
E is the electric field strength $(E=-\nabla\phi)$

The electrophoretic mobility is available for many particles in the literature, and it is generally best to use reported values since μ_{ep}, rather than being constant for each particle, can vary strongly depending on the environmental conditions (e.g., solution pH and ionic strength). It can, however, also be estimated for simple molecules using the Nernst–Einstein relation to give $\mu_{ep}=Dze/k_bT$, where D is the diffusion coefficient of the molecule, z is the net valence, e is the charge on an electron, k_b is the Boltzmann constant, and T is the temperature. As before for electroosmosis, we can also occasionally make the "thin double layer" approximation when the particle is much larger than the surrounding charge field. The Helmholtz–Smoluchowski approximation in such a case yields $\mu_{ep}=\varepsilon\varepsilon_o\zeta/\eta$, where ε_o is the permittivity of a vacuum ($\varepsilon_o=8.854\times10^{-12}$ C/V m), ε is the dielectric constant of the particle, and η is the fluid viscosity. Alternatively, when the particle is surrounded by a "thick double layer," it is more appropriate to use $\mu_{ep}=2\varepsilon\varepsilon_o\zeta/3\eta$. For details on the electrophoretic mobility of different materials, readers are referred to Hunter (1981) or Lyklema (1991, 1995).

1.2.4.4 Dielectrophoresis

The third term in Equations 1.4 is the dielectrophoretic mobility, and unlike electroosmosis and electrophoresis, it acts on a dielectric particle rather than a charged surface. Dielectrophoresis is the motion of a dielectric particle due to the polarization of the particle induced by a nonuniform electric field. It causes the particle to be attracted either to the region where the gradient is the strongest (positive dielectrophoresis) or to the region where it is the weakest (negative dielectrophoresis). The dielectrophoretic velocity of a particle is given by

$$v_{dep} = \mu_{dep}\nabla(E\cdot E) \tag{1.9}$$

where μ_{dep} is the dielectrophoretic mobility. It is given by $\mu_{dep}=a^2\varepsilon_w\varepsilon_o Re[f_{CM}]/6\eta$ where a is the particle radius, ε_w is the dielectric constant of the liquid, η is the fluid viscosity, and f_{CM} is the Clausius–Mossotti factor. The Clausius–Mossotti factor is a complex frequency-dependent term that describes the effect of the displacement current (from the polarization of the particle) and the ohmic current (from conduction). It is given by $f_{CM}=(\varepsilon_p-\varepsilon_w)/(\varepsilon_p+2\varepsilon_w)$, where ε is the complex permittivity, the subscript "p" refers to the particle, and the subscript "w" refers to the liquid. The complex permittivity is defined as $\varepsilon=\varepsilon+\sigma/j\varepsilon_o\omega$, where σ is the conductivity and ω is the frequency of the applied potential. The real part of the Clausius–Mossotti factor lies between −0.5 and 1, with negative values corresponding to negative dielectrophoresis and positive values indicating positive dielectrophoresis. Unlike electroosmosis and electrophoresis, the dielecrophoretic velocity is a function of the electric field gradient instead of the electric field itself, and, thus, dielectrophoresis only occurs in a channel with a nonuniform electric field. Additionally, dielectrophoresis can occur under the application of both an AC field and a DC field, unlike the other two electrokinetic effects described above.

In practice, an AC field is often applied to a fluidic channel to eliminate these other electrokinetic effects, while observing dielectrophoresis.

1.3 Optofluidic Transport

As is described throughout this book, the field of optofluidics emerged from a series of efforts in the mid-2000s in trying to fuse advanced planar optics with micro- and nanofluidics (Psaltis et al. 2006; Monat et al. 2007; Erickson et al. 2008). One current focus of this field is the development of optical devices that have new functionalities enabled by microfluidic elements. The advantages of these devices, which again are described primarily in the second part of this book, are associated with the ability to exploit microfluidic transport phenomena (as described above) in order to change optical properties like refractive index, gain, and nonlinearity, over very small length scales.

The opposite is of course also possible in that optical effects can be used to enhance microfluidic transport. Such techniques range from traditional optical tweezing (Ashkin 1970; Ashkin et al. 1986; Curtis et al. 2002; Grier 2003; Wang et al. 2005), rotational manipulation of components based on form birefringence (Neale et al. 2005), to more recent electro-optic approaches such as that by Chiou et al. (2005). Imasaka and coworkers (Hatano et al. 1997; Kaneta et al. 1997; Imasaka 1998; Makihara et al. 1999) provided the initial foundations for radiation pressure-driven separation techniques, which they termed optical chromatography (Hart and Terray 2003; Hart et al. 2004, 2006; Terray et al. 2005; Zhao et al. 2006). These and similar implementations tend to rely on scattering forces or forces that act along the direction of optical propagation, rather than trapping forces that tend to act normal to it. These methods therefore represent something much closer to transport in the way the term is usually applied (and described above), and are therefore more relevant to this chapter. For more information on optical trapping, please refer Chapter 14 by Chiou and Chapter 16 by Schmidt, respectively.

The precision with which particles or fluids can be transported and separated with these optical techniques makes them particularly useful for biomedical analysis devices. In terms of transport, however, these systems are practically limited by the fundamentals of free-space optics. Specifically, the light–particle interaction length is limited by the focal depth of the objective lens used to focus the light, to usually a few hundred microns (as shown in Figure 1.7a). To get around this limitation, a number of researchers have investigated the use of waveguides to perform such transport (as shown in Figure 1.7b). In recent papers (Mandal and Erickson 2007; Schmidt et al. 2007; Yang et al. 2009b), we have been referring to this form of photonically enabled transport as "optofluidic transport."

The first clear demonstrations of long-distance optical transport on waveguides focused on the use of solid-core, fluid-clad structures, which relied on the evanescent field of the waveguide to both capture and transport suspended particles (similar to what is shown in Figure 1.7). Readers interested in more details on planar waveguides and waveguiding should consult Chapter 3 by Janz and Chapter 16 by Schmidt, respectively. Kawata and Sugiura (1992) first demonstrated the use of an evanescent field–based optical trapping technique. This was further refined by Tanaka and Yamamoto (2000), who showed the propulsion of polystyrene spheres on a channel waveguide. Others such as Gaugiran et al. (2005) have demonstrated the use of silicon nitride waveguides for trapping and propulsion of yeast and red blood cells. The advantage of using silicon nitride waveguides is their ability to guide wavelengths of light at 1064 nm. One limitation of these devices is that the majority of the guided optical energy is confined within the solid core of the waveguide, and the particles only interact with the 10%–20% of the energy that is accessible in the evanescent field. As such, a number of recent works have investigated the possibility of using "liquid-core" waveguiding structures for optical transport. As an example, Mandal and Erickson (2007) recently demonstrated the use of a specially tailored hollow-core photonic crystal fiber (HCPCF) to propagate light within a liquid-core environment and levitate/transport dielectric particles. In a more chip-friendly format, Measor et al. (2008) demonstrated the use of particle transport within a planar, liquid-core anti-resonant reflective optical waveguide (ARROW) as a means of characterizing the optical performance of the waveguide. Yang et al. (2009b) also recently extended

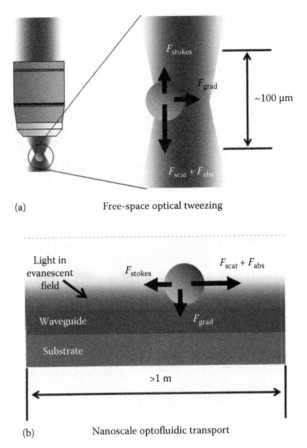

(a)   Free-space optical tweezing

(b)   Nanoscale optofluidic transport

FIGURE 1.7 Comparison between traditional and photonic optical transport. (a) Forces on a particle in a free-space optical trap and (b) forces on a particle trapped within the evanescent field of a photonic waveguide.

this by demonstrating the trapping and transport of nanoparticles and DNA in liquid-core solid-slot waveguides.

In this section of this chapter, we drift away from our general discussion of microfluidic transport and focus now on methods by which this transport can be uniquely enhanced through the use of optical forces. We begin by presenting the fundamental force equations describing the coupling between electromagnetics and hydrodynamics, as described above. The first part of this section describes the transport and the manipulation of particles directly using optical forces. This is then followed by a description of how the same basic principles have been used to manipulate fluids directly. As we describe, the work related to this latter area is particularly interesting as it has the possibility to enable methods by which flows can be routed in microfluidic devices without the need for on-chip pumps and valves.

1.3.1 Fundamental Force Equations

Optical forces are usually defined using a stress tensor (similar to the stress tensor used to describe fluidic forces) (Landau and Lifshitz 1960). The force on a system, F, can be defined as the sum of all the forces, f_V, on each of the volume elements, dV, that comprise the system. Thus,

$$F = \int f_V(r) dV \tag{1.10}$$

If f_V can be expressed as the divergence of a second-order tensor, T, then this volume integral can be written as a surface integral, and the total force can be written as

$$F = \int \nabla \cdot T \, dV = \oint T \cdot dS \qquad (1.11)$$

where dS is an infinitesimal surface element. This formulation applies for any stress tensor. In the special case of electromagnetics, this tensor is called the Maxwell stress tensor (MST) and is given by

$$T = \varepsilon_o \varepsilon_m EE + \mu_o \mu_m HH - \frac{1}{2} I \left(\varepsilon_o \varepsilon_m E \cdot E + \mu_o \mu_m H \cdot H \right) \qquad (1.12)$$

where

E and H represent the electric and magnetic fields, respectively
ε_o is the permittivity of a vacuum ($\varepsilon_o = 8.854 \times 10^{-12}$ C/V m)
ε_m is the dielectric constant of the medium
μ_o is the permeability of a vacuum ($\mu_o = 4\pi \times 10^{-7}$ N/A^2)
μ_m is the relative permeability of the medium
I is the identity tensor

and the dyadic notation is implied in the first two terms above.

Thus, the optical force can be calculated by first finding the electromagnetic field (or the optical field) in the region of interest, to give E and H, followed by using the MST to find the force.

1.3.2 Particle Manipulation and Transport

As mentioned above, the vast majority of the ways in which optical forces have been used to perform transport in microfluidic devices is through direct action on the transported particle itself. This is the case for both traditional optical tweezers and emerging near-field methods. In the simplest case, the net force acting on a moving particle in a fluid medium is given by the summation of Equation 1.10 with the fluid drag force, F_{flow}, as described by Equation 1.13:

$$F_{MST} + F_{flow} = m_p \frac{dv_p}{dt} \qquad (1.13)$$

where

m_p is the particle mass
dv_p/dt is the particle acceleration
F_{MST} is the electromagnetic force computed from the integration of Equation 1.11 using Equation 1.12

Note that at steady state (when the particle is not accelerating) this summation must equal zero. For the case of a stationary particle in a quiescent medium, F_{flow} must equal zero, and thus the integration of Equation 1.11 must also be zero.

In the most general case, F_{flow} is described by the integral of the normal component of the total flow stress tensor, T_{flow}, around the particle boundary, S, given by

$$F_{flow} = \oint_S \left(T_{flow} \cdot n \right) dS = \oint_S \left(-pI \cdot n + \mu \left(\nabla v + \nabla v^T \right) \cdot n \right) dS \qquad (1.14)$$

where the values for p and v are obtained from a solution to either Equation 1.1 or 1.2. While the above form of the equation is appropriate for the calculation of the drag force in numerical analysis or complex flow fields, it can be simplified for analytical approximations by using the Stokes drag approximation, which, for the purposes of this discussion, takes the form given by Equation 1.15:

$$F_{\text{flow}} = \frac{6\pi\eta a v_o}{g(a/h)} \tag{1.15}$$

where
$\quad a$ is the particle radius
$\quad v_o$ is the velocity of the flow relative to the particle
$\quad g(a/h)$ is a function of the particle radius divided by the distance from the center of the particle to the surface, h, and represents a correction for near-wall effects

When the particle is in an infinite medium, which is often the case for a traditional tweezer, $g(a/h)$ is equal to 1. When the particle is traveling near a surface (as is the case in many near-field manipulation methods), $g(a/h)$ is given by Equation 1.16 (Happel and Brenner 1983):

$$g(a/h) = \left[1 - \frac{9}{16}\left(\frac{a}{h}\right) + \frac{1}{8}\left(\frac{a}{h}\right)^3 - \frac{45}{256}\left(\frac{a}{h}\right)^4 - \frac{1}{16}\left(\frac{a}{h}\right)^5 \right] \tag{1.16}$$

For more complex geometries or nonuniform flow fields, the above approximations are less accurate, and thus the use of numerical simulations to determine F_{flow} is required. Although the approximation given by Equation 1.16 is only valid in the low Reynolds number regime, this is where the vast majority of optical transport occurs, and this limitation is not considered too restrictive. Extensive details on the numerical methods (including the use of commercial software packages) for solving the above equations in the context of optofluidic transport are available in Yang et al. (2009a) and Yang and Erickson (2008).

Generally speaking, two transport regimes exist for optically driven transport: (1) when the transported particle radius, a, is much smaller than the wavelength of light, λ, and (2) when the particle radius is approximately the same or much larger than λ. The first regime is referred to as the Rayleigh regime and is defined by the assumption that the electromagnetic field is uniform as it impinges on the particle (hence the limitation that $a = \lambda$). For this case, F_{MST} can be broken down into scattering, absorption, and trapping forces exerted on a particle (Svoboda and Block 1994; Happel and Brenner 1983; Ng et al. 2000a,b; Neuman and Block 2004), which take the form

$$F_{\text{scat}} = \frac{8\pi^3 I_o \alpha^2 \varepsilon_m}{3c\lambda^4} \tag{1.17a}$$

$$F_{\text{abs}} = \frac{2\pi\varepsilon_m I_o}{c\lambda} \text{Im}(\alpha) \tag{1.17b}$$

$$F_{\text{trap}} = \frac{2\pi\nabla I_o \alpha}{c} \tag{1.17c}$$

where

$\alpha = 3V(\varepsilon - \varepsilon_m)/(\varepsilon + 2\varepsilon_m)$ and is referred to as the polarizability, where V is the particle volume

c is the speed of light

ε and ε_m are the dielectric constants of the particle and the material, respectively

I_o is the optical intensity

Equating F_{scat}, F_{abs}, and F_{trap}, with Equation 1.15, we obtain

$$v_o = \frac{g(a/h)}{6ac\eta}\left(\frac{8\pi^2 \varepsilon_m \alpha^2 I_o}{3\lambda^4} + \frac{2\varepsilon_m \, \mathrm{Im}(\alpha)I_o}{\lambda} + 2\pi\alpha \nabla I_o \right) \qquad (1.18)$$

which is descriptive of the particle transport velocity in the Rayleigh regime. In the Mie regime ($a > \lambda$), F_{MST} must be computed directly from Equation 1.11.

As an illustrative example of this form of transport, consider that shown in Figure 1.8, as originally reported in Schmidt et al. (2007). In this experiment, a standard microfluidic channel is overlaid on a glass substrate with polymeric (SU-8) waveguides (Figure 1.8c), excited at 975 nm. When a particle floats by the waveguide within the evanescent field (Figure 1.8b), it is trapped (pulled down towards it) and then pushed along the waveguide. Frame captures for a polystyrene particle trapped and transported along the waveguide are shown in Figure 1.8d through f. In this case, the microfluidic flow is actually moving in the opposite direction of the optical transport, a qualitative indication of the general strength of transport achievable with this approach. For more experimental details, readers are referred to Schmidt et al. (2007) and Yang et al. (2009b).

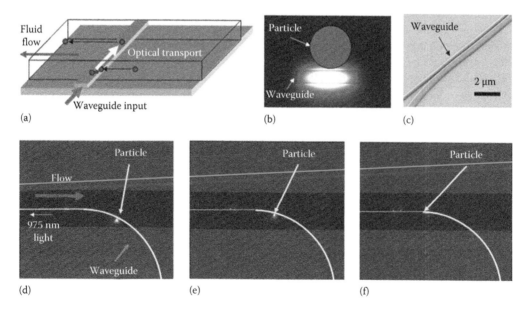

(a)   (b)   (c)

(d)   (e)   (f)

FIGURE 1.8 Optical trapping and transport in the evanescent field of an optical waveguide. (a,b) A particle flowing in a microchannel becomes captured in the evanescent field of the excited waveguide. (c) SEM of two waveguides. (d,e,f) Time-step images showing the transport of 3 μm polystyrene particles on a waveguide.

1.3.3 Direct Manipulation of Fluids Using Optical Methods

We end this section with a discussion of some of the physics behind how optical forces can act directly on a liquid in a microfluidic environment, as opposed to on particles within it. This is particularly interesting because it shows promise in being able to apply all the mature and highly parallel optical tweezing technologies described above, to direct fluid manipulation in microfluidic environments, potentially eliminating the need for on-chip valves, pumps, and other microfabricated flow control elements. Optical actuation is particularly attractive since it is a noncontact means of actuation and can also be dynamically reconfigurable in that one need not predefine a microfluidic network beforehand.

Generally speaking, there are two methods by which one can use light to manipulate fluids on the microscale. The first of these is the use of light to control the path of a fluid that is actuated by some other transport mechanism, and the second is the direct actuation of the fluid itself either directly through radiation pressure or indirectly through, for example, photothermal effects. An example of the latter of these is the work of Shirasaki et al. (2006), who used laser-induced infrared heating to directly gel and valve a thermally responsive polymer solution flowing in a channel demonstrating rapid flow switching times without the need for prefabricated valve elements. In another approach, Sugiura et al. (2009) developed a technique that utilizes light irradiation of a photoresponsive hydrogel sheet to dynamically define channel geometries and valves on the fly. We have recently developed a technique that uses photothermal conversion of light on an absorbing substrate to gel a thermo-rheological fluid creating reversible valves. The advantage of this technique is that it has relatively low optical power requirements (Krishnan et al. 2009).

The direct actuation of fluids using optical effects has also been demonstrated by a number of authors. For example, Wunenburger et al. (2006a,b) used laser light impinging on an interface between two fluids with different refractive indices, resulting in a deformation of the interface, as shown in Figure 1.9b. It has been shown that similar light-induced interfacial deformations can lead to fluid dynamical instabilities and create liquid jets of microdroplets (Casner and Delville 2003). By tuning the laser power, it is possible to use these "opto-hydrodynamic" instabilities to form large-aspect-ratio liquid channels and guide light through these channels (Brasselet and Delville 2008; Brasselet et al. 2008). Laser actuation has also been used to drive the motion of droplets using the thermocapillary effect, where a droplet in a carrier fluid placed in a thermal gradient experiences a surface tension gradient due to the dependence of surface tension on temperature (Ohta et al. 2007). This in turn induces a viscous stress in both fluids, resulting in interfacial flow. The absorption of laser light can be used

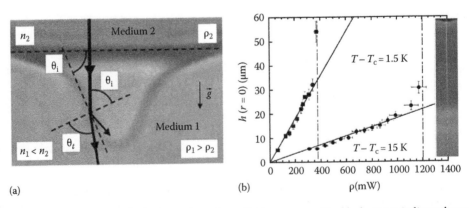

(a)   (b)

FIGURE 1.9 (a) Deformation of a fluid interface using radiation pressure. The black arrows indicate the applied laser beam. (Reprinted from Wunenburger, R. et al., *Phys. Rev. E*, 73, 036314, 2006a. With permission.) (b) Variation of $h(r)$ with beam power for different values of temperature above the critical temperature of the microemulsion. (Reprinted from Casner, A. and Delville, J.P., *Phys. Rev. Lett.*, 90, 144503, 2003. With permission.)

to create thermal gradients, and thus drive droplet flow. Laser-induced thermocapillary forces have been used to carry out mixing in microchannels, make optofluidic valves, and carry our sorting and merging of droplets (Baroud et al. 2007a,b). Finally, fluid actuation in a microchannel has also been carried out using photothermal nanoparticles suspended near the liquid–air interface in the channel (Liu et al. 2006). The photothermal nanoparticles convert optical energy from a laser beam of submilliwatt power, to thermal energy; the resulting liquid evaporation and condensation drives fluid flow along the path of the laser beam.

To provide a more detailed example of the physical analysis of these types of flows, we will now briefly describe one method to calculate optical forces on a fluid, using the change of photon momentum as a beam crosses an interface. The approach described above, involving the direct integration of the MST, can of course be applied to fluid elements instead of particles; however, such an approach can be extremely computationally intensive. This discussion is based on the work done by previous researchers (Wunenburger et al. 2006a).

Consider a fluid interface, as shown in Figure 1.9, where the refractive index of medium 1, n_1, is less than that of medium 2, n_2, to which a laser beam is applied deforming the interface. In this paper (see caption in Figure 1.9 for reference), the interface consists of a near-critical two-phase equilibrium emulsion state of a micellar phase of a microemulsion. The microemulsion has a critical temperature, T_c, above which the mixture separates into two micellar phases that form the two fluids across the interface. Figure 1.9a is a schematic showing deformation at the interface, while an experimental realization of such a system is shown in Figure 1.9b. The equilibrium shape of the interface, $h(r)$, can then be determined by a balance between the electromagnetic radiation, the hydrostatic pressure difference between the two fluids, and the surface tension or the Laplace pressure at the interface, giving the following equation:

$$(\rho_1 - \rho_2)gh(r) - \sigma\kappa(r) = \Pi(r, z, \theta_i), \qquad \kappa(r) = \frac{1}{r}\frac{d}{dr}\left[r\sin\theta_i(r)\right] = \frac{1}{r}\frac{d}{dr}\left(\frac{rh'(r)}{\sqrt{1 + h'^2(r)}}\right) \qquad (1.19)$$

where
the subscripts 1 and 2 refer to the two fluids
ρ is the fluid density (such that $\rho_1 > \rho_2$)
g is the acceleration due to gravity
σ is the surface tension

The two terms on the left-hand side of the equation represent the hydrostatic pressure difference between the two fluids and the Laplace pressure, respectively. The term on the right-hand side is the radiation pressure due to the laser beam, and is given by the expression:

$$\Pi(r, z, \theta_i) = -\frac{n_2}{c}\cos^2\theta_i\left(1 + R - \frac{\tan\theta_i}{\tan\theta_t}T\right)I \qquad (1.20)$$

where
θ_i is the angle of incidence
θ_t is the angle of refraction
R and T are the reflection and transmission coefficients, respectively
c is the speed of light in vacuum
I is the intensity of the incident beam

For a beam striking the interface at normal incidence ($\theta_i = 0$), the above expression reduces to

$$\Pi\left(r, z, \theta_i = 0\right) = -\frac{2n_2}{c} \frac{n_2 - n_1}{n_1 + n_2} I \tag{1.21a}$$

For a Gaussian beam, such as that of a laser, the intensity can be written as $I(r) = (2P/\pi\omega_0^2)\exp(-2r^2/2\omega_0^2)$, where P is the beampower, and ω_0 is the beam waist, such that

$$\Pi\left(r, z, \theta_i = 0\right) = -\frac{n_2}{c} \frac{n_2 - n_1}{n_1 + n_2} \frac{4P}{\pi\omega_0^2} \exp\left(-\frac{2r^2}{\omega_0^2}\right) \tag{1.21b}$$

For steady deformations of small amplitudes, Equation 1.19 can be linearized to give

$$\left(\rho_1 - \rho_2\right)gh\left(r\right) - \sigma\frac{1}{r}\frac{d}{dr}\left[rh'\left(r\right)\right] = -\frac{n_2}{c} \frac{n_2 - n_1}{n_2 + n_1} \frac{4P}{\pi\omega_0^2} \exp\left(-\frac{2r^2}{\omega_0^2}\right) \tag{1.22}$$

Figure 1.9b shows a typical plot describing the variation of $h(r=0)$ with the beam power, for two different temperatures, T, above the critical temperature. As can be seen from this figure, there is a threshold power above which the interface is no longer stable, leading to a dramatic increase in the value of $h(r)$ and indicating the formation of a long, cylindrical jet. The inset image here shows the cylindrical jet formed at this power. The threshold power was found to increase with the increase of temperature above the critical temperature of the microemulsion. In another experiment, the authors also found that the threshold power varied linearly with the laser beam waist. Figure 1.9b is for a beam waist of 3.5 µm. In the experiments described in this paper (see caption in Figure 1.9 for reference), the threshold power was found to vary between a few hundred milliwatts to on the order of one watt for different beam waists. Our description here only briefly discusses the balance between fluid forces and radiation pressure at an interface. For more details the reader is referred to the following works (Casner and Delville 2001; Wunenburger et al. 2006a,b; Brasselet et al. 2008).

1.4 Summary and Conclusions

In this chapter, we have introduced some of the fundamental and practical aspects of microfluidic flow and transport in a context relevant to optofluidics. In Section 1.1, we discussed the advantages of both electrokinetic and pressure-driven transport in general terms. The primary advantage of pressure-driven flow was its robustness and general applicability to a wide variety of fluids. This is particularly important for optofluidics in that we are usually interested in using a wide variety of different fluids to obtain a large range of optical properties like refractive index, nonlinearity, and gain. The disadvantage is that integrated systems must also include a pressure source, which is an undesirable complexity, particularly for the development of portable devices. Additionally, as we have discussed, pressure-driven flow does not scale well with decreasing length and is difficult to apply to nanoscale transport. Electrokinetic transport is in principle much simpler and requires less infrastructure in that only an external voltage needs to be applied to drive the flow, but the severe limitations on the types of fluids that can be used, limit its applicability to optical devices. In optofluidically enabled biomolecular or cellular analysis devices, however, electrokinetics can have significant advantages in terms of minimizing the tumbling and rotation of flowing targets and enabling faster switching of flow streams. In the second half of this chapter, we introduced some of the physics of optically driven microfluidics with a focus initially on waveguide-based particle transport and later on direct manipulation of fluids using optical forces. The advantage of the former of these areas is that optical forces can be imparted over longer distances than can be achieved in free space, opening the door to the possibility of developing new practical types of optical force–based bioanalytical devices, like the optical chromatograph. The use of optical forces to manipulate fluids directly is perhaps less well developed, but does show some very unique potential in terms of being able to reduce the complexity of microfluidic devices by eliminating the need for constructing on-chip valves and pumps.

References

Acar, Y. B., R. J. Gale, A. N. Alshawabkeh et al. 1995. Electrokinetic remediation—Basics and technology status. *Journal of Hazardous Materials* 40:117–137.

Ashkin, A. 1970. Acceleration and trapping of particles by radiation pressure. *Physical Review Letters* 24:156–159.

Ashkin, A., J. M. Dziedzic, J. E. Bjorkholm, and S. Chu. 1986. Observation of a single-beam gradient force optical trap for dielectric particles. *Optics Letters* 11:288–290.

Baroud, C. N., M. R. de Saint Vincent, and J. P. Delville. 2007a. An optical toolbox for total control of droplet microfluidics. *Lab on a Chip* 7:1029–1033.

Baroud, C. N., J. P. Delville, F. Gallaire, and R. Wunenburger. 2007b. Thermocapillary valve for droplet production and sorting. *Physical Review E* 75:5.

Becker, H. and L. E. Locascio. 2002. Polymer microfluidic devices. *Talanta* 56:267–287.

Biddiss, E., D. Erickson, and D. Q. Li. 2004. Heterogeneous surface charge enhanced micromixing for electrokinetic flows. *Analytical Chemistry* 76:3208–3213.

Brasselet, E. and J. P. Delville. 2008. Liquid-core liquid-cladding optical fibers sustained by light radiation pressure: Electromagnetic model and geometrical analog. *Physical Review A* 78:7.

Brasselet, E., R. Wunenburger, and J. P. Delville. 2008. Liquid optical fibers with a multistable core actuated by light radiation pressure. *Physical Review Letters* 101:4.

Casner, A. and J. P. Delville. 2001. Giant deformations of a liquid-liquid interface induced by the optical radiation pressure. *Physical Review Letters* 8705:4.

Casner, A. and J. P. Delville. 2003. Laser-induced hydrodynamic instability of fluid interfaces. *Physical Review Letters* 90:144503.

Chiou, P. Y., A. T. Ohta, and M. C. Wu. 2005. Massively parallel manipulation of single cells and microparticles using optical images. *Nature* 436:370–372.

Cho, S. K., H. J. Moon, and C. J. Kim. 2003. Creating, transporting, cutting, and merging liquid droplets by electrowetting-based actuation for digital microfluidic circuits. *Journal of Microelectromechanical Systems* 12:70–80.

Curtis, J. E., B. A. Koss, and D. G. Grier. 2002. Dynamic holographic optical tweezers. *Optics Communications* 207:169–175.

Daiguji, H., P. D. Yang, A. J. Szeri, and A. Majumdar. 2004. Electrochemomechanical energy conversion in nanofluidic channels. *Nano Letters* 4:2315–2321.

Darhuber, A. A., J. P. Valentino, S. M. Troian, and S. Wagner. 2003. Thermocapillary actuation of droplets on chemically patterned surfaces by programmable microheater arrays. *Journal of Microelectromechanical Systems* 12:873–879.

Dittrich, P. S., K. Tachikawa, and A. Manz. 2006. Micro total analysis systems. Latest advancements and trends. *Analytical Chemistry* 78:3887–3907.

Erickson, D. and M. Krishnan. 2009. Introduction to electrokinetic transport in microfluidic systems. In *Lab-on-a-chip Technology: Fabrication and Microfluidics*, A. Rasooly and K. Herold, Eds. Norwich, U.K.: Horizon Scientific Press.

Erickson, D. and D. Q. Li. 2004. Integrated microfluidic devices. *Analytica Chimica Acta* 507:11–26.

Erickson, D., X. Z. Liu, R. Venditti, D. Q. Li, and U. J. Krull. 2005. Electrokinetically based approach for single-nucleotide polymorphism discrimination using a microfluidic device. *Analytical Chemistry* 77:4000–4007.

Erickson, D., D. Sinton, and D. Q. Li. 2003. Joule heating and heat transfer in poly(dimethylsiloxane) microfluidic systems. *Lab on a Chip* 3:141–149.

Erickson, D., C. H. Yang, and D. Psaltis. 2008. Optofluidics emerges from the laboratory. *Photonics Spectra* 42:74–78.

Gast, F. U. and H. Fiehn. 2003. The development of integrated microfluidic systems at GeSiM. *Lab on a Chip* 3:6N–10N.

Gaugiran, S., S. Getin, J. M. Fedeli et al. 2005. Optical manipulation of microparticles and cells on silicon nitride waveguides. *Optics Express* 13:6956–6963.

Grier, D. G. 2003. A revolution in optical manipulation. *Nature* 424:810–816.

Happel, J. and H. Brenner. 1983. *Low Reynolds Number Hydrodynamics: With Special Applications to Particulate Media*. Boston, MA: Kluwer.

Hart, S. J. and A. V. Terray. 2003. Refractive-index-driven separation of colloidal polymer particles using optical chromatography. *Applied Physics Letters* 83:5316–5318.

Hart, S. J., A. Terray, K. L. Kuhn, J. Arnold, and T. A. Leski. 2004. Optical chromatography of biological particles. *American Laboratory* 36:13–17.

Hart, S. J., A. Terray, T. A. Leski, J. Arnold, and R. Stroud. 2006. Discovery of a significant optical chromatographic difference between spores of *Bacillus anthracis* and its close relative, *Bacillus thuringiensis*. *Analytical Chemistry* 78:3221–3225.

Hatano, T., T. Kaneta, and T. Imasaka. 1997. Application of optical chromatography to immunoassay. *Analytical Chemistry* 69:2711–2715.

Hunter, R. J. 1981. *Zeta potential in Colloid Science: Principles and Applications*. London, U.K.: Academic Press.

Imasaka, T. 1998. Optical chromatography. A new tool for separation of particles. *Analusis* 26:53–55.

Israelachvili, J. N. 1986. Measurement of the viscosity of liquids in very thin-films. *Journal of Colloid and Interface Science* 110:263–271.

Kaneta, T., Y. Ishidzu, N. Mishima, and T. Imasaka. 1997. Theory of optical chromatography. *Analytical Chemistry* 69:2701–2710.

Kawata, S. and T. Sugiura. 1992. Movement of micrometer-sized particles in the evanescent field of a laser-beam. *Optics Letters* 17:772–774.

Kock, M., A. Evans, and A. Brunnschweiler. 2000. *Microfluidic Technology and Applications*. Hertfordshire, U.K.: Research Studies Press.

Krishnan, M., J. Park, and D. Erickson. 2009. Opto-thermorheological flow manipulation. *Optics Letters* 34:1976–1978.

Kuiper, S. and B. H. W. Hendriks. 2004. Variable-focus liquid lens for miniature cameras. *Applied Physics Letters* 85:1128–1130.

Landau, L. D. and E. M. Lifshitz. 1960. *Electrodynamics of Continuous Media*. New York: Pergamon Press.

Laser, D. J. and J. G. Santiago. 2004. A review of micropumps. *Journal of Micromechanics and Microengineering* 14:35–64.

Li, D. Q. 2004. *Electrokinetics in Microfluidics*. Boston, MA: Elsevier Academic.

Liu, G. L., J. Kim, Y. Lu, and L. P. Lee. 2006. Optofluidic control using photothermal nanoparticles. *Nature Materials* 5:27–32.

Lyklema, J. 1991. *Fundamentals of Interface and Colloid Science, Volume 1: Fundamentals*. London, U.K.: Academic Press.

Lyklema, J. 1995. *Fundamentals of Interface and Colloid Science, Volume 2: Solid Liquid Interfaces*. London, U.K.: Academic Press.

Makihara, J., T. Kaneta, and T. Imasaka. 1999. Optical chromatography: Size determination by eluting particles. *Talanta* 48:551–557.

Mandal, S. and D. Erickson. 2007. Optofluidic transport in liquid core waveguiding structures. *Applied Physics Letters* 90:184103.

McDonald, J. C., D. C. Duffy, J. R. Anderson et al. 2000. Fabrication of microfluidic systems in poly(dimethylsiloxane). *Electrophoresis* 21:27–40.

Measor, P., S. Kuehn, E. J. Lunt, B. S. Phillips, A. R. Hawkins, and H. Schmidt. 2008. Hollow-core waveguide characterization by optically induced particle transport. *Optics Letters* 33:672–674.

Monat, C., P. Domachuk, and B. J. Eggleton. 2007. Integrated optofluidics: A new river of light. *Nature Photonics* 1:106–114.

Neale, S. L., M. P. Macdonald, K. Dholakia, and T. F. Krauss. 2005. All-optical control of microfluidic components using form birefringence. *Nature Materials* 4:530–533.

Neuman, K. C. and S. M. Block. 2004. Optical trapping. *Review of Scientific Instruments* 75:2787–2809.

Ng, J. M. K., I. Gitlin, A. D. Stroock, and G. M. Whitesides. 2002. Components for integrated poly(dimethylsiloxane) microfluidic systems. *Electrophoresis* 23:3461–3473.

Ng, L. N., B. J. Luf, M. N. Zervas, and J. S. Wilkinson. 2000a. Forces on a Rayleigh particle in the cover region of a planar waveguide. *Journal of Lightwave Technology* 18:388–400.

Ng, L. N., M. N. Zervas, J. S. Wilkinson, and B. J. Luff. 2000b. Manipulation of colloidal gold nanoparticles in the evanescent field of a channel waveguide. *Applied Physics Letters* 76:1993–1995.

Ohta, A. T., A. Jamshidi, J. K. Valley, H. Y. Hsu, and M. C. Wu. 2007. Optically actuated thermocapillary movement of gas bubbles on an absorbing substrate. *Applied Physics Letters* 91:3.

Pollack, M. G., A. D. Shenderov, and R. B. Fair. 2002. Electrowetting-based actuation of droplets for integrated microfluidics. *Lab on a Chip* 2:96–101.

Psaltis, D., S. R. Quake, and C. H. Yang. 2006. Developing optofluidic technology through the fusion of microfluidics and optics. *Nature* 442:381–386.

Quake, S. R. and A. Scherer. 2000. From micro- to nanofabrication with soft materials. *Science* 290:1536–1540.

Reichmuth, D. S., G. S. Chirica, and B. J. Kirby. 2003. Increasing the performance of high-pressure, high-efficiency electrokinetic micropumps using zwitterionic solute additives. *Sensors and Actuators B: Chemical* 92:37–43.

Schmidt, B. S., A. H. J. Yang, D. Erickson, and M. Lipson. 2007. Optofluidic trapping and transport on solid core waveguides within a microfluidic device. *Optics Express* 15:14322–14334.

Shirasaki, Y., J. Tanaka, H. Makazu et al. 2006. On-chip cell sorting system using laser-induced heating of a thermoreversible gelation polymer to control flow. *Analytical Chemistry* 78:695–701.

Stone, H. A., A. D. Stroock, and A. Ajdari. 2004. Engineering flows in small devices: Microfluidics toward a lab-on-a-chip. *Annual Review of Fluid Mechanics* 36:381–411.

Stroock, A. D., S. K. W. Dertinger, A. Ajdari, I. Mezic, H. A. Stone, and G. M. Whitesides. 2002. Chaotic mixer for microchannels. *Science* 295:647–651.

Sugiura, S., A. Szilagyi, K. Sumaru et al. 2009. On-demand microfluidic control by micropatterned light irradiation of a photoresponsive hydrogel sheet. *Lab on a Chip* 9:196–198.

Svoboda, K. and S. M. Block. 1994. Optical trapping of metallic Rayleigh particles. *Optics Letters* 19:930–932.

Tanaka, T. and S. Yamamoto. 2000. Optically induced propulsion of small particles in an evenescent field of higher propagation mode in a multimode, channeled waveguide. *Applied Physics Letters* 77:3131–3133.

Terray, A., J. Arnold, and S. J. Hart. 2005. Enhanced optical chromatography in a PDMS microfluidic system. *Optics Express* 13:10406–10415.

Thorsen, T., S. J. Maerkl, and S. R. Quake. 2002. Microfluidic large-scale integration. *Science* 298:580–584.

Unger, M. A., H. P. Chou, T. Thorsen, A. Scherer, and S. R. Quake. 2000. Monolithic microfabricated valves and pumps by multilayer soft lithography. *Science* 288:113–116.

Wang, P., Z. L. Chen, and H. C. Chang. 2006. A new electro-osmotic pump based on silica monoliths. *Sensors and Actuators B: Chemical* 113:500–509.

Wang, M. M., E. Tu, D. E. Raymond et al. 2005. Microfluidic sorting of mammalian cells by optical force switching. *Nature Biotechnology* 23:83–87.

Whitesides, G. M. 2006. The origins and the future of microfluidics. *Nature* 442:368–373.

Wunenburger, R., A. Casner, and J. P. Delville. 2006a. Light-induced deformation and instability of a liquid interface. I. Statics. *Physical Review E* 73:036314.

Wunenburger, R., A. Casner, and J. P. Delville. 2006b. Light-induced deformation and instability of a liquid interface. II. Dynamics. *Physical Review E* 73:036314.

Yang, A. J. H. and D. Erickson. 2008. Stability analysis of optofluidic transport on solid-core waveguiding structures. *Nanotechnology* 19:045704.

Yang, A. H. J., T. Lerdsuchatawanich, and D. Erickson. 2009a. Forces and transport velocities for a particle in a slot waveguide. *Nano Letters* 9:1182–1188.

Yang, J., F. Z. Lu, L. W. Kostiuk, and D. Y. Kwok. 2003. Electrokinetic microchannel battery by means of electrokinetic and microfluidic phenomena. *Journal of Micromechanics and Microengineering* 13:963–970.

Yang, A. H. J., S. D. Moore, B. S. Schmidt, M. Klug, M. Lipson, and D. Erickson. 2009b. Optical manipulation of nanoparticles and biomolecules in sub-wavelength slot waveguides. *Nature* 457:71–75.

Zeng, S. L., C. H. Chen, J. C. Mikkelsen, and J. G. Santiago. 2001. Fabrication and characterization of electroosmotic micropumps. *Sensors and Actuators B: Chemical* 79:107–114.

Zhao, B. S., Y. M. Koo, and D. S. Chung. 2006. Separations based on the mechanical forces of light. *Analytica Chimica Acta* 556:97–103.

2

Microfabrication

Aaron R. Hawkins
Matthew R. Holmes
Tao Shang
Yue Zhao

2.1 Introduction

2.1.1 Historical Context

Chip-based optofluidics has been made possible by the micro- and nanofabrication techniques developed for semiconductors by the microelectronics industry. The first semiconductor devices consisted of semiconductor slabs and connecting metal wires, put together one at a time. Semiconductor chips, as we now know them, originated when inventors at Fairchild Semiconductor formed multiple transistors simultaneously on a flat piece of silicon. These transistors were linked together with metal lines to form an "integrated circuit," and the modern electronic era was born (Riordan 2007).

The key to making integrated circuits (ICs) has been "planar" processing, meaning modification of the entire surface of a round silicon wafer during each manufacturing step. More traditional methods, like sawing, milling, drilling, or lathing, modify objects in a serial fashion, removing material at a single machining point. Planar processing is more akin to printing and has been an especially good match for ICs because it scales well in terms of components per area—the same steps and total fabrication time can be used to make one hundred or one million transistors per square centimeter. Planar processing has also been critical for component scaling, with transistors steadily decreasing in size from tens of microns in the 1960s to tens of nanometers today. Modern ICs are mind-boggling in their complexity, squeezing billions of nanoscale elements onto every chip, with an estimated 10^{18} total transistors produced annually. No other manufactured product comes close in terms of uniformity and cost per component.

The profitability and enormity of IC production has spawned an entire support industry, supplying machinery and infrastructure for processing silicon substrates. It was not long before researchers began asking themselves, "What else can be made with these machines and techniques? Can the success of silicon transistors be duplicated with different materials and components?" Part of the motivation for exploring alternatives was the very rapid pace of silicon development. Smaller research labs and universities could not afford the very latest machinery necessary for state-of-the-art circuit production. Still, facilities considered obsolete by silicon's standards were capable of producing innovative, microscale structures, given the right creativity.

Perhaps the first distinct field to widely benefit from the IC groundwork was photonics, and specifically optoelectronics. Efficient photodiodes for light detection could be readily made from silicon substrates even in the early years of IC development. Planar processes were soon adapted to semiconductors like gallium arsenide, and light-emitting diodes (LEDs), based on these materials, flooded the market (Schubert 2003). Low-cost, planar-processed semiconductor lasers made possible the fiber optic boom in the 1990s (Matthews 2000), and today's "integrated photonics" field can even be considered mature. Today, lasers, detectors, and interconnecting glass waveguides can routinely be built on chips with yields that rival those for silicon ICs. Specialty lines of processing equipment have even evolved with the field to handle the unique requirements of optoelectronic materials.

While the benefits of planar processing for electronic components seemed obvious, it was not until the 1970s that it was applied to microscale mechanical components—things with moving parts. Silicon, especially, became the building block for an array of devices beginning with pressure sensors and ink-jet printer nozzles (Peterson 1979). Soon mechanical elements were tied to electrical ones and the field of microelectromechanical systems (MEMS) began a rapid expansion (Fan et al. 1988). By the 1990s, MEMS claimed microaccelerometers (Eddy and Sparks 1998), and Texas Instrument's Digital Light Processing chip (Van Kessel et al. 1998) as part of its family of successes. MEMS development continues today with component geometries decades behind what are achievable with silicon ICs.

A relatively recent application of planar processing has occurred in the field of fluid analysis and manipulation. Its most widely acknowledged beginning dates to 1990 with the call for chip-based miniaturization (Manz et al. 1990). Researchers postulated that micron-scale tubes or channels could be fabricated on a planar surface with high precision and with several performance advantages when applied to chemical analysis. Before this point, fluid flow was analyzed on the micron scale in capillary tubes, but "microfluidics" was quickly adapted to planar or "on chip" channels. Analysis methods common in capillary tubes, like chromatography and electrophoresis, were also adapted, and the development objective became squeezing as many measurement steps as possible onto a microplatform—made from silicon, glass, or plastic starting materials. The terms "lab-on-a-chip" and "micro total analysis system—μTAS" are now synonymous with these efforts (Harrison et al. 1992).

2.1.2 Content and Chapter Organization

Optofluidics fabrication lies at the nexus of the three major offshoots from IC manufacturing: integrated photonics, MEMS, and labs-on-a-chip. Almost by definition, optofluidic devices will manipulate fluid and light on the micro- or nanoscale. Fluid-handling elements have many analogies in the lab-on-a-chip and MEMS worlds, while light generation, detection, modulation, and routing can be achieved through integrated photonics. Optimal fabrication methods draw from each of these disciplines, with branches stretching back to planar IC manufacturing.

In selecting content for this chapter, we concentrated on micro- and nanofabrication techniques that were most applicable to optofluidic devices as they exist today and are likely to exist in the future. We have tried to emphasize more established methods, especially those that can be done using commercially available equipment or with commercially available materials. It is impossible to be completely comprehensive, given that there are many volumes (Madou 2002) and even small libraries devoted to the broader topic. Our intention is to provide an introduction and reference source for readers new to microfabrication while giving more experienced readers a snapshot of processes that are most readily being applied in optofluidics.

We have divided micro- and nanofabrication, as it relates to optofluidics, into five classes of processes: (1) film formation, (2) micro- and nanopatterning, (3) forming impressions, (4) encapsulation, and (5) polishing and packaging. We will begin the substantive part of this chapter with an illustrative example that outlines the fabrication of a hypothetical optofluidic device, with necessary steps drawn from each of our five process classes. This will be followed by a section on common substrate materials and then sections dedicated to specific techniques within our five identified processing classes. We will conclude with additional examples of real devices and with some thoughts on future work in the field.

2.2 An Illustrative Example

Before delving into the details of specific micro- and nanofabrication techniques, we begin by outlining the fabrication of an imagined optofluidic device—a fluid analyzer employing the evanescent coupling of light into a fluid-filled microchannel. This example is meant to provide some visual context on how steps would progress from a blank substrate to a finished device. To our knowledge, the example analyzer we illustrate has not been actually fabricated by any development group.

We begin with a round glass substrate (wafer), typically about 1 mm thick. Our first fabrication step will be to deposit a thin film of glass over the entire surface of our wafer as shown in Figure 2.1A. The designation "thin" is somewhat subjective, but typically this will mean less than 5 μm in thickness. This film will have a refractive index higher than that of the wafer ($n = 1.45$). The relative refractive indexes are important because we intend to form total-internally-reflecting waveguides in the film. The most common deposition methods will be highlighted in Section 2.4.

After film application, the next major step involves transferring a pattern into a photosensitive polymer (photoresist) coated over the substrate. Microscopic windows of photoresist are simultaneously removed from the wafer's surface using a process called photolithography. Figure 2.1B shows a diagram of our example wafer after a layer of photoresist has been patterned. More details on the photolithography process, as well as other patterning procedures, are included in Section 2.5.

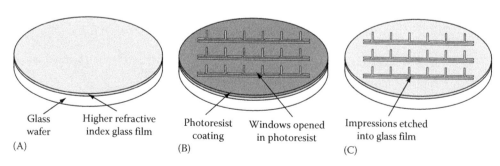

| Glass wafer | Higher refractive index glass film | Photoresist coating | Windows opened in photoresist | Impressions etched into glass film |
| (A) | | (B) | | (C) |

FIGURE 2.1 Processing steps representing: (A) film formation; (B) micro- and nanopatterning; (C) forming impressions.

The windows in the photoresist coating allow the next type of processing step—the forming of impressions. In our example, these impressions will be formed in the thin glass film we deposited over the substrate. This will be done by etching the glass wafer in a chemical that will remove the glass exposed by the windows while leaving the photoresist intact. The pattern of the etched impressions will match the original photoresist pattern, as illustrated in Figure 2.1C. Common etching methods will be expanded upon in Section 2.6, along with other methods like stamping and milling that are relevant to optofluidics.

A necessary component for optofluidic devices is a chamber to hold fluids. This is often a long, enclosed channel, with an open cross section of micron-scale dimensions. To form these types of structures, we must use an encapsulation procedure. For our example device, we will rely on wafer bonding as a means of encapsulation. This is done by bringing two clean wafers into physical contact and forming a chemical bond between them. In the illustration shown in Figure 2.2A, a second glass wafer is placed over the patterned one. The impressions formed in the first wafer have been transformed into enclosed channels. The second wafer has holes drilled through it and the holes aligned to specific points on the bottom wafer. The holes are intended to serve as small fluid reservoirs sitting above the opening to the channels. Bonding will be expounded upon, along with other encapsulation methods, in Section 2.7.

We could say that at this point the "simultaneous" processing is complete. Our particular micropattern has produced nine devices. What remains is the extraction of individual "die" from the bonded wafer along with edge polishing, as shown in Figure 2.2B. A section in this chapter is devoted to a more thorough discussion of polishing and packaging.

While we were mainly interested in the fabrication procedures used to produce our sample device, it is also instructive to point out its principles of operation. Figure 2.3 shows how the device could be integrated with optical fibers and a test fluid to create a sensor. Light from the fiber would be launched into the total-internal-reflection waveguide sandwiched between the two glass wafers. In order for light to

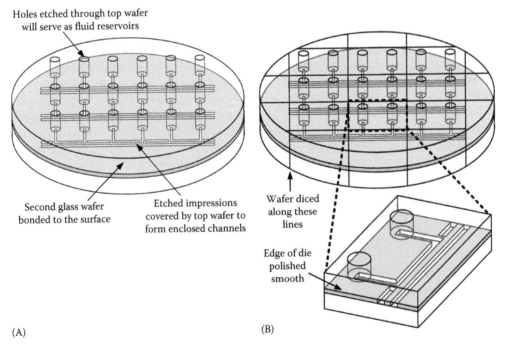

Holes etched through top wafer
will serve as fluid reservoirs

Second glass wafer
bonded to the surface

Etched impressions
covered by top wafer to
form enclosed channels

Wafer diced
along these
lines

Edge of die
polished
smooth

(A)   (B)

FIGURE 2.2 (A) Processing step representing encapsulation—the patterned bottom wafer is bonded to a second wafer containing through holes. (B) The wafer is then diced and the edges of each die are polished for optical interfacing.

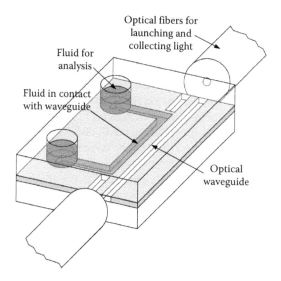

FIGURE 2.3 Operating principle of example device. Light is coupled into the waveguide formed between the glass wafers. The light interacts with fluid inside the channel that is in intimate contact with one of the waveguide's boundaries.

remain confined in the waveguide, it is critical that its refractive index be greater than any surrounding material. During fabrication, we ensured that the glass above and below the waveguide were of lower index. One side of the waveguide will always be surrounded by air ($n = 1$) and the other side will be surrounded by air and sample fluid as the waveguide passes by our enclosed channel—ensuring guiding because most liquids have refractive indexes below 1.45. The intimate contact between the fluid and the waveguide will allow light to be used to probe the fluid over a fairly long interaction length. While most interaction will necessarily take place through evanescent coupling of the optical mode into the fluid (a very small portion of the mode), significant measurements can still take place. For instance, if fluorescent particles have bound to the fluid channel's wall, light propagating through the waveguide could excite a fluorescent signal. A portion of this fluorescence could be collected by the waveguide and transmitted to the edge of the chip.

2.3 Substrates

The choice of a starting substrate material for an optofluidic device dictates the processes that can be used for microfabrication. Factors to consider when choosing a substrate include temperature limits (many standard processes involve heat), chemical resistance (etching relies on caustic solutions and gases), and mechanical strength (bonding and substrate handling are often required). Optical properties must also be considered, such as whether the substrate is transparent or naturally fluorescent for light wavelengths we intend to use. Interactions between test fluids and the substrate are also important as chip dissolution can change the properties of a fluid we are analyzing or manipulating. In this section, we describe the most important substrates for optofluidic fabrication along with details that can assist designers with substrate choice.

2.3.1 Silicon

Silicon is by far the most widespread used material in semiconductor electronic devices. It withstands very high temperatures (melting point 1414°C), forms a native oxide (SiO_2) that is easily grown in a furnace, and forms a better semiconductor/insulator interface than any other material. SiO_2 layers can act as etching windows and as an isolation for integrated metallization. Both silicon and SiO$_2$ are inert to most common acids and interact little with biological fluids or molecules.

The technology for silicon crystal growth was rapidly developed and the steady increase in wafer size has continued until today, with 300 mm diameter wafers now in production (Fischer et al. 2000). Wafers smaller than 300 mm are still widely used and available from many different vendors (http://www.ece. byu.edu/cleanroom/EW_Purchase.phtml). The vast majority of processing equipment is designed with silicon substrates in mind so there are infrastructure benefits to using silicon whenever possible. When fabricating labs-on-a-chip from silicon, many designers will make the argument that silicon electronics could be integrated with fluid analysis structures on the same chip. This is an appealing prospect, but there have been few working chips demonstrating this thus far.

Silicon is one of the most popular photodetector materials in the visible and near-IR regions (Huang et al. 2006). The bandgap of silicon is about 1.12 eV, so silicon can strongly absorb light at wavelengths <1.1 μm with a high index of refraction ($n > 3.5$). Silicon photodetectors achieve quantum efficiencies up to 80% for wavelengths between 0.8 and 0.9 μm, but silicon is transparent at the important telecommunication wavelengths of 1.3 and 1.55 μm. The field of silicon photonics focuses on using silicon for waveguiding at these wavelengths.

2.3.2 Silicon-on-Insulator

Silicon-on-insulator (SOI) is a semiconductor wafer technology that produces higher-performing, lower-power electronic devices than traditional bulk silicon techniques (DeJule 2009). SOI works by placing a thin, insulating layer, such as silicon oxide, between a thin layer of silicon and the silicon substrate. The cross section of an SOI wafer is illustrated in Figure 2.4.

The unique optical properties of SOI structures offer the ability to integrate photonic devices into CMOS IC technology (Schmidt et al. 2007). For example, low-loss SOI waveguides (Settle et al. 2006, Grillot et al. 2008) have been developed based on the large refractive index step between the silicon ($n > 3.5$) and the insulator (if SiO$_2$, $n = 1.46$). SOI substrates are compatible with most conventional silicon fabrication processes and can be used in the same process equipment.

2.3.3 Compound Semiconductors

Compound semiconductors are made up of two or more elements, usually combinations from the III-V or II-VI columns of the periodic table (Jackson 1998). The cost of developing the knowledge and technology to process raw materials into device quality semiconductors is enormous. As a consequence, there are only a few, highly developed compound semiconductor substrates: InSb, GaAs, InP, GaP, and CdTe.

Compound semiconductors are widely applied in optoelectronic and communication devices. A large segment of compound semiconductor production goes into the illumination market, including LEDs. Substrates suitable for LED applications are materials with direct bandgaps such as GaAs, and recent advances in controlling the crystal growth of nitrogen compounds have made it possible to fabricate short-wavelength (blue) light emitters based on GaN (Winser et al. 2001, Lee et al. 2007). Another large

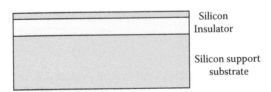

FIGURE 2.4   Cross-section structure of an SOI wafer.

application area is wireless communications. Other applications include photovoltaics (Alferov et al. 2004) and power semiconductors.

Compound semiconductor substrates are, in general, much more expensive than silicon, are and only available in smaller sizes (50–150 mm diameters). Most are compatible with temperatures up to 400°C, and every individual semiconductor has a different family of etch chemicals it is compatible with. Most compound semiconductors absorb visible light like silicon. The main motivation for using them for substrates instead of silicon in an optofluidic application would be their ability to produce on-chip light emitters, either LEDs or semiconductor lasers (Fasol 1996, Someya et al. 1999).

2.3.4 Glass

Quartz and borofloat glass wafers (Ahuja and Scypinski 2001) are widely used in the area of integrated photonics and lab-on-a-chip since they are compatible with many established microfabrication processes. Both materials are hard, insulating, and have low thermal expansion coefficients. They are commercially available in round wafer form, with 100 mm diameter being a popular size.

Quartz, the crystalline form of silicon dioxide, can transmit light from the UV range to the IR, which makes it suitable for both fluorescence detection and UV-absorption detection. Optical quality quartz is relatively expensive, and wafer bonding of quartz to silicon or glass is quite difficult because of quartz's high melting point (1650°C) (Geschke et al. 2004). In contrast, borofloat glass is much cheaper and can be bonded to silicon and other borofloat wafers at lower temperatures (~400°C). Borofloat can be used for some fluorescence applications since it is transparent to visible light (although it has some natural fluorescence), but it is not suitable for transmission of UV light less than 300 nm in wavelength. The refractive index of borofloat is about 1.472, which is slightly higher than that of quartz (1.458).

2.3.5 Plastics

Since plastics (polymers) are flexible, lightweight, inexpensive, and can be made transparent and biocompatible, they have increasingly been used as substrate materials for microfabricated devices, such as organic electronic circuits, biomedical devices, microfluidic flow networks, and flexible displays (He and Kanicki 2000). Typical plastic materials for substrates include polyethylene terephthalate (PET) (Zyung et al. 2005, Sierros and Kukureka 2007), polymethyl methacrylate (PMMA) (Horng et al. 2005), and polyimide (Xiao et al. 2008). In many ways, fabricating on plastics represents a radical departure from what has been developed on silicon substrates, but it is pursued vigorously because of the potential for low-cost manufacturing. Plastic substrates usually come in large flat sheets, but they can easily be formed into workable sizes using a saw or laser cutter.

PMMA is probably the most common plastic substrate encountered in optofluidics and is classified as a thermoplastic and transparent plastic. It is commonly called acrylic glass, or plexiglass. It is viewed as a potential substitute for glass and silicon-glass substrates for fabricating microfluidics, although it has higher optical absorption and more natural fluorescence for most wavelengths in and around the visible range. PMMA, like other plastics, must be processed at lower temperatures than semiconductor or glass wafers. Many solvents, like acetone, used for cleaning and photoresist removal are incompatible with PMMA.

2.4 Film Formation

Creating films on the surface of substrates allows us to create devices out of multiple materials instead of just the substrate itself. Films are usually formed uniformly over an entire wafer's surface and kept thin (less than a few microns) so not much of the film material is required. Semiconductors, metals, glasses, and polymers are all routinely deposited as films, and sophisticated devices may utilize multiple film layers of varied material composition. This section will discuss popular film deposition techniques encountered in the optofluidics field. Note that films formed using these techniques usually adhere well to semiconductor or glass substrates but can be challenging to attach to plastic substrates.

2.4.1 Thermal Oxide Growth on Silicon

Thermal oxidation (Hu 1984) is a technique that forces an oxidizing agent to diffuse into a silicon wafer and react with it at extremely high temperatures to promote the growth of SiO_2 layers. Thermal growth on silicon is usually accomplished at a temperature between 800°C and 1200°C.

A schematic diagram of a typical wafer oxidation furnace is shown in Figure 2.5. Two basic schemes are used, wet and dry oxidation (Hopper et al. 1975, Massoud and Plummer 1985), so correspondingly, either water vapor (steam) or molecular oxygen may be used as the oxidant. In wet oxidation, hydrogen and oxygen gases are introduced into a torch chamber where they react to form water molecules (high-purity steam at high temperature), which are piped into the furnace tube where they diffuse toward the wafers. Water molecules react with the silicon to produce the oxide and a byproduct, hydrogen gas. During dry oxidation, silicon wafers react with the ambient oxygen in the furnace tube, forming a layer of silicon dioxide on their surface. Thermal oxidation is straightforward, requiring only heat and an oxygen source, and produces a very robust SiO_2 layer. This type of film formation is only possible with a silicon substrate, however, and because the growth rate varies as (growth time)$^{1/2}$, creating thick (>2 µm) SiO_2 layers this way is impractical.

2.4.2 Chemical Vapor Deposition

Chemical vapor deposition (CVD) is a method that is widely practiced for depositing thin films on substrates. In a CVD process, a gas-phase chemical reaction is used to produce reactants that deposit on wafers to form a solid film. Deposition rates for these processes do not change over time like they do during diffusion-based SiO_2 growth, with typical deposition rates ranging between 0.1 and 3 nm/s.

2.4.2.1 Low-Pressure Chemical Vapor Deposition

The CVD process carried out under subatmospheric pressures is commonly referred to as low-pressure chemical vapor deposition (LPCVD). Reduced pressures tend to reduce unwanted gas-phase reactions

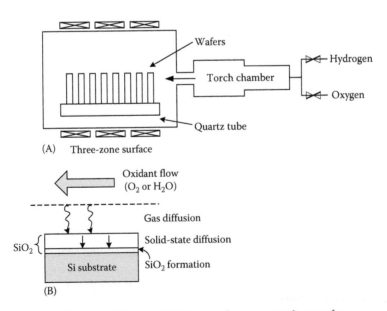

FIGURE 2.5   (A) Thermal SiO_2 growth furnace. (B) SiO_2 growth process on silicon wafers.

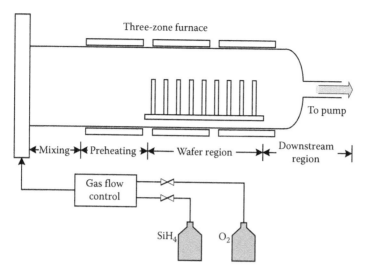

FIGURE 2.6 LPCD growth chamber plumbed for the growth of SiO_2 layers.

and improve film uniformity across a wafer. A typical LPCVD growth chamber is illustrated in Figure 2.6. The chamber is filled with flowing process gasses, and wafers are supported concentrically and perpendicularly to the flow within a quartz tube. The tube is heated (up to 1000°C), evacuated, and then refilled with a process gas at pressures ranging between 50 and 5000 mTorr. The LPCVD method is very successfully applied to the deposition of polysilicon thin films (Hatails and Greve 1988) from SiH_4 gas in the temperature range of 600°C–660°C and SiO_2 layers from SiH_4 gas and oxygen at 900°C.

2.4.2.2 Plasma-Enhanced Chemical Vapor Deposition

Plasma-enhanced chemical vapor deposition (PECVD) is a process that utilizes plasma to enhance chemical reaction rates of process gases. Figure 2.7 illustrates a typical plasma reactor system for PECVD. The plasma is created by applying RF (AC) frequency or DC discharge between two electrodes while the space between them is filled with reacting gases. Different from LPCVD, the energy for the PECVD process is provided by plasma power rather than heat, so that PECVD processing allows deposition at lower

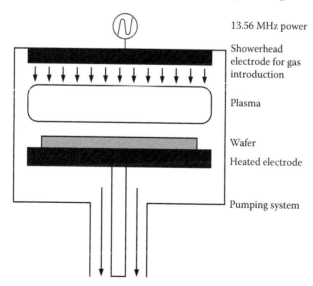

FIGURE 2.7 Elements of a PECVD system.

temperatures (typically between 200°C and 300°C). PECVD is often used in semiconductor manufacturing to deposit films onto wafers containing metal layers or other temperature-sensitive structures.

The most common films (Schuegraf 1988) formed by PECVD are insulating, silicon-based dielectric materials, such as SiO_2 (Adams et al. 1981, Devine 1990) and Si_3N_4, amorphous silicon, and polycrystalline silicon. SiO_2 can be deposited from dichlorosilane or silane and oxygen, typically at pressures from a few hundred mTorr to a few Torr, or from tetraethylorthosilicate (TEOS) in oxygen or oxygen–argon plasma (Kim 2000). Plasma-deposited silicon nitride, formed from silane and ammonia or nitrogen, is also widely used.

While PECVD films can be made at relatively low temperatures, they incorporate more defects than those of LPCVD or thermal oxide growth. These defects manifest themselves as scattering sites during optical transmission and lead to faster, more uneven etching when films are exposed to etch chemicals.

2.4.2.3 Atomic Layer Deposition

Atomic layer deposition (ALD) is a self-limiting process that uses pulses of gas to produce films with atomic scale precision. The majority of ALD reactions use precursors. At the beginning of a deposition, precursor molecules are pulsed into a chamber and cover the surface of a substrate. The precursor gas is selected so that it does not react with itself, and only a monolayer can reside on the substrate. To remove unreacted precursor from the chamber, a purge gas is introduced. The next step of deposition uses a pulse of oxidant that deposits oxygen and removes the precursor ligands. This is followed by another purge step. This series of gas pulses (precursor→purge→oxidant→purge) is repeated to build up a film. Figure 2.8 shows an example of the deposition of an Al_2O_3 film (Ghiraldelli et al. 2008).

ALD is known for its ability to produce uniform, conformal films with precise control of thickness. Film quality is influenced by the precursor and oxidant reactivity, precursor and oxidant pressure and time in the chamber, and substrate temperature. Ideally, film thickness scales with the number of growth cycles, and this process is most applicable when a high precision, very thin film is needed.

2.4.3 Sputtering

Sputtering (Fraser 1978) is a physical vapor deposition process by which atoms are dislodged from the surface of a solid cathode as a result of collisions with high-energy particles. A basic set-up for a sputtering system (Rastogi et al. 1987) is illustrated in Figure 2.9. Both target and substrate are planar plates. Ions are generated from an inert gas discharge and directed at the target material, the subsequent collisions liberating atoms from the target. The sputtered atoms are then transported to the substrate through a region of reduced pressure, and condense on the substrate, forming a thin film. Sputtering has become one of the most widely used techniques for depositing various films, including aluminum, aluminum alloys, gold, TiW, Si, SiO_2, and Si_3N_4 (Kominiak 1975, Class 1979). While substrates are often heated during deposition, sputtering can be done at temperatures lower than most CVD processes. Sputtering deposition rates do not change with film thickness and can be as high as several nanometers per second. The process can also be very scalable, with huge sputtering chambers and targets capable of coating hundreds of wafers at a time.

In the simplest case, a sputtering system can be operated with DC voltage as shown in Figure 2.9. An electric field is applied between two electrodes. The electric field accelerates electrons, causing them to collide with neutral atoms. These collisions result in ionization and the generation of ions which in turn bombard the cathode. Atoms from the cathode will be sputtered if the energy of the bombarding ions is large enough. DC sputtering systems work effectively for metals (Wilson and Weiss 1991), but not insulators (SiO_2, Al_2O_3, Si_3N_4, etc.). If we attempt to use a DC system with an insulating target, positive charge builds up on the target and effectively stops any sputtering.

Another technique, RF sputtering (Vossen 1971), enables us to sputter insulators more effectively. High frequency alternating current is applied to an electrode so that the target is alternatively

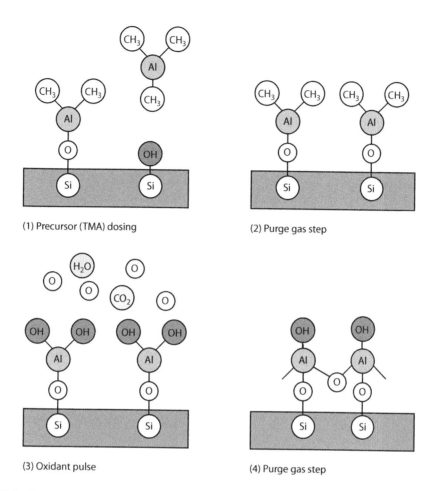

FIGURE 2.8 Deposition sequence for an ALD process forming films of Al_2O_3.

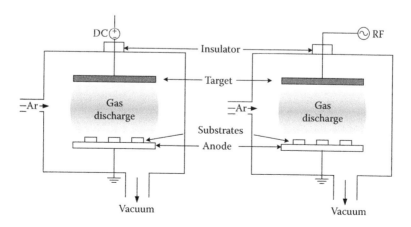

FIGURE 2.9 Elements in a sputtering system. Left: DC sputtering. Right: RF sputtering.

bombarded by positive ions and then negative electrons so as to neutralize charge buildup. It is common to use generators operating at 13.56 MHz for RF sputtering systems.

2.4.4 Evaporation

Evaporation (Harper and Advisor 1998) is perhaps the simplest film deposition technology and can be used to deposit most solids. As shown in Figure 2.10, the source material is heated in high vacuum until it is evaporated. Evaporated atoms rise from the source in a cone-shaped vapor cloud and then condense back to a solid state when they encounter a surface. Substrate wafers are typically placed at specific angles so that condensing vapor deposits almost uniformly across their surface. To increase film uniformity, a planetary system can also be used that spins the wafers with two degrees of rotation within the evaporant cloud.

There are two primary types of evaporation: thermal evaporation and e-beam evaporation (Andrew 1982). In a thermal evaporation system, Joule heating via a refractory metal element melts the material source. Low-melting-point metals, such as gold and aluminum can easily be deposited with a thermal evaporation source. Electron beam sources are somewhat more flexible and can be used with a larger

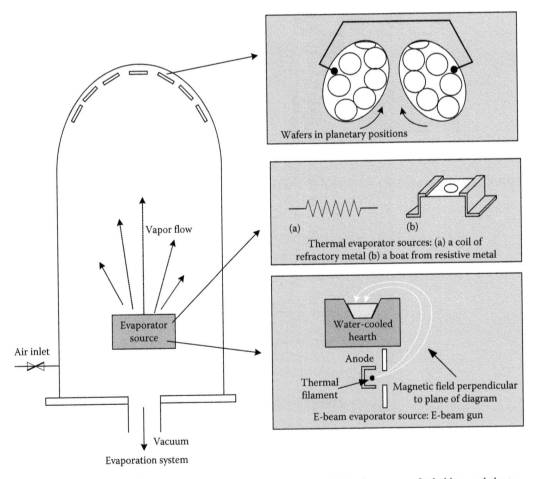

FIGURE 2.10 Evaporation deposition. Left: evaporation chamber. Right: planetary wafer holders and the two different types of evaporation sources.

number of materials. This type of source heats the evaporant by using a high energy beam of electrons, so that heating is not limited by the melting point of a heater element. Even high melting-point materials such as refractory metals can be evaporated.

Advantages of evaporation include its simplicity and low cost to implement. Deposition rates do not depend on film thickness and are typically between 0.2 and 5 nm/s. Most elemental metals can be evaporated along with a select number of insulators (Al_2O_3 and SiO for example). Parylene is a commonly employed insulating polymer that must be deposited in an evaporation system. Downsides to evaporation relate to the melting of the source material. At the high temperatures required, the source is easily contaminated. Alloys are also challenging to evaporate from a single melted source because one element will invariably have a higher vapor pressure than another at a given temperature. The result is a deposited alloy different in composition than the source alloy.

2.4.5 Spin-On Polymers and Glasses

In recent years, polymers (Becker and Gärtner 2000) have become very promising materials for microsystem technology, for substrates as well as films. Much of their appeal stems from their use in low-cost and batch-style fabrication such as injection molding and hot embossing. Polymers can also be engineered for specific applications by tailoring their thermal, electrical, and mechanical properties.

A standard method for preparing thin films of soluble polymers is centrifugal spinning. "Spin-coated" films begin as liquids and when spun, centrifugal forces thin out the liquid to reproducible thicknesses based on rotation speed. The liquid film transitions to a solid after subsequent baking or UV-curing steps.

A common example of a spin-on polymer is spin-on-glass (SOG) (Liu et al. 2003, Kim et al. 2008), which is an interlayer insulating material applied in liquid form to fill narrow gaps between conducting structures. SOG was found insoluble to solvents after cross-linking reaction from energy sources including UV or deep UV, ion beam, e-beam lithography, focused ion beam (Suzuki et al. 1994), and x-ray lithography. After exposure, the structure of ladder silicon SOG is changed to a silicon-dioxide-like structure and is an alternative to SiO_2 deposited by CVD. Another important family of polymers is the polyimide (Ghosh and Mittal 1996, Bhattacharya and Bhosale 2000), which are widely used in microelectronics to form insulating layers, sacrificial layers, adhesive films, sensors, and scanning probes. Photosensitive polymides (PSPI), which have essentially the same composition and properties as conventional polymides after being fully cured, can be patterned by direct exposure and development without the use of photoresist. Other available spin-on polymers include liquid crystal polymer (LCP), SU-8 (which is also photosensitive), and polydimethylsiloxane (PDMS). Teflon AF is another notable polymer used for optofluidics because it can produce films with refractive indices ($n = 1.29$) lower than water ($n = 1.33$).

2.4.6 Conformality

The conformality of a film deposited over a feature is closely related to the concept of step coverage. Step coverage is defined as the ratio of film thickness along the sidewalls of a step feature to the film thickness at the bottom of the step. Good step coverage is critical when connecting conducting lines up and over a feature or trying to encapsulate a feature.

The step coverage of evaporated films can be poor due to the shadowing effect. As shown in Figure 2.11, when evaporant atoms approach the substrate at a large incident angle θ, they will deposit on the outermost part of the surface while the walls and the lower part of the surface is shadowed by neighboring material. Heating (resulting in surface diffusion) and rotating the substrates (minimizing the shadowing) help improve the step coverage, but evaporation cannot form continuous films for aspect ratios (AR = step height/step width) greater than 1 (see Figure 2.12).

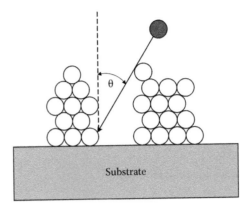

FIGURE 2.11 Illustration of the shadowing effect that can occur during film deposition. As angle θ increases, incoming atoms collide with the top of a trench and cannot create films on the trench's bottom and sides.

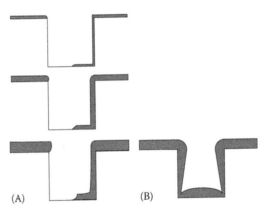

FIGURE 2.12 (A) Time evolution of the evaporative coating of a feature with aspect ratio of 1.0, with little surface atom mobility (i.e., low substrate temperature) and no rotation—assuming vapor enters feature from the left. (B) Final profile of deposition on rotated and heated substrates.

The conformality of sputtered film is also limited by the shadowing effect, but the working pressure of sputtering is higher than that of evaporation, which causes more random angled delivery (Blech and Vander Plas 1983). As a consequence, sputtering produces thin films with better conformality. CVD tends to have better conformality than either evaporation or sputtering (Cheng et al. 1991), especially for high aspect-ratio features. Conformality in a CVD process is highly dependent on gas pressure, substrate temperature, plasma power, and the flow rate ratios between process gases.

2.5 Micro- and Nanopatterning

A suitable method for micro- or nanopatterning is critical for any type of micro- or nanofabrication. Whatever method chosen will be the basis for transferring a design of optical, electrical, or fluid elements from a computer drawing onto real surfaces. In this section we describe the most common patterning methods in the field, including descriptions of their resolution capabilities and whether they are serial (slow) or highly parallel (fast). All of these techniques (except laser engraving) involve transferring a pattern into a masking layer, a thin film of material deposited on a substrate. Transferring the pattern into the substrate itself by modifying its surface is described in Section 2.6.

2.5.1 Photolithography

Photolithography is the process of transferring patterns from a photomask to a thin layer of radiation-sensitive material (called photoresist) spread over the surface of a substrate. Typically, UV light is used to chemically alter the photoresist, making it susceptible to removal in a chemical solution (development) (Chang and Sze 1996). Photolithography is the pattern transfer process used in semiconductor manufacturing and for a large fraction of MEMS, lab-on-a-chip, and optofluidic devices.

2.5.1.1 Example of Photolithography Procedure

Perhaps the best way to begin a discussion on photolithography is by describing a step by step photolithography procedure (Van Zant 2000, Jaeger 2002, Madou 2002, Liu 2006). This is illustrated in Figure 2.13 and consists of the following sequence steps: (1) Cleaning and preparation of wafer, (2) Application of adhesion promoter hexamethyldisilazane (HMDS) and photoresist to wafer, (3) Soft bake to solidify photoresist, (4) Mask and UV exposure, (5) Developing. In a "positive" lithography process, the portions of the photoresist that were exposed to light become soluble in the developer and are removed from the wafer surface. In a "negative" process, exposed resist becomes insoluble in the developer while the unexposed resist is removed. Figure 2.14 shows the pattern transfer of positive and negative photoresist.

2.5.1.2 Photomasks

Photomasks consist of a glass plate (either fused silica or soda lime glass) covered with a chrome layer. Patterns on masks are normally designed by computer layout programs (e.g. Cadence, L-Edit). The digital data from design files then drives a pattern generator. These machines often employ electron-beam writing (see section below) to open up windows in an electron sensitive chemical coated over the photomask's chrome layer. Exposed portions of chrome are then chemically removed to reproduce the pattern drawn on the computer (Reynolds 1979). Photomasks can be commercially obtained from a number of vendors with a partial list available at the following link: http://www.ee.byu.edu/cleanroom/commercial.phtml.

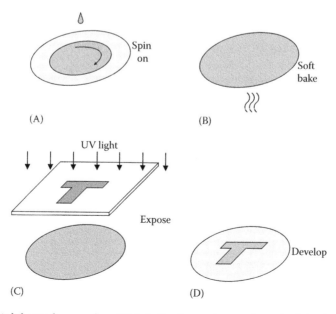

FIGURE 2.13 Basic lithography procedure. (A) Spin the photoresist onto the wafer; (B) soft bake; (C) expose; (D) develop pattern in differentiating chemical.

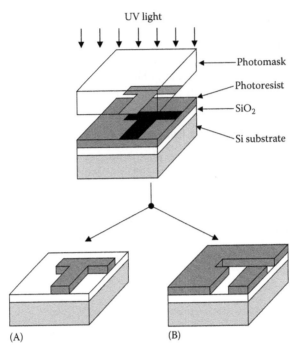

FIGURE 2.14 Details of lithographic pattern transfer process. Pattern transfer with (A) positive photoresist and (B) negative photoresist.

2.5.1.3 UV Exposure Systems

Exposure systems perform two functions: positioning a photomask above a wafer and illuminating the wafer with UV light. The light shines through the photomask, which blocks it in some areas and lets it pass in others. UV exposure systems can be classified by the optics that transfer the image from the mask to the wafer. Contact exposure, the simplest type of exposure system, puts a photomask in direct contact with the wafer and exposes it to uniform, collimated light. Proximity exposure puts a small gap between the photomask and wafer. In both cases, the mask covers the entire wafer, and simultaneously patterns multiple device designs or "die."

Projection exposure systems are common in high-volume microelectronic manufacturing. Projection masks (known as "reticles") contain the design for only one die or a small array of die (known as a "field"). The pattern on the reticle is reduced through a lens (often by 10 times) so that the features projected on the wafer are smaller than those on the photomask. Steppers are a sort of projection exposure system that repeats this exposure many times across a wafer's surface.

For most contact and proximity aligners, alignment between a wafer and a photomask is done manually using microscopes and positioning micrometers. Experienced operators can align patterns from a photomask to patterns on a wafer with around $0.25\,\mu m$ accuracy across a wafer. Most steppers provide automated alignment of wafer and mask using interferometers and machine vision. Alignment accuracy in this case can exceed $0.10\,\mu m$.

2.5.1.4 Achievable Linewidth

The ability of a lithography process to resolve narrow lines is always limited by the wavelength of the light used for illumination. For contact aligners, the smallest feature, or critical dimension (CD), that can be achieved is given by the equation

$$CD = \sqrt{\lambda g},\qquad(2.1)$$

where λ is the wavelength of light and g the spacing between the photomask and wafer. This limits CD for these types of aligners to about 0.5 μm, although in practice 1.0 μm is usually more typical.

For projection aligners, CD follows the equation

$$CD = k\frac{\lambda}{NA},$$ (2.2)

where

  k is a process dependent factor
  λ is the wavelength of the light
  NA is the numerical aperture of the lenses used for image reduction

Obviously, lowering the wavelength of light used in the illumination system increases the resolving power of the stepper. Lines as low as 32 nm are being resolved by current mass-production capable steppers using argon-fluoride (ArF) excimer lasers that emit light with a wavelength of 193 nm.

2.5.2 E-Beam Lithography

Electron beam (e-beam) lithography is the practice of scanning a beam of electrons in a patterned fashion across a wafer or mask surface covered with electron resist. A schematic of an e-beam lithography system is shown in Figure 2.15 (Sze 2001). The primary advantage of e-beam lithography is that it can achieve higher resolution and greater depth of focus than optical lithography because it can beat the diffraction limit of light and directly write features in the nanometer size range. The disadvantage of e-beam lithography is low throughput, since patterns are written serially across a wafer's surface. An approximate writing speed for an e-beam system is 10 wafers/h with 0.25 μm line resolution.

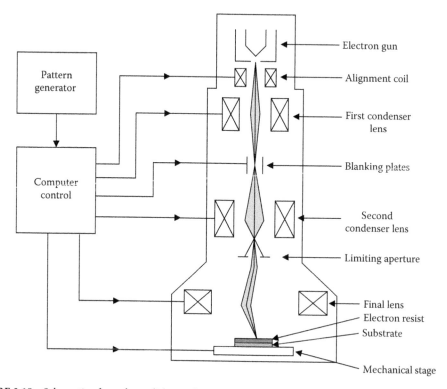

FIGURE 2.15 Schematic of an e-beam lithography system.

Condenser lenses are used to focus the e-beam to a spot size of 10 nm or smaller in diameter, consequently e-beam lithography resolution can approach 10 nm. The scan field of e-beam lithography is typically 1 cm by 1 cm. For research applications, it is very common to convert a scanning electron microscope (SEM) into an e-beam lithography system using a relatively low-cost accessory. Such converted systems can produce linewidths on the order of 30 nm or smaller.

2.5.3 Direct Laser Patterning

Direct laser patterning (DLP) is a way to structure photosensitive materials in three dimensions (Deubel et al. 2004, Christopher et al. 2006). It is a very popular form of optical maskless lithography and makes use of multiphoton absorption (MPA). MPA occurs when two or more photons, none of which is energetic enough to cause an electronic excitation on its own, are absorbed simultaneously to cause such an excitation. The absorption probability is proportional to the laser intensity and the number of photons absorbed. Thus, for a focused laser beam, efficient MPA occurs only in the focal spot, leading to a local polymerization of the targeted layer material. 3D structures can be written into photosensitive material like SU-8, Ormocere, PDMS, and chalcogenide glass.

Laser engraving is a practice that uses lasers to engrave or mark an object. A computer system is often used to control the movements of the laser beam and very precise engravings can be achieved at high rates on materials such as plastics and metals. Because light is utilized, laser engraving is a non-contact engraving method. In most systems, the workpiece is stationary and the laser moves around in the X and Y directions drawing vectors (Figure 2.16).

2.5.4 Nanoimprint Lithography

Nanoimprint lithography was recently introduced as a cost effective way to do very high resolution pattern transfer (Chou et al. 1996). It has two basic steps. The first is the imprint step in which a mold with nanostructures on its surface is pressed under a controlled pressure into a thin resist cast on a substrate. This step duplicates the nanostructures on the mold in the resist film. Silicon dioxide, silicon, or other materials such as metal and ceramics can all be used as mold materials. When heated up above the glass transition temperature of the polymer, the pattern on the mold forms into the softened polymer film. PMMA is one of the primary films used for this process. After being cooled down, the mold is separated from the sample and the pattern resist is left on the substrate (Figure 2.17).

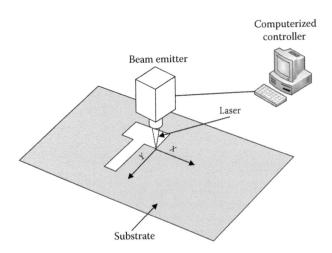

FIGURE 2.16   Laser engraving on a flat sheet.

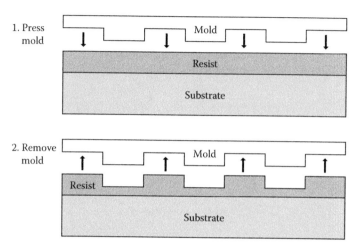

FIGURE 2.17 Basic steps of nanoimprinting.

Electrochemical nanoimprinting can also be achieved using a stamp made from a superionic conductor such as silver sulfide (Hsu et al. 2007). When the stamp is contacted with metal, electrochemical etching can be carried out with an applied voltage. The electrochemical reaction generates metal ions that move from the original film into the stamp. Eventually all the metal is removed and the complementary stamp pattern is transferred to the remaining metal. To ensure the uniformities of pressure and pattern for full wafer nanoimprint processes and to prolong the mold lifetime, a pressing method utilizing isotropic fluid pressure, named air cushion press (ACP) by its inventors (Gao et al. 2006), was developed and is being used by commercial nanoimprint systems.

2.6 Forming Impressions: Etching and Casting

Creating impressions—grooves and holes—in a surface is how physical changes are made to a substrate. These impressions may later be filled with fluid or be used to help guide light in an optofluidic device. Much of impression forming is done through etching of one form or another. Of the examples of etching discussed in this section, all but focused ion beam etching require a previously patterned etch mask if we want the etching to create specific patterns in the surface. Etching using an etch mask is simultaneous, meaning an entire wafer's surface is transformed at once. This section also discusses casting, another very popular way to form impressions that does not modify a wafer directly, but instead places a membrane containing patterned impressions over the wafer's surface.

2.6.1 Wet Chemical Etching

Etching is used to chemically remove layers from the surface of a wafer, typically after an etch "mask" is coated over the wafer and windows opened up in the mask using a micropatterning technique. Etching is critically important to changing the physical structure of the wafer surface, and wafers may undergo many etching steps in the production of a sophisticated microdevice (Chang and Sze 1996).

Wet chemical etching is the simplest type of etching, and essentially consists of immersing a wafer in a chemical solution. The mechanism for wet chemical etching involves three basic steps: (1) transporting reactants by diffusion to the reacting surface, (2) chemical reactions occur at the surface, and (3) the products from the surface are removed by diffusion (Figure 2.18). In many cases, patterned photoresist works well as an etch mask. Other situations require a more durable mask, such as SiO_2 or silicon nitride.

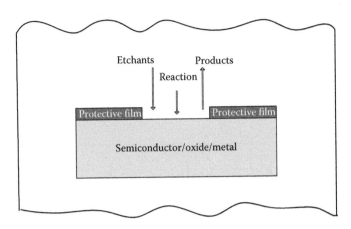

FIGURE 2.18 Basic mechanisms in wet chemical etching with wafer immersed in a solution.

Etchants encountered in microfabrication are usually acidic. For example, a common etch for silicon is a solution of $HF/HNO_3/CH_3COOH$. Glasses like SiO_2 are etched using HF. A solution of $HNO_3/H_3PO_4/HCl$ is used to etch aluminum. Every material has a unique set of etchants it will be susceptible to. A collection of commonly used materials and etches is found at this link: http://www.ece.byu.edu/cleanroom/wet_etch.phtml.

Except when etching crystals, wet chemical etching will be isotropic. This means that etchant removes material at the same rate in all directions, leading to etching underneath a mask edge and a rounded etch profile. Anisotropic etching means the etchant removes material at different rates for different directions in relation to the wafer surface. Some anisotropic etch profiles can look perpendicular to the surface (Sze 2001). Figure 2.19 shows the comparison between an isotropic and an anisotropic etching profile.

2.6.2 KOH Etching of Silicon

The mostly widely known example of anisotropic wet etching occurs when KOH is used to etch silicon. This is often done in the production of MEMS structures, because the etch produces smooth surfaces

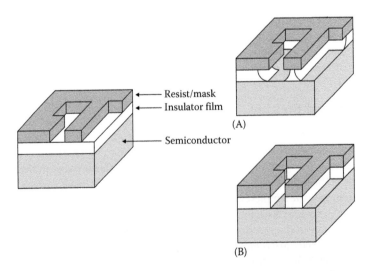

FIGURE 2.19 Comparison of etching profiles. (A) Isotropic etching; (B) anisotropic etching.

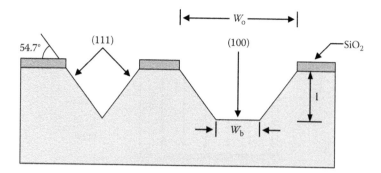

FIGURE 2.20 KOH etching of silicon through window patterns on (100) plane.

and can be very fast. The thru-wafer holes used to produce ink-jet heads are made using KOH etching. The unique etching profiles are made possible because in the silicon crystal lattice, the (111)-plane has more available bonds per unit area than the (110)- and (100)-planes; therefore the etch rate is slower for the (111)-plane. Orientation dependent etching of <100>-oriented silicon through a patterned silicon dioxide mask creates precise V-shaped grooves, the edges being (111)-planes at an angle of 54.7° from the (100)-surface. If the window of the mask is large enough or the etching time is short, a U-shaped groove will be formed (Figure 2.20). The width of the bottom surface (W_b) is given by

$$W_b = W_0 - \sqrt{2}l, \tag{2.3}$$

where
  W_0 is the width of the window on the wafer surface
  l is the etched depth (Bean 1978)

2.6.3 Plasma Etching

Plasma etching involves a high-speed stream of plasma of an appropriate process gas mixture impacting the surface of a sample. During the process, a wafer or die is placed in the plasma etcher, and the air is evacuated from the process chamber. A process gas is introduced at low pressure, and is excited into plasma using an RF voltage at 13.56 MHz and a few hundred watts. The types and amount of gas used depends upon the etch process. Low pressures (1–1000 mTorr) are maintained in the process chamber. Energetic free radicals in the plasma react with a wafer's surface, effectively removing exposed material.

Reactive ion etching (RIE) is one form of plasma etching (Sze 2001). RIE relies on specific electrode geometries to cause self-biasing of the plasma. Self-biasing sets up an electric field that accelerates ions toward a process wafer. Due to the mostly vertical delivery of reactive ions, RIE can produce very anisotropic etch profiles, which contrast with the typically isotropic profiles of wet chemical etching. Figure 2.21 shows a schematic of an RIE chamber.

In most RIE processes, gases are chosen so that the ion radicals chemically react with the etch surface without reacting significantly with an etch mask. Popular process plasmas often contain small molecules rich in chlorine or fluorine. For instance, carbon tetrachloride (CCl_4) etches silicon and aluminum, and trifluoromethane (CF_4) and sulfur hexafluoride (SF_6) etch silicon dioxide and silicon nitride. A plasma containing oxygen is used to oxidize photoresist and other polymers and facilitate their removal.

Advancements to basic RIE systems have been developed to decrease etch times or increase the anisotropic nature of the etching. One example is inductively coupled plasma (ICP) RIE. In this type of system, the plasma is generated through electromagnetic induction with an RF powered magnetic field.

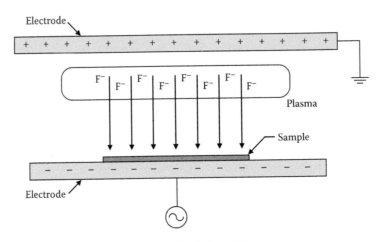

FIGURE 2.21   Schematic of RIE utilizing fluorine radicals for etching.

Very high plasma densities can be achieved, increasing etch rates. Electron cyclotron resonance (ECR) plasma etching is another type of RIE that also provides higher ion density using a separately controlled RF power source that controls the ion energy independent of the ion density.

Ion milling, or sputter etching, is another method for plasma etching. It bombards a wafer with energetic ions of noble gases (working at low pressures ~10^{-4} Torr), often Ar^+, which knock atoms from the substrate by transferring momentum. Because the etching is performed by ions, which approach the wafer approximately from one direction, this process is highly anisotropic. On the other hand, it tends to display poor selectivity, meaning etch mask and substrate material are both etched at equal rates during the process.

2.6.4 Focused Ion Beam Etching

Focused ion beam (FIB) etching is a technique used particularly in the semiconductor and materials science fields for site-specific analysis, deposition, and ablation of materials (Seliger et al. 1979, Orloff et al. 2003, Giannuzzi and Stevens 2004). An FIB resembles an SEM. However, while an SEM uses a focused beam of electrons for imaging, an FIB instead uses a focused beam of ions (Figure 2.22) for milling (Shul et al. 1997). Gallium ions are most widely used because it is easy to build gallium ion sources.

FIB is often used for maskless patterning to patch or modify an existing device. For example, in an integrated circuit, the gallium beam could be used to cut unwanted electrical connections, or to deposit conductive material in order to make a connection. FIB is also very useful for milling nanoscale holes or building up nanoscale structures. FIB is usually done in the presence of an SEM beam so that milling or deposition operations can be imaged in real time. The process is serial in nature, however, so is typically reserved for prototyping operations.

2.6.5 Casting

The casting of silicone-based elastomers has been widely used in optofluidics and lab-on-a-chip research (Kim et al. 1995). Polymer casting for a miniaturized separation device was reported as early as 1990 (Becker and Gärtner 2000) when a cast silicone rubber was used to form microchannels. This elastomeric layer was then placed between two glass plates for channel sealing and mechanical support. Casting generally offers flexible and low-cost access to planar microchannel structures. PDMS (Jo et al. 2000) is the most used material. Figure 2.23 shows a schematic of the casting process for PDMS. In this procedure, a mixture of elastomer precursor and its curing agent is poured over a molding template.

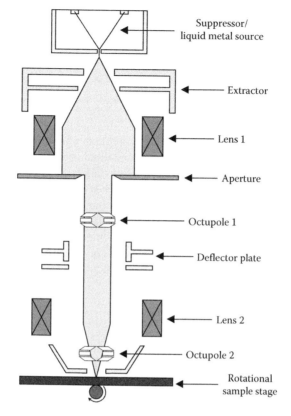

FIGURE 2.22 Schematic of a focused ion beam system.

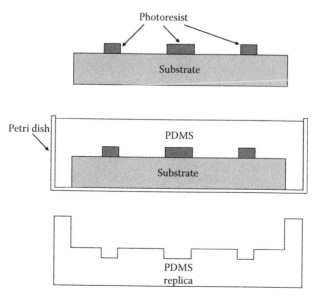

FIGURE 2.23 Schematic of the casting process using PDMS.

Templates are typically made by lithographically patterning a photoresist layer (Hosokawa et al. 1998), by silicon surface micromachining (Effenhauser et al. 1997, 1996) or by photostructuring a printed circuit board (Fielden et al. 1998, Baldock et al. 1998). These templates may be surface modified for a better mold release. After curing, the soft elastomer microstructures can be simply pulled from the mold and placed on a planar surface (glass, plastic, printed circuit board) to form closed channels.

2.7 Encapsulation

Encapsulation is a microfabrication topic uniquely important when fluid-handling chips are being made. To properly handle fluids, we must create enclosed tubes with lengths significantly longer than their diameter. Encapsulation processes used so far can be classified into two major types: sandwiched channels and surface machined channels. To form sandwiched channels, grooves are formed in a thick solid substrate or flexible polymer. The grooved substrate or polymer is then covered by a solid substrate and a bond is created. The grooves acquire a roof and a tube is formed. Wafer bonding and PDMS casting can be classified as sandwiching processes. Surface machined channels typically rely on sacrificial etching and/or conformal film deposition to form channels on the surface of a substrate. All forms of encapsulation have been demonstrated to form complex fluidic networks, and this section provides particulars on how they are executed.

2.7.1 Wafer Bonding

The general wafer bonding method is a versatile technique that allows many different materials with different surface profiles and functional characteristics to be joined together to form unique structures. The wafer bonding process can be performed in many different ways by varying temperatures, the presence of adhesion layers, the amount of pressure, and the presence of an electric field. For example, wafer bonding can be done at room temperature, low temperature (<100°C), or high temperature (>100°C). Wafer bonding can be done directly, where no intermediate adhesion layer is used, or it can be done indirectly, where an intermediate adhesion layer is used. Bonding can be started by applying pressure, mechanical forces, molecular attractive forces, or electrostatic forces. Figure 2.24 shows that generally the procedure starts with two wafers: a device wafer and a handle wafer. Several factors related to the surface condition will determine whether bonding produces a good seal: surface roughness, surface flatness, particles on the surface, localized surface protrusions, localized absence of a sufficient density of bonding species, and trapped air pockets.

Surface roughness and flatness are a function of how the wafers were polished. Pure mechanical polishing allows a high degree of smoothness (mean roughness <1 Å) and flatness (~0.05 μm) for silicon and silica glass substrates up to 200 mm in diameter (Haisma et al. 1989), but it produces a significant amount of subsurface damage (defects in the silicon lattice ~20 μm in depth).

Chemomechanical polishing often follows mechanical polishing to remove surface damage. Wafers are mounted on a flat plate and an external pressure is applied to press the wafer against a polishing pad that is moved over the wafer surface. A polishing slurry composed of a colloidal dispersion of silica (SiO_2) in an aqueous solution of potassium hydroxide is dropped onto the pad (Kern and Puotinen 1970, Pietsch et al. 1994). The process is illustrated in Figure 2.25. On silicon wafers, the polishing first removes native oxide from the surface and then layers of silicon. Typically, a silicon layer of ~25 μm has to be removed by the polishing process to create a damage-free, smooth surface. The mean microroughness of the resultant silicon surface is typically in the range of 1–2 Å.

Hydrogen-peroxide-based "RCA" wet cleans are most commonly used in wafer bonding of silicon and glass wafers to remove particles on surfaces (Meerakker and Straaten 1990). Oxygen plasma and UV ozone cleanings are also effective in removing organic contaminants (Tong and Gosele 1999). Both cleanings are particularly attractive when components such as metal electrodes are present since metals will be removed in RCA cleaning solutions.

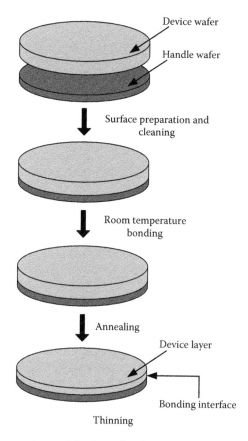

FIGURE 2.24 Step by step processing used for thermal wafer bonding.

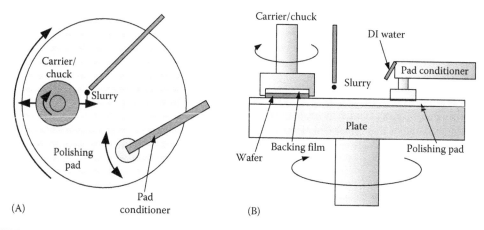

FIGURE 2.25 Schematic of the chemomechanical polishing process. (A) Top view of a polishing machine. (B) Side view of a polishing machine.

After the surface is sufficiently cleaned, bonding proceeds when the two wafers are brought into physical contact by applying a small amount of mechanical force or pressure (Lasky 1986). At this point, the two wafers will bond to each other due to the presence of molecular attractive forces. To increase the bonding strength, several techniques can be used as described below.

2.7.1.1 Thermal Bonding

In the thermal bonding process, wafers are annealed at temperatures sufficient to cause significant molecular diffusion so that covalent bonds can be created at the wafer-to-wafer interface. When working with silicon wafers, bonding usually occurs between SiO_2 layers grown on both surfaces. Thermal bonding of silicon typically takes place at over 800°C and temperatures can be much higher when bonding quartz wafers. Borofloat glass wafers are attractive for thermal bonding because they require much less heat (~400°C). One important consideration with this method is matching thermal expansion coefficients between materials. If this is not done, there is danger of stress cracking and bowing once the bonded wafers have cooled.

2.7.1.2 Solvent Bonding

In solvent bonding, wafers are placed together in the presence of a solvent that dissolves or melts the material. As the solvent evaporates, the surfaces of the wafers solidify together, forming a strong bond. An example of a solvent bonded system would be PMMA substrates with surfaces melted by acetone.

2.7.1.3 Epoxy and Eutectic Bonding

Epoxy and eutectic bonding uses an intermediate layer as glue between the substrates. For epoxy bonding, a layer of epoxy that may be patterned is applied to a substrate. The epoxy is heated and pressure is applied as the epoxy and second substrate are brought into contact. The wafer is allowed to cool and the epoxy bonds both substrates. A number of epoxies (Weckwerth et al. 1996), UV-curable epoxies (Nguyen et al. 2000), and photoresists can be used for bonding. SU-8 is used in many microfluidic applications for both spacer and adhesive layers (Ko 1985). Using polymers as an intermediate layer is advantageous where low process temperatures are required, such as devices that have aluminum structures in them or where substrate materials other than silicon are used.

Eutectic bonding is a common packaging technique in electronics. Eutectic bonding is commonly done using gold as preform or intermediate layer. Gold forms a eutectic alloy with silicon (composition 18.6 ± 0.5% Si in Au) at the relatively low temperature of 363°C. In order to bond wafers, one wafer is coated with a thin film of gold. After the wafers are sandwiched together, they are heated to a temperature a little higher than 363°C. At the eutectic point, the liquid phase of the eutectic Au/Si alloy is formed, gold dissolves in silicon, the eutectic point shifts to a higher temperature, and the solid-state phase of the gold–silicon bond is formed (Wolfenbuttel 1997).

2.7.1.4 Anodic Bonding

Anodic bonding is a simple and effective process to permanently bond a silicon wafer to a glass substrate. The bonding process is performed at a temperature ranging from 300°C to 500°C, with an applied voltage of 500–1000 V that creates an electric field that penetrates the substrates (Nguyen and Wereley 2002). The anode of the power supply is connected to the silicon wafer and the cathode to the glass wafer as illustrated in Figure 2.26. The bonding duration is dependent on the temperature and the voltage but is usually on the order of 10–15 min (Obermeier 1995).

The physical mechanism behind anodic bonding relies on the sodium ions contained within the glass, which at elevated temperature are displaced from the bonding surface of the glass by the applied electrical field. The depletion of sodium ions near the surface of the glass makes the surface highly reactive with the silicon surface, forming a solid chemical bond (Knowles and Helvoort 2006).

A good match of the thermal expansion coefficients between the silicon and glass is necessary to avoid buckling after bonding and cooling. Glasses such as Corning 7740 (Pyrex), Corning 7750, Schott 8392, Schott 8330, and Hoya SD-2 (Cheng 2001) are suitable matches with silicon. Because of the relatively low temperatures employed in anodic bonding, the bonding interface is usually free of gas bubbles as gas entrapment does not occur readily at these temperatures (Albaugh and Cade 1988).

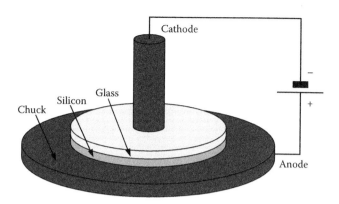

FIGURE 2.26 Illustration of the anodic bonding processing between glass and silicon substrates.

2.7.2 Sacrificial Etching

Sacrificial etching has become an attractive method for making micro- and optofluidic devices because they are easily fabricated using readily available silicon processing technology, and the technique works with a wide variety of substrates. The process involves a sacrificial material defined by photolithography and patterned on a substrate as shown in Figure 2.27A, then it is followed by the deposition and patterning of a conformal overcoat thin film shown in Figure 2.27B. The sacrificial material should ideally provide a mechanically rigid and chemically reliable support for the conformal overcoat layer under many different processing conditions (Barber et al. 2005). Many different polymers work well as sacrificial materials such as negative and positive photoresist, and even metals if the channels are thin. A partially removed sacrificial material layer is shown in Figure 2.27C. The sacrificial material is removed selectively, usually using either an acid or solvent solution. The disadvantage of the sacrificial etching method is that it can be time consuming. Because fresh etchant must diffuse to the sacrificial material, etch rates decrease with distance etched. The microfluidic channel with the sacrificial material layer completely removed is shown in Figure 2.27D.

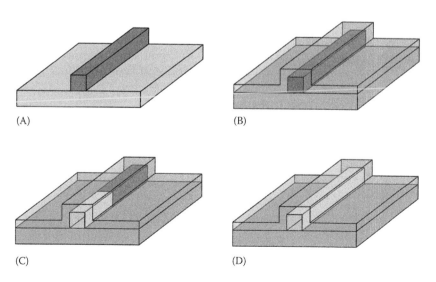

FIGURE 2.27 Sacrificial etching process used to create microfluidic channels. (A) Photoresist patterned on substrate; (B) conformal coating over the photoresist; (C) partially removed sacrificial photoresist core; (D) fully removed sacrificial photoresist core.

2.7.3 Shadow Deposition

Shadow deposition is the process whereby microfluidic channels may be created and sealed using CVD methods. Figure 2.28A shows two beams that have been etched to form the sides of a trench. Material is conformally deposited over the beams using PECVD or LPCVD (Liu and Tai 1999) as shown in Figure 2.28B. Finally, as the CVD material reaches a critical thickness the channel is sealed as shown in Figure 2.28C. This procedure can work for channels of many dimensions, but generally cavities with the smallest gaps are more easily and reliably sealed.

2.7.4 PDMS Covers

Encapsulated channels made from PDMS are very popular because of the easy accessibility of material, rapid fabrication, and desirable performance aspects. For example, PDMS has a low interfacial free energy, which avoids molecules of most polymers sticking on or reacting with its surface, is stable against humidity and temperature, is optically transparent and can be cured by UV light, can attach on nonplanar surfaces, and is mechanically durable (Xia and Whitesides 1998).

PDMS material can be obtained in viscous liquid precursor form from many vendors under various trade names, such as Sylgard Silicone Elastomer from Dow Corning and RTV silicone from GE Silicones.

The precursor materials consist of two parts: the base and curing agent. The two parts are mixed and then cured at room temperature, in a vacuum, or under elevated temperatures for a rapid cure. Under the recommended mixing ratio, this results in a thermoset, transparent elastomeric solid (Duffy et al. 1999).

To create three-dimensional features using PDMS, a casting process is used, as described in Section 6.5. The uncured precursor is poured over molds with three-dimensional patterned features that can be made in many different ways, for example in Figure 2.29A and B, photoresist is used. Once the PDMS is removed, it forms an inverted duplicate of the features that it was covering as shown in Figure 2.29C. The PDMS materials can then be bonded to a substrate to form an enclosed channel as shown in Figure 2.29D. The substrate can be silicon, glass, polyimide sheet, or another piece of PDMS. It is possible to integrate more than two layers of PDMS and form three-dimensional microfluidic circuits with complex three-dimensional channel geometries (Chiou et al. 2002). The term "soft lithography" has become synonymous with the PDMS molding and bonding process.

PDMS fabrication also has challenges, such as volume change and elastic deformation. When making PDMS parts it should be taken into consideration that the PDMS will shrink a certain amount depending on the materials used, materials poured, and the curing technique. Alternately, many organic solvents can cause PDMS to swell (Juncker et al. 2001). Elastic deformation can also limit the aspect ratio of PDMS structures. If the aspect ratio is too high, sagging and bowing results in regions that do not contact the substrate, which makes additional soft lithography steps impossible. The recommended aspect ratios for PDMS structures are between 0.2 and 2 (Xia and Whitesides 1998).

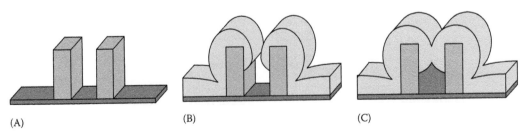

(A)   (B)   (C)

FIGURE 2.28 Shadow deposition process for encapsulating channels. (A) Narrow trenched is formed in a substrate's surface; (B) CVD SiO$_2$ is deposited conformally over the trenches; (C) SiO$_2$ deposition creates a channel by sealing the trench.

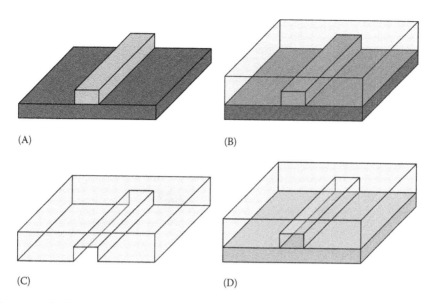

(A)　　　　　　　　　(B)

(C)　　　　　　　　　(D)

FIGURE 2.29 Soft lithography process using PDMS. (A) Photoresist is patterned on substrate; (B) PDMS is poured over the photoresist and cast; (C) PDMS mold is pulled from the substrate; (D) PDMS mold is bonded to alternate substrate.

2.8 Polishing and Packaging

Most optofluidic devices are still in the embryonic stage, therefore, they do not need the sophisticated packaging required for microelectronic chips. Even in test and development, though, some degree of packaging is required. This may be as simple as removing single die from a wafer, but may also include reservoir attachment and die polishing—the topics expanded upon in this section. Of course as the field develops further and optofluidics are pushed into the consumer market, more aspects of packaging will have to be considered such as providing compact interfaces between electronic, optical, and fluid sources.

2.8.1 Dicing

Wafer dicing is the process whereby individual silicon chips or ICs are separated from a larger silicon wafer upon completion of processing. Dicing is done with a dicing saw containing a diamond-coated blade. Materials that can be diced include glass, alumina, silicon, ceramics, and compound semiconductors. The process is accomplished by mounting a wafer to dicing tape that is sticky on the back and holds the wafer to a thin sheet-metal frame. The diamond saw then cuts through the wafer at a depth so that it does not penetrate through the dicing tape. Straight line cuts are made over the entire wafer, separating individual devices or die as shown in Figure 2.30. Die can be created in any shape generated by straight lines, and are typically rectangular or square shaped. The size of the die left on the tape may range from 35 to 0.5 mm².

2.8.2 Cleaving

Crystalline substrates can be cleaved along crystal planes to section up wafers into die. Usually wafers are cleaved along the crystalline direction (110) by placing the wafer on sticky mounting tape and scribing indentations on the edges of the wafer with a scriber. By pressing down firmly on the edge of the wafer, it should break in a straight line along the crystalline direction. For (100) wafers, cleaving produces

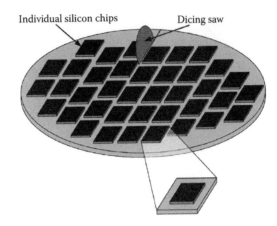

FIGURE 2.30   Wafer dicing to extract die from a substrate.

square or rectangular die. The cleaved facets are usually smooth and have the qualities of a mirror with a reflectivity of approximately 30%, which is determined by the index of the substrate material and air. Having a smooth edge (that does not need subsequent polishing) is very advantageous if the die is to be interfaced with optical elements (optical fiber launching light into on-chip waveguides).

2.8.3 Edge Polishing

Edge polishing has a significant effect on the optical performance of waveguides on die that do not have mirror smooth edges. Materials such as fused silica, silicon, silicon/oxide hybrids, sapphire, pyrex, phosphate glass, LiN, and LiTaO$_3$ can be used as substrates and can be polished at many different angles. Normally, edges that are diced (and even some that are cleaved), are left with rough or chipped edges. Some of the challenges of optical edge polishing involve poor flatness, excessive rounding, and compensating for the height differential that occurs during polishing hybrids of silicon and oxide layers and also waveguide structures filled with epoxy or photoresist. However, by using edge polishing it is possible to achieve an optical quality edge surface finish with 1–1.5 nm variation and an edge plane orientation of ±0.1°. Edge polishing is a serial process, meaning each die face must be polished individually, which can require a great deal of time and handling, downsides that are not very attractive for mass production.

2.8.4 Reservoir Attachment

For optofluidic devices that will handle liquids, it is often desirable to have an on-chip fluid reservoir capable of holding a significant volume of liquid (on the order of microliters). The small amounts of liquids contained in microchannels can evaporate quickly if they are not replenished. When channels are made using wafer bonding or PDMS soft lithography, reasonably sized reservoirs can be made by etching or punching through the encapsulating material. If these "thru hole" reservoirs are not available, or a larger storage volume is desired, an external reservoir is often attached to chips. The easiest means of attachment is gluing or epoxying a cylindrical tube to the chip's surface over the exit of a microchannel. Epoxies offer the advantage of a good seal by filling the gap between the external reservoirs and the device opening. The surface where the reservoir is to be attached can be roughened to improve the adhesion. More sophisticated reservoirs can also be purchased from companies like Upchurch and LabSmith. These reservoirs are glued to the chip and have threaded attachments so that capillary tubing can be directly interfaced with them.

Temporary fluid reservoirs can also be provided by microfluidic probes, like the Microports offered by Cascade Microtech. These probes allow a quick and easy interface for delivering fluids and electrical fields to microfluidic devices. Precise positioners allow for positioning temporary sealing reservoirs above a channel opening, and directly interface with capillary tubing. These types of reservoirs are intended for test and evaluation rather than a permanent optofluidic system.

2.9 Additional Microfabrication Examples

In this section we present two representative examples of microfabricated optofluidic devices. They are both more sophisticated than the conceptual device described in Section 2.2 and are presented so that readers can appreciate the diversity of fabrication methods and how they are applied. For each example, we will point out how individual processes fit within our five defined classes that were described in Sections 2.4 through 2.8: (1) film formation, (2) micro- and nanopatterning, (3) forming impressions, (4) encapsulation, and (5) polishing and packaging.

2.9.1 Fluorescence Sensor Utilizing Anti-Resonant Reflective Optical Waveguides

Hollow waveguides, capable of guiding light through air or liquids, can be constructed on-chip using the anti-resonant reflective optical waveguides (ARROW) principle. For more details regarding the physics of ARROWs, see Chapter 16. ARROWs manifest themselves as hollow microchannels surrounded by alternating layers of dielectrics (oxides or nitrides) with different refractive indices. ARROWs can be constructed into fluidic networks and be interfaced with solid waveguides to direct light and liquids across a chip. In the example presented here, ARROWs are constructed on a silicon wafer and the intended application is a fluorescence sensor capable of exciting and collecting light from particles suspended in a test fluid (Yin et al. 2007).

Figure 2.31 illustrates the first steps in the fabrication process. Intact, 100 mm diameter silicon wafers were used, but the figure shows the view of what happens to a single die. Alternating films of SiO_2 and SiN are deposited over the silicon using PECVD. Film thickness and refractive indices are controlled precisely in order to achieve optical guiding. This step represents Film Formation. The illustration is not meant to be to scale. For instance the silicon substrate is actually 500 µm thick while the PECVD films are on the order of 200 nm thick.

Figure 2.31 A also shows a line of SU8 patterned over the wafer's surface. SU8 is spun on and is photosensitive, so patterning utilized photolithography (micro- and nanopatterning step). Real SU8 features were 12 µm wide and 5 µm tall. The entire length of the Z-shaped line was approximately 1 cm. Again, given the real dimensions, the drawing of the SU8 feature is not meant to be to scale and is very compressed compared to the real length.

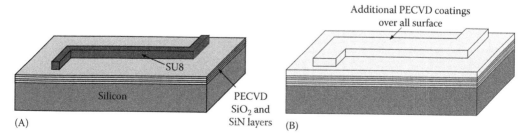

FIGURE 2.31 First steps of ARROW based sensor fabrication. (A) PECVD film deposition and lithographic patterning of SU8. (B) Additional SiO_2 and SiN films are added to the wafer's surface, coating all features. Topmost layer is SiO_2.

Further PECVD layers are then applied over the chip's surface, again alternating between SiO_2 and SiN. These films coat the top and sides of the SU8 line. The process relies on the fact that the films are conformal so there is good coverage of all surfaces. SU8 was used for creating features because it can withstand the PECVD temperatures (250°C) without liquefying or changing shape. This is the beginning of the encapsulation process used to create fluidic channels, which will be completed when the SU8 line is removed through sacrificial etching. Figure 2.31B shows how the surface of the wafer is transformed with the conformal top coating.

A photolithography step follows (Figure 2.32A), in which photoresist is spun on the wafer and then patterned to form windows adjacent to the original SU8 line. The patterned wafer is then placed in an RIE system that etches impressions into the SiO_2 on top of the wafer. Figure 2.32B shows the wafer after the photoresist etch mask is removed. This step falls into the forming impressions processing category, and its purpose is to form ridge waveguides. The SiO_2 ridges can guide light through total internal reflection at the SiO_2/air interfaces at their top and sides, and through ARROW waveguiding due to the dielectric stack below them.

Another photolithography and RIE etching step is used to create holes in the SiO_2 and SiN layers coating the ends of the original SU8 line. Opening these holes allows access to the SU8 so that it can be chemically removed. This is accomplished by putting the wafer in a hot bath of a sulfuric acid and hydrogen peroxide solution. The SU8 is dissolved slowly, leaving a hollow interior surrounded by dielectric layers, and thus completing the encapsulation process. Figure 2.33 illustrates the formation of a via hole and subsequent sacrificial etching of SU8.

A sensor chip is further prepared for testing by cleaving its edges so it can be separated from other die on the wafer. Cleaving leaves the edges optically smooth and ready for interfacing with optical fiber or lenses. Simple reservoirs are also added by epoxying cylinders over the access holes at the end of the hollowed-out, Z-shaped line. Figure 2.34 shows a diagram of the completed chip, indicating how fluid can enter the hollowed chamber that will act as a waveguide due to the surrounding ARROW layers. The ridge waveguides can be used to guide light on and off the chip, including to optical fiber as indicated in the drawing. In one application of this sensor, the ridge waveguide intersecting the hollow ARROW channel is used to deliver excitation light for fluorescence measurements. Fluorescing molecules flowing past the waveguide intersection emit signal that is carried by the hollow ARROW toward another ridge waveguide that leads off chip. This configuration is so sensitive even one particle passing the waveguide intersection can be detected (Yin et al. 2007).

This fabrication example relies upon only one substrate (no bonding) and for the most part standard microelectronic processes. One could even conceive of it being done in an existing silicon manufacturing facility. Its challenging aspects include the conformal coating needed to create the sacrificial etched channel (high stress films or bad step coverage produce broken channels) and the long sacrificial etch times required.

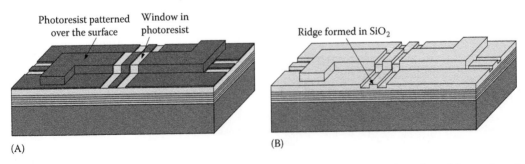

(A)   (B)

FIGURE 2.32 Patterning of ridge waveguide into surface. (A) Photoresist is spun on the wafer and then patterned using photolithography. (B) Impressions are formed in SiO_2 layer using RIE etching. Photoresist etch mask is then removed.

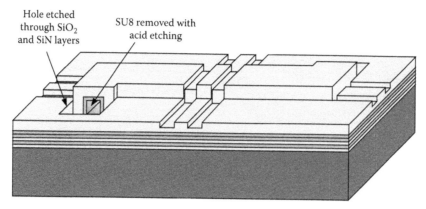

FIGURE 2.33 Photolithography and RIE etching are again used to open holes at the end of the Z-shaped structure. The exposed SU8 is removed using acid etching, producing a hollow structure surrounded by SiN and SiO_2 layers.

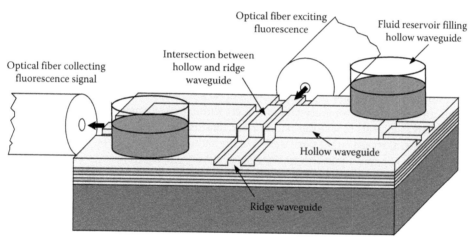

FIGURE 2.34 Implementation of ARROW-based sensor. Hollow waveguides are filled with fluid from attached reservoirs. Light is launched into and collected from the chip using optical fibers.

2.9.2 Fluidically Tuned Photonic Crystal

Planar photonic crystals are a relatively recent invention that takes advantage of the availability of SOI wafers and new designs for photonic crystal waveguiding. More information on the physics of photonic crystals can be found in Chapter 4. Because photonic crystals have nano-structured holes that can be filled with liquids or gases, they are a natural match to optofluidics. Here we describe the fabrication of a photonic crystal device that interfaces with fluids that dynamically change its optical properties (Erickson et al. 2006). The feature sizes necessary to create this device are an order of magnitude smaller than in our earlier example, so different techniques must be employed. E-beam lithography and PDMS casting are utilized for patterning, RIE etching for forming impressions, and soft lithography for encapsulation.

The first fabrication step involves the nanopatterning of an SOI substrate. E-beam resist is spun on the wafer and then e-beam lithography performed. RIE etching is then done through the wafer's top silicon layer, stopping at the SiO_2 layer below. The holes of the photonic crystal are approximately 250 nm in

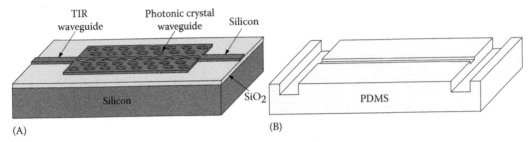

(A)   (B)

FIGURE 2.35 Photonics and fluidics portion of tunable photonic crystal. (A) The photonics layer—photonic crystal waveguide is formed into an SOI wafer using e-beam lithography and RIE etching. (B) The fluid layer— PDMS is cast to form deep and shallow trenches in its surface.

diameter and 200 nm deep. Ridge waveguides approximately 500 nm wide are also formed in the silicon layer. Both types of structures are illustrated after etching in Figure 2.35A, and make up what is termed the photonics layer of the device.

The fluid flow layer of the device is constructed from PDMS using soft lithography. Both shallow and deep trenches are cast into the PDMS as shown in Figure 2.35B. The trenches are intended to direct liquid from an external source over the center line of holes that make up the photonic crystal. The shallow trench meant to fit over the center holes is made 200 nm tall and 350 nm wide.

The cast PDMS is aligned over the silicon wafer using a modified photolithography aligner. The e-beam resist used for patterning is left on the wafer to enhance adhesion to the PDMS layer. Positioning the shallow trenches and photonic crystal center holes within tens of nanometers of each other is nearly impossible with an aligner. To take this into account, there is varied spacing between the multiple photonic crystal patterns etched on the same wafer and the multiple trenches cast on the same PDMS membrane. Within the alignment tolerance of the aligner, at least one set of trenches will match one set of photonic crystal holes. Figure 2.36 illustrates the PDMS layer positioned correctly and attached to the SOI substrate. The trenches in the PDMS become fluid channels and can be filled with liquids of varying refractive index. Index modulations alter the allowed wavelengths that can be guided through the photonic crystal. Additional fluid-handling PDMS layers can also be added to the structure that pump fluid past the crystal, effectively creating a fluid controlled optical switch.

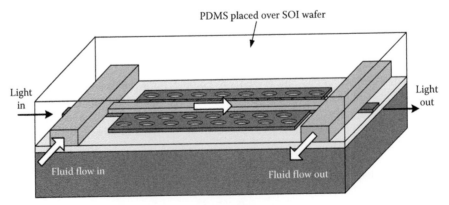

FIGURE 2.36 Liquid tunable photonic crystal device. PDMS is placed over the photonic crystal so that a fluid channel is formed only over the photonic crystal's center holes. Fluid passing over the holes modifies the surrounding refractive index, modulating the wavelength of lights allowed to pass through the waveguide.

2.10 Conclusion and Outlook

Micro- and nanofabrication of on-chip optofluidics can trace its origin to the manufacturing of silicon microelectronics. Although the planar fabrication principles are similar, optofluidics has fostered implementations all its own. A similar trend has taken place with the related fields of MEMS and integrated photonics. These new implementations are often related to new nonsilicon material system being used for devices. As optofluidics advances, we expect more fabrication innovation through research creativity and further adaptation of techniques developed by the silicon industry.

All chip-based optofluidic devices are still in a developmental phase, with most of them being conceived of and made in university laboratories. As promising devices leave the lab for commercial production, manufacturing cost constraints will become a large factor. In the academic world, these considerations are often relegated to some future roadblock to overcome. Applying interesting physics comes first. But low cost is one of the promises and motivations for optofluidics. As the field matures, devices that get deployed will be the ones that can be made reproducibly at low cost. Those costing thousands of dollars per chip will be remembered only in research papers. As such, we predict that the most enduring microfabrication techniques will also be the ones that lend themselves to low-cost manufacturing. Those resembling IC assembly follow a proven model in this regard.

Another important consideration for future optofluidics is packaging. In this chapter, we touched on packaging only briefly because most devices are only being evaluated in laboratories where packaging considerations are minimal. If these devices are to be a part of commercial analysis or communications systems, however, they cannot rely on *XYZ* positioning stages and handheld syringes. Light sources and detectors must be integrated, perhaps using techniques borrowed from the optoelectronics industry. Fluid interfaces too must be automated and robust. As in the optics industry, we expect packaging to account for the majority of manufacturing costs for on-chip optofluidics.

For all its challenges, one of the inherent advantages of wafer-scale microfabrication is the possibility of integrating directly with silicon circuitry. Combining optofluidics with silicon transistors brings together electronics, optics, and fluids in a tiny package with a long list of performance advantages. So far, work in this area has been modest, perhaps because of the huge gap between massive state-of-the-art silicon foundries and smaller optofluidic research labs. Simply getting transistor populated wafers can be a costly challenge. Wafer scale integration is an area we predict will make important strides, however, especially as more effort is put into introducing marketable μTAS chips.

One effort that may aid the advancement of optofluidics is coordination between research labs as the field matures. This may start with loose affiliations and move to standardizing a set of processes and device designs. Perhaps optofluidics foundries, similar to those established for microelectronics and MEMS, will be created. These facilities have the advantage of consolidating "start-up" costs for device development, although they do limit the scope of fabrication to a set of established steps and materials. Many of these efforts will likely hinge on the commercial success of a set of optofluidic chips. If we can produce the optofluidic equivalent of the MEMS DLP or accelerometer, the future will be very bright.

Acknowledgments

This work was enabled by support from the National Institutes of Health (grants R21EB003430 and R01EB006097) and the National Science Foundation (grant ECS-0528714).

References

Adams, A. C., Alexander, F. B., Capio, C. D., and Smith, T. E. 1981. Characterization of plasma deposited silicon dioxide. *Journal of Electrochemical Society* 128: 1545–1551.

Ahuja, S. and Scypinski, S. 2001. An innovative separation platform: Electrophoretic microchip technology. In *Handbook of Modern Pharmaceutical Analysis*, Jin, L. J., Ferrance, J., Huang, Z., and Landers, J. P., Eds. New York: Academic Press, p. 533.

Albaugh, K. B., and Cade, P. E. 1988. Mechanism of anodic bonding of silicon to Pyrex glass. In *Technical Digest IEEE Solid-State Sensors and Actuator Workshop*, New York, pp. 109–110.

Alferov, Zh. I., Andreev, V. M., and Rumyantsev, V. D. 2004. Solar photovoltaics: Trends and prospects. *Semiconductors* 38: 899–908.

Andrew, R. 1982. A simply constructed high-performance e-beam evaporation source. *Vacuum* 32: 376–377.

Baldock, S. J., Bektas, N., Fielden, P. R. et al. 1998. Isotachophoresis on planar polymeric substrates. In *Proceedings of Micro-TAS'98*, Banff, Canada, pp. 359–362.

Barber, J. P., Lunt, E. J., George, Z. A., Yin, D., Schmidt, H., and Hawkins, A. R. 2005. Fabrication of hollow waveguides with sacrificial aluminum cores. *IEEE Photonics Technology Letter* 17: 363–365.

Bean, Kenneth E. 1978. Anisotropic etching in silicon. *IEEE Transaction on Electron Devices* 25(10): 1185–1193.

Becker, H. and Gärtner, C. 2000. Polymer microfabrication methods for microfluidic analytical applications. *Electrophoresis* 21: 12–26.

Bhattacharya, P. K. and Bhosale, K. S. 2000. Relaxation of mechanical stress in polyimide films by soft-baking. *Thin Solid Films* 290–291: 74–79.

Blech, I. A. and Vander Plas, H. A. 1983. Step coverages simulation and measurements in a DC planar magnetron sputtering systems. *Journal of Applied Physics* 54: 3489–3496.

Chang, C. Y. and Sze, S. M., Eds. 1996. *ULSI Technology*. New York: McGraw-Hill.

Cheng, J. 2001. *Biochip Technology*. Philadelphia, PA: Harwood Academic Publishers.

Cheng, L. Y., Mcvittie, J. P., and Saraswat, C. 1991. New test structure to identify step coverage mechanisms in chemical vapor deposition of silicon dioxide. *Applied Physics Letters* 58: 2147–2149.

Chiou, C. H., Lee, G. B., Hsu, H. T., Chen, P. W., and Liao, P. C. 2002. Micro devices integrated with microchannels and electrospray nozzles using PDMS casting techniques. *Sensors and Actuators B* 86: 280–286.

Chou, S. Y., Krauss, P. R., and Renstrom, P. J. 1996. Imprint lithography with 25-nanometer resolution. *Science* 272: 85–87.

Christopher, N. L., Lim, D., O'Malley, K., Baldacchini, T., and Fourkas, J. T. 2006. Direct laser patterning of conductive wires on three-dimensional polymeric microstructures. *Chemistry of Materials* 18: 2038–2042.

Class, W. H. 1979. Deposition and characterization of magnetron sputtered aluminum and aluminum alloy films. *Solid State Technology* 22: 61–68.

Devine, R. 1990. On the structure of low-temperature PECVD-silicon-dioxide films. *Journal of Electronic Materials* 19: 1299–1301.

DeJule, R. 2009. SOI technology goes mainstream. *Semiconductor International*. March 1, 2009.

Deubel, M., Von Freymann, G., Wegener, M., Pereira, S., Busch, K., and Soukoulis, C. M. 2004. Direct laser writing of three-dimensional photonic-crystal templates for telecommunications. *Nature Materials* 3: 444–447.

Duffy, D. C., Schueller, O. J. A., Brittain, S. T., and Whitesides, G. M. 1999. Rapid prototyping of microfluidic switches in poly (dimethyl siloxane) and their actuation by electro-osmotic flow. *Journal of Micromechanics and Microengineering* 9: 211–217.

Eddy, D. S. and Sparks, D. R. 1998. Applications of MEMS technology in automotive sensors and actuators. *Proceedings of the IEEE* 86: 1747–1755.

Effenhauser, C. S., Bruin, G. I., Paulus, A., and Ehrat, M. 1996. *Analytical Methods and Instrumentation*. In Special issue μTAS'96, Basel, Switzerland, pp. 124–125.

Effenhauser, C. S., Bruin, G. I., Paulus, A., and Ehrat, M. 1997. Integrated capillary electrophoresis on flexible silicone microdevices: Analysis of DNA restriction fragments and detection of single DNA molecules on microchips. *Analytical Chemistry* 69: 3451–3457.

Erickson, D., Rockwood, T., Emery, T., Scherer, A., and Psaltis, D. 2006. Nanofluidic tuning of photonic crystal circuits. *Optics Letters* 31: 59–61.

Fan, L.-S., Tai, Y.-C., and Muller, R. S. 1988. IC-processed electrostatic micro-motors. In *IEEE International Electronic Devices Meeting*, San Francisco, CA, pp. 666–669.

Fasol, G. 1996. Room-temperature blue gallium nitride laser diode. *Science* 272: 1751–1752.

Fielden, P. R., Baldock, S. J., Goddard, N. J. et al. 1998. A miniaturized planar isotachophoresis separation device for transition metals with integrated conductivity detection. In *Proceedings of Micro-TAS'98*, Banff, Canada, pp. 323–326.

Fischer, A., Kurner, H., Kurner, W., and Kurner, P. 2000. Slip-free processing of 300 mm silicon batch wafers. *Journal of Applied Physics* 87: 1543–1549.

Fraser, D. B. 1978. The sputter and s-gun magnetrons. In *Thin film Processes*, Vossen, J. L. and Kern, W., Eds. New York: Academic Press, p. 115.

Gao, H., Tan, H., Zhang, W., Morton, K., and Chou, S. Y. 2006. Air cushion press for excellent uniformity, high yield, and fast nanoimprint across a 100 mm field. *Nano Letters* 6: 2438–2441.

Geschke, O., Klank, H., and Tellman, P. 2004. Glass micromachining. In *Microsystem Engineering of Lab-on-A-Chip Devices*, Petersen, D., Bo Mogensen, K., and Klank, H., Eds. Weinheim, Germany: Wiley-VCH Verlag GmbH & Co. KGaA, pp. 161–162.

Ghiraldelli, E., Pelosi, C., Gombia, E., Friger, C., Vanzetti, L., and Sevda, A. 2008. Growth of dielectric Al_2O_3 films by atomic layer deposition. *Japanese Journal of Applied Physics* 47: 8147–8177.

Ghosh, M. K. and Mittal, K. L. 1996. *Polyimides: Fundamentals and Applications*. Boca Raton, FL: Taylor & Francis.

Giannuzzi, L. A. and Stevens, F. A. 2004. *Introduction to Focused Ion Beams: Instrumentation, Theory, Techniques and Practice*. New York: Springer Press.

Grillot, F., Vivien, L., Cassan, E., and Laval, S. 2008. Influence of waveguide geometry on scattering loss effects in submicron strip silicon-on-insulator waveguides. *IET Optoelectronics* 2: 1–5.

Haisma, J., Spierings, G. A. C. M., Biermann, U. K. P., and Pals, J. A. 1989. Silicon-on-insulator wafer bonding-wafer thinning: Technological evaluations. *Japanese Journal of Applied Physics* 28: 1426–1443.

Harper, C. A. and Advisor, S. 1998. Film deposition techniques and processes. In *Thin Film Technology Handbook*, Barlow, F., Elshabini-Riad, A., and Brown, R., Eds. New York: McGraw-Hill, pp. 1–14.

Harrison, D. J., Manz, A., Fan, Z., Ludi, H., and Widmer, H. M. 1992. Capillary electrophoresis and sample injection systems integrated on a planar glass chip. *Analytical Chemistry* 64: 1926–1932.

Hatails, M. K. and Greve, D. W. 1988. Large grain polycrystalline silicon by low-temperature annealing of low pressure chemical vapor deposited amorphous silicon films. *Journal of Applied Physics* 63: 2260–2266.

He, Y. and Kanicki, J. 2000. High-efficiency organic polymer light-emitting heterostructure devices on flexible plastic substrates. *Applied Physics Letters* 76: 661–663.

Hopper, M. A., Clarke, R. A., and Young, L. 1975. Thermal oxidation of silicon. *Journal of the Electrochemical Society* 122: 1216–1222.

Horng, R. Y., Han, P., Chen, H. Y., Lin, K. W., Tsai, T. M., and Zen, J. M. 2005. PMMA-based capillary electrophoresis electrochemical detection microchip fabrication. *Journal of Micromechanics and Microengineering* 15: 6–10.

Hosokawa, K., Fujii, T., and Endo, I. 1998. Hydrophobic microcapillary vent for pneumatic manipulation of liquid in μ-TAS. In *Proceedings of Micro-TAS'98*, Banff, Canada, pp. 307–310.

Hsu, K. H., Schultz, P. L., Ferreira, P. M., and Fang, N. X. 2007. Electrochemical nanoimprinting with solid-state superionic stamps. *Nano Letters* 7: 446–451.

Hu, S. M. 1984. Thermal oxidation of silicon. *Journal of Applied Physics* 55: 4095–4105.

Huang, Z., Carey, J. E., Liu, M., Liu, M., Guo, X., Mazur, E., and Campbell, J. C. 2006. Microstructured silicon photodetector. *Applied Physics Letters* 89: 033506.

Jackson, K. Y. 1998. *Compound Semiconductor Devices*. Weinheim, Germany: Wiley-VCH Verlag GmbH & Co. KGaA.

Jaeger, R. C. 2002. *Introduction to Microelectronic Fabrication*, 2nd ed. *Modular Series on Solid-State Devices*, Vol. V, G. W. Neudeck and R. F. Pierret, Eds. Upper Saddle River, NJ: Prentice Hall.

Jo, B. H., Van Lerberghe, L. M., Motsegood, K. M., and Beebe, D. J. 2000. Three-dimensional microchannel fabrication in polydimethylsiloxane (PDMS) elastomer. *Journal of Microelectromechanical Systems.* 9: 76–81.

Kern, W. and Puotinen, D. A. 1970. Cleaning solution based on hydrogen peroxide for use in silicon semiconductor technology. *RCA Review* 31: 187–206.

Kim, E., Xia, Y., and Whitesides, G. M. 1995. Polymer microstructures formed by molding in capillaries. *Nature.* 376: 581–584.

Kim, J., Kim, H., and Lee, J. H. 2008. Thin layer laser bonding using spin-on-glass materials. *Applied Surface Science* 254: 6842–6848.

Kim, M. T. 2000. Deposition kinetics of silicon dioxide from tetraethylorthosillicate by PECVD. *Thin Solid Films* 360: 60–68.

Knowles, K. M. and Helvoort A. T. J. 2006. Anodic bonding. *International Materials Reviews* 51: 273–311.

Ko, W. H. 1985. Bonding techniques for microsensors. In *Micromachining and Micropackaging of Transducers*, Fung, C. D., Ed. Amsterdam, the Netherlands: Elsevier, pp. 41–61.

Kominiak, G. J. 1975. Silicon nitride by direct RF sputter deposition. *Journal of the Electrochemical Society* 122: 1271–1273.

Lasky, J. B. 1986. Wafer bonding for silicon-on-insulator technologies. *Applied Physics Letters* 48: 78–80.

Lee, S. K., Kim, T. H., Lee, S. Y., Choi, K. C., and Yang, P. 2007. High-brightness gallium nitride nanowire UV-blue light emitting diodes. *Philosophical Magazine* 87: 2105–2115.

Liu, C. 2006. *Foundations of MEMS.* Upper Saddle River, NJ: Pearson Prentice Hall.

Liu, C. and Tai, Y. C. 1999. Sealing of micromachined cavities using chemical vapor deposition methods: Characterization and optimization. *IEEE Journal of Microelectromechanical Systems* 8: 135–145.

Liu, Y. X., Cui, T. H., Sunkam, R. K., Coane, P. J., Vasile, M. J., and Geoettert, J. 2003. Novel approach to form and pattern sol-gel polymethylsilsesquioxane-based spin-on glass thin and thick films. *Sensors and Actuators B: Chemical* 88: 75–79.

Madou, M. J. 2002. *Fundamentals of Microfabrication: The Science of Miniaturization*, 2nd edn. Boca Raton, FL: Taylor & Francis.

Manz, A., Graber, N., and Widmer, H. M. 1990. Miniaturized total chemical analysis systems: A novel concept for chemical sensing. *Sensors and Actuators B: Chemical* 1: 244–248.

Massoud, H. Z. and Plummer, D. 1985. Thermal oxidation of silicon in dry oxygen growth-rate enhancement in the thin regime. *Journal of the Electrochemical Society* 132: 2685–2693.

Matthews, S. J. 2000. Semiconductor Lasers 2000: The early years: Promise and problems. *Laser Focus World* 36(4): 81–88.

Meerakker, J. E. and Straaten, M. H. 1990. A mechanistic study of silicon etching in NH_3/H_2O_2 cleaning solutions. *Journal of the Electrochemical Society* 137: 1239–1241.

Nguyen, N. and S. T. Wereley. 2002. *Fundamentals and Applications of Microfluidics.* Norwood, MA: Artech House.

Nguyen, H., Patterson, P., Toshiyoshi, H., and Wu, M. C. 2000. A substrate-independent wafer transfer technique for surface-micromachined devices. In *Proceedings of MEMS'00, 13th IEEE International Workshop Micro Electromechanical System*, Miyazaci, Japan, January 23–27, pp. 628–632.

Obermeier, E. 1995. Anodic wafer bonding. In *Proceedings of the Third International Symposium on Semiconductor Wafer Bonding: Science, Technology and Applications*, Hunt, C. E., Baumgart, H., Iyer, S. S., Abe, T., and Gosele, U., Eds. London, U.K.: Electrochemistry Society, pp. 212–220.

Orloff, J., Utlaut, M., and Swanson, L. 2003. *High Resolution Focused Ion Beams: FIB and Its Applications.* New York: Springer Press.

Peterson, K. E. 1979. Fabrication of an integrated planar silicon ink-jet structure. *IEEE Transaction on Electron Devices* ED-26: 1918–1920.

Pietsch, G. J., Higashi, G. S., and Chabal, Y. J. 1994. Chemomechanical polishing of silicon: Surface termination and mechanism of removal. *Applied Physics Letters* 64: 3115–3117.

Rastogi, R. S., Vankar, V. D., and Chopra, K. L. 1987. Simple planar magnetron sputtering source. *Review of Scientific Instruments* 58: 1505–1506.

Reynolds, J. A. 1979. An overview of E-beam mask-making. *Solid State Technology* 22: 87–94.

Riordan, M. 2007. The silicon dioxide solution. *IEEE Spectrum* December: 51–56.

Schmidt, B., Xu, Q. F., Shakya, J., Manipatruni, S., and Lipson, M. 2007. Compact electro-optic modulator on silicon-on-insulator substrates using cavities with ultra-small modal volumes. *Optical Express* 15: 3140–3148.

Schubert, E. F. 2003. *Light-Emitting Diodes*. New York: Cambridge University Press.

Schuegraf, K. K. 1988. Plasma-assisted chemical vapor deposition. In *Handbook of Thin-Film Deposition Processes and Techniques*, Nguyen, V. S., Ed. Park Ridge, NJ: Noyes Publications, pp. 124–132.

Seliger, R., Ward, J. W., Wang, V., and Kubena, R. L. 1979. A high-intensity scanning ion probe with sub-micrometer spot size. *Applied Physics Letters* 34: 310–312.

Settle, M., Salib, M., Michaeli, A., and Krauss, T. F. 2006. Low loss silicon on insulator photonic crystal waveguides made by 193 nm optical lithography. *Optical Express* 14: 2440–2445.

Shul, R. J., Lovejoy, M. L., Word, J. C., Howard, A. J., Rieger, D. J., and Kravitz, S. H. 1997. High rate reactive ion etch and electron cyclotron resonance etching of GaAs via holes using thick polyimide and photoresist masks. *Journal of Vacuum Science Technology B* 15: 657–664.

Sierros, K. A. and Kukureka, S. N. 2007. Tribological investigation of thin polyester substrate for displays. *Wear* 263: 992–999.

Someya, T., Werner, R., Forchel, A., Catalano, M., Cingolani, R., and Arakawa, Y. 1999. Room temperature lasing at blue wavelengths in gallium nitride microcavities. *Science* 285: 1905–1906.

Suzuki, K., Yamashita, M., Ueda, H., and Nakaue, A. 1994. Focused ion beam lithography using ladder silicon spin-on glass. *Journal of Applied Physics* 33: 7033–7036.

Sze, S. M. 2001. *Semiconductor Devices: Physics and Technology*, 2nd edn. New York: John Wiley & Sons.

Tong, Q.-Y. and Gosele, U. 1999. *Semiconductor Wafer Bonding: Science and Technology*. New York: John Wiley & Sons.

Van Kessel, P. F., Hornbeck, L. J., Meier, R. E., and Douglass, M. R. 1998. A MEMS-based projection display. *Proceedings of the IEEE* 86: 1687–1704.

Van Zant, P. 2000. *Microchip Fabrication*, 4th edn. New York: McGraw-Hill.

Vossen, J. L. 1971. Control of film properties by RF-sputtering technologies. *The Journal of Vacuum Science Technology* 8: S12–S30.

Weckwerth, M. V., Simmons, J. A., Harff, N. E. et al. 1996. Epoxy bond and stop-etch (EBASE) techniques enabling backside processing of (Al)GaAs heterostructures. *Superlattices Microstructures* 20: 561–567.

Wilson, R. J. and Weiss, B. L. 1991. The structure of DC magnetron sputtered Al-1%-Si films. *Thin Solid Films* 203: 147–159.

Winser, A. J., Harrison, I., Novikov, S. V. et al. 2001. Blue emission from arsenic doped gallium nitride. *Journal of Crystal Growth* 230: 527–532.

Wolfenbuttel, D. F. 1997. Low-temperature intermediate Au-Si wafer bonding: Electric or silicide bond. *Sensors and Actuators A* 62: 680–686.

Xia, Y. and Whitesides, G. M. 1998. Soft lithography. *Annual Review of Material Sciences* 28: 153–194.

Xiao, S. Y., Che, L. F., Li, X. X., and Wang, Y. L. 2008. A novel fabrication process of MEMS devices on polyimide flexible substrates. *Microelectronic Engineering* 85: 452–457.

Yin, D., Lunt, E. J., Rudenko, M. I., Deamer, D. W., Hawkins, A. R., and Schmidt, H. 2007. Planar optofluidic chip for single particle detection, manipulation, and analysis. *Lab on a Chip* 7: 1171–1175.

Zyung, T., Kim, S. H., Chu, H. Y. et al. 2005. Flexible organic LED and organic thin-film transistor. *Proceedings of the IEEE* 93: 1265–1272.

3
Passive Integrated Optics

Siegfried Janz

3.1 Introduction

Integrated optics seeks to emulate the architecture and varied functionality of integrated electronics on an optical platform. As such, this field encompasses research on all individual optical devices and optical circuits that can be fabricated monolithically on a single chip to carry out complex optical processing functions. The terms "integrated optics" and "waveguide optics" often seem interchangeable, since most integrated optical chips use optical waveguides as the basic building block for both devices and as the optical paths linking different devices on the chip. Many commercially successful devices such as charge-coupled device (CCD) arrays and vertical cavity surface emitting laser (VCSEL) arrays do not use waveguides, yet are obviously highly integrated multicomponent optical devices. Nevertheless, the purpose of this chapter is to provide a brief overview of only waveguide-based integrated optics. Passive waveguide devices are those that require no external power or control signals. An integrated optical chip may contain passive waveguide-based optical splitters and combiners, interferometers, spectrometers, as well as active devices such as modulators, photodetectors, and lasers.

Integrated optics evolved into a well-defined field of its own approximately 30 years ago, its growth and development being paralleled by the development and eventual adoption of optical fiber communication systems. Telecommunications has been the main application and driver for research in integrated optics, and much research is still motivated by applications in long-range communications and short-range data interconnects. Now that the physical principles and fabrication knowledge have been well established by a generation of researchers, integrated optical technology is also creeping into other application areas that include chemical and biological sensing (Lukosz 1991, Luff et al. 1996, 1998, Weisser et al. 1999, Boyd and Heebner 2001, Prieto et al. 2003, Horvath et al. 2005, Densmore et al. 2008, Xu et al. 2008), astronomy (Malbet et al. 1999, Haguenauer et al. 2000), and spectroscopy (Baldwin et al. 1996, Waterbury et al. 1998, Florjañczyk et al. 2007).

Most of the classical optical devices and systems we are familiar with are based on bulk optical elements such as glass lenses, mirrors, and diffraction gratings. Light travels between these elements following the well-known rules of wave propagation through free space or bulk materials. Integrated optics has some compelling advantages over these classical optical systems. First is the small size of even very complex integrated optical components, which are confined to a two-dimensional plane on a chip surface that occupies an area of a few square centimeters, or in some cases even less than a square millimeter in area (Sasaki et al. 2005, Dumon et al. 2006). Second, integrated optical devices are manufactured using the same mass production techniques employed in the electronics industry, so manufacturing costs for complex optical devices can be very low. For example, every DVD and CD drive contains a semiconductor laser—an integrated optical device that costs only a few cents per unit when manufactured in large quantities for the consumer electronics market.

A third important consideration is the degree of flexibility and control the integrated optics platform gives the optical device designer. In classical optics, the designer works with the laws of wave propagation in bulk media or free space to obtain the desired phase and amplitude of the light wave at any point within the system. In integrated optics, however, light can be delivered with any desired amplitude and phase to any point on the chip, limited only by the topological constraints in laying out waveguides on the chip surface. Optical design is more akin to laying out an electrical circuit, since waveguides carry light from one point on the chip to the other, much as metal lines carry electrical current throughout an electronic integrated circuit.

This chapter will attempt to give an overview of the basic waveguide components, operating principles, materials, and applications of passive integrated optics, beginning with a short overview of optical waveguide theory. The subsequent section covers material platforms and outline fabrication processes used for optical waveguides. The remainder of the chapter will focus on basic passive waveguide components that are the building blocks for an integrated optical device, and touch on difficulties and solutions involved in combining many devices on one chip. Where appropriate, approaches to effectively combining waveguide elements with fluids will be highlighted.

3.2 Background

3.2.1 Waveguides

An optical waveguide confines light in one or two transverse dimensions (e.g., x and y) but allows it to propagate along a third axis, which we denote as z. Figure 3.1 shows the simplest example: a three-layer slab waveguide. In a classical ray optics description (Hunsperger 1991), light rays are coupled into one end facet of the waveguide. If the core layer has a higher refractive index than the adjacent cladding layers, light incident on the core-cladding boundary at angles beyond the critical angle will undergo total internal reflection (TIR). One can then imagine that a ray of light undergoing successive TIR at the core

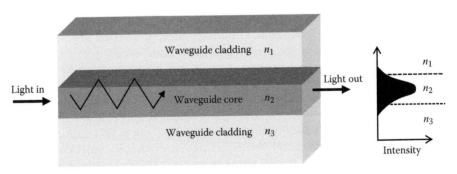

FIGURE 3.1 A three-layer slab waveguide. To form a waveguide, the core index of refraction, n_2, must be larger than both the upper and lower cladding indices, n_1 and n_3. The plot at right shows a typical optical intensity distributions across the three regions of the waveguide.

boundaries will propagate forever along the core layer with no loss. This TIR mechanism is the basis for optical confinement in most optical waveguides, including optical fibers.

It may seem that simple confinement of light to the core region is a sufficient condition for wave-guiding. Nevertheless, a predictive model of waveguide propagation requires the introduction of the waveguide mode concept. As is the case for any wave excitation in a confined system, only discrete solutions or eigenmodes of the wave equation can exist in the waveguide. In a material with isotropic index of refraction n, the Helmholtz wave equations for the optical electric field, **E**, and magnetic field, **H**, vectors

$$\nabla^2 \mathbf{E} = \frac{n^2}{c^2} \frac{\partial^2 \mathbf{E}}{\partial t^2}$$

$$\nabla^2 \mathbf{H} = \frac{n^2}{c^2} \frac{\partial^2 \mathbf{H}}{\partial t^2} \tag{3.1}$$

are obtained directly from Maxwell's equations. Here c is the speed of light in vacuum. Solving these waveguide equations in the multilayer waveguide structure of Figure 3.1 and applying the appropriate electromagnetic boundary conditions yields a set of eigenmode solutions, with each mode (denoted by mode number i) having a unique electric profile, $\mathbf{E}_i(x, y)$ (and corresponding magnetic field, $\mathbf{H}_i(x, y)$, as required by Maxwell's equations) and a corresponding propagation wave vector, β_i, that describes propagation along the waveguide axis:

$$\mathbf{E}_i = \mathbf{E}_i(x, y) e^{i(\beta_i z - \omega t)} \tag{3.2}$$

For the simplest case of the three-layer slab waveguide model in Figure 3.1, the mode profile solutions in each of the three regions have the following general form (Hunsperger 1991):

$$\mathbf{E}_i(x) = A\, e^{-\kappa_1 x} \qquad\qquad \text{upper cladding}$$

$$\mathbf{E}_i(x) = B \cos(hx) + C \sin(hx) \qquad\qquad \text{waveguide core}$$

$$\mathbf{E}_i(x) = D\, e^{\kappa_3 x} \qquad\qquad \text{lower cladding} \tag{3.3}$$

Here the amplitudes A, B, C, and D are determined by the electromagnetic boundary conditions at the various layer interfaces. In the core layer, the electric field profile of mode i has a sinusoidal variation determined by the coefficient

$$h_i = \sqrt{n_2^2 k^2 - \beta_i^2} \tag{3.4}$$

where $k = 2\pi/\lambda$ is the vacuum wave vector of light of wavelength λ. In the cladding layers, the electric fields decrease exponentially with distance from the core layer, with decay constants

$$\kappa_1 = \sqrt{\beta_i^2 - n_1^2 k^2}$$

$$\kappa_3 = \sqrt{\beta_i^2 - n_3^2 k^2} \tag{3.5}$$

The eigenmode solutions defined by Equations 3.3 through 3.5 are distinguished by the possible values that the wave vector β_i can take. These discrete eigenvalues arise in solving Equation 3.3 for the amplitude

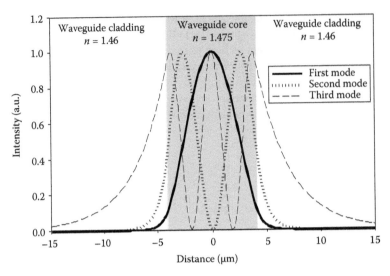

FIGURE 3.2 The intensity profiles of the first three transverse electric (TE) modes of a three-layer slab waveguide with an 8 μm thick core layer. For TE modes, the electric field vector of the light is polarized in the waveguide plane.

coefficients. A few of the lowest-order modes of a three-layer slab waveguide are shown in Figure 3.2. The lowest-order (i.e., $i=0$) mode field has one lobe in the waveguide core, and higher-order modes have $i+1$ lobes with the even modes being symmetric and the odd modes antisymmetric. The wave vector for each mode can be expressed in terms of a modal effective index N_{eff} defined such that $\beta=N_{eff}(2\pi/\lambda)$. This effective index plays a similar role to the index n in bulk materials, since the phase velocity of a waveguide mode is given by $v=c/N_{eff}$. Hence N_{eff} is often used instead of β when discussing waveguide propagation.

While solving the full electromagnetic wave equation seems well removed from the classical ray optics picture sketched in Figure 3.1, for simple cases the classical ray optic model can give the same waveguide modes as a rigorous solution of the electromagnetic wave equation (Hunsperger 1991). The key is to recognize that the ray in Figure 3.1 represents a plane wave being reflected back and forth between the two interfaces. The round trip phase change of the plane wave traveling from one side of the core to the other and back must be a multiple of 2π for a coherent superposition to occur between successive reflections. Since this can only occur for a set of discrete incident angles, θ_i, of the plane wave on the core-cladding interface, the discrete eigenmodes emerge naturally in the ray optics picture. Furthermore, by taking into account the phase change that occurs upon reflection from the core-cladding interfaces, one can obtain the correct values of the propagation wave vector β_i, which in the ray picture is just the projection $\beta_i=k\sin\theta_i$ of the ray wave vector k along the z-axis.

The three-dimensional waveguides used in integrated optical structures are inherently more complex and variable than the three-layer slab waveguide example of Figure 3.1, since lateral confinement in both the x and y transverse directions can be provided using a variety of different geometries. As shown in Figure 3.3, waveguides can take on the shape of cylinders as for optical fiber, rectangular channel waveguides, and ridge waveguides. However, the basic principles remain the same; the key design parameters for any application being the allowed number of waveguide modes, the mode field profile and polarization, and the propagation wave vector.

Even in the two-dimensional slab waveguide model of Equation 3.3, the exact mode solutions can only be found using numerical methods, and the same is true for three-dimensional waveguides. A vast amount of work has been done over the years to develop numerical techniques and approximations to solve for waveguide modes and model waveguide propagation in complex structures (Marcuse 1974,

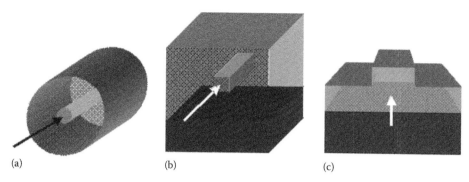

FIGURE 3.3 Three-dimensional waveguide geometries. (a) A cylindrical waveguide (e.g., optical fiber), (b) a buried channel waveguide, and (c) a ridge waveguide.

Kogelnik 1990, März 1995, Scarmozzino et al. 2000, Okamoto 2005). This large body of work is well beyond the scope of this chapter.

In addition to the large body of literature, there are many commercial software packages available for integrated optical design, based on a number of different approaches among which are the finite element method (FEM) and finite difference mode solvers, the beam propagation method (BPM), and finite difference time domain (FDTD) calculations.

The number of modes that can propagate in a waveguide is an important design factor in any integrated optical system. In a multimode waveguide, each mode has a different spatial electric field distribution (e.g., see Figure 3.2) and a unique wave vector, β_i. Each mode, therefore, propagates with a different phase velocity, and the total field amplitude at any point along the waveguide is a result of the coherent superposition of all the modes that are carrying optical power. The light in the waveguide may, therefore, have a virtually unpredictable phase and electric field profile at any time and position in a multimode waveguide. Nevertheless, if enough modes are excited simultaneously, the overall optical power distribution in a multimode waveguide can be quasi-uniform. Such a waveguide may even be desirable if the goal is simply to provide a light pipe to transport optical power from one point to another. Multimode waveguides usually have large core regions, and, therefore, it is also relatively easy to couple light in and out. The most common example is the multimode optical fiber, which may have a guiding core between several tens to hundreds of micrometers in diameter. Multimode fibers can be used to collect light and feed it to a device such as a spectrometer, or feed light to specific location as in the case of fiber optic lamps and illuminators, and the familiar fiber optic decorations and toys.

Most optical signal processing operations require that light at any point in the waveguide circuit have a well-defined and controllable phase and electric field amplitude. For this reason, even the simplest integrated optical devices usually use optical waveguides that support only one mode. The number of modes that a waveguide supports increases with waveguide size and with the core-cladding index step, and is also determined by the wavelength of the propagating light. On the other hand, if the waveguide cross section is too small, all modes may be cut-off and light is not guided at all. The single-mode requirement, therefore, places constraints on the possible waveguide geometry, the size, and the refractive index of the core and cladding materials. That being said, it is important to recognize that the waveguide mode is a mathematical concept. In an ideal waveguide, modes propagate forever with no loss or change in mode profile. However, in some cases light can also propagate considerable distances along a waveguide in "modes" that mathematically speaking are not confined, but still have relatively low propagation loss. The effect of power in these leaky modes, and the stray light generated as they radiate away, must also be considered in practical waveguide circuits. Furthermore, in real fabricated waveguides there are always optical loss mechanisms due to material absorption (Grand et al. 1990, Bazylenko et al. 1995) and scattering from interface roughness (Payne and Lacey 1994, Barwicz and Haus 2005, Schmid et al. 2008), so even a true waveguide mode may undergo significant attenuation as it propagates.

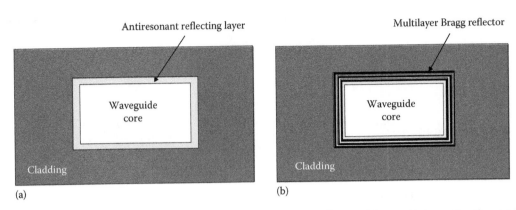

FIGURE 3.4 Waveguide cross section of two waveguides that use single or multilayer reflectors rather than total internal reflection to confine light to the waveguide core. (a) ARROW waveguide with a single reflecting layer. (b) A Bragg waveguide that uses a reflective multilayer coating.

Although TIR is necessary to implement a truly lossless waveguide, "leaky" but usable waveguides can be made by surrounding the core by reflecting structures. Examples include waveguide cores bounded by metal films (Jenkins et al. 2003), or reflecting dielectric layers in the case of Bragg (Cho et al. 1977, Yi et al. 2006) and ARROW waveguides (Duguay et al. 1986), shown schematically in Figure 3.4. When only a single dielectric layer is inserted between the core and cladding, the reflecting layer structure behaves like an antiresonant Fabry–Perot (FP) cavity at the operating wavelength. Since an FP cavity is close to 100% reflecting far from its resonance wavelength or incident angle, even a single layer structure can effectively confine light within the enclosed core region. The latter is referred to as an antiresonant reflecting optical waveguide, or ARROW.

Such waveguides are not commonly used in telecommunications since they are inherently more difficult to make than TIR waveguides, and do have a small but unavoidable intrinsic loss. They are however very useful when a very wide waveguide that supports only one mode is desired, or where the core index may be smaller than that of the surrounding materials. This is the case in many applications in optofluidics where the object is to guide light within a fluid layer (Bernini et al. 2004, Campopiano et al. 2004, Yin et al. 2005). Since many of the liquids of interest are water based, the index of refraction will be around $n=1.33$, which is lower than most dielectrics (e.g., in glass, $n=1.44$–1.5), polymers, and semiconductors. The only readily available cladding material with index lower than water is PTFE (Teflon), which has an index slightly lower than water (Datta et al. 2003). Hence implementing the TIR mechanism is often impractical when fluid waveguide core layers are desired.

3.2.2 Materials

Waveguides have been fabricated using many different materials ranging from glasses, semiconductors, polymers (Eldada and Shacklette 2000), and ferroelectrics like lithium niobate (Wooten et al. 2000). Given the space limitations and intent of this chapter, this section will focus on only a small subset of the more important and commonly encountered waveguide material systems. Some of the material and design issues specific to these systems will be discussed, since these issues are representative of the considerations common to every waveguide platform. Most of the waveguides discussed here are designed for operation with $\lambda=1550\,nm$ light, since most waveguide technologies were developed with telecommunications in mind. However, the physical principles discussed in the previous section apply equally well to waveguiding at shorter or longer wavelengths, although the semiconductor waveguide systems are opaque to light with photon energy higher than the energy gap. Waveguides in materials such as glass, silicon nitride, and polymers can be designed for wavelengths from the visible to the near-infrared, while silicon and III–V semiconductor waveguides can be transparent from their absorption edge (usually in the near-infrared) out to the mid-infrared.

Figure 3.5 shows representative examples of single-mode waveguide designs in glass, III–V semiconductors, and silicon. These designs are typical of waveguides with similar index of refraction profiles and size range. For example, while a discussion of organic and polymer waveguides is omitted here, they often have a refractive index profile similar to that of glass waveguides.

3.2.2.1 Glass Waveguides

The glass waveguide of Figure 3.5a is typical of structures used in passive waveguide devices for telecommunications applications. The fabrication sequence entails the deposition of the lower cladding glass layer on a silicon substrate wafer, followed by the higher index core layer using plasma-enhanced chemical vapor deposition (PECVD) or flame hydrolysis. At this point the waveguide patterns are masked by photolithography, and the core layer is removed everywhere except at the location of the channel waveguide, using some form of reactive ion etching (RIE). Finally the entire structure is covered by a lower index material similar to the lower cladding. More detail on the microfabrication of glass waveguides is given in Chapter 4.

The waveguide in Figure 3.5a consists of a $5\,\mu m$ square core of refractive index $n = 1.47$, with a cladding of $n = 1.46$. The small refractive index difference of $\Delta n = 0.01$ (~0.7%) is typical of glass waveguides, and is determined by varying the concentration of dopants such as Ge, P, and B in the constituent glass regions (Lee et al. 1988, Kominato et al. 1990, Kilian et al. 2000). It is possible to grow glass waveguides with higher refractive index cores by increasing the doping levels, or by using silicon nitride (Si_3N_4) or silicon oxynitrides (Si_xN_yO) as the waveguide core material. For the latter, refractive indices ranging between $n = 1.5$ and $n = 2.0$ can be obtained (Henry et al. 1987, Wörhoff et al. 1999a,b).

As drawn, the symmetrical waveguide of Figure 3.5a will support two identical but orthogonally polarized modes, with the electric field polarized horizontally for the transverse electric (TE) mode,

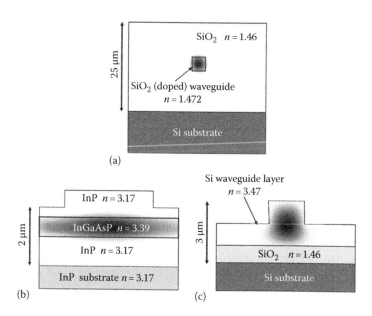

FIGURE 3.5 Cross section of the structures and mode intensity profiles of (a) a glass waveguide with a $5\,\mu m$ square core, (b) an $InP/In_{1-x}Ga_xAs_{1-y}P_y$ ridge waveguide with core composition corresponding to a band gap of $0.95\,eV$ ($Q = 1.3$), and (c) a silicon-on-insulator ridge waveguide. The TE mode intensity profiles are shown as grayscale contour plots.

and vertically for the transverse magnetic (TM) mode. The core-cladding refractive index step sets the range of practical waveguide dimensions. If the glass core is much larger than the 5 μm dimension shown, higher-order modes will also exist. If on the other hand the waveguide core is reduced in size, the waveguide mode will begin to expand into the cladding layer, and be more susceptible to bend losses at curved waveguide paths. Glass waveguides usually have optical modes similar to that of glass optical fiber, and by using index matching fluid to eliminate reflections, an almost seamless coupling can be achieved between glass waveguide facet and optical fiber.

Glass and oxynitride waveguides have some common materials and fabrication related issues that must be considered in waveguide circuit design and fabrication. The stress in the waveguide layers is perhaps the most problematic for telecommunication applications. These glass or silicon oxynitride layers are deposited on a silicon substrate at temperatures of 400°C or more, and often annealed at even higher temperatures. Upon cooling, a strong in-plane compressive stress evolves in the films, because the thermal expansion coefficient of the silicon substrate is much larger than that of SiO_2 (Janz 2004). Through the stress-optic effect, the stress fields in turn cause a refractive index anisotropy such that the light polarized in the wafer plane (TE) experiences a different index of refraction than light polarized perpendicular to the wafer (TM) (Janz 2004). This birefringence increases with annealing temperature up to the softening temperature above which glass becomes viscous and no longer supports internal strain (Janz et al. 2003). The resulting birefringence can be as large as $\Delta n = n_x - n_y \sim 10^{-3}$. This number is small relative to the refractive index of the glass, but it leads to a polarization-dependent shift of the order of $\delta\lambda = \lambda(\Delta n/n)$ in the wavelength output of any filter, resonator, or wavelength routing devices. While this may not be a problem when linearly polarized light is used, light delivered from optical fiber is usually unpolarized. As a result, a number of methods to reduce polarization birefringence have been developed to reduce the polarization dependence of waveguides to a level where the birefringence Δn is below 10^{-5}, particularly those that carry out telecommunications signal processing, routing, and receiving functions. For example, grooves can be etched beside the waveguide core to relieve stress (Nadler et al. 1999). The upper cladding layers can be doped with B and P in order to modify the thermal expansion coefficients to match that of the silicon substrate (Chun et al. 1996, Suzuki et al. 1997, Ojha et al. 1998, Kilian et al. 2000), resulting in an isotropic stress distribution. In certain cases, it is also possible to design the waveguide itself so that the birefringence arising purely from the geometrical anisotropy of the waveguide cancels out the stress birefringence. This approach has been successful for echelle grating demultiplexers (Janz et al. 2004) that employ slab waveguides, for which it is not possible to create an isotropic stress field in the core layer by other means.

Optical absorption by molecular bonds can also occur in glass and silicon nitride waveguides (Grand et al. 1990, Bazylenko et al. 1995). While the strongest optical absorption occurs at the OH bond overtones near a wavelength $\lambda = 1400$ nm, absorption by Si–N and Si–H bonds can cause significant loss in the telecommunication C-band ($\lambda = 1530$–1560 nm). For SiO_2 near stoichiometric composition, annealing is usually sufficient to reduce absorption to the point where the waveguide loss is less than 0.1 dB/cm, which is essentially undetectable over the short propagation lengths on a waveguide chip. However, losses can be persistent in silicon oxynitride and silicon nitride waveguides, and some effort in process optimization is required to suppress absorption to an acceptable level (Wörhoff et al. 1999a,b). In the context of optofluidics and waveguide biosensor applications, similar absorption features in water from $\lambda \sim 1480$ to 1600 nm can limit propagation distance of light to only a few millimeters (Curcio and Petty 1951) in waveguides where light interacts with a water layer.

3.2.2.2 III–V Semiconductor Waveguides

While glass is excellent material for passive waveguides, being an insulator with no significant electro-optic effects it offers limited potential for making active devices. On the other hand, the direct gap III–V semiconductors based on InP and GaAs are used for almost all waveguide-based lasers, and are also used extensively for high-speed photodetectors and modulators. Active devices and optoelectronics will be covered in other chapters of this book. However, integration of active and passive devices on a single semiconductor chip has long been a goal of much integrated optics research. This quest has given rise to an enormous body of scientific literature on passive III–V semiconductor waveguide devices, despite the fact that III–V waveguides are small and usually suffer from higher coupling and material losses than glass waveguides.

The III–V semiconductor compounds grown on the GaAs or InP substrate wafers (e.g., InGaAs, GaAlAs, InGaAsP, among others) can be used for electrically pumped lasers, modulators, switches, and photodetectors. Since these semiconductors have a direct energy band gap (i.e., no momentum change is involved in the recombination of an excited electron and a hole state), electrons and holes injected into the semiconductor recombine very efficiently by emitting a photon with the energy corresponding to the valence–conduction band energy difference. Hence III–V semiconductors have a high stimulated emission rate, an essential property of any lasing gain medium. The quantum efficiency for creating an electron and hole pair by photon absorption is also very high. The use of very thin (~50–200 nm) quantum well layers to modify the electron and hole states can further improve the gain and absorption of III–V semiconductor waveguides. As a result, the III–V semiconductors can be used to make both excellent lasers and photodetectors. These materials also have useful electro-optic effects that can be used to vary the local index of refraction or optical absorption by applying either a voltage across the semiconductor waveguide (Wakita 1998, Li and Yu 2003), or injecting electrons and holes into the waveguide (Ng et al. 2007).

Epitaxial growth of these materials on GaAs and InP wafers has been developed over many decades so that it is now possible to build III–V waveguides capable of absorbing or emitting light at wavelengths from $\lambda = 600$–3000 nm, and even beyond if intersubband quantum well transitions are used. The two most commonly used growth techniques are molecular beam epitaxy (MBE) and metal-organic chemical vapor deposition (MOCVD). Figure 3.5b shows a waveguide fabricated using InGaAsP alloy layer structure on an InP substrate. This particular material system is of particular importance in telecommunications, since InP/InGaAsP materials are used in most semiconductor lasers and photodetectors operating in the 1300–1700 nm wavelength range.

GaAs wafers can be used as substrates for layers of AlGaAs, InGasAs, and related compounds. The wavelength ranges covered by lasers and detectors built on the GaAs substrates usually range from about 600–1000 nm. However, recent work (Gupta et al. 2008) on growing alloy layers on GaAs wafers that contain N and Sb has led to the successful demonstration of lasers on GaAs substrates operating at wavelengths in the 1500 nm range.

The refractive index of commonly used III–V semiconductor alloys ranges from 3.0 to about 3.5. For example, Figure 3.6 shows the index of refraction of $In_{1-x}Ga_xAs_{1-y}P_y$ layers grown on InP. Here, the alloy compositions are chosen to create layers with a lattice constant that matches the underlying InP substrate, to avoid high film stress and formation of crystal defects. The core-cladding index steps in III–V waveguides can be anywhere from $\Delta n = 0$ to $\Delta n = 0.5$ depending on the application and material system. The core-cladding index steps are, therefore, significantly higher than in glass waveguides. Therefore, waveguide dimensions in III–V devices must be much smaller than for glass, to achieve single-mode operation. The mode sizes are similar to that shown in Figure 3.5b, typically being a few microns wide and one micron or less in vertical extent.

The motivation for using passive III–V waveguides is the ability to combine the active detectors and lasers on the same chip as passive waveguides and waveguide devices. Examples of active–passive III–V semiconductor integration include waveguide spectrometers monolithically integrated with photodetectors to make channel monitors (Tolstikin et al. 2003, 2004) or spectrometers integrated with semiconductor lasers to create tunable wavelength sources (Soole et al. 1992). Both detection and lasing require materials with electron-hole transition energies (i.e., the band gap energy) that correspond to the desired photon energy or wavelength. Unfortunately this means the active material will absorb light at the operating wavelength and cannot be used for the passive waveguide devices on the same chip, since these must be transparent. The problem of combining materials that are transparent to the operating wavelength with materials that can emit or detect light has been and remains a central problem that drives much III–V waveguide research.

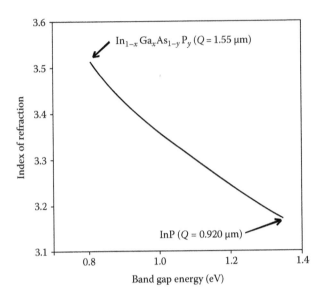

FIGURE 3.6 Index of refraction for $In_{1-x}Ga_xAs_{1-y}P_y$ alloys lattice matched to InP, over a composition range from InP (absorption edge at $\lambda = 920\,nm$) to an alloy with a band gap energy of 0.8 eV (absorption edge at $\lambda = 1550\,nm$). The index values are for light at a wavelength of $\lambda = 1550\,nm$.

Many solutions to this fundamental integration problem have been proposed and demonstrated. Figure 3.7a illustrates the use of epitaxial regrowth, to align active (i.e., absorbing) and passive (transparent) waveguide layers on a single semiconductor wafer (Soole et al. 1992, Zirngibl et al. 1994). In such an approach, the passive waveguide may be grown first on the substrate wafer. Then the passive waveguide is etched away in where the active device is to be located, and the active (i.e., absorbing) waveguide layers are grown.

It is also possible to grow multilevel waveguide layer structures with both transparent and absorbing layers as in Figure 3.7b. In the as-grown structure guided light is coupled to active layers. These active absorbing layers are arranged so they can simply be etched away in areas where passive, transparent waveguides are needed. In the example shown in Figure 3.7b, light couples from a transparent waveguide core layer to a second absorbing layer that forms the active layer of a waveguide photodetector (Tolstikin et al. 2003, 2004). Although this approach does not offer the design freedom of a regrowth methodology, it avoids some difficulty in growing high-quality crystalline material over surfaces with preexisting topology.

Another approach is to use post-growth processing to change the composition of the waveguide layers in selected areas so that the passive waveguide regions become transparent (i.e., the absorption edge is blue-shifted) while the active sections retain their original composition. This has been demonstrated using local ion-implantation in the passive waveguide region (Haysom et al. 1999), or annealing of the semiconductor wafer with patterned capping layers over the passive sections (McDougall et al. 1998). Both these methods change the alloy composition by intermixing the alloy components at the boundary of two dissimilar layers, as shown in Figure 3.7c. Even when interdiffusion is enhanced by ion-implantation or other methods, the effective diffusion length is very short. Therefore, to be effective, the layers must be extremely thin. Hence, intermixing works well for waveguides consisting of quantum wells, but not for waveguides with individual layers much thicker than 100 nm.

3.2.2.3 Silicon Waveguides

The silicon-on-insulator (SOI) waveguide platform is the most recent material systems to be widely adopted by the integrated optics community (Pavesi and Lockwood 2004, Reed and Knights 2004). The

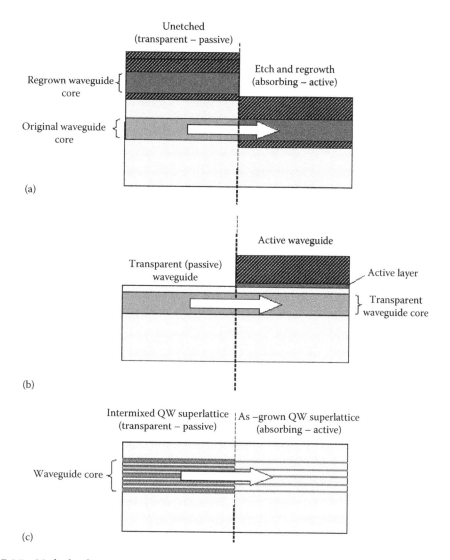

FIGURE 3.7 Methods of integrating passive (transparent) waveguides and active (absorbing) waveguides on a single semiconductor chip: (a) patterned etch and regrowth of active layer, (b) patterned etch to remove active layer, and (c) quantum well intermixing.

typical SOI waveguide shown in Figure 3.8 consists of a thin single crystal silicon guiding layer over a silicon dioxide layer (the BOx, or buried oxide layer). Both of these layers are supported by a silicon wafer substrate that is usually 500–1000 μm thick. The exact thickness of the silicon waveguide layer and the BOx layer can be anywhere between a few tens of nanometers to several microns thick, depending on the application.

The use of silicon for integrated optics may seem surprising, since it has neither the strong electro-optic effects nor direct optical transitions of the III–V semiconductors that enable high-speed modulation, photodetection, and lasing; nor the high transparency and good fiber-to-waveguide mode match of glass waveguides. Interest in silicon photonics first arose because it is the same material used by the electronics industry. It therefore seemed natural that if one wanted to integrate optical processing elements with electronics, the waveguides should be fabricated on the same chip and made of the same silicon-based materials as the electronics. Although much of the early work in this area concentrated on waveguides

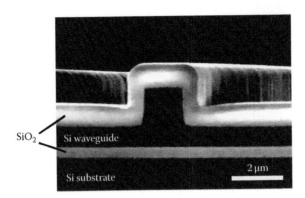

FIGURE 3.8   Cross-sectional electron microscope view of a silicon-on-insulator ridge waveguide.

made from silicon germanium alloys (Soref et al. 1991, Soref 1993), growth of SiGe waveguide layers on silicon is difficult because of the lattice constant mismatch, low refractive index contrast with silicon, and stress induced birefringence (Janz et al. 1998). It gradually became evident that SOI was a more generally usable waveguide system (Rickman et al. 1991, Fisher et al. 1996, Vivien et al. 2002) particularly as the quality of SOI wafers improved. Very high-quality SOI wafers have become readily available within the last 10 years, as SOI has been widely adopted as an electronic chip substrate by the electronics industry. As a result, waveguide losses in Si are almost entirely due to scattering from sidewall roughness of the fabricated waveguides, rather than from intrinsic material or wafer manufacturing defects. While producing sufficiently smooth waveguides can be challenging, waveguide losses of less than 3 dB/cm or less are obtained using standard fabrication procedures (Lardenois et al. 2003, Dumon et al. 2004).

As the field of silicon photonics has emerged, it has become evident that the reasons to use silicon go well beyond the need to combine electronic and optics on a simple CMOS compatible platform. Silicon has a high refractive index ($n=3.47$ at $\lambda=1550$ nm), and in particular there is a large refractive index step of Δn ~2 between the Si waveguide core and the SiO_2 or similar claddings below and above the waveguide. Although III–V semiconductors also have a high index of refraction, the index step between the waveguide core layers and the underlying substrate is usually only a few percent, as is the case of the InGaAsP/InP ridge waveguide in Figure 3.5b. As a consequence, the III–V waveguides have a much weaker optical confinement than the SOI waveguide of Figure 3.5c.

The high refractive index and core-cladding index contrast of silicon have enabled a whole new family of integrated optical devices based on submicron silicon waveguides with cross-sectional dimensions of the order of 0.2–0.5 μm (Ahmad et al. 2002, Bogaerts et al. 2004, Dumon et al. 2004, 2006, Sasaki et al. 2005, Xu et al. 2005, Dulkeith et al. 2006, De Vos et al. 2007, Densmore et al. 2008, Xu et al. 2008). These photonic wire waveguides are not only small themselves, but they can support virtually lossless waveguide bends with radii of only a few microns. As a result, complex integrated optical circuits formed from silicon photonic wire waveguides will fit in surprising small chip areas. Similar thin Si waveguide layers are also commonly used as the starting material for most two-dimensional photonic crystal devices (McNab et al. 2003).

3.3 Integrated Optical Circuit Design Elements

The individual waveguide paths in an integrated optic chip play the same role as the metal interconnections in an electronic device, but there are fundamental differences. In electronic chips the quantum mechanical wavelength of electrons in the metal conductors is small compared to the metal thickness. Therefore the precise geometrical details of a wire or metal conductor are not critical, at least to a first approximation. The cross-sectional dimensions of a single-mode optical waveguide, however, are comparable to or often smaller than the wavelength of light. Ensuring the efficient transmission of light through optical circuit elements as trivial as a waveguide bend or a splitter requires careful analysis of how waves will propagate around or through these structures.

3.3.1 Waveguide Bends

The advantages of integrated optics stem in part from the ability to manipulate the optical path at will, thereby freeing the designer from the constraints of diffraction and line-of-sight propagation imposed by free-space optics. This presupposes that the waveguides making up the optical circuit can be laid out in sufficiently compact curved paths to be useful. Thus the waveguide bend is the first fundamental building block of an optoelectronic circuit.

The important parameters of the waveguide bend shown in Figure 3.9 are the bend radius of curvature R, and the associated bend loss. In designing an integrated optical device, the overall chip size is usually determined simply by how sharply light can be forced to change direction. If the waveguide bend radius is too small, light will radiate away from the waveguide in the region of the bend introducing losses (Marcuse 1974, Hunsperger 1991). There are two intuitive pictures for understanding the source of this loss. In the ray optics picture, the ray of light traveling along the waveguide strikes the waveguide wall in the curved section. At the bend, its angle of incidence can exceed the critical angle for TIR and reflection losses accrue as the ray is guided around the bend by successive reflections. A more complex picture looks at the evanescent tail of the waveguide mode in the bend region. As light travels around the bend, the field in the evanescent tail at the outer edge of the waveguide is forced to travel faster than the light at the centre of the waveguide. If the bend is too sharp, this light would be forced to travel faster that the speed of light in the cladding material, and instead separates from the waveguide mode and is radiated away. This latter model leads to a semiquantitative analysis of bend loss (Hunsperger 1991) that yields a formula for loss coefficient

$$\alpha = C_1 e^{-C_2 R} \tag{3.6}$$

defined such that the attenuation of mode intensity after propagating through a curved section of length L is just $I = I_0 e^{-\alpha L}$. A detailed derivation and discussion of the coefficients C_1 and C_2 is far beyond the scope of this chapter, and in any case an analytical solution is only available for the simplest two-dimensional waveguide models. The important point is that the bend loss varies exponentially with bend radius R, so even small variations in bend radius can cause large changes in the total transmitted power.

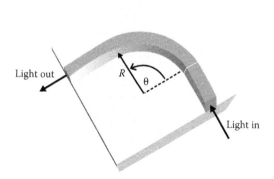

FIGURE 3.9 A schematic waveguide bend of radius R through angle θ, as viewed from above.

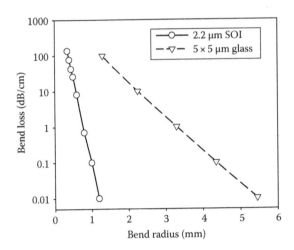

FIGURE 3.10 A comparison of calculated bend loss for a 5×5 μm core glass waveguide similar to that in Figure 3.5a with index step $\Delta n = 0.011$, and for a 2.2 μm thick silicon-on-insulator ridge waveguide similar to those in Figures 3.5c and 3.8.

In the two-dimensional waveguide system, the coefficients in Equation 3.6 depend on the dimensions of the waveguide, the waveguide effective index N_{eff}, and the index of refraction for the surrounding cladding. In particular the C_2 coefficient varies with the index step Δn between waveguide core and cladding, and this holds true for real three-dimensional waveguides as well. This is evident in Figure 3.10, which compares the calculated bend loss for a glass waveguide with index step $\Delta n = 0.01$ as in Figure 3.5a, and a 2.4 μm wide silicon ridge waveguide as in Figure 3.8. This graph clearly shows the advantages of a high index contrast system such as silicon. Assuming an acceptable loss must be less than 0.1 dB per 90° bend, the minimum bend radius is less than 1 mm for a silicon waveguide and about 5 mm for glass waveguides. For silicon photonic wire waveguides with submicron cross-sectional dimensions, a 90° bend can have a radius less than 10 μm. Given that for comparable devices the chip size will scale with bend radius, silicon photonic wire optical chips can be four orders of magnitude smaller than integrated optical chips using low contrast systems based on glass waveguides.

3.3.2 Optical Splitters and Couplers

In building an optical waveguide circuit, the next level in complexity is to take a single optical input and send it to two or more destinations via separate waveguides. The inverse operation combines the light from two or more waveguides into one output waveguide. Since Maxwell's equations are reversible in time, both the splitting and combination function can be performed by the same structure.

3.3.2.1 Y-Junction Splitters

The simplest method of splitting or combining light is to use the 1×2 Y-junction splitter of Figure 3.11. The operation of this splitter is self-evident. The waveguide mode propagates along the waveguide until it nears the junction region where it evolves into two identical modes that travel separately down the two output waveguides. If the two output beams were reversed, they would travel down the waveguides toward the junction and recombine to produce a replica of the input mode traveling in the opposite direction. However, this is only true for a lossless splitter, and in particular when the two beams are precisely in phase as they recombine. Simply sending two beams of arbitrary relative phase down each branch will result in power loss in the form of radiation away from the Y-junction. Similarly, if only one of the two waveguide branches carries light toward the Y-junction, half the light will be lost to radiation at the Y-junction vertex. Although it may seem counterintuitive, applying energy conservation along

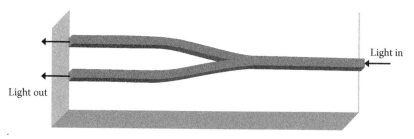

FIGURE 3.11 A schematic top view of waveguide Y-junction splitter.

with time reversal of the fields imposes this condition on any splitter with a 50% (i.e., 3 dB) power splitting ratio (Izutsu et al. 1982).

Y-junction splitters are conceptually simple, and they have the advantage that they are relatively wavelength independent. As was the case for a waveguide bend, one would like to minimize overall chip size by designing the splitter to cleanly divide the beam into two separate propagating beams in as short a distance as possible. On the other hand, low splitter loss requires that the junction region vary slowly enough along the propagation direction that the mode field evolves adiabatically from the fundamental waveguide mode of the input guide to the two separate fundamental modes in the two output waveguides. The length of the splitter is also determined by the waveguide curvature of the two output branches, which sets the distance over which the two output waveguides can achieve the desired separation. Hence, here again, bend loss plays a critical role. The length of a splitter in a low index contrast system such as glass can be several millimeters long, while in a high index contrast system such as silicon, the splitting can be accomplished in distances of the order of a few tens of microns. Splitting loss also arises from unavoidable light scattering in the vicinity of the Y-junction vertex. All types of lithography have a spatial resolution limit, and the etching process itself limits the smallest feature size that can be reliably transferred from a lithographic mask to the waveguide material. No fabrication technique will produce a perfectly sharp vertex at the centre line of the Y-junction. The remaining blunt vertex tip must inevitably scatter some light. In general, the performance of a well-designed Y-junction is limited mainly by a combination of factors related to fabrication and material refractive index. Nevertheless overall splitting losses of less than 1 dB can be routinely achieved.

3.3.2.2 Directional Couplers

The directional coupler of Figure 3.12 consists of two waveguides that are so close to each other that the evanescent fields of the waveguide modes extend into the adjacent waveguide. If light is initially coupled into one of the waveguides, it will gradually transfer into the other waveguide as it propagates, and then back again to the first, and so on, with a periodicity known as the coupling length. The directional coupler can therefore be used to tap some or all of the light from one waveguide into a second waveguide, with the desired splitting ratio being determined by the waveguide separation and coupler length.

The simplest picture to envisage this power transfer process is to consider the two waveguides in the coupler as one system that supports a set of waveguide supermodes. In particular, there will be a symmetric mode and an antisymmetric mode as shown Figure 3.13, with propagation wave vectors β_s and β_a.

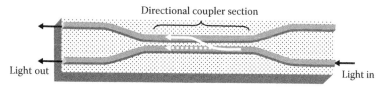

FIGURE 3.12 A schematic top view of a waveguide directional coupler.

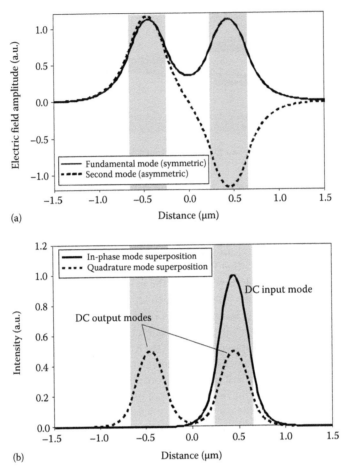

FIGURE 3.13 (a) The electric field amplitude for the symmetric and antisymmetric supermodes of a directional coupler (DC) as shown in Figure 3.12. (b) Total intensity when the two modes are in phase (at the input of the DC) and when they are in quadrature (at the output of a DC designed to give a 3 dB splitting ratio).

As the waveguide separation is increased, the values of these propagation constants will both tend to approach the value of the fundamental waveguide mode of an isolated waveguide. Conversely, the difference between β_s and β_a will grow as the waveguides approach each other, and tend toward the wave vectors of the fundamental and first mode of a waveguide that is twice the width of the single waveguide, when the gap between the waveguides goes to zero.

In the weak coupling limit (i.e., the waveguides are sufficiently far apart) the lobes of the two modes will be very similar in shape to the fundamental mode of an isolated waveguide. If light is coupled into one of the two waveguides, the resulting electric field distribution in the waveguides can be constructed by a coherent superposition of the symmetric and antisymmetric modes. As the light propagates down the waveguide pair, the two modes propagate with different phase velocity and their relative phase changes. After a distance L_c, the relative phase change is $(\beta_s - \beta_a)L_c = \pi$, and almost all the light appears in the adjacent waveguide. As light continues to propagate, the power will oscillate back and forth between the two waveguides with a period of the coupling length L_c. By separating the two waveguides after an appropriate distance, one can design a 2×2 optical splitter with any desired power splitting ratio.

The phenomenology and theory of optical directional couplers has been discussed in many texts and scientific articles (Somekh et al. 1972, Hunsperger 1991, Trinh et al. 1995, Okamoto 2005). The design issues revolve around ensuring that the waveguides are close enough together that the coupling lengths required to achieve 1×2 splitting, or full transfer of power to the second waveguide (i.e., $L_c = \pi/(\beta_s - \beta_a)$) are acceptable. The caveat is that loss occurs in the transition region between the coupler sections where the waveguides are parallel, and where they are far enough apart to be considered to be two independent waveguides. This loss arises again from bend loss, and also the difference in local mode shape of the electric field profile in each waveguide for the waveguide pair and a single isolated waveguide. The loss can be minimized by ensuring that the transition region is long enough that the changes in mode profiles occur adiabatically with no induced loss.

The advantage of the directional coupler is that any splitting ratio can be achieved simply by choosing the correct coupling length. Power loss is determined largely by how long the transition regions between the coupler section and separate waveguides can be made, the goal being to make the transition as adiabatic as possible to minimize mode conversion loss, and also minimize bend loss. As usual, in high index contrast systems such as SOI, directional couplers can be a few tens of microns or less in length, depending on the desired splitting ratio. In glass they tend to be several millimeters long. Unlike the simple Y-junction vertex, the directional coupler is intrinsically wavelength dependent, since the propagation wave vectors are explicitly wavelength dependent (i.e., since $\beta = N_{eff} 2\pi/\lambda$), and also depend on the wavelength dispersion of the waveguide effective index. This means that a directional coupler will impose an additional and usually unwanted wavelength dependence on any device where it is used, such as a ring resonator (Delâge et al. 2009).

The design and fabrication of direction couplers is relatively straightforward, with the critical parameter being the waveguide separation which determines the coupling length L_c and therefore the splitting ratio. As one moves to smaller devices with submicron waveguide cross sections such as silicon photonic wires, meeting the tolerances on waveguide separation to achieve a target splitting ratio begins to challenge the limits of even state-of-the-art lithography and etching tools.

3.3.2.3 Multimode Interference Couplers

The simple 2×2 multimode interference (MMI) coupler pictured in Figure 3.14 is similar in operating principle to the direction coupler discussed above, since it relies on coherent interference between the modes propagating in the wide MMI waveguide section of the coupler. In the 2×2 MMI, the main difference from the directional coupler is that the fundamental and first lateral modes of the MMI take the place of the symmetric and antisymmetric modes of the directional coupler. Higher-order modes are also involved in MMI couplers, which in their most general form can be considered to be an imaging

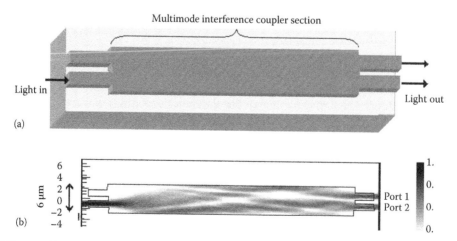

FIGURE 3.14 (a) A schematic view of a waveguide multimode interference coupler, viewed from above. (b) The calculated electric field inside a silicon ridge waveguide MMI design for 3 dB splitting, when light is launched into one of the input ports.

system capable of projecting multiple images of light launched from an input waveguide onto N output waveguides at the output plane of the MMI.

The MMI coupler in Figure 3.14 consists of two single-mode input waveguides and two single-mode output waveguides connected by an MMI section, which is simply a much wider multimode waveguide. When light from one of the single-mode waveguides is launched into the MMI section, it will couple power into a number of waveguide modes of the MMI section such that the superposition of the field from all the excited modes at the input plane forms a close approximation to the input waveguide mode field shape. The more modes that are supported by the MMI, the closer the true input waveguide mode shape can be approximated by a superposition of MMI modes. Any optical power not coupled into the MMI modes is radiated away from the MMI as unconfined radiation. As the excited modes propagate along the MMI section, their relative phases change because of the different phase velocity of each mode. One can show that, given appropriate constraints on the MMI dimensions and MMI mode wave vectors β_j (Bachmann et al. 1994, Soldano and Pennings 1995), the modes will line up at certain planes along the MMI and recreate single or multiple images of the input waveguide mode. For example, a splitter can be made by choosing an MMI length such that two images appear at the output plane of the MMI waveguide, so that light is coupled equally into the two output waveguides, as in Figure 3.14. Alternatively one may choose a plane where a single image of the input mode field is formed, but reflected in the centre axis of the MMI waveguide. The 2×2 MMI coupler design can therefore be adapted to be either a splitter or to transfer light from one waveguide to another.

The 2×2 MMI is just one example of a class of multimode imaging devices that can split and recombine light in various ways (Bachmann et al. 1994, Soldano and Pennings 1995). MMIs can be implemented in any of the common waveguide material systems. As in previous examples, MMI devices in glass will be much larger than their counterparts in III–V semiconductor and SOI system, but designing and fabricating the MMI in semiconductors can be more demanding because of the much smaller length scales involved. Nevertheless, silicon MMI couplers may be easier to make than other couplers because they rely on imaging phenomenon. Obtaining the correct splitting ratio of the MMI coupler relies on the precise values of the width and length of the MMI section. This is however primarily a challenge in metrology and reproducibility of fabrication processes. Extremely small, difficult to fabricate features are absent, such as the gap in a directional coupler or the sharp vertex of Y-junction. Obtaining optimal performance from an MMI coupler, including minimal over all insertion loss, can still involve careful consideration of a number of design variables so that the power in the input and output waveguide modes efficiently coupled to the modes of the MMI section. In the case of silicon waveguides, many of the key design parameters have been identified and analyzed in detail by Halir et al. (2008).

The wavelength dependence and polarization dependence for MMIs can be less than for directional couplers (Xu et al. 2006, 2007, Delâge et al. 2009). MMIs are therefore a useful alternative to directional couplers in photonic wire ring resonators, enabling the design of polarization independent ring resonator filters, with wavelength independent resonance depths and widths. The one disadvantage of a 2×2 MMI coupler for simple coupling applications is that the available splitting ratios are limited to discrete ratios such as 0.5/05 and 0.85/0.15, rather than being continuously variable as is the case for the directional coupler. At the cost of additional complexity, other splitting ratios may be achieved by forming a coupler by connecting two identical rectangular MMIs in series, and adjusting the lengths of the interconnecting waveguides (Le and Cahill 2008), or by using specially shaped MMIs (Besse 1996).

3.3.2.4 Star Couplers

In some integrated optic devices, it is necessary to transfer light from a single waveguide to many waveguides. The most common occurrence is in the arrayed waveguide grating (AWG) spectrometer used to make wavelength multiplexers and demultiplexers. MMI couplers can be designed to achieve $1 \times N$ splitting, but the phase across the N outputs is not uniform. Splitting into multiple waveguides can also be achieved using successive 1×2 splitters, but a star coupler makes more efficient use of chip area

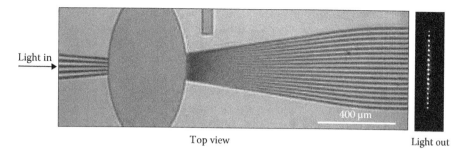

Top view   Light out

FIGURE 3.15 Top view of an SOI ridge waveguide star coupler. Inset at right shows an image of the output facet when λ=1550 nm light is coupled into an input waveguide.

(Dragone et al. 1989). In its simplest form, a star coupler such as the example in Figure 3.15 consists of one or more input waveguides opening onto a wide slab waveguide section. The light spreads outward as it propagates according to the two-dimensional diffraction equation, to form a circular wave front in the far field. The *N* output waveguides are placed along a circular arc that matches a line of constant phase in the far field, so that far field radiation is coupled into each of the *N* waveguides with the same phase—an essential requirement for AWG demultiplexers.

While the design and fabrication of a simple star coupler is straightforward, considerable effort can go into optimization of insertion loss and other parameters that may be specific to certain applications. For example, the output waveguides can be widened as they approach the slab waveguide array boundaries so they adiabatically merge into the slab section, leaving little dead space between waveguide apertures (Dragone 1990, Dumon et al. 2006). The objective is to ensure all of the light is transferred into the output waveguides, and eliminate back reflections that can contribute to return loss and unwanted signal from stray light. In some waveguide platforms the mode structure of the slab waveguide and the input and output waveguides may be fundamentally different. For example, in SOI waveguides thicker than a few hundred nanometers, a slab waveguide may support several vertical waveguide modes, while the ridge waveguides can be designed to guide only one fundamental mode. In the slab waveguide section, the fundamental mode and higher-order modes will have different phase velocities (i.e., different effective index values). Inadvertent coupling back and forth from higher-order slab modes to the fundamental mode of the input and output ridge waveguides not only causes power loss, but introduces a randomly varying noise signal onto the transmitted light. Again the solution is to design the junctions between slab and input and output waveguides to minimize unwanted coupling to and from higher-order slab modes (Schmid et al. 2007, Cheben et al. 2007).

3.3.3 Input and Output Coupling Structures

The potential of integrated optics for optics and optical processing can only be realized if light can be coupled onto the chip by practical, robust, and cost-effective means, and the optical output can be effectively coupled off the chip (Matthews et al. 1990). The physical problem is that of coupling light beams with diameters of about ten microns or larger from single-mode optical fiber or free-space beams into waveguides ranging from several micrometers across to a few hundred nanometers for silicon photonic wire waveguides. In most integrated optics configurations, light is delivered to the chip by optical fiber, and coupled from the chip to an output fiber. The power coupling efficiency, η, depends on the mode overlap integral between the incident beam field profile, $E(x, y)_{inc}$, and the waveguide mode profile, $E(x, y)_{mode}$, at the waveguide facet (Hunsperger 1991):

$$\eta = \left| \int E(x, y)_{mode} E^*(x, y)_{inc} \, dx \, dy \right|^2 \qquad (3.7)$$

Here it is assumed field profiles are normalized to give unity optical power in both the incident beam and waveguide. Effective coupling therefore requires that the incident beam spot and waveguide mode are closely matched in size and shape, and also spatially aligned within tolerances much less than the waveguide cross-sectional dimensions. Furthermore, this alignment must be maintained for the anticipated range of operating environments. With free-space optics, light can only be focused down to spot sizes of the order of the wavelength of light. Thus, there can be a particularly large mismatch between the optical spot size in free space and the mode size of small semiconductor waveguides.

The simplest cases are those where free-space beams or large diameter multimode optical fibers are coupled to large multimode waveguides. As long as the incident beam diameter and the waveguide cores are tens of microns or larger, physical alignment to within several micron tolerances is usually sufficient. Efficient coupling of light can be achieved using simple lenses, or choosing a fiber and/or the waveguide design such that the guiding cores are of comparable dimensions, and if necessary minimizing reflections using index matching material between the fiber and waveguides.

Next in level of difficulty is coupling of single-mode glass fiber to single-mode glass planar waveguides, as is required in telecommunications applications. Here the optical fiber and glass waveguide are made of similar materials and have similar optical mode sizes. Since the optical waveguide mode is now approximately 5–10 μm in diameter, alignment tolerances of less than 1 μm are required. This places additional mechanical demands on the apparatus for aligning the fiber and waveguide chip during the fiber attach process. However, since the mode sizes are similar, the fiber and waveguide facet often only need to be brought into close proximity and be well aligned. Special coupling structures are not essential for achieving good mode overlap. In telecommunication applications involving single-mode light propagation, back reflections (i.e., return loss), and multiple reflection interference effects can be critical parameters affecting overall system performance. Therefore, index matching gels or epoxies are commonly used to attach the fiber to the waveguide chip, while ensuring that the optical path through the fiber facet, across the gap, and into the waveguide has an almost uniform index of refraction.

Coupling of III–V and silicon semiconductor waveguide devices to optical fiber is more demanding in terms of mode transformation and alignment. SOI waveguide dimensions can range from ridge waveguides with width and heights near 5 μm, to silicon photonic wire waveguides with cross-sectional dimensions of a few hundred nanometers. The III–V semiconductor waveguides such as those used for lasers have mode dimension of the order of 1 μm. For such a waveguide the coupling loss to a standard telecommunication fiber can be of the order of −20 dB (1%) even if only mode size mismatch is considered (cf. Equation 3.7), while for silicon photonic wire this loss can be less than −30 dB (0.1%). In these situations, additional means to improve input and output coupling need to be implemented.

External to the chip, fiber-to-waveguide coupling can be improved in a number of ways. One approach is to insert a short focal length lens between the waveguide chip and the input and output fibers (Sumida and Takemoto 1984). This lens could be a ball lens or a gradient index (GRIN) lens, which simply focuses the output mode of the fiber down to match the waveguide mode. Rather than using a ball lens, the fiber can also be machined to create a rounded or pointed tip which can compress or focus the guided light exiting the fiber tip (Alder et al. 2000). Fiber tips for this purpose are commercially available. Both these methods are constrained by diffraction to generate spot diameters no smaller than approximately 1 μm, assuming the vacuum wavelength is about $\lambda = 1.55$ μm.

Coupling to small waveguides can be facilitated by modifying the on-chip waveguide itself in the region near the facet. Intuitively, one may expect that simply adiabatically increasing the size of the waveguide (and thereby the mode diameter) as it approaches the facet would be effective. If the waveguide is only expanded in the horizontal plane, the mode remains tightly confined in the vertical plane and coupling improvement is limited. Given that conventional waveguide fabrication is done by a uniform layer deposition and two-dimensional lithography, varying waveguide vertical thickness along the propagation direction can add considerable complexity to the fabrication process (Sure et al. 2003). Rather than varying the waveguide thickness, one approach is to grow a GRIN structure on the waveguide surface (Delâge et al. 2006). This structure accepts the large incident mode size, but the light focuses down and couples into the smaller waveguide mode size as it propagates. However, this approach still requires the growth and patterning of special GRIN lenses near the waveguide facet, which adds several steps to the fabrication procedure.

Rather than widening the waveguide at the facet, it can be more effective to taper the waveguide down to very small dimensions as low as 100 nm across. As the waveguide cross section shrinks, the optical confinement weakens and eventually the mode expands out into the surrounding cladding materials, both horizontally and vertically as shown in Figure 3.16. By choosing the final inverse taper tip width and cladding material properties, it is possible to control the mode diameter in both the horizontal and vertical directions as it exits the waveguide. Tapering a typical silicon photonic wire waveguide mode out to a full 10 μm mode diameter is difficult to achieve because the necessary tip dimensions are less than 100 nm. However, inverse tapers can be used to achieve excellent coupling of waveguides to somewhat smaller input and output modes, as encountered when tapered fiber tips or high numerical aperture input and output lenses are used. Originally developed to facilitate coupling to semiconductor lasers (Shani et al. 1989) inverse tapers have been adapted to the silicon waveguide platform (Shoji et al. 2002, Almeida et al. 2003, Lee et al. 2005), and are now a standard element in the silicon photonics tool kit, particularly for coupling to small cross section photonics wire waveguides. These inverse tapers can increase the mode area of an Si photonic wire waveguide (~220 nm diameter) by a factor of 100, and measured coupling efficiencies from tapered fiber to waveguide are improved by an order of magnitude (Almeida et al. 2003). A variation of the inverse taper uses segmented waveguide structures along the tapered waveguide to reduce confinement without the need for such narrow tips. The coupling performance of these segmented tips can be comparable to the standard continuous inverse taper, while fabrication tolerance may be somewhat relaxed (Chou et al. 1996, Cheben et al. 2006).

Gratings can be used to couple light into waveguides from the surface of the waveguide, eliminating facet coupling altogether, and allowing waveguide devices to be tested at the wafer level, and even packaged without facet cleaving and polishing. Although not critical for the final application of a chip, eliminating the need for accurately cleaved and/or polished facets enables wafer level testing which may significantly reduce testing and manufacturing costs. A grating coupler in its simplest form is just a series of grooves or ridges written onto the waveguide surface. The grating line spacing or period Λ is precisely chosen such that the grating will diffract the incident beam into the waveguide.

In other words, the grating wave vector, $K_g = 2\pi/\Lambda$, is determined by the difference (modulo 2π) between the waveguide mode wave vector, β, and the projection of the incident beam wave vector on the waveguide surface, $k = k_0 \sin(\theta)$.

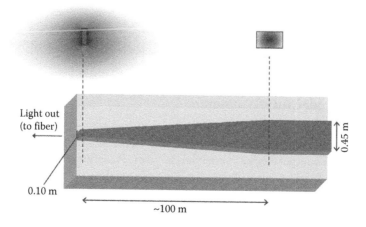

FIGURE 3.16 A schematic view of an inverse taper coupler on a 250 nm thick Si photonic wire waveguide. The calculated waveguide TE mode intensity profiles are shown at the start of the taper and at the waveguide facet.

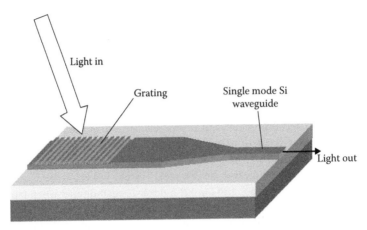

FIGURE 3.17   A schematic view of a grating waveguide coupler for coupling light from an external beam to a single-mode silicon photonic wire waveguide.

Figure 3.17 shows a schematic of a grating coupler in a silicon photonic wire platform. The optical fiber is positioned a few microns above the chip surface and illuminates the grating, which in turn couples the light into the waveguide mode. Since the grating must be wide enough to capture all the light emitted from the fiber, the waveguide at this point is very wide and can support many lateral modes. Therefore, the coupler is immediately followed by an adiabatic taper section over which the waveguide width is narrowed down to the final single-mode waveguide width. There are many variants of grating couplers published in the literature (Taillert et al. 2003, 2004, Roelkens et al. 2006, 2007, Masanovic et al. 2003, Van Laere et al. 2007). A simple grating can achieve up to 30% coupling from standard fiber mode (~10 μm mode diameter) to a 220 nm thick Si photonic wire waveguide, while a more elaborate grating structure including an underlying reflecting layer has achieved coupling efficiency of 69% (Van Laere et al. 2007). For practical implementation, the challenge is to design gratings which can be fabricated on chip with few or no extra fabrication steps beyond that required for the on-chip device itself. While many grating coupler designs are effective, they usually require multiple etch depths, angled etches, or additional layer depositions.

3.4 Dispersive Waveguide Devices

The previously described waveguide elements address the elementary problems of coupling light on and off chip, and distributing light within the chip. The following section reviews waveguide circuit elements with dispersive properties that can be used to implement fundamental optical functions such as wavelength filtering, dispersion control, and converting optical phase shifts into measurable intensity changes.

3.4.1 Mach–Zehnder Interferometers

The Mach–Zehnder interferometer (MZI) is the first example of a device that combines basic waveguide elements into a circuit that performs a specific function. Figure 3.18 shows a schematic diagram of a waveguide MZI, and the output intensity variation. Here two 1×2 splitters are aligned back to back, and are connected by simple waveguide sections. The input waveguide is split into two arms, using either a simple Y-junction, MMI coupler or direction coupler. After propagating through the two arms, the light is recombined into a single output waveguide by a second identical splitter. At the second splitter the light from the two arms will interfere, and the output intensity will vary with the relative phase difference $\phi_1 - \phi_2$ of the light arriving from the two arms. The phase accumulation of light traveling through a waveguide of length L will have the form $\phi = N_{eff}(2\pi/\lambda)L$, where N_{eff} is the effective index of

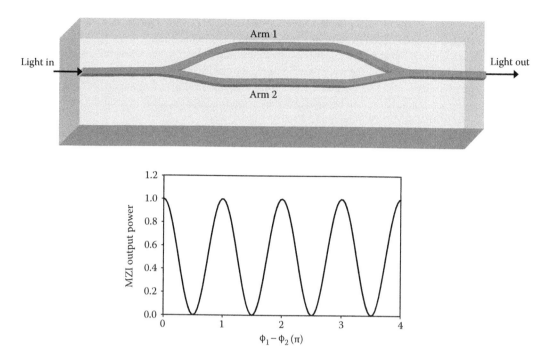

FIGURE 3.18 A schematic top view of a channel waveguide MZI, and a plot if the output power variation with the phase difference between light traveling in the two arms.

the waveguides. In the simplest case where the two waveguide arms are identical except for a length difference ΔL, the MZI acts as purely passive wavelength filter with an oscillating output spectrum of the form:

$$I = \cos^2\left(N_{\text{eff}}\frac{2\pi}{\lambda}\Delta L\right) \tag{3.8}$$

Therefore MZI devices can be used as passive or tunable optical filter elements. Although the wavelength response of a single MZI is a simple sinusoidal spectrum as given by Equation 8, by combining a series of MZI devices an arbitrarily complex filter function can be built up (Offrein et al. 2000).

If the effective index itself is changed by an increment δN_{eff} over a section of length L_s in one arm of the MZI, the output intensity from Equation 3.8 can be rewritten as

$$I = \cos^2\left(\delta N_{\text{eff}}\frac{2\pi}{\lambda}L_s + \varphi_0\right) \tag{3.9}$$

where the background phase φ_0 now includes the phase difference $N_{\text{eff}}(2\pi/\lambda)\Delta L$ that originated from the physical length difference between the two MZI arms. In this case, it becomes possible to control the output of the MZI by changing δN_{eff}. Since the MZI thereby provides a means of converting refractive index change into a measurable intensity change, the MZI forms the basis for high-speed optical modulators. The ability to transform phase into intensity response makes MZIs one of the essential configurations in implementing waveguide-based sensors of fluid properties and molecular detection.

In telecommunications, the local effective index of one of the MZI arms can be modulated through a number of electro-optical effects. In the case of silicon and III–V semiconductor waveguides, the index modulation can be achieved through injection of electrons and holes into the waveguide core (Xu et al. 2005, Ng et al. 2007), or through the use of electric field induced refractive index changes based on the quantum confined Stark effect or Pockels effect (Wakita 1998). Similarly, high-speed lithium niobate (LiNbO$_3$) waveguide modulators can be implemented using the Pockels effect to modulate one or both arms of a LiNbO$_3$ MZI device (Wooten et al. 2000).

MZIs can also be used to interrogate waveguide-based molecular and liquid sensors. For example, evanescent field molecular sensors are based on measuring the change in the waveguide effective index N_{eff} as molecules of a particular species become attached to the waveguide surface. Similar devices can be used to detect changes in refractive index of liquid above the waveguide surface. Although these effective index changes are usually less than $\delta N_{eff} \sim 10^{-3}$, such small changes are easily measured using an MZI circuit, in which the sensor element and MZI are both monolithically integrated on a chip (Luff et al. 1998, Densmore et al. 2008). Figure 3.19 shows one example in which a spiral silicon photonic wire waveguide sensor is inserted in one arm of the MZI. The spiral waveguide geometry is used to amplify the sensor response by making the sensor waveguide length (i.e., L_s in Equation 3.9) as long as possible. When molecules bind to the waveguide surface, the effective index changes and the transmitted power through the MZI undergoes a series of oscillations as described by Equation 3.9. The detection limit of such MZI based sensors has been shown to be much less than 1% of a monolayer of protein molecules (Densmore et al. 2008), largely because of the strong interaction of the propagating waveguide mode with surface molecules that occur in high index contrast systems such as SOI. A particularly useful feature of MZI interferometers is that the overall optical path length of the two arms can be balanced, effectively setting the ϕ_0 term in Equation 3.9 to zero. By doing this the temperature dependence and wavelength dependence of the MZI output, which arise largely through variation of the product $N_{eff}\Delta L$, can be reduced or eliminated without any loss of sensitivity to molecular binding. This considerably simplifies the performance requirements of the light source and packaging of the sensor chips.

In the case of MZI biosensors and modulators, the advantages of an integrated optics approach become obvious. The MZI is the waveguide analog of the two-beam interference geometry (e.g., Michelson interferometer) in free space. In the free space version, the incident beam is split into two beams by partially reflecting mirrors; beam paths must be carefully aligned, and then the two beams must be recombined. The entire arrangement can take a considerable amount of space, and the output is very sensitive to misalignment and thermal/mechanical drift. The waveguide MZI accomplishes the same task, but since the

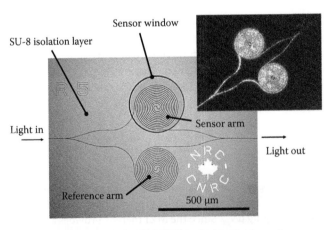

FIGURE 3.19 A top view of an MZI with folded spiral arms, where the arm in the sensor window is exposed to the overlying fluid for optical sensing. The inset shows an IR camera image when $\lambda = 1550\,nm$ light is coupled into the device. Note that because of destructive interference there is no light in the output waveguide.

splitting, propagation, and recombination of light are all accomplished on a monolithic chip, the entire device is impervious to most disturbances short of a catastrophic mechanical shock.

3.4.2 Ring Resonators

Optical resonators can be used as optical filters, to convert a local refractive index change into a power variation (e.g., for modulators), or to enhance light–matter interactions by amplifying the optical power within the resonator cavity (e.g., lasers). In free-space optics variations of the Fabry-Perot (FP) resonator, a cavity formed between two partially reflecting mirrors, are used to implement these functions. The waveguide ring resonators shown in Figure 3.20 are an integrated optic analog of the FP cavity. The resonator consists of a circular or oval waveguide resonator cavity that is coupled through a 2×2 coupler to a straight bus or drop waveguide, and possibly a second output waveguide as in Figure 3.20b, though this is not essential for every application. The couplers depicted in Figure 3.20 are directional couplers, but the exact coupling mechanism itself is not important. However, the resonator properties are determined by the coupler performance. Figure 3.21 shows a schematic of generalized 2×2 couplers. The coupler performance is characterized by the transmission coefficients t_{12} and t_{34} describing the modal field amplitude transmitted through the coupler to the same waveguide, and the coupling coefficients κ_{14} and κ_{32} describing the mode field transferred to the adjacent waveguide, such that the modal fields at the output ports 2 and 3 are given by $E_2=t_{12}E_1+\kappa_{32}E_3$ and $E_4=t_{34}E_3+\kappa_{14}E_1$. For a symmetric coupler with identical waveguides at each port, one can take the magnitudes of the transmission coefficients to be the same, $|t_{12}|=|t_{34}|=|t|$, and the same applies for the coupling coefficients so that $|\kappa_{14}|=|\kappa_{32}|=|\kappa|$.

In operation, a power fraction $|\kappa|^2$ of the light propagating along the input waveguide (port 1) in Figure 3.21 will couple into the ring waveguide. The remainder is transmitted through the coupler to the drop waveguide output with a transmitted power $|t|^2$. Similarly, if a second "add" waveguide is present as in Figure 3.20b, light will also couple from the ring to this waveguide with the field coupling and transmission coefficient denoted by κ' and t'. After one trip around the ring the remaining light will be coupled back to the drop waveguide again, while the rest is transmitted back into the ring. The output

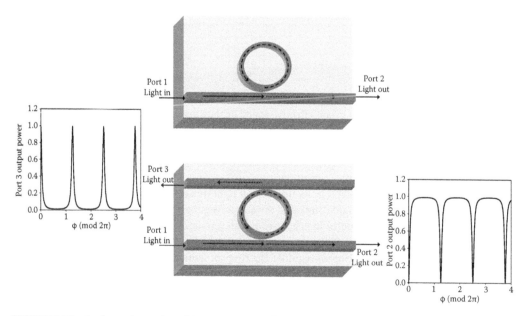

FIGURE 3.20 A schematic top view of ring resonators with one and two coupled waveguides. The plot shows typical output spectra for the drop port (port 2) and port (port 3).

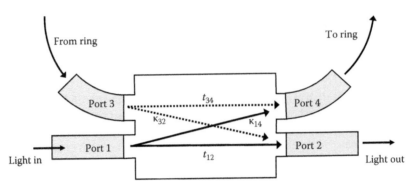

FIGURE 3.21 Generalized 2×2 coupler schematic showing the transmission coupling coefficients corresponding to the various beam paths.

intensities I_1 and I_2 at the outputs of the drop waveguide and the add waveguide respectively, as in Figure 3.20, are given by the equations (Yariv 2000)

$$I_1 = I_0 \cdot \frac{\alpha^2 t'^2 - 2\alpha|t'||t|\cos\varphi + |t|^2}{1 - 2\alpha|t'||t|\cos\varphi + \alpha^2|t'|^2|t|^2} \tag{3.10a}$$

$$I_2 = I_0 \cdot \frac{\left(1 - |t|^2\right)\left(1 - |t'|^2\right)\alpha}{1 - 2\alpha|t||t'|\cos\varphi + \alpha^2|t|^2|t'|^2} \tag{3.10b}$$

The purely dissipative loss (i.e., scattering and absorption) in the ring is represented by α (i.e., $\alpha = 1$ corresponds to no loss). The phase $\phi = N_{eff}L\,(2\pi/\lambda) + \phi_0$ in Equation 3.10 includes the propagative phase accumulated as light makes one complete circuit of the ring of length L, as well as any additional phase ϕ_0 arising in the couplers.

The resonances are spaced at wavelength intervals of the free spectral range $\Delta\lambda_{FSR}$:

$$\Delta\lambda_{FSR} = \frac{\lambda^2}{N_g L} \tag{3.11}$$

Here, $N_g = N_{eff} - \lambda(\partial N_{eff}/\partial\lambda)$ is the group effective index of the ring waveguide, a parameter that takes into account the wavelength dependence of the effective index. In the absence of dispersion, $N_g = N_{eff}$, but some waveguides of the group effective index and phase effective index can be very different. For example, in silicon photonic wire waveguides used in biosensors (Delâge et al. 2009), $N_g = 4.8$ is more than twice as large as $N_{eff} = 2.2$ of the waveguide.

The calculated transmission spectra in Figure 3.20 for light transmitted through the drop waveguide and past the ring resonator, and light coupled to the add waveguide exhibit multiple resonances whenever an integer number of wavelengths matches the overall resonator round trip path length. The width and depth of the resonances depends on the coupling between input and output waveguides and the ring, and the internal loss of the ring waveguide. The ring configuration of Figure 3.20b acts as a filter or wavelength router, since it removes light at the resonance wavelengths from the input or drop waveguide, and adds them to the add waveguide. In the case of the resonator of Figure 3.20a which has no add waveguide, Equations 3.10 still apply but with $t' = 1$. The light removed from the drop waveguide is simply absorbed or scattered away in the ring waveguide. This configuration is useful for modulators and optical sensors. If the ring loss $\alpha = 1$ (i.e., no loss), the device of Figure 3.20a will exhibit no resonances in the output intensity spectrum, thus forming an all-pass filter. However, the phase of the output light will still show a strong variation near the nominal resonance wavelength, making these all-pass filter devices useful for dispersion compensation (Madsen et al. 1999).

Since the phase ϕ shift in Equation 3.10 scales with optical path length $N_{eff}L$, any change in N_{eff} shifts the resonance wavelength, thereby changing the output power at any fixed wavelength near a resonance. Therefore, in addition to being a simple optical filter, the ring resonator can be used as the basis of an optical modulator or a sensor. The ring resonator behaves much the same as a classical FP resonator, except that the waveguide couplers take the place of mirrors. The coupler transmission coefficients, t, play the role of FP mirror reflectivity while the coupling coefficients, κ, correspond to mirror transmission. For the ring with only one bus waveguide as in Figure 3.20a, the classical analog is an FP cavity with one mirror having 100% reflectivity.

The width of the resonances depends on the total round trip resonator loss, αt, which is the product of waveguide losses within the ring (α) and coupling loss (t) at the coupler. The ratio of the 3 dB resonance line width, $\Delta\lambda_{3dB}$, to the free spectral range is known as the finesse of the resonator:

$$F = \frac{\Delta\lambda_{FSR}}{\Delta\lambda_{3dB}} = \pi \left[\arccos\left(\frac{2\alpha|t|}{1 + \alpha^2|t|^2} \right) \right]^{-1} \qquad (3.12)$$

The finesse is related to the resonator quality factor Q,

$$Q = \frac{\lambda}{\Delta\lambda_{3dB}} = \frac{\lambda}{\lambda_{FSR}} F \qquad (3.13)$$

which is often used to describe resonators in optics and electronics. In optics, the quality factor is the ratio of the energy stored in the resonator to the energy dissipated per optical cycle. However, taking Q as the ratio of wavelength and resonance line width, as implicit in Equations 3.12 and 3.13, is a more practical working definition, especially when dealing with very lossy resonators.

Rings are often used for the transmission or filtering of very narrow wavelength bands, or to translate an effective index modulation into a large intensity modulation. For these applications rings are designed to have both a high finesse (or Q) and a large extinction ratio. The resonance depth or extinction ratio is determined by the coupling conditions. At critical coupling where the transmission and loss coefficients are equal, the transmission past the ring element at resonance will be exactly zero. A useful example is the case where the ring waveguide has no loss, and the two couplers are identical. In this ideal case, all of the light is transmitted to the output waveguide on resonance, resulting in lossless wavelength filter.

Several different ring resonator configurations and applications have been explored by many groups, with the references mentioned here representing only a very small sample of the available literature. Rings have been used for wavelength selective filtering and routing (Little et al. 1997), optical modulation (Rabiei et al. 2002, Xu et al. 2005), dispersion management (Madsen et al. 1999) and optical buffer memory (Xia et al. 2007), as well as optical sensing (Ksendzov et al. 2004, Yalcin et al. 2006, De Vos et al. 2007, Xu et al. 2008).

The ring resonator can be a more complicated circuit element to implement than the simple MZI, but is often used to perform similar functions. In the case of wavelength routing and filtering the advantage of the ring is clear, since rings can be designed to give extremely sharp resonances. In sensing and modulator applications, the advantage of a resonator over simpler interferometer configurations is that light circulating repeatedly around the resonator cavity interacts with the waveguide more than once, amplifying the phase shift induced by modulation of the waveguide effective index. In comparing ring resonators and MZI, one may consider the effective index change δN_{eff} required to cause a 3 dB (50%)

change in intensity for each. The effective index change needed to cause ring resonance wavelength shift of half the resonance line width, $\Delta\lambda_{3dB}$, is

$$\delta N_{eff} = \frac{\lambda}{2LF} \tag{3.14}$$

Near critical coupling, this corresponds to a 3 dB change in output intensity. Hence the response of a ring resonator of any given length can also be enhanced by increasing the cavity finesse, thus providing an additional degree of design freedom. A comparable intensity change in an MZI modulator or sensor requires an MZI sensor or modulator arm length equivalent to the ring length L scaled by a factor $F/4$ (Janz et al. 2009). A more detailed description of ring resonators in sensor and optofluidic applications will be given by Suter and Fan in Chapter 11.

3.4.3 Waveguide Spectrometers

The development of waveguide spectrometer devices was greatly accelerated by the advent of wavelength division multiplexing (WDM) in optical telecommunications during the early 1990s, and the consequent need for devices that multiplex and demultiplex signals on many different wavelength channels to and from a single optical fiber. These functions can be carried out using discrete optical filters, or free-space grating spectrometers, but building such devices requires rather complex opto-mechanical assembly and alignment of many discrete optical elements. Waveguide spectrometers offer a much smaller and robust package that is entirely monolithic from input waveguide to the individual output wavelength channels.

3.4.3.1 Echelle Grating Spectrometers

A waveguide spectrometer can be designed by adapting a conventional free-space grating spectrometer design into the planar waveguide. This approach was in fact taken by the first researchers reporting waveguide-based spectrometers (Watanabe and Nosu 1980, Yen et al. 1981, Fujii and Minowa 1983), and this work eventually evolved into echelle grating devices fabricated and demonstrated on III–V semiconductor (He et al. 1998, Soole et al. 1991, Kremer et al. 1991) and glass (Janz 2004) waveguide platforms. The basic echelle grating device layout in Figure 3.22 consists of an input waveguide coupled to a slab section through which the light propagates and diverges to illuminate a reflection grating formed by a corrugated wall etched vertically into the waveguide surface. This curved grating diffracts and focuses light back toward an array of output waveguides positioned along an arc known as the Rowland circle. The concave grating will form an image of the input waveguide at the Rowland circle (März 1995). The operation of a waveguide echelle grating spectrometer is essentially identical to that of its free space counterpart. As the wavelength of light changes, the angle of diffraction θ (as measured relative to the grating normal) changes according to the grating equation (Delâge et al. 2004):

$$\sin\theta = \frac{m\lambda}{dN_{eff}} - \sin\varphi \tag{3.15}$$

where φ is the incident angle and m is the diffraction order. The diffracted focal spot shifts along the Rowland circle. The output waveguides are arrayed along this arc at positions and spacing chosen to give the desired channel wavelength and channel separation.

Waveguide echelle grating spectrometers have been designed and fabricated to give state-of-the-art performance for WDM applications. The glass echelle waveguide structure described in Janz et al. (2004) had 48-wavelength channels with 100 GHz (~0.8 nm) channel spacing. It was fabricated using a 5 µm thick phosphorous doped silica glass (PSG) core surrounded by upper and lower cladding glass

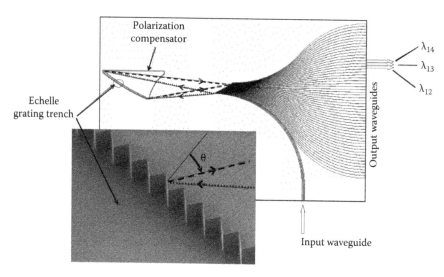

FIGURE 3.22 The layout of an echelle grating waveguide spectrometer. The inset shows an electron microscope image of a detail of the fabricated diffraction grating (Janz 2004). The indicated angle, θ, is the diffraction angle as measured from the normal to the grating line.

layers, with an index step of approximately 1% between the core and cladding. As shown in Figure 3.20, the diffraction grating is formed by etching a 10 μm deep trench through the core layer. The wall of the trench facing the input and output waveguides on the Rowland circle is stepped to form a 20th order echelle grating. After etching the grating facet is metal coated to enhance reflectivity. The fiber-to-fiber insertion loss for the 48-channel devices is approximately 4 dB, with a measured adjacent channel crosstalk better than −35 dB. The overall die size was $18 \times 17 \, \text{mm}^2$ for the 48-channel device.

Nevertheless, echelle gratings present unique design challenges for both the waveguide designer and the fabrication process engineer. The grating diffraction efficiency depends critically on the verticality of the echelle grating facets. In the case of a glass waveguide echelle grating device, a grating wall tilt of only 1° from the vertical can create a coupling loss of more than 0.5 dB from the incident fundamental mode to the reflected fundamental mode (Delâge et al. 2004). The diffraction efficiency of a metal coated grating is also polarization dependent due to the different electromagnetic boundary conditions for the components of the optical electric field polarized parallel and perpendicular to the grating surfaces. To suppress this polarization dependence and avoid a strong polarization dependent loss in the channel outputs, metal was selectively applied to only one face of each grating tooth (Janz 2004).

A polarization dependent channel wavelength shift arises from the differences in waveguide effective index for the TE and TM polarized light, largely arising from stress in the case of glass waveguides as discussed in Section 3.2.2.1. Since the dispersive section is a slab waveguide, birefringence reduction methods developed for ridge or channel waveguides cannot be used. An alternative polarization correction method has been developed that uses a triangular shaped compensator section in the slab waveguide between the grating and the output waveguides (He et al. 1999). By modifying the birefringence and adjusting the shape of this section, paths of the TE and TM diffracted beams are bent by slightly different angles so that light of both polarizations arrives at the same location on the Rowland circle. Birefringence adjustment can be achieved simply by thinning the waveguide in the compensator section, as was done in III–V devices reported by He et al. (1998). Another approach is to add a thin high index layer to the compensator section to modify the birefringence. This method was implemented in a glass 40-channel echelle grating device using silicon nitride as the high index layer (Janz 2004). In this device the polarization induced wavelength shift was less than 10 pm, or about 1% of the 0.8 nm channel spacing.

Despite the difficulties faced in designing and fabricating echelle grating waveguide devices, echelle gratings offer some interesting advantages. The echelle configuration is several times smaller than the equivalent AWG spectrometer as described in the following section, since the incident beam is diffracted back on itself to give a folded beam path. An echelle spectrometer can also easily be scaled up to larger channel counts of more than 160 channels and smaller channel spacing. Similar scaling of AWG devices quickly leads to components of unmanageable size, because of the length increment conditions on the individual array waveguides. Nevertheless, in recent years the waveguide spectrometer field has come to be dominated by AWG technology.

3.4.3.2 Arrayed Waveguide Grating Demultiplexer

The AWG spectrometer replaces the diffraction grating of an echelle device with an array of waveguides of constantly incrementing length. The fundamental concept was first proposed by Smit (1988), and rapidly evolved into the typical AWG configuration shown in Figure 3.23. Here light is coupled into a single input guide leading to a slab waveguide section. This section is simply a star coupler as described in Section 3.3.2.4, with the input light spreading by diffraction and illuminating an array of many waveguides. The length of each successive waveguide in the array increases by a constant increment ΔL. The array waveguides all converge to a second slab coupler which is often identical to the initial star coupler, but now operating in reverse as a combiner that brings light to a focus at its opposite end where the output channel waveguides are located. Because of the length increment between each array waveguide, the corresponding phase increment between light emerging from adjacent waveguides varies linearly across the array with a phase increment $\Delta\phi = N_{eff}\pi\Delta L/\lambda$. Here N_{eff} is again the effective index of the array waveguide mode. As a result, light from all the array waveguide outputs combines coherently to form a converging beam propagating at an angle θ relative to the combiner center line. At a certain center wavelength λ_0 the phase increment is a multiple of 2π, so $\theta=0$ and the beam is focused on the center channel at the far side of the combiner.

As the wavelength shifts by $\Delta\lambda = \lambda - \lambda_0$, the phase increment at the array waveguide outputs is no longer an exact multiple of 2π, and the beam is directed at an angle θ away from the center waveguide. This diffraction angle can be found from the equation

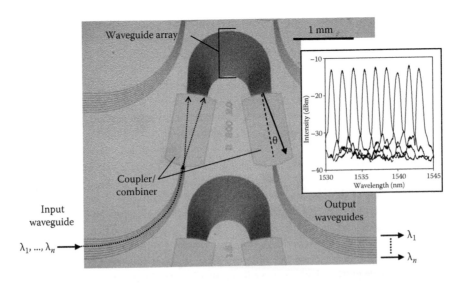

FIGURE 3.23 A microscope image of a nine-channel arrayed waveguide grating spectrometer on a silicon-on-insulator chip. The inset shows the measured channel wavelength spectra for each of the nine output waveguides.

$$\sin\theta = \frac{\Delta\lambda m}{dn_c} \cdot \frac{n_g}{N_{eff}} \qquad (3.16)$$

Here

d is the separation of the array waveguide outputs at the combiner star coupler
n_c is the effective index of the combiner slab waveguide
the order m is defined in terms of the center wavelength λ_0 as $m = N_{eff}\Delta L/\lambda_0$
n_g is the group effective index of the array waveguide

Equation 3.16 has the same form as the grating equation that also describes diffraction from a conventional ruled grating in Littrow configuration, when illuminated at normal incidence. Thus from a mathematical point of view the operation of a conventional grating, echelle grating, and AWG are essentially the same. The great advantage of the AWG is that fabrication requires only the ability to make single-mode waveguides with sufficient reproducibility and uniformity that the relevant phase relationships between the optical paths through the array section are maintained. Since the first proposal by Smit, AWGs have evolved into a mature technology that displaces the dielectric filter demultiplexer in the commercial telecommunications market (Smit and van Dam 1996).

We have presented the two most established approaches to planar waveguide demultilpexers in this section. The AWG is a mature technology that has been commercialized for a number of years, and is based on standard waveguide technology. While proposed and demonstrated first, diffraction grating configuration with acceptable performance for WDM has only been recently achieved on glass waveguides, and a number of unique challenges needed to be overcome to successfully fabricate this device. There are also other approaches to planar waveguide spectrometers and wavelength demultiplexing being explored which cannot be included here, such as the use of superprism effects in two-dimensional photonic crystals (Kosaka et al. 1999), and assemblies of ring resonators as wavelength filters.

3.5 Waveguides and Optofluidics

Optofluidics and the use of waveguides to facilitate the interaction of light with fluids will be covered in detail in various chapters of this book. This section will only attempt to highlight how some of the passive integrated optics concepts outlined in this chapter are applied in probing fluids, with examples taken from the area of refractive index-based molecular sensing.

The application of integrated optics to optofluidics can be divided into two categories. The first covers the use of waveguides to simply deliver light to a fluid, or collect light from a fluid. For example, waveguides can be used to deliver pump light to excite fluorescence in a liquid, and collect fluorescent light emitted by an optically excited fluid (Ruano et al. 2003, Yin et al. 2005, 2007). Applications involving fluorescence include molecular detection by fluorescent tagging and excitation of target molecules. The second category encompasses those devices where the measurement of interaction of light with the fluid depends on determining precise values of the optical phase. This class of devices includes most waveguide refractive index sensors, since the transduction signal is always a change in optical phase that must be measured using interferometric methods. Evanescent field molecular sensors are one example. In these devices the waveguide must be single mode so that the optical phase along the entire waveguide length is well defined.

Waveguides that are primarily intended for collection and delivery of light (e.g., in fluorescence sensing) can be multimode and have large cross sections, since the optical phase of the propagating light is not a critical parameter. Such waveguides can also be useful in measuring optical absorption of a fluid. However, these waveguides must enclose and confine a fluid, and at the same time guide light. Such waveguide may have a hollow core that acts as the fluidic channel, and there are a number of strategies for accomplishing this. The waveguide cladding may be deposited over a sacrificial core that is subsequently etched away leaving a hollow channel in the center of the waveguide (Hawkins and Schmidt

2007, Decorby et al. 2007). The channel may be formed by natural enclosed void that is created between two closely spaced glass ridges deposited by PECVD (Dumais et al. 2005). Finally the channels may be formed by a combination of micromachining and layer deposition (Lo et al. 2004). Optical waveguides with liquid cores can take many forms, including hollow-core photonic crystal fiber, Bragg waveguides, ARROWs, and slot and liquid–liquid waveguides. Their characteristics and applications to biosensing are covered in detail in Chapters 15 and 16.

A different approach to waveguide-based optofluidic device is to let the fluid form the cladding of the waveguide, and light interacts with the fluid through the evanescent tail of the waveguide mode. Every waveguide mode has some portion of its electric field distribution that extends outside the high index waveguide core layer into the cladding, as described in Section 2.1. This field decays exponentially to zero with distance into the cladding, with a decay length δ determined from Equation 3.5:

$$\frac{1}{\delta} = \left(\frac{2\pi}{\lambda}\right)\sqrt{N_{\text{eff}}^2 - n_{\text{clad}}^2} \tag{3.17}$$

This length depends on the difference between the waveguide effective index N_{eff} and the cladding material index. The evanescent field-cladding interaction can be used to deliver light that excites fluorescence in a fluid that surrounds the waveguide, or measure its optical absorption. However, another useful capability of evanescent field probing is the sensitivity to fluid refractive index near the waveguide surface. A change in refractive index manifests itself through a change in the phase velocity of the waveguide mode $v = c/N_{\text{eff}}$. Hence a waveguide-based refractive index measurement must employ a single-mode waveguide so that the phase is well defined, and can be measured using interferometric techniques such as the MZI or ring resonators described previously.

Optical perturbation theory can be used to obtain a general expression for the effective index change δN_{eff} induced by a small localized refractive index change (Kogelnik 1990). If the function $\delta n(x, y)$ describes the refractive index perturbation at any point in the waveguide, the corresponding change in waveguide mode effective index is given by

$$\delta N_{\text{eff}} = c \int \Delta \varepsilon E E^* \, dy \, dx = 2c\varepsilon_0 \int n \, \Delta n E E^* \, dy \, dx \tag{3.18}$$

Here E is the electric field amplitude of the waveguide mode, normalized such that guide power is unity, and the integral is over the cross section of the waveguide. This equation predicts that the shift in mode effective index is proportional to the overlap of the modal intensity ($I \sim EE^*$) with the perturbed refractive index distribution. In most waveguides the portion of the waveguide mode extending into the cladding is very small, and hence interaction of the guided light with the cladding materials is very weak. However, waveguides can be engineered to enhance the evanescent field-cladding overlap (Tiefenthaler and Lukosz 1989, Parriaux and Veldhuis 1998). For example, the waveguide mode can be forced to expand into the waveguide cladding simply by reducing the size of the waveguide core to the point where most of the modal field is propagating in the cladding material. A complementary strategy is to choose a substrate cladding with a much lower refractive index than the sample liquid forming the upper cladding. This reverse symmetry index profile (Horvath et al. 2002) causes the waveguide mode evanescent tail to extend further into the liquid. Such reverse symmetry waveguides can be used to study not just bulk liquid properties but also the attachment of and dynamics of living cells on the waveguide surface (Horvath et al. 2005).

Evanescent field waveguide refractive index sensors can be used to monitor changes of fluid chemical composition since the liquid refractive index will change with dissolved molecular concentration. Measuring fluid glucose concentration in water is one example (Chao and Guo 2003). However, in biomolecular studies, a refractive index measurement alone is of limited value because there is no specificity.

In other words, a simple refractive index measurement cannot be used to identify molecular species in solution, but only mark a change in concentration of any one of possibly many fluid components.

The more important application for evanescent field waveguide sensor is to monitor refractive index changes occurring within a few nanometers of the waveguide surface, in particular those caused by the attachment of molecules to the waveguide surface (Lukosz 1991, Luff et al. 1996, 1998, Weisser et al. 1999, Boyd and Heebner 2001, Prieto et al. 2003, Densmore et al. 2008, Xu et al. 2008). These molecular sensor surfaces are usually functionalized by the prior attachment of a receptor molecule, that will ideally form a bond only with a specific target molecule of interest. When target molecules are present in the overlying fluid, they will bind to the receptor and change the refractive index in the near surface region. As in the case of a bulk liquid refractive index sensor, the interaction of the evanescent tail of the waveguide mode with the region of modified refractive index causes the mode effective index to shift. Molecular binding thereby produces a phase change $\Delta\phi = \delta N_{eff} L$ in the light transmitted through a sensor waveguide of length L, while molecular identification is provided through the specificity of the receptor–target molecule binding reaction.

Surface affinity binding sensors differ in one fundamental aspect from bulk fluid refractive index sensors. In a surface sensor the region of perturbed refractive index extends only a few nanometers above the exposed waveguide surface. In the integral of Equation 3.18, the molecular layer thickness is usually much thinner than the evanescent field decay length. The sensor response therefore depends on the localization of the mode field at the waveguide surface. In bulk sensors, the response depends only on the fraction of the mode propagating in the liquid, but the distance the evanescent tail extends into the liquid makes little difference to the overall response. On the other hand, optimizing a surface affinity binding sensor also requires the evanescent field to be tightly confined to the near surface region. It is evident from Equation 3.16 that the latter requirement is best met by choosing a waveguide with a very large difference between the waveguide mode effective index and the cladding liquid refractive index. Therefore, the optimal surface affinity binding sensor should have a high core refractive index to give a very short evanescent field decay length, and also be as thin as possible so that a large fraction of the mode power is contained in the evanescent tail. Finally, a significant enhancement is surface sensor response is obtained by choosing the waveguide mode polarization to be perpendicular to the waveguide surface. The electromagnetic boundary conditions require the normal vector component of the displacement $\mathbf{D} = \varepsilon \mathbf{E}$ to be continuous across an interface, forcing the electric field amplitude at the sensor surface to increase discontinuously across the interface by a factor of $\varepsilon_{core}/\varepsilon_{clad} = (n_{core}/n_{clad})^2$. In the case of a silicon waveguide, this can result in a factor of three increase in sensor response for the TM polarized mode, relative to the TE polarized mode in the same waveguide (Densmore et al. 2006).

Due to the much higher refractive index of Si ($n = 3.5$) relative to the surroundings, silicon waveguides with a thickness between 200 and 300 nm have the highest response to surface binding of any of the common waveguide material platforms (Densmore et al. 2006). This has been confirmed by recent work (Densmore et al. 2008, Xu et al. 2008) on silicon photonic wire waveguide sensors. Figure 3.24 shows the TM mode profile for a silicon photonic wire waveguide as used in sensor applications, for $\lambda = 1550$ nm light. This waveguide is 0.22 µm thick and 0.45 µm wide. The thickness corresponds to the point of maximum response to molecular binding on the waveguide surface (Densmore et al. 2006), and the width is the maximum possible while ensuring the waveguide can only support a single mode. Here the light propagates mostly in the cladding region within approximately 100 nm of the surface, resulting in the high sensitivity to any near surface perturbations.

The engineering of waveguide size, refractive index, and electromagnetic boundary conditions has also led to a new class of slot waveguides in which light is confined in an extremely small channel separating two silicon photonic wire waveguides (Almeida et al. 2004). Figure 3.25 shows a cross section of such a waveguide as well as the waveguide mode intensity profile. Here the two Si waveguides are separated by a 60 nm gap. The slot waveguide mode overlap with the liquid is almost unity, so that such waveguides can make very efficient probes of bulk liquid properties in the narrow channel. In addition, the overlap integral of the waveguide mode and the near surface region at the inner waveguide

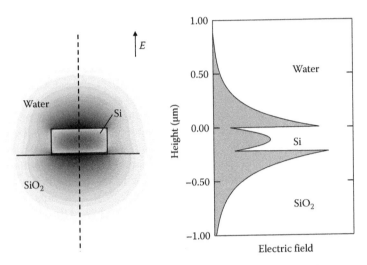

FIGURE 3.24 The waveguide mode profile of a silicon photonic wire sensor shown in the cross section at left, and as an amplitude plot along the dashed line.

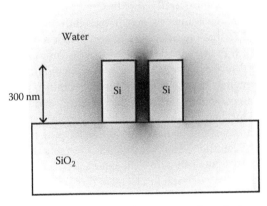

FIGURE 3.25 A cross section of a silicon slot waveguide consisting of two Si channel waveguides separated by a 60 nm slot. The horizontally polarized component of the electric field for the TE mode profile is shown as a grayscale contour plot (dark=high field amplitude).

sidewalls can be several times higher than for the simple photonic wire sensor described above, since the evanescent field is now completely localized to the gap spacing of 60 nm rather than the natural decay length as calculated by Equation 3.17. However, due to the high aspect ratio, functionalizing the narrow gap surface with appropriate molecular layers, and transporting and exchanging fluids within these extremely narrow gaps requires more care in surface preparation and fluidic design. Recent work has shown that in addition to sensing, this high electric field in slot waveguides can be used to create optical traps for particles and large molecules such as DNA (Yang et al. 2009). The properties and applications of slot waveguides are discussed in detail in Part III of this book.

The fundamental response of most evanescent field waveguide sensors is a change in the effective index of the waveguide mode, which is equivalent to a change in phase velocity. As a result, the transduction signal for light propagating through a sensor is a cumulative phase change $\phi = \delta N_{eff}L(2\pi/\lambda)$, that is proportional to both the effective index shift δN_{eff} (which scales with the surface density of bound molecules) and the sensor length L. This phase change can be converted into a directly measurable

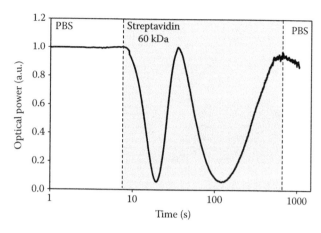

FIGURE 3.26 The measured output intensity of an MZI incorporating a silicon photonic wire sensor as Figure 3.19, as streptavidin molecules form a monolayer on the sensor surface (Densmore 2008). The shade area indicates the time interval over which the sensor is exposed to the streptavidin molecules in the solution. Before and after exposure, the sensor is immersed in PBS solution.

intensity change by incorporating the waveguide sensor into the ring resonator or MZIs as described in previous sections. Figure 3.26 shows the output intensity of MZI incorporating a silicon photonic wire sensor in one arm. As a monolayer of streptavidin protein binds to the surface, the output signal undergoes several full intensity swings, suggesting that the sensor resolution is significantly less than 1% of a monolayer.

It is also possible to measure the molecular binding induced effective index shift using a grating embedded in the waveguide. Light will be coupled into a waveguide, or out of a waveguide, using a grating with periodicity chosen such that the grating wave vector matches the difference between the waveguide propagation wave vector $\beta = N_{\text{eff}}(2\pi/\lambda)$ and the parallel component of a free-space wave vector $k_p = (2\pi/\lambda) \sin \theta$, where θ is the incident or emitted beam angle with respect to the waveguide surface normal, as shown schematically in Figure 3.27. Since β changes as the effective index, molecules binding to the surface of the waveguide will change the optimal coupling angle θ, or the optimal coupling wavelength if θ is fixed. For example, some of the earliest waveguide sensors were based on a grating coupler that coupled light into a waveguide, and the measured transduction signal was the change in coupled intensity as molecules bound to the waveguide grating surface (Tiefenthaler and Lukosz 1989). A closely related configuration uses light reflected from a waveguide grating structure. In this configuration incident light is coupled into the sensor waveguide by the grating, but light in the waveguide is coupled out into the specular reflected beam direction. As a result the reflected power in a coherent superposition of specularly reflected light and light emitted in the same direction by the grating waveguide structure. Near the grating waveguide coupling resonance, this reflected beam shows large resonant peak. As before, shifts in the resonance angle or wavelength can be used to detect changes in the waveguide effective index. Such waveguide grating sensors have been implemented recently in 0.22 μm silicon slab waveguides (Sinclair 2009) using the backside illumination configuration of Figure 3.27a. Figure 3.27b shows the calculated resonance shift for such a sensor, as a 2 nm thick layer with refractive index $n = 1.5$ is deposited on the surface. In this example the grating is formed by lines of 180 nm thick SiO_2 patterned on the silicon waveguide surface. The ambient fluid is assumed to be water.

The advantage of the grating configurations is that the sensor may be interrogated in a reflection geometry using simple bulk optics. No fiber-to-waveguide mode coupling and attachment is needed. However, the response of such grating sensors depends on a number of factors. The mean free path (or photon lifetime) within the grating waveguide structure and the incident beam diameter should be matched. The longer the photon travels within the waveguide, the sharper the grating resonance will be

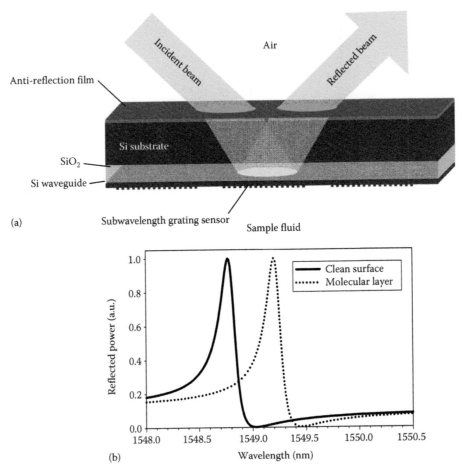

FIGURE 3.27 (a) A schematic diagram of a grating-based silicon evanescent field sensor. (b) The calculated grating reflectivity spectra for a clean surface and with a 2 nm thick layer with index $n = 1.5$ deposited on the sensor.

and the bigger the intensity change upon molecular adsorption. The grating must therefore be carefully designed to give the desired photon path length. Achieving very sharp grating resonances requires that the incident beam diameter be large (>5 mm) and very well collimated. This begins to place demands on the quality and alignment of the integration optics.

On the other hand, in simple waveguide sensors the sensor length can be as long as the achievable waveguide loss permits. Since the phase response scales with sensor length, the response to molecular binding can be very high. While an extremely long straight waveguide can result in an impractically large sensor size, silicon photonic wire waveguide sensors can be folded into extremely compact spiral and serpentine geometries such that several millimeter long waveguides are contained in a total chip area less than a few hundred micrometers across (Densmore et al. 2008, Xu et al. 2008). Such structures can be incorporated into both ring resonator and the Mach–Zehnder interrogation circuit shown in Figure 3.19. These folded structures are made possible by the high refractive index of silicon which allows lossless waveguide bends with radii less than 10 μm. While the waveguide length gives these sensors a higher response to molecular binding, the small chip area of each sensor allows them to be functionalized using inkjet-like spotting machines that dispense small microdroplets of solutions at precisely determined locations. Such sensor geometries are also much more compatible with liquid sample delivery by microfluidic channels, since the sensor fits well within typical microchannel sizes of a few hundred micrometers across. Since the channel floor is covered by a dense network of sensor waveguides, the sensor-fluid cross section is high leading to a much faster response and smaller liquid volumes.

3.6 Summary

This chapter has presented an overview of some of the basic materials and passive waveguide building blocks that are used to construct passive integrated optical circuits. In combination with active materials and device geometries, these passive elements enable much of the modern telecommunications infrastructure, and are also used for new optical technologies in optofluidics, biosensing, and short range optical data interconnects. There is an obvious analogy between waveguides carrying light in an optical circuit and the metal lines carrying electrons in an electronic circuit. Nevertheless, unlike metal interconnects, the design of even the most elementary waveguide circuit elements described in this chapter requires the understanding and analysis of wave propagation through the structures. Design considerations for the fundamental elements such as waveguide bends, 1×2 splitters, star couplers, interferometers, and resonators have been reviewed, albeit only superficially. Despite the added complexity, waveguide optics offers an optical device designer the ability to make extremely compact, robust, monolithic optical devices that perform many basic optical processing functions including signal distribution, wavelength routing, and filtering.

Some of the material systems commonly used in integrated optics have been introduced here as well. The III–V semiconductor platform provides the lasers, photodetectors, and modulators used in optical telecommunications. Glass waveguides are another well-established waveguide platform that is used extensively in telecommunications, for example in AWG multiplexers and demultiplexers to combine and separate wavelength channels in WDM systems. Finally, recent work on the silicon photonics platform has been reviewed. This last material system is of great current research interest in part because of its potential for CMOS compatible optical interconnects. However, it also provides a platform that allows optical devices to be scaled down to the subwavelength dimensions, so that light and its interaction with matter can be manipulated at the nanometer scale.

The ability to manipulate the phase, amplitude, and spatial distribution of light at microscopic length scales enables researchers and engineers to exploit the interaction of light with fluids in ways that are not possible with bulk optics. Hollow waveguides can be fabricated that enclose both fluid and guided light in the same channel so it becomes possible to carry out optical monitoring of very small volumes of liquids and the molecules they contain through measurements of fluorescence, absorption, and refractive index. The alternate approach uses solid core nanometer size waveguides such as silicon photonic wires. Here a significant fraction of the guide optical mode propagates in the liquid around the waveguide, in the evanescent tail light that extends about 100 nm beyond the waveguide core. Therefore, complex waveguide circuits can be built up that are designed for probing the fluid above the waveguide, monitoring the properties of the liquid or detecting trace amounts of molecules through techniques such as affinity binding measurements. Finally, recent work is pointing the way to using waveguides to create intense local electric fields that not only detect but actually trap and manipulate particles and molecules. By incorporating the basic integrated optic circuit elements reviewed in this chapter and elsewhere, it also becomes possible to incorporate the sensors and optical actuators with basic optical signal processing functions on monolithic integrated optical chips.

References

Ahmad, R.U., Pizzuto, F., Camarda, G.S., Espinola, R.L., Rao, H., and Osgoode, R.M. 2002. Ultracompact corner-mirrors and T-branches in silicon-on-insulator. *IEEE Phot. Technol. Lett.* 14: 65–67.

Alder, T., Stöhr, A., Heinzelmann, R., and Jäger, D. 2000. High efficiency fiber-to-chip coupling using low-loss tapered single-mode fiber. *IEEE Phot. Technol. Lett.* 12: 1016–1018.

Almeida, V.R., Panepucci, R.R., and Lipson, M. 2003. Nanotaper for compact mode conversion. *Opt. Lett.* 28: 1302–1304.

Almeida, V.R., Xu, Q., Barrios, C.A., and Lipson, M. 2004. Guiding and confining light in void nanostructure. *Opt. Lett.* 29: 1209–1211.

Bachmann, M., Besse, P., and Melchior, H. 1994. General self-imaging properties in N×N multimode interference couplers including phase relations. *Appl. Opt.* 33: 3905–3911.

Baldwin, J., Schüler, N., Butler, I.S., and Andrews, M.P. 1996. Integrated optics evanescent wave surface enhanced Raman scattering (IO-EWSERS) of mercaptopyridines on a planar optical chemical bench: Binding of hydrogen and copper ion. *Langmuir* 12: 6389–6398.

Barwicz, T. and Haus, H.A. 2005. Three-dimensional analysis of scattering losses due to sidewall roughness in microphotonic waveguides. *J. Lightwave Technol.* 23: 2719–2732.

Bazylenko, M.V., Gross, M., Allen, P.M., and Chu, P.L. 1995. Fabrication of low-temperature PECVD channel waveguides with significantly improved loss in the 1.50–1.55 mm wavelength range. *IEEE Phot. Technol. Lett.* 7: 774–776.

Bernini, R., Campopiano, S., Zeni, L., and Sarro, P.M. 2004. ARROW optical waveguide based sensors. *Sens. Actuators B* 100: 143–146.

Besse, P.A., Gini. E., Bachmann, M., and Melchior, H. 1996. New 2×2 and 1×3 multimode interference couplers with free selection of power splitting ratios. *J. Lightwave Technol.* 14: 2286–2293.

Bogaerts, W., Tailaert, D., Luyssaert, B. et al. 2004. Basic structures for photonic integrated circuits in silicon-on-insulator. *Opt. Express* 12: 1583–1591.

Boyd, R.W. and Heebner, J.E. 2001. Sensitive disk resonator photonic biosensor. *Appl. Opt.* 40: 5742–5747.

Campopiano, S., Bernini, R., Zeni, L., and Sarro, P.M. 2004. Microfluidic sensor based on integrated optical hollow waveguides. *Opt. Lett.* 1894–1896.

Chao, C.Y. and Guo, L.J. 2003. Biochemical sensors based on polymer microrings with sharp asymmetrical resonance. *Appl. Phys. Lett.* 83: 1527–1529.

Cheben, P., Xu, D.-X., Janz, S., and Densmore, A. 2006. Subwavelength waveguide grating for mode conversion and light coupling in integrated optics. *Opt. Express* 14: 4695–4702.

Cheben, P., Schmid, J.H., Delâge, A. et al. 2007. A high-resolution silicon-on-insulator arrayed waveguide grating microspectrometer with sub-micrometer aperture waveguides. *Opt. Express* 15: 2299–2306.

Cho, A.Y., Yariv, A., and Yeh, P. 1977. Observation of confined propagation in Bragg waveguides. *Appl. Phys. Lett.* 30: 471–472.

Chou, M.H., Arbore, M.A., and Fejer, M.M. 1996. Adiabatically tapered periodic segmentation of channel waveguides for mode-size transformation and fundamental mode excitation. *Opt. Lett.* 21: 794–796.

Chun, Y.Y., Lee, Y.T., Lee, H.J., and Chung, S.J. 1996. Birefringence reduction in a high core boron-doped silica-on-silicon planar optical waveguide. *J. Korean Phys. Soc.* 29, 140.

Curcio, J.A. and Petty, C.C. 1951. The near infrared absorption spectrum of liquid water. *J. Opt. Soc. Am.* 41: 302–304.

Datta, A., Eom, I., Dhar, A. et al. 2003. Microfabrication and characterization of Teflon coated AF-coated liquid core waveguide channels in silicon. *IEEE Sens. J.* 3: 788–795.

Decorby, R.G., Ponnampalam, N., Nguyen, H.T., Pai, M.M., and Clement, T.J. 2007. Guided self-assembly of integrated hollow Bragg waveguides. *Opt. Express* 15: 3902–3915.

Delâge, A., Bidnyk, S., Cheben, P. et al. 2004. Recent developments in integrated spectrometers. In *Proceedings of the 2004 International Conference on Telecommunications and Optical Networking (ICTON)*. Toronto, Canada: Microstruct. Sci., Nat. Res. Council of Canada, pp. 78–83.

Delâge, A., Janz, S., Lamontagne, B. et al. 2006. Monolithically integrated asymmetric graded index and step-index couplers for microphotonic waveguides. *Opt. Express* 14: 148–161.

Delâge, A., Xu, D.-X., McKinnnon, R.W. et al. 2009. Wavelength dependent model of a ring resonator sensor excited by a directional coupler. *J. Lightwave Technol.* 27: 1172–1180.

Densmore, A., Xu, D.-X., Waldron, P. et al. 2006. A silicon-on-insulator photonic wire based evanescent field sensor. *IEEE Phot. Technol. Lett.* 18: 2520–2522.

Densmore, A., Xu, D.-X., Janz, S. et al. 2008. Spiral-path high-sensitivity silicon photonic wire molecular sensor with temperature-independent response. *Opt. Lett.* 6: 596–598.

De Vos, K., Bartolozzi, I., Schacht, E., Bienstman, P., and Baets, R. 2007. Silicon-on-insulator microring resonator for sensitive and label-free biosensing. *Opt. Express* 15: 7610–7615.

Dragone, C. 1990. Optimum design of a planar array of tapered waveguides. *J. Opt. Soc. Am. A* 7: 2081–2093.

Dragone, C., Henry, C.H., Kaminow, I.P., and Kistler, R.C. 1989. Efficient multichannel integrated optics star coupler on silicon. *IEEE Phot. Technol. Lett.* 1: 241–243.

Duguay, M.A., Kokubun, Y., Koch, T., and Pfeiffer, L. 1986. Antiresonant reflecting optical waveguides in SiO_2-Si multilayer structures. *Appl. Phys. Lett.* 49: 13–15.

Dulkeith, E., Xia, F., Schares, L., Green, W.M.J., and Vlasov, Y.A. 2006. Group index and group velocity dispersion in silicon-on-insulator photonic wires. *Opt. Express* 14: 3853–3863.

Dumais, P., Callender, C.L., Noad, J.P., and Ledderhof, C.J. 2005. Silica-on-silicon optical sensor based on integrated waveguides and microchannels. *IEEE Phot. Technol. Lett.* 17: 441–443.

Dumon, P., Bogaerts, W., Wiaux, V. et al. 2004. Low-loss SOI photonic wires and ring resonators fabricated with deep UV lithography. *IEEE Phot. Technol. Lett.* 16: 1328–1330.

Dumon, P., Bogaerts, W., Van Thourhout, D., Tailllaert, D., and Baets, R. 2006. Compact wavelength router based on a silicon-on-insulator arrayed waveguide grating pigtailed to a fiber array. *Opt. Express* 14: 664–669.

Eldada, L. and Shacklette, L.W. 2000. Advances in polymer integrated optics. *IEEE J. Sel. Top. Quantum Electron.* 6: 54–68.

Fisher, U., Zinke, T., Knopf, J.R., Arndt, F., and Petermann, K. 1996. 0.1 dB/cm waveguide losses in single mode SOI rib waveguides. *IEEE Phot. Technol. Lett.* 8: 647.

Florjañczyk, M., Cheben, P., Janz, S., Scott, A., Solheim, B., and Xu, D.-X. 2007. Multiaperture planar waveguide spectrometer formed by arrayed Mach-Zehnder interferometers. *Opt. Express* 15: 18176–18189.

Fujii, Y. and Minowa, J. 1983. Optical demultiplexer using a silicon concave diffraction grating. *Appl. Opt.* 22: 974–978.

Grand, G., Jadot, J.P., Denis, H., Valette, S., Fournier, A., and Grouillet, A.M. 1990. Low loss PECVD silica channel waveguides for optical communications. *Electron. Lett.* 26: 2135–2136.

Gupta, J.A., Barrios, P.J., Aers, G.C., and Lapointe, J. 2008. 1550 nm GaInNAsSb distributed feedback laser diodes on GaAs. *Electron. Lett.* 44: 578–579.

Haguenauer, P., Berger, J.-P., Rousselet-Perrault, K. et al. 2000. Integrated optics for astronomical interferometry. III. Optical validation of a planar optics two telescope beam combiner. *Appl. Opt.* 38: 2130–2139.

Halir, R., Molina-Fernandez, I., Ortega-Monux, A. et al. 2008. A design procedure for high-performance rib-waveguide-based multimode interference couplers on silicon-on-insulator. *J. Lightwave Technol.* 26: 2928–2936.

Hawkins, R.H. and Schmidt, H. 2007. Optofluidic waveguides: II. Fabrication and structures. *Microfluid Nanofluid* 4: 17–32.

Haysom, J.E., Delâge, A., He, J.-J. et al. 1999. Experimental analysis and modeling of buried waveguides fabricated by quantum well intermixing. *IEEE J. Quantum Electron.* 35: 1354–1363.

He, J.-J., Lamontagne, B., Delage, A., Erickson, L., Davies, M., and Koteles, E.S. 1998. Monolithic integrated wavelength demultiplexer based on waveguide Rowland circle grating in InGaAsP/InP. *J. Lightwave Technol.* 16: 631–638.

He, J.-J., Koteles, E.S., Lamontagne, B., Erickson, L., Delâge, A., and Davies, M. 1999. Integrated polarization compensator for WDM waveguide demultiplexers. *IEEE Phot. Technol. Lett.* 11: 224–226.

Henry, C.H., Kazarinov, R.F., Lee, H.J., Ostrowsky, K.J., and Katz, L.E. 1987. Low loss Si_3N_4-SiO_2 optical waveguides on Si. *Appl. Opt.* 26: 2621–2624.

Horvath, R., Lindvold, L.R., and Larsen, N.B. 2002. Reverse symmetry waveguides: Theory and fabrication. *Appl. Phys. B* 74: 383–393.

Horvath, R., Pedersen, H.C., Skivesen, N., Selmeczi, D., and Larsen, N.B. 2005. Monitoring living cell attachment and spreading using reverse symmetry waveguide sensing. *Appl. Phys. Lett.* 86: 071101–071103.

Hunsperger, R.G. 1991. *Integrated Optics: Theory and Technology*, 3rd edition. Berlin, Germany: Springer-Verlag.

Izutsu, M., Nakai, Y., and Sueta, T. 1982. Operation mechanism of the single-mode optical-waveguide Y junction. *Opt. Lett.* 7: 136–138.

Janz, S. 2004. Silicon-based waveguide technology for WDM. In *Silicon Photonics*. Berlin, Germany: Springer-Verlag, pp. 323–360.

Janz, S., Baribeau, J.-M., Delâge, A. et al. 1998. Optical properties of pseudomorphic SiGe for Si based waveguides at the λ = 1300 and 1550 nm telecommunications wavelength bands. *IEEE J. Select. Top. Quantum Electron.* 4: 990–996.

Janz, S., Cheben, P., Dayan, H., and Deakos, R. 2003. Measurement of birefringence in thin film waveguides by Rayleigh scattering. *Opt. Lett.* 28: 1778–1780.

Janz, S., Balakrishnan, A., S. Charbonneau, S. et al. 2004. Planar waveguide echelle gratings in silica-on-silicon. *IEEE Phot. Technol. Lett.* 16: 503–505.

Janz, S., Densmore, A., Xu, D.-X. et al. 2009. Silicon photonic wire waveguide sensors. In *Advanced Photonic Structures for Chemical and Biological Sensing*. Fan, X., Ed., Berlin, Germany: Springer-Verlag.

Jenkins, R.M., McNie, M.E., Blockley, A.F., Price, N., and McQuillan, J. 2003. Hollow waveguides for integrated optics. In *Proceedings of European Conference on Optical Communications (ECOC)*. Rimini, Italy.

Kilian, A., Kirchoff, J., Kuhlow, B., Przyrembel, G., and Wischmann W. 2000. Birefringence free planar optical waveguide made by flame hydrolysis deposition (FHD) through tailoring of the overcladding. *J. Lightwave Technol.* 18: 193.

Kogelnik, H. 1990. Theory of optical waveguides. In *Guided Wave Optoelectronics*, 2nd edition. Tamir, T., Ed., Berlin, Germany: Springer-Verlag, pp. 7–87.

Kominato, T., Ohmori, Y., and Okazaki, H. 1990. Very low loss GeO_2 doped silica waveguides fabricated by flame hydrolysis deposition method. *Electron. Lett.* 26: 327.

Kosaka, H., Kawashima, T., Tomita, A. et al. 1999. Superprism phenomena in photonic crystals: Toward microscale lightwave circuits. *J. Lightwave Technol.* 17: 2032.

Kremer, C., Ebbinghaus, G., Heise, G., Muller-Nawrath, R., Schienle, M., and Stoll, L. 1991. Grating spectrograph in InGaAsP/InP for dense wavelength division multiplexing. *Appl. Phys. Lett.* 59: 627.

Ksendzov, A., Homer, M.L., and Manfreda, A.M. 2004. Integrated optics ring-resonator chemical sensor with polymer transduction layer. *Electron. Lett.* 40: 63–65.

Lardenois, S., Pascal, D., Vivien, L. et al. 2003. Low-loss submicrometer silicon-on-insulator rib waveguides and corner mirrors. *Opt. Lett.* 28: 1150–1152.

Le, T.L. and Cahill, L.W. 2008. The design of multimode interference couplers with arbitrary power splitting ratios on an SOI platform. In *Proceedings of the 21st IEEE Laser and Electro-optical Society Annual Meeting*. Piscatawy, NJ: IEEE, pp. 378–379.

Lee, H.J., Henry, C.H., Orlowsky, K.J., Karazinov, R.F., and Kometani, T.Y. 1988. Refractive-index dispersion of phosphosilicate glass, thermal oxide, and silicon nitride films on silicon. *Appl. Opt.* 27: 4104–4109.

Lee, K.K., Lim, D.R., Pan, D. et al. 2005. Mode transformer for miniaturized optical circuits. *Opt. Lett.* 30: 498–500.

Li, G.L. and Yu, P.K.L. 2003. Optical intensity modulators for digital and analog applications. *J. Lightwave Technol.* 21: 2010–2030.

Little, B.E., Chu, S.T., Haus, H.A., Foresi, J.S., and Laine, J.-P. 1997. Microring resonator channel dropping filters. *J. Lightwave Technol.* 15: 998–1005.

Lo, S., Wang, M., and Chen, C. 2004. Semiconductor hollow waveguides formed by omni-directional reflectors. *Opt. Express* 12: 6589–6593.

Luff, B.J., Harris, R.D., Wilkinson, J.S., Wilson, R., and Schiffrin, D.J. 1996. Integrated-optical directional coupler biosensor. *Opt. Lett.* 21: 618–620.

Luff, B.J., Wilkinson, J.S., Piehler, J., Hollenback, U., Ingenhoff, J., and Fabricius, N. 1998. Integrated optical Mach-Zehnder biosensor. *J. Lightwave Technol.* 16: 583–591.

Lukosz, W. 1991. Principles and sensitivities of integrated optical and surface plasmon sensors for direct affinity sensing and immunosensing. *Biosens. Bioelectron.* 6: 215–225.

Madsen, C.K., Lenz, G., Bruce, A.J., Cappuzo, M.A., Gomez, L.T., and Scotti, R.E. 1999. Integrated all-pass filters for tunable dispersion and dispersion slope compensation. *IEEE Phot. Technol. Lett.* 11: 1623–1625.

Malbet, F., Kern, P., Schanen-Dupont, I., Berger, J.P., Rousselet-Perrault, K., and Benech, P. 1999. Integrated optics for astronomical interfereometry. I. Concept and astronomical applications. *Astrom. Astrophys. Suppl. Ser.* 138: 135–145.

Marcuse, D. 1974. *Theory of Dielectric Optical Waveguides*. New York: Academic Press.

März, R. 1995. *Integrated Optics*. Boston, MA: Artech House.

Masanovic, G.Z., Passaro, V.M.N., and Reed, G.T. 2003. Dual grating-assisted directional coupling between fibers and thin semiconductor waveguides. *IEEE Phot. Technol. Lett.* 15: 1395–1397.

Matthews, M.R., MacDonald, B.M., and Preston, K.R. 1990. Optical components—The new challenge in packaging. In *Proceedings of the 40th Electronic Components and Technology Conference*. Las Vegas, NV, pp. 206–212.

McDougall, S.D., Kowalski, O.P., Hamilton, C.J. et al. 1998. Monolithic integration via a universal damage enhanced quantum well intermixing technique. *IEEE J. Select. Top. Quantum Electron.* 4: 636–646.

McNab, S., Moll, N., and Vlasov, Y. 2003. Ultra-low loss photonic integrated circuit with membrane-type photonic crystal waveguides. *Opt. Express* 11: 2927–2939.

Nadler, C.K., Wildermuth, E.K., Lanker, M., Hunziker, W., and Melchior, H. 1999. Polarization insensitive, low-loss, low cross-talk, wavelength multiplexer modules. *IEEE J. Select. Top. Quantum Electron.* 5: 1407–1412.

Ng, S., Janz, S., McKinnon, W.R., Barrios, P., Delage, A., and Syrett, B.A. 2007. Performance optimization of a reconfigurable waveguide digital optical switch on InGaAsP-InP: Design, materials, and carrier dynamics. *IEEE J. Quantum Electron.* 43: 1147–1158.

Offrein, B.J., Horst, F., Bona, G.L., Germann, R., Salemink, H.W.M., and Beyeler, R. 2000. Adaptive gain equalizer in high-index-contrast SiON technology. *IEEE Phot. Technol. Lett.* 12: 504–506.

Ojha, S.M., Cureton, C., Bricheno, T., Day, S., Moule, D., Bell, A.J., and Taylor, J. 1998. Simple method of fabricating polarization insensitive and very low crosstalk AWG grating devices. *Electron. Lett.* 34, 78–79.

Okamoto, K., 2005. *Fundamentals of Optical Waveguides*. New York: Academic Press.

Parriaux, O. and Veldhuis, G.J. 1998. Normalized analysis for the sensitivity optimization of integrated optical evanescent-wave sensors. *J. Lightwave Technol.* 16, 573–582.

Pavesi, L. and Lockwood, D.J. 2004. *Silicon Photonics*. Berlin, Germany: Springer-Verlag.

Payne, F.P. and Lacey, J.P.R. 1994. A theoretical analysis of scattering loss from planar optical waveguides. *Opt. Quantum Electron.* 26: 977–986.

Prieto, F., Sepulveda, B., Calle, A. et al. 2003. An integrated optical interferometric nanodevice based on silicon technology for biosensor applications. *Nanotechnology* 14: 907–912.

Rabiei, P., Steier, W., Zhang, C., and Dalton, L.R. 2002. Polymer micro-ring filters and modulators. *J. Lightwave Technol.* 20: 1968–1975.

Reed, G.T. and Knights, A.P. 2004. *Silicon Photonics: An Introduction.* John Wiley & Sons: Chichester, U.K.

Rickman, A.G., Reed, G.T., and Namavar, F. 1991. Silicon-on-insulator optical rib waveguide loss and mode characteristics. *IEEE J. Lightwave Technol.* 12: 1771.

Roelkens, G., Van Thourhout, D., and Baets, R. 2006. High efficiency silicon-on-insulator grating coupler based on a poly-silicon overlay. *Opt. Express, OSA* 14: 11622–11630.

Roelkens, G., Thourhout, D., and Baets, R. 2007. High efficiency grating coupler between silicon-on-insulator waveguides and perfectly vertical optical fibers. *Opt. Lett., OSA* 32: 1495–1497.

Ruano, J.M., Glidle, A., Cleary, A., Walmsley, A., Aitcheson, J.S., and Cooper, J.M. 2003. Design and fabrication of a silica on silicon integrated optical biochip as a fluorescence microarray platform. *Biosens. Bioelectron.* 18: 175–184.

Sasaki, K., Ohno, F., Motegi, A., and Baba, T. 2005. Arrayed waveguide grating of $70 \times 60 \, mm^2$ size based on Si photonic wire waveguides. *Electron. Lett.* 41: 801–802.

Scarmozzino, R., Gopinath, A., Pregla, R., and Helfert, S. 2000. Numerical techniques for modeling guided-wave photonic devices. *IEEE J. Select. Top. Quantum Electron.* 6: 150–162.

Schmid, J.H., Lamontagne, B., Cheben, P. et al. 2007. Mode converters for coupling to high aspect ratio silicon-on-insulator channel waveguides. *IEEE Phot. Technol. Lett.* 19: 855–857.

Schmid, J.H., Delage, A., Lamontagne, B. et al. 2008. Interference effect in scattering loss of high index contrast planar waveguides caused by boundary reflections. *Opt. Lett.* 33: 1479–1481.

Shani, Y., Henry, C.H., Kistler, K.J., Orlowsky, K.J., and Ackerman, D.A. 1989. Efficient coupling of a semiconductor laser to an optical fibre by means of a tapered waveguide on silicon. *Appl. Phys. Lett.* 55: 2389–2391.

Shoji, T., Tsuchizawa, T., Watanabe, T., Yamada, K., and Morita, H. 2002. Low loss mode size converter from 0.3 mm square Si wire waveguides to single mode fibers. *Electron. Lett.* 38: 1669–1670.

Sinclair, W., Schmid, J.H., Waldron, P. et al. 2009. Silicon photonic evanescent field molecular sensor using resonant grating interrogation. In *Conference on Lasers and Electro-Optics (CLEO).* Washington, DC: Optical Society of America, Paper JTUD89.

Smit, M.K. 1988. New focussing and dispersive planar waveguide component based on optical phased array. *Electron. Lett.* 24: 385–386.

Smit, M.K. and van Dam, C. 1996. PHASAR-based WDM-devices: Principles, design and applications. *IEEE J. Select. Top. Quantum Electron.* 2: 236–250.

Soldano, L.B. and Pennings, E.C.M. 1995. Optical multi-mode interference devices based on self imaging: Principles and applications. *J. Lightwave Technol.* 13: 615–627.

Somekh, S., Garmire, E., Yariv, A., Garvin, H.L., and Hunsperger, R.G. 1972. Channel optical waveguide directional couplers. *Appl. Phys. Lett.* 22: 46–47.

Soole, J.B.D., Scherer, A., Leblanc, H.P., Andreakis, N.C., Bhat, R., and Koza, M.A. 1991. Monolithic InP/InGaAsP/InP grating spectrometer for the 1.48–1.56 µm wavelength range. *Appl. Phys. Lett.* 58: 1949.

Soole, J.B.D., Poguntke, K.R., Scherer, A. et al. 1992. Wavelength selectable laser emission from a multistripe array grating integrated cavity laser. *Appl. Phys. Lett.* 61: 2750–2752.

Soref, R.A. 1993. Silicon-based optoelectronics. *Proc. IEEE* 81: 1687–1706.

Soref, R.A., Schmidtchen, J., and Petermann, K. 1991. Large single mode ridge waveguides in GeSi-Si and Si-on SiO₂. *IEEE J. Quantum Electron.* 27: 1971.

Sumida, M. and Takemoto, K.J. 1984. Lens coupling of laser diodes to single mode fibres. *J. Lightwave Technol.* 2: 305–311.

Sure, A., Dillon, T., Murakowski, J., Lin, C., Pustai, D., and Prather, D. 2003. Fabrication and characterization of three-dimensional silicon tapers. *Opt. Express* 11: 3555–3561.

Suzuki, S., S. Sumida, Inoue, Y., Ishii, M., and Ohmori, Y. 1997. Polarisation insensitive arrayed waveguide gratings using dopant-rich silica-based glass with thermal expansion adjusted to Si substrate. *Electron. Lett.* 33: 1173–1174.

Taillaert, D., Chong, H., Borel, P.I., Fransden, H., de la Rue, R.M., and Baets, R. 2003. A compact two-dimensional grating coupler used as a polarization splitter. *IEEE Phot. Technol. Lett.* 15: 1249–1251.

Taillaert, D., Bienstman, P., and Baets, R. 2004. Compact efficient broadband grating coupler for silicon-on-insulator waveguides. *Opt. Lett., OSA* 29: 2749–2751.

Tiefenthaler, K. and Lukosz, W. 1989. Sensitivity of grating couplers as integrated-optical chemical sensors. *J. Opt. Soc. Am. B* 6: 209–220.

Tolstikin, V.I., Densmore, A., Pimenov, K., and Laframboise, S. 2003. Single-mode vertical integration of p-i-n photodetectors with optical waveguides for monitoring in WDM transmission systems. *IEEE Phot. Technol. Lett.* 15: 843–845.

Tolstikin, V.I., Densmore, A., Pimenov, K. et al. 2004. Monolithically integrated optical channel monitor for DWDM transmission systems. *J. Lightwave Technol.* 22: 146–152.

Trinh, P.D., Yegnanarayanan, S., and Jalali, B. 1995. Integrated optical directional couplers in silicon-on-insulator. *Electron. Lett.* 31: 2097–2098.

Van Laere, F., Roelkens, G., Ayre, M. et al. 2007. Compact and highly efficient grating couplers between optical fiber and nanophotonic waveguides. *J. Lightwave Technol.* 25: 151–156.

Vivien, L., Laval, S., Dumont, B., Lardenois, S., Koster, A., and Cassan, E. 2002. Polarization-independent single-mode rib waveguides on silicon on insulator for telecommunications wavelengths. *Opt. Commun.* 210: 43–49.

Wakita, K. 1998. *Semiconductor Optical Modulators*. Boston, MA: Kluwer Academic Publishers.

Watanabe, R. and Nosu, K. 1980. Slab waveguide demultiplexer for multimode optical transmission in the 1.0–1.4 µm wavelength region. *Appl. Opt.* 19: 3588.

Waterbury, R.D., Wensheng, Y., and Byrne, R.H. 1998. Long pathlength absorbance spectroscopy: trace analysis of Fe(II) using a 4.5 µm liquid core waveguide. *Anal. Chim. Acta* 357: 99–102.

Weisser, M., Tovar, G., Mittler-Neher, S. et al. 1999. Specific biorecognition reactions observed with and integrated Mach-Zehnder interferometer. *Biosens. Bioelectron.* 14: 405–411.

Wooten, E.L., Kissa, K.M., Yi-Yan, A. et al. 2000. A review of lithium niobate modulators for fiber-optic communication systems. *IEEE J. Select. Top. Quantum Electron.* 6: 69–81.

Wörhoff, K., Driessen, A., Lambeck, P.V., Hilderink, L.T.H., Linders, P.W.C., and Popma, Th.J.A. 1999a. Plasma enhanced chemical vapour deposition silicon oxynitride optimized for application in integrated optics. *Sens. Actuators* 74: 9–12.

Wörhoff, K., Lambeck, P.V., and Driessen, A. 1999b. Design, tolerance analysis and fabrication of silicon oxynitride based planar optical waveguides for communication devices. *J. Lightwave Technol.* 17: 1401.

Xia, F., Sekaric, L., and Vlasov, Y. 2007. Ultracompact optical buffers on a silicon chip. *Nat. Phot.* 1: 65–71.

Xu, Q., Schmidt, B., Pradhan, S., and Lipson, M. 2005. Micrometre-scale silicon electro-optic modulator. *Nature* 435: 325–327.

Xu, D.-X., Janz, S., and Cheben, P. 2006. Design of polarization-insensitive ring resonators in silicon-on-insulator using MMI couplers and cladding stress engineering. *IEEE Phot. Technol. Lett.* 18: 343–345.

Xu, D.-X., Densmore, A., Waldron, P. et al. 2007. High bandwidth SOI photonic wire ring resonators using MMI couplers. *Opt. Express* 15: 3149–3155.

Xu, D.-X., Densmore, A., Delâge, A. et al. 2008. Folded cavity SOI microring resonator sensors for high sensitivity and real time measurement of biomolecular binding. *Opt. Express* 16: 15137–15148.

Yalcin, A., Popat, K.C., Aldridge, J.C. et al. 2006. Optical sensing of biomolecules using microring resonators. *IEEE J. Select. Top. Quantum Electron.* 12: 148–155.

Yang, A.H.J., Moore, S.D., Schmidt, B.S., Klug, K., Lipson, M., and Erickson, D. 2009. Optical manipulation of nanoparticles and biomolecules in sub-wavelength slot waveguides. *Nature* 457: 71–75.

Yariv, A. 2000. Universal relations for coupling of optical power between microresonators and dielectric waveguides. *Electron. Lett.* 36: 321–323.

Yen, H.W., Friedrich, H.R., Morrison, R.J., and Tangonan, G.L. 1981. Planar Rowland spectrograph for fibre optic demultiplexing. *Opt. Lett.* 6: 639.

Yi, Y., Akiyama, S., Bernel, P., Duan, X., and Kimmerling, L.C. 2006. Sharp bending of on-chip silicon Bragg cladding waveguide with light guiding in low index core materials. *IEEE J. Select. Top. Quantum Electron.* 12: 1345–1349.

Yin, D., Barber, J.P., Hawkins, A.R., and Schmidt, H. 2005. Highly efficient fluorescence detection in picoliter volume liquid core waveguides. *Appl. Phys. Lett.* 87: 211111–211113.

Yin, D., Lunt, E.J., Barman, A., Hawkins, A.R., and Schmidt, H. 2007. Microphotonic control of single molecule fluorescence correlation spectroscopy using planar opto fluidics. *Opt. Express* 15: 7290–7295.

Zirngibl, M., Joyner, C.H., Stulz, L.W. et al. 1994. Digitally tunable laser based on the integration of a waveguide grating multiplexer and an optical amplifier. *IEEE Phot. Technol. Lett.* 6: 516–518.

4

Photonic Crystal Hollow Waveguides

Fetah Benabid
P. John Roberts

4.1 Introduction

Since the advent of the laser, a huge effort has been underway to guide this light in a confined manner over distances that are much larger than the Rayleigh range. Guiding light within air or other gases in a confined way over length scales of many kilometers has proven to be particularly challenging until recently. This is because total internal reflection (TIR), which enabled confined optical guidance to be realized in fibers comprising solid materials (mainly glass materials), cannot be used to confine light in a gas or vacuum. Although several methods have been attempted to guide light in air, it is the advent of the hollow-core photonic crystal fiber (HC-PCF) that has finally made the confined low-loss guidance of light in air possible.

Among the early attempts at developing diffraction-free light guidance in air is the capillary waveguide. However, this form of guidance has been limited to a propagation loss of the order of 1 dB/m and mode areas many thousands of λ^2, with λ being the optical wavelength. This is due to the inherent confinement loss characteristics of the capillary modes and to the sensitivity of such modes to bend-induced mode coupling. Indeed, the capillary modes should be thought of as leaky modes, with small but significant radiation field components, rather than true guided modes.

The motivation for guiding light in a hollow waveguide stems from a number of application areas. From a telecommunications perspective, guiding light in air or vacuum largely removes limitations due to Rayleigh scattering (Marcatili and Schmeltzer 1964) and intrinsic material absorption. Rayleigh scattering will be dramatically reduced together with material absorption, potentially allowing lower loss guidance over a broader spectral range. A further advantage of air guidance is the ability to convey high-power pulses before the onset of deleterious nonlinear effects, material breakdown, or heating-related sensitivity.

Further motivation for the development of hollow waveguides comes from the ability to fill the voids with gaseous or liquid media (i.e., fluidic media). If the light is tightly confined in the waveguide, high intensities can be attained even at relatively low powers. If the attenuation is low, an effective interaction between the light and the medium in the waveguide may occur over many meters or kilometers. Such a co-confinement of light and fluidic media opens up new areas of exploration in nonlinear and quantum optics, as well as in other applications such as gas sensing or microscale optofluidics. Rather than attempt to review the applications of the fibers, this chapter focuses on a description of the mechanisms responsible for the optical guidance in various forms of fluid-filled fiber and factors that affect propagation loss.

Section 4.2 discusses the historical attempts at achieving confined guidance of light in air, leading to the development of HC-PCF. Section 4.3 describes different types of HC-PCF and the waveguiding mechanisms they rely upon. Section 4.4 concentrates on liquid-filled fibers, for which both TIR guidance and bandgap guidance are possible. TIR guidance can occur if just the core of the fiber is liquid filled, whereas the cladding may contain air holes to ensure the cladding effective index is below that of the liquid core. Bandgap guidance can arise if all the holes in the cladding are filled with liquid that is of higher refractive index than the solid host medium. The index contrast scaling laws associated with all forms of guidance, which become rigorous at weak contrast, will also be reviewed.

4.2 A History of Hollow-Core Guidance

4.2.1 Hollow Capillaries

The development of the laser in early 1960s led to much interest in the guided propagation of light. Among the earliest proposed waveguides were hollow metal and dielectric tube waveguides. The metal waveguide had been successfully used at microwave wavelengths, but due to metals being lossy at optical frequencies, the guidance in the optical domain is greatly compromised. Marcatili developed a theory applicable to this form of waveguide that involves nonperfect metals or dielectrics (Marcatili and Schmeltzer 1964).

Assuming electric and magnetic fields of the form $\mathbf{E}(\mathbf{r}_\perp;\omega)e^{i(\beta z-\omega t)}$ and $\mathbf{H}(\mathbf{r}_\perp;\omega)e^{i(\beta z-\omega t)}$, respectively, with Oz the fiber axis direction and the subscript $\perp$ denoting vector components in the transverse x–y plane (see Figure 4.1), the guided modes can be found as solutions of the following vector wave equations derived from the source-free Maxwell equations (Snyder and Love 1983):

$$\left[\nabla_\perp^2 + k^2 n^2\right]\mathbf{E}_\perp + \nabla_\perp\left[\left(\nabla_\perp \ln n^2\right)\cdot\mathbf{E}_\perp\right] = \beta^2\mathbf{E}_\perp$$

$$\left[\nabla_\perp^2 + k^2 n^2 + \left(\nabla_\perp \ln n^2\right)\times\nabla_\perp\times\right]\mathbf{H}_\perp = \beta^2\mathbf{H}_\perp \qquad (4.1)$$

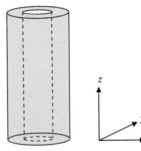

FIGURE 4.1 A hollow capillary showing the adopted Cartesian coordinate system.

where
  $\nabla_\perp$ is the transverse gradient operator
  n is the refractive index distribution
  β is the complex propagation constant
  k is the magnitude of the free-space wave number $k=\omega/c=2\pi/\lambda$,
    where ω the light frequency, c the speed of light, and λ the
    vacuum wavelength

The real part of the propagation constant determines the spatial-propagation contribution to the field phase of the mode. This is also often represented by an effective index of the mode $n_{\text{eff}} = \text{Re}(\beta)/k$. The imaginary part of β is directly related to the leakage rate or propagation loss of the propagating mode.

For a structure with cylindrical symmetry, the modes decouple according to their azimuthal variation, indexed by an integer m, such that the z-components of the E and H fields vary as $E_z(\mathbf{r}_\perp; \omega) = E_z^{(m)}(\rho; \omega) \exp(im\phi)$ and $H_z(\mathbf{r}_\perp; \omega) = H_z^{(m)}(\rho; \omega) \exp(im\phi)$, respectively. Here, ρ is the radial coordinate measured from the centre of the fiber and ϕ the azimuthal angle. The remaining E and H field components can be found from E_z and H_z using Maxwell's equations. Within locally homogeneous media, all the terms involving $\nabla_t(\ln n^2)$ in Equation 4.1 vanish so that $E_z^{(m)}(\rho; \omega)$ and $H_z^{(m)}(\rho; \omega)$ can be expressed in terms of Bessel functions of order m. Within the guiding core of refractive index n_g, these field components vary simply as a Bessel function, $J_m(K_g\rho)$, with $K_g = \sqrt{n_g^2 k^2 - \beta^2}$. Outside the core, in an assumed semi-infinite medium of index n_h, the components for a guided or leaky mode vary as $H_m^{(1)}(K_h\rho)$, with $H_m^{(1)}$ the outward-going Hankel function of order m and $K_h = \sqrt{n_h^2 k^2 - \beta^2}$. Demanding that the components of Y_m (the modified Bessel function of order m) vanish in the core and that the components of the inward traveling Hankel waves outside the core vanish, a determinental equation for guided and leaky modes is derived. Guided solutions are found at (k,β) values satisfying this equation, which will be written $D_m(\omega, \beta) = 0$. By analytic continuation of β into the complex plane, leaky mode solutions are found when the outside index n_h is such that $n_{\text{eff}} < n_h$, which is the situation for the hollow glass capillary.

As well as decoupling according to the azimuthal variation, the modes also decouple according to polarization. Two classes of modes labeled TE_{0n} and TM_{0n} ($n = 1, 2, 3, \ldots$) are associated with $m = 0$ azimuthal variation. The former mode type is characterized by a vanishing E_z component whereas the latter has a vanishing H_z component. The mode class HE_{mn} ($m = 1, 2, \ldots, n = 1, 2, \ldots$), with m indexing the azimuthal variation, is characterized by a nonvanishing of both E_z and H_z components. The n subscript represents the order of the root of $D_m(\omega, \beta) = 0$. The HE_{mn} modes come in degenerate pairs since the positive, definite m references both positive and negative azimuthal variations, $\exp(im\phi)$ and $\exp(-im\phi)$. Within this mode classification, the determinental equation is simplified to the following equation (Marcatili and Schmeltzer 1964):

$$\left(\frac{J_m'(K_g\rho)}{J_m(K_g\rho)} - \frac{K_g H_m^{(1)'}(K_h\rho)}{K_h H_m^{(1)'}(K_h\rho)} \right) \left(\frac{J_m'(K_g\rho)}{J_m(K_g\rho)} - n_h^2 \frac{K_g H_m^{(1)'}(K_h\rho)}{K_h H_m^{(1)'}(K_h\rho)} \right) - \left(\frac{2\pi m}{k^2(K_g\rho)} \right)^2 \left(1 - \frac{K_g}{K_h} \right)^2 = 0, \qquad (4.2)$$

where the prime on a Bessel function denotes differentiation with respect to its argument.

In the limit of short wavelengths (i.e., $\lambda \ll \rho$), the determinental equation, $D_m(\omega, \beta) = 0$, determining the leaky core modes of a cylindrical capillary of core radius ρ further simplifies so that closed form expressions can be derived. At a wave number k (related to ω by $ck = \omega$) and an azimuthal variation of the dominant field component governed by $\exp(im\phi)$, solutions are found at β values determined by the real and imaginary parts:

$$\text{Re}(\beta_{m,n}) = n_{\text{eff}} k = k \left[n_g^2 - \left(\frac{u_{m,n}}{k\rho} \right)^2 \right]^{1/2} \quad \text{and} \quad \text{Im}(\beta_{m,n}) = \kappa k = A \left(\frac{u_{m,n}}{2\pi} \right)^2, \qquad (4.3)$$

where

$u_{m,n}$ is the nth root of the Bessel function of order $m-1$, i.e., $J_{m-1}(u_{m,n}) = 0$, with $0 < u_{m,1} < u_{m,2} < \ldots$
κ is the imaginary part of the refractive index

the factor A is determined from

$$A = \frac{1}{\sqrt{n_r^2 - 1}} \times \begin{cases} 1, & \text{for TE}_{0n} \text{ modes} \\ n_r^2, & \text{for TM}_{0n} \text{ modes} \\ (n_r^2 + 1)/2, & \text{for HE}_{mn} \text{ modes} \end{cases} \qquad (4.4)$$

where $n_r = n_h/n_g$ is the ratio of the refractive index of the medium surrounding the core to that of the core.

Figure 4.2 shows the numerically computed $-\log |D_m(\omega, \beta)|$ plotted as a function of k and $n_{eff} = \beta/k$ for a glass capillary of core radius $\rho = 4.1204\,\mu m$ and $m = 1, 5$, and 10, respectively. Only real values of β are considered in the plot so that leaky modes appear as broadened lines, the width of which gives information about the leaky mode decay rate $\kappa = \text{Im}(\beta)$. The core is assumed to have a refractive index $n_g = 1$ and the glassy region is made of silica (with index taken as $n_h = 1.45$) that is sufficiently thick to be assumed infinitely extended. As a consequence of choosing to plot $-\log |D_m(\omega, \beta)|$, the "guided" modes are located at maximal values, which are represented by a white region in the figure. Leaky mode solutions for the silica glass capillary thus have n_{eff} versus k trajectories, located near the centers of the broadened white lines. Regions in the n_{eff}-k plane where no states exist are identified by the black regions.

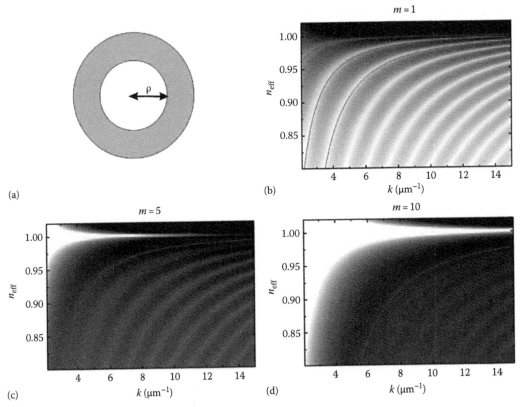

(a)

(b)

(c)

(d)

FIGURE 4.2 (a) A glass capillary with a hollow core of radius ρ. Plots of $-\log |D_m(\omega, \beta)|$ as a function of wave number k and effective index n_{eff}, when $\rho = 4.1204\,\mu m$, for azimuthal variation (b) $m = 1$, (c) $m = 5$, and (d) $m = 10$. The dotted lines show the dispersion curves of HE_{11}, HE_{12}, HE_{51}, $HE_{10,1}$ found using Equation 4.3.

The highest (in n_{eff}) solution for $m=1$ corresponds to the HE_{11} core mode. Figure 4.2 also shows the dispersion (n_{eff} vs. k) curve of representative modes using the simplified analytical expression given by Equation 4.3. The figures clearly show that the analytical expressions of Equation 4.3 are a very good approximation in determining the dispersion properties of the capillary waveguide when the core is at least a few wavelengths in extent.

The lowest loss mode for a 1 mm core diameter silica capillary waveguide, the HE_{11} mode, is found to suffer a confinement loss of 1.85 dB/km at a wavelength of 1.0 μm. This figure was, however, found to be extremely sensitive to bending of the waveguide, with a bend radius of as much as 10 km calculated to double the loss. Furthermore, the difficulty in coupling light efficiently to the low-loss mode in a waveguide of such a large diameter makes this waveguide impractical. Using a smaller diameter capillary improved the coupling, but the waveguide loss for a hollow dielectric waveguide scales as λ^2/r_0^3, so the loss increases dramatically as the core radius is reduced. For example, the calculated loss for a 10 μm diameter core is ~1 dB/mm, even before considering bend loss, thus rendering its implementation in applications of very limited impact.

4.2.2 Bragg Reflection Waveguides

Historically, it was Yeh and Yariv who showed that, in principle, lossless guidance is possible in a low-index slab bounded by periodic layers of alternating layers of higher refractive index, see Figure 4.3 (Yeh et al. 1977). In this structure, light is confined to the central region of index n_g, bound by alternating planes of higher-index (n_2) and lower-index (n_1) media. The various refractive indices obey the inequalities $n_g < n_1 < n_2$. The use of planar Bragg reflectors was at the time already commonplace in distributed feedback and distributed Bragg reflection lasers, but Yeh and Yariv were the first to consider in detail the waveguiding capabilities afforded by a pair of these reflectors.

Neglecting material absorption and interface roughness, a waveguide of this type, formed with an infinite number of periodic layers, would be expected to provide lossless guidance. Due to the rapid evanescent decay of the field through the periodic layers, the number of layers required to provide a good approximation to loss-free guidance is small. The optimal layer thickness, which provides the minimum loss for a given number of layers, is one-quarter of a wavelength (Yeh 1998). Each layer then acts as an antiresonator, which improves the light exclusion from the cladding layers.

The planar Bragg system is easily analyzed using a transfer matrix formalism. The mode fields for this geometry decouple according to polarization, with TE-polarized modes having no **E**-field component normal to the layers and TM polarization having no **H**-field component normal to the layers. The

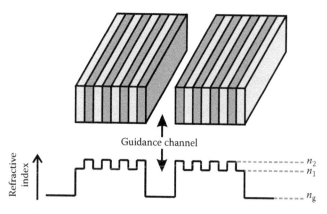

FIGURE 4.3 Schematic diagram of a slab Bragg reflection waveguide (top) and its refractive index profile (bottom). In this structure, light can be guided in the central low-index channel.

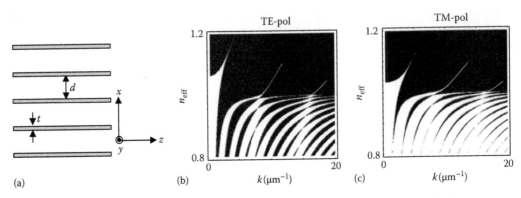

FIGURE 4.4 (a) A Bragg mirror comprising parallel planar silica layers of thickness *t* in air. The layers are separated by distance *d*. Pass-bands of Bragg mirror arrangement when $t=0.3658\,\mu\text{m}$ and $d=2.7469\,\mu\text{m}$ for (b) TE polarization and (c) TM polarization.

adopted coordinate system has the Ox direction normal to the layers, as shown in Figure 4.4a. Without loss of generality, the mode field will be assumed not to vary along Oy, but varies as $\exp(i\beta z)$ along Oz. For polarization $\lambda(=\text{TE or TM})$, the field in the *j*th layer with index n_j is written as a superposition of plane-wave components:

$$\mathbf{E} = \exp(i\beta z)\left[A_\lambda^{(j)}\hat{\mathbf{e}}_\lambda^{(+)}\exp(iK_j x)+B_\lambda^{(j)}\hat{\mathbf{e}}_\lambda^{(-)}\exp(-iK_j x)\right]$$

$$\mathbf{H} = \exp(i\beta z)\left[A_\lambda^{(j)}\mathbf{k}_j\times\hat{\mathbf{e}}_\lambda^{(+)}\exp(iK_j x)+B_\lambda^{(j)}\mathbf{k}_j\times\hat{\mathbf{e}}_\lambda^{(-)}\exp(-iK_j x)\right]\Big/k_j, \qquad (4.5)$$

where
$$K_j = \sqrt{n_j^2 k^2 - \beta^2}$$
$\mathbf{k}_j=(K_j,0,\beta)$ is the wavevector in the medium
$\hat{\mathbf{e}}_\lambda^{(+/-)}$ are the polarization directions for positive and negative going (along Ox) waves

The electromagnetic boundary conditions are used to relate $\left(A_\lambda^{(j)}, B_\lambda^{(j)}\right)$ in adjacent layers by a relationship of the form

$$\begin{pmatrix} A_\lambda^{(j)} \\ B_\lambda^{(j)} \end{pmatrix} = \mathbf{M}_\lambda^{(j,j-1)}\begin{pmatrix} A_\lambda^{(j-1)} \\ B_\lambda^{(j-1)} \end{pmatrix}. \qquad (4.6)$$

The transfer matrix formulation can be used to relate fields across a unit cell of a 1-D periodic system. Bloch solutions are obtained when the associated transfer matrix has eigenvalues that are complex conjugate pairs with unit modulus, whereas bandgaps are identified as regions of (ω,β) where this transfer matrix has eigenvalues that do not have unit modulus. Within the bandgaps, incident light decays when entering the crystal (now considered of finite extent) with an exponential envelope. Such 1-D Bragg stacks thus behave as mirrors.

The transfer matrix methodology was used to find the allowed bands for the example 1-D crystal arrangement shown in Figure 4.4a. The technique has been extensively used in the design and the analysis of multilayer dielectric mirrors.

Figure 4.4 shows band diagrams for the mode propagating along the z-axis obtained with this method. Instead of the usual choice of indices larger than 2, the structure chosen here comprises silica layers of a given thickness and pitch because of its relevance to the air-guiding optical fibers described below. Each layer has a thickness $t=0.3658\,\mu\text{m}$, immersed in air, and separated by a distance $d=2.7469\,\mu\text{m}$ (i.e., the pitch $\Lambda=d+t\approx3.1\,\mu\text{m}$). The bands are shown in Figure 4.4b and c for TE and TM polarization, respectively, in the wave number (k) – effective index (n_{eff}) plane. The white regions show where electromagnetic modes of this 1-D periodic structure (assumed infinite) exist, and the black regions indicate bandgaps. The results show that air guidance is possible for both polarizations.

The transfer matrix formalism is readily extended to deal with a cavity introduced between two Bragg stack mirrors. The cavity allows electromagnetic waveguide modes to form. Recently, such a design has been adopted as the basis of a hollow waveguide integrated in a CMOS compatible chip (Yang et al. 2007). Rather than give a detailed analysis of such waveguides, which only prevent diffraction in one spatial direction, we will next discuss waveguides that truly offer diffraction-free propagation. This will also elucidate the connection between the so-called antiresonant reflection optical waveguide (ARROW) and bandgap guidance.

Yeh and Yariv extended their theory to deal with a cylindrically symmetric Bragg reflector structure, or Bragg fiber (Yeh et al. 1978), illustrated in Figure 4.5. This study showed that such a structure should provide guidance similar to the periodic slab waveguide. The structure is analyzed using an extension of the analysis used for the capillary that incorporates the same transfer matrix methodology that was used for the planar Bragg system. The fields in the core and outside the final cladding layer (in an assumed semi-infinite medium) are of the same form as of the capillary. Within the jth cladding layer of refractive index n_j, $E_z^{(m)}$ and $E_z^{(m)}$ are expressed as a superposition of Bessel (J_m) and Neumann (Y_m) wave components:

$$E_z^{(m)}(\rho;\omega)= A_m^{(j)}J_m\left(K_j\rho\right)+B_m^{(j)}Y_m\left(K_j\rho\right)$$

$$H_z^{(m)}(\rho;\omega)= C_m^{(j)}J_m\left(K_j\rho\right)+D_m^{(j)}Y_m\left(K_j\rho\right), \tag{4.7}$$

where $K_j=\sqrt{n_j^2k^2-\beta^2}$. The electromagnetic boundary conditions allow the expansion coefficients within layer j to be related to the coefficients in an adjacent layer via a transfer matrix. Demanding the components of Y_m in the core to vanish and the components of the inward traveling Hankel waves outside the cladding to vanish, a determinental equation for guided modes, $D_m(\omega,\beta)=0$, is derived. Again, leaky modes are found by analytic continuation of the determinental equation to complex β when $\text{Re}[\beta/k]<n_h$. The confinement loss of such a mode is determined by the imaginary part of the complex β solution.

To provide a useful comparative measure for the broadband guiding HC-PCFs described in the next section, a Bragg fiber comprising one or more concentric layers of silica in air will be analyzed. The

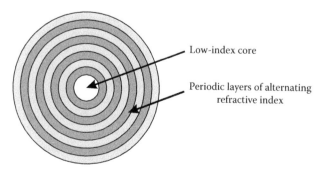

FIGURE 4.5 Diagram of a cylindrical Bragg reflector optical fiber. This fiber consists of a low-index core (white), and alternating layers of higher but different refractive indexes.

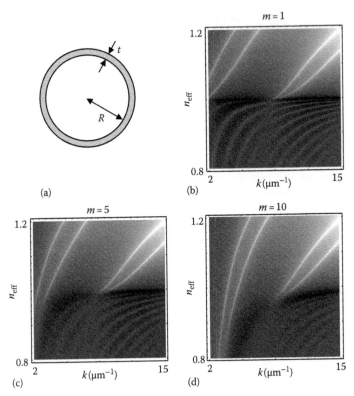

FIGURE 4.6 (a) A thin silica capillary with core radius R and thickness t. Plots of $-\log|D_m(\omega, \beta)|$ for this structure, with $R=4.1204\,\mu$m and $t=0.3658\,\mu$m, as a function of wave number k and effective index n_{eff} for azimuthal variation (b) $m=1$, (c) $m=5$, and (d) $m=10$.

structure consisting of a single layer of thickness, t, surrounding a core of radius, R, is shown in Figure 4.6. This corresponds to a cylindrical ARROW (Duguay et al. 1986). Also shown is $-\log|D_m(\omega, \beta)|$ for $m=1$, 5, and 10 plotted as a function of the wave number $k=\omega/c$ and the effective index $n_{\mathrm{eff}}=\beta/k$. Comparing with the corresponding capillary results shown in Figure 4.2, the single silica layer gives rise to strong perturbations of the core guided modes near transverse resonances given by $n_{\mathrm{eff}}=\sqrt{n_1^2-\left[\pi j/(kt)\right]^2}$, with $j=1,2,\ldots$. Near these resonances of the silica layer, the core modes show an enhanced confinement loss. The resonances emerge above the air light line, $n_{\mathrm{eff}}=1$, as guided modes of the ring.

A structure comprising three concentric layers of the same thickness t is shown in Figure 4.7 for the same core radius $R=4.1204\,\mu$m, together with $-\log|D_m(\omega, \beta)|$ for $m=1$, 5, and 10, respectively. This structure is not experimentally practical, since the silica layers reside in air and are not mechanically connected to one another, but it provides a useful comparison benchmark for the PCF designs described later. Again, core modes are perturbed close to transverse resonances of the layers. The resonances associated with each of the layers, which become guided modes above the air light line $n_{\mathrm{eff}}=1$, interact with one another except well above the air light line.

With the advent of the photonic bandgap (PBG) concept, the early proposals of Yeh and Yariv weret revisited by Fink et al. (1999) who designed and fabricated a short (10 cm) multimode cylindrical Bragg guide operating at a central wavelength of $10\,\mu$m. The guide was formed by alternating layers of tellurium and polystyrene deposited on the inside of a ~2 mm silica tube. The layers were calculated to provide effectively 100% reflection for a broad wavelength range at near grazing incidence (from air), and better than 90% reflection over a similar range at normal incidence. Transmission of 80% was obtained across a 2 μm bandwidth with a 90° bend in the 10 cm waveguide.

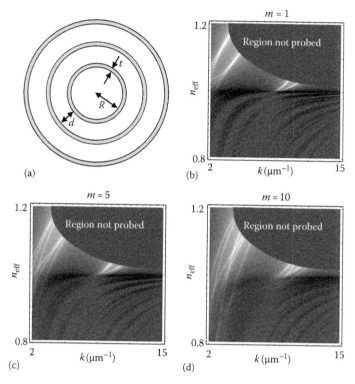

FIGURE 4.7 (a) A cylindrical Bragg reflector optical fiber comprising silica and air, with core radius R and concentric silica layers all of thickness t. Plots of $-\log |D_m(\omega, \beta)|$ for this structure with $R = 4.1204\,\mu m$, $t = 0.3658\,\mu m$, as a function of wave number k and effective index n_{eff} for azimuthal variation (b) $m = 1$, (c) $m = 5$, and (d) $m = 10$.

Due to the weak dependence of reflection on the incidence angle, the Bragg waveguide fabricated by Fink et al. was termed an "OmniGuide" waveguide. This class of waveguide was theoretically studied in detail by Johnson et al. (2001), and there is now a wealth of theoretical work on OmniGuide waveguides. Experimental work is more limited, and mainly concentrated on guidance in the mid-infrared region. OmniGuide fibers with an attenuation of ~1 dB/m have been demonstrated at 10.6 μm (Temelkuran et al. 2002), with higher losses of ~5 dB/m in the near-infrared (Kuriki et al. 2004), measured over several-meter-long lengths of fiber.

4.3 Hollow-Core Photonic Crystal Fiber

The hollow waveguide that has attracted the most interest over the past 10 years is the HC-PCF. This fiber consists of introducing a hollow core (defect) into a periodic two-dimensional photonic crystal structure, as shown schematically in Figure 4.8. The ability of the structure to confine and guide light was proposed by Russell in 1991 (Russell 2003), and later analyzed theoretically by Birks et al. (1995). While the existence of in-plane-of-periodicity PBGs in photonic crystal structures was then well known (John 1987, Yablonovitch 1987), the HC-PCF relies on an out-of-plane bandgap (see section below). Epistemologically speaking, the development of HC-PCF did not stem from the earlier work of Yariv and Yeh, who showed the possibility of optical guidance in air. It was instead triggered by the emergence of the PBG physics (John 1987, Yablonovitch 1987) and from Russell's earlier work on propagating optical Bloch modes (Russell 2003).

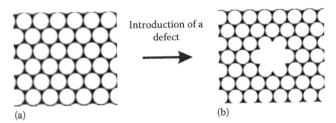

Introduction of a defect

(a)   (b)

FIGURE 4.8 An example of a two-dimensional photonic crystal structure (a) without a defect, and (b) with the introduction of a core defect. This example consists of a triangular arrangement of perfectly circular air holes (white) in a silica background (black). The introduction of a defect leads to the formation of guided modes within the defect, provided that the defect exhibits supported modes within the k-n_{eff} region of the cladding bandgap.

As a result of subsequent developments, we now divide HC-PCF into two distinct classes. These will be treated separately in the following subsections. The first of these is the PBG HC-PCF introduced above, in which the cladding structure provides a PBG. These fibers exhibit very low loss but a limited bandwidth of guidance. The guidance properties of these fibers will be considered further in Section 4.3.1. A following section elucidates the formation of the PBG by using a photonic analogue to the tight-binding model in solid-state physics.

The second class of HC-PCF has a photonic crystal cladding that does not possess a PBG capable of giving guidance in air. This class of HC-PCF, first reported by Benabid et al. (2002), provides broadband guidance (over the visible and near-infrared) with relatively low attenuation, and a comparable loss to PBG HC-PCF at short wavelengths. The guidance of this class of fiber is due to inhibited coupling between the cladding and the core modes (Benabid and Roberts 2008), and will be explored in Section 4.3.3.

4.3.1 Photonic Bandgap HC-PCF

Here we consider the guidance of the class of HC-PCF that relies on the formation of an out-of-plane PBG in the photonic crystal cladding structure. The term out-of-plane refers to the presence of a PBG for a nonzero propagation constant β along the fiber axis direction, such that no optical modes (or photonic state) exist with **E** and **H** field dependences, $\mathbf{E}(\mathbf{r}_t)\exp(i\beta z - \omega t)$ and $\mathbf{H}(\mathbf{r}_t)\exp(i\beta z - \omega t)$, respectively, over a given range of wavelengths. Equivalently, the bandgap implies that light cannot propagate in such photonic crystals over the associated wavelength and effective index range; light entering the crystal with k-n_{eff} values within the bandgap is exponentially attenuated. The crystal, therefore, behaves as a mirror to such light fields.

The HC-PCF bandgaps are due to a 2-D periodicity in the cladding structure, which contrasts with the 1-D variations associated with the Bragg fibers described in Section 4.2. The bandgaps for the 2-D periodic structures are complete over their specified regions in the k-n_{eff} plane, whereas for the lower dimensional Bragg structures, cladding modes can always be found near the air light line. The Bragg fibers can guide because the cylindrical symmetry ensures that only a restricted subset of these modes can interact with the core guided mode. This subset of modes, which share the same azimuthal variation (indexed by an integer m) as the core mode, does exhibit a bandgap.

It was recognized that the attainment of a complete bandgap would lead to guidance more robust to structural and external perturbations, and perhaps even more importantly, could be attained using a singly connected arrangement of a single material of relatively low refractive index such as silica. The latter aspect makes the structure appropriate for fabrication in a conventional fiber-draw tower, enabling the advantageous mechanical properties and low intrinsic loss of a material such as silica to be exploited to the full. In order to understand the possibility of an out-of plane PBG opening up even at extremely low-index contrast (Argyros et al. 2005), where an in-plane ($n_{\mathrm{eff}}=0$) bandgap is not attainable, one needs to consider the transverse index contrast defined as $n_{gl,\perp}/n_{air,\perp} = K_{gl}/K_{air}$, where $K_{gl} = \sqrt{\left(n_{gl}k\right)^2 - \beta^2}$ and $K_{air} = \sqrt{\left(n_{air}k\right)^2 - \beta^2}$ are the transverse **k**-vector components in glass and air, respectively. This is the relevant contrast ratio in bandgap formation. Inspection of Equations 4.2 and 4.7 describing the simpler 1-D systems will at least make this assertion plausible. The transverse index contrast can reach much larger values than the silica/air refractive index contrast, $n_{gl}/n_{air} \approx 1.46$. Equivalently, while the 1.46 silica/air refractive index contrast is insufficient for the scattering from any geometrical arrangement of these constituents to be strong enough to open up a bandgap for in-plane propagation (i.e., $\beta = kn_{eff} = 0$), the scattering strength increases as β increases. As β increases toward $n_{air}k$, the transverse index ratio reaches sufficiently high values to enable gaps to open up below the air light line.

The rigorous analysis of the 2-D periodic systems is more complicated than for the 1-D Bragg systems, and necessitates the use of heavy numerical simulations. Early computational schemes were taken over from solid-state physics, but due care to properly include vector effects associated with field polarization needs to be incorporated. The initial calculations were performed using either a plane-wave expansion technique or a transfer matrix approach based on discretization of the Maxwell equations. The existence of out-of-plane bandgaps in a silica/air structure was theoretically demonstrated by Birks et al. (1995) using these computational approaches. Two examples of cladding structures for bandgap HC-PCF, fabricated at Bath University, are shown in Figure 4.9. The first one is the cladding structure of the so-called triangular-lattice HC-PCF (Couny et al. 2007a), which is the standard PBG HC-PCF. The cladding consists of essentially identical air holes, and we will explore its properties below. The second cladding example entails an arrangement with two periods.

The location of PBGs for a given periodic structure is illustrated in a plot of the density of photonic states (DOPS). Historically, the DOPS was typically plotted against both normalized frequency ($k\Lambda$) and normalized axial wavevector ($\beta\Lambda$), with Λ being the cladding pitch (see Figure 4.10). This diagram is essentially equivalent to the ones shown in the previous section for capillary and Bragg fibers which indicate the position of modes in the $k\Lambda$ −effective refractive index (n_{eff}) plane. The schematic diagram in Figure 4.10 illustrates the position of full PBGs (black regions) for a silica/air structure. The bandgaps are full (or complete) in the sense that no states of any symmetry or polarization are present in these regions. The diagram also shows three dispersion lines which are key to understanding the propagation regimes for bandgap HC-PCF. These mark the boundary between allowed and forbidden propagation in a particular medium. The first of these is the air (or vacuum) light line, $k = \beta$. Below this line, i.e., the region $k < \beta$, propagation in air is forbidden (i.e., propagation is necessarily evanescent in character). Below the second line, given by, $n_{SiO_2}k = \beta$, no states of the structure can be supported. This is because

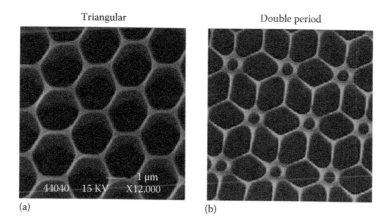

Triangular　　　　　　Double period

(a)　　　　　　(b)

FIGURE 4.9 Electron micrographs of two types of bandgap HC-PCF fiber cladding. (a) Standard triangular-lattice cladding for low-loss HC-PCF and (b) "double-period" cladding, formed using two thicknesses of capillary.

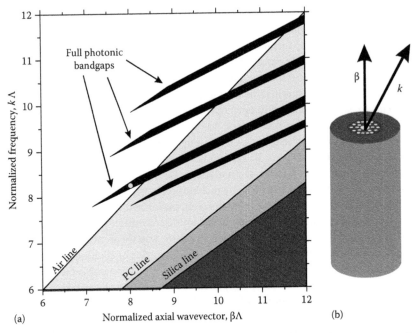

FIGURE 4.10 (a) A band-edge diagram or "finger-plot" illustrating the position of PBGs in terms of the normalized frequency and normalized axial wavevector. The "light lines" are boundaries below which the light is evanescent and unable to propagate within the indicated medium. (b) The wavevector k and the propagation constant, β. The wavevector k has a length equal to the free-space wave number $k = 2\pi/\lambda$, and its projection along the fiber axis direction is β. It should be remembered that due to the inhomogeneous nature of the fiber cross section, the fiber modes are distinct from plane waves, so states are not labeled by k.

SiO$_2$ is the highest index material within the fiber structure, The third line, labeled the PC-line, set by $n_{FSM}k = \beta$, represents the cutoff of the cladding photonic crystalline structure. No states of the cladding exist below this line. In the absence of defects which have index higher than n_{FSM}, this line also gives the true cutoff of the entire cladding structure of the HC-PCF, and thus n_{FSM} represents the effective refractive index of the structure. The subscript FSM stands for fundamental space-filling mode (Birks et al. 1997). This mode represents the fundamental mode (i.e., the mode with the highest effective index) of the cladding structure when considered as a waveguide. The FSM transverse profile fills the entire cladding structure, and in the case of triangular-lattice HC-PCF, the field of this mode is localized in the interstitial nodes of the cladding holes. The physics behind the formation of this mode will be discussed later.

It can be seen in Figure 4.10 that the PBGs extend into the white region, where propagation is allowed in air ($k\Lambda = \beta\Lambda$). It is in these regions, at points such as the one indicated by the dot in the figure, that bandgap guidance in a hollow core is possible; light is free to propagate in the air core, but unable to escape into the cladding since there are no supported photonic states.

Examples of numerically calculated DOPS plots are shown in Figure 4.11. The photonic structure considered here is that of a triangular arrangement of perfect circular air holes imbedded in silica background. The gray scale is linear from light gray representing a low DOPS to black for a high density of states. The regions with zero photonic states, which demark the bandgaps and correspond to DOPS=0, are shown as white-colored regions. Normalized frequency $k\Lambda$ is plotted against $(\beta - k)\Lambda$, such that the air-line is vertical as indicated in the plots.

The plots in this figure are calculated for different ratios between hole diameter and pitch (D/Λ). In practice, the air-filling fraction (*aff*) per unit cell is more often quoted than D/Λ. For the present

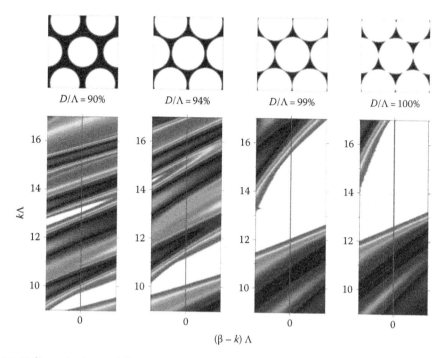

FIGURE 4.11 (See color insert following page 11-20.) Index profiles (top) and DOPS plots for four cladding structures (bottom), with ratios of hole diameter to pitch 0.90, 0.94, 0.99, and 1 from left to right. The color scale represents the photonic density-of-states from low (light gray) to high (black). Zero density is highlighted by white color, and shows, for $(\beta - k)\Lambda < 0$, the permissible transmission band location for an HC-PCF possessing such a cladding structure. The vertical line is the air-line, $\beta = k$. (From Benabid, F., *Philos. Tran. R. Soc. A*, 364, 3439, 2006.)

structure, the *aff* is related to D/Λ by $aff = \pi\left(D/\Lambda\right)^2/\left(2\sqrt{3}\right)$. The four plots correspond to a ratio D/Λ of 0.90, 0.94, 0.99, and 1, respectively. The code used to perform these calculations is based on the plane-wave expansion method and the DOPS was normalized to that of silica bulk material (Hedley 2006).

For $D/\Lambda = 0.90$ and below, the band diagram exhibits several narrow bandgaps that extend above the air-line (i.e., $(\beta - k)\Lambda < 0$) in corroboration with the results of Birks et al. (1995). As D/Λ increases, the higher-order bandgaps (those located at $k\Lambda > 9$ for $D/\Lambda = 0.90$) tend to shrink and eventually disappear while the fundamental PBG follows a clear trend: it becomes wider and wider and its location shifts to higher frequencies. The width of the fundamental bandgap as a function of D/Λ is shown in Figure 4.12. For example, at $D/\Lambda \sim 0.98$, i.e., an *aff* of 88%, only a narrow bandwidth of 45 THz is expected when the pitch is $\Lambda = 2\,\mu m$. The bandwidth displays an almost exponential dependence on *aff*, which has recently been elucidated theoretically through the development of a photonic model akin to the solid-state tight-binding model (Couny et al. 2007b). This model will be introduced later in the section.

The geometry considered in Figure 4.11 does not perfectly match with that of fabricated fibers whose structure becomes slightly distorted during the fiber-draw process. Fabricated fibers exhibit air holes of a hexagonal shape with curved corners instead of circular ones (Benabid 2006) (see the inset of Figure 4.13). However, the trend shown by the PBG width and location remains valid. Furthermore, because of the change in structure, fabricated HC-PCF can exhibit much higher *aff* than the upper limit of 91% of a triangular lattice of perfectly circular holes. Consequently, as will become clearer below, the PBG of fabricated HC-PCF can be slightly wider and centered at larger $k\Lambda$ relative to the structure considered above.

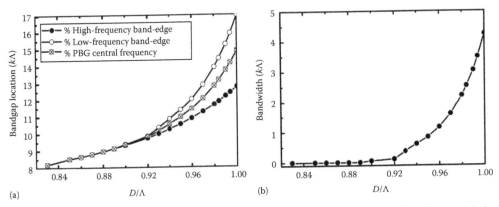

(a)

(b)

FIGURE 4.12 PBG evolution with D/Λ for the lattice structure shown in Figure 4.11. (a) The evolution of the band edges and the central frequency of the fundamental PBG as D/Λ increases. The frequencies were taken at the air-line (i.e., $n_{\text{eff}}=1$ or $(\beta-k)\Lambda=0$). (b) The evolution of the PBG bandwidth with D/Λ. (From Benabid, F., *Philos. Tran. R. Soc. A*, 364, 3439, 2006.)

The introduction of a defect into the photonic crystal cladding structure (e.g., the core of an HC-PCF) introduces guided modes centered within or at the periphery of the defect. Figure 4.13a shows a DOPS diagram for a periodic photonic structure inferred from the cladding of a fabricated HC-PCF (see inset for the structure unit cell). The HC-PCF has a hollow core that corresponds in size to approximately seven missing cladding cells. Indeed, the fiber was fabricated by omitting seven capillaries in the stack which was drawn down to form the fiber. The fiber's cladding structure constitutes a triangular arrangement of nearly hexagonally shaped air holes with an *aff* of ~93% and a pitch of 2.2 μm. Figure 4.13b shows calculated mode trajectories for this fiber. These are shown superimposed on a DOPS plot. We distinguish two classes of modes that are guided by the hollow-core defect: (1) core guided modes and (2) surface modes. The fields of the core guided modes are mostly localized in the hollow region of the defect (i.e., they are "air modes") and the modes exhibit dispersion curves with a shallow gradient relative to the air-line. This is exemplified by the mode branches for the fundamental (HE$_{11}$) core mode and the first quartet of higher-order core modes indicated on the diagram in Figure 4.13b. Here the adjective

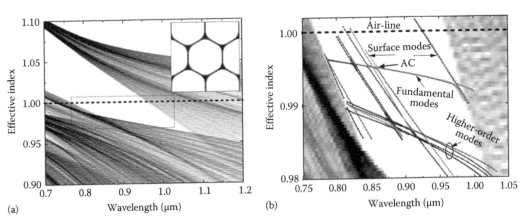

(a)

(b)

FIGURE 4.13 (See color insert following page 11-20.) (a) The calculated DOPS for an HC-PCF and (b) focuses in on the bandgap region near the air light line and shows calculated mode trajectories superimposed. The extension of the guided modes out of the bandgap is due to the DOPS being calculated for the average cladding hole diameter, while the modes are calculated for the actual structure with slight variations in hole diameter across the cladding.

"fundamental" used to label the degenerate HE_{11} mode was adopted to follow convention. Rigorously speaking, the fundamental mode corresponds to the one with the highest effective index which is, in the case of HC-PCF, either the cladding fundamental space-filling mode (see below) or a TIR-guided mode associated with a locally high concentration of silica (typically residing near the core surround).

The surface modes within the bandgap are other guided modes that exist at the interface between the hollow core and the cladding, with their field distributions concentrated at the silica core surround. These are also called core interface modes or core surround modes. Core interface modes appear as the more vertical trajectories in the figure with gradients that are closer to those of the band edges.

The presence of core interface modes provides a loss mechanism for the fundamental-like (and other core guided) modes. At anticrossings between core and core interface modes, such as the point marked AC in Figure 4.13b, power can couple from the fundamental mode to the lossy-surface mode to which it is nearly phase-matched (i.e., nearly the same β value). The surface mode is far lossier than a core mode since it has much higher field intensity at the glass/air interfaces. These interfaces inevitably have a degree of roughness, as will be described in Section 4.3.4, which leads to a higher scattering loss suffered by surface modes. A typical loss figure for a core surround mode is of order 1 dB/m, whereas the core mode, when free from coupling or hybridization with a core surround mode, can show a loss of order 1 dB/km. Furthermore, the presence of the surface modes has the effect of reducing the usable bandwidth of the fiber.

The experimentally measured spatial intensity profile of core guided and surface modes of an HC-PCF whose core approximately corresponds to 19 missing unit cells (a "19-cell core" fiber) are shown in Figure 4.14. In this figure, Figure 4.14a is the fundamental core mode, and Figure 4.14b and c are higher-order core modes. To a good approximation, within the core itself, the intensity profile of HC-PCF core modes are the same as those of a hollow capillary of the same core diameter (Marcatili and Schmeltzer 2006). In addition, the figure shows the intensity profile of the transmitted light near an anticrossing between the HE_{11} core mode and a surface mode (Figure 4.14d). Figure 4.14e relates to the transmitted intensity profile near an anticrossing between a higher-order mode (HOM) and a surface mode. Finally, away from any anticrossing, and thus where there is no hybridization with the core modes, the typical intensity profile of a surface mode is shown in Figure 4.14f.

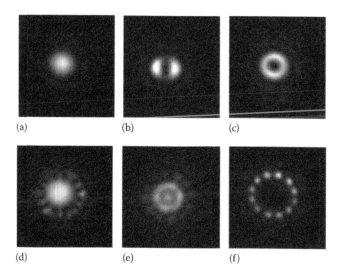

(a)   (b)   (c)

(d)   (e)   (f)

FIGURE 4.14 Measured guided mode intensity profiles for a 19-cell core HC-PCF with a bandgap centered at 1550 nm. The intensity profile of (a) the fundamental and (b and c) higher-order core modes excited at a single wavelength within the PBG by changing the input coupling conditions and corresponding to the different core modes. Intensity profile of transmitted modes near an anticrossing between (d) a surface mode and a fundamental mode and (e) a surface mode and higher-order core mode. (f) Intensity profile of a surface mode excited at a wavelength outside the air-guidance transmission window.

Since the bandgap position is normalized to the pitch of the fiber, the spectral location can be chosen by appropriate scaling of the cladding structure. Thus the core size of a particular fiber depends not only on the number of cells removed to form the core defect, but also on the wavelength at which the fiber guides.

Although we have seen above that it is the presence of an out-of-plane bandgap in the cladding structure that allows HC-PCF to guide light in a hollow core, the numerical simulations provide no real physical insights into the bandgap's origin. There are several works that have considered the nature of cladding states, dating back to Yeh and Yariv's Bragg reflection waveguide (Yeh et al. 1977). In that work, it was qualitatively observed that the allowed bands in the cladding structure were formed from several modes, predominantly confined in the higher-index layers, the number of modes being determined by the number of layers. This observation indicated the role of the resonance properties of the high-index layers in bandgap formation.

Later, the ARROW model was introduced by Duguay (Duguay et al. 1986). This model was originally applied to simple planar waveguides by considering the resonance properties that depend on the transverse component of the wavevector in the high-index confining layers. The layers act as Fabry–Perot resonators, and as such, high reflection is obtained for antiresonant wavelengths. The ARROW model has more recently been applied to Bragg reflection fibers and a particular class of PCF (Litchinitser 2002, 2003). In the case of PCF, the studies considered high-index cylindrical inclusions in a low-index background, and show that the transmission bands and bandgaps are predominantly determined by the modal properties of the individual inclusions, which themselves act as waveguides. In the short wavelength limit, $\lambda \ll \Lambda$, the band structure is independent of the lattice constant, and dependents solely on the geometry of the inclusions. The ARROW concept extends this idea to longer wavelengths, and is based on the assumption that bandgaps will still occur between the isolated inclusion mode lines, even though the mode lines become broadened due to inter-inclusion electromagnetic coupling. The model further assumes that the gap will extend slightly below the refractive index of the background material to enable guidance within a core. While the ARROW model provides important insights, it is not able to accurately model PCFs.

Birks et al. introduced a method for calculating band structure that is an intermediate approach between full numerical simulation and the ARROW model (Birks et al. 2006), and provides further insight into the guidance mechanism of Bragg fibers and all-solid PCF which guide by virtue of a bandgap. The proposed model is analogous to the cellular method used in solid-state physics, and unlike the ARROW model, this model does take account of the low-index regions that separate the high-index rods. For this reason it is able to provide good approximations to the true band structure.

Couny et al. (2007b) have extended the analogy with solid-state physics to the tight-binding model, which gives additional information about the guidance mechanism, and of the origin of the PBG formation in complex structures such as that of triangular-lattice HC-PCF. The central ideas are summarized in the following subsection.

4.3.2 Photonic Tight-Binding Model Description of HC-PCF Cladding

The tight-binding model has been successfully used in atomic and solid-state physics to give an approximate description of the origin of electronic bands (allowed and forbidden) (Ashcroft and Mermin 1976). In this picture, electronic bands in a solid (e.g., a crystal) are considered to be the result of bringing together isolated and identical atomic sites that constitute the solid. This approach has the double advantage of holding a high degree of predictive power as well as being simple. The model simplifies the complex problem of finding the allowed energy band of a crystal to that of treating the Hamiltonian of identical atomic constituting elements of the crystal, each with well-defined energy states and wave functions. Only atomic sites that are close to one another interact, leading to a limited number of off-diagonal (or hopping) terms in the Hamiltonian of the system. As these atomic sites become closer, each single electronic state, whose wave functions (orbitals) were well localized within the potential well created by the single atom, splits to form a band of energy levels with delocalized wave functions (Bloch functions). The wave functions are all a linear superposition of the isolated atomic orbitals. Furthermore, for typical crystal configurations, the upper energy level of the formed band corresponds to an antisymmetric wave function, and the lower energy level corresponds to a symmetric and tightly bound state. It can be expected that a simple transposition of the tight-binding model to a periodic photonic structure will provide a wealth of information on the nature of the modes in the allowed bands and of the formation of bandgap. In the photonic structure, the electronic bands are substituted by bands of photonic states which are, in the case of a PCF, optical modes propagating along the axis of the fiber. The photon binding is locally provided by TIR, which is made possible by the index contrast between the structural features of the photonic structure. The analogue to an atom in a crystal is a dielectric wave-guiding feature within the unit cell of the photonic structure. A proof of principle of this picture was reported using a toy model in Benabid and Roberts (2008). As a result, the mode structure of a PCF cladding consists of photonic bands, each of which contains optical (Bloch) modes which in the tight-binding picture are considered to be a linear superposition of modes associated with individual guiding elements within the cladding. Consequently, once the identification of these guiding elements is elucidated, all the rules of thumb deduced from the tight-binding model apply.

For the triangular-lattice PBG HC-PCF, it was shown experimentally and theoretically that the structural features behind the photonic states of the fiber cladding are the interstitial silica apexes and the silica struts that form the silica web (see Figure 4.15b) on one hand and the air hole on the other hand (Couny et al. 2007b). Figure 4.15d through g presents the numerically calculated DOPS and the modal

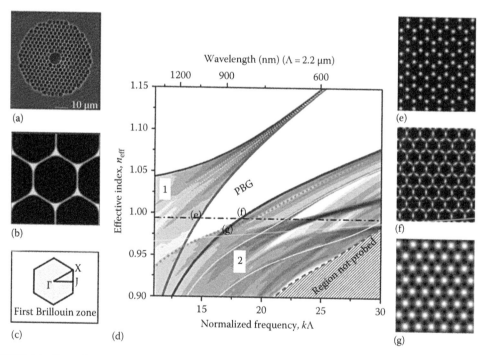

FIGURE 4.15 (See color insert following page 11-20.) (a) SEM of an HC-PCF with core guidance at 1064 nm. (b) Details of the cladding structure used for the numerical modeling. (c) Brillouin zone symmetry point nomenclature. (d) Propagation diagram for the HC-PCF cladding lattice. White represents zero DOS and black maximum DOS. The upper x-axis shows the corresponding wavelengths for an HC-PCF guiding at 800 nm (pitch $\Lambda = 2.15\,\mu m$). The solid and dotted lines represent the Γ-point and J-point mode trajectories, respectively. The trajectories of the cladding modes on the edges of the PBG are represented (e) in red for the interstitial "apex" mode with the near field, (f) in blue for the silica "strut" mode, and (g) in green for the "air-hole" mode. The first two modes are shown at an effective index of 0.995 (dash-dotted white line), whereas the air-hole mode is shown at $k\Lambda = 15.5$ and $n_{eff} = 0.973$.

properties of the HC-PCF cladding structure shown in Figure 4.15b. Either side of the out-of-plane PBG region (shown in black in Figure 4.15d) are allowed bands where the cladding can support propagating modes (regions 1 and 2 in Figure 4.15d). These bands follow the photonic tight-binding model (PTBM) in the sense that the bands shrink as $k\Lambda$ increases. Of particular interest is the mode forming the lower-index edge of region 1, since it determines the lower-frequency edge of the PBG of the cladding. The dispersion of this mode is represented in red in Figure 4.15d, and its near field (NF) is shown in Figure 4.15e at a representative normalized wave number $k\Lambda$. This confirms that the light is predominantly guided in the interstitial apexes. The apexes are thus identified as the most important optical resonators associated with the bandgap edge of higher effective index (Benabid and Russell 2005). The nature of the modes at the upper-frequency edge of the PBG is more complicated than those at the lower-frequency edge. Indeed, the modes near the bandgap edge with lower effective index are formed from overlapping and interacting allowed photonic bands. This results in the bandgap edge itself being determined by a combination of the trajectories of two cladding modes of different symmetry. At frequencies (i.e., $k\Lambda$) below $k\Lambda = 16.9$, this edge is due to a mode located at the J-point of the Brillouin zone (Figure 4.15g, dotted green curve in Figure 4.15d) which guides predominantly in the air holes ("air-hole" mode). Above $k\Lambda = 16.9$, the edge is due to a mode located at the Γ-point (Figure 4.15f, blue traces in Figure 4.15d) guiding predominantly within and close to the silica struts ("strut" mode) that join neighboring apexes, with little light penetrating into the apexes. Consequently, the silica struts directly limit the upper frequency of the fiber transmission band.

These observations were corroborated experimentally by measuring spectrally resolved NF images and spatially resolved transmission spectra of 3 mm long HC-PCFs using a scanning near-field optical microscope (SNOM). Both the NF and spectra were taken by exciting the fiber with super-continuum light over a narrow angular range near the air light line. The transmission spectra of the relevant modes show distinctive cutoffs relating to the fiber PBG location. Figure 4.16a shows the

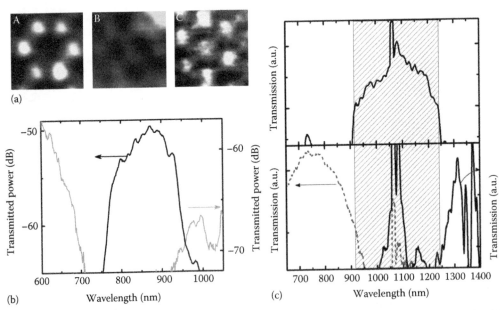

FIGURE 4.16 (a) SNOM images of (A) the "apex" mode, (B) the "strut" mode, and (C) the "air-hole" mode of the fiber cladding. (b) Optical spectrum of the HC-PCF guiding around 800 nm taken with the SNOM tip aligned with the core (black line) and near an air hole of the cladding (gray line). (c) Optical spectrum of the HC-PCF guiding around 1064 nm taken with the SNOM tip aligned (top) with the core, (bottom) with an interstitial apex (black solid line), and with an air hole of the cladding (gray dotted line). The peaks around 1064 nm are due to the residual super-continuum pump. (From Couny, F. et al., *Opt. Express*, 15, 325, 2007.)

typical measured NF profiles when the fiber is excited by light near the lower-frequency bandgap edge and the upper-frequency edge (Figure 4.16a(A) and 4.16a(B and C), respectively). The NF is obtained by scanning the SNOM tip, with 0.15 μm step, over an area covering a few unit cells of the HC-PCF cladding structure. Figure 4.16a(A) clearly shows that the imaged mode corresponds to that of the apex resonator, while Figure 4.16a(B and C) shows light confined predominantly in the struts and air, respectively. Moreover, when the tip is aligned with an interstitial apex, the transmission spectra of the two different fibers show a cutoff around the lower-frequency edge of the PBG (see the solid black line on the lower graph of Figure 4.16c for the 1060 nm HC-PCF). This corresponds to the apex mode frequency cutoff near the air-line in accordance with the numerical simulation (see Figure 4.15d). Similarly, for strut and air-hole modes, the transmission spectra show a clear cutoff at the short wavelength side of the HC-PCF transmission bandwidth (see the gray line of Figure 4.16b for the 800 nm HC-PCF and the gray dotted line of the lower graph of Figure 4.16c for the 1060 nm HC-PCF). However, due to the limited spatial resolution of the SNOM and to the hybridization between the two constituent resonators, the transmission spectra collected when the tip was aligned on top of a strut or in an air hole do not show a measurable difference in their frequency cutoff. Nevertheless, the above results using the SNOM do confirm that the PBG is formed by the interplay of three distinct resonators.

4.3.3 Broadband Hollow-Core Guidance

Bandgap guidance enables low propagation loss to be attained, but the bandwidth for guidance is restricted by the achievable width of the bandgap close to the air light line. Since the fiber must be mechanically sound, the glass nodes that can act as antiresonant reflection optical waveguides are compromised by the necessary presence of connecting glass struts, which restricts the frequency range of the broadest bandgap in practical glass/air structures to about 15% of the bandgap central frequency. The bandgap fibers with high air-fill fraction claddings, that have recently become accessible to fabrication, enable higher-order gaps to open up. These gaps, however, are relatively narrow and separated by wide spectral regions where attenuation within the core will be high. Many applications either require or would benefit from an increased bandwidth of guidance. A class of fibers that provides broad frequency ranges of relatively low-loss attenuation is based on cladding structures which entail a connected network of nearly constant thickness glass struts. An example of such a fiber is shown in Figure 4.17b.

The guidance within the core shown by these fibers does not rely on the formation of cladding bandgaps. Instead, the cladding structure is designed in such a way that the cladding modes that are present close to the air light line do not interact strongly with the core mode of interest.

The strategy behind the design of broadband guiding fibers is, in many respects, opposite to that adopted for bandgap fibers. The latter rely on sizeable glass nodes where the struts cross to form the bandgap, and the glass struts which are necessary to connect them act as distinct resonators that tend to narrow the bandgaps. The broadband guiding fiber designs, on the other hand, require that the glass nodes are maintained small, since these possess resonances which can exist close to the air light line at wavelengths distinct from the transverse resonances associated with the struts, thus narrowing the guiding bandwidth available. A common theme behind the design of hollow-core fibers is to minimize, as far as possible, the number of distinct forms of resonators that operate close to the core-mode effective index (which is very close to the air light line, $n_{eff} = 1$).

Aspects of the broadband HC-PCF guidance can be understood by considering a hypothetical glass/air Bragg fiber, which comprises concentric layers of silica in air such that the geometry possesses cylindrical symmetry. This geometry has already been considered in Section 4.2.2. Due to the symmetry, the modes of this fiber geometry decouple according to an azimuthal field variation of the form $\exp(im\phi)$ in the z-components of their **E** and **H** fields; modes with different m do not interact. The confinement loss of the HE_{11} core guided mode, which belongs to the $m = 1$ mode class, can be reduced arbitrarily by incorporation of a sufficient number of glass shells, except close to transverse

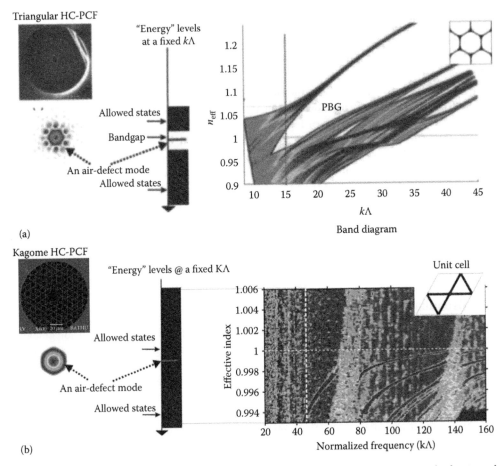

(a)

(b)

FIGURE 4.17 (See color insert following page 11-20.) (a) Top left: Scanning electron micrograph of a triangular-lattice HC-PCF. Bottom left: near-field profile of the fundamental (HE_{11}-like) air-guided core mode lying within a bandgap. Center: schematic representation of the mode spectrum when an air defect mode is guided. Right: DOPS diagram showing the presence of the PBG. (b) Same as (a) but for Kagome-lattice HC-PCF. The fundamental core mode lies within a continuum of cladding modes and the band diagram does not exhibit a PBG.

resonances given by the condition, $kt\sqrt{n_{gl}^2 - 1} = \pi j$ ($j = 1, 2, 3, \ldots$), with n_{gl} the refractive index of glass, and t the thickness of the glass layer. The broadband HC-PCF also shows low transmission at wavelengths near where this resonance is fulfilled. The Bragg fiber does possess cladding modes that can phase-match with the core mode, but symmetry precludes interaction since the former are from a different m mode class. The singly connected nature of broadband HC-PCF dramatically lowers the symmetry from C_∞, so symmetry-induced decoupling is not relied upon to give good core guidance. Instead, mode incompatibility primarily associated with very different phase variations of the modes inhibits the coupling.

A variety of broadband guiding single-material HC-PCFs has now been experimentally demonstrated. These show propagation loss which can be below 1 dB/m over sizeable wavebands, far lower than a capillary guide at a comparable core size. As well as the Kagome-based structure shown in Figure 4.17b, fibers based on a square-lattice arrangement of air holes have recently shown favorable propagation properties. Numerical simulations show that, if the fiber structure can be better controlled, a loss as low as 10 dB/km can be maintained over quite broad wavebands without increasing the size of the core (Couny et al. 2008).

Other than the size of the glass nodes within the microstructure, the main structural difference between bandgap HC-PCF and broadband HC-PCF is the pitch. In contrast to bandgap fibers, the wavelength ranges of operation for a given broadband guiding geometry is not set by the cladding pitch, but rather by the strut thicknesses. The deleterious effects of the strut intersections are weakened as the cladding pitch is made larger, but the practical limit to pitch enlargement is set by micro- or macrobend-induced coupling between the core mode of interest and other core modes and cladding modes that have a high field content in the air holes. The detuning in effective index between the core mode and these other modes decreases as Λ increases. The (relatively) low-loss broadband fibers fabricated to date have pitches in the range 10–20 μm. At a pitch value within this range, a core size that extends over less than about two cladding pitches is found to give quite robust guidance within a single core guided mode if care is taken to excite only the required HE_{11}-like mode at the input. The fibers can be designed so that HOMs in the core nearly phase-match with air dominated cladding modes; the consequent field hybridization renders the core HOMs more subject to leakage loss and the quasi-single mode operation of the fibers is enhanced.

The wide bandwidths of propagation that involve only weak interaction of the core mode with cladding modes results in the fibers showing very low chromatic dispersion, which is highly beneficial in many applications, particularly those that involve short pulse propagation (e.g., sub-100 fs pulses) (Couny et al. 2007b).

4.3.4 Loss Mechanisms

4.3.4.1 Confinement

The bandgap fibers can be rendered free of confinement loss over the bandgap range by ensuring a sufficient number of cladding periods surround the guiding core. In practice, for a cladding with a structure corresponding to Figure 4.18a, 10 layers are sufficient to confine the HE_{11}-like core mode over the majority of the bandgap such that confinement loss is below 1 dB/km. For a cladding of the form shown in Figure 4.18b, just eight periods are required, and for the form shown in Figure 4.18c, only six are needed. The requisite number of cladding periods for all these cladding types, which in *aff* range from 85% to 96%, is readily incorporated into fabricated fibers, which leaves other loss mechanisms responsible for attenuation in bandgap HC-PCF. Chief among these is scattering loss due to roughness at the glass/air interfaces, which will be treated in Section 4.3.4.2.

The main loss mechanism in broadband guiding HC-PCF is confinement; the absence of a PBG, and the presence of weak residual interaction between core and cladding mode constituents, implies that incorporating more and more cladding periods does not lead to progressive loss reduction. The number of cladding periods beyond which further loss reduction does not occur depends on the details of the cladding structure, but for the example structure shown in Figure 4.17b, this threshold is found to be just two periods. The realization that most of the guided power resides outside the solid fiber constituent,

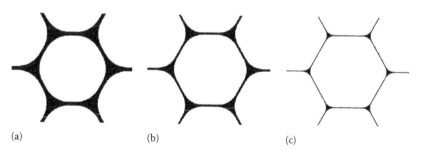

(a)   (b)   (c)

FIGURE 4.18 Bandgap cladding structures with (a) 85%, (b) 92%, and (c) 96% air-filling fractions. All these cladding forms are amenable to fabrication.

and that confinement loss is constrained in practical designs to be of order 0.1 dB/m, has given impetus to the development of polymer-based broadband guiding HC-PCF (Argyros and Pla 2007). If the fiber draw can be sufficiently well controlled when using such a material, the loss is expected to be similar to that of the silica-based broadband HC-PCF. In fact, it appears that the broadband HC-PCF are easier to fabricate when using polymers than the bandgap HC-PCF forms, and the former have so far shown lower overall loss than the latter with use of this material.

4.3.4.2 Scattering Loss

Since most of the guided light in bandgap HC-PCF resides in the core or the cladding holes, mode attenuation due to Rayleigh scattering loss within the silica is much reduced compared to all-solid fibers. The Rayleigh scattering coefficient of air at standard temperature and pressure is many times below that of silica, so it was hoped that the loss of the bandgap HC-PCF could be reduced below the figure of 0.16 dB/km attainable with the solid fibers. This would open up exciting possibilities for deployment of the bandgap HC-PCF as long-haul communication fiber. Fabricated bandgap HC-PCF has showed considerably higher attenuation than this, the lowest reported loss being 1.2 dB/km, yet the confinement loss was confirmed by calculation to be negligible. It was clear that the presence of the glass/air interfaces contributed to the loss, as had been inferred from earlier studies of small solid-core TIR-guiding PCF. The roughness at the interfaces is primarily due to surface capillary waves (SCWs) that become frozen in as the glass (or polymer) moves through the glass transition during the fiber draw (Roberts et al. 2005a). The timescales associated with the glass-forming process are such that equilibrium thermodynamics are believed to be sufficient to describe the static SCWs that remain after the transition; the mean energy of each SCW component is $k_B T_{tr}/2$, where k_B is the Boltzmann constant and T_{tr} is the "fictitious" glass transition temperature (Jäckle and Kawasaki 1995). The relative phase of each thermally excited SCW is random, resulting in a rough surface that must be described statistically.

Given the seemingly immutable nature of the interface roughness, the strategy employed to reduce the loss has been to design the fibers such that they exclude light as much as possible from the interfaces. This has either involved introducing features to the glass ring that surrounds the air core to render it antiresonant within the wavelength range of the bandgap, or thinning this core surround ring which has the effect of dispelling unwanted core surround related guided modes from the bandgap close to the air light line. The former approach ultimately enables lower loss to be achieved, but suffers from a reduced useable bandwidth due to an increase in the number of core surround related guided modes that reside in the bandgap near $n_{eff}=1$.

Conventional wisdom specifies that dispelling the unwanted core surround related guided modes from the bandgap close to the air light line, so their trajectories in n_{eff}-λ space do not come close to the trajectory of the wanted core mode, will lead to a reduction of the field intensity of the latter mode in the proximity of the core surround, including its interfaces. The idea is that interaction and consequent hybridization will decrease with increasing detuning. If the core surround modes are rendered leaky by being pushed into the continuum of cladding modes, this will also tend to decrease the core-mode field intensity within the core surround region due to any residual interaction. Such a core surround mode expulsion can be achieved in a practical design by appropriately thinning the core surround ring (West et al. 2004). Fabricated examples of fibers with thinned core surrounds and a high *aff* within the cladding have indeed shown reduced loss, with sub-10 dB/km guidance having been reported. These fibers also have the benefit of a broad low-loss operating wavelength range. This aspect also leads to a lower and more slowly varying dispersion, which is often beneficial in applications. It is noteworthy that, in the attainment of ultimate low loss, the favorable-field exclusion properties of antiresonance can prevail over the detrimental effects associated with the encroachment of core surround modes into the bandgap.

Alternative glasses or polymers can be considered in connection with reducing roughness scattering loss but, at least for glasses for which T_{tr} and surface tension γ have been tabulated, these all appear to have a larger T_{tr}/γ ratio than silica, which leads to a larger roughness. Nevertheless, a possible route to loss reduction involves using glasses that transmit at longer wavelengths, since the mode propagation loss at a given roughness amplitude decreases rapidly with wavelength (Roberts et al. 2005b). Glasses with low refractive index are preferred in connection with loss reduction, since the loss scales with $\left(n_{gl}^2 - 1\right)^2$. Of the glasses with tabulated properties, ZBLAN glass, which transmits with low attenuation in the mid-IR, has been identified as a candidate material from which to fabricate bandgap HC-PCF that shows lower loss than can be attained using silica. It remains to be seen whether the mechanical properties of ZBLAN glass allow the intricate HC-PCF geometry to be realized during a fiber draw.

4.4 Guidance within a Liquid Infiltrated Fiber

The previous sections concentrated on guidance in a hollow core and by extension to gas-filled HC-PCF given the insignificant index change introduced by filling the fiber with a gas-phase material. The same ideas can be applied to guidance in a liquid with a refractive index below that of the solid material component of a fiber. Given the wide range of refractive indices of liquids, the guidance mechanism and the transmission window strongly depend on the filling liquid and on the filling configuration. We can distinguish two main filling design options. The first one consists of filling all the holes in the fiber (i.e., cladding holes and the core hole) and the second one consists of filling only the core while keeping the cladding holes air filled.

In the first option, guidance will result but the scale of the structure must be altered to take account of the reduced index contrast. A simple scaling rule applies in the scalar field approximation, which becomes rigorous at weak index contrast. This scaling will be explored below. For the case where just the core of an HC-PCF is filled with liquid, depending on the liquid index, the guidance can be PBG-based, TIR-based, or a combination of both guidance mechanisms. Unless the dispersion control associated with bandgap guidance is central to the application, the TIR form of liquid-core guidance is generally preferred because the guidance is usually more robust, less sensitive to the details of cladding microstructure, and wider transmission bandwidth is available.

4.4.1 Guidance in HC-PCF with Just the Core Filled with Liquid

For this design option, just the core of an HC-PCF is filled with liquid and the cladding holes remain air filled. Since the cladding is substantially composed of air, its effective refractive index is well below the index of the solid material component. This cladding effective index is simply the index of the FSM, n_{FSM}. This corresponds to the PC-line shown in Figure 4.10. Figure 4.19b shows the DOPS of a fabricated PBG HC-PCF whose SEM is shown in Figure 4.19a and is the same as the one shown in Figure 4.15. However, in Figure 4.19, the plotted frequency range extends to larger values, and one can clearly see the photonic band which is due to the splitting of the fundamental modes of the silica apexes. It narrows as $k\Lambda$ increases to ultimately form a single dispersion line. At the high $k\Lambda$ limit, this line coincides with the effective index trajectory of a single-apex mode, and the ARROW approximation applies. This behavior is also in accord with the PTBM.

The FSM is the mode of this apex-related photonic band with highest effective index; and corresponds to the "tightly bound" mode in the picture of the PTBM mentioned above, and shows no phase variation within the fiber cross section. The index n_{FSM} of this mode is the most important factor that determines which guidance mechanism operates when the core of the PBG HC-PCF is filled with the liquid.

For liquids with a refractive index $n_l < n_{FSM}$, only PBG guidance is possible (provided that the cladding structure exhibits a PBG below n_l). Given the dispersive nature of n_{FSM}, the inequality $n_l < n_{FSM}$ is not necessarily fulfilled over the whole spectrum. Figure 4.19b shows that for the case of the considered HC-PCF the minimum value of n_{FSM} is ~1.05, which occurs for normalized frequencies $k\Lambda$ of about 12. Consequently, for fluids with indices less than about 1.05, the guidance can only take place via PBG irrespective of the wavelength. Since liquids have refractive indices larger than 1.05, in practice this regime

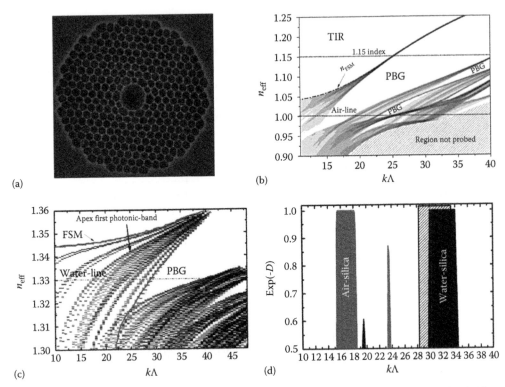

FIGURE 4.19 (a) SEM of a typical bandgap HC-PCF. (b) Density of photonic states (DOPS) for the cladding of the fiber shown in (a) with holes filled with air, plotted as a function of the normalized frequency and the effective index. The PBG for air guidance occurs around $k\Lambda \sim 16$. When the core of the fiber is filled by a liquid with index higher than n_{FSM} at a particular optical frequency, one or more TIR-guided modes with effective index values above n_{FSM} will result. Other guided modes confined primarily in the core are possible. These are HOMs residing within the bandgap effective index range. If the liquid (or gas) index of the core is lower than n_{FSM}, only bandgap guidance is possible. (c) DOPS for the cladding of the fiber shown in (a) but the air holes are filled with water (index ~1.33). The bandgap region shifts approximately according to the scaling rules described in Section 4.4. Guidance within a water-filled core is possible in the bandgap at effective index values below the water line $n_{eff} = 1.33$, which is centered around $k\Lambda \sim 16$. (d) Numerically calculated spectral location of the PBG and expressed as e^{-D} (D being the DOPS function appropriately normalized) for the case of a PBG HC-PCF filled with (gray) air at an effective index of $n_{eff} = 1$, and (black) water at an effective index of 1.33. The hatched region is the location of the water-filled PBG HC-PCF as predicted by the index-scaling law (Equation 4.10).

is not encountered when liquid filling the core of a standard HC-PCF design, but it is the regime of relevance when dealing with a gas-filled core. In this configuration, because of the increase of the defect effect index, the PBG-induced transmission windows of the liquid-filled fiber exhibits a typical trend; the fiber transmission window is blue-shifted relative to that of air-guiding fiber, and its bandwidth gets wider in accord with the PTBM.

Another possible scenario is that of a liquid with a higher index than n_{FSM} in one spectral region and lower in another region. Figure 4.19b illustrates this with a liquid index of $n_l = 1.15$. For normalized frequencies $k\Lambda < 25$, the guidance can occur via TIR, while for $k\Lambda > 25$, only PBG guidance takes place. In fact, even when $n_l > n_{fsm}$, bandgap guidance is still possible. Such guidance necessarily corresponds to high-order core modes that have effective index values within the bandgap range. With careful coupling optics, it is possible to selectively excite the fundamental core guided mode, which is TIR guided, if single mode operation is required.

Both TIR and bandgap guidance of a "fundamental" liquid-core mode show similar properties. The main difference between TIR and PBG guidance manifests in the dispersion characteristic in each transmission spectral window.

If the liquid of interest is of higher refractive index than the solid material (here silica) component of the PCF, TIR guidance becomes possible at all wavelengths. At such index values, the liquid can of course also be used within the core of a standard glass capillary to attain TIR guidance.

In the TIR-guiding regime, it is often advantageous to ensure that the effective index step between the core and cladding is relatively low so that guidance is not greatly multimode. The HC-PCF geometry into which the liquid is introduced can be designed with this in mind. For example, filling an HC-PCF such as the one shown in Figure 4.19a with water and setting an operating wavelength to be 1.06 μm requires n_{FSM} to be close to the index of water, 1.33. This in turn requires an operating normalized frequency $k\Lambda$ around 44, and consequently a pitch of about 7.5 μm for this fiber design, which can be commercially available.

Although it may be attractive to use an "off-the-shelf" HC-PCF in liquid-core guiding applications such as the above TIR-based example, it may be preferred to use a fiber with smaller air holes in the cladding (i.e., low *aff*) so that the pitch and core size can be kept relatively small. Note that the initial unfilled fiber does not need to guide well or even at all when air filled, so that a relatively cheap below-spec HC-PCF may be commercially attainable for liquid-core applications.

4.4.2 Guidance in an HC-PCF with Core and Cladding Holes Filled with Liquid

Bandgap guidance may be preferred to TIR guidance either due to the associated advantageous dispersion characteristic or to the wavelength filtering possibilities. Bandgap guidance predominantly within an HE_{11}-like guided mode of the liquid core can result if the liquid index is below that of the solid material of the fiber. An important example of this involves aqueous solutions introduced into the holes and core of an HC-PCF. The guidance properties that result will be described in Section 4.4.4 in connection with scaling rules obeyed by the mode spectrum.

If all the holes, including the core, are filled with liquid with refractive index above the background solid component, both TIR and bandgap guidance are possible. Bandgap guidance necessarily involves higher-order core modes; the lowest-order (HE_{11}-like) core mode will guide by TIR. The TIR-guided mode of an HC-PCF with all holes filled with high-index liquid may be preferred to simply filling a capillary, since the effective index step between core and cladding can be adjusted such that single or quasi-single mode guiding within the core ensues.

4.4.3 Designs with Lower Mode Field Overlap with the Liquid

The designs described above ensure that most of the guided field resides in the liquid of interest. Some applications can tolerate or require a smaller overlap of the mode field with the liquid. If the liquid has lower refractive index than the solid background material, it can be placed in the cladding holes of a standard TIR-guiding solid-core PCF, which if being purchased commercially, is a cheaper option than a hollow-core PCF. The field is evanescent within the liquid, but still the overlap may be large enough, particularly if a fiber with a small core and relatively large cladding holes is selected for infiltration. If the liquid has higher index than the host material, filling the holes of a standard TIR-guiding solid-core PCF results in bandgap guidance. An example of this form of fiber is when the filling liquid is a liquid crystal (LC) (Larsen et al. 2003). By applying a voltage or heat source, the LC enables a reasonable degree of tunability of the bandgap position and so may be used as the basis of a tunable filter. In a sensing application, the overlap of the mode field with a high-index liquid introduced in the holes of a standard PCF can be controlled by appropriate choice of the hole size in the PCF being filled. Larger holes (relative to the pitch and the core) lead to a larger overlap, but also to less robust guidance, with confinement loss being a potential issue.

4.4.4 Guiding and Scaling Laws in Liquid-Filled HC-PCF

When a liquid is introduced into the core and the cladding holes of an HC-PCF, the modal spectrum and the associated dispersion changes. In order to rigorously work out the optical properties which result, Maxwell's equations need to be solved afresh each time a new liquid with a different refractive index is used. Figure 4.19c shows the DOPS diagram for the cladding of the HC-PCF shown in Figure 4.19a when filled with water. Just below the water cutoff index, the bandgap is centered around $k\Lambda \approx 32$. This represents a substantial blue-shift from the air-filled HC-PCF PBG location where the bandgap was centered around $k\Lambda \approx 17$. Furthermore, the water-filled cladding structure exhibits a wider normalized PBG width (bandgap width divided by central frequency) relative to the air-filled one. This is so because as the index contrast is lowered, the deleterious vector coupling effects associated with mode formation become weaker and the bandgaps actually become broader when expressed in terms of the appropriate dimensionless parameters.

Although only strictly valid for low-index contrast, very useful scaling laws can be invoked to predict the manner in which the photonic mode structure changes with index contrast without the need for further heavy computation (Birks et al. 2003). Within the scalar-wave approximation, the vectorial wave equations (Equation 4.1) simplify to the following equation:

$$\nabla_\perp^2 \Psi(x,y) + (k^2 n^2 - \beta^2)\Psi(x,y) = 0 \qquad (4.8)$$

Here $\Psi(x,y)$ is a scalar function that describes the transverse distribution of any component of the electric or magnetic field in the photonic structure of the fiber. In the case of a PCF structure consisting of a material with high index n_1 and a material with low index n_2 with a pitch Λ, Birks et al. (2003) found that the photonic states scale so that the following quantities remain invariant under a change of frequency, material index, or length scale Λ:

$$v^2 = k^2 \Lambda^2 (n_1^2 - n_2^2)$$

$$w^2 = \Lambda^2 (\beta^2 - k^2 n_2^2) \qquad (4.9)$$

These invariants can thus be used to yield useful scaling laws that can describe the shift in frequency of the photonic states (and hence bandgap locations) of the fiber cladding when the index contrast of the latter is altered. In particular, from the expression of the invariant v, when the low-index material n_2 of the PCF is varied while the high index n_1 remains unchanged, so that the initial index contrast $N_0 = n_1/n_2$ becomes N, any bandgap found originally at a wavelength λ_0 will shift to a new wavelength λ given by (Johnson et al. 2001):

$$\lambda \approx \lambda_0 \sqrt{\frac{1 - N^{-2}}{1 - N_0^{-2}}} \qquad (4.10)$$

This scaling law holds for any wavelength within the bandgap and provides a very useful tool to have an approximate location of the bandgap of a fluid-filled HC-PCF. The hatched region in Figure 4.19d shows the location of the shifted-PBG as deduced from the scaling law (Equation 4.10) and compared to the numerical deduced from the results shown in Figure 4.19b and c for air-filled and water-filled, respectively.

The scaling law is of relevance to any application that requires filling the entire air region (core and cladding) of HC-PCF with gases or liquids. The frequency shift of the bandgap was experimentally demonstrated by Antonopoulos et al. (2006) using HC-PCFs first in their air-filled form and then with the air replaced with liquid deuterium oxide or heavy water (D_2O). D_2O has a refractive index of about 1.33 in the visible part of the spectrum. By replacing the air in the HC-PCF with heavy water, the index contrast of the fiber structure was reduced from about 1.46 to 1.10. D_2O was preferred to ordinary water because it is much less lossy in the near-infrared (Tam and Patel 1979). White light and supercontinuum transmission spectra were taken before and after the fibers were filled with D_2O. It was then possible to record the changes in the transmitted spectra due to the change in the index contrast alone, while using the same piece of fiber and so keeping the pitch and symmetry of the lattice unchanged. This eliminated any spectral changes that would be due to structural differences between different fiber samples of varying index contrast.

The scaling law can also be applied to changing the index of the glass host while keeping the liquid or gas infiltrate within the holes at the same. This can be useful if the required operating wavelength is substantially outside the transmission band of silica, e.g., for wavelengths longer than about 3.5 μm. If the index contrast becomes larger than about 1.7, however, the scaling law breaks down such that bandgap in operation in standard HC-PCF effectively closes up (Pottage et al. 2003). This is due to the strengthening of vectorial effects. Fortunately, a new and quite robust bandgap opens up to replace it; this can be used for probing gases at long wavelengths when a chalcogenide glass forms the fiber host material.

4.5 Conclusions

In this chapter we have endeavored to describe the optical transmission properties and guidance mechanisms of hollow waveguides. We have seen how these are impacted when the waveguide is filled with a fluidic material, be it gaseous or liquid. An emphasis was given to the HC-PCF because of its current widespread use, its low transmission loss and relatively small mode area. The last aspect enables quasi-single mode operation essentially free from bend-loss constraints, and facilitates nonlinear interactions due to high field intensities being possible even at low optical powers.

A historical and theoretical approach was taken in this chapter so as to trace the development relevant to HC-PCF through the understanding of the different underlying guidance mechanisms. The salient features of each type of hollow waveguide were highlighted. Rather than reviewing the computational approaches that have been used in the study of complex waveguides such as HC-PCF, an introduction to the relatively simple analysis of planar and cylindrical Bragg waveguides was offered. The conceptually useful PTBM was introduced to shed light on bandgap formation and the nature of the photonic modes. These simple models give a sound basis for understanding the guidance of the more complex waveguides and elucidate how reported asymptotic regimes such as ARROW guidance relate to PBG.

This chapter made no attempt to survey all forms of hollow waveguides or the applications for which they have been used, but aimed to provide the theoretical backbone for the optofluidic applications using PCFs that are described in more detail in later chapters. It is also hoped that it provides the reader with a useful foundation for choosing or designing an appropriate waveguide for use in microfluidic applications.

Acknowledgments

The authors are grateful to Francois Couny and Phil Light for providing some of the chapter material.

References

Antonopoulos, G., Benabid, F., Birks, T.A., Bird, T.M., Knight, J.C., and Russell, P.S.J. 2006. Experimental demonstration of the frequency shift of bandgaps in photonic crystal fibers due to refractive index scaling. *Opt. Express* 14:3000–3006.

Argyros, A. and Pla, J. 2007. Hollow-core polymer fibres with a kagome lattice: Potential for transmission in the infrared. *Opt. Express* 15:7713–7719.

Argyros, A., Birks, T., Leon-Saval, S., Cordeiro, C.M.B., and Russell, P.S.J. 2005. Guidance properties of low-contrast photonic bandgap fibres. *Opt. Express* 13:2503–2511.

Ashcroft, N.W. and Mermin, N.D. 1976. *Solid State Physics*. Saunders College, Philadelphia, PA.

Benabid, F. 2006. Hollow-core photonic bandgap fibres: New guidance for new science and technology. *Philos. Tran. R. Soc. A* 364:3439–3462.

Benabid, F. and Roberts, P.J. 2008. Guidance mechanisms in hollow-core photonic crystal fiber. *Proc. SPIE* 6901:69010U.

Benabid, F. and Russell, P.S.J. 2005. Hollow-core PCF: Progress and prospects. *Proc. SPIE* 5733:176–190.

Benabid, F., Knight, J.C., Antonopoulos, G., and Russell, P.S.J. 2002. Stimulated Raman scattering in hydrogen-filled hollow-core photonic crystal fiber. *Science* 298:399–402.

Birks, T.A., Roberts, P.J., Russell, P.S.J., Atkin, D.M., and Shepherd, T.J. 1995. Full 2D photonic bandgaps in silica/air structures. *Electron. Lett.* 31:1941–1943.

Birks, T.A., Knight, J.C., and Russell, P.S.J., 1997. Endlessly single-mode photonic crystal fiber. *Opt. Lett.* 22:961–963.

Birks, T., Bird, D., Hedley, T., Pottage, J., and Russell, P. 2003. Scaling laws and vector effects in bandgap-guiding fibres. *Opt. Express* 12:69–74.

Birks, T.A., Pearce, G.J., and Bird, D.M. 2006. Approximate band structure calculation for photonic band-gap fibres. *Opt. Express* 14:9483–9490.

Couny, F., Benabid, F., Roberts, P.J., Burnett, M.J., and Maier, S.A. 2007a. Identification of Bloch-modes in hollow-core photonic crystal fiber cladding. *Opt. Express* 15:325–338.

Couny, F., Benabid, F., Roberts, P.J., Light, P.S., and Raymer, M.G. 2007b. Generation and photonic guidance of multi-octave optical-frequency combs. *Science* 318:1118–1121.

Couny, F., Roberts, P.J., Birks, T.A., and Benabid, F. 2008. Square-lattice large-pitch hollow-core photonic crystal fiber. *Opt. Express* 16:20626–20636.

Duguay, M.A., Kukubun, Y., Koch, T.L., and Pfeiffer, L.N. 1986. Antiresonant reflecting optical wave-guides in SiO_2-Si multilayer structures. *Appl. Phys. Lett.* 49:13–15.

Fink, Y., Ripin, D.J., Fan, S.H., Chen, C.P., Joannopoulos, J.D., and Thomas, E.L. 1999. Guiding optical light in air using an all-dielectric structure. *J. Lightwave Technol.* 17:2039–2041.

Hedley, T.D. 2006. Modelling of photonic crystal fibre, in *Physics*. University of Bath, Bath, U.K.

Jäckle, J. and Kawasaki, K. 1995. Intrinsic roughness of glass surfaces. *J. Phys.: Condens. Matter* 7:4351–4358.

John, S. 1987. Strong localization of photons in certain disordered dielectric superlattices. *Phys. Rev. Lett.* 58:2486–2489.

Johnson, S.G., Ibanescu, M., Skorobogatiy, M., Weisberg, O., Engeness, T.D., Soljacic, M., Jacobs, S.A., Joannopoulos, J.D., and Fink, Y. 2001. Low-loss asymptotically single-mode propagation in large-core OmniGuide fibers. *Opt. Express* 9:748–779.

Kuriki, K., Shapira, O., Hart, S.D., Benoit, G., Kuriki, Y., Viens, J.F., Bayindir, M., Joannopoulos, J.D., and Fink, Y. 2004. Hollow multilayer photonic bandgap fibers for NIR applications. *Opt. Express* 12:1510–1517.

Larsen, T., Bjarklev, A., Hermann, D., and Broeng, J. 2003. Optical devices based on liquid crystal photonic bandgap fibres. *Opt. Express* 11:2589–2596.

Litchinitser, N.M., Abeeluck, A.K., Headley, C., and Eggleton, B.J. 2002. Antiresonant reflecting photonic crystal optical waveguides. *Opt. Lett.* 27:1592–1594.

Litchinitser, N.M., Dunn, S.C., Usner, B., Eggleton, B.J., White, T.P., McPhedran, R.C., and de Sterke, C.M. 2003. Resonances in microstructured optical waveguides. *Opt. Express* 11:1243–1251.

Marcatili, E.A.J. and Schmeltzer, R.A. 1964. Hollow metallic and dielectric waveguides for long distance optical transmission and lasers. *Bell Syst. Tech. J.* 43:1783.

Pottage, J.M., Bird, D.M., Hedley, T.D., Birks, T.A., Knight, J.C., Russell, P.S.J., and Roberts, P.J. 2003. Robust photonic band gaps for hollow core guidance in PCF made from high index glass. *Opt. Express* 11:2854–2861.

Roberts, P.J., Couny, F., Sabert, H., Mangan, B.J., Williams, D.P., Farr, L., Mason, M.W. et al. 2005a. Ultimate low loss of hollow-core photonic crystal fibres. *Opt. Express* 13:236–244.

Roberts, P.J., Couny, F., Sabert, H., Mangan, B., Birks, T., Knight, J., and Russell, P. 2005b. Loss in solid-core photonic crystal fibers due to interface roughness scattering. *Opt. Express* 13:7779–7793.

Russell, P. 2003. Photonic crystal fibers. *Science* 299:358–362.

Snyder, A.W. and Love, J.D. 1983. *Optical Waveguide Theory*. Springer, New York.

Tam, A.C. and Patel, K.N. 1979. Optical absorption of light and heavy water by laser optoacoustic spectroscopy. *Appl. Opt.* 18:3348–3358.

Temelkuran, B., Hart, S.D., Benoit, G., Joannopoulos, J.D., and Fink, Y. 2002. Wavelength-scalable hollow optical fibres with large photonic bandgaps for CO_2 laser transmission. *Nature* 420:650–653.

West, J., Smith, C., Borrelli, N., Allan, D., and Koch, K. 2004. Surface modes in air-core photonic band-gap fibers. *Opt. Express* 12:1485–1496.

Yablonovitch, E. 1987. Inhibited spontaneous emission in solid-state physics and electronics. *Phys. Rev. Lett.* 58:2059–2062.

Yang, W., Conkey, D.B., Wu, B., Yin, D., Hawkins, A.R., and Schmidt, H. 2007. Atomic spectroscopy on a chip. *Nat. Photon.* 1:331–335.

Yeh, P. 1998. *Optical Waves in Layered Media*. Wiley-Interscience, Hoboken, NJ.

Yeh, P., Yariv, A., and Hong, C.S. 1977. Electromagnetic propagation in periodic stratified media. I. General theory. *J. Opt. Soc. Am.* 67:423–438.

Yeh, P., Yariv, A., and Marom, E. 1978. Theory of Bragg fiber. *J. Opt. Soc. Am.* 68:1196–1201.

5

Optoelectronics

Romeo Bernini
Luigi Zeni

The term "optoelectronics" covers a wide range of scientific and technical fields related to the generation, the detection, and the control of light with electronic devices. Examples of optoelectronic devices are light-emitting diodes (LED) and laser diodes (LD), useful to convert an electrical signal into an optical one, or photodiodes and photomultipliers, able to convert an optical signal into an electrical one. Optoelectronics is essentially based on a few principles that are well founded in electronics and optics. Optoelectronic devices are fundamental for the fabrication and characterization of optofluidic devices, also considering that, in all practical applications, the input and output signals of a device are electrical currents or voltages. LDs and photodiodes can be used, as an example, to detect, control, and analyze small particles in optofluidic devices. The integration of optoelectronics with optofluidics not only facilitates the realization of complex devices as "lab-on-a-chip" systems, but also allows the implementation of novel functionalities, which are not achievable with solid-state optical materials taken alone. Recent examples of this fruitful interaction are the microfluidic tuning of distributed feedback quantum cascade lasers and the fabrication of integrated optofluidic distributed feedback dye lasers.

In this chapter, we first introduce the basic concepts of light–matter interaction and laser principles. Next, we discuss the theory of semiconductor pn junctions and its main properties. Finally, the operating principles and the basic structures of commonly used optoelectronic devices, such as LEDs, LDs, photodetectors, and photomultipliers, are presented.

5.1 Light–Matter Interaction

5.1.1 Matter's Aggregation States and Their Properties

Matter consists of atoms. When an atom comes close enough to interact with neighboring atoms, they can constitute molecules, and matter in the liquid or solid state is formed. Isolated atoms and molecules can exist only in discrete energy levels, according to the rules of quantum mechanics. These energy levels can be arranged in the order of ascending values: E_0, E_1, E_2,..., E_m (where m is an integer), and each sequence is characteristic of a particular atom or molecule (Figure 5.1a). For solids, however, identical

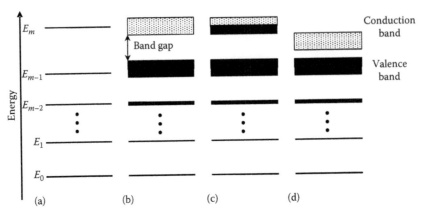

FIGURE 5.1   Energy levels of (a) isolated atoms, (b) insulators, (c) conductors, and (d) intrinsic semiconductors.

atoms and molecules, with their characteristic sets of energy levels, lie in close proximity to each other. In this case, the particular energy level splits into a new set of energy levels, some lower and some higher than the original values, and becomes a band of closely spaced levels. Electrons will fill the available energy states in ascending order.

Depending on the band structure, solids exhibit different electronic properties and can be divided into different types: insulators, conductors, and semiconductors.

In an insulator (Figure 5.1b), the valence band (VB, the one containing the electrons responsible for the chemical forces between atoms, also known as valence electrons) is filled with electrons whereas the next allowable band (the conduction band [CB]) is empty. Because there is a large energy gap (band gap) between the bands (typically >3 eV), very few electrons can attain sufficient thermal energy to jump the energy gap and contribute to conductivity. Therefore, if an electric field is applied to this solid, it would have very little effect on the electrons, since they are unable to move in response to the force exerted by the field.

In conducting solids, such as metals, the CB is only partially filled with electrons (Figure 5.1c). The abundance of unfilled states in this band means that electrons can readily gain energy from external agents, such as an applied electric field, and can move about easily.

A pure or intrinsic semiconductor (at $T = 0\,K$) has a filled VB and an empty CB, and looks like an insulator (Figure 5.1d). However, in this case, the energy gap is much smaller, and as the temperature increases, the conductivity rises since some electrons are thermally excited into the empty CB and, once there, they can then move freely in response to an applied electric field. Furthermore, when an electron is excited into the CB, it leaves behind an empty state, or a hole, in the VB, allowing one of the remaining neighboring electrons in the VB to move into the empty state under the effect of an electric field. This action fills up the hole but creates a new hole. This process may repeat itself with the result that the holes appear to drift in the opposite direction to the electrons when an electric field is applied, and thus they behave like positive charges. The charge of a hole is equal in magnitude to the charge of an electron. Hence, in a semiconductor, the excitation of an electron to the CB can be viewed as a process whereby an electron–hole pair (EHP) is created, with each particle contributing to the current flowing in response to an applied voltage.

The intrinsic semiconductor described above has equal concentrations of thermally generated electrons and holes. However, this balance can be altered by doping the semiconductor. Doped semiconductors in which the majority charge carriers are electrons are called n-type (negative carrier) while doped semiconductors in which the majority charge carriers are holes are called p-type (positive carrier). Doping an intrinsic semiconductor is achieved by introducing a small number of impurity atoms. These impurity atoms may have five valence electrons (e.g., phosphorous), thus achieving an n-type semiconductor, or three valence electrons (e.g., boron), thus achieving a p-type semiconductor. Such impurity-doped materials are called extrinsic semiconductors (Sze and Ng 2006).

5.1.2 Light–Matter Interaction Processes

Matter contains electric charges and dipoles that interact with the electromagnetic field of light. In particular, an electromagnetic wave forces charged particles in gases, liquids, and solids, to oscillate or accelerate. Conversely, an oscillating charge emits an electromagnetic wave.

The light of frequency (v) may interact with an atom only if its energy (hv) corresponds to the difference between two of the discrete energy levels, in accordance with the relation

$$h\nu = E_i - E_j \tag{5.1}$$

where h is Planck's quantum constant (6.626×10^{-34} J s). This emphasizes that when light interacts with matter, it exhibits a quantized nature that can be described in terms of particles. These particles or "quanta" of light are called photons.

As showed by Einstein, three forms of interaction are possible between atoms and photons. All of them involve a quantum jump of energy within the atom; typically, in the visible region, this is between 1.77 and 3 eV. Figure 5.2 illustrates these three basic processes.

As depicted in Figure 5.2a, the first process is the absorption of a photon, which can occur when the atom initially is in the lower state and the quantum energy, hv, of the photon equals the energy difference between the two levels. The photon may be absorbed inducing the atom to rise to the upper energy level. The absorption rate (number of absorptions per unit time) caused by an external source with a photon density, n_{ph}, is given by

$$\left.\frac{dN_2(t)}{dt}\right|_{Absorption} = K_{12} n_{ph}(t) N_1(t) \tag{5.2}$$

where

N_1 is the lower-level population density
K_{12} is a constant

The second process is spontaneous emission (Figure 5.2b): When an atom is initially in the upper energy level, it may drop spontaneously to the lower level emitting a photon, losing a quantum of energy in the process. This process is called spontaneous because it is independent of the presence of other photons. The spontaneous emission rate from level 2 to level 1 can be written as

$$\left.\frac{dN_2(t)}{dt}\right|_{Spontaneous} = -\gamma_{21} N_2(t) \tag{5.3}$$

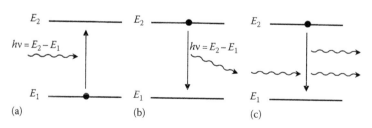

FIGURE 5.2 Atom–photon interaction processes: (a) absorption, (b) spontaneous emission, and (c) stimulated emission.

where

N_2 is the upper-level population density
γ_{21} indicates the total spontaneous transition rate or decay rate from level 2 to level 1

The third process is called stimulated emission (Figure 5.2c), in which an atom in the upper energy level is stimulated to emit a photon by the arrival of another resonant photon ($h\nu = E_2 - E_1$). The stimulated emission rate caused by an external source with a photon density, n_{ph}, is given by

$$\left.\frac{dN_2(t)}{dt}\right|_{\substack{\text{Stimulated} \\ \text{emission}}} = -K_{21}n_{ph}(t)N_2(t) \tag{5.4}$$

where K_{21} is a constant. The proportionality constant of the downward transition has exactly the same value as the constant of the upward transition, $K_{21} = K_{12}$. The proportionality constants, K_{21}, K_{12}, and γ_{21}, can also be expressed as $\gamma_{21} = A$ and $K_{21} = K_{12} = h\nu B$, where A and B are the so-called Einstein coefficients.

The stimulated emission process is the inverse of absorption and has two important characteristics. First, the emitted photon is an exact copy of the incident photon with the same properties: frequency, direction, phase, and polarization. Second, it is multiplicative, i.e., one photon becomes two. If these two photons interact with two other excited atoms, this will yield a total of four photons, and so on. However, stimulated emission and absorption occur simultaneously, and the net result is that light may be absorbed or amplified, depending on the relative populations of the atoms in the upper and lower states. Because in a material close to thermodynamic equilibrium, the population of the lower state will exceed that of the upper state, the material will absorb light. In order to amplify light, it is necessary that the number of atoms in the higher energy level exceeds the number of atoms in the lower energy level, a condition known as population inversion. This condition may be obtained using an external energy supply that selectively excites the atoms in the upper state.

Let us consider a two-energy-level system, as shown in Figure 5.2. At equilibrium ($dN_i/dt = 0$), we can write

$$K_{12}n_{ph}(t)N_1(t) = K_{21}n_{ph}(t)N_2(t) + \gamma_{21}N_2(t) \tag{5.5}$$

since the total upward and downward transition rates must be equal. When n_{ph} is large, spontaneous emission can be neglected, and we have $N_1 = N_2$. Hence, we cannot achieve population inversion in a two-energy-level system by using direct optical pumping between the two levels.

The above processes are also called radiative transitions because they result in photon absorption and emission. In addition, atoms can undergo transitions between energy levels also by nonradiative means. Nonradiative transitions permit energy transfer by mechanisms such as molecular vibrations and inelastic collisions between constituent atoms, and inelastic collisions with the walls of the vessel. When solids are concerned, the transitions occur between or within energy bands and the excess energy is converted into lattice vibrations (phonons) or transferred to impurity atoms.

According to the rules of quantum mechanics, semiconductor materials can be divided into two categories: direct band gap semiconductors and indirect band gap semiconductors. In the first category, the intra-band transitions can occur with a mere exchange of energy because the momentum conservation law is always fulfilled due to the particular shape of the band diagram in the energy–momentum plane (radiative transition), while in the second category, the intra-band transitions require energy and momentum exchange (radiative and nonradiative transitions).

In most materials, absorption of light merely causes heating of the absorbing material, with all the absorbed energy being converted to internal kinetic energy. However, in some materials, only part of the energy of an optically excited state is dissipated as heat, and one or more photons having lower energies than the incident one are radiated. This most commonly involves an internal process, which generates a photon and one or more phonons, with a total energy equal to that of the absorbed incident photon. An example of such a process is fluorescence emission. Fluorescence (or photoluminescence) is an inelastic energy relaxation process, by which light is absorbed at one wavelength and then reemitted at another, longer wavelength (lower energy). Typically, the excited state generated by the absorption relaxes rapidly by nonradiative processes to a lower, metastable energy state, before relaxing to another lower energy state by emitting a photon. The emission of light by fluorescence has no preferred direction and is also randomly polarized.

5.2 Laser

5.2.1 Laser Principles

Laser is the acronym of "light amplification by stimulated emission of radiation." The laser is a light source and like other light sources, it is a device that converts energy into light. An input energy is used to excite atoms or molecules, which then emit light. However, differently from conventional light sources that are based on spontaneous emission, the laser is a light source based on stimulated emission of radiation. This imparts several unique properties like high directionality, monochromaticity, coherence, and brightness.

The laser is essentially an optical oscillator and can be viewed as the optical equivalent of an electrical oscillator. An oscillator is composed of a resonant amplifier with a positive feedback (Figure 5.3a). If the gain becomes greater than the losses, the system becomes unstable and oscillation begins. The oscillation amplitude grows until the amplifier saturates. The effective gain then decreases and a stable condition is obtained when the gain compensates the losses. The oscillation frequency depends on the frequency characteristics of the gain and the feedback. In Figure 5.3b, a very general scheme of a laser is given. The essential elements of a laser, independent of its type, are

1. An active medium, which can be an appropriate solid, gas, or liquid
2. A pumping process to excite the atoms (molecules, etc.) to higher energy levels
3. An optical feedback that permits the light to bounce back and forth through the laser medium

The active medium, with an adequate pumping process, acts as an amplifier. The optical feedback is usually supplied by mirrors aligned to form an optical resonator so that the light travels back and forth between these mirrors with very small losses.

In order to obtain optical gain, it is necessary that the pumping process produces a condition of population inversion in the active medium. Because in two-energy-level systems it is impossible to achieve inversion, laser materials have at least three energy levels. Let us consider the four-energy-level system

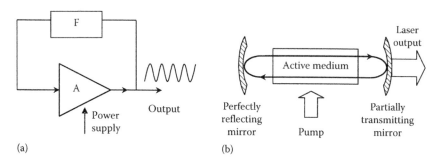

FIGURE 5.3 (a) Electrical oscillator and (b) laser (optical oscillator).

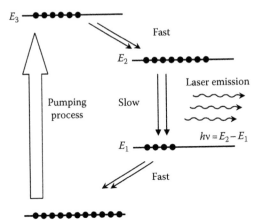

FIGURE 5.4 Four-energy-level system with a population inversion between levels 1 and 2. The black circles indicate the atom population of the levels.

shown in Figure 5.4. This is an ideal model but representative of many real laser systems. The pumping process causes the excitation of atoms (or molecules) from the ground level, E_0, to the energy level, E_3. This excitation may be produced by electron impact in gas discharge, by absorption from the light of a lamp, or by several other mechanisms.

A spontaneous radiative or nonradiative decay will occur from E_3 to E_2 and also from E_3 to E_1; but if the $E_3 \rightarrow E_2$ decay is fast enough, atoms will accumulate in level E_2. If the levels are chosen appropriately, the $E_2 \rightarrow E_1$ decay can be slow and the $E_1 \rightarrow E_0$ decay relatively faster. Clearly, the result of this will be an inversion of population between levels 2 and 1.

This inverted population can now be exploited to give optical amplification at frequency ν.

Let us quantify this amplification. We define an effective pumping rate, $R = \eta R_0$, where R_0 is the pumping rate from E_0 to E_3, and η is the fraction of the atoms that from level E_3 decays to level E_2. This quantity is a measure of the pumping efficiency of the laser system.

Using Equations 5.2 through 5.4, we can write the total rate equation for the energy level E_2, including stimulated and spontaneous transitions:

$$\frac{dN_2(t)}{dt} \approx R - \gamma_{21} N_2(t) \tag{5.6}$$

And, analogously, for level E_1,

$$\frac{dN_1(t)}{dt} \approx \gamma_{21} N_2(t) - \gamma_{10} N_1(t) \tag{5.7}$$

If the pumping process is constant, the system reaches a steady state condition in which $dN_1/dt = dN_2/dt = 0$, and these equations can be solved to obtain

$$N_2 = R/\gamma_{21} \quad \text{and} \quad N_1 = \left(\gamma_{21}/\gamma_{10}\right)N_2 \tag{5.8}$$

So, the population difference on the laser transition can be written as

$$N_2 - N_1 = R\frac{\left(\gamma_{10} - \gamma_{21}\right)}{\gamma_{10}\gamma_{21}} \tag{5.9}$$

This equation shows that if the decay rate, γ_{10}, is fast compared to γ_{21} ($\gamma_{10} > \gamma_{21}$), a population inversion ($N_2 > N_1$) between levels 1 and 2 can occur.

Now, we can define an optical gain coefficient (gain per unit length) of the medium between levels 2 and 1 as

$$g(v) = (N_2 - N_1)\sigma(v) \tag{5.10}$$

where $\sigma(v)$ is the stimulated emission cross section, which is a characteristic parameter of the laser media. The dependence of the gain g on the frequency is related to the fact that, normally, the emission and absorption processes do not occur at a well-defined frequency corresponding to the laser transition, but they are distributed over a frequency interval with a central peak.

The variation of the optical gain with the frequency is called the optical gain lineshape. This spectral broadening depends on a number of factors. In gas lasers, it is primarily due to the Doppler shift caused by the thermal motion of atoms and molecules.

Let us now consider the active media in an optical cavity resonator (Figure 5.5). As an example, this can be done by placing mirrors at each end of the medium, to form a Fabry–Perot resonator. Suppose that a small amount of electromagnetic radiation is generated by spontaneous emission at the laser transition frequency and travels back and forth between the cavity mirrors. If the round-trip overall gain, G, including laser gain and total losses, is greater than unity then this noise radiation is amplified on each successive round trip by stimulated emission building up the laser oscillation. Hence, the condition for the initiation of laser oscillation can be written as

$$G = \frac{I_f}{I_i} = R_1 R_2 \exp\left[(g-\alpha)2L\right] > 1 \tag{5.11}$$

where

I_f and I_i are the final and starting intensities for a complete round trip of a path $2L$
R_1 and R_2 are the reflectivities of the two mirrors
α is the loss per unit length in the medium (due to absorption, scattering, wall losses, etc.)

From Equation 5.11, we can define a gain threshold, g_{th}, for which $G = 1$, i.e., the minimum optical gain necessary to achieve a continuous laser emission:

$$g_{th} = \alpha + \frac{1}{2L}\ln\left(\frac{1}{R_1 R_2}\right) = \alpha_t \tag{5.12}$$

where α_t is the total loss coefficient that describes the overall losses in one round trip.

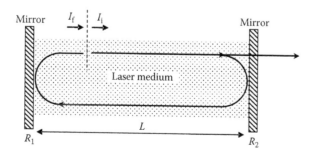

FIGURE 5.5 Optical cavity resonator.

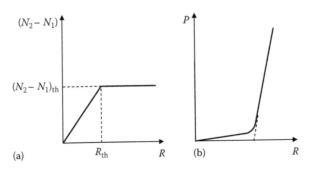

FIGURE 5.6 (a) Laser population inversion and (b) laser output power, P, versus the pumping rate, R.

Initially, the gain medium must have a gain coefficient, g, greater than g_{th} according to Equation 5.11, but as the radiation intensity inside the cavity increases the optical amplifier saturates. The gain then decreases until a steady state condition is reached when $g = g_{th}$. This saturation occurs when the atoms of the upper laser level are involved in the stimulated emissions at a rate that begins to match the pumping rate at that level. The threshold gain, g_{th}, corresponds to a threshold population inversion achieved at a threshold pumping rate Rth (Figure 5.6a):

$$(N_2 - N_1)_{th} = \frac{\sigma}{g_{th}} \tag{5.13}$$

If we plot the coherent laser output power, P, as a function of the pumping rate, R (Figure 5.6b), we observe that as far as the pumping rate induces a population inversion $(N_2 - N_1)$ below the threshold, the laser output is small; any increase in R gives rise only to an increase in $(N_2 - N_1)$ and is manifested as an increase in the spontaneous emission radiation. When R increases above a threshold value, R_{th}, $(N_2 - N_1)$ remains clamped to $(N_2 - N_1)_{th}$ because g must remain equal to g_{th}. Any additional increase in the pumping rate increases the rate of stimulated emission, and so increases the laser output power.

The spectral properties of laser light are primarily determined by two factors: the gain bandwidth of the lasing medium and the longitudinal modes of the resonator.

The simplest basic optical resonator is a pair of shaped mirrors at each end of the laser medium, to form a Fabry–Perot resonator. The resonator sustains only some frequencies called longitudinal modes. In particular, a frequency is admitted only if the total phase shift in a single round trip is a multiple of 2π. The allowed resonant frequencies, ν_m, of the cavity are

$$\nu_m = m \frac{c}{2L} \tag{5.14}$$

where
   m is an integer called mode number
   c is the velocity of light in the laser medium

Equation 5.14 gives the resonant frequencies at which the laser may oscillate, but the oscillating frequencies are only those falling within that portion of the gain profile of the laser transition exceeding the loss threshold. Hence, the output spectrum of laser light is the result of combining the two conditions given by Equations 5.12 and 5.14, as shown in Figure 5.7. In Figure 5.7c, we can see that the oscillating laser frequencies exhibit the Fabry–Perot mode structure enveloped by the optical gain lineshape of the laser transition.

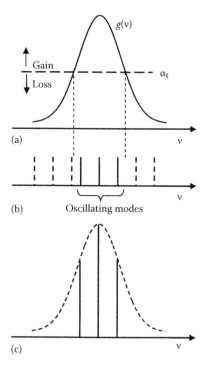

FIGURE 5.7 Gain profile and resonant frequencies in a cavity laser: (a) optical gain lineshape of the laser transition, (b) allowed resonances of the Fabry–Perot cavity, and (c) oscillating laser frequencies.

So far, we have dealt only with longitudinal modes and tacitly assumed that a plane wave travels inside the cavity. However, all real laser cavities have a finite transverse size, and different electrical field patterns, called transversal modes, are admitted inside the cavity. A spatial field distribution can be a cavity mode only if it reproduces itself after one round trip. Hence, the mode of an optical cavity can be described by three integers, p, q, and m. The integers p and q define a transverse mode with an associated spatial distribution. For given p and q, the number m defines the longitudinal modes with the same spatial distribution but with different frequencies. Moreover, two different transverse modes do not necessarily have the same longitudinal frequency implied by Equation 5.14.

Examples of the transverse mode supported by a laser cavity are the Hermite–Gauss resonator modes depicted in Figure 5.8.

When only one longitudinal and one transverse mode are selected, the bandwidth of the laser light is very narrow.

Lasers can be divided into two main categories: continuous wave (CW) and pulsed. In the CW operation, highly coherent, high-intensity, continuous, and uninterrupted output light is produced. In the

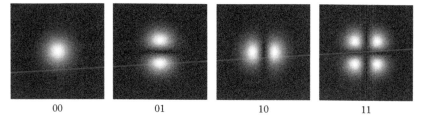

FIGURE 5.8 Low-order Hermite–Gauss resonator modes.

pulsed mode, light pulses with high peak intensities are generated. A widely used approach to produce pulsed lasers is the Q-switching method. By properly controlling the resonator losses (i.e., by mechanically or electrically spoiling the resonator quality factor, Q), the feedback and so the laser action can be allowed for just a short time interval, thus letting the huge amount of energy stored in the laser gain medium to be released in the form of an intense light pulse. A different technique used to achieve a pulsed emission is the so-called mode locking. It basically consists of locking a large number (M) of longitudinal modes in a laser oscillator, so giving rise to their coherent superpositions resulting in a single pulse circulating inside the cavity whose duration is inversely proportional to M and whose intensity is directly proportional to M^2. Thus, the output of a mode-locked laser consists of a train of pulses with a fixed period equal to the round-trip time of the cavity and intensity depending on the transmission coefficient of the output mirror. The simplest way to achieve mode locking in a laser, provided that a sufficient number of longitudinal modes are able to oscillate, is to insert an amplitude modulator inside the cavity with a modulation frequency equal to the inverse of the round-trip time.

5.2.2 Gas Lasers

Gas lasers are one of the most commonly used laser types. In a gas laser, the active medium is a gas or a mixture of gases in a transparent tube. Generally, population inversion is created by an electrical discharge in which excitation of gas atoms, ions, or molecules is obtained by collisions with energetic electrons. An electrical discharge is relatively simple to construct and operate by using a high voltage source. In Figure 5.9, the basic scheme of a gas laser is shown.

The He-Ne laser is one of the best-known gas lasers. Typically, it operates at a wavelength of 632.8 nm. The gain medium consists of a mixture of two noble gases, He and Ne, with a population ratio of about 10 to 1. The amplifying medium is neon, which is pumped into a state of population inversion by collision with excited helium atoms. The He atoms are first excited to higher energy levels by the electrical discharge and then they pass energy to the Ne atoms through inelastic collisions. He-Ne lasers can also radiate at wavelengths of 0.543, 1.55, and 3.39 μm. The wavelength selection is achieved by a filtering element inserted into the cavity. Excimer lasers are gas-based lasers that generate intense, short ultraviolet pulses. The laser tube contains a mixture of an inert gas and a halogen. The excimer molecules (KrF, ArF, etc.) exist only in an excited state since the constituents are repulsive in the ground state. From its upper state, the excimer can undergo laser transitions to its ground state. This results in a huge, effective population inversion and very high gains, which allow simple resonator designs, especially advantageous in medical and industrial applications. In recent years, they have also gained great attention in micromachining and microlithography fields.

Table 5.1 reports some of the most common gas lasers.

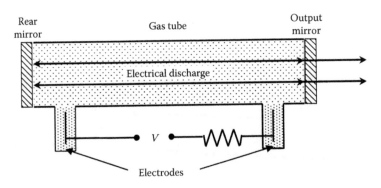

FIGURE 5.9   Basic scheme of a gas laser.

TABLE 5.1 Most Commonly Used Lasers

Laser Type	Wavelength	Power or Pulse Energy	Pulsed or CW
Gas Lasers			
He–Ne	632.8 nm, 0.543 μm, 1.55 μm, 3.39 μm	1–50 mW	CW
Ar⁺	488 nm, 515 nm	2–20 W	CW
He–Cd	441.6 nm, 325 nm	50–200 mW	CW
CO_2	10.6 μm	10^2–10^4 W	CW
ArF	193 nm	500 mJ	Pulsed
KrF	248 nm	500 mJ	Pulsed
F	157 nm	500 mJ	Pulsed
Solid-State Lasers			
Nd:YAG	1.064 μm	1.5 J/700 W	Pulsed or CW
Nd:glass	1.06 μm	50 J	Pulsed
Titanium sapphire ($Ti:Al_2O_3$)	670 nm–1070 nm	50 W	Pulsed or CW
Dye Lasers			
Rhodamine 6 G	560–640 nm	200 mJ/100 mW	Pulsed or CW
Coumarin	400–500 nm		

5.2.3 Solid-State Lasers

In a solid-state laser, the gain medium is a crystalline or an amorphous (glass) solid in the form of a transparent rod, doped with active ions. The active ions may represent substitution of the crystal lattice atoms or may be dispersed as an impurity into the glass host. A wide range of laser wavelengths may be obtained by combining different dopant ions and host materials. An important group of dopant elements are the rare earth ions (Nd, Er, and Yb). Their peculiar electron configurations make them very suitable for laser operation.

In most commercially available solid-state lasers, population inversion is created by optical excitation from continuous or pulsed lamps or from semiconductor diode lasers.

One of the most used solid-state lasers is the Nd:YAG. In this case, the yttrium aluminum garnet (YAG) crystal is doped with neodymium ions (Nd^{3+}). The laser operating wavelength is 1064 nm.

Another important solid-state laser is titanium–sapphire (Ti–Sapphire). The active medium is sapphire (Al_2O_3) doped with titanium ions (Ti^{3+}). This material exhibits a very broad gain bandwidth band with a peak wavelength near 800 nm and produces lasers with an operating wavelength tunable over a wide range (670–1070 nm). It is used both in CW and in pulsed modes; a pulsed Ti–Sapphire laser can produce ultrashort pulses (down to several femtoseconds) using mode locking. A listing of some of the most common solid-state lasers is given in Table 5.1.

5.2.4 Dye Lasers

The active medium of a dye laser is an organic molecule dissolved in a solvent (alcohol, water, etc.). The characteristic molecular structure of these organic molecules results in broad absorption and emission bands. The laser can be pumped by flashlamps or by another laser (see Figure 5.10).

Thanks to the broad emission band, tunable laser radiation can be achieved over a bandwidth of about 30 nm from a single molecule and over wavelengths ranging from 320 to 1500 nm, if a mixture of

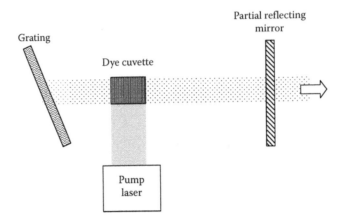

FIGURE 5.10 Basic scheme of a dye laser.

molecules is used. For example, coumarin dyes give rise to oscillation in the blue-green spectra (400–500 nm), whereas rhodamine-6 G dyes can be used to generate radiation ranging from 560 to 640 nm.

The wavelength selection is achieved by a filtering element (prism or diffraction grating) inserted into the cavity.

However, dye molecules have the tendency to become trapped in excited states, not involved in laser transition and with a very slow decay rate. As a long time is required before they can take part again in the laser process, typically, in dye lasers, the active medium is contained in a separate tank and is continuously injected through a thin cuvette where the lasing process, in always fresh dyes, actually occurs. This configuration permits the use of dye molecules only for a short time while passing through the cuvette, and gives them enough time to return to the ground state while passing through the tank, before they are used again.

5.3 Semiconductor-Based Devices

Semiconductor-based devices are of great interest in the development of optofluidics because they allow, in principle, the integration, on the same substrate, of light sources, photodetectors, and processing electronics, while the microfluidics channels can be realized, according to the particular case, on the same substrate or on a polymeric layer purposely deposited on the top. Furthermore, whenever hybrid integration is possible (i.e., layers of different semiconductor materials can be grown on a silicon wafer to allow a proper realization of light sources and detectors), the full potential of the well-established silicon-based processing electronics can be exploited.

5.3.1 pn Junction—The Diode Equation

A pn junction, in its simplest realization, can be regarded as an abrupt discontinuity between p-type and n-type semiconductor regions of the same material (e.g., silicon) called a metallurgic junction. A schematic of the pn junction is depicted in Figure 5.11a, where fixed ionized donors and free electrons can be recognized in the n region and fixed ionized acceptors and free holes can be seen in the p region. Due to the differences in the concentration of free carriers (electrons and holes) in the two sides of the junction, they diffuse one toward each other and recombine, so leaving exposed positive ions on the right side and exposed negative ions on the left side. These fixed charges of opposite signs give rise to an electric field that opposes the diffusion of free carriers until equilibrium is reached and a space charge layer (SCL), or depletion region, is created across the junction (see Figure 5.11b). The

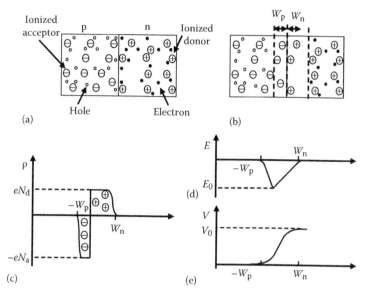

FIGURE 5.11 pn junction properties. (a) p and n materials before the junction formation, (b) charge repositioning after the junction formation, (c) charge density profile across the junction, (d) electric field profile across the junction, and (e) potential profile across the junction.

semiconductor regions away from the SCL are called neutral regions because no net charge exists. The concentration of positive ions equals the donor's concentration, N_d, while the concentration of negative ions equals the acceptor's concentration, N_a. The amount of these fixed charges and their spatial distribution across the junction is shown in Figure 5.11c. Of course, the extent of the SCL on the p side, W_p, and that on the n side, W_n, depends on N_d and N_a, so that the total positive charge equals the total negative charge: $N_a W_p = N_d W_n$. The electric field profile across the junction can be easily derived from Gauss's law, and is depicted in Figure 5.11d. According to the electric field profile, a potential, $V(x)$, develops across the junction and can be found by integrating the electric field. The result is shown in Figure 5.11e, where the reference has been set to zero away from the p side of the junction. The overall voltage across the junction, V_o, is called built-in potential, and it can be shown to be related to the dopant's concentration:

$$V_o = \frac{eN_aN_d\left(W_n + W_p\right)^2}{2\varepsilon\left(N_a + N_d\right)} = \frac{k_BT}{e}\ln\left(\frac{N_aN_d}{n_i^2}\right) \tag{5.15}$$

where
  k_B is Boltzmann's constant
  e is the electron charge
  ε is the material dielectric constant
  n_i is the intrinsic carriers' concentration
  T is the temperature

This voltage represents a potential barrier to the free carriers' diffusion.

5.3.1.1 Forward Bias

In an open-circuit pn junction, no current flows (otherwise, the energy conservation law would be violated), but a current can flow if the junction is inserted into a circuit and forward biased. Forward biasing a junction means to apply an external battery, with a voltage V, whose positive pole is attached to the p side and the negative pole to the n side. The applied voltage, V, reduces the potential barrier to $V_o - V$, and this reduction allows the holes and electrons to diffuse toward the n side and the p side, respectively. This process is called minority carrier injection because holes are injected in the n region and electrons are injected in the p region. The injected minority carriers recombine with the majority carriers and these lost carriers are continuously replenished by the battery. In other words, the battery injects electrons in the n side of the junction and extracts electrons (so, equivalently, injects holes) from the p side. The net result is current flowing into the circuit where the junction is inserted. It can be shown that the current density is

$$J = J_{so} \left[\exp\left(\frac{eV}{k_B T} \right) - 1 \right] \tag{5.16}$$

where J_{so}, called the reverse saturation current density, depends on the dopant's concentration, on the electron and hole diffusion constants and lifetimes, and on n_i, which in turn is strongly temperature dependent. This equation is also known as the Shockley equation or the diode equation. Depending on the junction characteristics, the diode equation may change, and this is usually taken into account by inserting a "diode ideality factor," η, that can be 1 for diffusion-controlled and 2 for SCL recombination–controlled characteristics. The diode equation thus becomes

$$J = J_{so} \left[\exp\left(\frac{eV}{\eta k_B T} \right) - 1 \right] \tag{5.17}$$

For further details, the reader is referred to Sze and Ng (2006).

5.3.1.2 Reverse Bias

A pn junction is said to be reverse biased when the sign of the external battery is reversed (i.e., the positive pole is attached to the n side and the negative pole to the p side). In this case, the potential barrier is increased by the external potential, let us say $V_r = -V$, and the free carriers are pushed away from the SCL and its width increases. A small current flows in a reverse-biased pn junction, as long as the reverse bias, V_r, is kept below a certain threshold (breakdown voltage). From a physical point of view, the total reverse current is due to the EHPs thermally generated in the neutral regions and able to reach the SCL, and those thermally generated directly inside the SCL. In a reverse-biased junction, the Shockley equation still holds true and contributes to the total reverse current density, J_{rev}, with $-J_{so}$ (as it can be easily understood by observing that the exponential term in the Shockley equation becomes negligible for $V_r > k_B T/e$). Nevertheless, an additional term taking into account the thermal generation inside the SCL has to be added, so, $J_{rev} = -(J_{so} + eWn_i/\tau_g)$ where $W = W_p + W_n$ is the total extent of the SCL and τ_g is the mean time to generate an EHP by virtue of the thermal vibrations of the lattice. Of course, the actual diode current, I, can be calculated by multiplying the current density by the junction cross-sectional area, A. Figure 5.12 shows the diode I–V characteristic.

5.3.1.3 Avalanche Breakdown

When the reverse voltage in a pn junction is increased above a certain threshold, a breakdown phenomenon can occur and the reverse current undergoes a sudden increase. This phenomenon can be understood considering that the thermally generated EHPs inside the depletion layer are accelerated by the electric field and, provided that the SCL length is sufficiently long, each free carrier can reach a kinetic energy so high as to generate other EHPs by impact with the lattice atoms. If the process is repeated a number of times, a very high number of free carriers are generated within the SCL (starting from the

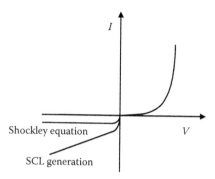

FIGURE 5.12 Diode *I–V* characteristic. (The discontinuity in the origin is due to the different scales used on the positive and negative axes.)

few initial, thermally generated EHPs). If this avalanche effect is properly controlled by the external circuit, it does not result in junction destruction and can be exploited for a number of applications, including the realization of extremely sensitive photodetectors (see Section 5.3.3.4). These photodetectors are especially important in optofluidics applications that typically require low-light detection. In single-molecule fluorescence-correlation spectroscopy for example, they can be used for the detection and measurement of extremely weak light signals on the level of single photons emitted by a single or a few molecules.

5.3.1.4 Depletion Layer Capacitance

A very important aspect affecting the switching speed of a pn junction is represented by the depletion layer capacitance. In fact, a pn junction has positive and negative charges separated over a distance, W, so resembling a parallel-plate capacitor, even if the total charge, $Q = eN_dW_nA = -eN_aW_pA$, does not depend linearly on the voltage across the device. As a change in the applied voltage gives rise to a change in the SCL width and, accordingly, to a change in the total charge, it can be shown that the depletion layer capacitance is expressed as

$$C_{dep} = \frac{\varepsilon A}{W} = \frac{A}{\left(V_o - V\right)^{1/2}}\left[\frac{e\varepsilon\left(N_aN_d\right)}{2\left(N_a + N_d\right)}\right]^{1/2} \tag{5.18}$$

This capacitance rapidly decreases under reverse-bias conditions as the reverse voltage increases.

5.3.1.5 Recombination Lifetime

The behavior of a pn junction, as previously stated, is dependent on the recombination of EHPs. The recombination process can be conveniently described by the recombination lifetime. This important parameter depends, usually in a very complicated manner, on the semiconductor type (direct or indirect band gap), on the presence of recombination centers, on the dopant concentration, and on injection conditions. Nevertheless, in the case of direct band gap semiconductors (those used, e.g., to realize semiconductor light sources) and under weak injection conditions, it can be shown that the recombination lifetime is inversely proportional to the dopant concentration (and hence constant for a fixed junction). On the other hand, if a strong injection regime is considered, the recombination lifetime is inversely proportional to the instantaneous carriers' concentration. This is particularly important when directly modulated light-emitting diodes (LEDs) or laser diodes (LDs) have to be realized because the changes in the junction characteristics as a consequence of the carriers' lifetime changes, is one of the factors leading to a nonlinear distortion of the emitted light.

5.3.2 Integrated Light Sources

5.3.2.1 Light-Emitting Diodes

An LED is basically a pn junction (diode), realized using a direct band gap semiconductor (such as GaAs, even if the addition of proper impurities also allows the use of certain indirect band gap semiconductor alloys), under forward-bias conditions. In such a junction, the EHPs recombination, occurring primarily in the depletion layers as a consequence of the forward bias, gives rise to energy release in the form of photons that are allowed to escape from the junction through a properly designed transparent window. The recombination zone in light-emitting devices is usually called the active region. Typical LED structures are depicted in Figure 5.13.

As a rule of thumb, the emitted photons energy is approximately equal to the semiconductor band gap energy, $h\nu \approx E_g$. The actual emitted photons' energy distribution depends on the distribution of the electrons and the holes within the CB and the VB, respectively. So the emitted spectrum appears like a bell-shaped curve, as shown in Figure 5.14a.

Figure 5.14b and c shows the relative light intensity as a function of diode current, and the diode *I–V* characteristics, respectively. The output spectrum of an LED is temperature dependent because of temperature variations of the semiconductor band gap. In particular, for GaAs materials, the band gap decreases with increasing temperature and this leads to an increase in the peak wavelength of emission of about 3 nm per 10°C. A large variety of LEDs are now available, whose emission spectra depend both on the employed materials and on the device structure and doping levels. An overview of the emitted wavelengths as functions of the materials is reported in Table 5.2. As it can be seen, the available wavelengths range from ultraviolet to mid-infrared.

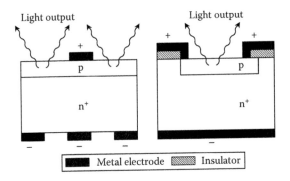

FIGURE 5.13   Typical LED structures.

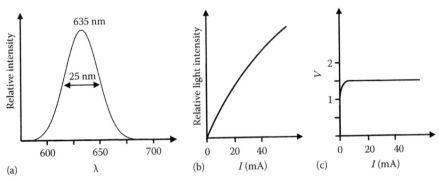

FIGURE 5.14   (a) Typical output spectrum of an LED, (b) typical relative light intensity as a function of diode current, and (c) typical diode *I–V* characteristics.

TABLE 5.2 Selected LED Semiconductor Materials

Semiconductor	Wavelengths (nm)	Direct Band Gap or Indirect Band Gap
GaAs	870–900	D
AlGaAs	740–870	D
InGaAsP	1–1.6 μm	D
InGaN	430–530	D
SiC	560–470	I
InAlGaAsP	590–630	D
GaAs$_{1-y}$P$_y$ ($y < 0.45$)	630–870	D
GaAs$_{1-y}$P$_y$ ($y > 0.45$)	560–700	I

Note: D, direct band gap; I, indirect band gap; DH, double heterostructure.

The actual available power is quantified by the external efficiency, $\eta_{external}$, which is a measure of the device's ability to convert electrical energy into optical energy, defined as

$$\eta_{external} = \frac{P_{out}(\text{Optical})}{IV} \times 100\% \tag{5.19}$$

where the product IV is just the current–voltage product, representing the electrical power delivered to the LED by power supply. Typical external efficiencies range from extremely low values (0.02% for indirect band gap materials) to relatively high values (20% for special devices realized with direct band gap materials). The highest external efficiencies are usually obtained by means of so-called double heterostructure (DH) diodes, i.e., diodes where the junction is realized using different semiconductors with different band gaps. The DH diodes allow optimizing the external efficiency because they make it possible to separate the actual light-emitting junction (realized with a very thin layer of a material with a certain band gap) from the rest of the structure (realized with a different material exhibiting a wider band gap). In this way, the photons emitted by the junction are not absorbed by the surrounding material due to its wider band gap, so they easily find their way out of the LED. Figure 5.15 shows a typical

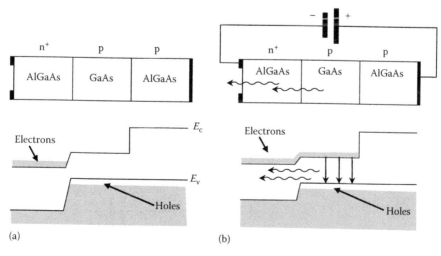

FIGURE 5.15 DH-LED. (a) Unbiased band diagram and (b) band diagram under forward bias.

double heterostructure light-emitting diode (DH-LED) realized with AlGaAs ($E_g \cong 2\,\text{eV}$) and GaAs ($E_g \cong 1.4\,\text{eV}$). Note that the n-type AlGaAs is heavily doped; as a consequence, the depletion layer extends almost entirely in the thin GaAs layer (so representing the active region) and the surrounding n-doped and p-doped AlGaAs layers are unable to absorb the emitted photons due to their wider band gaps.

An overview of the most commonly used LED materials, along with their characteristics, is reported in Table 5.2.

LEDs are very cheap and easy-to-use light sources, covering a wide range of wavelengths and available in many shapes and sizes. Thus, they represent extremely attractive devices for optofluidics applications whenever no high spectral purity is required.

5.3.2.2 Fabry–Perot Laser Diodes

Even if LEDs represent very versatile light sources and are employed in a large variety of applications, they nonetheless suffer from a number of limitations that make them less useful for applications where extremely narrow-band (and so, highly coherent) and high-power light sources are needed. Light sources exhibiting an extremely high coherence and able to deliver very high optical powers are represented by the lasers introduced in Section 5.2. The integrated version of a laser is represented by the LD, which can be regarded as an evolution of the LED. Keeping in mind the principle of laser emission, i.e., the simultaneous occurrences of optical gain, achieved via stimulated emission of radiation, and optical feedback, it is clear that an LED can become an LD if the condition of population inversion between the CB and the VB is reached and a form of optical feedback is somehow established. It can be shown that a pn junction between heavily doped (degenerate) semiconductors, under a proper forward bias, gives rise to an inversion region where optical gain is achieved via stimulated emission. This occurs for a definite photon energy interval: $E_g < h\nu < E_{Fn} - E_{Fp}$, where E_g is the semiconductor band gap while E_{Fn} and E_{Fp} are the Fermi energy levels in the n side and the p side of the junction, respectively. Of course, photons whose energy falls outside the above interval are either absorbed ($h\nu > E_{Fn} - E_{Fp}$) or not affected at all, i.e., for $h\nu < E_g$ the material is transparent. Figure 5.16a and b represents, respectively, the energy band diagrams of such a junction with no external bias and with an external bias sufficient to cause population inversion.

From an inspection of the above figure, it is clear that the population inversion occurs in the energy interval where an electron-filled CB corresponds to a hole-filled VB. An incoming photon whose energy

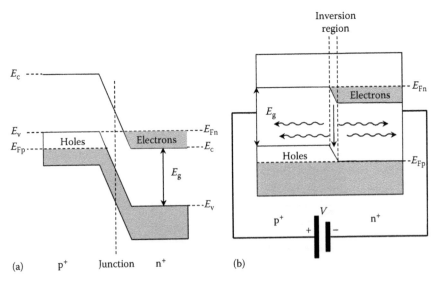

FIGURE 5.16 (a) Unbiased and (b) forward-biased pn junction between degenerate semiconductors.

belongs to the inversion interval is more likely to cause the emission of another photon than being absorbed. As the population inversion is caused by forward bias, the pumping mechanism is the forward diode current and it is usually referred to as current injection pumping. Now that the conditions for achieving optical gain are established, a form of optical feedback is required to obtain laser emission. It is easily achievable considering that the semiconductor crystal can be cleaved along the crystalline planes to obtain a perfectly smooth surface able to act as a mirror (reflectivity ~30%), so realizing a Fabry–Perot resonator. Figure 5.17 shows the simplest realization of a Fabry–Perot LD, which, being realized with only one type of semiconductor material (GaAs, in this case), is called a homojunction laser diode.

The Fabry–Perot resonators support a discrete number of longitudinal modes whose wavelengths obey the resonance condition, as given by Equation 5.14. The laser action (lasing) starts from spontaneous emission when the diode current reaches a threshold, I_{th}, such that the corresponding optical gain overcomes the losses, as illustrated in Figure 5.18. From this figure, it also evident how the number of oscillating modes (peaks in the output spectrum) depends on the injected current.

Unfortunately, in homojunction LDs, optical losses are too high due to the insufficient guiding effect of the active region. This leads to excessively high threshold currents, which prevent the possibility to

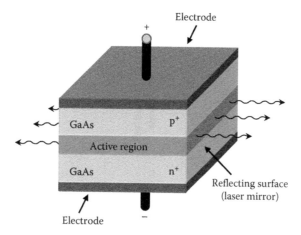

FIGURE 5.17 Simplest realization of a Fabry–Perot homojunction LD.

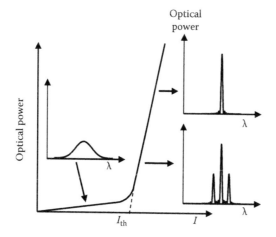

FIGURE 5.18 LD's optical power output versus diode current.

operate at room temperature. However, the threshold current can be reduced by orders of magnitude by using heterostructure LDs. In fact, heterostructure laser diodes and, in particular, double heterostructure laser diodes (DH-LDs)on one hand exhibit, just like DH-LEDs, an extremely thin active region where it is much easier to reach the high current density needed for population inversion, and, on the other hand, realize a much more efficient optical guide because the active region is sandwiched between different materials with lower refractive indices. A typical DH-LD scheme is depicted in Figure 5.19 along with the band diagram, refractive index profile, and the photon density spatial distribution, while a possible practical realization of such a device is shown in Figure 5.20.

The output spectrum of a Fabry–Perot LD is affected by several factors. First of all, several longitudinal and lateral (transverse) modes can exist, depending on the resonator length, its transverse cross section, and on the optical gain spectral distribution. Furthermore, the output spectrum depends on the actual current density and on temperature. This means that even if an LD is designed to operate on a single longitudinal mode and a single lateral mode, changing the case temperature leads to so-called mode hopping and the output is seen to jump between adjacent longitudinal modes (and the output frequency changes accordingly), as shown in Figure 5.21. To avoid the above effects, a careful temperature control of the device is mandatory.

5.3.2.3 DBR and DFB Laser Diodes

When an extremely narrow-band output is required from an LD, the simple Fabry–Perot resonator is no longer exploitable. This occurs because it cannot assure sufficiently narrow resonance peaks due to

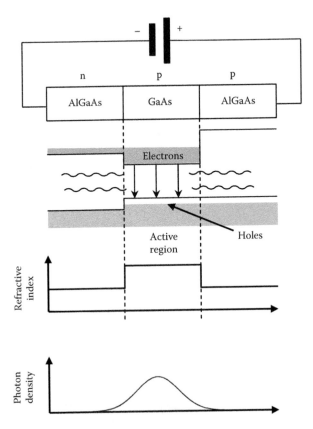

FIGURE 5.19   DH-LD, along with the band diagram, refractive index profile, and output photon density distribution across the junction.

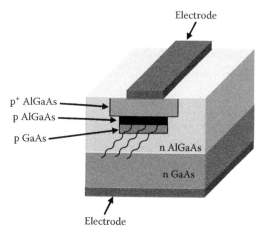

FIGURE 5.20 A possible practical realization of a DH-LD.

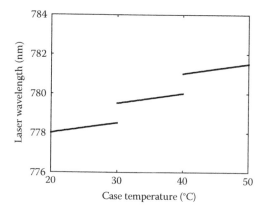

FIGURE 5.21 Mode hopping in single-mode LD.

low reflectivities of the cleaved surfaces and this, together with the fact that the material gain spectrum is relatively wide because the mechanism of population inversion occurs between the CB and the VB of the semiconductor, leads to a relatively wide output spectrum. To assure a narrow-band single-mode operation, highly selective mirrors are employed to replace the cleaved semiconductor surfaces. These mirrors are usually realized in the form of Bragg reflectors. Bragg reflectors are diffraction gratings made up of corrugated periodic structures where the partial reflections of waves from the corrugations interfere constructively to give a reflected wave only when the wavelength, λ_B, corresponds to twice the corrugation periodicity, Λ:

$$l\frac{\lambda_B}{n} = 2\Lambda \tag{5.20}$$

where
  n is the refractive index
  l is an integer called the diffraction order

The structure of a distributed Bragg reflector laser diode (DBR-LD) is shown in Figure 5.22.

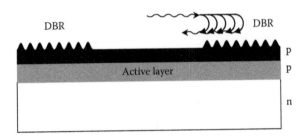

FIGURE 5.22 Structure of a DBR-LD.

A different possibility is offered by so-called distributed feedback laser diodes (DFB-LDs), in which the necessary optical feedback for laser operation is no longer provided by the reflectors placed at opposite sides of the cavity, but it is distributed all along the cavity length. Figure 5.23 shows the structure of a DFB-LD. It can be shown that the allowed modes for a DFB structure do not correspond exactly to the Bragg wavelengths but are placed symmetrically around λ_B:

$$\lambda_m = \lambda_B \pm \frac{\lambda_B^2}{2nL}(m+1) \tag{5.21}$$

where
   L is the corrugation length
   m is an integer

In DBR- and DFB-LDs, the complexity of the fabrication process is balanced by an extremely narrow-band output and relatively high power. Nevertheless, the stability of the output characteristics requires a precise control of the device temperature and bias current, which presents significant challenges for integration in optofluidics chips.

5.3.2.4 Quantum Well Lasers

Starting from a DH-LD, if the thickness of the narrow band gap semiconductor (such as GaAs) sandwiched between the wider band gap semiconductors (AlGaAs) is less than ~50 nm, quantum effects prevail and the band structure assumes a stepwise profile, thus somehow returning to a discrete-level structure typical of classical lasers. These effects are depicted in Figure 5.24 and occur because the electrons are confined, in the x direction, to a small length, d, so their energy is quantized.

Similar considerations hold true for holes. A quantum well (QW) laser is a device exploiting the above-described structure and presents some definite advantages over the "bulk" DH semiconductor

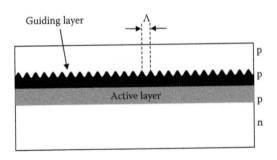

FIGURE 5.23 Structure of a DFB-LD.

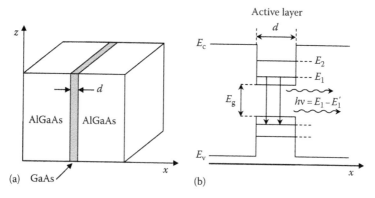

FIGURE 5.24 Quantum well LD: (a) structure and (b) energy levels.

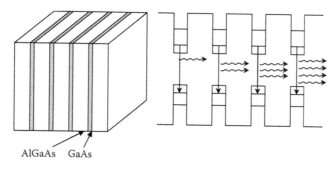

FIGURE 5.25 MQW structure.

lasers. In fact, population inversion is more easily achieved because the lower energy level is rapidly populated by electrons (while in bulk semiconductors, the electrons are spread over the CB), so the threshold current is markedly reduced. Furthermore, the spread of the output spectrum is substantially narrower because lasing occurs between discrete levels instead of energy bands.

The advantages of QW lasers can be extended over a larger volume of semiconductor materials by using multiple quantum well (MQW) devices that provide better overlap with the propagating optical mode. In MQW lasers, the structure presents alternating ultrathin layers of wide and narrow band gap semiconductors, as shown in Figure 5.25, and this allows to achieve more optical power. At present, many commercially available LDs are MQW devices with a distributed feedback structure to obtain a single-mode operation.

5.3.2.5 Vertical-Cavity Surface-Emitting Lasers

The LDs described so far are edge-emitting devices and their output beams have elliptical cross sections. Taking once again advantage of DBRs, it is possible to realize LDs in which the resonant cavity axis is orthogonal to the semiconductor wafer. Their basic structure is shown in Figure 5.26a, where dielectric mirrors are actually narrow-band Bragg reflectors realized with alternating low and high refractive index material layers. The output beam has a circular cross section because it does not emerge from the side of the active layer, which acts as a rectangular diffraction aperture, but from a circular aperture in the contact layer. Vertical-cavity surface-emitting lasers (VCSELs) can also be easily arranged in arrays (see Figure 5.26b), so realizing a matrix emitter with important applications in optical interconnects and optical computing technologies. They represent interesting devices in optofluidics applications because

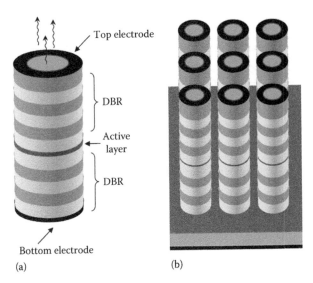

FIGURE 5.26   (a) Structure of a VCSEL and (b) VCSEL's array.

they allow an easy coupling of the light inside the fluidic channels that can be realized on a different substrate and conveniently bonded to the semiconductor wafer containing the VCSELs. Recently, the use of VCSELs as laser sources for the realization of an integrated optical trap has attracted particular interest since VCSELs are much smaller and less expensive than commonly used solid-state lasers, and have a low beam divergence and a more symmetric beam profile compared to edge-emitting LDs. This makes it easy to collimate or focus the output beam with a simple lens, which does not have to have a very high numerical aperture.

5.3.2.6 Organic LEDs

In recent years, many efforts have been devoted to the development of light sources based on organic materials because of their versatility (they can be realized by ink-jet printing, spinning, or other polymer layer–deposition techniques) and integrability in polymeric chips for photonics processing and sensing applications.

In organic LEDs (OLEDs), electricity is directly converted into light. Therefore, the evaluation of the overall light output and its relation to the driving current are of fundamental interest, though understanding and tailoring of their emission color also falls in central device physics and practical application problems. A large amount of effort has been expended preparing various material compositions, particularly of the binary and ternary systems, and measuring their emission spectra with the hope of finding new and useful OLED emitters.

The construction of OLEDs consists of laminated organic thin films laid on top of a substrate, as shown in Figure 5.27. Put simply, it features a thin, simple construction where organic materials are sandwiched between two electrodes and laid on a glass or plastic substrate.

To be more specific, the organic device is normally constructed of three layers. An emissive layer is sandwiched between two transport layers that come into contact with the cathode and the anode, respectively. The function of the transport layers is to carry the electric charge smoothly from the respective electrodes to the emissive layer.

When a voltage is applied to OLEDs (see Figure 5.28), the holes and the electrons are generated from each of the two electrodes, which have a positive and a negative electric charge, respectively. When they recombine in the emissive layer, organic materials make the emissive layer turn into a high energy state termed "exciton." Light is emitted when the layer returns to its original stability.

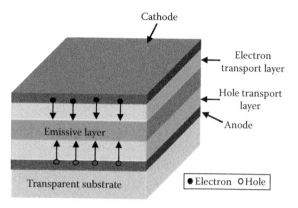

FIGURE 5.27 OLED structure.

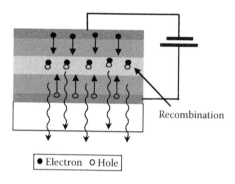

FIGURE 5.28 Biased OLED.

The combinations of molecular structures in organic materials are virtually limitless. Within these combinations, identifying organic materials that provide high efficiency and long life will determine their practical application.

Published emission spectra of OLEDs cover a spectral range from infrared to ultraviolet and show either a shape characteristic of molecules dispersed in an electronically neutral medium, broad maxima from disordered emitters involving two molecules underlying excited states, narrow lines reflecting the excitation of metal ions in their complexes with organic compounds, or microcavity and lasing effects in the layered structures with strongly injecting electrodes. Furthermore, since the composition of the population of emitting excited states produced in multilayer electroluminescence devices varies with their operating conditions, the emission spectra are dependent on the voltage evolution.

5.3.2.7 Integration Aspects

The above-discussed integrated light sources offer different potentials for integration in optofluidic chips. In fact, as already mentioned throughout the previous sections (Sections 5.3.2.1 through 5.3.2.6), different levels of integration can be envisaged depending on the material hosting the fluidic channels. If the host material is a direct band gap semiconductor, light sources can be directly realized on the chip provided that no special requirements in terms of temperature stability are needed for their proper operation, and the technological steps for channels realization are compatible with the other steps. On the other hand, if the fluidic card is polymeric, two ways are, in principle, possible: the use of VCSELs on a separate chip conveniently bonded to the fluidic card and the direct realization of OLEDs inside the fluidic card. The latter solution is undoubtedly the most fascinating, but it can be implemented only if the spectral and power requirements of the light source are compatible with OLEDs. Whenever high spectral purity, high power light sources are required for an application, a fiber-coupled external LD is probably the most suitable solution.

5.3.3 Integrated Light Detectors

5.3.3.1 Photoconductors

The ability of semiconductor materials to convert photons, whose energy is greater than the band gap, into EHPs is the key phenomenon allowing the realization of semiconductor-based photodetectors. In fact, the above process, known as photogeneration, causes an increase in the material electrical conductivity (photoconductivity) that can be easily converted into a current or voltage change by means of a suitable circuit. The increase in conductivity is due to the photogenerated EHP's concentration, $\Delta n = \Delta p$. The simplest photodetector can be realized by considering a semiconductor material with a couple of electrodes and an external battery, as shown in Figure 5.29.

Assuming that the dark conductivity, i.e., the conductivity when no light impinges on the material, is negligibly small, the illumination causes a current, I_{ph}, to flow into the circuit. This current is known as photocurrent. Of course, if the dark conductivity is not negligible, a background current, known as dark current, will always flow in the circuit and has to be added to the photocurrent. An important parameter in photodetectors is the external quantum efficiency, defined as

$$\eta = \frac{I_{ph}/e}{P_i/h\nu} \tag{5.22}$$

where
  P_i is the incident optical power
  e is the electron charge
  $h\nu$ is the photon energy

It represents the ratio between the number of photogenerated EHPs and the number of incident photons. The parameter that usually appears in a photodetector's data sheet is the responsivity, measured in amperes per watt and defined as

$$R = \frac{I_{ph}}{P_i} = \eta \frac{e\lambda}{hc} \tag{5.23}$$

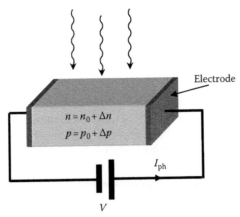

FIGURE 5.29 Schematic representation of a photoconductor-based photodetector.

where
λ is the wavelength
c is the speed of light

For a more detailed physical insight, it is interesting to derive a simple expression for the photocurrent. Writing the number of photogenerated EHPs per unit volume as

$$\Delta n = \Delta p = \frac{P_i \tau \eta}{h\nu(w \cdot d \cdot l)} = N \tag{5.24}$$

where τ is the recombination lifetime of the EHPs, the photocurrent can be written as

$$I_{ph} = e\left(\mu_e + \mu_h\right)N\frac{V}{l}(w \cdot d) = e\left(\mu_e + \mu_h\right)\tau\eta\frac{P_i}{h\nu}\frac{E}{l} \tag{5.25}$$

where
μ_e and μ_h are the mobilities of electrons and holes, respectively
$E = V/l$ is the electric field between the electrodes

Taking into account the definition of mobility, after some algebra, the final expression for the photocurrent becomes

$$I_{ph} = e\tau\eta\frac{P_i}{h\nu}\frac{v_e\left(1 + \mu_h/\mu_e\right)}{l} = I_0\tau\frac{v_e}{l} = I_0\frac{\tau}{t_e} \tag{5.26}$$

where
v_e is the electrons' velocity
t_e is the electrons' transit time between the electrodes.

The above expression highlights some interesting aspects, useful for the optimization of photodetectors. First, the photocurrent, apart from a constant factor, I_0, is directly proportional to the EHP's recombination lifetime. Second, it is inversely proportional to the electrons' transit time. This means that to maximize I_{ph}, a material with long recombination lifetime is required, and the electrode distance should be as short as possible. On the other hand, reducing the electrode distance causes an increase in the device capacitance, thus increasing the switching time. It is clear that a trade-off exists and a careful optimization of the device is required, according to the specific requirements. An extremely versatile solution is represented by a reverse-biased pn junction, as described in Section 5.3.3.2.

5.3.3.2 Photodiodes

A photodiode in its simplest form is a reverse-biased pn junction. It exhibits an extremely low dark current, represented by the diode reverse current, and allows the optimization of the performances according to the required specifications. The schematic structure of a photodiode is represented in Figure 5.30a. The p side is heavily doped and very thin (less than a micron), so the SCL extends deeply in the n side, usually for a few microns. The illuminated side has a window delimited by an annular electrode, to allow the photons to enter the device and generate EHPs. The electric field in the SCL separates the electrons and the holes and drifts them in opposite directions until they reach the neutral regions. This generates the photocurrent in the external circuit that provides a voltage on the load resistor. The actual spatial distribution of the EHPs inside the device depends on the incident wavelengths, as each

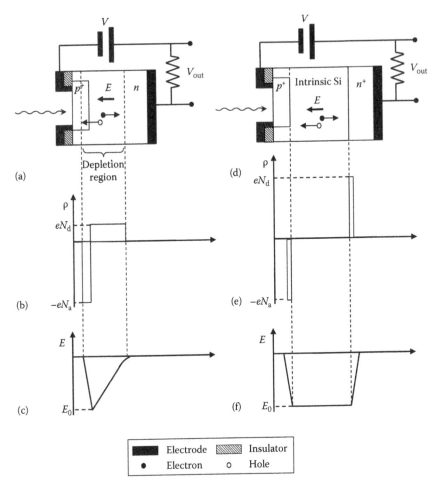

FIGURE 5.30 (a) pn junction photodiode structure, (b) charge distribution, and (c) electric field profile, (d) pin photodiode structure, (e) charge distribution, and (f) electric field profile.

wavelength has its definite absorption coefficient and hence its penetration depth in the material. Figure 5.31 shows the absorption coefficients of different semiconductor materials, as functions of the wavelengths. Note that both direct and indirect band gap materials can be used for photodetectors because photogeneration is always an allowed process, provided that the photons' energy is greater than the semiconductor band gap.

5.3.3.3 pin Photodiode

The above-described photodiode is based on the simple pn junction but, even if useful for some applications, presents two drawbacks caused by the short length of the SCL. The first drawback is the relatively high value of the depletion layer capacitance, which causes problems at high modulation frequencies. The second drawback is the limited extension of the layer where the electric field is nonzero along with its nonuniform profile (again the SCL). This means that many EHPs happen to be generated outside the SCL, i.e., in the neutral regions, especially for photons with high penetration depths. These EHPs, in the absence of the electric field, do not separate and so are more easily subject to recombination, thus becoming useless for the photocurrent and reducing the quantum efficiency. Fortunately, these drawbacks can be easily eliminated by slightly modifying the photodiode structure. The modified structure

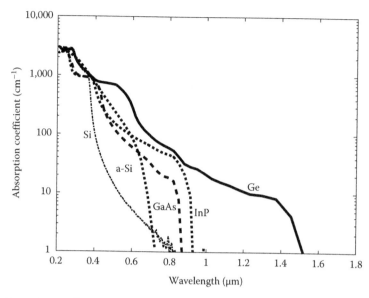

FIGURE 5.31 Absorption coefficients versus wavelengths for various semiconductor materials.

is shown in Figure 5.30d and is known as a pin photodiode because a layer of an intrinsic semiconductor (silicon in the figure) is inserted between the heavily doped p⁺ and n⁺ layers. The intrinsic layer is completely depleted and can be made up to 50 μm wide, according to the applications. Such a layer also exhibits a uniform electric field and gives rise to a small depletion layer capacitance, C_{dep} (typically of the order of a picofarad), which is also practically independent of the reverse bias applied to the junction, as long as the width W is dominated by the intrinsic layer extension:

$$C_{\text{dep}} = \frac{\varepsilon A}{W} \tag{5.27}$$

Of course, increasing W also increases the transit time, thus slowing down the photodetector's speed of response. Again a trade-off exists: The extension of the intrinsic layer has to be established according to the wavelength range to be detected and the required speed. The typical performance parameters of a pin photodiode are reported in Table 5.3.

TABLE 5.3 Characteristics of Some Photodetectors

Photodiode	Wavelength Range (nm)	Peak Wavelength (nm)	Responsivity at Peak Wavelength (A/W)	Gain	Response Time (ns)	Dark Current (nA)
Si pn junction	200–1100	600–900	0.5–0.6	<1	0.5	0.001–0.1
Si pin	300–1100	800–900	0.5–0.6	<1	0.03–0.05	0.01–0.1
Si APD	400–1100	830–900	40–130	10–100	0.1	1–10
Ge APD	700–1700	1500–1600	4–14	10–20	0.1	10^3–10^4
InGaAs–InP pin	800–1700	1500–1600	0.7–0.9	<1	0.03–0.1	0.1–10
InGaAs–InP APD	800–1700	1500–1600	7–18	10–20	0.07–0.1	10–100
PMT	115–1700	135–1550	0.01–0.09	10^4–10^7	0.1–10	0.05–5

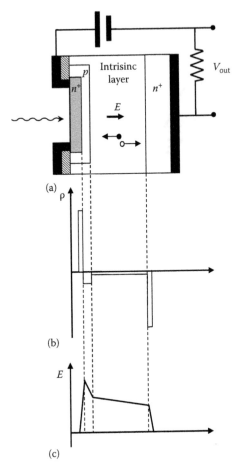

FIGURE 5.32 (a) Avalanche photodiode structure, (b) electric charge distribution, and (c) electric field profile.

5.3.3.4 Avalanche Photodiode

It is often required in many applications, such as the detection of fluorescence emission from small numbers of molecules, to measure extremely low optical powers. In these cases, a photodetector with a form of internal amplification for the photocurrent can be exploited. Such a device is known as an avalanche photodiode (APD). The basic idea is to take advantage of the avalanche breakdown, occurring in purposely designed pn junctions, triggered by photogenerated EHPs. A schematic diagram of an APD is shown in Figure 5.32a. It looks like a modified pin photodetector where a further p-doped layer is added. Figure 5.32b shows the net charge distribution across the diode due to exposed dopant ions. Under adequate reverse bias conditions, the depletion layer extends across the intrinsic layer, thus giving rise to the electric field profile shown in Figure 5.32c. The absorption of photons and hence the photogeneration take place mainly in the π layer, where the nearly uniform low electric field separates the EHPs and pulls them toward the n$^+$ and p$^+$ sides, respectively. As the drifting electrons reach the p layer, they experience a greater field and, therefore, acquire sufficient kinetic energy to impact-ionize the material, releasing more EHPs, as shown in Figure 5.33. If the p layer (avalanche region) is sufficiently long, the new EHPs cause further impact ionization and so on. Thus, from a single electron entering the avalanche region, a large number of EHPs can be generated, all of which contribute to the photocurrent. The reason for keeping the avalanche region and the absorption region separate is that avalanche multiplication is a statistical process and so leads to carrier generation fluctuations eventually resulting in

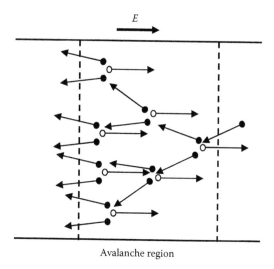

FIGURE 5.33 Electron-driven impact ionization process. View of the charge carriers path in the avalanche region.

excess noise (known as avalanche noise). This noise is minimized if the impact ionization is restricted to carriers with the highest ionization coefficient (electrons, if silicon is used). An avalanche multiplication factor, M, is defined as

$$M = \frac{\text{Multiplied photocurrent}}{\text{Primary photocurrent}} = \frac{I_{\text{ph}}}{I_{\text{pho}}} \tag{5.28}$$

The primary photocurrent is the photocurrent measured in the absence of multiplication, for example, under a small reverse bias. As regards the dynamic response of APDs, it depends on three factors: the time required for the photogenerated carriers to cross the absorption region, the time required for the avalanche process to build up, and the time it takes for the last hole released in the avalanche process to transit through the absorption region. This means that the speed of an APD is somewhat lower than that of the corresponding pin photodiode. A further optimization of APDs can be obtained by using heterojunction photodiodes. The use of different semiconductor materials arranged to form MQWs allows, in fact, precisely tailoring the different regions of an APD, so increasing the efficiency and reducing the noise. A summary of the characteristics of commercially available photodiodes is reported in Table 5.3. It should be mentioned that properly designed APDs can be used for single photon counting.

5.3.3.5 Image Sensors: CCD Cameras

When spatially resolved light detection is required, an image sensor can solve the problem. The simplest and most popular presently available image sensor is represented by the charge coupled device (CCD)-based camera. A CCD camera is a device realized starting from a metal-oxide-semiconductor (MOS) capacitors array. An MOS capacitor consists of a semiconductor substrate covered by a thin oxide layer with a metal electrode on the top. When properly biased, it can attract electrons (or holes) generated in the semiconductor material by an impinging light source with adequate photon energy, and hold them under the metal electrode. If the metal electrode, rather than being realized by a single metal layer, is composed of three closely spaced metal layers able to be biased with independent voltages, a pixel is formed. An array of such pixels can be used to capture the light intensity distribution hitting its surface and store it as a distribution of charges (say electrons) in the semiconductor substrate. This is easily achieved by setting the central electrode of each pixel to a voltage higher than those of the two adjacent electrodes. To extract the information from the CCD array, its rows are interrogated one by one exploiting a suitable clock signal able to change the bias of the electrodes, so letting the charge packet, stored under each pixel, move step by step toward the end of the row. At the end of each row, a synchronized charge amplifier turns the charge packet into a voltage level, which is converted into a digital form and stored in a memory chip. In this way, a digital map of the light intensity is stored in the memory chip.

CCD cameras are cheap devices but they usually do not exhibit high sensitivity. Thus, they can be used in optofluidic applications to image the light distribution in a fluidic channel if needed, when the light intensity is sufficiently high. They can also be easily placed on the top or the bottom of a polymeric fluidic card.

5.3.3.6 Integration Aspects

All the above-discussed light detectors offer the possibility to be directly integrated into an optofluidic chip, provided that the latter is realized on a direct or an indirect band gap semiconductor material substrate. In fact, light detectors can be fabricated, differently from the light sources, on both direct and indirect band gap semiconductors. This aspect is extremely important when the integration of processing electronics (amplifiers, filters, analog-to-digital converters, etc.) is also considered, because it allows the use of silicon substrates.

5.4 Photomultiplier Tubes

Photomultiplier tubes (PMTs) are detectors that also permit optical measurements at extremely low light levels down to single photons. These devices are based on the external photoelectric effect.

Let us consider a thin metal film kept in vacuum; when light strikes on it, in some cases, if the photon energy is sufficiently high, bound electrons inside the metal thin film are released into the vacuum. This phenomenon is called external photoelectric effect, and the metal thin films with such a photoemissive surface are usually called photocathodes. The minimum photon energy for photoemission depends on the material; it is called the work function and it strongly depends on the material. Most photocathodes are made of a compound material mostly consisting of alkali metals, and changing the photocathode materials makes it possible to achieve sensitivity at various wavelengths of the light spectrum.

PMTs are vacuum tubes with a glass envelope and consist of a photocathode, an electron multiplier system, and an electron collector (anode). In Figure 5.34, a schematic of a PMT is represented. Light passes through the input window and strikes on the photocathode. The electrons emitted into the vacuum are accelerated and focused by the focusing electrode in the electron multiplier section. The physical process responsible for the electron multiplication is a secondary emission process. It occurs when

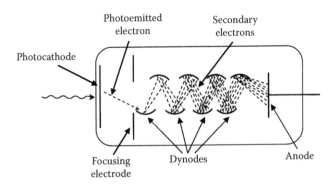

FIGURE 5.34   Schematic of a PMT.

an electron collides with a metallic target. If the electron energy is sufficient, a number of other electrons (secondary electrons) are emitted into the vacuum.

In the electron multiplying section of a PMT, this secondary emission is repeated several times by means of successive, properly polarized electrodes (dynodes). Usually, in a PMT, 6–14 stages of dynodes are used in order to obtain a multiplying factor of 10^5 to 10^8. Finally, secondary electrons emitted from the last dynode are ultimately collected by the anode and outputted to an external circuit as an electrical signal.

Photomultipliers are widely used for their high sensitivity and time resolution, which they achieve at room temperature.

Since PMTs are a kind of electron tubes, they are relatively large in size, but are superior in sensitivity and response speed with respect to ADPs, making them useful as optical detectors in a variety of applications, such as analytical instruments, medical equipment, and industrial measurement systems. However, it must be underlined that semiconductor detectors are the most likely candidates for integrated applications while PMTs are almost impossible to integrate.

5.5 Future Trends

In this chapter, we have introduced the optoelectronic basic concepts and devices. Optoelectronic devices play a fundamental role in the fabrication and characterization of optofluidic devices, and integrated semiconductor optoelectronics will be a fundamental tool in optofluidic applications. Semiconductor technology allows the integration of passive and active optical components (emitters, detectors, waveguides, microchannels, and related devices) onto the same substrate, allowing simple and flexible development of complex optofluidic devices. This integration offers several additional advantages such as miniaturization, robustness, reliability, high accuracy, and simplicity in the optical alignment of different devices.

In particular, among the several available technologies, silicon technology seems to be the most promising one. Silicon, the dominant material for microelectronics fabrication, is nowadays largely used in sensing applications, and recently several silicon photonic components also have been developed. The driving force behind the development of silicon photonics is the telecommunications field, but this technology will be successfully applied for the development of optofluidics.

Reference

Sze, S.M. and Ng, K.K. 2006. *Physics of Semiconductor Devices*, Wiley, Hoboken, NJ.

6

Spectroscopic Methods

Jin Z. Zhang

Optical spectroscopic methods are widely used in the study of optical and other properties of different materials. The various techniques are usually based on measuring absorption, emission, or scattering of light that contains information about physical properties of the materials. Commonly used techniques include electronic absorption (UV–vis), photoluminescence (PL), infrared (IR) absorption, and Raman scattering. Other more specialized techniques include fluorescence correlation spectroscopy (FCS), dynamic light scattering (DLS), fluorescence lifetime imaging microscopy (FLIM), total internal reflection fluorescence (TIRF), surface-enhanced Raman scattering (SERS), and single-molecular spectroscopy (SMS). These various techniques can provide different information about the molecular properties of interest. In this chapter, several common spectroscopic techniques are reviewed with emphasis on their principles of operation as well as spectral interpretation. The main objective is to explain how one can obtain useful physical information about the materials under study from the optical spectra measured experimentally.

Spectroscopic methods are an integral and important part of optofluidics since the measurement in optofluidic devices for (bio)-sensing applications are in one way or another based on different types of spectroscopy that rely on absorption, emission, or scattering of light. In this chapter, we will review some of the common spectroscopic techniques with emphasis on those closely associated with or potentially useful for optofluidic applications. Some of these spectroscopic methods have already been applied successfully in conjunction with optofluidic techniques, while others are expected to be implemented in the near future.

6.1 UV–Visible Electronic Absorption Spectroscopy

6.1.1 Operating Principle: Beer's Law

The basic operating principle of electronic absorption spectroscopy is based on measurement of light absorption due to electronic transitions in a sample. Since the wavelength of light required for electronic transitions is typically in the UV and visible region of the electromagnetic (EM) radiation spectrum, electronic absorption spectroscopy is often called "UV–visible" or "UV–vis spectroscopy."

Based on Beer's law, the absorbance (A) or optical density (OD) is related to the incident light intensity, I_0, transmitted light intensity, I, concentration of a liquid sample, c, path length of the sample, l, absorption (or extinction) coefficient, α, and molar absorptivity, ε, by the following equation:

$$A = OD = \log I_0/I = -\log T = \alpha c = \varepsilon l c \qquad (6.1)$$

where T is the transmittance, defined as I/I_0. In an experiment, both I_0 and I can be measured and thus A can be determined experimentally. If l and c are known, the absorption coefficient can be determined by Equation 6.1. The concentration should be kept reasonably low to avoid saturation that would distort the spectrum. Similar equations can be derived for gas or solid samples. The solid sample needs to be thin enough to avoid saturation of absorption.

The absorption coefficient is wavelength dependent and a plot of ε as a function of wavelength, λ, is the spectrum of interest. The spectrum is characteristic of a given sample and reflects the fundamental electronic properties of the molecules in the sample. The absorption coefficient is related to the imaginary part of the complex refractive index (RI) of the sample.

6.1.2 Instrument: UV–Visible Spectrometer

Electronic absorption or UV–visible spectroscopy is one of the simplest and yet most useful optical techniques for studying optical and electronic properties of molecules. UV–vis spectra are typically measured using commercially available spectrometers at reasonable cost (e.g., Hewlett-Packard, Ocean Optics, and Shimadzu). As illustrated in Figure 6.1, the intensity of light from a light source, e.g., a lamp, is measured by a light detector, photodiode, photomultiplier tube (PMT), or charge-coupled device (CCD) detector, with a sample between the light source and the detector. If the sample absorbs light at a specific wavelength, the transmitted light will be attenuated. The intensity of the transmitted

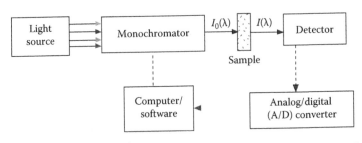

FIGURE 6.1 Schematic diagram of key components of a typical UV–vis spectrometer that include a light source, e.g., a lamp; a monochromator to disperse the incident light; a sample cuvette and holder; a detector, e.g., PMT, photodiodes, or CCD; analog-to-digital (A/D) converters; and a computer with software to control the scan of the monochromator and data acquisition.

light plotted as a function of light wavelength will give an absorption spectrum of the sample. Typical spectrometers cover the wavelength range from about 200 to 800 nm. Extending the measurement beyond 800 nm is possible but usually requires different light source, optics, and detector.

A UV–visible spectrum is relatively easy to measure. However, there are still some common mistakes that one should avoid making. First, as mentioned earlier, the sample concentration cannot be too high. Too high a concentration or OD leads to saturated absorption and a distorted, unreal spectrum. Second, a proper blank or background measurement needs to be taken before the sample spectrum is measured. Ideally, one should use the same cuvette and solvent. If there are other species in the sample that have absorption in the region measured but are of no interest, they should be part of the "blank" or background measurement. Third, one needs to make sure that the sample is clear and does not contain floaters that are visible to the eye. Visible floaters will cause significant scattering and distort the measured spectrum. If the sample contains only a small amount of floaters such as aggregates, they can usually be filtered away or precipitated out by centrifugation before a measurement is made. Solid samples need to be thin to avoid saturation and clear to avoid scattering.

6.1.3 Spectrum and Interpretation

The electronic absorption spectrum contains information about the fundamental electronic properties of the sample, e.g., density-of-states (DOS), energy levels, and electronic dipole moment. An in-depth discussion of electronic absorption theory is beyond the scope of this chapter but can be found in most quantum chemistry or spectroscopy books (McHale 1998). Suffice to say here, at the most basic level, it is the initial state electronic wave function, final state wave function, and electrical dipole moment that determine an electronic transition. The DOS plays an important role for samples that involve multiple electronic transitions due to multiple initial electronic states and/or final states, which often result in broad spectral features.

An atomic spectrum results from electronic transition from a lower atomic energy level, usually the ground electronic state, to a higher energy level though interaction of the atom with an EM field such as light. The transitions are governed by spectroscopic selection rules originating from the electrical dipole moment.

Likewise, for molecules, the electronic absorption spectrum is a result of electronic transitions from a lower electronic energy level or state to a higher energy level. In most cases, the initial state is the ground electronic state. Based on quantum mechanical perturbation theory for spectroscopy, the electronic transition probability and intensity is determined by the matrix element

$$\langle \psi_f / \boldsymbol{\mu} \cdot \boldsymbol{E} / \psi_i \rangle \tag{6.2}$$

where
  ψ_i and ψ_f are the wave functions of the initial and final states
  $\boldsymbol{\mu}$ is the electrical dipole moment operator (a vector)
  $\boldsymbol{E}$ is the applied external electrical field (also a vector), e.g., light

The bracket notation implies complete integration over the entire space in all dimensions involved by the wave functions and dipole moment operator. Since $\boldsymbol{E}$ is usually considered as space independent on the scale of a typical molecule, its absolute value can be taken out of the integration or bracket. The evaluation of Equation 6.2 determines not only if there can be an electrically allowed transition (if the integration does not result in zero in value) but also how strong the transition is, i.e., the spectral intensity (proportional to the square of the value calculated by Equation 6.2).

Figure 6.2 illustrates schematically two typical situations involved in molecular electronic transitions. The left panel shows a transition from a bound, stable ground electronic state to a bound, meta-stable excited electronic state, while the right panel shows a transition from a similar ground state to an unbound,

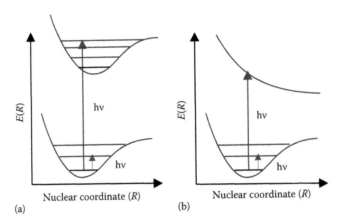

FIGURE 6.2 Schematic illustration of electronic transitions in a diatomic molecule with two different excited state potential energy surfaces: bound (a) and repulsive (b). The long vertical arrows indicate electronic transition while the shorter arrows indicate vibrational transitions or IR absorption in the ground electronic state.

unstable excited electronic state that is repulsive or dissociative. In this figure, $E(R)$, the nuclear potential energy is plotted as a function of the nuclear coordinate or the distance between two nuclei or atoms, R, which is usually called a "potential energy curve." It is so called because $E(R)$ represents the potential energy in the Hamiltonian for the nuclear Schrödinger equation and it is the total energy for the electronic Schrödinger equation. The electronic energy levels have different functional dependence on the nuclear coordinate. The illustration is for a diatomic molecule. For molecules containing N atoms ($N > 2$), there are $3N - 6$ (for nonlinear molecules) or $3N - 5$ (for linear molecules) nuclear coordinates or modes. Thus, the electronic energy is a function of these multiple nuclear coordinates. A plot of the electronic energy as a function of these nuclear coordinates in more than one coordinate is often called a "potential energy surface or hypersurface." Figure 6.2 illustrates a simple way to think about electronic transitions in molecules.

The examples shown in Figure 6.2 are the simplest possible scenarios, two states involved in a simple diatomic molecule with one vibrational mode and no interaction with other molecules. The electronic spectrum can be predicted or accurately calculated once the ground state and excited state potential energy curves and the dipole moment operator are known or obtained from electronic structure calculation. In reality, there are many possible factors that can make the spectrum as well as interpretation more complicated. First, all polyatomic molecules have more than one vibrational mode, as mentioned earlier. The nuclear potential energy is thus a function of all these modes. Some modes can be coupled or strongly interacting with each other. Second, there are many excited electronic states and some are coupled. Third, molecules can interact with each other, especially in liquids and solids, or with their environment or embedding medium. All these factors will make the spectrum more complicated, usually broader and with less resolved features due to homogeneous (intrinsic lifetime) and inhomogeneous (different environments of individual molecules) broadening.

6.1.4 Application and Optofluidics

Optical absorption is a fundamental property of molecules. Thus, measurement of the electronic absorption spectrum is essential to studying and understanding the optical properties and applications of molecules or materials. The absorption will determine not only the color of the materials but also other important optical properties such PL, lasing, electroluminescence, photovoltaic, photocatalytic, and photoelectrochemical properties.

In terms of optofluidics applications, UV–vis absorption has been applied in the form of transmission measurement in microfluidic Fabry–Perot (FP) intracavity spectroscopy (Shao et al. 2006). Presence of samples such as cells inside the cavity modified the resonant conditions of the microfluidic cavity and thereby the transmission spectra in terms of the number of transverse modes, mode shape, and the relative mode spacing. For objects that are much smaller than the wavelength of light, scattering is insignificant and absorption is simply equal to the total incident light intensity minus the transmitted light intensity. Thus, measuring transmission is essentially the same as measuring absorption. Using FP intracavity spectroscopy, label-free detection of single biological cells, including yeast cells and human blood cells, as well as polystyrene spheres, has been successfully demonstrated (Shao et al. 2006). Because of the relatively large size of the cells (8–15 μm), light scattering is substantial. Therefore, the reduced transmission of light is a result of the combination of absorption and scattering. This technique has recently been extended to study canine lymphoma cells with spectra distinct from single-mode spectra of normal lymphocytes (Shao et al. 2008). In a somewhat related study, a novel variable optical attenuator has been designed and demonstrated based on optofluidic technology with the modulation of attenuation achieved by changing the optical path length through a light-absorbing liquid brought about by the deformation of a membrane actuated by the air pressure (Yu et al. 2008). Such a device structure could conceivably be useful for analyzing liquid samples.

In general, optofluidics devices based on absorption or transmission measurements are somewhat limited, likely due to relatively low sensitivity because such measurements are not zero-background experiments. This is in contrast to fluorescence-based techniques that are zero-background experiments and thereby much more sensitive, as discussed next.

6.2 Fluorescence Spectroscopy

6.2.1 Operating Principle

At the fundamental level, the principle underlying fluorescence (also often termed PL) spectroscopy is very similar to that of electronic absorption spectroscopy. They both involve electronic transition of initial and final states coupled by the electrical dipole operator. The main difference is that the transition involved in fluorescence is from a higher energy level or state to a lower energy level. There is also an important practical difference between the two techniques in that PL is a zero-background experiment, i.e., no signal detected when there is no PL, which is in contrast to absorption spectroscopy that is a nonzero background experiment. Zero-background experiments are intrinsically more sensitive than nonzero background experiments. Therefore, PL is typically more sensitive than electronic absorption measurement.

A typical PL spectrum is just a plot of the PL intensity as a function of wavelength for a fixed excitation wavelength. A photoluminescence excitation (PLE) spectrum, however, is a measure of PL at a fixed emission wavelength as a function of excitation wavelength. To a good approximation, PLE is similar to the electronic absorption spectrum as long as no complications are involved, e.g., involvement of multiple overlapping excited states or formation of excimers (excited dimers). PLE is useful for studying samples for which an electronic absorption spectrum is challenging to obtain, for example, if there is low transmission as a result of thickness or high concentration of the sample.

6.2.2 Instrumentation: Spectrofluorometer

Figure 6.3 shows a schematic of the key components of a typical spectrofluorometer used for PL measurements. While several components are similar to that in a UV–vis spectrometer, including light source, sample cuvette, and detector, the detection scheme is different in that emitted light from the sample is detected, rather than transmitted light in UV–vis spectroscopy. Rayleigh scattering, which has the same wavelength as the excitation light, should be avoided in the detection of PL. Raman scattering can show up in a PL spectrum, especially when the Raman signal is strong while PL is relatively weak. The nature of Raman scattering and Raman spectroscopy will be discussed later.

In a typical PL measurement, a specific wavelength of light is selected from a light source by a monochromator and directed at the sample of interest. Light emitted from the sample is collected through lenses, dispersed by another monochromator, and detected by a photodetector. The analog electrical

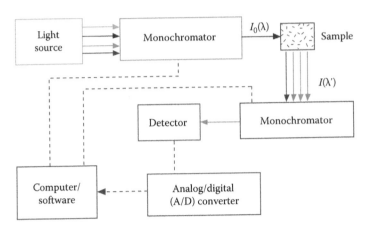

FIGURE 6.3 Schematic illustration of key components of a typical spectrofluorometer that include a light source, e.g., a lamp; monochromator to disperse the incident light; sample cuvette and holder; a second monochromator to disperse the emitted light; detector, e.g., PMT, photodiodes, or CCD; analog-to-digital (A/D) converters; and computer with software to control the scan of the monochromators and data acquisition.

signal generated by the photodetector is converted into a digital signal by an A/D (analog-to-digital) convertor and processed by software on a computer. The spectrum is displayed in terms of the intensity of emitted PL light (proportional to the electrical signal generated) as a function of the wavelength of emitted light.

PL is usually red-shifted with respect to the incident excitation light, i.e., appearing at longer wavelength. Unwanted Rayleigh scattering is at the same wavelength of the incident light (elastic), so PL can be easily distinguished from Rayleigh scattering that is usually blocked by optical filters and excluded in the spectral range scanned. Another important scattering process is inelastic Raman scattering with wavelength longer (Stokes Raman) or shorter (anti-Stokes Raman) than the incident light due to energy loss or again caused by vibration (or rotation) molecules (Ferraro and Nakamoto 1994). Raman scattering is usually much weaker than PL so it does not present a problem. However, when PL is weak and/or Raman scattering is strong, e.g., from solvent molecules, the Raman signal can be readily observed. For Stokes Raman scattering, to be explained in more detail later, the signal is also red-shifted with respect to the incident light, similar to PL. Thus, it is sometimes not easy to tell if the signal is Raman scattering or true PL from the sample of interest.

There are a couple of practical ways to distinguish PL from Raman scattering. First, Raman spectral features are usually much narrower than those of PL, especially for samples in liquid or solid forms. Second, and more reliably, Raman peaks should shift with changes in the excitation wavelength while PL usually does not, especially when the change in the excitation wavelength is small. The reason that the Raman peaks shift with excitation wavelength is that the energy difference between the Raman scattered light and the incident light is a constant, equal to the vibrational frequency of molecules, and therefore, shift in the excitation wavelength will result in a shift in the Raman frequencies, in the same direction and with the same amount of energy or frequency. Static PL peaks do not shift with excitation wavelength because the PL originates from the same initial vibrational state (usually ground vibrational state in the excited electronic state) and end up in the same final vibrational state in the ground electronic state as long as the excitation is within the same electronic absorption band.

6.2.3 Spectrum and Interpretation

In order to extract useful physical information from the measured PL or Raman spectrum, it is necessary to understand the basic principles behind PL and Raman and their connection to properties of the

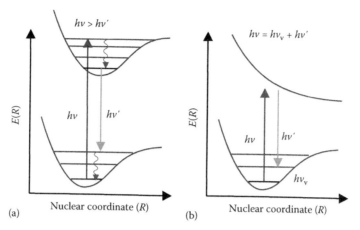

FIGURE 6.4 Illustration of PL (a) from a bound, excited state and (Stokes) Raman scattering involving a dissociative excited state (b), where hv, hv', and hv_v are the incident light energy, scattered (PL or Raman) light energy, and molecular vibrational energy (fundamental mode), respectively.

sample. We mention both Raman and PL here because of some common features they share. Figure 6.4 shows a simple illustration of the PL and Raman scattering processes in a simple diatomic molecule with two different excited state potential energy curves. A more detailed discussion of Raman scattering will be given later. The focus is primarily on PL in this section. The data interpretation of PL (and Raman) is more complex for large molecules because of the many more nuclear degrees of freedom. In principle, one needs to consider all the possible nuclear degrees of freedom. In practice, however, only some modes are active and need to be considered when interpreting PL or Raman spectra.

The situation is even more complicated for liquids and solids due to strong intermolecular interactions that have effects on the PL (and Raman) spectra. As a result, PL and Raman can be used, in turn, to probe properties of individual molecules as well as interactions between molecules in a liquid or solid. Besides PL spectrum, the PL quantum yield (QY), measured as the intensity ratio of the emitted light over the absorbed light, provides important indications about the properties of the molecules. The PL yield (Φ_{PL}) is related to the excited electronic state lifetime (τ_{ob}) via the following relations:

$$\Phi_{PL} = \tau_{ob}/\tau_r \tag{6.3}$$

$$1/\tau_{ob} = 1/\tau_r + 1/\tau_{nr} \tag{6.4}$$

where τ_r and τ_{nr} are the radiative and non-radiative lifetimes of the excited state. The observed lifetime (τ_{ob}) is also called "fluorescence lifetime" when measured using time-resolved PL techniques. This should not be confused with the radiative lifetime. Any lifetime measured experimentally based on time-resolved PL or other measurements is just the observed lifetime τ_{ob}, which contains contributions from both τ_r and τ_{nr}. When one of these lifetimes is very long, the observed lifetime is dominated by the other one.

Here we wish to offer a few words of caution for practical PL measurements. In PL measurements, one needs to be careful about the possible appearance of Raman signal, as mentioned before and to be discussed in more detail later. One also needs to be careful about high order Rayleigh scattering from the strong incident light. Mistaking Rayleigh or Raman scattering as PL signals can easily occur, and such mistakes have made it into the literature. For example, if 300 nm light is used for excitation, multiples of 300 nm can appear in the measured PL spectrum at 600 or 900 nm due to higher orders of the grating in a spectrometer, when the 300 nm Rayleigh scattered light is not completely blocked by optical filters. This is also why the

PL spectrum is usually measured by starting the scan to the red or lower energy compared to the incident light, to avoid the first order Rayleigh scattering of the incident light. However, this cannot eliminate the higher order scattering of the incident light if the scan covers that spectral range (e.g., 600 nm). Likewise, besides first order, higher grating order Raman signal can also appear in a measurement. Therefore, care must be taken to identify and avoid such potential problems or artifacts in PL studies.

6.2.4 Applications

Due to its high sensitivity, PL is one of the most widely used spectroscopic techniques for analytical applications in various fields. It is used for both identifying and quantifying species. PL can be molecular specific for gas phase species with high frequency resolution. For species in liquid or solid, the PL spectra are usually broad due to homogeneous and, more importantly, inhomogeneous broadening, and thereby usually lack molecular specificity. Most applications thus take advantage of the high sensitivity of PL for quantification of the amount of material. PL-based spectroscopic techniques have been successfully integrated with micro- or nanofluidics for optofluidic applications. A number of specific examples will be given next, including those based on Förster (or fluorescence) resonance energy transfer (FRET), FCS, FLIM, and TIRF. Some of these techniques are used in conjunction for certain applications. We will discuss some of them in the next few sections.

6.3 Förster Resonance Energy Transfer

In this section, we will discuss one of the most commonly encountered techniques based on PL—FRET. FRET is often used to determine distance between two functional groups in a molecule (Lakowicz 1999). FRET involves non-radiative transfer of energy from a donor molecule (or nanoparticle) to an acceptor molecule (or nanoparticle). Therefore, the signature of FRET is quenching of the donor fluorescence followed by lower energy or longer wavelength fluorescence of the acceptor. FRET efficiency, E_{FRET}, defined as the fraction of energy (in photons) absorbed by the donor that was subsequently transferred to the acceptor, is expressed as

$$E_{FRET} = \frac{R_0^6}{\left(R_0^6 + R^6\right)} = \frac{1}{1 + (R/R_0)^1} \tag{6.5}$$

where
    R_0 is the so-called Förster distance
    R is the distance between the center of the donor and the acceptable fluorophore dipole moments

From this equation, it is clear that, at the Förster distance, the FRET efficiency is 50%. The Förster distance, R_0, is determined by the spectral overlap between the donor fluorescence and acceptor absorption, fluorescence QY of the donor in the absence of the acceptor, the RI of the medium, and the dipole orientation factor.

More relevant to experimental measurement is the expression

$$E_{FRET} = 1 - \frac{F_{DA}}{F_D} = 1 - \frac{\tau_D}{\tau_{DA}} \tag{6.6}$$

where
    F_{DA} and τ_{DA} are, respectively, the fluorescence intensity and lifetime of the donor in the presence of the acceptor
    F_D and τ_D are the fluorescence intensity and lifetime of the donor when the acceptor is far away or absent

Equation 6.6 shows that the FRET efficiency ranges from 0 (when the acceptor is far away and $F_{DA} = F_D$) to 1 when the acceptor is very close to the donor and F_{DA} is near 0 (complete quenching). By attaching appropriate donors and acceptors to functional groups of interest in a molecule and measuring the FRET efficiency, one can determine the distance between the functional groups. For FRET to work effectively, the donor fluorescence spectrum must overlap with the absorption spectrum of the acceptor. This is to ensure effective coupling or interaction between the donor and acceptor dipoles. In addition, the distance between the donor and acceptor cannot be too far, usually within 5 nm.

While most earlier studies of FRET are based on molecular systems, usually dye molecules (Medintz et al. 2003, 2004), recent work has been reported on using semiconductor nanoparticles (NPs) or quantum dots (QDs) for FRET successfully, often as donors (Willard et al. 2001; Medintz et al. 2003; Zhou et al. 2005; Zhang et al. 2007a) and, to a lesser degree, as acceptors (Kagan et al. 1996a,b; Huang et al. 2006), or both as donors and acceptors (Wang et al. 2002). QDs are less ideal as acceptors due to their typically broad absorption that easily results in absorption of the light used to excite the donor. One limitation for QDs in FRET is their relatively large size compared to molecular systems. Their advantages include tunable absorption and emission with size as well as enhanced photostability when properly passivated, as mentioned before.

So far, limited research has been done on combining FRET with optofluidic devices. In one recent report, an optofluidic ring resonator dye laser has been successfully demonstrated using highly efficient energy transfer (Shopova et al. 2007). The active lasing material consists of a donor and acceptor mixture and flows in a fused silica capillary whose circular cross section forms a ring resonator and supports whispering gallery modes (WGMs) of high Q-factors (>10^7). The excited states are created in the donor and transferred to the acceptor through the FRET process, whose emission is coupled into the WGM. FRET lasers using cascade energy transfer and using QDs as the donor have also been demonstrated (Shopova et al. 2007). This work has shown the good potential of developing novel microfluidic lasers with low lasing thresholds and excitation/emission flexibility, as well as the possibility for laser intracavity biological and chemical sensing. Given the versatility and sensitivity of FRET and compact nature of optofluidic devices, further development is expected on combining the two techniques in the future.

6.4 Fluorescence Correlation Spectroscopy

FCS is based on measurement of fluctuation of fluorescence (or concentration) of molecules. It was first demonstrated in the early 1970s (Magde et al. 1972). Since then significant progress has been made both on the theoretical and instrumentation fronts. Due to the high sensitivity of fluorescence to changes in molecular structure, chemistry, and local environment of molecules, FCS has become a powerful and popular analytical method for studying chemical kinetics and dynamics (Hess et al. 2002).

In a typical FCS setup, light is focused on a sample and the measured fluorescence intensity fluctuations, due to some kinetic or dynamic processes such as diffusion, chemical reactions, or aggregation, are analyzed using the temporal autocorrelation. The setup usually consists of a laser beam that is reflected into a microscope objective by a dichroic mirror. The laser beam is then focused in the sample, which contains fluorescent molecules at such a low concentration that only few are within the focal spot (usually 1–100 molecules in 1 fL). When the particles cross the focal volume, they fluoresce, and this fluorescence light is collected by the same objective and, because it is red-shifted with respect to the excitation light, it passes the dichroic reaching a detector, typically a PMT or avalanche photodiode (APD). The resulting electronic signal can be stored either directly as intensity versus time trace to be analyzed at a later point, or, computed to generate the autocorrelation directly (which requires special acquisition cards).

Appropriate models and equations need to be developed for the corresponding system in order to effectively extract useful physical information from the measured fluorescence autocorrelation function. For example, FCS can be used to obtain quantitative information such as diffusion coefficients, hydrodynamic radii, average concentrations, kinetic chemical reaction rates, and singlet–triplet dynamics. FCS is commonly used to characterize the dynamics of fluorescent species, e.g., fluorescent nanostructures (Dong et al. 2006; Tsay et al. 2006; Kyoung and Sheets 2007) and autofluorescent proteins in living cells (Hink and Visser 2002; Hink et al. 2003; Ohsugi et al. 2006). The model and equations are usually more complicated when there are multiple fluorescent species and/or multiple dynamic processes involved (Hess et al. 2002).

The temporal evolution of concentration fluctuations can be quantified via the corresponding fluctuation in $\delta F(t)$, in the fluorescence signal $F(t)$ around its mean value $\langle F(t) \rangle$, where $\delta F(t) = F(t) - \langle F(t) \rangle$ and t is time. The normalized autocorrelation function $G(t)$ of a fluorescence fluctuation at a given time, $\delta F(t)$, and at a later time, $\delta F(t + \tau)$, is given by

$$G(\tau) = \langle \delta F(t) \delta F(t + \tau) \rangle / \langle F(t) \rangle^2 \qquad (6.7)$$

Thus, $G(\tau)$ contains information about concentration fluctuations of molecules observed, reflecting the dynamic or kinetic processes involved. The functional form of $G(\tau)$ is thereby system dependent.

For a simple case that involves single diffusing species in a prolate ellipsoidal Gaussian observation volume without chemical reactions, the autocorrelation function is given by (Aragon and Pecora 1976; Eigen and Rigler 1994)

$$G_D(\tau) = 1/[N(1 + \tau/\tau_D)(1 + \tau/\omega^2\tau_D)^{1/2}] \qquad (6.8)$$

where

τ_D is the characteristic diffusion time during which a molecule resides in the observation volume with an axial (z_0) to lateral (r_0) dimension ratio $\omega = z_0/r_0$

N is the mean number of fluorescent molecules in the observation volume

As N increases, $G_D(\tau)$ decreases. This can be rationalized based on the fact that the relative effect of a single molecule on the total fluorescence signal is small when N is large. In other words, when N is too large, the overall fluctuations are small in comparison to the total signal and may become difficult to resolve. On the other hand, if N is too small, the individual fluctuation events are too sparse in time, it may take too long to do the measurement. Therefore in practice, there is an optimal value for N, usually ~1–10 for $C \sim 10^{-9}$ M (Hess et al. 2002).

From Equation 6.8, one can see that by measuring the autocorrelation function and fitting with the function given in the equation, one can obtain τ_D. In a calibrated observation volume with a Gaussian profile, $1/e^2$ radial waist r_0 and known ω, the diffusion coefficient, D, can thus be obtained from $D = r_0^2/4\tau_D$. As mentioned above, the data analysis becomes more complicated for systems that involve multiple fluorescent species and/or kinetic processes.

Figure 6.5 shows an example of a typical FCS setup and representative fluorescence data and the resulting autocorrelation function (Hess et al. 2002).

FCS has been applied to optofluidic applications (Foquet et al. 2004; Bi et al. 2006; Petrasek et al. 2007; Yin et al. 2007), usually with single-molecule detection capability. In one case, FCS has been successfully implemented in conjunction with planar optofluidics on a chip (Yin et al. 2007). Full planar integration was achieved by lithographic definition of subpicoliter excitation volumes using intersecting solid and liquid-core optical antiresonant reflecting optical waveguides (ARROWs). Figure 6.6 shows a schematic of an optofluidic FCS ARROW chip. Several steps are involved in implementing FCS on an ARROW chip, including lithographic definition of a subpicoliter fluorescence excitation volume, optical waveguides to guide light to the excitation region, and an efficient way to collect the fluorescence and guide it to the photodetector.

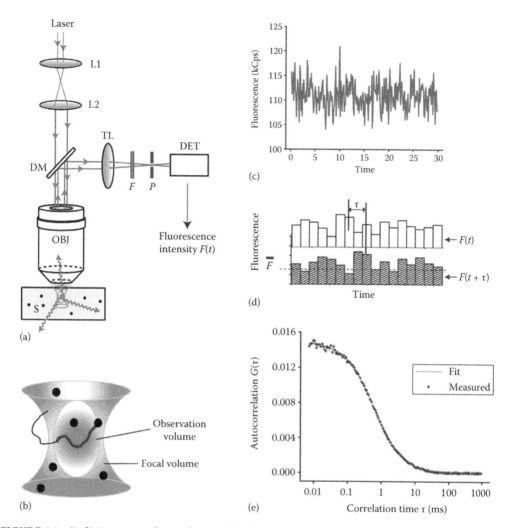

FIGURE 6.5 (Left) Experimental setup for FCS. (a) A laser beam is first expanded by a telescope (L1 and L2), and then focused by a high-NA objective lens (OBJ) on a fluorescent sample (S). The epifluorescence is collected by the same objective, reflected by a dichroic mirror (DM), focused by a tube lens (TL), filtered (F), and passed through a confocal aperture (P) onto the detector (DET). (b) Magnified focal volume (light grey) within which the sample particles (black circles) are illuminated. (Right) (c) A typical fluorescence signal, as a function of time, measured for rhodamine green (RG) with a λ_x of 488 nm. (d) Portion of the same signal in panel c, binned, with an expanded time axis and average fluorescence. The signal is correlated with itself at a later time $(t + \tau)$ to produce the autocorrelation $G(\tau)$. (e) Measured $G(\tau)$ describing the fluorescence fluctuation of RG molecules due to diffusion only as observed by FCS. (Reproduced from Hess, S.T. et al., *Biochemistry*, 41, 697, 2002. With permission.)

Based on this FCS chip architecture, single-molecule detectivity was demonstrated using Alexa 647 dye molecules (Yin et al. 2007). Figure 6.7 shows the concentration-dependent FCS autocorrelation signal (a) and the average number of molecules as a function of concentration (b). The results suggested a resolution of less than one molecule on average (0.35) in the integrated device, in good agreement with estimates of the average molecule number obtained using the nominal concentration and effective excitation volume in time-integrated fluorescence. Due to the advantages afforded by SMS and micro- and nano-optofluidics, one could anticipate further research in integrating these two techniques for high sensitivity analytical applications.

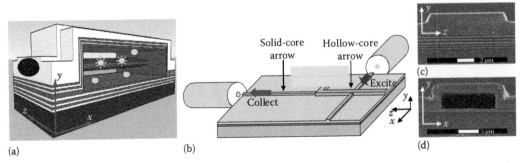

(a)   (b)   (d)

FIGURE 6.6 (a) Excitation geometry for FCS in an optofluidic ARROW chip. The excitation beam enters the liquid-core waveguide along the x-direction. Light is confined in the liquid and solid cores via dielectric multilayers shown in different shades of gray. Molecules within the excitation beam emit fluorescence that is collected in the z-direction. (b) Schematic view of optofluidic FCS chip. Liquid-core waveguide (pink) is connected to fluidic reservoir and optically coupled to solid-core waveguides (green). (c,d) SEM images of solid- and hollow-core cross sections. (Reproduced from Yin, D. et al., *Opt. Express*, 15, 7290, 2007. With permission.)

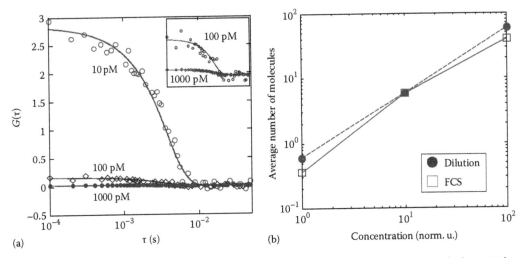

(a)   (b)

FIGURE 6.7 (a) FCS autocorrelation signal on ARROW chip for different concentrations of Alexa 647 dye. Symbols: experiment. Lines: model. Inset: magnification of 100 and 1000 pm data. (b) Number of molecules in excitation volume versus dye concentration. Squares: FCS data. Circles: integrated fluorescence detection. (Reproduced from Yin, D. et al., *Opt. Express*, 15, 7290–7295, 2007. With permission.)

6.5 Fluorescence Lifetime Imaging Microscopy

FLIM is a technique for determining the spatial distribution of excited state lifetimes in microscopic samples (van Munster et al. 2005). Typically, FLIM instruments are designed to measure lifetimes in the nanosecond (ns) range, since the lifetimes of most fluorophores used in modern fluorescence microscopy are within this range. FLIM has root in the fields of microscopy and fluorescence spectroscopy. Its key advantage is the combination of spatial resolution and high signal sensitivity. FLIM measurement can also be conducted in either frequency domain or time domain. In recent years, FLIM has become increasingly popular in the fields of biological and biomedical sciences.

The main principle behind FLIM is to combine microscopy with fluorescence lifetime measurement to form images that contain spatial and lifetime information about the sample. In terms of the techniques used to obtain lifetime information, FLIM was originally developed based on frequency domain measurement and subsequently evolved into the time domain, which is easier to understand conceptually. The microscopy methods are based on wide-field, confocal, multiphoton, structural illumination, or total internal reflection (TIR). In frequency domain FLIM measurements, the intensity of the excitation light is continuously modulated. Due to the finite lifetime of the fluorescence decay, the fluorescence emission will display a phase shift and a decrease in modulation with respect to the excitation light. The fluorescence lifetime, τ, can be determined by measuring the phase shift, denoted as τ_φ, or modulation decrease, denoted as τ_M. It can be shown that these two lifetimes can be expressed as (van Munster and Gadella 2005)

$$\tau_\varphi = \frac{\tan(\varphi\varphi_{em} - \varphi_{ex})}{\omega} \tag{6.9}$$

$$\tau_M = \frac{1}{\omega}\sqrt{\frac{1}{(M_{em}/M_{ex})^2 - 1}} \tag{6.10}$$

where

φ_{em} and φ_{ex} represent phase, and

M_{em} and M_{ex} represent modulation depth of the fluorescence emission and excitation light, respectively

ω is the angular frequency of modulation

In the case of a single exponential decay for the fluorescence, τ_φ and τ_M are equal. If the fluorescence decay contains multiple exponentials with different lifetime components, $\tau_\varphi < \tau_M$ (Gadella et al. 1993). Thus, it is easy to distinguish single from multiple exponential decays in frequency domain FLIM.

In time-domain FLIM, a short (relative to the fluorescence lifetime) pulse of light, usually from a pulsed laser with pulse duration shorter than the fluorescence lifetime, is used to excite the sample. The emitted fluorescence is measured directly in a time-resolved manner and the measured result is presented in terms of fluorescence intensity as a function of real time following the excitation (Lakowicz 1999). For example, time-domain measurements can be done using time-correlated single-photon counting (TCSPC) whereby the arrival time of the first photon after each pulse is monitored at very high time resolution (O'connor and Philips 1984). By recording the arrival times of a large number of photons, a representation of the decay curve is obtained. Another time-domain technique is the collection of photons in a fixed number of discrete time intervals using gated detection. For the case of a single exponential decay, provided the excitation light pulse width or duration is much shorter than the lifetime to be measured, τ, the emitted fluorescence intensity, I, as a function of time (t) is given by

$$I(t) = I(0)\exp\left(\frac{-t}{\tau}\right) \tag{6.11}$$

where $I(0)$ is the fluorescence intensity at $t=0$, defined as the time of the excitation. By fitting the measured data with Equation 6.11, one can obtain the fluorescence lifetime τ. If the excitation pulse is long, deconvolution is necessary to extract the fluorescence lifetime of interest from measured data. In this case, the excitation pulse shape needs to be determined first experimentally or assumed theoretically before deconvolution can be performed. If the dynamics are more complex than a simple monoexponential decay, a more complex function or multiple exponentials need to be used to fit the measured data and to extract the lifetimes desired. In measurements discussed so far, the time resolution is determined by the photodetector, which is usually much lower than the laser pulse duration, typically

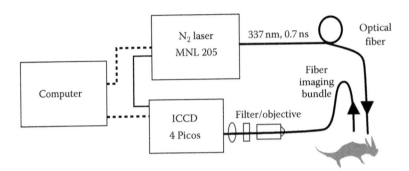

FIGURE 6.8 Schematic of a portable fiber-based FLIM microscope system developed at the Center for Biophotonics Science and Technology (CBST) in the University of California, Davis. GRIN for gradient index and ICCD for intensified charge-coupled device. (Adapted from the University of California, Davis Web site http://cbst. ucdavis.edu/research/summaries/fiber-based-flim-microendoscope/.)

on the order of tens to hundreds of picoseconds. Shorter time resolution can be achieved, close to that of the laser pulse width, by using more sophisticated techniques such as pump probe, where two pulses, usually originating from the same laser source, with controllable delay between them are used as excitation light (Fleming 1986). The fluorescence intensity is measured as a function of delay time between the two pulses and the time resolution is thereby determined by the cross-correlation between them, rather than by the photodetector. The time resolution can thus be easily on the subpicosecond time scale when femtosecond lasers are used. The limitation with such a pump-probe approach is the use of high power and expensive lasers and the likelihood of photo bleaching and photochemical reactions.

Almost all FLIM instruments have been custom-built, in whole or in part, for specific applications of interest. As a result, there is a large variety of FLIM instruments. What all FLIM instruments have in common include the excitation light, usually laser light intensity-modulated or pulsed in time, emitted fluorescence light measured in a time-resolved manner, and an optical microscope with optical detectors that allows for sample manipulation and image formation. Figure 6.8 shows a schematic diagram of a fiber-based FLIM microendoscope system developed at the University of California, Davis that can be used in the operating suite to characterize tissue in vivo, without the need for excisional biopsy.

FLIM has been successfully applied in optofluidics applications as demonstrated in probing the behavior of fluorescent molecules diffusing in submicrometer SiO_2 channels (Petrasek et al. 2007). Measurements reveal differences in fluorescence lifetimes compared to bulk solution that result from the effects of confinement and the presence of interfaces. Figure 6.9a shows fluorescence intensity and lifetime images of 2.5 µm wide channels filled with 10 nM Alexa 546 dye solution. The gradual change in the shape and dimensions of a bubble is reflected in the blurred edges of the bubbles. In the intensity images the bubbles appear 15–30 times brighter than the liquid-filled regions in the channels. The FLIM image and the histogram of lifetime values show that the fluorescence lifetime in the liquid-filled channels is constant within the precision of the measurement and has a mean value of 3.7 ± 0.3 ns, which is comparable with the value of 3.72 ± 0.1 ns obtained in the bulk solution. The lifetime in bubble regions is longer: $\tau_F = 4.2 \pm 0.2$ ns. Figure 6.9b shows fluorescence intensity and fluorescence lifetime images of 5 µm wide channels filled originally with 10 µM Alexa 546 solution. The brighter portions of the channels are bubble regions, while the darker zones are filled with liquid. The intensity ratio of bubble to liquid-filled region is approximately 3 in the vertical channel and 1.5 in the horizontal feed channel. The fluorescence lifetime in the liquid is 3.6 ± 0.1 ns, in agreement with the bulk value. The lifetime in the bubble region is 4.0 ± 0.1 ns and is independent of the bubble brightness.

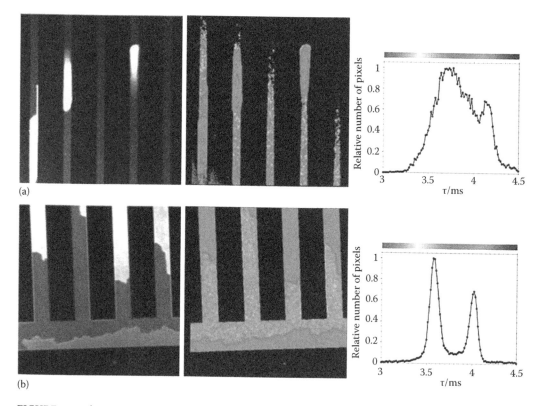

FIGURE 6.9 Fluorescence intensity (left) and lifetime (middle) images of 2.5 μm wide channels filled with 10 nM (a) and 10 μM (b) solution of Alexa 546 in water; (right) lifetime distribution in the lifetime image. Field of view: 58 μm. (Reproduced from Petrasek, Z. et al., *Microsc. Res. Tech.*, 70, 459, 2007. With permission.)

6.6 Total Internal Reflection Fluorescence

TIRF is another fluorescence-based spectroscopic technique proven to be useful for the study of surface and interface behavior of molecules, especially biological molecules, and their aggregates (Bright 1993; Tschmelak et al. 2005). It can provide *in situ*, real-time, nondestructive, and highly sensitive detection in microliter quantities (~10 nL minimum). This is particularly useful for expensive biological samples that are available in small quantities.

The key principle behind TIRF is total internal reflection of light at the interface of two media with different indexes of refraction when the incidence angle (θ_i) is greater than the critical angle (θ_c). When a beam of light propagating within a medium of RI n_1 encounters an interface with another medium of lower RI n_2, it can undergo TIR for $\theta_i > \theta_c$. While most of the incident light totally reflects at the interface, a small portion of the EM radiation penetrates through the interface into the medium with lower RI, usually lower density like a liquid. This is illustrated in Figure 6.10 (Rockhold et al. 1983). The intensity of this interfacial field, usually called the "evanescent wave," decays exponentially with distance from the interface.

The critical angle is given by

$$\theta_c = \sin^{-1}(n_2/n_1) \tag{6.12}$$

The penetration depth (d_p) of the interfacial evanescent wave is given by

$$d_p = \lambda_i / (2\pi n_1 [\sin^2 \theta_i - (n_2/n_1)^2]^{1/2}) \tag{6.13}$$

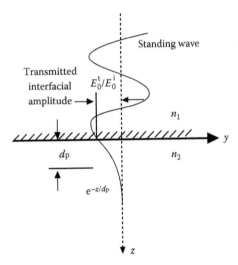

FIGURE 6.10 Schematic illustration of the surface wave in the y–z plane shows the standing wave pattern and exponential decay of the interfacial electric field amplitude into the optically less dense medium 2. E_0^i is the electrical field amplitudes at the interface in medium 2, and E_0^t is the transmitted electrical field amplitude. (Adapted from Rockhold, S.A. et al., *J. Electroanal. Chem.*, 150, 261, 1983.)

where λ_i is the incident light wavelength. If fluorophores are present in the lower RI medium, the evanescent wave can cause electronic excitation of the fluorophore when the wavelength is appropriate. Since the penetration depth is usually very small, on the order of wavelength of light and depending on the incident angle, only a small sample volume is excited and the excitation only occurs for molecules at or near the interface. This minimizes primary absorption effects. Also, the optical geometry can be set up in such a way that the fluorescence signal passes though the waveguide rather than the bulk solution, reducing any secondary absorption effects.

Commercial TIRF sample cells are available for easy use and for studying kinetics in a flow system. Such cells are usually compatible with standard fluorometers and convenient for adjusting the incident angle of light. Applications of TIRF include measurement of protein adsorption (kinetics ad isotherms, conformational changes, and surface mobility) (Rockhold et al. 1983; Shibata and Lenhoff 1992; Lassen and Malmsten 1996; Jennissen et al. 1999), immunoassay systems (e.g., antibody–antigen interaction) (Coille et al. 2002; Willard et al. 2003; Albrecht et al. 2008), and cell adhesion to surfaces (Fuhr et al. 1998; Patel et al. 2008).

While TIRF has been used widely as a sensitive analytical technique in various applications, to date, it has not yet been used directly in conjunction with optofluidic devices. Of course, the principle of TIR has been employed in optofluidic applications already. It can be anticipated that TIRF itself will be combined more directly with optofluidics in the future. In fact, it has been proposed theoretically that an immiscible fluid vesicle (e.g., air bubble) be used to generate nanoscopic fluid films of ca. 200 nm thickness on the surface of an interfacial TIRF sensor (Jennissen and Zumbrink 2004). The thickness of the liquid film can be easily probed and measured by evanescent wave technology. This nanofilm technique increases the mass transport coefficient and thereby eliminates the mass transport limitation associated with macromolecules in conventional sensors, making the binding rates reaction-rate limited. This proposed technique is yet to be demonstrated experimentally.

6.7 Infrared Vibrational Spectroscopy

IR and Raman are two common vibrational spectroscopy techniques useful for characterizing structural properties such as vibrational frequencies of molecules and phonons as well as crystal structures of solids. Since they often have different selection rules for transitions, they are usually complementary to each other. For IR absorption to occur, the electronic transition dipole moment must change with vibration, i.e., $(\partial\mu/\partial Q)_0 \neq 0$, where μ is the electrical dipole moment and Q represents the coordinate or displacement for a specific vibrational mode. This mode is called "IR active." Otherwise, the mode is IR inactive and no absorption will occur at the vibrational frequency of that mode. For Raman transitions to occur, the polarizability, α, must change with vibration, i.e., $(\partial\alpha/\partial Q)_0 \neq 0$, and the mode is called "Raman active." Otherwise, the mode is Raman inactive and no Raman scattering can take place at the corresponding vibrational frequency. Raman spectroscopy will be discussed later.

IR spectroscopy is based on the measurement of transmitted IR light through a sample. The absorbance measured as a function of frequency contains information about the vibration or phonon modes or frequencies of the sample. The key components for an IR spectrometer are similar to that of UV–visible spectrometer except that the light is in the IR and the detector and optical components, such as gratings and mirrors, all need to be appropriate for IR light. Various commercial spectrometers, including Fourier transform IR (FTIR), are available in many research institutions.

The sample for IR spectroscopy measurement needs to be thin or dilute enough so saturation can be avoided and Beer's law is valid, similar to UV–visible spectroscopy. For molecules, the IR spectrum reflects their vibrational modes, determined by selection rules. For solids, the IR spectrum measures the photon modes, also governed by appropriate selection rules that are determined by their crystal symmetry properties.

IR absorption has been used in conjunction with optofluidic applications. For example, microfluidic laser intracavity absorption spectroscopy has been demonstrated with mid-IR ($\lambda \approx 9\,\mu m$) quantum cascade lasers (Belkin et al. 2007). A deep-etched narrow ridge waveguide laser is placed in a microfluidic chamber, with the evanescent tails of the laser mode penetrating into a liquid on both sides of the ridge. The absorption lines of the liquid modify the laser waveguide loss and result in significant changes in the laser emission spectrum and the threshold current.

In another study, IR spectra have been measured based on multiple internal reflection IR waveguides embedded with a parallel array of nanofluidic channels (Oh et al. 2008). The channel width is maintained substantially below the mid-IR wavelength to minimize IR scattering from the channel structure and to ensure TIR at the channel bottom. The flow rate of the liquid in the channels is controlled by a DC gate potential on a field effect transistor (FET). Manipulation of the surface charge on the SiO_2 channel bottom and sidewalls changes their ζ-potential. Using the IR technique, one can probe the vibrational modes of dye molecules, as well as those of the solvent. The observed IR absorbance accounts for the amount of dye molecules advancing or retracting in the nanochannels, as well as adsorbing to and desorbing from the channel bottom and sidewalls. Figure 6.11 shows sample spectra taken every 5 min with $2\,cm^{-1}$ resolution while the Rhodamine B dye molecules flow into the nanochannels due to electroosmosis and respond to the gate bias.

6.8 Raman Spectroscopy and SERS

6.8.1 Normal Raman Scattering

As mentioned earlier, Raman spectroscopy, named after the Indian scientist Raman who first discovered it in 1928, is based on the phenomenon of inelastic light scattering. In a Raman scattering measurement, a single frequency light, usually from a single-mode laser source, shines on the sample and scattered light is measured off angle with respect to the incident light to minimize Rayleigh scattering. The inelastically scattered light with lower (Stokes scattering) or higher (anti-Stokes scattering) frequencies can be measured with a photodetector. The energy difference between the scattered and the incident light, the so-called Raman shift (usually given in wave number $cm^{-1} = 1/\lambda$ with wavelength, λ, expressed in centimeters), equals the vibrational or phonon frequencies of the sample, as long as selection rules allow. The spectrum is usually presented in terms of intensity of the Raman scattered light as a function of Raman shift.

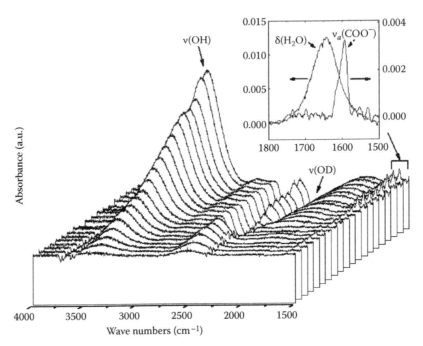

FIGURE 6.11 A time-series IR absorbance spectra of Rhodamine B in D_2O taken during FET flow control. The background spectrum is taken after filling the nanochannels with a D_2O buffer solution. The sample spectra are taken every 5 min, while Rhodamine B flows into the nanochannels by electroosmosis. The O–D stretching vibrational modes of D_2O at 2800–2200 cm^{-1}, as well as O–D and O–H stretching vibrational modes of HDO at 2600–2500 and 3600–3200 cm^{-1} are observed. The inset shows that vibrational mode associated with the COO_2 functional group of Rhodamine B at 1590–1600 cm^{-1} is distinguishable from H_2O scissoring mode at 1640 cm^{-1}. (Reproduced from Oh, Y.J. et al., *Lab Chip*, 8, 251, 2008. With permission.)

Raman spectrometers are commercially available from several sources, e.g., Thermal, Horiba, Bruker, Jobin-Yvon, and Renishaw. The basic setup is similar to that of a PL spectrometer in that Raman scattered light is detected when a sample is illuminated. It should be noted that even though the Raman shift typically appears in the region of a few hundred to a few thousand cm^{-1}, similar to that of IR measurement (a few to a few tens of microns), the signal detected is usually in the visible if the incident light is in the visible. If UV light were used, the Raman signal would be in the UV or near UV. The Stokes Raman signal is most often detected and it always appears lower in energy or frequency compared to the incident light frequency. Since visible detectors are usually more sensitive than IR detectors, Raman detection is therefore considered more sensitive than IR detection. In addition, because Raman is a zero-background measurement while IR is not, Raman should be more sensitive than IR in this regard as well. However, the Raman scattering cross section or QY is often very low, on the order of 10^{-6} to 10^{-8}, a relatively high intensity light source, such as a laser, is usually needed and the signal detection require long data acquisition time.

The Raman signal can be enhanced by using light frequency that is on resonance with an electronic transition of the sample, namely "resonance Raman scattering." While the signal can be enhanced by a factor of 100–1000, sample degradation often occurs due to the electronic absorption of the incident light and subsequent photochemical reactions. The resonance enhancement is due to the fact that the Raman signal depends on the "detuning" or energy difference between the incident photon frequency and the energy difference between the electronic and ground electronic state. As this energy difference becomes smaller (more on resonance), the Raman signal becomes larger (McHale 1998).

As mentioned earlier, Raman signal can often be observed in practical PL measurements, especially when the PL signal is weak. The Raman signal from the embedding medium, such as solvent, can also be easily seen due to the very high concentration. Raman peaks are often narrower than PL peaks, especially for most NPs. A simple way to verify if a sharp peak is Raman or PL is to change the excitation light wavelength slightly, say 10–20 nm, and observe if the peak position changes accordingly.

To date, there has been only limited studies reported on using normal Raman spectroscopy in conjunction with microfluidic devices (Leung et al. 2005). This is possibly due to the small signal of normal Raman scattering and the small sample volume in microfluidic devices. However, there have been several reports on combining SERS with microfluidic devices, as discussed next. In this one study reported, a microfluidic reaction system operating in continuous flow is used in conjunction with confocal Raman microscopy to afford rapid molecule synthesis and product quantification. As a result, the approach allows for rapid reaction optimization within a continuous flow system. This was demonstrated by the catalytic oxidation of isopropyl alcohol to acetone using tetra-N-propylammonium perruthanate/N-methylmorpholine N-oxide in a radial interdigitated micromixer as a model reaction system. A more detailed discussion of Raman detection in microchips and microchannels will be given in Chapter 17 by Benford et al.

6.8.2 Surface-Enhanced Raman Scattering

SERS is based on the enhancement of Raman scattering of an analyte molecule near or on a roughened metal substrate surface. It is an extremely useful method for chemical and biochemical analysis and detection. The origin of the Raman enhancement is largely due to an enhanced EM field at the metal surface due to increased absorption of the incident light. This is related to several other surface enhancement phenomena, including surface plasmon resonance (SPR) and surface-enhanced fluorescence (Matveeva et al. 2005, 2007; Haes et al. 2006; Sherry et al. 2006; Fu et al. 2007; Meli and Lennox 2007; Ray et al. 2007; Stancu et al. 2007; Toderas et al. 2007; Willets and Van Duyne 2007; Zhang et al. 2007b,c).

Normal Raman scattering, as discussed in Section 6.8.1, is a widely used spectroscopic technique for chemical analysis, detection, and imaging. Its key advantage is molecular specificity or selectivity that allows for unique identification of samples. The major limitation of normal Raman is the very small signal or low QY of scattering (on the order of 10^{-7}) (Ferraro and Nakamoto 1994). SERS, as discovered in the middle 1970s, has made Raman or SERS more popular for practical applications due to the combined advantages of molecular specificity and large signal (Fleischmann et al. 1974; Albrecht and Creighton 1977; Jeanmaire and Vanduyne 1977; Moskovits 1978; Tian et al. 2007; Wang et al. 2008). Since the initial discovery, extensive experimental and theoretical work has been conducted on both understanding the fundamentals of SERS and its practical applications for chemical and biochemical analysis. With enhancement of many orders of magnitude compared to normal Raman scattering, SERS allows measurements with very low analyte concentration, low laser intensity, and short data collection time, with detection of single molecules suggested as a possibility (Kneipp et al. 1997, 1999; Nie and Emery 1997; Michaels et al. 2000; Jiang et al. 2003). However, due to dependence of the SERS activity on the metal substrate structure (e.g., distance from the surface, orientation and conformation of the molecules, and strength of interaction between the molecule and substrate), a SERS spectrum can differ substantially from normal Raman spectra for the same molecule in terms of peak intensity distribution and particular vibrational modes detected. Therefore, care needs to be taken to interpret SERS spectra (Seballos et al. 2006, 2007).

In practical SERS measurements, several critical factors need to be taken into consideration. The first and most important is the distance between the molecule and the SERS substrate, usually silver or gold, and, to a lesser degree, other metals (Moskovits 1985). The closer the molecule is to the surface, the stronger the SERS intensity, given all other factors being the same. To date, there is not a simple analytical mathematical equation to relate SERS intensity to distance. As a matter of fact, different distance dependences of SERS have been reported experimentally and rationalized theoretically using different models. This is in part because it is actually challenging to unambiguously determine the distance dependence of SERS due to complications caused by changes in other factors when distance is varied on the atomic scale (Seballos et al. 2006, 2007). One theoretical model based on the mechanism of EM enhancement has suggested a SERS distance dependence approximately as (Kennedy et al. 1999)

$$I = \left(1 + \frac{r}{a}\right)^{-10} \tag{6.14}$$

where
   I is the intensity of the Raman mode
   a is the average size of the field enhancing features on the surface
   r is the distance from the surface to the adsorbate

This theoretical prediction has been corroborated by recent experimental SERS study of pyridine adsorbed on Ag film over nanospheres (AgFON) using Al_2O_3 films as a spacer to vary the distance between the molecules and the AgFON substrate surface (Dieringer et al. 2006).

A second important factor is the location of the analyte molecule on the substrate surface. As an example, for spherical particles with incident laser light polarized in a particular direction, the EM field enhancement is anisotropic. Therefore, the location of the molecule will significantly affect how strong the SERS will be for a given molecule. The third factor is the orientation of the molecule with respect to the surface normal of the metal substrate. Given the anisotropic EM field distribution on the substrate surface, the SERS activity depends strongly on the orientation of the molecule. The strongest enhancement will occur when the polarizability or, to a good approximation, the dipole moment of the molecule is parallel to the electrical field vector of the enhanced light field. Finally, the strength of interaction between the molecule and the substrate is also important to SERS. Stronger interaction usually results in shorter distance between the molecule and the substrate, and thereby stronger SERS. When the interaction is very strong, charge transfer between the molecule and metal substrate can take place, which is often used to explain the "chemical enhancement."

In addition, SERS spectra are clearly sensitive to the structure of the metal substrate. The size and shape as well as their distribution will affect the sensitivity as well as consistency of SERS measurement. Much effort has been made to improve SERS spectra consistency in practical measurement by developing metal substrates that are uniform or monodisperse in size and shape. For example, recently a small SERS probe using hollow gold nanospheres (HGNs or HAuNSs) has been developed (Schwartzberg et al. 2006a). The unique optical and structural properties of HGNs afford several advantages for SERS applications, including small size, spherical shape, tunability, and narrow absorption from the visible to near IR (Schwartzberg et al. 2006a). Furthermore, compared to aggregates or NPs with inhomogeneous distribution in size and shape (Michaels et al. 2000; Jiang et al. 2003; Talley et al. 2004), the HGNs afford SERS spectra that are highly consistent measured by the peak ratio of SERS spectra of single HGNs. Figure 6.12 shows a comparison of SERS spectra of mercapto benzoic acid (MBA) obtained using HGNs vs. normal silver nanoparticle aggregates. The inset shows a histogram detailing the distribution of the peak ratio of two SERS peaks for the two different substrates. The HGNs show much narrower and thus better distribution (lighter bars in the middle), indicating highly uniform structure and narrow SPR, and thereby consistent SERS for single-nanostructure SERS (Schwartzberg et al. 2006b). The high consistency is a direct result of the narrow plasmon absorption band and high structural uniformity.

One of the most intriguing issues in SERS studies is whether one can observe SERS from a single molecule. While it is reasonably easy to establish single-nanoparticle SERS, it is often challenging to prove beyond doubt single-molecule SERS (Kneipp et al. 1997; Doering and Nie 2002; Otto 2002; Kneipp and Kneipp 2005). A recent experiment based on isotope substitution is perhaps the most elegant to show that it is possible to observe SERS from a single molecule. This is demonstrated using two isotopologues of Rhodamine 6G (R6G) that offer unique vibrational signatures. When an average of one molecule was

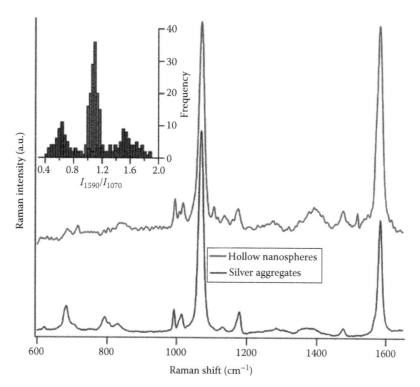

FIGURE 6.12 Single-particle SERS spectrum of MBA on HGSs (top trace) and silver aggregates (bottom trace). The inset is a histogram of the relative intensity of the two most intense peaks of MBA at 1070 and 1590 cm^{-1} of 150 HGNs (lighter bars in the middle) and 150 silver aggregates (darker bars on the two sides). (Reproduced from Schwartzberg, A.M. et al., *Anal. Chem.*, 78, 4732, 2006b. With permission.)

adsorbed per silver nanoparticle, only one isotopologue was typically observed (Dieringer et al. 2007). This was observed even when the original sample was a mixture of unsubstituted and substituted R6G. This is essentially a proof that only one R6G molecule was detected by SERS.

SERS detection based on an optofluidic chip has been successfully demonstrated in several studies (Keir et al. 2002; de Mello and Beard 2003; Connatser et al. 2004; Song et al. 2006; Measor et al. 2007; Piorek et al. 2007; Wang et al. 2007; Connatser et al. 2008). In an early study, surface-enhanced resonance Raman scattering (SERRS) of a model derivative of TNT was detected using a microflow cell designed within the framework of the lab-on-a-chip concept (Keir et al. 2002). The SERRS substrate, silver colloid, was prepared *in situ*, on-chip, by borohydride reduction of silver nitrate. Quantitative behavior was obtained over 4 orders of magnitude with a detection limit of 10 fmol, which is between 1 and 2 orders of magnitude better than that achieved using a macroflow SERRS cell.

In a recent study, interconnected solid and liquid-core ARROWs form a planar beam geometry that allows for high mode intensities along microfluidic channels containing molecules optimized for SERS (Measor et al. 2007). The excitation power and concentration dependence of SERS from R6G molecules adsorbed to silver NPs was systematically studied. Representative data are shown in Figure 6.13. The data can be described by a model that takes into account the microphotonic structure. Detection sensitivity to a minimum concentration of 30 nM was found, demonstrating the suitability of ARROW-based optofluidic chips for high sensitivity detection with molecular specificity. Related SERS sensors or probes have been developed based on D-shaped and photonic crystal fibers recently (Zhang et al. 2007d; Shi et al. 2008a,b). The fibers provide a convenient, compact, and low-cost alternative platform for SERS sensors.

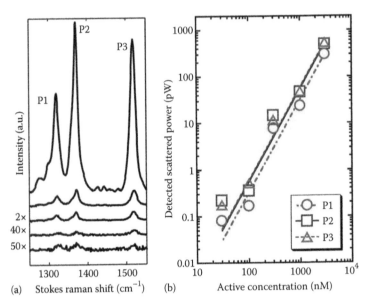

FIGURE 6.13 (a) Detected SERS signals at different active concentrations corrected to a nominal input excitation irradiance level of 15.2 kW/cm² for differing alignments between measurements. The lowest three concentrations are multiplied by factors of 2, 40, and 50 as denoted. (b) Symbols: detected SERS power versus concentration. Lines: quadratic fits. (Reproduced from Measor, P. et al., *Appl. Phys. Lett.*, 90, 211107, 2007. With permission.)

6.9 Surface Plasmon Resonance

SPR refers to an analytical technique based on SPR absorption of metal films or nanostructures. SPR absorption occurs when the collective electron oscillation frequency of a metal surface or nanostructure is resonant with the EM radiation, e.g., light. This will be discussed in more detail in Chapter 18 by Sinton et al. SPR as an analytical method has been used for optical detection of biomolecules such as DNA and proteins. It involves measuring spectral shifts or changes in SPR due to binding or conjugation of biomolecules onto a metal nanoparticle surface. These changes can be used to detect or image biomolecules (Heaton et al. 1995; Rella et al. 2004; Li et al. 2006a; Stancu et al. 2007; Beusink et al. 2008; Yuk et al. 2008). For example, SPR change has been used in a homogeneous assay for DNA detection based on reducing the interparticle distance through binding of DNA (Elghanian et al. 1997; Storhoff et al. 2002). In this approach, two batches of gold NPs, each modified with a different oligonucleotide were mixed and single-stranded DNA was added that was complementary to both oligonucleotides immobilized on the gold NPs. The gold particles were connected into a network and the interparticle distance was decreased, resulting in color change. Interestingly, this color change is reversible by heating the solution above the melting temperature of the formed DNA duplex. Different variations of such approach have been developed for detecting DNA, including single-base mismatch discrimination (Li and Rothberg 2004; Li et al. 2005). Other spectral changes such as appearance of new absorption peaks due to strong aggregation caused by binding of biomolecules can also be used to determine the biomolecule of interest. SPR-based techniques are usually not molecule specific and rely on prior information for identifying the molecules detected.

A typical SPR instrument measures changes in the surface plasmon absorption band of a thin metal layer or metal nanostructures caused by interaction with a target analyte of interest. When polarized light hits a prism covered by a thin metal, usually gold, layer and under certain conditions, including wavelength, polarization, and incidence angle, free electrons in the metal absorb the incident light photons and convert them into surface plasmon waves. A dip in reflectivity of the light is observed under these SPR conditions. Perturbations to the metal surface, such as an interaction with target analyte molecules of interest, induce a modification of resonance conditions, which in turn results in a change in reflectivity. Measurement of this change in reflectivity is the basis for the SPR technique, which is used widely in the form of biochips for quantification and analysis of biological molecules (Abiko et al. 2007; Chen et al. 2007; Du and Wang 2007; Dong et al. 2008). Improvements in these techniques using metal nanostructures, instead of thin films, have been explored (Petrova et al. 2007a). SPR has been further developed in an imaging technique that is sensitive and label-free for visualizing the whole of the biochip via a video CCD camera (Wilkop et al. 2004; Li et al. 2006b; Yuk et al. 2006). The biochips can be prepared in an array format with each active site providing SPR information simultaneously. It captures all of the local changes at the surface of the biochip and provides detailed information about molecular binding, biomolecular interactions, and kinetic processes (Petrova et al. 2007b).

SPR has been used in conjunction with micro-optical fluidic systems in a number of ways. For instance, since SPR can be generated by evanescent waves, bent polished fibers (Slavik et al. 1999), tapered or cladding-stripped fibers (Tubb et al. 1997), or integrated optical waveguides (Weiss et al. 1996; Mouvet et al. 1997; Dostalek et al. 2001; Wang et al. 2004) with metal coatings on their surface have been demonstrated as very compact biosensors. The incident light and detected light are both transmitted through the fiber, with the SPR triggered by the incident light at the gold-coated tapered surface. Near-field scanning optical microscopy (NSOM) using nanosized apertures based on SPR has been used for the detection of DNA sequences in conjunction with a specially designed microfluidic system based on an array of solid-state nanopores (Chansin et al. 2007a).

6.10 Refractometry and Polarimetry

"Refractometry" is the method of measuring the index of refraction of substances, denoted as n. The index of refraction of a substance is defined as the speed of light in vacuum divided by the speed of light in the substance of medium, which is a dimensionless quality larger than 1. Commercial instruments, called refractometers, are available for measuring RI of mostly liquids, and to a lesser degree, gases and solids. Its main application is to determine the composition, concentration, or purity of a material. The RI of a substance is strongly dependent on temperature and the wavelength of light. The variation with wavelength is called dispersion, which is what causes separation of different wavelengths of light by a prism. For most liquids, the RI decreases with increasing temperature. Thus, RI measurements are usually reported at a reference temperature of 20°C and reference wavelength of 589.3 nm (the sodium D line). For example, the RI of water at 20°C taken at the sodium D Line is 1.3330 and is often reported as $1.3330n_D^{20}$ or $1.3330n_D^{20}$, with n for RI, D for sodium D line, and 20 for 20°C.

Refractometry has been applied to study live cells by measuring the average RI associated with live cells confined in microfluidic channels (Lue et al. 2006). Similarly, a nano-optical probe has been used for real-time RI measurement of single live cell (Lee et al. 2008). The probe was made of a tapered fiber tip coated with 13 nm diameter gold NPs. The evanescent wave near the tip excited localized SPR of gold NPs, and the RI of the cell can be determined by measuring the localized plasmonic effect of the probe (Lee et al. 2008). In another recent study, a diffraction grating under TIR was used to study monodisperse particle dispersions passing through an integrated microfluidic channel (Sarov et al. 2008). The device has been used to accurately determine in real time the specific RI for the nanoparticle suspension and the nanoparticle concentration. The effective RI can be determined with an accuracy of 7×10^{-4} for the polymeric and 2×10^{-4} for the metallic and ferromagnetic dispersions.

"Polarimetry" is the method for measuring the polarization of transverse waves, usually EM waves such as light (Purdie and Swallows 1989). The measurement is typically done on EM waves that have traveled through or reflected, refracted, or diffracted from some material as a way to characterize that material. The instrument used to perform the polarization measurement is called a "polarimeter," which is commercially available (e.g., Rudolph and Anton Paar). Polarimetry of thin films and surfaces is commonly referred to as ellipsometry. Similar to RI, many of the properties measured by a polarimeter are temperature dependent, and the effect of temperature thus needs to be taken into account. Likewise, the measured results are dependent on the wavelength of light used. For example, the optical rotatory dispersion (ORD) is a measure of the optical rotation as a function of wavelength.

Anisotropic crystalline solids, and samples containing an excess of one enantiomer of a chiral molecule, can rotate the orientation of plane-polarized light. Such substances are said to have "optical activity." Measurement of this change in polarization orientation by polarimetry is useful for studying the structure of anisotropic materials and for determining the purity of chiral mixtures. Optical rotation occurs because optically active samples have different refractive indices for left- and right-circularly polarized light. In other words, left- and right-circularly polarized light travel through an optically active sample at different velocities.

Linearly or plane-polarized light is the superposition of equal intensities of left- and right-circularly polarized light. As plane-polarized light travels through an optically active sample, the left- and right-circularly polarized components travel at different velocities. This difference in velocities creates a phase shift between the two circularly polarized components when they exit the sample. Summing the two components still produces linearly polarized light, but at a different orientation from the light entering the sample.

The simplest polarimeter consists of a monochromatic light source, a polarizer, a sample cell, a second polarizer, which is called the analyzer, and a light detector. The analyzer is oriented 90° to the polarizer so that no light reaches the detector. When an optically active substance is present in the beam, it rotates the polarization of the light reaching the analyzer so that there is a component that reaches the detector. The angle that the analyzer must be rotated to return to the minimum detector signal is the optical rotation, α. The amount of optical rotation depends on the number of optically active species through which the light passes, and thus depends on both the sample path length and the analyte concentration.

The application of polarimetry in optofluidics has been limited so far. A somewhat related example is a micropolarimeter with a 40 nL probe volume based on a capillary (Swinney et al. 2001). The optical configuration consists of two polarizing optics, a capillary, a laser source, and a photodetector. This unique polarimeter is based upon the interaction of a linearly polarized laser beam and a capillary tube, with an inner diameter of 250 μm. Side illumination of the tube results in a 360° fan of scattered light, which contains a set of interference fringes that change in response to optically active solutes. Solutes that exhibit optical activity can be quantified and detected by analyzing the polarization state of the backscattered light.

6.11 Single-Molecule Spectroscopy

Most spectroscopy studies of samples have been carried out on large numbers of molecules. The properties measured are thus ensemble averages or sums of the properties of individual molecules. Due to heterogeneous distributions in size, shape, environment or embedding medium, and surface properties, the spectrum measured is thus inhomogeneously broadened. This results in loss of spectral information and complications in the interpretation of the results. One way to solve the above problem is to remove the heterogeneity by conducting the measurement on single molecules (Moerner 1994; Basche et al. 1997). In this approach, signal is detected only from one molecule at a time, and such measurement based on spectroscopy is termed "single-molecule spectroscopy" (SMS). Most SMS studies have been based on detection of emission or scattering from single molecules even though absorption measurements have also been demonstrated (Gerhardt et al. 2007).

To date, there are several demonstrations on the use of SMS in conjunction with optofluidic devices, particular with FCS (Foquet al. 2004; Bi et al. 2006; Yin et al. 2007). As discussed in Section 6.4, FCS combined with the optofluidic liquid-core (LC)-ARROW platform has been suggested to be able to detect, analyze, and manipulate molecules and biological particles with single-molecule sensitivity (Yin et al. 2006, 2007; Schmidt and Hawkins 2008). In a similar study, microfluidic channels with two lateral dimensions smaller than 1 μm were fabricated in fused silica for high sensitivity single-molecule detection and FCS (Foquet et al. 2004). Driven by electrokinetic forces, analytes could be flowed rapidly through the observation volume, drastically increasing the rate of detection events and reducing data acquisition times. Bursts of fluorescence from single fluorophores could easily be observed in these nanochannels. Figure 6.14a shows fluorescence time traces from single molecules moving past the observation volume, positioned inside the narrow channel, for four different applied electric fields. Figure 6.14b compares burst size histograms corresponding to single-molecule detection recordings at three different electric fields shown in Figure 6.14a. An increase in the burst frequency by almost two orders of magnitude was achieved while the average burst size was reduced only slightly, attributed to excitation saturation.

As another example of SMS, DNA translocation events have been detected through an array of solid-state nanoholes on an aluminum/silicon nitride membrane (Chansin et al. 2007b). The DNA strands are electrokinetically driven through the holes and during the translocation process, the molecules are confined to the walls of the nanofluidic channels, allowing 100% detection efficiency. Figure 6.15 shows a schematic of the experimental setup used for the measurement. With the opaque aluminum layer acting as an optical barrier between the illuminated region and the analyte reservoir, high-contrast imaging of single-molecule events can be performed. The efficiency of the approach was demonstrated using a 10 pM fluorescently labeled λ-DNA solution as a model system to detect simultaneous translocation events. Single-pore translocation events were also successfully detected using single-point confocal spectroscopy. Other SMS methods in relation to optofluidics, e.g., SMS based on zero-mode waveguides, will be discussed in Chapter 16. A more detailed discussion of SMS will be given in Chapter 13.

6.12 Optical Resonators

Optical resonators, including microcavity resonators and FP resonators, have been commonly used for biomolecule analysis applications (Armani et al. 2003; White et al. 2006; Liu et al. 2008). In an optical ring resonator, light propagates along a curved surface in WGMs as a result of TIR. The guided light interacts repeatedly with the sample on the surface of the ring through the evanescent field. Since the WGM resonant wavelength (λ) is determined by the effective index of refraction, n_{eff}, that depends on the sample on the surface of the ring resonator, the change in n_{eff} can thus be used to detect the sample:

$$\lambda = \frac{2\pi r n_{eff}}{m} \tag{6.15}$$

where r is the ring outer radius and is an integer related to the WGM angular momentum. Even though the size of the microcavity ring is usually small, it provides sensing capabilities often better than an optical waveguide.

For example, label-free single-molecule detection of interleukin-2 in serum has been demonstrated with a dynamic range of 10^{12} in concentration using a microtoroid resonator with a silica surface for binding and ultrahigh quality factor ($>10^8$) whispering gallery microcavity (Armani et al. 2007). Similar to microsphere or planar ring resonators, liquid-core optical ring resonator (LCORR) can also be considered as a ring resonator in which a micron-sized glass capillary is used as a ring resonator for the sample while an optical fiber positioned perpendicular to the capillary is used to couple light into the LCORR wall through evanescent field (White et al. 2007). LCORR has been used successfully in the detection of bovine serum albumin absorbed onto the surface of the LCORR and activated with 3-aminopropyl-trimethosiloxane (Zhu et al. 2007). LCORRs are discussed in more detail in Chapter 11 by Suter and Fan.

An FP resonator based on a pair of gold-coated optical fibers has been used in a micro-optical fluidic system for measuring the RI of a single Mardin-Darby canine kidney (MDCK) cell trapped by a micropipette (Song et al. 2006). Significant improvement in precision and stability was subsequently made by using a pair of FP gratings without any coating for optically trapping the cell inside the cavity by laser force, since the cell can be perfectly aligned along the cores of the fibers (Barer et al. 1953; Liang et al. 2007).

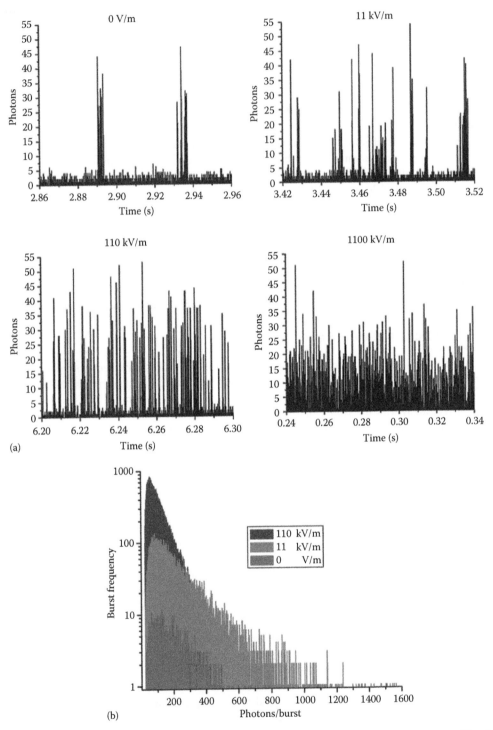

FIGURE 6.14 Single-fluorophore detection in nanochannels. (a) Time traces of fluorescence intensity (50 μs bin size) from 2 nM Alexa Fluor 488 5-dUTP (molecular probes) for four different applied electric fields and (b) corresponding histograms of burst size distributions (2 min acquisition), showing the large increase in burst frequency with increasing flow speed or higher driving voltage. (Reproduced from Foquet, M. et al., *Anal. Chem.*, 76, 1618, 2004. With permission.)

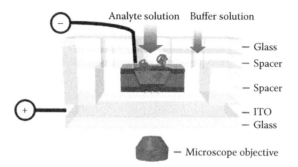

FIGURE 6.15 Schematic of experimental setup used to monitor DNA translocation events using nanopore membranes. Excitation light from a microscope objective is blocked by the membrane with the fluorescence signal from a translocating molecule collected by the same objective and directed to the electron multiplying CCD camera. The cover slip is coated with a conductive layer of ITO. Solid-state nanopores have an average diameter of 300 nm. (Reproduced from Chansin, G.A.T. et al., Towards ultra-fast parallel DNA analysis: Sub-wavelength metallic nanopore arrays for high-throughput single molecule spectroscopy, *MicroTAS*, Paris, 2007b. With permission.)

In addition, photonic crystal resonators have been explored as a platform for sensing applications (Loncar et al. 2003; Lee and Fauchet 2007). In photonic crystals, the EM field is localized in the low RI region, i.e., the air pores. When molecules are introduced onto the pore walls, the resonant peak will shift as a result of change in the effective central pore diameter or index of refraction. This shift is highly sensitive to molecules immobilized on the pore wall and can thus be used to detect the molecules of interest.

6.13 Dynamic Light Scattering

DLS is a relatively simple spectroscopic technique based on Rayleigh scattering to determine the size of nanoscale objects such as particles, aggregates, and large molecules, e.g., polymers and proteins (BarZiv et al. 1997; Kamyshny et al. 2002; Shukla et al. 2004; Chakraborty et al. 2005; Xie et al. 2007). When light hits particles that are small compared to the wavelength of light, a major scattering mechanism is elastic Rayleigh scattering. DLS is based on measuring the speed at which particles move under Brownian motion by monitoring the intensity of light scattered by the sample at some fixed angle. Brownian motion causes constructive and destructive interference of the scattered light.

Technically, DLS uses the method of autocorrelation to uncover information contained in the light intensity fluctuations. Autocorrelation measures how well a signal matches a time-delayed version of itself, as a function of the amount of time delay. Faster decay in the autocorrelation function indicates faster motion of the particles. The autocorrelation function can be related to the hydrodynamic radius (R_H) of the particles, which is the information of interest, by the Stokes–Einstein equation (Wuelfing et al. 1999). Larger particles correspond to slower Brownian motion and slower decays of the autocorrelation function.

One way to quantify the light intensity fluctuation is to use a second order correlation function defined as

$$g^{(2)}(\tau) = \frac{\langle I(t)I(t+\tau) \rangle}{\langle I(t) \rangle^2} \tag{6.16}$$

where

$I(t)$ is the intensity of scattered light at time t, and the brackets indicate averaging over all time t

τ is the delay time between two intensity measurements, $I(t)$ and $I(t+\tau)$

For a monodisperse sample, the second order correlation function can be analyzed based on the following equation (Berne and Pecora 2000):

$$g^{(2)}(\tau) = B + \beta \exp(-2\Gamma\tau) \tag{6.17}$$

where
  B is the baseline at infinite delay
  β is the correction function amplitude at zero delay minus the baseline
  Γ is the decay rate or constant of the correlation function

By fitting the experimentally measured correlation function using Equation 6.15, these parameters, including Γ, can be obtained. One can then calculate the diffusion constant or coefficient, D, of the particle via the relation

$$D = \frac{r}{q^2} \tag{6.18}$$

where q is the magnitude of the scattering vector given by

$$q = \frac{4\pi n_0}{\lambda_0} \sin\left(\frac{\theta}{2}\right) \tag{6.19}$$

where
  n_0 is the solvent index of refraction
  λ_0 is the vacuum wavelength of the incident light
  θ is the scattering angle

For a given experimental setup, q can be calculated, and thus D can be obtained based on Equation 6.16. Finally, the hydrodynamic radius of the particle, R_H, of a diffusing spherical particle can be calculated from the Stokes–Einstein relation:

$$R_H = \frac{kT}{6\pi\eta D} \tag{6.20}$$

where
  k is Boltzmann's constant
  T is the temperature in K
  η is the solvent viscosity

DLS has been used frequently to determine the size of inorganic, organic (polymer), or biological NPs and aggregates (Khlebtsov et al. 2003; Dickson et al. 2006; Xie et al. 2007). To date, however, there has been no report on integrating DLS with optofluidic devices. Such integration is conceivable in the future since optofluidic devices can conceivably offer a convenient and compact platform for DLS measurement. Other applications of light scattering include flow cytometry for which optofluidic devices exist. Flow cytometry is discussed in more detail in Chapter 19 by Chen et al.

6.14 Summary

In this chapter, we have discussed some of the most commonly used optical spectroscopy techniques including electronic absorption, luminescence (PL, FCS, TIRF, and FLIM), IR, SPR, Raman/SERS, SMS, and DLS. Many of these techniques, e.g., fluorescence and SERS, have extremely high sensitivity and allow for single-molecule detection. Most of them have been successfully exploited for optofluidics applications, albeit using mostly bulk optics. The current trend is moving toward a more complete optical integration, as discussed in several chapters in this book, especially chapters in Part II. Selection of these different techniques for specific applications depends on the information of interest. Both the fundamental principle of operation and application examples have been presented. Some of the spectroscopic methods presented are yet to be applied to optofluidic devices; however, it is expected to be simply a matter of time before they will merge. The noninvasive nature and typically high sensitivity of most optical spectroscopy methods, when combined with the compactness and flexibility of optofluidic platforms, are expected to significantly change the way chemical and biochemical analyses will be conducted in the future.

References

Abiko, F., K. Tomoo, A. Mizuno et al. (2007). Binding preference of eIF4E for 4E-binding protein isoform and function of eIF4E N-terminal flexible region for interaction, studied by SPR analysis. *Biochemical and Biophysical Research Communications* 355(3): 667–672.

Albrecht, M. G. and J. A. Creighton (1977). Anomalously intense Raman-spectra of pyridine at a silver electrode. *Journal of the American Chemical Society* 99(15): 5215–5217.

Albrecht, C., N. Kaeppel, and G. Gauglitz (2008). Two immunoassay formats for fully automated CRP detection in human serum. *Analytical and Bioanalytical Chemistry* 391(5): 1845–1852.

Aragon, S. R. and R. Pecora (1976). Fluorescence correlation spectroscopy as a probe of molecular-dynamics. *Journal of chemical Physics* 64(4): 1791–1803.

Armani, D. K., T. J. Kippenberg, S. M. Spillane, and K. J. Vahala (2003). Ultra-high-Q toroid microcavity on a chip. *Nature* 421(6926): 925–928.

Armani, A. M., R. P. Kulkarni, S. E. Fraser, R. C. Flagan, and K. J. Vahala (2007). Label-free, single-molecule detection with optical microcavities. *Science* 317(5839): 783–787.

Barer, R., K. F. A. Ross, and S. Tkaczyk (1953). Refractometry of living cells. *Nature* 171(4356): 720–724.

BarZiv, R., A. Meller, T. Tlusty et al. (1997). Localized dynamic light scattering: Probing single particle dynamics at the nanoscale. *Physical Review Letters* 78(1): 154–157.

Basche, T., W. E. Moerner, M. Orrit, and U. P. Wild, Eds. (1997). *Single-Molecule Optical Detection, Imaging and Spectroscopy*. Weinheim, Germany: VCH.

Belkin, M. A., M. Loncar, B. G. Lee et al. (2007). Intra-cavity absorption spectroscopy with narrow-ridge microfluidic quantum cascade lasers. *Optics Express* 15(18): 11262–11271.

Berne, B. J. and R. Pecora (2000). *Dynamic Light Scattering: With Applications to Chemistry, Biology, and Physics*. New York: Dover Publications.

Beusink, J. B., A. M. C. Lokate, G. A. J. Besselink, G. J. M. Pruijn, and R. B. M. Schasfoort (2008). Angle-scanning SPR imaging for detection of biomolecular interactions on microarrays. *Biosensors & Bioelectronics* 23(6): 839–844.

Bi, R., P. D. Zhang, C. Q. Dong, and J. C. Ren (2006). Combination of micro-fluidic chip with fluorescence correlation spectroscopy for single molecule detection. *Chinese Chemical Letters* 17(4): 521–524.

Bright, F. V. (1993). Probing biosensor interfaces by multifrequency phase and modulation total internal-reflection fluorescence (MPM-TIRF). *Applied Spectroscopy* 47(8): 1152–1160.

Chakraborty, S., B. Sahoo, I. Teraoka, and R. A. Gross (2005). Solution properties of starch nanoparticles in water and DMSO as studied by dynamic light scattering. *Carbohydrate Polymers* 60(4): 475–481.

Chansin, G. A. T., R. Mulero, J. Hong et al. (2007a). Single-molecule spectroscopy using nanoporous membranes. *Nano Letters* 7(9): 2901–2906.

Chansin, G. A. T., R. Mulero, J. Hong et al. (2007b). Towards ultra-fast parallel DNA analysis: Sub-wavelength metallic nanopore arrays for high-throughput single molecule spectroscopy. *MicroTAS*, Paris.

Chen, L. L., L. Deng, L. L. Liu, and Z. H. Peng (2007). Immunomagnetic separation and MS/SPR end-detection combined procedure for rapid detection of *Staphylococcus aureus* and protein A. *Biosensors & Bioelectronics* 22(7): 1487–1492.

Coille, I., S. Reder, S. Bucher, and G. Gauglitz (2002). Comparison of two fluorescence immunoassay methods for the detection of endocrine disrupting chemicals in water. *Biomolecular Engineering* 18(6): 273–280.

Connatser, R. M., L. A. Riddle, and M. J. Sepaniak (2004). Metal-polymer nanocomposites for integrated microfluidic separations and surface enhanced Raman spectroscopic detection. *Journal of Separation Science* 27(17–18): 1545–1550.

Connatser, R. M., M. Cochran, R. J. Harrison, and M. J. Sepaniak (2008). Analytical optimization of nano-composite surface-enhanced Raman spectroscopy/scattering detection in microfluidic separation devices. *Electrophoresis* 29(7): 1441–1450.

de Mello, A. J. and N. Beard (2003). Dealing with 'real' samples: Sample pre-treatment in microfluidic systems. *Lab on a Chip* 3(1): 11N–19N.

Dickson, J. L., S. S. Adkins, T. Cao, S. E. Webber, and K. P. Johnston (2006). Interactions of core-shell silica nanoparticles in liquid carbon dioxide measured by dynamic light scattering. *Industrial & Engineering Chemistry Research* 45(16): 5603–5613.

Dieringer, J. A., A. D. McFarland, N. C. Shah et al. (2006). Surface enhanced Raman spectroscopy: New materials, concepts, characterization tools, and applications. *Faraday Discussions* 132: 9–26.

Dieringer, J. A., R. B. Lettan, K. A. Scheidt, and R. P. Van Duyne (2007). A frequency domain existence proof of single-molecule surface-enhanced Raman spectroscopy. *Journal of the American Chemical Society* 129(51): 16249–16256.

Doering, W. E. and S. M. Nie (2002). Single-molecule and single-nanoparticle SERS: Examining the roles of surface active sites and chemical enhancement. *Journal of Physical Chemistry B* 106(2): 311–317.

Dong, C. Q., H. F. Qian, N. H. Fang, and J. C. Ren (2006). Study of fluorescence quenching and dialysis process of CdTe quantum dots, using ensemble techniques and fluorescence correlation spectroscopy. *Journal of Physical Chemistry B* 110(23): 11069–11075.

Dong, H., X. D. Cao, C. M. Li, and W. H. Hu (2008). An in situ electrochemical surface plasmon resonance immunosensor with polypyrrole propylic acid film: Comparison between SPR and electrochemical responses from polymer formation to protein immunosensing. *Biosensors & Bioelectronics* 23(7): 1055–1062.

Dostalek, J., J. Ctyroky, J. Homola et al. (2001). Surface plasmon resonance biosensor based on integrated optical waveguide. *Sensors and Actuators B* 76: 8–12.

Du, X. Z. and Y. C. Wang (2007). Directed assembly of binary monolayers with a high protein affinity: Infrared reflection absorption spectroscopy (IRRAS) and surface plasmon resonance (SPR). *Journal of Physical Chemistry B* 111(9): 2347–2356.

Eigen, M. and R. Rigler (1994). Sorting single molecules - application to diagnostics and evolutionary bio-technology. *Proceedings of the National Academy of Sciences of the United States of America* 91(13): 5740–5747.

Elghanian, R., J. J. Storhoff, R. C. Mucic, R. L. Letsinger, and C. A. Mirkin (1997). Selective colorimetric detection of polynucleotides based on the distance-dependent optical properties of gold nanoparticles. *Science* 277(5329): 1078–1081.

Ferraro, J. R. and K. Nakamoto (1994). *Introductory Raman Spectroscopy*. Boston, MA: Academic Press.

Fleischmann, M., P. J. Hendra, and A. J. McQuillan (1974). Raman spectra of pyridine adsorbed at a silver electrode. *Chemical Physics Letters* 26(2): 163–166.

Fleming, G. R. (1986). *Chemical Applications of Ultrafast Spectroscopy*. New York, Oxford University Press.

Foquet, M., J. Korlach, W. R. Zipfel, W. W. Webb, and H. G. Craighead (2004). Focal volume confinement by submicrometer-sized fluidic channels. *Analytical Chemistry* 76(6): 1618–1626.

Fu, E., T. Chinowsky, K. Nelson et al. (2007). SPR imaging-based salivary diagnostics system for the detection of small molecule analytes. *Oral-Based Diagnostics* 1098: 335–344.

Fuhr, G., E. Richter, H. Zimmermann et al. (1998). Cell traces—Footprints of individual cells during locomotion and adhesion. *Biological Chemistry* 379(8–9): 1161–1173.

Gadella, T. W. J., T. M. Jovin, and R. M. Clegg (1993). Fluorescence lifetime imaging microscopy (film)—Spatial-resolution of microstructures on the nanosecond time-scale. *Biophysical Chemistry* 48(2): 221–239.

Gerhardt, I. et al. (2007). Scanning near-field optical coherent spectroscopy of single molecules at 1.4 K. *Optics Letters* 32(11): 1420–1422.

Haes, A. J., S. L. Zou, J. Zhao, G. C. Schatz, and R. P. Van Duyne (2006). Localized surface plasmon resonance spectroscopy near molecular resonances. *Journal of the American Chemical Society* 128(33): 10905–10914.

Heaton, R. J., P. I. Haris, J. C. Russell, and D. Chapman (1995). Application of SPR & FTIR spectroscopy to the study of protein-biomaterial interactions. *Biochemical Society Transactions* 23(4): S502–S502.

Hess, S. T., S. H. Huang, A. A. Heikal, and W. W. Webb (2002). Biological and chemical applications of fluorescence correlation spectroscopy: A review. *Biochemistry* 41(3): 697–705.

Hink, M. A. and A. Visser (2002). Fluorescence cross-correlation spectroscopy is able to monitor interaction and dynamics of GFP-fused proteins diffusing in the plasma membrane of live cells. *Biophysical Journal* 82(1): 1670.

Hink, M. A., J. W. Borst, and A. Visser (2003). Fluorescence correlation spectroscopy of GFP fusion proteins in living plant cells. *Biophotonics* 361(Pt B): 93–112.

Huang, X. Y., L. Li, H. F. Qian, C. Q. Dong, and J. C. Ren (2006). A resonance energy transfer between chemiluminescent donors and luminescent quantum-dots as acceptors (CRET). *Angewandte Chemie-International Edition* 45(31): 5140–5143.

Jeanmaire, D. L. and R. P. Vanduyne (1977). Surface Raman spectroelectrochemistry. 1. Heterocyclic, aromatic, and aliphatic-amines adsorbed on anodized silver electrode. *Journal of Electroanalytical Chemistry* 84(1): 1–20.

Jennissen, H. P. and T. Zumbrink (2004). A novel nanolayer biosensor principle. *Biosensors & Bioelectronics* 19(9): 987–997.

Jennissen, H. P., A. Sanders, H. J. Schnittler, and V. Hlady (1999). TIRF-rheometer for measuring protein adsorption under high shear rates: Constructional and fluid dynamic aspects. *Materialwissenschaft und Werkstofftechnik* 30(12): 850–861.

Jiang, J., K. Bosnick, M. Maillard, and L. Brus (2003). Single molecule Raman spectroscopy at the junctions of large Ag nanocrystals. *Journal of Physical Chemistry B* 107(37): 9964–9972.

Kagan, C. R., C. B. Murray, and M. G. Bawendi (1996a). Long-range resonance transfer of electronic excitations in close-packed CdSe quantum-dot solids. *Physical Review B-Condensed Matter* 54(12): 8633–8643.

Kagan, C. R., C. B. Murray, M. Nirmal, and M. G. Bawendi (1996b). Electronic energy transfer in CdSe quantum dot solids. *Physical Review Letters* 76(9): 1517–1520.

Kamyshny, A., D. Danino, S. Magdassi, and Y. Talmon (2002). Transmission electron microscopy at cryogenic temperatures and dynamic light scattering studies of glucose oxidase molecules and self-aggregated nanoparticles. *Langmuir* 18(8): 3390–3391.

Keir, R., E. Igata, M. Arundell et al. (2002). SERRS. In situ substrate formation and improved detection using microfluidics. *Analytical Chemistry* 74(7): 1503–1508.

Kennedy, B. J., S. Spaeth, M. Dickey, and K. T. Carron (1999). Determination of the distance dependence and experimental effects for modified SERS substrates based on self-assembled monolayers formed using alkanethiols. *Journal of Physical Chemistry B* 103(18): 3640–3646.

Khlebtsov, N. G., V. A. Bogatyrev, B. N. Khlebtsov, L. A. Dykman, and P. Englebienne (2003). A multilayer model for gold nanoparticle bioconjugates: Application to study of gelatin and human IgG adsorption using extinction and light scattering spectra and the dynamic light scattering method. *Colloid Journal* 65(5): 622–635.

Kneipp, K. and H. Kneipp (2005). Detection, identification, and tracking of biomolecules at the single molecule level using SERS. *Biophysical Journal* 88(1): 365A.

Kneipp, K., Y. Wang, H. Kneipp et al. (1997). Single molecule detection using surface-enhanced Raman scattering (SERS). *Physical Review Letters* 78(9): 1667–1670.

Kneipp, K., G. R. Harrison, S. R. Emory, and S. M. Nie (1999). Single-molecule Raman spectroscopy - Fact or fiction? *Chimia* 53(1–2): 35–37.

Kyoung, M. J. and E. D. Sheets (2007). TIR-FCS/optical trapping for investigating nanoparticle dynamics near surfaces. *Biphysical Journal* 159A.

Lakowicz, J. R. (1999). *Principles of Fluorescence Spectroscopy*. New York: Kluwer.

Lassen, B. and M. Malmsten (1996). Competitive protein adsorption studied with TIRF and ellipsometry. *Journal of Colloid and Interface Science* 179(2): 470–477.

Lee, M. and P. M. Fauchet (2007). Two-dimensional silicon photonic crystal based biosensing platform for protein detection. *Optics Express* 15(8): 4530–4535.

Lee, J. Y., C. W. Lee, E. H. Lin, and P. K. Wei (2008). Single live cell refractometer using nanoparticle coated fiber tip. *Applied Physics Letters* 93(17): 3.

Leung, S. A., R. F. Winkle, R. C. R. Wootton, and A. J. deMello (2005). A method for rapid reaction optimisation in continuous-flow microfluidic reactors using online Raman spectroscopic detection. *Analyst* 130(1): 46–51.

Li, H. X. and L. Rothberg (2004). Colorimetric detection of DNA sequences based on electrostatic interactions with unmodified gold nanoparticles. *Proceedings of the National Academy of Sciences of the United States of America* 101(39): 14036–14039.

Li, J. H., X. Chu, Y. L. Liu et al. (2005). A colorimetric method for point mutation detection using high-fidelity DNA ligase. *Nucleic Acids Research* 33(19): 168.

Li, H., D. F. Cui, J. Q. Liang, H. Y. Cai, and Y. J. Wang (2006a). Imaging array SPR biosensor immunoassays for sulfamethoxazole and sulfamethazine. *Chinese Chemical Letters* 17(11): 1481–1484.

Li, Y., H. J. Lee, and R. M. Corn (2006b). Fabrication and characterization of RNA aptamer microarrays for the study of protein-aptamer interactions with SPR imaging. *Nucleic Acids Research* 34(22): 6416–6424.

Liang, X. J., A. Q. Liu, C. S. Lim, T. C. Ayi, and P. H. Yap (2007). Determining refractive index of single living cell using an integrated microchip. *Sensors and Actuators A: Physical* 133(2): 349–254.

Liu, A. Q., H. J. Huang, L. K. Chin, Y. F. Yu, and X. C. Li (2008). Label-free detection with micro optical fluidic systems (MOFS): A review. *Analytical and Bioanalytical Chemistry* 391(7): 2443–2452.

Loncar, M., A. Scherer, and Y. M. Qiu (2003). Photonic crystal laser sources for chemical detection. *Applied Physics Letters* 82(26): 4648–4650.

Lue, N., G. Popescu, T. Ikeda et al. (2006). Live cell refractometry using microfluidic devices. *Optics Letters* 31(18): 2759–2761.

Magde, D., W. W. Webb, and E. Elson (1972). Thermodynamic fluctuations in a reacting system - Measurement by fluorescence correlation spectroscopy. *Physical Review Letters* 29(11): 705–708.

Matveeva, E. G., Z. Gryczynski, J. Malicka et al. (2005). Directional surface plasmon-coupled emission: Application for an immunoassay in whole blood. *Analytical Biochemistry* 344(2): 161–167.

Matveeva, E. G., I. Gryczynski, A. Barnett et al. (2007). Metal particle-enhanced fluorescent immunoassays on metal mirrors. *Analytical Biochemistry* 363(2): 239–245.

McHale, J. L. (1998). *Moleculqar Spectroscopy*. Upper Saddle River, NJ: Prentice Hall.

Measor, P., L. Seballos, D. L. Yin et al. (2007). On-chip surface-enhanced Raman scattering detection using integrated liquid-core waveguides. *Applied Physics Letters* 90(21): 211107–211109.

Medintz, I. L., A. R. Clapp, H. Mattoussi et al. (2003). Self-assembled nanoscale biosensors based on quantum dot FRET donors. *Nature Materials* 2(9): 630–638.

Medintz, I. L., E. R. Goldman, M. E. Lassman, and J. M. Mauro (2003). A fluorescence resonance energy transfer sensor based on maltose binding protein. *Bioconjugate Chemistry* 14(5): 909–918.

Medintz, I. L., G. P. Anderson, M. E. Lassman et al. (2004). General strategy for biosensor design and construction employing multifunctional surface-tethered components. *Analytical Chemistry* 76(19): 5620–5629.

Meli, M. V. and R. B. Lennox (2007). Surface plasmon resonance of gold nanoparticle arrays partially embedded in quartz substrates. *Journal of Physical Chemistry C* 111(9): 3658–3664.

Michaels, A. M., J. Jiang, and L. Brus (2000). Ag nanocrystal junctions as the site for surface-enhanced Raman scattering of single Rhodamine 6G molecules. *Journal of Physical Chemistry B* 104(50): 11965–11971.

Moerner, W. E. (1994). Examining nanoenvironments in solids on the scale of a single, isolated impurity molecule. *Science* 265(5168): 46–53.

Moskovits, M. (1978). Surface-roughness and enhanced intensity of Raman-scattering by molecules adsorbed on metals. *Journal of Chemical Physics* 69(9): 4159–4161.

Moskovits, M. (1985). Surface-enhanced spectroscopy. *Reviews of Modern Physics* 57(3 Pt. 1): 783–826.

Mouvet, C., R. D. Harris, C. Maciag et al. (1997). Determination of simazine in water samples by waveguide surface plasmon resonance. *Analytica Chimica Acta* 338(1–2): 109–117.

Nie, S. M. and S. R. Emery (1997). Probing single molecules and single nanoparticles by surface-enhanced Raman scattering. *Science* 275(5303): 1102–1106.

O'Connor, D. V. and D. Philips (1984). *Time-Correlated Single Photon Counting*. New York: Academic Press.

Oh, Y. J., T. C. Gamble, D. Leonhardt et al. (2008). Monitoring FET flow control and wall adsorption of charged fluorescent dye molecules in nanochannels integrated into a multiple internal reflection infrared waveguide. *Lab on a Chip* 8(2): 251–258.

Ohsugi, Y., K. Saito, M. Tamura, and M. Kinjo (2006). Lateral mobility of membrane-binding proteins in living cells measured by total internal reflection fluorescence correlation spectroscopy. *Biophysical Journal* 91(9): 3456–3464.

Otto, A. (2002). What is observed in single molecule SERS, and why? *Journal of Raman Spectroscopy* 33(8): 593–598.

Patel, H., I. Konig, M. Tsujioka et al. (2008). The multi-FERM-domain-containing protein FrmA is required for turnover of paxillin-adhesion sites during cell migration of *Dictyostelium*. *Journal of Cell Science* 121(8): 1159–1164.

Petrasek, Z., M. Krishnan, I. Monch, and P. Schwille (2007). Simultaneous two-photon fluorescence correlation spectroscopy and lifetime imaging of dye molecules in submicrometer fluidic structures. *Microscopy Research and Techniques* 70: 459–466.

Petrova, H., C. H. Lin, M. Hu et al. (2007b). Vibrational response of Au-Ag nanoboxes and nanocages to ultrafast laser-induced heating. *Nano Letters* 7(4): 1059–1063.

Petrova, H., C. H. Lin, S. de Liejer et al. (2007a). Time-resolved spectroscopy of silver nanocubes: Observation and assignment of coherently excited vibrational modes. *Journal of Chemical Physics* 126(9): 094709.

Piorek, B. D., S. J. Lee, J. G. Santiago et al. (2007). Free-surface microfluidic control of surface-enhanced Raman spectroscopy for the optimized detection of airborne molecules. *Proceedings of the National Academy of Sciences of the United States of America* 104(48): 18898–18901.

Purdie, N. and K. A. Swallows (1989). Analytical applications of polarimetry, optical-rotatory dispersion, and circular-dichroism. *Analytical Chemistry* 61(2): 77A–89A.

Ray, K., H. Szmacinski, J. Enderlein, and J. R. Lakowicz (2007). Distance dependence of surface plasmon-coupled emission observed using Langmuir-Blodgett films. *Applied Physics Letters* 90(25): 251116.

Rella, R., J. Spadavecchia, M. G. Manera et al. (2004). Liquid phase SPR imaging experiments for biosensors applications. *Biosensors & Bioelectronics* 20(6): 1140–1148.

Rockhold, S. A., R. D. Quinn, R. A. Vanwagenen, J. D. Andrade, and M. Reichert (1983). Total internal-reflection fluorescence (TIRF) as a quantitative probe of protein adsorption. *Journal of Electroanalytical Chemistry* 150(1–2): 261–275.

Sarov, Y. E., I. Capek, T. B. Ivanov et al. (2008). On total internal reflection investigation of nanoparticles by integrated micro-fluidic system. *Nano Letters* 8(2): 375–381.

Schmidt, H. and A.R. Hawkins (2008). Optofluidic waveguides: I. Concepts and implementations. *Microfluidics and Nanofluidics* 4(1–2): 3–16.

Schwartzberg, A. M., T. Y. Olson, C. E. Talley, and J. Z. Zhang (2006a). Synthesis, characterization, and tunable optical properties of hollow gold nanospheres. *Journal of Physical Chemistry B* 110(40): 19935–19944.

Schwartzberg, A. M., T. Y. Oshiro, J. Z. Zhang, T. Huser, and C. E. Talley (2006b). Improving nanoprobes using surface-enhanced Raman scattering from 30-nm hollow gold particles. *Analytical Chemistry* 78(13): 4732–4736.

Seballos, L., T. Y. Olson, and J. Z. Zhang (2006). Effects of chromophore orientation and molecule conformation on surface-enhanced Raman scattering studied with alkanoic acids and colloidal silver nanoparticles. *Journal of Chemical Physics* 125(23): 234706.

Seballos, L., N. Richards, D. J. Stevens et al. (2007). Competitive binding effects on surface-enhanced Raman scattering of peptide molecules. *Chemical Physics Letters* 447: 335–339.

Shao, H., D. Kumar, and K. L. Lear (2006). Single-cell detection using optofluidic intracavity spectroscopy. *IEEE Sensors Journal* 6(6): 1543–1550.

Shao, H., W. N. Wang, S. E. Lana, and K. L. Lear (2008). Optofluidic intracavity spectroscopy of canine lymphoma and lymphocytes. *IEEE Photonics Technology Letters* 20(5–8): 493–495.

Sherry, L. J., R. C. Jin, C. A. Mirkin, G. C. Schatz, and R. P. Van Duyn (2006). Localized surface plasmon resonance spectroscopy of single silver triangular nanoprisms. *Nano Letters* 6(9): 2060–2065.

Shi, C., C. Lu, C. Gu et al. (2008a). Inner wall coated hollow core waveguide sensor based on double substrate surface enhanced Raman scattering. *Applied Physics Letters* 93(15): 150101.

Shi, C., H. Yan, C. Gu et al. (2008b). A double substrate "sandwich" structure for fiber surface enhanced Raman scattering detection. *Applied Physics Letters* 92(10): 103107.

Shibata, C. T. and A. M. Lenhoff (1992). TIRF of salt and surface effects on protein adsorption. 1. Equilibrium. *Journal of Colloid and Interface Science* 148(2): 469–484.

Shopova, S. I., J. M. Cupps, P. Zhang et al. (2007). Opto-fluidic ring resonator lasers based on highly efficient resonant energy transfer. *Optics Express* 15(20): 12735–12742.

Shukla, A., M. A. Kiselev, A. Hoell, and R. H. H. Neubert (2004). Characterization of nanoparticles of lidocaine in w/o microemulsions using small-angle neutron scattering and dynamic light scattering. *Pramana-Journal of Physics* 63(2): 291–295.

Slavik, R., J. Homola, and J. Ctyroky (1999). Single-mode optical fiber surface plasmon resonance sensor. *Sensors and Actuators B: Chemical* 54(1–2): 74–79.

Song, W. Z., X. M. Zhang, A. Q. Liu et al. (2006). Refractive index measurement of single living cells using on-chip Fabry-Perot cavity. *Applied Physics Letters* 89(20): 203901.

Stancu, I. C., A. Fernandez-Gonzalez, and R. Salzer (2007). SPR imaging antimucin-mucin bioaffinity based biosensor as label-free tool for early cancer diagnosis. Design and detection principle. *Journal of Optoelectronics and Advanced Materials* 9(6): 1883–1889.

Storhoff, J. J., R. Elghanian, C. A. Mirkin, and R. L. Letsinger (2002). Sequence-dependent stability of DNA-modified gold nanoparticles. *Langmuir* 18(17): 6666–6670.

Swinney, K., J. Nodorft, and D. J. Bornhop (2001). Capillary-scale polarimetry for flowing streams. *Analyst* 126(5): 673–675.

Talley, C. E., L. Jusinski, C. W. Hollars, S. M. Lane, and T. Huser (2004). Intracellular pH sensors based on surface-enhanced Raman scattering. *Analytical Chemistry* 76(23): 7064–7068.

Tian, Z. Q., B. Ren, J. F. Li, and Z. L. Yang (2007). Expanding generality of surface-enhanced Raman spectroscopy with borrowing SERS activity strategy. *Chemical Communications* (34): 3514–3534.

Toderas, F., M. Baia, L. Baia, and S. Astilean (2007). Controlling gold nanoparticle assemblies for efficient surface-enhanced Raman scattering and localized surface plasmon resonance sensors. *Nanotechnology* 18(25).

Tsay, J. M., S. Doose, and S. Weiss (2006). Rotational and translational diffusion of peptide-coated CdSe/CdS/ZnS nanorods studied by fluorescence correlation spectroscopy. *Journal of the American Chemical Society* 128(5): 1639–1647.

Tschmelak, J., N. Kappel, and G. Gauglitz (2005). TIRF-based biosensor for sensitive detection of progesterone in milk based on ultra-sensitive progesterone detection in water. *Analytical and Bioanalytical Chemistry* 382(8): 1895–1903.

Tubb, A. J. C., F. P. Payne, R. B. Millington, and C. R. Lowe (1997). Single-mode optical fibre surface plasma wave chemical sensor. *Sensors and Actuators B: Chemical* 41(1–3): 71–79.

van Munster, E. B. and T. W. J. Gadella (2005). Fluorescence lifetime imaging microscopy (FLIM). *Microscopy Techniques* 95: 143–175.

Wang, S. P., N. Mamedova, N. A. Kotov, W. Chen, and J. Studer (2002). Antigen/antibody immunocomplex from CdTe nanoparticle bioconjugates. *Nano Letters* 2(8): 817–822.

Wang, T. J., C. W. Tu, F. K. Liu, and H. L. Chen (2004). Surface plasmon resonance waveguide biosensor by bipolarization wavelength interrogation. *IEEE Photonics Technology Letters* 16(7): 1715–1717.

Wang, M., N. Jing, I. H. Chou, G. L. Cote, and J. Kameoka (2007). An optofluidic device for surface enhanced Raman spectroscopy. *Lab on a Chip* 7(5): 630–632.

Wang, T., X. G. Hu, and S. J. Dong (2008). A renewable SERS substrate prepared by cyclic depositing and stripping of silver shells on gold nanoparticle microtubes. *Small* 4(6): 781–786.

Weiss, M. N., R. Srivastava, and H. Groger (1996). Experimental investigation of a surface plasmon-based integrated-optic humidity sensor. *Electronics Letters* 32(9): 842–843.

White, I. M., H. Oveys, X. Fan, T. L. Smith, and J. Y. Zhang (2006). Integrated multiplexed biosensors based on liquid core optical ring resonators and antiresonant reflecting optical waveguides. *Applied Physics Letters* 89(19).

White, I. M., H. Y. Zhu, J. D. Suter et al. (2007). Refractometric sensors for lab-on-a-chip based on optical ring resonators. *IEEE Sensors Journal* 7(1–2): 28–35.

Wilkop, T., Z. Z. Wang, and Q. Cheng (2004). Analysis of mu-contact printed protein patterns by SPR imaging with a LED light source. *Langmuir* 20(25): 11141–11148.

Willard, D. M., L. L. Carillo, J. Jung, and A. Van Orden (2001). CdSe-ZnS quantum dots as resonance energy transfer donors in a model protein-protein binding assay. *Nano Letters* 1(9): 469–474.

Willard, D., G. Proll, S. Reder, and G. Gauglitz (2003). New and versatile optical-immunoassay instrumentation for water monitoring. *Environmental Science and Pollution Research* 10(3): 188–191.

Willets, K. A. and R. P. Van Duyne (2007). Localized surface plasmon resonance spectroscopy and sensing. *Annual Review of Physical Chemistry* 58: 267–297.

Wuelfing, W. P., A. C. Templeton, J. F. Hicks, and R. W. Murray (1999). Taylor dispersion measurements of monolayer protected clusters: A physicochemical determination of nanoparticle size. *Analytical Chemistry* 71(18): 4069–4074.

Xie, H., K. L. Gill-Sharp, and P. O'Neal (2007). Quantitative estimation of gold nanoshell concentrations in whole blood using dynamic light scattering. *Nanomedicine-Nanotechnology Biology and Medicine* 3(1): 89–94.

Yin D., J.P. Barber, D.W. Deamer, A.R. Hawkins, H. Schmidt (2006). Single-molecule detection using planar integrated optics on a chip. *Optics Letters* 31: 2136–2138.

Yin, D., E. J. Lunt, A. Barman, A. R. Hawkins, and H. Schmidt (2007). Microphotonic control of single molecule fluorescence correlation spectroscopy using planar optofluidics. *Optics Express* 15(12): 7290–7295.

Yu, H. B., G. Y. Zhou, C. F. Siong, and L. Feiwen (2008). A variable optical attenuator based on optofluidic technology. *Journal of Micromechanics and Microengineering* 18(11): 115016.1–115016.5.

Yuk, J. S., D. G. Hong, H. I. Jung, and K. S. Ha (2006). Application of spectral SPR imaging for the surface analysis of C-reactive protein binding. *Sensors and Actuators B: Chemical* 119(2): 673–675.

Yuk, J. S., J. W. Jung, Y. M. Mm, and K. S. Ha (2008). Analysis of protein arrays with a dual-function SPR biosensor composed of surface plasmon microscopy and SPR spectroscopy based on white light. *Sensors and Actuators B: Chemical* 129(1): 113–119.

Zhang, J., Y. Fu, M. H. Chowdhury, and J. R. Lakowicz (2007a). Enhanced Forster resonance energy transfer on single metal particle. 2. Dependence on donor-acceptor separation distance, particle size, and distance from metal surface. *Journal of Physical Chemistry C* 111(32): 11784–11792.

Zhang, J., Y. Fu, M. H. Chowdhury, and J. R. Lakowicz (2007b). Metal-enhanced single-molecule fluorescence on silver particle monomer and dimer: Coupling effect between metal particles. *Nano Letters* 7(7): 2101–2107.

Zhang, J. M., Z. Dai, N. Guo et al. (2007c). Fluorescence resonance energy transfer between CdTe quantum dot donors and Au nanoparticles labeled DNA acceptors. *Chemical Journal of Chinese Universities-Chinese* 28(2): 254–257.

Zhang, Y., C. Shi, G. Claire et al. (2007d). Liquid core photonic crystal fiber sensor based on surface enhanced Raman scattering. *Applied Physics Letters* 90(19): 193504–193506.

Zhou, D. J., J. D. Piper, C. Abell et al. (2005). Fluorescence resonance energy transfer between a quantum dot donor and a dye acceptor attached to DNA. *Chemical Communications* (38): 4807–4809.

Zhu, H. Y., I. M. White, J. D. Suter, P. S. Dale, and X. D. Fan (2007). Analysis of biomolecule detection with optofluidic ring resonator sensors. *Optics Express* 15(15): 9139–9146.

7

Lab-on-a-Chip

Su Eun Chung
Wook Park
Seung Ah Lee
Sung Eun Choi
Jisung Jang
Sung Hoon Lee
Sunghoon Kwon

7.1 Introduction

7.1.1 Concept and Motivation for a Lab-on-a-Chip

A lab-on-a-chip (LOC) refers to a device or system that performs macroscale laboratory processes at small scales, typically from a few millimeters to centimeters in size. LOCs started in the field of biology and chemistry, where the improvement of analysis throughput and sensitivity was sought through miniaturization and advanced instrumentation. In a similar context, micro total analysis systems (μTAS), which integrate the steps necessary to perform a chemical analysis on one chip, were first described in 1990 (Manz et al., 1990). An LOC describes a wide range of chip-format devices that miniaturize a laboratory process, and they include both microfluidic chips and non-fluidic biochips, which have been demonstrated with microarray technology (Dittrich and Manz, 2006).

Since LOCs deal with extremely small volumes of liquid or gas samples, understanding and utilizing the physics of fluids at small scales, a field known as "microfluidics," has been one of the most important milestones in the development of LOCs. Microfluidics covers the motion of fluids at low Reynolds numbers, solution phase reactions in nano- to picoliter systems, and the interaction between the fluids themselves or the contents in the fluids with respect to the external forces applied, such as pressure or electromagnetic force. Microfluidic techniques that take into account such phenomena are essential for sample preparation, transport, reaction, and detection in LOC devices as discussed in Section 7.2.

The miniaturization of chemical processes was facilitated in the late 1990s by the prevalence of microfabrication technology. In the early stages of development, LOC devices were made of silicon and glass using fabrication techniques employed by the semiconductor industry and in the manufacturing of microelectromechanical systems (MEMS). As the field expanded, the demand for inexpensive materials and simpler fabrication methods grew, and LOCs built from polymers appeared. Soft lithography with polydimethylsiloxane (PDMS) has become a very popular technique for the fabrication of microfluidic channels, along with techniques such as rapid prototyping and micromolding. PDMS is a soft elastomer that has been one of the key materials in the development of LOCs because of the following properties: optically transparent, inexpensive, chemically inert, and easy to mold. The use of soft lithography popularized the research of microfluidics at the laboratory level, because microfluidic devices could be rapidly fabricated at a relatively low manufacturing cost. However, for specialized systems or commercialized devices, glass and silicon are still favored because of their chemical and mechanical stability. Various nano-/microfabrication techniques are also widely used for the integration of external tools other than fluidic channels, such as surface modification, electrodes, and optical components as discussed in Chapter 2.

Over the past few decades, many LOC devices and systems have been used for various applications in chemical analysis and biological research. Specific examples include analyte detection for clinical diagnostics, high-throughput genomic sequencing platforms, chemical screening for drug discovery, chemical synthesis, development of a chip-scale platform for cell biology, and tissue engineering. For these purposes, on-chip devices and control schemes for each stage of analysis, including sample preparation, injection, fluid and particle handling, reactors and mixers, separation, and detection have been demonstrated (Auroux et al., 2002). Academia and industry have driven many efforts to demonstrate system-level devices that perform entire analysis processes on-chip, some of which have succeeded in commercialization. DNA microarray technology was one of the first commercialized LOC systems to perform bioassays on-chip (Affymetrix; Schena et al., 1995). Many companies such as Agilent Technology, Caliper Life Sciences, Fluidigm, and Invitrogen have successfully launched commercial LOC products for microarrays, miniaturized polymerase chain reactions (PCR), electrophoreses, and sample preparations, thus making LOCs an attractive field for biotech startups (Clayton, 2005). In academia, biological/chemical research based on LOC platforms and clinical demonstrations of LOC technology as well as novel LOC devices is actively progressing.

Typical LOC assay processes include the following steps: sample preparation, injection, sorting, reaction, and detection (Figure 7.1a). Biological or chemical samples need to be analyzed and prepared in an appropriate manner before being injected into the device. Flow must then be actuated and guided to deliver the analytes or particles to a specific location at the desired time. After the controlled reaction of

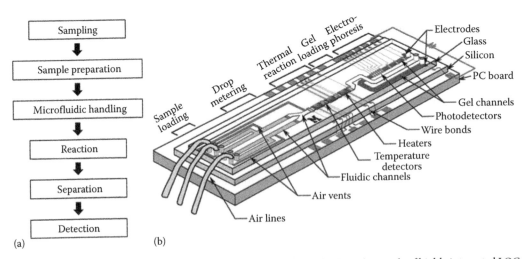

FIGURE 7.1 (a) General flow chart of chemical analysis in an LOC. (b) Typical example of highly integrated LOC systems for DNA analysis. The device integrates microfluidic channels, heaters, temperature sensors, and photodetectors to analyze nanoliter-sized DNA samples. (Reprinted from Burns, M.A. et al., *Science*, 282, 484, 1998. With permission from AAAS.)

analytes, the user has to interpret the products to analyze the results. Performing each action on a chip requires many microfluidic components that represent microscale versions of laboratory equipment. The implementation of such tools on a chip has been extensively studied. In addition to implementing various laboratory functions at the component level, it is important to integrate various LOC components in a chip to automate the assay—a typical example for DNA sequencing is shown in Figure 7.1b (Burns et al., 1998). Miniaturization and integration are thus two important aspects for successful LOC systems.

Miniaturization of chemical and biological processes provides a number of advantages over macroscale laboratory processes. In particular, LOCs offer an advanced platform for miniaturization by integrating multiple tools and processes. Miniaturized systems deal with small volumes of sample, which reduces reagent cost and generates less waste. The ability to perform analyses in small volumes is particularly advantageous in diagnostics and forensics, where less sample consumption is preferred. Moreover, by fractioning a bulk sample into smaller volumes, the surface-to-volume ratio is increased, thereby providing a smaller heat capacity that leads to lower energy consumption and faster heating or cooling. In early stages, capillary-based microreactors were used to take advantage of the large surface-to-volume ratio and low Reynolds number. Also capillary-based electrophoresis was used to speed up conventional slab-gel electrophoresis. While these simple one-dimensional capillaries easily implement the benefits of scaling down, their shape is limited to a one-dimensional tube. Based on lithography and microprocessing technologies, the LOC approach enables the design of networks of capillaries in 3D with flexibility in their shape and dimension. This not only produces capillaries in large volume with great control but also enables new microfluidic devices. Also LOCs offer an advanced platform for miniaturization by integrating multiple tools and processes on a single chip. Miniaturized LOC devices allow for faster mass transport, further increasing the speed and quality of the reaction and analysis.

The strength of LOC devices lies in the integration of multiple functionalities in a single chip. Different functional devices, for example, a flow controller or a microheater, can be embedded in one fluidic channel so that the reaction conditions within the device, such as local temperature, local electric field, and compound concentration, can be precisely regulated and monitored. The precision that LOC devices provide is very high compared to conventional laboratory equipment. This reduces error, increases uniformity, and allows for precise experiments that are not possible with bulk-scale equipment. Integrated microdevices also allow for automated control that can minimize user dependency. In addition, multiple steps in the analysis, such as sample preparation, reaction, and detection, can be integrated in one device, allowing for massively parallel analysis. Parallelization greatly increases throughput compared to serial macroscale laboratory processes.

Each biological or chemical procedure that utilizes LOCs benefits from miniaturization and integration. For example, on-chip chemical synthesis takes advantage of various aspects of LOCs, such as the combination of multiple processes on one chip via continuous-flow synthesis (Chung et al., 2005; Shestopalov et al., 2004), on-chip combinatorial chemistry in a network of microfluidic channels (Kikutani et al., 2002; Watts and Haswell, 2003), and the reaction of very small amount of reagents using droplet-based microfluidics (Chan et al., 2005; Hung et al., 2006). Droplet-based assays are also convenient in chemical screening, where compartmentalized droplets containing different chemical compositions can be used in the optimization of drug candidates and protein crystallization conditions (Dittrich et al., 2005; Zheng et al., 2003, 2005). Genome sequencing is another procedure that takes advantage of the LOC integration process, which allows a series of processes required for sequencing, such as PCR (Khandurina et al., 2000; Kopp et al., 1998), hybridization (Fan et al., 1999; Lenigk et al., 2002), and electrophoresis (Lagally et al., 2000), to be performed in one chip (Paegel et al., 2003). In addition, LOC techniques provide microscale spatiotemporal stimulation to cells or other microorganisms, thus overcoming the precision limitations of macroscale laboratory equipment for single-cell research (Carlo et al., 2006; Wheeler et al., 2001), neuroscience (Chung et al., 2005), stem cell research (Kim et al., 2006), and tissue engineering (El-Ali et al., 2006).

Because of their numerous advantages over macroscale processing, LOCs are expected to revolutionize chemistry and biology laboratories in the near future. Additionally, in commercial products, cheap and disposable LOC devices can be manufactured in mass, which could revolutionize the drug discovery process by high-throughput screening and provide low-cost point-of-care devices for Third World applications. Since the LOC technology is not yet a fully developed field, it is still being improved for successful popularization among nonscientific users and even the general public. Major improvements are required for the interface between the microchip and the macroworld, such as automated fluid injection and driving schemes without additional external devices and improved imaging and detection techniques. LOC devices for research laboratories and pharmaceutical companies are currently being commercialized, and hospital diagnostics would be the next big market for LOC systems. Eventually, LOC systems can have a large social impact when they are applied to home diagnostics.

7.1.2 Optofluidic Lab-on-a-Chip

LOC devices usually comprise a fluidic domain and an external control apparatus. The processing and reaction of liquid sample occur in the fluidic domain. Supplemental parts provide spatial and temporal control of the fluidic domain for flow control, particle manipulation, and detection. Many supplemental parts such as mechanical, electrical, magnetic, and optical schemes have been developed and utilized in many LOC devices. Recently, optofluidic LOC devices have been introduced to merge optics and microfluidics for providing control, imaging, and detection on the devices. Optics has strong advantages over other methods because it is noninvasive and highly precise. Transparent microfluidic devices based on glass or PDMS provide easier optical access into the fluidic domain. Detection or control using optical beams minimizes mechanical disturbance to the flow compared to invasive methods such as mechanical actuation. Optical energy from laser beams is converted into mechanical or thermal energy at a small spot in the fluid. This energy conversion is used in various ways: manipulating cells with optical tweezers and optical trapping (Chiou et al., 2005; Enger et al., 2004), sorting cells (Wang and Lee, 2005), fluid pumping (Ladavac and Grier, 2004; Leach et al., 2006), or guiding the flow (Liu et al., 2006). In addition to high-power laser beams, the incorporation of photolithography in microfluidics developed a new concept of optofluidic lithography for *in situ* high-throughput fabrication of particle-based biosensors on a chip (Chung et al., 2007; Dendukuri et al., 2006; Pregibon et al., 2007) (Section 7.3.2). Imaging and detection are the applications in which optics has been crucial. Various optical detection schemes have been implemented on-chip, with the integration of optical components, such as LED, lens, waveguides, and filters, to microfluidic devices (Chediak et al., 2004). True on-chip optical imaging has been achieved via optofluidic microscopes, where integration of complementary metal-oxide-semiconductor (CMOS) sensors to a microfluidic device allows for real-time imaging of live samples without a lens (Cui et al., 2008). Detailed descriptions of each optofluidic LOC device are discussed in the following sections.

7.2 Standard Lab-on-a-Chip Operation Process

The ultimate goal of an LOC system is on-chip performance of all functions performed in a conventional laboratory. A fully operational LOC system for a specific application requires integration of many separate component-level modules. In this section, fundamental component-level LOC components such as valves, mixers, and pumps are described with example literature in the order of functional work flow shown in Figure 7.1b: sample preparation, microfluidic handling, reaction, separation, and detection.

7.2.1 Sample Preparation

The first step in standard LOC operational work flow is the appropriate preparation of a sample. Various kinds of samples are used in the LOC system, including DNA, cells, viruses, tissues from humans or other living organisms, soil, food, or chemical reagents. Because most raw materials contain a variety of unnecessary components forming a complex matrix, an appropriate pretreatment that isolates necessary components is often required to prepare samples. Sample pretreatment includes filtration, purification, and preconcentration of the sample. Performance of the LOC device depends on the quality of the sample preparation and the detection efficiency of the analytical system, because high concentration of the introduced sample increases detection sensitivity and enhances further analysis (Mello and Beard, 2003). Various microfluidic devices for preconcentration (Kang et al., 2006; Lee et al., 2008a,b; Ying-Chih et al., 2005), concentration gradient generation (Jeon et al., 2000), and automatic filtration (Paegel et al., 2006) have been investigated.

Filtration sorts analytes of interest and removes unnecessary components from the sample. Commonly used methods for filtration in LOCs mimic conventional filters used in macroscale processes (Andersson et al., 2001a,b; He et al., 1999). For example, He et al. fabricated a lateral percolation system that retains analytical particles within a pillar of arrays shaped like a filter bed. The fluid retains particles introduced perpendicular to the filter bed, while particles smaller than the pitch of the pillars are trapped in a filter that rejects larger components laterally. The operating principle of this filter system is very simple, but its efficiency depends on the resolution of pillar arrays. In addition, particles with sizes smaller than the pitch, such as dust, might also be captured inside a filter network.

An integrated filtering system for immunomagnetic separation of cells from blood was presented by Furdui et al. This device consists of two pumps for a sample and a buffer, and a channel for a magnetic field that traps rare cells captured by antibody-coated magnetic beads (Furdui et al., 2003). The system has high specific-filtration efficiency due to the covalent binding of biotin or antibodies to the bead surface, but it requires sophisticated bead handling techniques. Since many filtration or extraction methods overlap with microfluidic particle handling technologies, more detailed examples are discussed in Section 7.2.2.4.

In addition to filtration or extraction of the sample, sonication may be used to speed up the reaction. The minisonicator system developed by Belgrader et al. has been used to extract, amplify by PCR, and detect DNA in a short period of time. This portable device has a short response time of less than 15 min for the entire process, including detection and monitoring of suspected agents in the sample (Belgrader et al., 1999). Samples that are pretreated with the various methods discussed above are injected into the LOC device or microfluidic chip for further analysis.

7.2.2 Fluid and Particle Handling

Fundamentally, LOC systems are designed to process small amounts of fluids in a continuous manner, as previously mentioned. To exploit the advantages of LOCs, various microscale apparatus for small and precise fluid and particle handling are required to manipulate samples on the chip. For example, pretreated and concentrated samples can be transferred to a designated area within a channel by pumping fluid into the chip. On some LOCs this process can be controlled with microvalves. Then, samples undergo a reaction process using micromixers. The final products harvested at the reaction step in the channel are then separated by a microfluidic sorting process.

7.2.2.1 Pumps

Micropumps are used to deliver reagents from macroscale reservoirs into small chips. A number of micropumps have been developed, from conventional pumps miniaturized by MEMS technology to pumps that employ new techniques that take advantage of physiochemical characteristics at the microscale. Here, we focus on fully integrated on-chip pumps and describe them in terms of their actuation mechanism.

Mechanically actuated pumps have moving parts—a diaphragm in most cases—that have direct contact with the fluid. The pumping volume of the fluid corresponds to the volume change in the chamber due to the deformation of the diaphragm. A pressure difference between the inside and the outside of

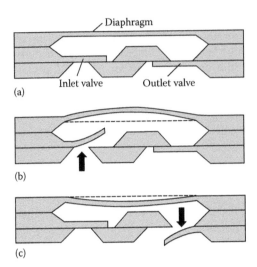

FIGURE 7.2 Micropump with diaphragm: (a) initial state, (b) chamber expansion due to the inflow, and (c) chamber contraction due to the outflow. (Reproduced from Springer Science + Business Media: Iverson, B.D. and Garimella, S.V., *Microfluid. Nanofluid.*, 5, 145, 2008. With permission.)

the chamber causes an actuation of the diaphragm in a micropump. To guide the flow through the inlet valve, the outside pressure should be higher than the pressure inside the chamber; check valves can be used to prohibit backflow for the opposite case (Figure 7.2) (Iverson and Garimella, 2008).

There are several actuation mechanisms for diaphragms: piezoelectric, pneumatic, electrostatic, or magnetic. Piezoelectric-based pumps induce internal stress that results from the diaphragm's vibration in response to an applied AC electric voltage. Relatively large displacement and force can be obtained using piezoelectric diaphragm-actuating pumps to control the flow of liquids (Iverson and Garimella, 2008).

Pneumatic diaphragm pumps combined with elastic materials are one of the most widely used micropumps in LOC devices. One side of the diaphragm is filled with fluid, and the other side is filled with gas to control the pressure of the fluid. The pressure is controlled by an external pressure source. The diaphragm tends to inflate to the lower pressure side as the pressure difference increases (Xia and Whitesides, 1998). Unger et al. prepared a monolithic diaphragm valve and pump made of elastomer by stacking membrane-embedded microfluidic channels. Sequential control of these pumps allows a peristaltic movement (Unger et al., 2000). They also showed the capability of integrated valves and pumps to have microfluidic random-access memory (Thorsen et al., 2002).

Electrostatic actuation of a diaphragm can be achieved by applying electrostatic forces between two embedded electrodes on a chip (Bourouina et al., 1997; Teymoori and Abbaspour-Sani, 2005). One electrode is embedded in the diaphragm and the other is placed on the fixed substrate. As the voltage between the two electrodes varies, the electrodes repel or attract each other. In addition to electrostatic actuation, electromagnetic pumps that use coils have been developed, because their capability for longer-distance actuation is greater than those pumps using the electrostatic mechanism (Gong et al., 2000).

Electroosmotic flow is a unique pumping strategy at the microscale. The surface charge of the channel walls can attract counter ions in a solution onto the channel surface, thereby forming an electric double layer transverse to the channel. When an electric field is applied along the channel in the axial direction, the counter ions are actuated by the electric field and drag adjacent fluids (Figure 7.3) (Chen and Santiago, 2002). With the integration of asymmetric electrodes inside the chip under the flow of applied AC voltage to the electrode, liquid flow is generated (Brown et al., 2001). This simple integration of electrodes has popularized electroosmotic pumps in the LOC field (McKnight et al., 2001; Studer et al., 2004; Zeng et al., 2001). Electroosmotic pumping is especially useful for microfluidic channels of very small diameter, where pneumatic actuation is difficult.

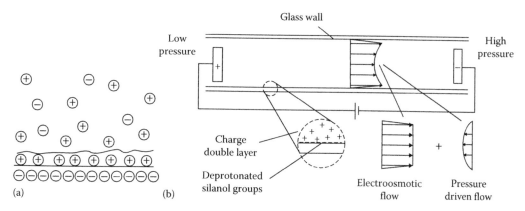

FIGURE 7.3 (a) When fixed charges are induced in the wall, counter ions are attracted and an electric double layer is formed. (b) Operation of a micropump by electroosmosis. Ions in an electric double layer are moved by an external electric field, experiencing drag with adjacent molecules. (Reprinted from Chen, C.H. and Santiago, J.G., *J. Microelectromech. Syst.*, 11, 672, 2002. With permission. © 2009 IEEE.)

Similar to the electroosmosis-based micropump system, a magnetic field can be used for pumping. When the magnetic field is applied perpendicular to the current, a Lorentz force is generated orthogonal to both the magnetic field and the current. Typically, AC current is used to prevent electrolysis of fluid contents that causes bubble formation. Lemoff et al. demonstrated a micropump that uses the Lorentz force as the driving force for the fluid (Lemoff and Lee, 2000).

Surface tension plays a significantly greater role in the micro world than it does in the macro world. Electric energy applied to a surface can change the wettability of the surface. This phenomenon, called electrowetting, can also be an effective pumping mechanism. Droplets on a hydrophobic surface can be actuated by electrowetting (Sung Kwon et al., 2003). By moving a droplet, a diaphragm can also be controlled to pump fluid (Yun et al., 2002). The surface tension change can also cause fluid to move by varying the electric potential of the solid surface containing the electrode or by using dielectric materials (Lee et al., 2002).

Heating a channel can locally change the pressure, density, or surface energy. By continuously changing these properties at an appropriate speed, fluid can be forced to migrate. Song and Zhao moved vapor slugs in a microchannel by moving the heating source, thereby causing the leading edge of the liquid phase to evaporate and the other side to be condensed. The high pressure in the newly evaporated sides and low pressure on the condensed side resulted in fluid movement (Song and Zhao, 2001). Heating fluid via laser is also a good approach to move fluid because it is not confined to a particular heating area and can be controlled in a noninvasive manner. Handique et al. used a laser to heat air bubbles inside a channel to move fluid. As the air bubble was heated, the volume of the bubble expanded and pushed the fluid ahead of it. When the bubble cooled down, it allowed another fluid slug to enter inside the channel (Handique et al., 2001). Liu et al. locally induced photothermal nanoparticles in the fluid to induce movement. The local temperature change in the fluid causes evaporation and formation of small droplets ahead of the surface. The fluid moves toward the droplets due to surface tension (Figure 7.4) (Liu et al., 2006).

As an application of the optofluidic concept to microfluidic pump systems, optical tweezers are introduced for their noninvasive and precise control of dielectric particles. The principle is that when light beam is focused to the dielectric particle forming strong electric field gradient nearby, the dielectric particles are attracted toward the center of the beam with the strongest electric field. Dielectric particles can be moved *in situ* to generate flow inside the channel (Ashkin et al., 1986). In addition, Terray et al. controlled silica particles using piezoelectric mirrors to control multiple particles with a single light source. They also showed that the rotation of silica particles can function as a rotary pump

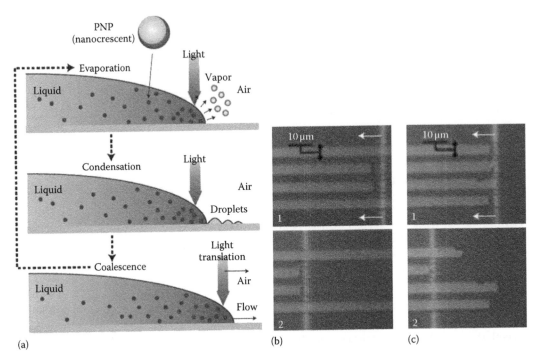

FIGURE 7.4 (a) Principle of guiding. When light heats the leading edge of the fluid, some of the fluid evaporates and condenses as droplets ahead of the leading edge. The front edge of the fluid then coalesces with the droplets and moves forward. (b) Simultaneous fluidic guiding. (c) Guiding different fluids. (Reprinted from Macmillan Publishers Ltd.: Dendukuri, D. et al., *Nat. Mater.*, 5, 365, 2006. With permission. Copyright 2009.)

in a given channel geometry. In another scheme, aligned silica particles can move in succession to function like a peristaltic pump, as shown in Figure 7.5 (Terray et al., 2002). Birefringent particles can get spin, i.e., angular momentum, from a circularly polarized light source. Leach et al. used this fact to rotate vaterite micro particles using circularly polarized light. As the particles rotate, the surrounding fluid flows along the direction of rotation (Leach et al., 2006). Optical tweezers are not limited to facile manipulation of spherical particles, but they can also manipulate microstructures for pumping in a channel. Naele et al. fabricated and rotated birefringent polymer microgears with electron-beam lithography and reactive ion etching. Similar to the study of Leach et al., circularly polarized light was used to drive microgear rotation (Neale et al., 2005). Maruo et al. prepared a polymer micro-robe via two-photon lithography in a confined region of the microchannel and rotated it with the time-divided laser scanning method (Maruo and Inoue, 2006).

Micropumps are divided into subcategories based on their actuating principle as summarized in Table 7.1. The choice of pumping mechanism in LOCs should be application specific, and one needs to consider pros and cons of each method. Generally, pumps with mechanical actuation require more involved fabrication to implement the moving parts. Nonmechanical pumps are relatively simple in their structures and therefore have low manufacturing cost. Certain optofluidic pumping often requires expensive external instruments such as lasers.

7.2.2.2 Valves

Microvalves, first reported in 1979, regulate the flow in a fluidic channel by opening or closing the liquid's path to control flow rate, direction, or volume as well as to mix several different liquids (Terry et al., 1979). According to their mechanism, microvalves can be classified into active valves that use external

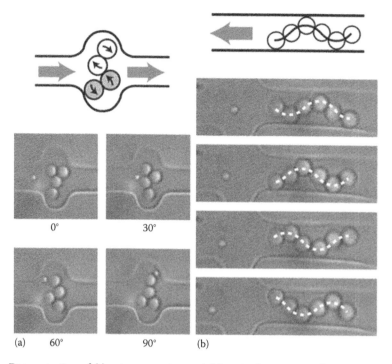

FIGURE 7.5 Demonstration of (a) rotary pumping and (b) peristaltic pumping by moving silica beads with optical tweezers. (Reprinted from Terray, A. et al., *Science*, 296, 1841, 2002. With permission from AAAS.)

forces, and passive valves that do not (Kwang and Chong, 2006). The number of valves in an LOC system often represents the system's complexity and functionality, just as the number of transistors does for an electronic circuit.

Magnetic, electric, piezoelectric, or thermal forces can be applied to actuate valves mechanically. Nonmechanical valves based on electrochemical or rheological methods can also be implemented (Qing et al., 2001; Suzuki and Yoneyama, 2003). An integrated, magnetic-induction-based valve is a kind of mechanically activated microvalve with an inductor capable of generating force to pull a silicon membrane up or down (Losey et al., 2002). The pressure difference at the opening of the valve causes liquid to flow from the inlet to the outlet. In contrast, a ferrofluid can be used as a valve by moving an external permanent magnet (Hartshorne et al., 2004).

Built-in modular or pneumatic valves are typical external-type active valves that directly actuate the flow via external forces such as gas. When gas is applied to the upper pneumatic channel in a two-layered pneumatic valve made through the PDMS rapid prototyping method, the membrane is deflected at the intersection with the bottom channel (Figure 7.6) (Unger et al., 2000). The opened valve is closed within several milliseconds by withdrawing the gas, and it returns to its initial state by its own restoring spring force. This pneumatic line-controlled valve is currently the most widely used active valve in microfluidic flow control.

Flows through passive microvalves are controlled by their special geometry instead of an externally applied force. For example, a "check valve" is a device that only opens to forward pressure, like a diode. In one implementation, a parylene membrane valve was constructed that was comprised of two parts, one for controlling flow in one direction and the other for controlling flow in the opposite direction (Figure 7.7a) (Guo-Hua and Eun Sok, 2004). Since each valve blocks unwanted fluid leakage during each half of the pumping cycle, there is no intermixing between the upstream and downstream flows.

TABLE 7.1   Categories of Micropump Systems: Advantages and Disadvantages of Each
Micropump System

	Categories	Advantages	Disadvantages
Mechanical actuation	Piezoelectric actuation	Large displacement and forces Absolute flow rate and pressure	High-cost process Small pumping rate
	Pneumatic actuation	Low power consumption Continuous displacement Low manufacturing cost	Requires closed-loop control (driving pressure varies over the time) Low flow rate
	Electric actuation	High pumping rate Low power consumption Full integration capability	Small actuation stroke Degradation of performance with time
	Magnetic actuation	Low power consumption Simple design of driver electronics Faster mechanical response	Small pumping rate
Nonmechanical actuation	Electroosmosis actuation	Simple design High backpressure	Requires high voltage Charged working liquid
	Electrowetting actuation	Effective for transporting finite quantities of fluid	Individual control of moving droplets
	Heating actuation	Sustains very high pressure Low input voltage High pumping rate and force	Power consumption for heat generation
	Optical actuation	Noninvasive Accurately transports small quantity of liquid	Requires expensive lasers

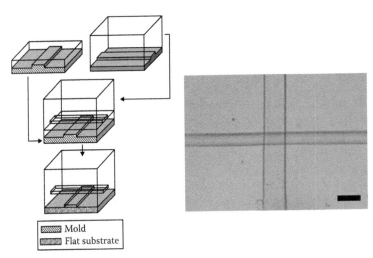

Mold
Flat substrate

FIGURE 7.6   Pneumatic valve using a PDMS elastomer membrane. Pneumatic pressure applied to the channel (top) blocks the flow in the bottom channel. (Reprinted from Unger, M.A. et al., *Science*, 288, 113, 2000. With permission from AAAS.)

A capillary-force-based valve, which is a kind of nonmechanically induced passive valve, utilizes the geometries or surface properties of a microchannel. In an example implementation, a PDMS microdispenser array system allows for sufficient pressure to enable liquids to enter a narrow hydrophobic microchannel (Figure 7.7b) (Yamada et al., 2004). When the applied pressure drops below a threshold, the liquid stops at the end of the broad channel. However, if the pressure is higher than the threshold value, the liquid is allowed to pass.

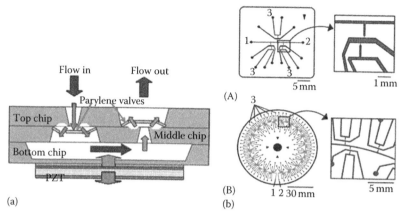

FIGURE 7.7 (a) Working principle of the parylene membrane valve. (Reproduced by permission of Institute of Physics Publishing. With permission.) (b) Liquid microdispenser array. (Reproduced by permission of American Chemical Society. With permission.)

7.2.2.3 Mixers

LOC systems usually deal with at least one dimension of channel in the microscale, resulting in a laminar flow with low Reynolds numbers. In this regime, molecules in the fluid usually mix with each other only by diffusion. The slow mixing by diffusion is an obstacle for fast reactions on a microchip. Rapid mixing using an active micromixer is important to speed up reaction times.

The simplest way to mix two fluids is to place two streams in one channel from two different branches for each inlet. A "T-sensor" (or T-mixer) or a Y-shaped mixer is one of the oldest and most popular schemes for microfluidic mixing. There are many theories and examples of these mixers (Gobby et al., 2001; Hatch et al., 2001; Ismagilov et al., 2000; Kamholz and Yager, 2002; Veenstra et al., 1999; Wong et al., 2004) (Figure 7.8).

Mixing by diffusion consumes time even though the traveling distance on the microchip is relatively small. Therefore, relying only on diffusion is not efficient for on-chip reactions. One of the strategies used to solve this problem is to split up and recombine fluids. Bessoth et al. split fluid into many small branches, mixing two fluids with each other over small segments before recombining them in the end (Bessoth et al., 1999). He et al. also mixed fluids simultaneously in many small channels by using a packed bed-like mixer (Andersson et al., 2001a,b).

Making the channel long and compact is one solution to lessen the traveling distance if diffusion consumes time. It also induces chaotic advection by efficiently mixing different fluids (Ottino, 1990). The geometry of the channel or the structures inside of it can also be designed to induce chaotic advection. Mengeaud et al. used a zigzag channel for mixing via recirculation of the fluid (Figure 7.9a) (Mengeaud et al., 2002). Other investigators fabricated 3D serpentine channels, exploiting their chaotic advection as well as shorter

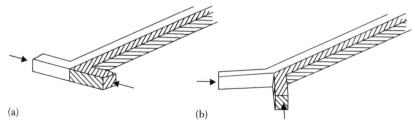

FIGURE 7.8 Basic mixers: (a) T-mixer and (b) Y-mixer. (Reproduced from Nguyen, N.T. and Wu, Z.G., *J. Micromech. Microeng.*, 15, R1, 2005. With permission.)

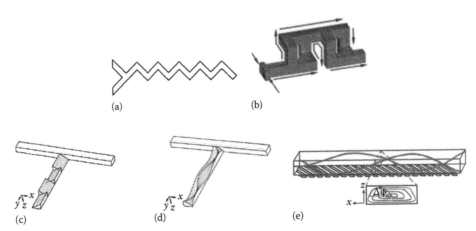

FIGURE 7.9 Geometries for chaotic advection: (a) zigzag, (Reproduced by permission of American Chemical Society. With permission.) (b) 3D serpentine, (Reproduced by permission of American Insitutute of Physics. With permission.) (c,d) twisted, (Reproduced by permission of the Royal Society of Chemistry. With permission.) (e) grooved microchannels. (Reprinted from Stroock, A.D. et al., *Science*, 295, 647, 2002. With permission from AAAS.)

time for diffusion (Figure 7.9b) (Chen and Meiners, 2004). Jen et al. showed the advantage of chaotic advection introduced by twisting the bottom of the channel over a conventional T-mixer (Figure 7.9c) (Jen et al., 2003), and Hong et al. demonstrated other useful geometries (Hong et al., 2004). Grooves on the channel (Figure 7.9c) (Stroock et al., 2002; Wang et al., 2003) or microstructures (Chang and Yang, 2004; Losey et al., 2002) or both (Guo-Hua and Eun Sok, 2004) are also useful for chaotic advection (Hessel et al., 2005).

External energy induces perturbations of solutions inside the fluidic channel, thereby allowing for rapid mixing. Since the integration of electrodes or magnetic coils on a chip is simple, electric and magnetic fields are widely used as a driving force in LOC devices. For example, heterogeneous surfaces induced by electrodes result in local recirculation of electroosmotic flow to enhance mixing (Erickson and Li, 2002). El Moctar et al. induced electric potential via electrodes perpendicular to the flow stream on the T-mixer. When two fluids with different electric properties such as permittivity and conductivity are used, they experience forces perpendicular to the flow and are mixed when AC voltage is applied (El Moctar et al., 2003). In addition, Bau et al. used the Lorentz force to move fluid perpendicular to the stream by applying an AC electric field and magnetic field in the same direction. Stretching and deformation of fluid makes mixing faster (Bau et al., 2001). Micromagnetic stirrers are another simple tool for rapid microfluidic mixing induced by magnetic force (Alaverdian et al., 2002).

Disturbance by perturbation of fluid pressure also facilitates mixing. Piezoelectric material is popular for generating pressure perturbation inside the channel because of its ease of integration and control. Ultrasonic waves resulting from the vibration of piezoelectric material have been employed by many groups (Gong et al., 2000; Lenigk et al., 2002; Nguyen and Huang, 2001; Rife et al., 2000; Yaralioglu et al., 2004) and used for DNA hybridization (Chronis et al., 2003). Pressure perturbation can also be induced from the fluid's source. Pulsing pressure (Niu and Lee, 2003) and the oscillatory change in pressure (Glasgow and Aubry, 2003) caused when two sources are out of phase also promote chaotic mixing.

Passive mixers with parallel lamination are very simple in the design and fabrication of the channels. However, it takes a long time for fluid mixing and requires bigger chip area due to the long mixing channel. To reduce mixing time and chip size, chaotic advection mixing can be used by utilizing smart fluidic channel geometries. But passive micromixers generally suffer from low mixing efficiency. In contrast, active micromixers using electric, magnetic field, or ultrasonic waves, enable rapid mixing and high mixing efficiency. Disturbing liquid flows using active mixers allows uniform concentration of mixed liquids, but those mixers have rather complex structures requiring high fabrication cost (Table 7.2).

TABLE 7.2 Categories of Micromixers Systems: Advantages and Disadvantages of Each Micromixer System

	Categories	Advantages	Disadvantages
Passive micromixer	Parallel lamination	Simple design	Time-consuming process
			Requires long channel
			Low mixing efficiency
	Chaotic advection	Enhanced mixing efficiency than parallel lamination	Low mixing efficiency
		Rapid mixing	
Active micromixer	Piezoelectric	Fast mixing	High fabrication cost
		Uniform concentration	
	Electrohydrodynamic	High mixing efficiency	Complex structure
		Rapid mixing	High fabrication cost
	Magnetohydrodynamic	High mixing efficiency	Complex structure
		Rapid mixing	

7.2.2.4 Sorting and Separation

The last step of the microfluidic flow and particle handling process in highly integrated LOCs is sorting and separation of reacted particles or solutions. To analyze material properties or characteristics of specific analytes, it is necessary to separate or sort out only the required analytes from the bulky mixed solution. Many microfluidic sorting and separation systems have been developed in the last few decades.

Hydrodynamic continuous particle separation can be achieved through asymmetric microchannels that contribute different forces to different-sized particles. Yamada et al. first developed the hydrodynamic asymmetric sorting method using fractionized flows with widths that differed in accordance with the particle size. This allowed for separation of particles from the original fluid as shown in Figure 7.10a. In this method, particles in a liquid reach a pinched segment in the microchannel, where they are aligned to a sidewall of the segment when pushed by a buffer flow. The particles are then successfully separated perpendicular to the direction of the buffer flow according to their sizes by spreading their flow profile inside the microchannel. Furthermore, higher separation performance can be achieved by forming a drain channel that can receive a large portion of the liquid flow from an asymmetric channel, as shown in Figure 7.10b (Takagi et al., 2005; Yamada and Seki, 2004).

Similar to the fractionized pinched-flow method, Yamada et al. also demonstrated a particle-sorting microfluidic method, i.e., filtration, with multiple side-branch channels to control flow rates near the walls at the branch points. Hydrodynamic filtration is one of the most common techniques used to sort out necessary organisms or particles from mixed materials (Takagi et al., 2005; Yamada and Seki, 2006). When the relative flow rates into side channels at a branch point are sufficiently low, only a small portion of the liquid near the sidewalls is withdrawn from the main stream. In this case, particles with a diameter closer to the width of the withdrawn stream do not flow to the side channels. In contrast, if the flow rates distributed into the side channels are increased, particles near the sidewalls flow into the side channels (Yamada and Seki, 2006).

As an example of a continuous particle separation method, laterally positioned obstacles within a channel can operate as filters by blocking the flow of specific-sized particles (Huang et al., 2004). Asymmetric bifurcation is caused around obstacles in a device as shown in Figure 7.10c and d. Because particles smaller than the lane width follow the streamlines, particles from any of the three lanes return to their original lane assignments. However, larger particles follow the center streams of the three lanes between obstacles. The geometric design of obstacles thus allows continuous-flow separation of different-sized particles (Figure 7.10c right).

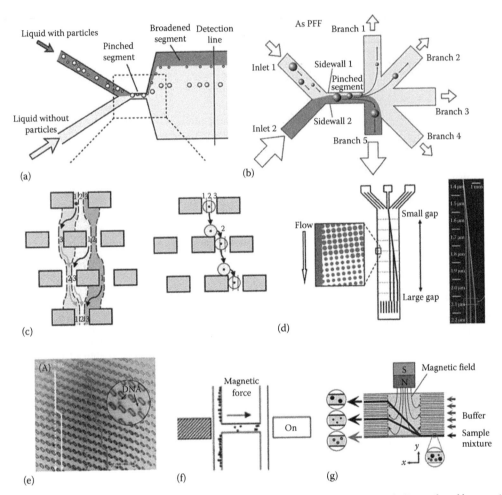

FIGURE 7.10 Various sorting methods. (a) Principle of pinched flow fractionation (PFF). (Reproduced by permission of American Chemical Society. With permission.) (b) Schematic diagrams of particle separation using asymmetric PFF. (Reproduced from Jen, C.-P. et al., *Lab Chip*, 3, 77, 2003. With permission of Royal Society of Chemistry.) (c,d) Continuous particle separation using an obstacle matrix. Particles can be separated according to their size by using an obstacle array. (Reprinted from Huang, L.R. et al., *Science*, 304, 987, 2004. With permission from AAAS.) (e) Electric-field-based sorting. (From Chou, C.F. et al., *Proc. Natl. Acad. Sci. USA*, 96, 13762, 1999. With permission. Copyright 2009, National Academy of Sciences.) (f) Magnetic-field-based sorting with H-filter. (Reprinted from Tergaard, S. et al., *J. Magn. Magn. Mater.*, 194, 156, 1999. With permission from Elsevier.) (g) Continuous-flow separation by magnetophoresis. (Reproduced by permission of American Chemical Society. With permission.)

Electrical or magnetic force can also be used for sorting and filtration. The construction of an electrical force-based sorting system is relatively easy since it only requires introduction of electrodes into channels. By using electrophoresis, a separation technique based on the mobility of ions in an electric field, and asymmetric obstacles, a DNA-sorting technique has been developed for the observation of the motion of different-sized DNA molecules (Chou et al., 1999). Chou et al. demonstrated separation of macromolecules using the principle of rectified Brownian motion and electrophoresis (Figure 7.10e). This technique is not limited to DNA analysis, but it can also be applied to the separation of proteins, colloidal particles, and cells. Dielectrophoresis (DEP), a phenomenon in which a force is exerted on a dielectric particle when it is subjected to a nonuniform electric field, has been successfully applied to manipulation and separation of a variety of biological cells, including bacteria, yeast, and mammalian cells. For example, DEP cell separation can be performed for label-free separations with a minor effect on viability and a limited stress on cells (Huang, 2003).

Magnetic sorting methods are commonly used to separate magnetically labeled particles or cells in a manner similar to electric-sorting techniques. The force on particles varies according to the field strength and the size and magnetism of the particles. These different properties of particles can be utilized as an effective particle-sorting mechanism in an LOC device. The basic principle of magnetic sorting can be explained using H-shaped channel networks (Figure 7.10f) (Tergaard et al., 1999). Two electromagnets positioned at each end of the H-shaped channel are controlled to sort and drag suspended magnetic particles that are mixed with other materials from the left side to the right side of the channel through the connecting branch in the middle. Furthermore, continuous-flow separation of magnetic particles can be performed using free-flow magnetophoresis (Pamme and Manz, 2004). When magnetic particles are pumped into a laminar flow chamber, a magnetic field is applied perpendicular to the direction of flow, and the particles are then separated from nonmagnetic material according to their size and magnetism (Figure 7.10g).

A variety of other methods can also be used for microfluidic sorting via LOCs, such as acoustic force using ultrasonic standing waves to manipulate microparticles and cells (Sergey Kapishnikov et al., 2006) or antigen–antibody reaction mechanisms to sort out biomaterials captured by a specific antibody. Furthermore, the combination of separate techniques can enhance separation efficiency, as demonstrated in research on detection of circulating tumor cells (CTCs) (Nagrath et al., 2007). CTCs serve as an index of cancer progression. However, they are rare, accounting for only one cell per 10^9 hematologic cells in the blood of a cancer patient. To increase the efficiency of sorting CTCs, Nagrath et al. fabricated a "CTC-chip" to detect CTCs with a microfluidic device with an appropriate number of pillars inside the channel. Each pillar is coated with an antibody that provides the specificity necessary for CTC capture. Therefore, the CTC-chip uses a bimodal sorting mechanism by combining obstacles like pillars and antibody–antigen reaction simultaneously. This method finally enabled the identification of CTCs from 99% of patients with approximately 50% purity.

Flow cytometry, an optofluidic sorting system, is a technique for measuring various characteristics of a single cell, including cell type and size, the presence of specific DNA and RNA, and a wide range of membrane-bound and intracellular proteins, by analyzing the optical properties of a single cell (Brown and Wittwer, 2000). Specific proteins and other cellular components can be stained using antibodies conjugated to fluorescent dyes. When the cells pass the detection spot where the excitation laser is illuminated, the fluorescent modules in the labeled cells are excited to a higher-energy state, as shown in Figure 7.11. The use of multiple fluorochromes on different antibodies allows for the detection of various characteristics in a single cell as discussed in Chapter 19.

7.2.3 Analyte Detection

The analyte detection process, another important process for LOCs, detects and analyzes pretreated, injected, and/or fluidically manipulated analytes. As stated, miniaturization of LOCs causes a decrease in analyte sample volume. A decrease in volume implies a decrease in the amount of molecules or ions available for detection. Therefore, scalability at small dimensions and the development of detection technologies with high sensitivity to analyze a very small amount of materials are the main issues in this LOC process.

Optical detection methods are the most common for LOCs as discussed in Chapter 6. The first example of optical detection is UV/visible absorption spectroscopy in which the peak of an absorption spectrum determines the composition or concentration of the sample by the Lambert–Beer law. Absorbance spectroscopy is simple and popular in macroscale analysis, but the decrease in detection volume results in a decrease in the optical path length for detection. Sensitivity is thus limited for microscale absorption spectroscopy. Therefore, absorbance detection technology in LOC devices has been uniquely developed to extend optical path length while maintaining small detection volume. Recently, there have been many trials to develop small-volume and long-path-length absorbance detectors. To lengthen the optical

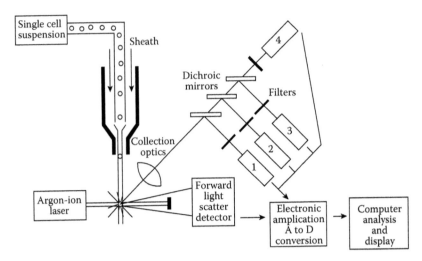

FIGURE 7.11 Schematic of flow cytometry. A single-cell suspension is hydrodynamically focused with a sheath fluid to intersect an argon-ion laser. Signals are collected by a forward angle light scatter detector, a side scatter detector (1), and multiple fluorescence emission detectors (2–4). The signals are amplified and converted to digital form for analysis and display on a computer screen. (Reproduced Reproduced from Brown, M. and Wittwer, C., *Clin. Chem.*, 46, 1221, 2000. With permission. Copyright 2009, American Association for Clinical Chemistry.)

pathway, U- and Z-shaped flow cells have been made for beam collections (Liang et al., 1996; Llobera et al., 2007; Ro et al., 2005; Verpoorte et al., 1992). In these flow cells, an optical beam is transmitted along the flow direction, and UV or visible light absorption is measured along the channel. Usually, optical fibers are used to transmit and collect beams for the detection of the analyte of interest (Liang et al., 1996). In U-cells, laser light launched at one end of the optical fiber passes through the analyte, and the absorbed light is detected at the collecting optical fiber positioned near the other end of the flow cell. In addition to the use of optical fibers in detection technologies, several other components can be used to aid the lengthening of the optical pathway, such as slits to block stray light (Ro et al., 2005) or air bubbles (Llobera et al., 2007). Recently, a completely new absorption spectroscopy using plasmonic resonance energy transfer (PRET)-enabled path-length-free absorption spectroscopy using plasmonic nanoparticles (Choi et al., 2009) as discussed in Chapter 6.

Fluorescence detection is another very common optical detection method with high sensitivity. In fluorescence detection, molecules are illuminated by excitation light, and fluorescence light emitted from the molecules is detected by photodetectors. Lasers and light-emitting diodes (LEDs) with narrow emission bands should be used to insure the spectrum of excitation light does not contain the emission band. Detection filters, which block excitation light and pass emitted light, should be used before the detector to increase signal-to-noise ratio. Many researchers have fabricated highly integrated fluorescence-detection systems by integrating various components such as lenses, waveguides, LEDs, filters, or photodetectors on a single chip. Seo et al. fabricated a self-aligned 2D compound lens to increase device sensitivity and reduce process time for aligning optical components (Seo and Lee, 2004). Chediak et al. assembled several detection components with directly reflecting light to a detector array for simpler structure to increase sensitivity (Chediak et al., 2004), and Schmidt et al. demonstrated a chip-sized spectrometer by combining a band-pass filter with a CMOS camera in a microfluidic channel. As shown in Figure 7.12a, real-time analyzing of the emission spectrum from flowing analytes is obtained with a chip-sized wavelength detector recording the fluorescence of a moving analyte, as shown in Figure 7.12b (Schmidt et al., 2007).

Organic LEDs can be used as excitation sources and organic photodiode as detectors to reduce manufacturing costs as much as possible. Chediak et al. used two crossed polarizers enclosing a single

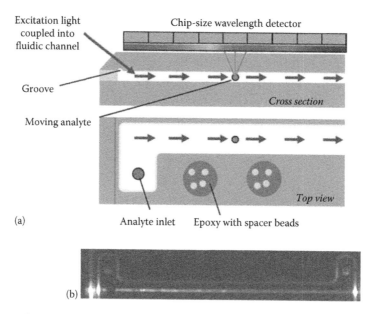

FIGURE 7.12 (See color insert following page 11-20.) (a) (Top) Cross-sectional view of a fluidic chamber. Light is guided within the fluid, thereby continuously exciting the analyte. The chip-sized wavelength detector records the fluorescence spectrum of the analyte while it moves along the detector. (Bottom) Top view of the fluidic chamber. (b) Photograph of the top view of the chamber, demonstrating large-volume fluorescence excitation. (Reproduced from Takagi, J. et al., *Lab Chip*, 5, 778, 2005. With permission of Royal Society of Chemistry.)

microfluidic device to reduce noise due to a leakage from excitation light. The polarizers filter out excitation light and only transmit emission signals to the photodiode detector. Utilization of organic optoelectronic devices is a cost-effective alternative for detection in microfluidic LOC systems (Chediak et al., 2004). For multiplexing applications, total microscopic systems can be integrated into chip-scale devices. Kwon et al. demonstrated integrating laser scanning microscopes on a chip using scanning microlenses (Kwon, 2004).

In contrast to absorbance or fluorescence detection, chemiluminescence detection does not require an excitation source. Instead, photochemical emission is achieved through using direct analyte binding or enzyme labeling to the sample. For example, a multiplexed chemiluminescent immunoassay sensor detects toxins, bacteria, and viruses by using a capillary flow biosensor, controlling a maximum of 10 different channels for 10 different immunoassays (Yacoub-George et al., 2007). The system has high sensitivity and short assay times because it is designed to have capillaries arranged in parallel.

Interferometric detection is a label-free detection method based on analyte binding in a sample with high sensitivity. Essentially, a single light source is divided into two light paths, namely, a reference path and an optical path. Refractive index change due to the analyte binding in the sample causes a phase shift in the optical path compared to the reference path and affects the interference patterns of the two beams. The binding event is observed through the imaging of these interference patterns. As shown in Figure 7.13, observation of interference patterns from three different analytes is carried out with four channels composed of three optical paths and one reference path in parallel. Guided beams from four channels form an interference pattern according to the number of analytes attached to each channel (Ymeti et al., 2007). Molecular binding interferometric detection technology can also be employed in free solution, instead of on the surface, by using back-scattering interferometry (Bornhop et al., 2007).

Surface plasmon resonance (SPR) detection measures refractive index change on a functionalized metal surface caused by the probe molecule's binding. The incident light passes through a prism at a

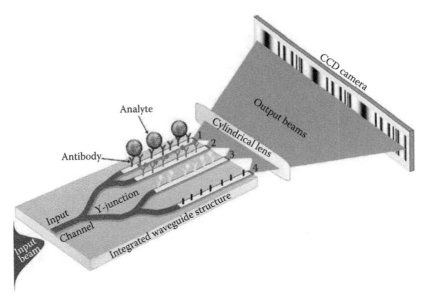

FIGURE 7.13 Schematic of the four-channel integrated optical Young interferometry detector. Channels 1, 2, and 3 are the measuring channels, and 4 is the reference channel. (Reprinted from Tergaard, S. et al., *J. Magn. Magn. Mater.*, 194, 156, 1999. With permission from Elsevier.)

specific angle with respect to the metal film, exciting the surface plasmon on the metal surface. The SPR absorbance causes the abrupt decline in light intensity measured by the system. The SPR detection system has been simplified and minimized from laboratory scale to microscale. Several commercial applications of this system are available, including the Sensata Spreeta SPR sensor, which observes biological binding reactions by detecting reflected light from an SPR chip (Chinowsky et al., 2003), or the briefcase-style SPR platform that combines optical elements and an LED light source (Chinowsky et al., 2007).

While optical detection requires rather bulky and sophisticated instrumentation, electrochemical detection (ECD) can greatly simplify detection instrumentation. Solution conductivity, current, and potential changes in a sample concentration can be measured with high sensitivity using only electrochemically active electrodes in microfluidic channels. This feature benefits the miniaturization of an LOC system for point-of-care devices (Myers and Lee, 2008; Verpoorte, 2003).

In all these detection schemes, the sample concentration required for detection is of importance. Various preconcentration schemes based on nanochannel or ion depletion may be incorporated for the detection of samples in very low concentrations (Kang et al., 2006; Lee et al., 2008a,b; Ying-Chih et al., 2005).

7.3 Applications of Optofluidic Lab-on-a-Chip Systems

Since optofluidic LOC devices can provide sample preparation, fluid and particle handling, and detection, they are promising solutions for the biosensing field, which will be thoroughly discussed in Part III of this book. In the previous section, we mainly described various components of LOC systems. In this section, we briefly introduce some examples of optofluidic LOC systems that integrate these components for different application areas such as DNA sequencing, particle-synthesis, encoded-particle-based biochips, and fluidic self-assembly. In the sequencing application, the optofluidic detection scheme is used to increase sequencing throughputs. Optics provides detection and fluidics provides a way to manipulate molecules in solution for separation and induction of reactions. Optofluidic concepts can be used in particle synthesis, generation of encoded particles, and fluidic self-assembly in nano- and microscale. These examples of optofluidic detection and fabrication should give a taste of important characteristics of optofluidic systems such as noninvasiveness and flexibility.

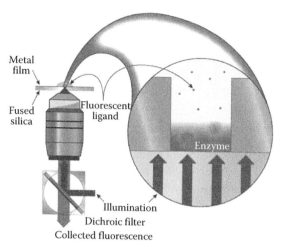

Metal film

Fused silica

Fluorescent ligand

Enzyme

Illumination
Dichroic filter
Collected fluorescence

FIGURE 7.14 An apparatus for single-molecule enzymology using zero-mode waveguides. (Reprinted from Levene, M.J. et al., *Science*, 299, 682, 2003. With permission from AAAS.)

7.3.1 LOC Systems for DNA Sequencing

In DNA sequencing, micro-total analysis devices have several advantages over capillary array electrophoresis with simple injection. Here, we briefly deal with sequencing methods that overcome DNA polymerase-dependent strategies through the use of LOC devices.

Sanger sequencing, also known as chain terminator sequencing, has been one of the most influential DNA sequencing methods (Sanger, 1988). Gel electrophoresis had been heavily used as a read-out method that sorts out fragments of labeled, single-strand DNA by length. However, running the gel was a very time-consuming process, usually taking several hours. The introduction of capillary electrophoresis greatly reduced the time needed for the electrophoresis by taking advantage of scaling. Microfluidic LOC systems based on a microfabricated array of capillaries and integrated PCR function have been successfully demonstrated to greatly increase the throughput of sequencing (Woolley et al., 1996). This chip-based technique was improved by further automating sample preparation steps into the microfluidic chips. Toriello et al. demonstrated complete genomic analysis based on automatic sample genomic preparation from the whole cell (Toriello et al., 2008).

Pyrosequencing is another technique based on the detection of nucleotide incorporation using primer-directed polymerase extension. This technique measures the release of inorganic pyrophosphate, which is proportionally converted into visible light by a series of enzymatic reactions, and the sequence can be deduced iteratively (Hyman, 1988; Ronaghi et al., 1998). Incorporation of the concept of pyrosequencing into state-of-the-art optofluidic instrumentation enabled sequence read-out from a single DNA molecule, thereby eliminating the need for PCR. The zero-mode waveguide method used nanofabricated subwavelength-sized holes in a metal film to observe single-molecule activity of DNA polymerase on a chip. By using this method, enzymes are absorbed onto the bottom of the waveguides in the presence of a solution containing the fluorescently tagged ligand molecules. The coverslip is illuminated through a microscope objective from below, and fluorescence is collected through the same objective (Figure 7.14) (Levene et al., 2003). The enzymatic synthesis of double-stranded DNA by DNA polymerase is observed by detecting light signals from the zero-mode waveguides. The subwavelength aperture greatly reduces optical detection volume to sense DNA synthesis action, thus enabling the study of single-molecular dynamics at much higher concentration than other single-molecular techniques. This single-molecule analysis at higher concentrations is possible by providing an optically efficient, highly parallel, simple platform (Levene et al., 2003).

7.3.2 Optofluidic LOC Systems for Particle Synthesis

An LOC provides a synthesis platform for making microparticles widely used in various applications such as microbeads for biological molecular handling, microcapsules for drug delivery, and microabrasives for chemical-mechanical polishing. Conventional fabrication of spherical microparticles is based on batch fabrication using emulsion of two immiscible fluids. Although this emulsion-based method can generate spherical particles in large volumes, fabricated particles vary significantly in size. The microfluidic concept has been recently applied to generate monodispersed microparticles. Compared with the emulsion method, particle generation based on LOCs can mean better controllability and uniformity with regard to particle size and materials. For example, the T-junction method and flow-focusing method enable the generation of uniform droplets. By using these methods, droplet size can be controlled through varying the flow rate and the channel dimension (Thorsen et al., 2001). Spherical particles have been fabricated using various approaches based on monodispersed droplet-generation methods (Oh et al., 2006). However, the shapes of the particle based on these fabrication methods are limited to the spherical type because they are derived from droplet formation.

Applying photolithography to transfer a geometric pattern in a photomask onto a photoresist-coated substrate provides greater flexibility in particle shape control. After filling photoresist into a microfluidic channel, various polymeric structures can be fabricated inside, using the microfluidic channel as a substrate for photolithography by *in situ* photopolymerization (Figure 7.15a). By using this technique, various microfluidic active components that are anchored in the channel, such as a hydrogel microvalve, which swells and shrinks in response to the solution's pH changes, can be formed (Beebe, 2000).

Microscope projection photolithography is a simple photolithography technique in which a transparent photomask is inserted in front of a microscope objective lens to pattern a photoresist-coated substrate. This method results in rapid prototyping of possible test structures quickly and inexpensively (Love et al., 2002). Dendukuri et al. employed this lithography technique to photopolymerize oligomeric diacrylate monomers flowing in a PDMS microfluidic device, a process known as high-throughput continuous-flow lithography (CFL). As shown in Figure 7.15b, free-floating microparticles of triangular, cubical, cylindrical, and other irregular shapes can be fabricated in a microfluidic channel. PDMS, a popular material for microfluidic channels, is highly oxygen permeable. The concentration of oxygen is very high near the PDMS channel surface, allowing for oxygen to take up the initiator radicals for photopolymerization. Polymerization is locally inhibited on the walls of the channel, thereby enabling fabrication of the free-floating particles (Dendukuri et al., 2006).

Instead of a static photomask, a dynamic mask scheme is applied by using a spatial light modulator, known as DMD (digital micromirror device), which generates arbitrary patterns using micromirror array patterns. Optofluidic-maskless lithography (OFML) is a technique that uniquely combines this dynamic mask scheme and CFL techniques in microfluidic channels (Figure 7.16). This provides real-time control of the *in situ* polymerization process to dynamically synthesize extruded polymeric microstructures with various two-dimensional shapes (Chung et al., 2007), in contrast to CFL, which uses physical masks. As shown in Figure 7.16(right), by continuously controlling exposure spot in real time, curved polymer wire and multicompositional structure are fabricated using the OFML technique.

For continuous high-throughput fabrication of a large number of distinctive microparticles, maskless-lithography techniques with programmable exposure patterns can significantly improve the performance and flexibility of fluidic lithography systems. In continuous *in situ* polymerization methods, the particle resolution and the maximum throughput are limited by the flow rate of the oligomer. Since the oligomer is moving in a fluidic channel, the particles are smeared during the synthesis of particles by photopolymerization. In order to obtain the high-resolution feature, the flow rate should be decreased. However, when the flow rate is decreased, the throughput of production also decreases and diffusion occurs in laminar flows. In order to overcome these limitations, stop-flow lithography has been developed. In stop-flow lithography, the flow of oligomer is controlled by a three-way valve. The flow of oligomer is stopped during synthesis of particles and flushed out after the fabrication step. During

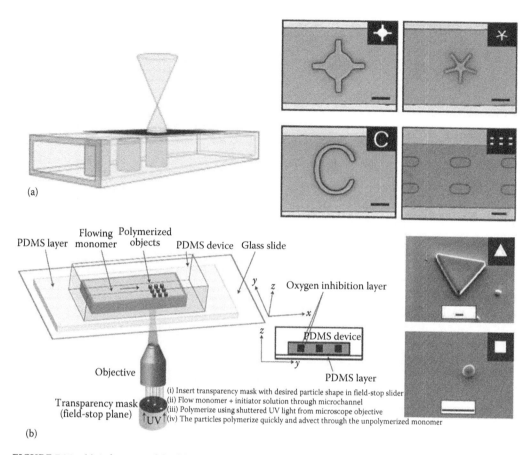

FIGURE 7.15 (a) A diagram of the fabrication method and images showing a variety of shapes that were polymerized. (Left) The fabrication method of hydrogel polymerization with various shapes. (b) Schematic depicting the experimental setup used in continuous-flow lithography (CFL) and various microstructures fabricated by CFL. (Reprinted from Macmillan Publishers Ltd: Braeckmans, K. et al., *Nat. Mater.*, 2, 169, 2003. With permission. Copyright 2009.)

photopolymerization, the particles are precisely fabricated without smear because the oligomer flow stops (Dendukuri et al., 2007).

Stop-flow lithography requires repetitive flow control for the fabrication and flushing out of microparticles. The moving-mask lithography technique using OFML is used to generate high-precision photopolymerized microstructures, without slowing the flow of the oligomer. Moving-mask lithography synchronously shifts the UV exposure pattern with the oligomer flow in order to prevent it from smearing during particle synthesis. This technique enables the generation of a dynamic mask moved in sync with the speed of the photopolymer flow rate as measured by a computer vision system (Figure 7.17a). The particles are precisely photopolymerized by moving the UV light pattern and floating along the oligomer flow. Moving-mask lithography improves the resolution of the lithography without interrupting the oligomer flow.

7.3.3 Optofluidic LOC Systems for Encoded-Particle-Based Biochips

Two-dimensional biochip array technologies are based on depositing probe molecules or cells on a 2D planar array. The location of the probe molecules in the array is used to identify the molecules. To increase the number of probes in a planar array, the size of each probe should be decreased, resulting

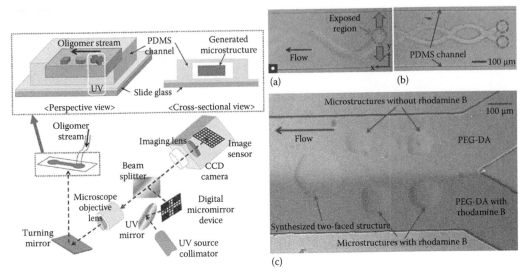

FIGURE 7.16 Schematic diagram of the proposed optofluidic maskless lithography system for dynamic control of the photopolymerization process in microfluidic devices. (a) Curved polymer wire fabricated by moving the circular exposure pattern radius. (b) Overlapped microwires synthesized simultaneously by moving two circular UV beams. (c) A microscope image of multimaterial microstructure synthesis at a Y-shaped microfluidic channel. (Reproduced from Chung, S.E. et al., *Appl. Phys. Lett.*, 91, 041106, 2007. With permission. Copyright 2009, American Institute of Physics.)

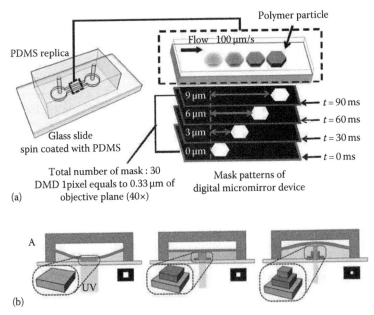

FIGURE 7.17 (a) Schematic diagram of moving-mask lithography. The mask is shifted at the same speed as the flow rate. (b) Schematic illustration of the 3D-OFML. The height of the bottom channel is controlled via deformation of the PDMS membrane under pneumatic pressure of the top chamber. (Reproduced from Lee, S.A. et al., *Lab Chip*, 9, 1670, 2009. With permission of Royal Society of Chemistry.)

in decreased manufacturing cost for the biochip array. Patterning large number of probes on a surface by photolithography or spotting takes up the majority of the manufacturing cost of the planar array. Another limitation of 2D planar arrays comes from the fabrication process. When the fabrication process of different chips changes, immobilization conditions of probe materials on the surface of those chips become unreliable. Thus, experimental results may not be reproducible even though the experiment is carried out with the same material in each biochip.

In an encoded particle-based approach, a specific biological molecule is attached to the surface of the particle with a code used to identify the attached molecule. The encoded- particle-based biochips have advantages over the planar array because of their high analysis throughput from high reaction kinetics of particles with probes and their low manufacturing cost. Each type of encoded particle can be fabricated in a batch process, thereby allowing cost-effective mass production under a single condition. The particles can be arranged into an array for scalable biochip applications, in which case simply adding particles from different batches can increase the number of probes without additional fabrication steps (Braeckmans et al., 2002).

Optical and fluidic methods have been used for encoding and detecting particles. Luminex Corporation has used organic color-coded microspheres (Figure 7.18a). The color codes are generated by the combination of different sizes and color intensities to identify an individual microsphere. The microspheres are coated with a reagent specific to a particular bioassay. These microspheres can capture and detect the

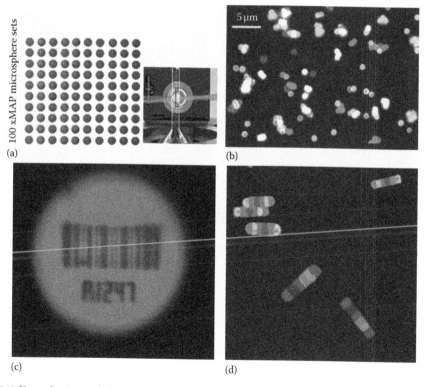

(a)   (b)   (c)   (d)

FIGURE 7.18 (See color insert following page 11-20.) (a) Color encoding (http://www.luminexcorp.com/index. html). (b) Quantum dots: fluorescence micrograph of a mixture of CdSe/ZnS QD-tagged beads emitting single color. (Reprinted from Macmillan Publishers Ltd: Han, M. et al., *Nat. Biotech.*, 19, 631, 2001. With permission. Copyright 2009.) (c) 2D graphical coding. Polybeads with a barcode patterned by photobleaching. (Reprinted from Macmillan Publishers Ltd: Liu, G. L. et al., *Nat. Mater.*, 5, 27, 2006. With permission. Copyright 2009.) (d) Color barcode using doping with rare-earth ions. (From Dejneka, M.J. et al., *Proc. Natl. Acad. Sci. USA*, 100, 389, 2003. With permission. Copyright 2009, National Academy of Sciences.)

specific analytes from a sample. They assist the multiplexing process by allowing 100 unique assays with a single sample in a rapid and precise manner. To analyze the results of the assay, the internal dyes in the microspheres are excited by laser and detected by an optical analyzer based on flow cytometry. The different emission spectra from two organic dyes require different wavelengths for excitation. However, photoluminescent nanoparticles, known as quantum dots, can be excited with the same wavelength. Their emission spectra are narrow enough to generate multiple codes resolved in the visible region. They also have more resistance to photobleaching compared to fluorescent dyes. Because of these characteristics, quantum dots are regarded as ideal color-coding material (Figure 7.18b) (Han et al., 2001).

The codes for microcarriers or microparticles can also be achieved with 2D graphical patterns such as barcodes. In order to generate the 2D graphical pattern as a code, optical and chemical approaches have been used. Photobleaching was used to generate unique graphical patterns on microbeads. Normally, fluorescent dyes gradually lose their fluorescent characteristics when they are excited by the laser. By using this phenomenon, a spatially selective focused laser beam using a confocal microscope was used for patterning unique graphical patterns on polystyrene microbeads (Figure 7.18c) (Braeckmans et al., 2003). However, this method has a limitation that the codes fade over time due to the diffusion of the fluorescent molecule. Microbarcodes composed of glass doped with rare-earth ions are resistant to photobleaching (Figure 7.18d). Rare-earth ions are excited with the same UV wavelength, and their emission spectra are in the visible region. However, the production time for one milliliter is about one month (Dejneka et al., 2003).

As described in the previous section, CFL generates a large number of polymeric microparticles of various shapes in a PDMS microfluidic device (Dendukuri et al., 2006). Multifunctional encoded particles were fabricated based on this CFL technique. Multifunctional encoded particles are composed of analyte-encoding parts and target capturing parts synthesized at the interface between two monomer streams that form laminar flows in a single step (Figure 7.19a through c). With this technique, a theoretically infinite number of codes can be generated by changing the photomask inserted into the optical system (Figure 7.19d). Additionally, various probes for capturing target molecules can be easily loaded by binding them with monomer and flowing them into a microfluidic channel (Pregibon et al., 2007). In this system, the optofluidic technique is used not only for synthesis of encoded particles but also for

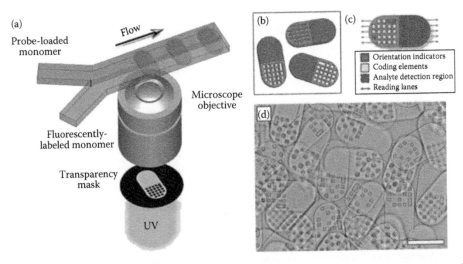

FIGURE 7.19 (a) Schematic diagram of dot-coded particle synthesis showing polymerization across two adjacent laminar streams to make a single probe that is half-fluorescent (shown in (b)). (c) Diagrammatic representation of particle features for encoding and analyte detection. Encoding scheme shown allows for the generation of 1,048,576 unique codes. (d) Differential interference contrast (DIC) image of particles generated by using the scheme shown in (a). (Reprinted from Pregibon, D.C. et al., *Science*, 315, 1393, 2007. With permission from AAAS.)

analysis of encoded particles on chip. Encoded particles after the hybridization experiment are flowed through a microfluidic device for detection. The hybridized and unhybridized encoded particles are aligned in a narrow channel whose width is slightly larger than the width of the encoded particle. In the flow-focused channel, an aligned particle can be observed on an inverted fluorescence microscope. The scheme provides high-throughput analysis among other simple, cost-effective advantages.

7.3.4 Optofluidic LOC Systems for Scalable Self-Assembly

Self-assembly is a fundamental process in which small components spontaneously join or link themselves to form a large organized structure in nature. This principle has been developed as a practical fabrication method of microstructures and nanostructures because it enables parallel and scalable production. In conventional micro-/nanofabrication approaches such as photolithography or nanoimprint lithography, the product scale and heterogeneity are determined by the lithography process; thus, it is difficult and expensive to scale up the product size and number. In contrast, in a self-assembly method, the product size and number are not determined by the device scale or the number of steps but by simple configuration of the assembly conditions such as the driving force, reaction time, and number of components. For example, in order to fabricate a hydrogel microstructure in an LOC device, the conventional lithography process includes deposition, alignment, and exposure for each material as shown in Figure 7.20a (Albrecht et al., 2005). For heterogeneous microstructures with N types of material, N times of the above processes are required. However, fluidic self-assembly can reduce the number of steps in the fabrication process (Stauth and Parviz, 2006). In the fluidic self-assembly process, after prefabricating different hydrogels at different sites, the various hydrogels are introduced into LOCs and fluidically assembled at the assembly site that anchors the structures (Figure 7.20b). This fluidic self-assembly technique is often massively parallel and faster; however, ordered control of assembly is difficult and the design of the assembled structures is limited.

In order to assemble more complex or highly ordered structures with fluidic self-assembly, shape-matching approaches have been developed. Assembly yield using shape-matching approaches is not as high as conventional robotic assembly owing to the probabilistic nature of self-assembly. In order to have a higher probability of matching and assembling, it is necessary to increase the chance of matching

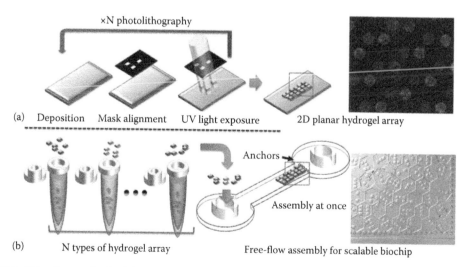

FIGURE 7.20 Two methods of fabricating biochips using photolithography and free-flow particle assembly. (a) Photolithography process for N number of materials for a 2D hydrogel array. (Reproduced from Schmidt, O. et al., *Lab Chip*, 7, 626, 2007. With permission of Royal Society of Chemistry.) (b) Fluidic self-assembly for N types of hydrogels.

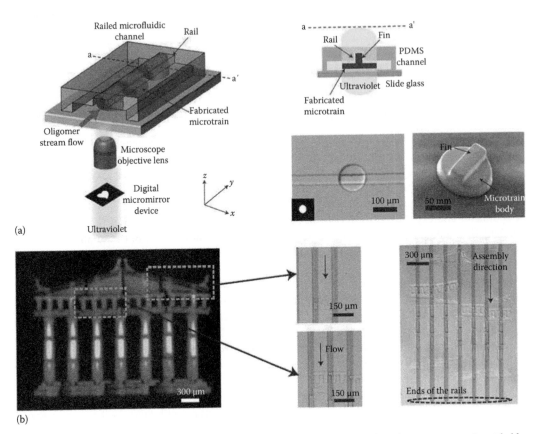

FIGURE 7.21 (a) Concept of railed microfluidics and a guiding mechanism. A finned microstructure is guided by a groove pattern on the top of the microchannel. (b) Complex self-assembly in railed microfluidics. (Reprinted from Macmillan Publishers Ltd: Beebe, D.J. et al., *Nature*, 404, 588, 2000. With permission. Copyright 2009.)

components by overloading the number of parts floating around the assembly sites, as demonstrated in fluidic self-assembly of RFID or assembly of DNA origami (Rothemund, 2006; Smith, 2003). Mass production of microcomponents is required in most fluidic self-assembly processes. However, if microstructures are accurately guided in a fluidic environment, an efficient fluidic self-assembly process combining the advantages of both high-yield robotic assembly and high-throughput fluidic self-assembly could be achieved.

Railed microfluidics is a deterministic high-yield and high-throughput fluidic assembly technique based on OFML. As shown in Figure 7.21, a groove, patterned by a two-step mold fabrication process, functions as a guided rail track inside the microfluidic channel. After the channel is filled with a UV-curable oligomer solution, a microtrain with a fin is created on the rail using *in situ* photopolymerization. The finned microtrains normally should follow the flow field in the microfluidic channel, but the matching of the fin and the rail enables microtrains to deviate from the flow field and instead follow the rail. In this manner, the movement of the particles can be controlled fluidically to allow assembly in the microfluidic channel. The real benefit of railed microfluidics over the conventional fluidic self-assembly lies in its capability to assemble complex systems made up of a large number of different parts (Figure 7.21b) (Chung et al., 2008).

An LOC provides a high-throughput production platform for making uniform microparticles. Combining optical devices such as a microscope and a DMD enables the generation of various shapes of microparticles. OFML that uses this method has potential as a fabrication technology to generate 3D structures with various configurations of high resolution and complexity.

7.4 Conclusion

This chapter described and discussed various optofluidic LOC systems. First, a general overview of an LOC system was provided, with an emphasis on the benefit of miniaturization and integration in Section 7.1. Next, common work flow and components of LOC systems were introduced with a few typical examples in Section 7.2. Finally, system-level implementations of LOCs for specific applications were described in Section 7.3.

The optofluidic implementation in the LOC field is obviously beneficial due to its noninvasive characteristics. This noninvasive access to the contents inside microfluidic channels can greatly simplify the fabrication of the devices and provides great flexibility. By integrating optics and microfluidics into one chip, various optical manipulation technologies using optical tweezers or flow cytometry and optical detection systems can be achieved. Based on this flexibility and integration, future chip-based LOC systems will continue to evolve as next generation clinical diagnostics.

References

Affymetrix, http://www.Affymetrix.Com/index.Affx.

Alaverdian, L., S. Alaverdian, O. Bilenko et al. 2002. A family of novel DNA sequencing instruments based on single-photon detection. *Electrophoresis* 23: 2804–2817.

Albrecht, D. R., V. L. Tsang, R. L. Sah et al. 2005. Photo- and electropatterning of hydrogel-encapsulated living cell arrays. *Lab on a Chip* 5: 111–118.

Andersson, H., W. van der Wijngaart, P. Nilsson et al. 2001a. A valve-less diffuser micropump for microfluidic analytical systems. *Sensors and Actuators B: Chemical* 72: 259–265.

Andersson, H., W. v. d. Wijngaart, and G. Stemme. 2001b. Micromachined filter-chamber array with passive valves for biochemical assays on beads. *Electrophoresis* 22: 249–257.

Ashkin, A., J. M. Dziedzic, J. E. Bjorkholm et al. 1986. Observation of a single-beam gradient force optical trap for dielectric particles. *Optics Letters* 11: 288–290.

Auroux, P.-A., D. Iossifidis, D. R. Reyes et al. 2002. Micro total analysis systems. 2. Analytical standard operations and applications. *Analytical Chemistry* 74: 2637–2652.

Bau, H. H., J. H. Zhong, and M. Q. Yi. 2001. A minute magneto hydro dynamic (mhd) mixer. *Sensors and Actuators B: Chemical* 79: 207–215.

Beebe, D. J. 2000. Functional hydrogel structures for autonomous flow control inside microfluidic channels. *Nature* 404: 588–590.

Belgrader, P., D. Hansford, G. T. Kovacs et al. 1999. A minisonicator to rapidly disrupt bacterial spores for DNA analysis. *Analytical Chemistry* 71: 4232–4236.

Bessoth, F. G., A. J. deMello, and A. Manz. 1999. Microstructure for efficient continuous flow mixing. *Analytical Communications* 36: 213–215.

Bornhop, D. J., J. C. Latham, A. Kussrow et al. 2007. Free-solution, label-free molecular interactions studied by back-scattering interferometry. *Science* 317: 1732–1736.

Bourouina, T., A. Bosseboeuf, and J. P. Grandchamp. 1997. Design and simulation of an electrostatic micropump for drug-delivery applications. *Journal of Micromechanics and Microengineering* 7: 186–188.

Braeckmans, K., S. C. De Smedt, M. Leblans et al. 2002. Encoding microcarriers: Present and future technologies. *Nature Reviews Drug Discovery* 1: 447–456.

Braeckmans, K., S. C. De Smedt, C. Roelant et al. 2003. Encoding microcarriers by spatial selective photobleaching. *Nature Materials* 2: 169–173.

Brown, M. and C. Wittwer. 2000. Flow cytometry: Principles and clinical applications in hematology. *Clinical Chemistry* 46: 1221–1229.

Brown, A. B. D., C. G. Smith, and A. R. Rennie. 2001. Pumping of water with ac electric fields applied to asymmetric pairs of microelectrodes. *Physical Review E* 6302: 016305.

Burns, M. A., B. N. Johnson, S. N. Brahmasandra et al. 1998. An integrated nanoliter DNA analysis device. *Science* 282: 484–487.

Carlo, D. D., L. Y. Wu, and L. P. Lee. 2006. Dynamic single cell culture array. *Lab on a Chip* 6: 1445–1449.

Chan, E. M., A. P. Alivisatos, and R. A. Mathies. 2005. High-temperature microfluidic synthesis of CdSe nanocrystals in nanoliter droplets. *Journal of American Chemical Society* 127: 13854–13861.

Chang, C. C. and R. J. Yang. 2004. Computational analysis of electrokinetically driven flow mixing in microchannels with patterned blocks. *Journal of Micromechanics and Microengineering* 14: 550–558.

Chediak, J. A., Z. Luo, J. Seo et al. 2004. Heterogeneous integration of CdS filters with GaN LEDs for fluorescence detection microsystems. *Sensors and Actuators: A* 111: 1–7.

Chen, H. and J. C. Meiners. 2004. Topologic mixing on a microfluidic chip. *Applied Physics Letters* 84: 2193–2195.

Chen, C. H. and J. G. Santiago. 2002. A planar electroosmotic micropump. *Journal of Microelectromechanical Systems* 11: 672–683.

Chinowsky, T. M., J. G. Quinn, D. U. Bartholomew et al. 2003. Performance of the Spreeta 2000 integrated surface plasmon resonance affinity sensor. *Sensors and Actuators B: Chemical* 91: 266–274.

Chinowsky, T. M., M. S. Grow, K. S. Johnston et al. 2007. Compact, high performance surface plasmon resonance imaging system. *Biosensors and Bioelectronics* 22: 2208–2215.

Chiou, P. Y., A. T. Ohta, and M. C. Wu. 2005. Massively parallel manipulation of single cells and microparticles using optical images. *Nature* 436: 370–372.

Choi, Y., T. Kang, and L. P. Lee. 2009. Plasmon resonance energy transfer (PRET)-based molecular imaging of cytochrome *c* in living cells. *Nano Letters* 9: 85–90.

Chou, C. F., O. Bakajin, S. W. P. Turner et al. 1999. Sorting by diffusion: An asymmetric obstacle course for continuous molecular separation. *Proceedings of the National Academy of Sciences of the United States of America* 96: 13762–13765.

Chronis, N., G. L. Liu, K. H. Jeong et al. 2003. Tunable liquid-filled microlens array integrated with microfluidic network. *Optics Express* 11: 2370–2378.

Chung, B. G., L. A. Flanagan, S. W. Rhee et al. 2005. Human neural stem cell growth and differentiation in a gradient-generating microfluidic device. *Lab on a Chip* 5: 401–406.

Chung, S. E., W. Park, H. Park et al. 2007. Optofluidic maskless lithography system for real-time synthesis of photopolymerized microstructures in microfluidic channels. *Applied Physics Letters* 91: 041106–041103.

Chung, S. E., W. Park, S. Shin et al. 2008. Guided and fluidic self-assembly of microstructures using railed microfluidic channels. *Nature Materials* 7: 581–587.

Clayton, J. 2005. Go with the microflow. *Nature Methods* 2: 621–625.

Cui, X., L. M. Lee, X. Heng et al. 2008. Lensless high-resolution on-chip optofluidic microscopes for *Caenorhabditis elegans* and cell imaging. *Proceedings of the National Academy of Sciences* 105: 10670.

Dejneka, M. J., A. Streltsov, S. Pal et al. 2003. Rare earth-doped glass microbarcodes. *Proceedings of the National Academy of Sciences of the United States of America* 100: 389–393.

Dendukuri, D., D. C. Pregibon, J. Collins et al. 2006. Continuous-flow lithography for high-throughput microparticle synthesis. *Nature Materials* 5: 365.

Dendukuri, D., S. S. Gu, D. C. Pregibon et al. 2007. Stop-flow lithography in a microfluidic device. *Lab on a Chip* 7: 818–828.

Dittrich, P. S. and A. Manz. 2006. Lab-on-a-chip: Microfluidics in drug discovery. *Nature Reviews Drug Discovery* 5: 210–218.

Dittrich, P. S., M. Jahnz, and P. Schwille. 2005. A new embedded process for compartmentalized cell-free protein expression and on-line detection in microfluidic devices. *ChemBioChem* 6: 811–814.

El-Ali, J., P. K. Sorger, and K. F. Jensen. 2006. Cells on chips. *Nature* 442: 403–411.

El Moctar, A. O., N. Aubry, and J. Batton. 2003. Electro-hydrodynamic micro-fluidic mixer. *Lab on a Chip* 3: 273–280.

Enger, J., M. Goksor, K. Ramser et al. 2004. Optical tweezers applied to a microfluidic system. *Lab on a Chip* 4: 196–200.

Erickson, D. and D. Q. Li. 2002. Influence of surface heterogeneity on electrokinetically driven microfluidic mixing. *Langmuir* 18: 1883–1892.

Fan, Z. H., S. Mangru, R. Granzow et al. 1999. Dynamic DNA hybridization on a chip using paramagnetic beads. *Analytical Chemistry* 71: 4851–4859.

Furdui, V. I., J. K. Kariuki, and D. J. Harrison. 2003. Microfabricated electrolysis pump system for isolating rare cells in blood. *Journal of Micromechanics and Microengineering* 13: S164–S170.

Glasgow, I. and N. Aubry. 2003. Enhancement of microfluidic mixing using time pulsing. *Lab on a Chip* 3: 114–120.

Gobby, D., P. Angeli, and A. Gavriilidis. 2001. Mixing characteristics of t-type microfluidic mixers. *Journal of Micromechanics and Microengineering* 11: 126–132.

Gong, Q. L., Z. Y. Zhou, Y. H. Yang et al. 2000. Design, optimization and simulation on microelectromagnetic pump. *Sensors and Actuators a-Physical* 83: 200–207.

Guo-Hua, F. and K. Eun Sok. 2004. Micropump based on PZT unimorph and one-way parylene valves. *Journal of Micromechanics and Microengineering* 14: 429–435.

Han, M., X. Gao, J. Z. Su et al. 2001. Quantum-dot-tagged microbeads for multiplexed optical coding of biomolecules. *Nature Biotechnology* 19: 631–635.

Handique, K., D. T. Burke, C. H. Mastrangelo et al. 2001. On-chip thermopneumatic pressure for discrete drop pumping. *Analytical Chemistry* 73: 1831–1838.

Hartshorne, H., C. J. Backhouse, and W. E. Lee. 2004. Ferrofluid-based microchip pump and valve. *Sensors and Actuators B: Chemical* 99: 592–600.

Hatch, A., A. E. Kamholz, K. R. Hawkins et al. 2001. A rapid diffusion immunoassay in a t-sensor. *Nature Biotechnology* 19: 461–465.

He, B., L. Tan, and F. Regnier. 1999. Microfabricated filters for microfluidic analytical systems. *Analytical Chemistry* 71: 1464–1468.

Hessel, V., H. Lowe, and F. Schonfeld. 2005. Micromixers—A review on passive and active mixing principles. *Chemical Engineering Science* 60: 2479–2501.

Hong, C. C., J. W. Choi, and C. H. Ahn. 2004. A novel in-plane passive microfluidic mixer with modified Tesla structures. *Lab on a Chip* 4: 109–113.

Huang, Y., S. Joo, M. Duhon et al. 2003. Dielectrophoretic cell separation and gene expression profiling on microelectronic chip arrays, *Analytical Chemistry* 74: 3362–3371.

Huang, L. R., E. C. Cox, R. H. Austin et al. 2004. Continuous particle separation through deterministic lateral displacement. *Science* 304: 987–990.

Hung, L. H., K. M. Choi, W. Y. Tseng et al. 2006. Alternating droplet generation and controlled dynamic droplet fusion in microfluidic device for CdS nanoparticle synthesis. *Lab on a Chip* 6: 174–178.

Hyman, E. D. 1988. A new method of sequencing DNA. *Analytical Biochemistry* 174: 423–436.

Ismagilov, R. F., A. D. Stroock, P. J. A. Kenis et al. 2000. Experimental and theoretical scaling laws for transverse diffusive broadening in two-phase laminar flows in microchannels. *Applied Physics Letters* 76: 2376–2378.

Iverson, B. D. and S. V. Garimella. 2008. Recent advances in microscale pumping technologies: A review and evaluation. *Microfluidics and Nanofluidics* 5: 145–174.

Jen, C. P., C. Y. Wu, Y. C. Lin et al. 2003. Design and simulation of the micromixer with chaotic advection in twisted microchannels. *Lab on a Chip* 3: 77–81.

Jeon, N. L., S. K. W. Dertinger, D. T. Chiu et al. 2000. Generation of solution and surface gradients using microfluidic systems. *Langmuir* 16: 8311–8316.

Kamholz, A. E. and P. Yager. 2002. Molecular diffusive scaling laws in pressure-driven microfluidic channels: Deviation from one-dimensional Einstein approximations. *Sensors and Actuators B: Chemical* 82: 117–121.

Kang, J., H. Chun, I. H. Shin et al. 2006. Preliminary evaluation of the use of a CDMA-based emergency telemedicine system. *Journal of Telemedicine and Telecare* 12: 422–427.

Kapishnikov, S., V. Kantsler and V. Steinberg. 2006. Continuous particle size separation and size sorting using ultrasound in a microchannel. *Journal of Statistical Mechanics* P01012: 1–15.

Khandurina, J., T. E. McKnight, S. C. Jacobson et al. 2000. Integrated system for rapid PCR-based DNA analysis in microfluidic devices. *Analytical Chemistry* 72: 2995–3000.

Kikutani, Y., T. Horiuchi, K. Uchiyama et al. 2002. Glass microchip with three-dimensional microchannel network for 2 × 2 parallel synthesis. *Lab on a Chip* 2: 188–192.

Kim, L., M. D. Vahey, H. Y. Lee et al. 2006. Microfluidic arrays for logarithmically perfused embryonic stem cell culture. *Lab on a Chip* 6: 394–406.

Kopp, M. U., A. J. Mello, and A. Manz. 1998. Chemical amplification: Continuous-flow PCR on a chip. *Science* 280: 1046.

Kwang, W. O. and H. A. Chong. 2006. A review of microvalves. *Journal of Micromechanics and Microengineering* 16: R13–R39.

Kwon, S. and L. P. Lee. 2004. Micromachined transmissive scanning confocal microscope. *Optics Letters* 29: 706–708.

Ladavac, K. and D. Grier. 2004. Microoptomechanical pumps assembled and driven by holographic optical vortex arrays. *Optics Express* 12: 1144–1149.

Lagally, E. T., P. C. Simpson, and R. A. Mathies. 2000. Monolithic integrated microfluidic DNA amplification and capillary electrophoresis analysis system. *Sensors and Actuators, B* 63: 138–146.

Leach, J., H. Mushfique, R. di Leonardo et al. 2006. An optically driven pump for microfluidics. *Lab on a Chip* 6: 735–739.

Lee, J., H. Moon, J. Fowler et al. 2002. Electrowetting and electrowetting-on-dielectric for microscale liquid handling. *Sensors and Actuators a-Physical* 95: 259–268.

Lee, J. H., Y.-A. Song, and J. Han. 2008a. Multiplexed proteomic sample preconcentration device using surface-patterned ion-selective membrane. *Lab on a Chip* 8: 596–601.

Lee, J. H., Y.-A. Song, S. R. Tannenbaum et al. 2008b. Increase of reaction rate and sensitivity of low-abundance enzyme assay using micro/nanofluidic preconcentration chip. *Analytical Chemistry* 80: 3198–3204.

Lee, S. A., S. E. Chung, W. Park et al. 2009. Three-dimensional fabrication of heterogeneous microstructures using soft membrane deformation and optofluidic maskless lithography. *Lab on a Chip* 9, 1670–1675 (DOI: 10.1039/b819999j).

Lemoff, A. V. and A. P. Lee. 2000. An ac magnetohydrodynamic micropump. *Sensors and Actuators B: Chemical* 63: 178–185.

Lenigk, R., R. H. Liu, M. Athavale et al. 2002. Plastic biochannel hybridization devices: A new concept for microfluidic DNA arrays. *Analytical Biochemistry* 311: 40–49.

Levene, M. J., J. Korlach, S. W. Turner et al. 2003. Zero-mode waveguides for single-molecule analysis at high concentrations. *Science* 299: 682–686.

Liang, Z., N. Chiem, G. Ocvirk et al. 1996. Microfabrication of a planar absorbance and fluorescence cell for integrated capillary electrophoresis devices. *Analytical Chemistry* 68: 1040–1046.

Liu, G. L., J. Kim, Y. Lu et al. 2006. Optofluidic control using photothermal nanoparticles. *Nature Materials* 5: 27.

Llobera, A., S. Demming, R. Wilke et al. 2007. Multiple internal reflection poly(dimethylsiloxane) systems for optical sensing. *Lab on a Chip* 7: 1560–1566.

Losey, M. W., R. J. Jackman, S. L. Firebaugh et al. 2002. Design and fabrication of microfluidic devices for multiphase mixing and reaction. *Journal of Microelectromechanical Systems* 11: 709–717.

Love, J. C., D. B. Wolfe, H. O. Jacob, and G. M. Whitesides. 2002. Microscope projection photolithography for rapid prototyping of masters with micron-scale features for use in soft lithography. *Langmuir* 17: 6005–6012.

Manz, A., N. Graber, and H. M. Widmer. 1990. Miniaturized total chemical analysis systems: A novel concept for chemical sensing. *Sensors and Actuators, B* 1: 244–248.

Maruo, S. and H. Inoue. 2006. Optically driven micropump produced by three-dimensional two-photon microfabrication. *Applied Physics Letters* 89: 144101.

McKnight, T. E., C. T. Culbertson, S. C. Jacobson et al. 2001. Electroosmotically induced hydraulic pumping with integrated electrodes on microfluidic devices. *Analytical Chemistry* 73: 4045–4049.

Mello, A. J. d. and N. Beard. 2003. Dealing with "real" samples: Sample pre-treatment in microfluidic system. *Lab on a Chip* 3: 11N–19M.

Mengeaud, V., J. Josserand, and H. H. Girault. 2002. Mixing processes in a zigzag microchannel: Finite element simulations and optical study. *Analytical Chemistry* 74: 4279–4286.

Myers, F. B. and L. P. Lee. 2008. Innovations in optical microfluidic technologies for point-of-care diagnostics. *Lab on a Chip* 8: 2015–2031.

Nagrath, S., L. V. Sequist, S. Maheswaran et al. 2007. Isolation of rare circulating tumour cells in cancer patients by microchip technology. *Nature* 450: 1235–1239.

Neale, S. L., M. P. Macdonald, K. Dholakia et al. 2005. All-optical control of microfluidic components using form birefringence. *Nature Materials* 4: 530–533.

Nguyen, N. T. and X. Y. Huang. 2001. Miniature valveless pumps based on printed circuit board technique. *Sensors and Actuators a-Physical* 88: 104–111.

Niu, X. Z. and Y. K. Lee. 2003. Efficient spatial-temporal chaotic mixing in microchannels. *Journal of Micromechanics and Microengineering* 13: 454–462.

Oh, H.-J., S.-H. Kim, J.-Y. Baek et al. 2006. Hydrodynamic micro-encapsulation of aqueous fluids and cells via "on the fly" photopolymerization. *Journal of Micromechanics and Microengineering* 16: 285–291.

Ottino, J. M. 1990. Mixing, chaotic advection, and turbulence. *Annual Review of Fluid Mechanics* 22: 207–253.

Paegel, B. M., R. G. Blazej, and R. A. Mathies. 2003. Microfluidic devices for DNA sequencing: Sample preparation and electrophoretic analysis. *Current Opinion in Biotechnology* 14: 42–50.

Paegel, B. M., W. H. Grover, A. M. Skelley et al. 2006. Microfluidic serial dilution circuit. *Analytical Chemistry* 78: 7522–7527.

Pamme, N. and A. Manz. 2004. On-chip free-flow magnetophoresis: continuous flow separation of magnetic particles and agglomerates. *Analytical Chemistry* 76: 7250–7256.

Pregibon, D. C., M. Toner, and P. S. Doyle. 2007. Multifunctional encoded particles for high-throughput biomolecule analysis. *Science* 315: 1393.

Qing, Y., M. B. Joseph, S. M. Jeffrey et al. 2001. Responsive biomimetic hydrogel valve for microfluidics. *Applied Physics Letters* 78: 2589–2591.

Rife, J. C., M. I. Bell, J. S. Horwitz et al. 2000. Miniature valveless ultrasonic pumps and mixers. *Sensors and Actuators a-Physical* 86: 135–140.

Ro, K. W., K. Lim, B. C. Shim et al. 2005. Integrated light collimating system for extended optical-path-length absorbance detection in microchip-based capillary electrophoresis. *Analytical Chemistry* 77: 5160–5166.

Ronaghi, M., M. Uhlen, and P. Nyren. 1998. A sequencing method based on real-time pyrophosphate. *Science* 281: 363–365.

Rothemund, P. W. K. 2006. Folding DNA to create nanoscale shapes and patterns. *Nature* 440: 297–302.

Sanger, F. 1988. Citation classic—DNA sequencing with chain-terminating inhibitors. *Current Contents/Life Sciences* 50: 23–23.

Schena, M., D. Shalon, R. W. Davis et al. 1995. Quantitative monitoring of gene expression patterns with a complementary DNA microarray. *Science* 270: 467–470.

Schmidt, O., M. Bassler, P. Kiesel et al. 2007. Fluorescence spectrometer-on-a-fluidic-chip. *Lab on a Chip* 7: 626–629.

Seo, J. and L. P. Lee. 2004. Disposable integrated microfluidics with self-aligned planar microlenses. *Sensors and Actuators, B* 99: 615–622.

Shestopalov, I., J. D. Tice, and R. F. Ismagilov. 2004. Multi-step synthesis of nanoparticles performed on millisecond time scale in a microfluidic droplet-based system. *Lab on a Chip* 4: 316–321.

Smith, J. S. 2003. Fluidic self-assembly of active antenna, in: Technology, A. (Ed.), The Regents of the University of California, US Patent.

Song, Y. J. and T. S. Zhao. 2001. Modelling and test of a thermally-driven phase-change nonmechanical micropump. *Journal of Micromechanics and Microengineering* 11: 713–719.

Stauth, S. A. and B. A. Parviz. 2006. Self-assembled single-crystal silicon circuits on plastic. *Proceedings of the National Academy of Sciences* 103: 13922.

Stroock, A. D., S. K. W. Dertinger, A. Ajdari et al. 2002. Chaotic mixer for microchannels. *Science* 295: 647–651.

Studer, V., G. Hang, A. Pandolfi et al. 2004. Scaling properties of a low-actuation pressure microfluidic valve. *Journal of Applied Physics* 95: 393–398.

Sung Kwon, C., M. Hyejin, and K. Chang-Jin. 2003. Creating, transporting, cutting, and merging liquid droplets by electrowetting-based actuation for digital microfluidic circuits. *Journal of Microelectromechanical Systems* 12: 70–80.

Suzuki, H. and R. Yoneyama. 2003. Integrated microfluidic system with electrochemically actuated on-chip pumps and valves. *Sensors and Actuators B: Chemical* 96: 38–45.

Takagi, J., M. Yamada, M. Yasuda et al. 2005. Continuous particle separation in a microchannel having asymmetrically arranged multiple branches. *Lab on a Chip* 5: 778–784.

Tergaard, S., G. Blankenstein, H. Dirac et al. 1999. A novel approach to the automation of clinical chemistry by controlled manipulation of magnetic particles. *Journal of Magnetism and Magnetic Materials* 194: 156–162.

Terray, A., J. Oakey, and D. W. M. Marr. 2002. Microfluidic control using colloidal devices. *Science* 296: 1841–1844.

Terry, S. C., J. H. Jerman, and J. B. Angell. 1979. A gas chromatographic air analyzer fabricated on a silicon wafer. *IEEE Transactions on Electron Devices* 26: 1880–1886.

Teymoori, M. M. and E. Abbaspour-Sani. 2005. Design and simulation of a novel electrostatic peristaltic micromachined pump for drug delivery applications. *Sensors and Actuators a-Physical* 117: 222–229.

Thorsen, T., R. W. Roberts, F. H. Arnold et al. 2001. Dynamic pattern formation in a vesicle-generating microfluidic device. *Physical Review Letters* 86: 4163.

Thorsen, T., S. J. Maerkl, and S. R. Quake. 2002. Microfluidic large-scale integration. *Science* 298: 580–584.

Toriello, N. M., E. S. Douglas, N. Thaitrong et al. 2008. Integrated microfluidic bioprocessor for single-cell gene expression analysis. *Proceedings of the National Academy of Sciences* 105: 20173–20178.

Unger, M. A., H. P. Chou, T. Thorsen et al. 2000. Monolithic microfabricated valves and pumps by multi-layer soft lithography. *Science* 288: 113–116.

Veenstra, T. T., T. S. J. Lammerink, M. C. Elwenspoek et al. 1999. Characterization method for a new diffusion mixer applicable in micro flow injection analysis systems. *Journal of Micromechanics and Microengineering* 9: 199–202.

Verpoorte, E. 2003. Chip vision-optics for microchips. *Lab on a Chip* 3: 42N–52N.

Verpoorte, E., A. Manz, H. Ludi et al. 1992. A silicon flow cell for optical detection in miniaturized total chemical analysis systems. *Sensors and Actuators B: Chemical* 6: 66–70.

Wang, C. H. and G. B. Lee. 2005. Automatic bio-sampling chips integrated with micro-pumps and micro-valves for disease detection. *Biosensors & Bioelectronics* 21: 419–425.

Wang, H. Z., P. Iovenitti, E. Harvey et al. 2003. Numerical investigation of mixing in microchannels with patterned grooves. *Journal of Micromechanics and Microengineering* 13: 801–808.

Watts, P. and S. J. Haswell. 2003. Microfluidic combinatorial chemistry. *Current Opinion in Chemical Biology* 7: 380–387.

Wheeler, A. R., W. R. Throndset, R. J. Whelan et al. 2001. Microfluidic device for single-cell analysis. *Electrophoresis* 22: 283–288.

Wong, S. H., M. C. L. Ward, and C. W. Wharton. 2004. Micro t-mixer as a rapid mixing micromixer. *Sensors and Actuators B: Chemical* 100: 359–379.

Woolley, A. T., D. Hadley, P. Landre et al. 1996. Functional integration of PCR amplification and capillary electrophoresis in a microfabricated DNA analysis device. *Analytical Chemistry* 68: 4081–4086.

Xia, Y. N. and G. M. Whitesides. 1998. Soft lithography. *Annual Review of Materials Science* 28: 153–184.

Yacoub-George, E., W. Hell, L. Meixner et al. 2007. Automated 10-channel capillary chip immunodetector for biological agents detection. *Biosensors and Bioelectronics* 22: 1368–1375.

Yamada, M. and M. Seki. 2004. Nanoliter-sized liquid dispenser array for multiple biochemical analysis in microfluidic devices. *Analytical Chemistry* 76: 895–899.

Yamada, M. and M. Seki. 2006. Microfluidic particle sorter employing flow splitting and recombining. *Analytical Chemistry* 78: 1357–1362.

Yamada, M., M. Nakashima, and M. Seki. 2004. Pinched flow fractionation: Continuous size separation of particles utilizing a laminar flow profile in a pinched microchannel. *Analytical Chemistry* 76: 5465–5471.

Yaralioglu, G. G., I. O. Wygant, T. C. Marentis et al. 2004. Ultrasonic mixing in microfluidic channels using integrated transducers. *Analytical Chemistry* 76: 3694–3698.

Ying-Chih, W., F. Jianping, M. Pan et al., Nanofluidic molecular filters for efficient protein separation and preconcentration, *The 13th International Conference on Solid-State Sensors, Actuators and Microsystems*, Seoul, Korea, 2005, pp. 352–355.

Ymeti, A., J. Greve, P. V. Lambeck et al. 2007. Fast, ultrasensitive virus detection using a young interferometer sensor. *Nano Letters* 7: 394–397.

Yun, K. S., I. J. Cho, J. U. Bu et al. 2002. A surface-tension driven micropump for low-voltage and low-power operations. *Journal of Microelectromechanical Systems* 11: 454–461.

Zeng, S. L., C. H. Chen, J. C. Mikkelsen et al. 2001. Fabrication and characterization of electroosmotic micropumps. *Sensors and Actuators B: Chemical* 79: 107–114.

Zheng, B., L. S. Roach, and R. F. Ismagilov. 2003. Screening of protein crystallization conditions on a microfluidic chip using nanoliter-size droplets. *Journal of American Chemical Society* 125: 11170–11171.

Zheng, B., C. J. Gerdts, and R. F. Ismagilov. 2005. Using nanoliter plugs in microfluidics to facilitate and understand protein crystallization. *Current Opinion in Chemical Biology* 15: 548–555.

II

Optical Elements and Devices

8

Fluid-Controlled Optical Elements

Christian Karnutsch
Benjamin J. Eggleton

8.1 Introduction

Modern optical components are typically precisely machined and highly optimized in their performance for a robust and durable operation. This often comes at the expense of their ability to tune and adapt—an asset that would be very beneficial for various applications. Examples of such application areas include dynamic imaging and displays, scanning, optical communications, sensing, spectroscopy, health care and life sciences, lab-on-a-chip technology, and optical information processing.

To overcome these limitations in adaptability, the emerging interdisciplinary research field of optofluidics—the marriage between microphotonics and microfluidics—provides enormous opportunities for realizing tunable, reconfigurable, and adaptive optical devices (Psaltis et al. 2006, Monat et al. 2007a). Optofluidics holds the promise for generating highly miniaturized and densely integrated optical elements and circuits (Monat et al. 2007b, 2008, Hunt and Wilkinson 2008, Grillet et al. forthcoming). To accomplish this, optofluidics is taking advantage of (1) the inherent mobility and flexibility of fluids and (2) a large variety of physical mechanisms, such as pressure, electrowetting, magnetism, and optical processes including thermo-optic, acousto-optic, and electro-optic phenomena (Levy and Shamai 2008).

This chapter is structured as follows: We begin by reviewing the properties and advantages of fluid-based optical systems, present an overview of previous work in this field (Section 8.2.1), and then introduce an ideal platform for fluid control of optical elements—planar photonic crystals (PhCs). We demonstrate that PhC structures can be infiltrated with liquids, leading, for example, to optofluidic microcavities (Section 8.2.2).

8.2 Fluid-Controlled Optical Elements

The use of fluids (liquids and gases) lends itself naturally to achieving dynamic control of optical elements. In the following, we will restrict our discussion to liquids, as these are most frequently used in practical implementations. There are many benefits of using liquids for achieving tunability and adaptability:

- Liquids offer a large variety of optical properties, such as refractive index, absorption spectrum, and gain.
- Liquids can be mixed directly on a "lab-on-a-chip," creating a specific desired refractive index or viscosity (Gennes et al. 2004, Galas et al. 2005, Huh et al. 2005, Bilenberg et al. 2006).
- Complex suspensions can be prepared, where suspended particles may include quantum dots (Fushman et al. 2005, Bose et al. 2007, Dorfner et al. 2007, Wu et al. 2007, Martiradonna et al. 2008), metallic nanoparticles (Cohen et al. 1973, Aslan et al. 2005), fluorescent dyes (Brackmann 2000), or nanodiamonds (McCutcheon and Loncar 2008, Fu et al. 2008). Through these particles, loss, gain, and luminescence can be introduced, allowing the control of these parameters in a device.
- Liquids form optically smooth interfaces, which is an important quality for low-loss optical devices. Optofluidic waveguides (liquid–liquid waveguides, liquid-core waveguides, and others) and interfaces between solid waveguides and microfluidic channels (Grillet et al. 2004, Domachuk et al. 2005, 2006a,b, Mahmud et al. 2008), all benefit from low scattering losses due to smooth surfaces.
- Liquids exhibit a relatively high thermo-optic coefficient of the order of 10^{-4} to 10^{-3} K^{-1} (Weber 2002), making them ideal candidates for the thermal tuning of the refractive index.

In order to fluidically control or alter the optical properties of a device, the key optical phenomena that one wishes to control are as follows:

- *Transmission* can be controlled via optical switches, interferometers, and resonant structures such as ring resonators or Fabry–Perot (FP) cavities.
- *Dispersion* is influenced by waveguides, resonators, and PhC structures.
- *Light generation and emission* can be manipulated by resonators, PhC structures, and the gain medium.
- *Propagation direction* and *light guiding* is influenced by waveguides, beam splitters, Y junctions, and PhC structures.

8.2.1 Optofluidic Implementations of Fluidically Controlled Photonic Devices

There are numerous reports on optofluidic implementations of controlling optical properties. In this section, we highlight some examples in the context of fluid control of photonic devices.

8.2.1.1 Tunable Lenses

Lenses constitute key elements of an optical imaging system. A standard lens, made of a solid material, is very difficult to tune or adapt. Tunable lenses, however, could be realized, for example, by using liquid lenses and altering either their shape or the refractive index of the liquid (Bains 2006). Huang gives more details in Chapter 9.

8.2.1.2 Resonators

Optofluidic resonators have been theoretically proposed and experimentally realized in our group at the University of Sydney (see Section 8.2.2.2). We suggested using a PhC waveguide structure and infiltrating selected air holes of the PhC lattice to create a resonator (see Figures 8.1 and 8.11). This approach offers the possibility to achieve resonators with high quality factors of the order of $Q \approx 60,000$ (Bog et al. 2008). It also enables the geometrical and spectral reconfiguration of these optical resonators, offering the potential for a broad range of all-optical applications integrated on a single chip. These include arbitrarily defined post-processed PhC structures and sensing architectures. The use of liquids offers an additional advantage: The cavities can be rendered temperature insensitive (Karnutsch et al. forthcoming); this aspect is elucidated in Section 8.2.2.7.

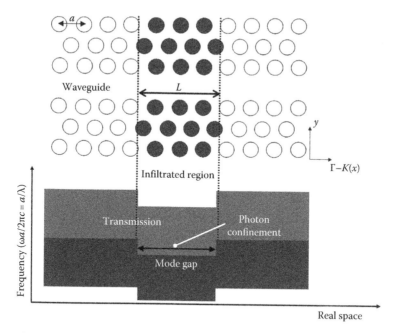

FIGURE 8.1 Schematic of an optofluidic double-heterostructure photonic crystal (DH-PhC) cavity. A PhC slab with a line-defect waveguide formed by omitting a row of air holes in the Γ–K direction serves as a platform. The central region of air holes is infused with a liquid, leading to a mode gap. Bottom: Schematic of the band diagram along the waveguide direction. The light-gray area indicates the transmission region where the propagation of photons is allowed through the waveguide, and the dark area indicates the mode gap region where propagation is inhibited. In other words, photons having frequency within the mode gap can only propagate in the infiltrated waveguide region.

8.2.1.3 Waveguides

Optofluidic waveguides are useful to achieve three distinct functionalities:

1. Confine and transmit light through fluid-filled cores. Examples include PhC fibers (see Chapter 15 by Barth et al.), slot waveguides (Di Falco et al. 2009, Yang et al. 2009), and antiresonant reflecting optical waveguides (Schmidt et al. 2005, Yin et al. 2007, Hawkins and Schmidt 2008) (see Chapter 16 by Schmidt).
2. Monolithically combine waveguides with microfluidic channels. Many optofluidic devices rely on interfacing optical waveguides with microfluidic channels. Examples include on-chip integrated flow cytometers (Lien et al. 2005) (see Chapter 19 by Chen et al.); low-cost, portable point-of-care diagnostic devices; and other lab-on-a-chip-type devices that rely on acquiring optical parameters from a liquid analyte.
3. Confine and guide light through 2D planar PhC circuits. Optofluidic waveguides can be created via selected infiltration of air holes in planar PhC structures (Leonard et al. 2000, Busch et al. 2004, Erickson et al. 2006). With this technique, complex, arbitrary waveguide structures are feasible, including "S-shaped" and 90° bends, beam splitters, and Y junctions (Mingaleev et al. 2004, Intonti et al. 2006, Kurt and Citrin 2008).

8.2.1.4 Slow-Light Structures

Slow-light planar PhC waveguides are of great interest in the context of optical delay lines and nonlinear optics (Krauss 2007, 2008, Baba 2008). It is generally assumed that light–matter interactions are enhanced in the slow-light regime, which would reduce the optical power needed to achieve a desired effect or could lead to a miniaturization of devices (Soljacic et al. 2002, Krauss 2007, Settle et al. 2007).

In PhC structures, the slow-light effect is caused by coherent backscattering and omnidirectional reflections of light at the PhC lattice (Krauss 2007). The slow-light regime is typically characterized by a high group velocity dispersion, which distorts optical pulses and, thereby, compromises the benefit of slow light (Engelen et al. 2006, Mok et al. 2006). Hence, it is important to realize slow-light structures with nearly dispersionless properties. It has recently been shown numerically that an optofluidic method can be employed to achieve slow-light PhC waveguides with low dispersion (Ebnali-Heidari et al. 2009). The proposed technique exploits the selective liquid infiltration of a uniform PhC waveguide. A large range of group refractive indices—ranging from 25 to 110—have been predicted with low dispersion from a single PhC waveguide geometry, achieved simply by altering the refractive index of the infused liquid.

8.2.1.5 Optical Switches

Microfluidic optical switches have been demonstrated based on controlling total internal reflection through the flow of two different miscible liquids using micron-sized channels made of a silicon elastomer, polydimethylsiloxane (PDMS) (Campbell et al. 2004, Groisman et al. 2008). The specific advantage of using PDMS for an optofluidic switch is its low refractive index of $n_{PDMS} = 1.41$, which can be matched by benign and easy-to-handle aqueous salt solutions. The proposed device has two layers of microchannels, separated by thin flexible membranes in regions where they overlap. The first layer—the "flow" layer—consists of a flat, parallel mirror channel that constitutes the main functional element of the device. It can be filled with either pure water or an index-matching salt solution. The channels connecting the inlets and the outlet with the mirror are completely sealed by the flexible membranes when a pressure is applied to the membranes through control lines in the second—"control"—layer of the chip. The membranes serve as integrated pressure-actuated valves. Optical switching is then performed by exchanging the liquid in the mirror channel, changing the angle of total internal reflection, and hence the transmission of a laser beam directed at the mirror channel. This and other types of optofluidic switches are described in more detail in Chapter 10 by Zamek et al.

8.2.1.6 Emission Control

Light sources are key components of many optical systems, where incoherent emission as well as lasing is often desired. Vignolini et al. (2008) have demonstrated an incoherent nanofluidic light source that was based on a silicon PhC microcavity by selectively infiltrating an air-pore cavity with a liquid containing colloidal PbS quantum dots. Their optofluidic light source is reconfigurable and emits in the telecom band at around $\lambda = 1.3\,\mu m$. This approach is potentially useful for integrated emitters, because it offers the possibility for local emitters in silicon PhC platforms and the opportunity to erase and rewrite them.

Kim et al. (2008) have demonstrated an optofluidic PhC nanolaser suitable for biological and chemical detection in lab-on-a-chip systems. It consists of a PhC nanocavity with an integrated microfluidic unit that improves the heat transfer and modulates the refractive index contrast. This system allows the dynamic modulation of the lasing wavelength and the far-field radiation pattern, resulting in efficient unidirectional emission.

Another type of tunable, optofluidic laser was demonstrated by Li et al. (2006). They presented a continuously tunable, optofluidic distributed feedback dye laser, where optical feedback was provided by a phase-shifted higher-order Bragg grating, embedded in the liquid core of a single-mode buried channel waveguide. The laser could be tuned mechanically by stretching the elastomeric PDMS substrate, resulting in a tuning range of 60 nm from a single laser chip. Optofluidic light sources are covered in more detail in Chapter 12 by Kristensen and Mortensen.

8.2.2 Case Studies of Liquid-Controlled Photonic Elements

Optofluidic devices exploit the characteristics of liquids to achieve a dynamic adaptation of their optical properties. The use of liquids allows for the functionalities of optical elements to be created, reconfigured, or tuned. In Sections 8.2.2.1 through 8.2.2.7, we highlight by way of case studies the benefits of an optofluidic approach, with specific reference to microfluidic optical cavities created in PhC platforms.

8.2.2.1 2D Planar Photonic Crystal Optofluidic Elements

PhC cavities represent a versatile building block for realizing micron-scale optical control (Akahane et al. 2003, 2005, Vahala 2003, 2005, Song et al. 2004, Asano et al. 2006), promoting their use in applications such as channel-drop filters (Song et al. 2003) and optical switches (Centeno and Felbacq 2000). In this section, we show that PhC microcavities can be formed by infusing a liquid into a selected section of a uniform PhC waveguide, and that the optical properties of these cavities can be tuned and adapted—reconfigured—by (1) altering the dimensions of the infiltrated region, or (2) removing the infiltrated liquid by using organic solvents.

In addition, by taking advantage of the negative thermo-optic coefficient of liquids, we introduce a concept to render PhC cavities insensitive to temperature changes in the environment. This is one example where the fluid control of optical elements results in a functionality that would be very hard to realize with other methods and techniques.

8.2.2.2 Optofluidic Microcavities

Figure 8.1 summarizes the principle of an optofluidic DH-PhC cavity. The central region of a PhC structure contains a lattice whose holes have been infused with a liquid of refractive index n_L, while its adjacent regions contain a regular PhC lattice of air holes. A waveguide across the DH experiences different dispersions associated with the different PhC sections. This results in the waveguide modes within the two sections existing at different frequencies, which causes a so-called mode gap, where light coupled to the waveguide mode of the infiltrated PhC section (with a frequency lying in this gap) does not propagate in the surrounding regular PhC section (Song et al. 2005, Tomljenovic-Hanic et al. 2006). Instead, the light remains localized within the infiltrated PhC area, decaying evanescently in the adjacent regions.

8.2.2.3 Photonic Crystal Platform

Our PhC structures are fabricated into a silicon-on-insulator substrate with a refractive index of $n_{Si} = 3.49$ (at $\lambda = 1400\,nm$). A triangular PhC lattice of air holes with periodicity $a = 410\,nm$ and hole diameter $\varnothing = 265\,nm$ $(0.646a)$ is defined by electron-beam lithography, and then transferred into the substrate by a dry etching process. (See Li et al. (2008) for a detailed description of the fabrication.) By wet etching with hydrofluoric acid, a suspended silicon membrane of thickness $T = 220\,nm$ $(0.537a)$ is created. A line-defect waveguide is introduced into these structures during the electron-beam lithography step by omitting one row of holes along the Γ–K direction of the lattice, resulting in a so-called W1 waveguide (see Figure 8.1). For our samples, the two PhC segments on either side of the waveguide are shifted closer together by 35 nm to reduce the guide width to 0.9 times the width of a standard W1 waveguide, creating a so-called W0.9 waveguide. This modification is done to increase the frequency range between the fundamental waveguide mode and the low-frequency edge of the PhC band gap (Smith et al. 2008, Karnutsch et al. 2009). The mode gap corresponding to the infiltrated W0.9 waveguide section does not overlap the photonic band gap edge and allows for a broad selection of liquid refractive indices to configure our devices, where an increase in the refractive index of the liquid extends the mode gap.

8.2.2.4 Infiltration Technique

Infiltrating air pores of a PhC lattice with diameters less than 300 nm is a challenging task. At these small dimensions, interface forces, such as surface tension and capillary forces, are the dominating phenomena governing the penetration of liquids into the holes. We use a glass micro-tip with an apex diameter of approximately 220 nm for the infiltration process, which is schematically illustrated in Figure 8.2.

The micro-tip is initially inserted within a meniscus of the infiltration liquid. The movement of the micro-tip during the infiltration process is controlled by a piezo-actuated translation stage with a positioning accuracy of ±20 nm. When the micro-tip is withdrawn from the liquid, droplets form along its

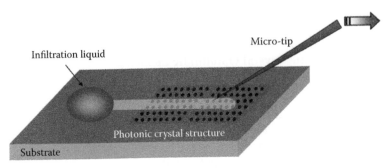

FIGURE 8.2 Schematic illustration of our liquid infiltration process: A micro-tip is immersed into the liquid and is drawn across a PhC structure to create a microfluidic optical element; in the depicted example, a microfluidic DH cavity. (From Bog, U. et al., *Opt. Lett.*, 33, 2206, 2008. With permission.)

length due to adhesive forces between the glass and the liquid. The droplets are subsequently deposited onto the substrate, in close proximity to the PhC structure of interest. The complete procedure is performed under an optical microscope equipped with an objective with an adequate working distance to allow for the insertion of the micro-tip during infiltration. Lastly, the micro-tip is used to maneuver a specific droplet across the PhC lattice to create infiltrated sections where the liquid enters the holes by capillary action. It is, in principle, possible to fill single holes when using a slightly modified technique, where the micro-tip is not moved across the PhC area while in contact with the surface, but instead is brought into contact solely with the intended hole to infiltrate.

8.2.2.5 Characterization Method—Evanescent Coupling

In order to optically characterize our microfluidic PhC components, we employ an evanescent coupling technique (Srinivasan et al., 2004, 2005, Grillet et al. 2007, Lee et al. 2008). Coupling to nanophotonic PhC components (such as waveguides and cavities) is a demanding task due to the small mode-field dimensions of these elements.

8.2.2.5.1 Tapered Fiber Fabrication

For evanescent coupling, we use a silica fiber that has a tapered region in which its diameter has been reduced to less than 1.5 µm. The fabrication process of this tapered fiber is highly complex. The basis is provided by a standard single-mode silica fiber (SMF28). The fiber is held under precise tension in a computer-driven taper rig, while a gas flame locally heats the fiber (Lizé et al. 2004). The fiber diameter is then adiabatically reduced from 125 µm to less than 1.5 µm (Birks and Li 1992) over a length of approximately 2 mm. The finished tapered fiber is attached to a microscope slide for mechanical support. Our tapered fibers have an induced "loop" shape with a radius of curvature of ≈30 µm to localize coupling to the nanoscale-infused sections of the PhC. After the looping procedure, the total transmission loss through the tapered fiber is ≈2 dB (at $\lambda = 1550$ nm).

8.2.2.5.2 Evanescent Coupling Technique

Because of the reduced dimensions of the tapered fiber, the electromagnetic field of the propagating mode extends significantly beyond the boundary of the fiber, allowing its evanescent field to interact with the PhC. Coupling between the tapered fiber and the PhC modes occurs when phase matching is achieved (Knight et al. 1997, Grillet et al. 2006). In our experimental setup (see Figure 8.3), light from a broadband source is launched into a single-mode silica fiber that is connected to the section of the tapered fiber. The transmitted signal through the tapered fiber is monitored with an optical spectrum analyzer. Examples of the resulting transmission spectra are given in Sections 8.2.2.6 and 8.2.2.7.

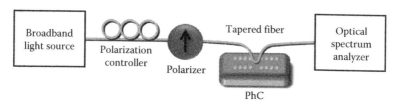

FIGURE 8.3 Schematic of our experimental setup for evanescent coupling. TE-polarized light is launched into a tapered fiber that couples light evanescently to the PhC sample. The transmitted light is monitored with an optical spectrum analyzer. (From Smith, C.L.C. et al., *Opt. Express*, 16, 15887, 2008. With permission.)

8.2.2.6 Reconfigurability

A complete reconfigurability of optofluidic devices is highly desirable, as it enables the creation and the tuning of optical functional elements from the same uniform PhC platform. In this section, we demonstrate that the spatial and the spectral reconfiguration of optofluidic PhC cavities can be achieved by controlling the size of a selectively liquid-infused region of air holes. The dimensions of the microfluidic cavity can be altered either by infiltrating a greater number of holes, i.e., gradually increasing the width of the infiltrated region, or by removing liquid from the infiltrated holes. Removal of liquid can be achieved by several methods. Our current approach is to wash the sample with an organic solvent, removing all infiltrated liquid in one simple processing step. If a more localized removal of the liquid is required, a potential method would be to use a tightly focused laser spot to locally evaporate the liquid.

In order to characterize the transmission spectra of our optofluidic cavities, we use the noninvasive evanescent coupling technique introduced in Section 8.2.2.5.

Our reconfigurability experiments comprise infiltrating a short DH cavity and then incrementally increasing the infiltrated region on the same PhC structure. We have first taken a reference measurement on an uninfiltrated waveguide (Figure 8.4a). The observed spectral signature around 1392 nm is associated with coupling to the fundamental TE-like waveguide mode. The discrete features in the spectrum are attributed to Fresnel reflections at the open ends of the PhC waveguide. We have then infiltrated a 2 μm long cavity, and subsequently increased the length of this cavity. Figure 8.5 illustrates a synopsis of measured transmission spectra for various cavity lengths, and Figure 8.6 shows a single measurement of a cavity of length 20.1 μm.

After the initial infiltration step (2 μm cavity; Figure 8.5), the spectral features appear at longer wavelengths compared to the reference measurement (Figure 8.4a). This is due to the increased effective refractive index of the modes due to the presence of the liquid. The observed fringe spectra in Figure 8.5 are attributed to Fabry–Pérot modes sustained by the microfluidic cavities. The coupling strength of the FP fringes with the probing fiber varies with the wavelength. It is at a maximum where phase matching between the fiber and the waveguide is the strongest.

As the cavity length is increased, the fringe spacing becomes smaller, consistent with an increased density of modes for larger cavities. For each investigated cavity length, the fringe spacing, $\Delta\lambda$, becomes smaller with the increasing wavelength, which is a result of the dispersive nature of the PhC waveguide. When the PhC region is completely filled (25 μm; Figure 8.5, bottom right), the majority of the fringes disappear as the mode gap effect no longer exists. In this case, there is only coupling to the fluid-filled PhC fundamental waveguide mode, containing features that again occur due to Fresnel reflections at the open ends of the PhC structure.

In order to reconfigure the PhC structure, we then cleaned the infiltrated PhC by immersing the sample in toluene for several minutes. This procedure completely removed the infiltrated liquid (see Figure 8.4b), which is confirmed by the spectral signature being nearly identical to the reference spectrum of the uninfiltrated case (Figure 8.4a).

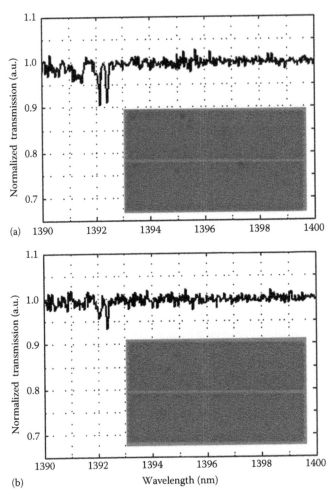

FIGURE 8.4 Normalized transmission through a tapered fiber when probing (a) an uninfiltrated PhC waveguide structure and (b) the reconfigured (cleaned) sample. (From Smith, C.L.C. et al., *Opt. Express*, 16, 15887, 2008. With permission.)

8.2.2.7 Temperature Stabilization

In this section, we introduce a method to temperature-stabilize optofluidic PhC cavities (Karnutsch et al. forthcoming). Temperature-stable cavities constitute a major building block in the development of a large suite of applications, ranging from high-sensitivity sensor systems for chemical and biomedical applications, to microlasers, optical filters, and switches. The performance of these devices shows a strong dependency on ambient temperature. As an example, refractive index sensors based on optical resonance techniques suffer from a temperature drift that introduces noise, and hence degrades the sensor's sensitivity (White and Fan 2008).

We, therefore, conducted an experimental and a theoretical study of the temperature dependence of optofluidic DH-PhC cavities (see Figure 8.1). We propose a method that enables a specific mode of the cavity to be made insensitive to changes in ambient temperature.

The key idea behind our optofluidic temperature stabilization concept is that a liquid with a negative thermo-optic coefficient can balance the positive thermal drift of the host PhC material. Figure 8.7 illustrates the temperature dependence of the refractive indices for silicon and a selected infiltration liquid, with the thermo-optic coefficients $\partial n_{Si}/\partial T = +2 \times 10^{-4}$ K^{-1} for silicon and $\partial n_L/\partial T = -3 \times 10^{-4}$ K^{-1}

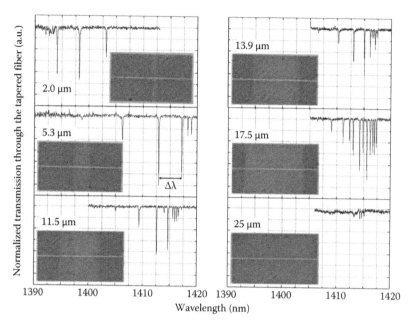

FIGURE 8.5 Normalized transmission through a tapered fiber when probing optofluidic PhC cavities of varying lengths. Numbers indicate the respective cavity length. Insets are microscope images (150×) of the corresponding cavities.

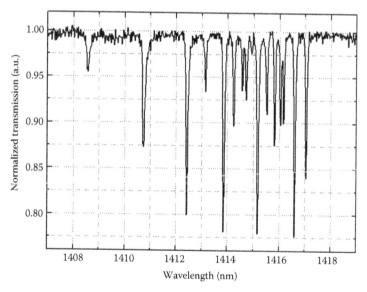

FIGURE 8.6 Normalized transmission through a tapered fiber when probing an optofluidic PhC cavity of length 20.1 μm.

for Cargille immersion oil type B. In other words, for a negative thermo-optic coefficient, a temperature increase causes a decrease in the refractive index.

For a PhC waveguide whose air holes have been infiltrated with a liquid, the effective refractive index experienced by a particular guided mode depends on the combination of the two refractive indices (PhC host material and infiltrated liquid), weighted by the filling fraction:

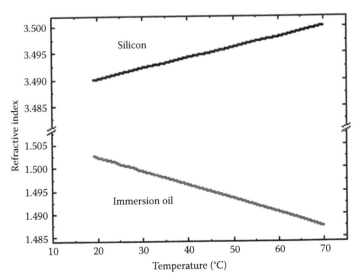

FIGURE 8.7 Refractive index as a function of temperature for silicon and Cargille immersion oil type B at $\lambda \approx 1410\,\text{nm}$. (From Karnutsch, C. et al., *Appl. Phys. Lett.*, 94, 231114, 2009. With permission.)

$$f_i = \frac{\int_i n_i^2 |E(r)|^2 \, dr}{\int n_i^2 |E(r)|^2 \, dr} \tag{8.1}$$

which gives the relative electric field, $E(r)$, overlap with the corresponding material of refractive index n_i. Given that the thermo-optic coefficient for most common PhC host materials—silicon, III–V compound semiconductors—is positive (Della Corte et al. 2000), while it is negative for most liquids (Kim and Su 2004) and polymers (Zhang et al. 2006), there is a large range of material combinations for which the effective index of the combination may be rendered temperature insensitive, if the guided mode has an appropriate fraction of electric field overlap in each material.

This concept can be expressed in mathematical terms by considering the electric field distribution in a resonant fluid-infused PhC cavity. To calculate the change in the resonance frequency with temperature, we apply the standard electromagnetic perturbation theory (Mortensen et al. 2008), linearizing the refractive index, n, around a reference temperature, T_0:

$$n(r,T) = n(r,T_0) + \left. \frac{\partial n(r,T)}{\partial T} \right|_{T=T_0} (T - T_0) \tag{8.2}$$

with r being a geometrical coordinate. Since

$$\frac{\partial \lambda}{\partial T} = \frac{\partial \lambda}{\partial n} \frac{\partial n}{\partial T} \tag{8.3}$$

the temperature-induced wavelength shift of a resonance can be expressed with the help of the perturbation theory as

$$\frac{\Delta \lambda}{\Delta T} \approx \sum_i \frac{\lambda(T_0)}{n_i(T_0)} \frac{\partial n_i}{\partial T} \cdot f_i \tag{8.4}$$

where

f_i is defined as given in Equation 8.1

$\sum_i f_i = 1$

For a two-component structure, such as a silicon membrane PhC with an infiltrated liquid, we find for the change of resonant wavelength, $\Delta\lambda$, for a given temperature change, ΔT:

$$\frac{\Delta\lambda}{\Delta T} \cong \lambda(T_0)\left[\frac{1}{n_1(T_0)}\frac{\partial n_1}{\partial T}f_1 + \frac{1}{n_2(T_0)}\frac{\partial n_2}{\partial T}(1-f_1)\right] \qquad (8.5)$$

where

$\lambda(T_0)$ is the free-space wavelength of the resonance at temperature T_0

n_i and $\partial n_i/\partial T$ are the refractive index and the thermo-optic coefficient of material i (silicon and infiltrated liquid), respectively

In Equation 8.5, we have implicitly assumed that $f_1 + f_2 = 1$, i.e., the mode only occupies the two materials, corresponding to complete hole filling and coverage of the PhC membrane structure. This is only strictly true in two dimensions, not in three dimensions, but the agreement between experimental and theoretical results shows that these assumptions are indeed justified. Comparing Equation 8.5 to the experimentally observed shift of a given resonance allows for the extraction of f_1 for that particular wavelength, which can then be compared to the value obtained through simulations. Deviations between these two values would to some extent shine light on the completeness of the liquid infiltration. From numerical simulations of the waveguide dispersion relation, $\omega(k)$, and the corresponding Bloch states, we can calculate f_1 and f_2 for different frequencies. In particular, from Equation 8.5 we can calculate the temperature-stable wavelength, λ_{Stable}, which separates resonances with a positive and a negative temperature-induced wavelength shift. The filling fraction, f_{Liquid}, at which this occurs is determined by

$$f_{\text{Liquid}}(\lambda_{\text{Stable}}) = \left[1 - \frac{n_{\text{Si}}}{n_{\text{L}}}\frac{\partial n_{\text{L}}/\partial T}{\partial n_{\text{Si}}/\partial T}\right]^{-1} \qquad (8.6)$$

For a resonance to be temperature independent, we must have

$$\sum_i \frac{1}{n_i(T_0)}\frac{\partial n_i}{\partial T}\cdot f_i = 0 \qquad (8.7)$$

which for a two-component structure—a membrane PhC infused with a liquid—means that the temperature derivatives must have opposite signs:

$$\frac{\partial n_1}{\partial T} = \frac{1-f_1}{f_1}\frac{n_1(T_0)}{n_2(T_0)}\left(-\frac{\partial n_2}{\partial T}\right) \qquad (8.8)$$

For a given PhC cavity design, λ_{Stable} can be derived from a numerical calculation of the filling fraction (Mortensen et al. 2008). At the wavelength λ_{Stable}, the mode profile is such that the negative thermo-optic coefficient of the infiltrated liquid compensates the positive thermo-optic coefficient of the host PhC. As an example, it follows from Equation 8.6 that temperature stabilization requires the filling fraction to be $f_{\text{Liquid}}(\lambda_{\text{Stable}}) = 22\%$, assuming $n_{\text{L}} = 1.5015$, $n_{\text{Si}} = 3.49$, and thermo-optic coefficients

$\partial n_{Si}/\partial T = +2 \cdot \times 10^{-4}$ K^{-1} for silicon and $\partial n_L/\partial T = -3 \cdot \times 10^{-4}$ K^{-1} for Cargille immersion oil type B. To experimentally achieve temperature stability for a given PhC structure, the interplay between the geometry of the cavity and the thermo-optic coefficient, $\partial n_L/\partial T$, of the infiltrated liquid must be optimized. This means that the optical cavity has to support a resonance at the wavelength λ_{Stable}, and, simultaneously, this specific mode must satisfy the condition for the filling fraction, $f_{Liquid}(\lambda_{Stable})$ (calculated according to Equation 8.6), such that the two temperature drifts are balanced.

The advantage of the presented optofluidic scheme is that a Fabry–Perot resonance can be aligned with λ_{Stable} by changing the cavity length, i.e., increasing or decreasing the size of the infiltrated region. Therefore, the resonant frequencies and the filling fractions can be tuned independently, a feature that cannot be achieved with standard PhC cavity designs that rely on photolithographic methods.

8.2.2.7.1 Experimental Results

We infiltrated Cargille immersion oil type B ($n_L = 1.5015$ at $\lambda = 1415$ nm, $T = 292$ K) into silicon PhC membranes (see Section 8.2.2.3 for detailed parameters). The infiltrated PhC sample is mounted on a Peltier module to vary the temperature from 25°C to 65°C (see Figure 8.8).

Figure 8.9 shows the transmission spectra as a function of temperature while probing an optofluidic cavity of length $L \approx 6.7$ µm. We observe five Fabry–Perot resonances sustained by the microfluidic cavity that exhibit moderate quality factors of the order of $Q \approx 15,000$ for resonance (1) and $Q \approx 20,000$ for resonance (2). The depth of the Fabry–Perot dips for the various modes varies with temperature due to the change of coupling efficiency between the tapered fiber and the optofluidic cavity. Note that this is a consequence of our chosen characterization method, and is not an inherent property of the investigated optofluidic cavities.

We observe from Figure 8.9 that the resonance wavelengths of the Fabry–Perot cavity shift with temperature with different gradients. In order to make this behavior clearer, Figure 8.10 displays the shift in wavelength as a function of temperature for the resonances labeled in Figure 8.9. In the investigated temperature range of 25°C–65°C, resonances (2) to (5) show a blueshift of the resonant wavelength between −0.03 nm K^{-1} (resonance (2)) and −0.06 nm K^{-1} (resonance (5)), while resonance (1) remains exceptionally stable at $\lambda = 1405$ nm, with an extremely low gradient of −0.003 nm K^{-1}. This represents a 20-fold reduction in temperature sensitivity compared to resonance (5), and a 27-fold reduction compared to a standard silicon PhC waveguide (Uenuma and Moooka 2009).

Applying Equation 8.5 to our experimental results and solving for f_1, we derive a filling fraction of $f_1 = 22.2\%$ for resonance (1), which is in excellent agreement with the theoretical prediction of 22% using Equation 8.6.

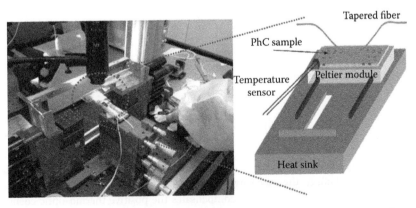

FIGURE 8.8 Experimental setup for temperature investigation. The PhC sample is mounted on an actively controlled Peltier module and probed with a tapered fiber.

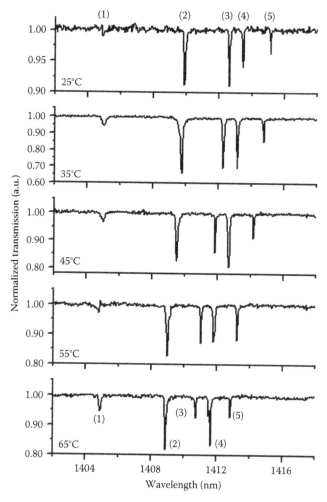

FIGURE 8.9 Normalized transmission spectra at various temperatures while probing an optofluidic PhC cavity of length $L \approx 6.7\,\mu m$. Resonance peaks are labeled for clarification.

The presented optofluidic temperature stabilization concept is not limited to a specific wavelength range. It can easily be extended to longer wavelengths, including telecommunication bands and mid-infrared, or into the visible spectral region, when a suitably transparent PhC host material is used, for example, gallium phosphide (Fu et al. 2008, Rivoire et al. 2008) or silicon nitride (McCutcheon and Loncar 2008). As many polymers exhibit a negative thermo-optic coefficient, it is also feasible to infiltrate a liquid polymer solution into the air holes and subsequently solidify the polymer, for example, by cross-linking it via temperature treatment or irradiation with UV light (Tay et al. 2007, El-Kallassi et al. 2008). This would lead to an all-solid-state temperature-stable PhC cavity.

8.3 Conclusions and Prospects

Optofluidics—marrying microfluidics with photonics—is a promising technology platform that offers innovative ways to render optical devices adaptable, tunable, and reconfigurable. To this end, planar PhC structures as well as microstructured optical fibers are well-suited platforms. Both contain easily accessible air voids that can be filled with liquids or gases, thereby offering the

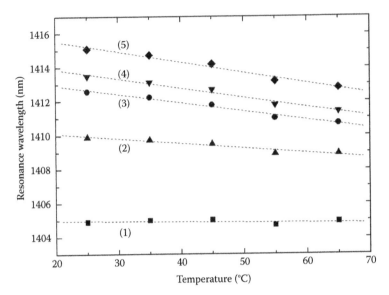

FIGURE 8.10 Wavelength shift as a function of temperature for the Fabry–Perot resonances labeled in Figure 8.9. Dashed lines are linear fits to the data. (From Karnutsch, C. et al., *Appl. Phys. Lett.*, 94, 231114, 2009. With permission.)

potential for strong light–matter interaction, and a large flexibility and tunability of their optical properties.

We have presented microfluidic optical cavities to exemplify the wide range of possibilities offered by fluid-controlled optical devices. These versatile cavities exhibit quality factors of up to 60,000; they can easily be configured and can be rendered temperature stable. In addition, adding active nanoparticles, such as quantum dots or fluorescent dyes, into the infiltration liquid could generate reconfigurable light sources that can be integrated into (silicon-based) lab-on-a-chip devices.

A PhC optofluidic microcavity is just one building block that will provide the basis for realizing more complex functions that could finally lead to a reconfigurable, integrated PhC circuit, as depicted in Figure 8.11. To achieve this challenging goal, novel technologies have to be developed that enable the controlled infiltration of precisely selected single holes in PhC lattices. One possible approach is using a system that facilitates a "pixel-by-pixel" infiltration, which could consist of a microinfiltration system that uses a micropipette to deposit droplets with a volume of less than 1 fL (Intonti et al. 2006), and a confocal laser scanning microscope in combination with a regular optical microscope to monitor the infiltration process.

Other selective writing methods could be established using, for example,

- Dip-pen nanolithography (Jaschke and Butt 1995, Piner et al. 1999, Salaita et al. 2007), where an atomic force microscope tip is used to transfer a molecular "ink" to a surface with a resolution below 20 nm.
- Microcontact printing, which uses an elastomeric stamp to deposit molecules directly onto substrates (Kim et al. 1995, Xia et al. 1996, Beh et al. 1999). This method is highly parallel, allowing the deposition of an entire pattern in one step.
- Hybrid microfluidic–photonic platforms (Adams et al. 2005, Diehl et al. 2006, Erickson et al. 2006, Mandal and Erickson 2008, Kim et al. 2008), where planar PhC components are integrated with a microfluidic network. In this situation, valves and pumps of a microfluidic circuitry could be remotely actuated to create a particular pattern in the underlying planar PhC structure, by filling and draining specific regions of the PhC lattice. This photonic circuit could then potentially be erased and reconfigured at a later stage.

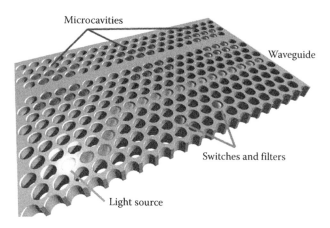

FIGURE 8.11 (See color insert following page 11-20.) Schematic of a reconfigurable, integrated optofluidic PhC circuit. The basic platform is a planar PhC with air holes. Numerous components can be integrated onto the same chip through selective infiltration of air holes.

When a reconfigurable optofluidic PhC-integrated circuit becomes reality, it will open the door to many exciting applications, including optical rapid prototyping—"optical circuits on demand." This would represent a leap forward in the turnaround time and the flexibility of optical engineering.

Acknowledgments

We acknowledge the support of the Australian Research Council through its Federation Fellow, Centre of Excellence and Discovery Grant programs, and the support of the School of Physics, University of Sydney, through its Denison Foundation and the International Science Linkages program by the ISL DEST grant. Christian Karnutsch acknowledges support from the University of Sydney through a Bridging Support Grant.

References

Adams, M. L., Loncar, M., Scherer, A., and Yueming, Q. 2005. Microfluidic integration of porous photonic crystal nanolasers for chemical sensing. *IEEE Journal on Selected Areas in Communications* 23: 1348–1354.

Akahane, Y., Asano, T., Song, B. S., and Noda, S. 2003. High-Q photonic nanocavity in a two-dimensional photonic crystal. *Nature* 425: 944–947.

Akahane, Y., Asano, T., Song, B. S., and Noda, S. 2005. Fine-tuned high-Q photonic-crystal nanocavity. *Optics Express* 13: 1202–1214.

Asano, T., Song, B. S., and Noda, S. 2006. Analysis of the experimental Q factors (~1 million) of photonic crystal nanocavities. *Optics Express* 14: 1996–2002.

Aslan, K., Lakowicz, J. R., and Geddes, C. D. 2005. Plasmon light scattering in biology and medicine: New sensing approaches, visions and perspectives. *Current Opinion in Chemical Biology* 9: 538–544.

Baba, T. 2008. Slow light in photonic crystals. *Nature Photonics* 2: 465–473.

Bains, S. 2006. Going with the flow. *IEE Review* 52: 42–45.

Beh, W. S., Kim, I. T., Qin, D., Xia, Y., and Whitesides, G. M. 1999. Formation of patterned microstructures of conducting polymers by soft lithography, and applications in microelectronic device fabrication. *Advanced Materials* 11: 1038–1041.

Bilenberg, B., Rasmussen, T., Balslev, S., and Kristensen, A. 2006. Real-time tunability of chip-based light source enabled by microfluidic mixing. *Journal of Applied Physics* 99: 0231021–0231025.

Birks, T. A. and Li, Y. W. 1992. The shape of fiber tapers. *IEEE Journal of Lightwave Technology* 10: 432–438.

Bog, U., Smith, C. L. C., Lee, M. W. et al. 2008. High-Q microfluidic cavities in silicon-based two-dimensional photonic crystal structures. *Optics Letters* 33: 2206–2208.

Bose, R., Yang, X., Chatterjee, R., Gao, J., and Wong, C. W. 2007. Weak coupling interactions of colloidal lead sulphide nanocrystals with silicon photonic crystal nanocavities near 1.55 μm at room temperature. *Applied Physics Letters* 90: 111117.

Brackmann, U. 2000. *Lambdachrome Laser Dyes*. Lambda Physik, Göttingen, Germany.

Busch, K., Lölkes, S., Wehrspohn, R. B., and Föll, H. 2004. *Photonic Crystals-Advances in Design, Fabrication, and Characterization*. Wiley-VCH, Berlin, Germany.

Campbell, K., Groisman, A., Levy, U. et al. 2004. A microfluidic 2 × 2 optical switch. *Applied Physics Letters* 85: 6119–6121.

Centeno, E. and Felbacq, D. 2000. Optical bistability in finite-size nonlinear bidimensional photonic crystals doped by a microcavity. *Physical Review B* 62: R7683.

Cohen, R. W., Cody, G. D., Coutts, M. D., and Abeles, B. 1973. Optical properties of granular silver and gold films. *Physical Review B* 8: 3689.

Della Corte, F. G., Cocorullo, G., Iodice, M., and Rendina, I. 2000. Temperature dependence of the thermo-optic coefficient of InP, GaAs, and SiC from room temperature to 600 K at the wavelength of 1.5 μm. *Applied Physics Letters* 77: 1614–1616.

Di Falco, A., O'Faolain, L., and Krauss, T. F. 2009. Chemical sensing in slotted photonic crystal heterostructure cavities. *Applied Physics Letters* 94: 063503.

Diehl, L., Lee, B. G., Behroozi, P. et al. 2006. Microfluidic tuning of distributed feedback quantum cascade lasers. *Optics Express* 14: 11660–11667.

Domachuk, P., Cronin-Golomb, M., Eggleton, B. J., Mutzenich, S., Rosengarten, G., and Mitchell, A. 2005. Application of optical trapping to beam manipulation in optofluidics. *Optics Express* 13: 7265–7275.

Domachuk, P., Littler, I. C. M., Cronin-Golomb, M., and Eggleton, B. J. 2006a. Compact resonant integrated microfluidic refractometer. *Applied Physics Letters* 88: 1111071–1111073.

Domachuk, P., Magi, E., Eggleton, B. J., and Cronin-Golomb, M. 2006b. Actuation of cantilevers by optical trapping. *Applied Physics Letters* 89: 0711061–0711062.

Dorfner, D. F., Hurlimann, T., Abstreiter, G., and Finley, J. J. 2007. Optical characterization of silicon on insulator photonic crystal nanocavities infiltrated with colloidal PbS quantum dots. *Applied Physics Letters* 91: 233111.

Ebnali-Heidari, M., Grillet, C., Monat, C., and Eggleton, B. J. 2009. Dispersion engineering of slow light photonic crystal waveguides using microfluidic infiltration. *Optics Express* 17: 1628–1635.

El-Kallassi, P., Balog, S., Houdré, R. et al. 2008. Local infiltration of planar photonic crystals with UV-curable polymers. *Journal of Optical Society of America B* 25: 1562–1567.

Engelen, R. J. P., Sugimoto, Y., Watanabe, Y. et al. 2006. The effect of higher-order dispersion on slow light propagation in photonic crystal waveguides. *Optics Express* 14: 1658–1672.

Erickson, D., Rockwood, T., Emery, T., Scherer, A., and Psaltis, D. 2006. Nanofluidic tuning of photonic crystal circuits. *Optics Letters* 31: 59–61.

Fu, K. M. C., Santori, C., Barclay, P. E. et al. 2008. Coupling of nitrogen-vacancy centers in diamond to a GaP waveguide. *Applied Physics Letters* 93: 234107.

Fushman, I., Englund, D., and VuCkovic, J. 2005. Coupling of PbS quantum dots to photonic crystal cavities at room temperature. *Applied Physics Letters* 87: 241102.

Galas, J. C., Torres, J., Belotti, M., Kou, Q., and Chen, Y. 2005. Microfluidic tunable dye laser with integrated mixer and ring resonator. *Applied Physics Letters* 86: 264101.

Gennes, P.-G. d., Brochard-Wyart, F., Quéré, D., and Reisinger, A. 2004. *Capillarity and Wetting Phenomena: Drops, Bubbles, Pearls, Waves*. Springer, New York.

Grillet, C., Domachuk, P., Ta'eed, V. et al. 2004. Compact tunable microfluidic interferometer. *Optics Express* 12: 5440–5447.

Grillet, C., Smith, C., Freeman, D. et al. 2006. Efficient coupling to chalcogenide glass photonic crystal waveguides via silica optical fiber nanowires. *Optics Express* 14: 1070–1078.

Grillet, C., Monat, C., Smith, C. L. C. et al. 2007. Nanowire coupling to photonic crystal nanocavities for single photon sources. *Optics Express* 15: 1267–1276.

Grillet, C., Monat, C., Smith, C. L. C. et al. Forthcoming. Reconfigurable photonic crystal circuits. *Laser & Photonics Review*.

Groisman, A., Zamek, S., Campbell, K., Pang, L., Levy, U., and Fainman, Y. 2008. Optofluidic 1×4 Switch. *Optics Express* 16: 13499–13508.

Hawkins, A. and Schmidt, H. 2008. Optofluidic waveguides: II. Fabrication and structures. *Microfluidics and Nanofluidics* 4: 17–32.

Huh, D., Gu, W., Kamotani, Y., Grotberg, J. B., and Takayama, S. 2005. Microfluidics for flow cytometric analysis of cells and particles. *Physiological Measurement* 26: R73–R98.

Hunt, H. and Wilkinson, J. 2008. Optofluidic integration for microanalysis. *Microfluidics and Nanofluidics* 4: 53–79.

Intonti, F., Vignolini, S., Turck, V. et al. 2006. Rewritable photonic circuits. *Applied Physics Letters* 89: 211117.

Jaschke, M. and Butt, H.-J. 1995. Deposition of organic material by the tip of a scanning force microscope. *Langmuir* 11: 1061–1064.

Karnutsch, C., Tomljenovic-Hanic, S., Monat, C., and Eggleton, B. J. 2009. Reconfigurable photonic crystal circuits using microfluidics. In *Optofluidics—Fundamentals, Devices, and Applications*, Y. Fainman, L. Lee, D. Psaltis and C. Yang, Eds. McGraw-Hill, New York.

Karnutsch, C., Smith, C. L. C., Graham, A. et al. 2009. Temperature stabilization of optofluidic photonic crystal cavities. *Applied Physics Letters* 94: 231114, DOI: 10.1063/1.3152998.

Kim, C. B. and Su, C. B. 2004. Measurement of the refractive index of liquids at 1.3 and 1.5 micron using a fibre optic Fresnel ratio meter. *Measurement Science and Technology* 15: 1683–1686.

Kim, E., Xia, Y., and Whitesides, G. M. 1995. Polymer microstructures formed by moulding in capillaries. *Nature* 376: 581–584.

Kim, S.-H., Choi, J.-H., Lee, S.-K. et al. 2008. Optofluidic integration of a photonic crystal nanolaser. *Optics Express* 16: 6515–6527.

Knight, J. C., Cheung, G., Jacques, F., and Birks, T. A. 1997. Phase-matched excitation of whispering-gallery-mode resonances by a fiber taper. *Optics Letters* 22: 1129–1131.

Krauss, T. F. 2007. Slow light in photonic crystal waveguides. *Journal of Physics D: Applied Physics* 40: 2666–2670.

Krauss, T. F. 2008. Why do we need slow light? *Nature Photonics* 2: 448–450.

Kurt, H. and Citrin, D. S. 2008. Reconfigurable multimode photonic-crystal waveguides. *Optics Express* 16: 11995–12001.

Lee, M. W., Grillet, C., Poulton, C. G. et al. 2008. Characterizing photonic crystal waveguides with an expanded k-space evanescent coupling technique. *Optics Express* 16: 13800–13808.

Leonard, S. W., Mondia, J. P., van Driel, H. M. et al. 2000. Tunable two-dimensional photonic crystals using liquid crystal infiltration. *Physical Review B* 61: R2389.

Levy, U. and Shamai, R. 2008. Tunable optofluidic devices. *Microfluidics and Nanofluidics* 4: 97–105.

Li, Z., Zhang, Z., Scherer, A., and Psaltis, D. 2006. Mechanically tunable optofluidic distributed feedback dye laser. *Optics Express* 14: 10494–10499.

Li, J., White, T. P., O'Faolain, L., Gomez-Iglesias, A., and Krauss, T. F. 2008. Systematic design of flat band slow light in photonic crystal waveguides. *Optics Express* 16: 6227–6232.

Lien, V., Zhao, K., and Lo, Y.-H. 2005. Fluidic photonic integrated circuit for in-line detection. *Applied Physics Letters* 87: 194106.

Lizé, Y., Mägi, E., Ta'eed, V., Bolger, J., Steinvurzel, P., and Eggleton, B. 2004. Microstructured optical fiber photonic wires with subwavelength core diameter. *Optics Express* 12: 3209–3217.

Mahmud, T., Zeller, E., Karnutsch, C., and Mitchell, A. (2008) Lithographically defined intersecting optical waveguides and fluidic channels. In *Micro- and Nanotechnology: Materials, Processes, Packaging, and Systems IV*. SPIE, Melbourne, Australia.

Mandal, S. and Erickson, D. 2008. Nanoscale optofluidic sensor arrays. *Optics Express* 16: 1623–1631.

Martiradonna, L., Carbone, L., Tandaechanurat, A. et al. 2008. Two-dimensional photonic crystal resist membrane nanocavity embedding colloidal dot-in-a-rod nanocrystals. *Nano Letters* 8: 260–264.

McCutcheon, M. W. and Loncar, M. 2008. Design of a silicon nitride photonic crystal nanocavity with a Quality factor of one million for coupling to a diamond nanocrystal. *Optics Express* 16: 19136–19145.

Mingaleev, S. F., Schillinger, M., Hermann, D., and Busch, K. 2004. Tunable photonic crystal circuits: Concepts and designs based on single-pore infiltration. *Optics Letters* 29: 2858–2860.

Mok, J. T., de Sterke, C. M., Littler, I. C. M., and Eggleton, B. J. 2006. Dispersionless slow light using gap solitons. *Nature Physics* 2: 775–780.

Monat, C., Domachuk, P., and Eggleton, B. 2007a. Integrated optofluidics: A new river of light. *Nature Photonics* 1: 106–114.

Monat, C., Grillet, C., Domachuk, R. et al. 2007b. Frontiers in microphotonics: Tunability and all-optical control. *Laser Physics Letters* 4: 177–186.

Monat, C., Domachuk, P., Grillet, C. et al. 2008. Optofluidics: A novel generation of reconfigurable and adaptive compact architectures. *Microfluidics and Nanofluidics* 4: 81–95.

Mortensen, N. A., Xiao, S., and Pedersen, J. 2008. Liquid-infiltrated photonic crystals: Enhanced light-matter interactions for lab-on-a-chip applications. *Microfluidics and Nanofluidics* 4: 117–127.

Piner, R. D., Zhu, J., Xu, F., Hong, S., and Mirkin, C. A. 1999. "Dip-Pen" nanolithography. *Science* 283: 661–663.

Psaltis, D., Quake, S. R., and Yang, C. H. 2006. Developing optofluidic technology through the fusion of microfluidics and optics. *Nature* 442: 381–386.

Rivoire, K., Faraon, A., and Vuckovic, J. 2008. Gallium phosphide photonic crystal nanocavities in the visible. *Applied Physics Letters* 93: 063103.

Salaita, K., Wang, Y., and Mirkin, C. A. 2007. Applications of dip-pen nanolithography. *Nature Nanotechnology* 2: 145–155.

Schmidt, H., Yin, D. L., Barber, J. P., and Hawkins, A. R. 2005. Hollow-core waveguides and 2-D waveguide arrays for integrated optics of gases and liquids. *IEEE Journal of Selected Topics in Quantum Electronics* 11: 519–527.

Settle, M. D., Engelen, R. J. P., Salib, M., Michaeli, A., Kuipers, L., and Krauss, T. F. 2007. Flatband slow light in photonic crystals featuring spatial pulse compression and terahertz bandwidth. *Optics Express* 15: 219–226.

Smith, C. L. C., Bog, U., Tomljenovic-Hanic, S. et al. 2008. Reconfigurable microfluidic photonic crystal slab cavities. *Optics Express* 16: 15887–15896.

Soljacic, M., Johnson, S. G., Fan, S., Ibanescu, M., Ippen, E., and Joannopoulos, J. D. 2002. Photonic-crystal slow-light enhancement of nonlinear phase sensitivity. *Journal of Optical Society of America B* 19: 2052–2059.

Song, B. S., Noda, S., and Asano, T. 2003. Photonic devices based on in-plane hetero photonic crystals. *Science* 300: 1537–1537.

Song, B. S., Asano, T., Akahane, Y., Tanaka, Y., and Noda, S. 2004. Transmission and reflection characteristics of in-plane hetero-photonic crystals. *Applied Physics Letters* 85: 4591–4593.

Song, B. S., Noda, S., Asano, T., and Akahane, Y. 2005. Ultra-high-Q photonic double-heterostructure nanocavity. *Nature Materials* 4: 207–210.

Srinivasan, K., Barclay, P. E., Borselli, M., and Painter, O. 2004. Optical-fiber-based measurement of an ultrasmall volume high-Q photonic crystal microcavity. *Physical Review B* 70: 081306.

Srinivasan, K., Barclay, P. E., Borselli, M., and Painter, O. J. 2005. An optical-fiber-based probe for photonic crystal microcavities. *IEEE Journal on Selected Areas in Communications* 23: 1321–1329.

Tay, S., Thomas, J., Momeni, B. et al. 2007. Planar photonic crystals infiltrated with nanoparticle/polymer composites. *Applied Physics Letters* 91: 221109.

Tomljenovic-Hanic, S., de Sterke, C. M., and Steel, M. J. 2006. Design of high-Q cavities in photonic crystal slab heterostructures by air-holes infiltration. *Optics Express* 14: 12451–12456.

Uenuma, M. and Moooka, T. 2009. Temperature-independent silicon waveguide optical filter. *Optics Letters* 34: 599–601.

Vahala, K. J. 2003. Optical microcavities. *Nature* 424: 839–846.

Vignolini, S., Riboli, F., Intonti, F. et al. 2008. Local nanofluidic light sources in silicon photonic crystal microcavities. *Physical Review E* 78: 045603(R).

Weber, M. J. 2002. *Handbook of Optical Materials*. CRC Press, Boca Raton, FL.

White, I. M. and Fan, X. 2008. On the performance quantification of resonant refractive index sensors. *Optics Express* 16: 1020–1028.

Wu, Z., Mi, Z., Bhattacharya, P., Zhu, T., and Xu, J. 2007. Enhanced spontaneous emission at 1.55 µm from colloidal PbSe quantum dots in a Si photonic crystal microcavity. *Applied Physics Letters* 90: 171105-1–171105-3.

Xia, Y., Kim, E., Zhao, X.-M., Rogers, J. A., Prentiss, M., and Whitesides, G. M. 1996. Complex optical surfaces formed by replica molding against elastomeric masters. *Science* 273: 347–349.

Yang, A. H. J., Moore, S. D., Schmidt, B. S., Klug, M., Lipson, M., and Erickson, D. 2009. Optical manipulation of nanoparticles and biomolecules in sub-wavelength slot waveguides. *Nature* 457: 71–75.

Yin, D. L., Lunt, E. J., Rudenko, M. I., Deamer, D. W., Hawkins, A. R., and Schmidt, H. 2007. Planar optofluidic chip for single particle detection, manipulation, and analysis. *Lab on a Chip* 7: 1171–1175.

Zhang, Z., Zhao, P., Lin, P., and Sun, F. 2006. Thermo-optic coefficients of polymers for optical waveguide applications. *Polymer* 47: 4893–4896.

9

Optofluidic Imaging Elements

Xiaole Mao
Zackary S. Stratton
Tony Jun Huang

9.1 Introduction

Imaging technology refers to techniques used to generate, collect, and preserve image data. It is one of the most thriving branches of modern technology, yet its origin can be dated back to 400 BC, when the first "pinhole experiment" was documented in ancient China. Optical imaging devices are pervasive in our modern lives, with applications ranging from optical communication to entertainment to consumer products to scientific research. Among a wide variety of imaging instruments, miniaturized imaging devices have received notable attention in recent years due to their applications in the rapidly developing markets of personal electronic devices (including cameras, cell phones, PDAs, and ultraportable laptop computers), in vivo bio-imaging devices such as endoscopes (Descour et al., 2002), surveillance/security systems, and miniaturized microscopes (Kwon and Lee, 2004). Development of miniaturized imaging systems is notoriously difficult as it requires precise fabrication, alignment, and actuation of various optical components—tasks that are difficult enough for traditional-scale systems—all within the confines of an extremely limited space.

In the past several years, the emergence of optofluidic technology (Psaltis et al., 2006; Monat et al., 2007), which combines microfluidic technology for fluid manipulation with optical components and methods, has provided a new technical route for the development of miniaturized imaging systems. Modern optofluidic technology has its roots in the ancient idea of using liquid as a mirror: Isaac Newton was reportedly among the first group of people to realize that the free surface of a rotating liquid could serve as a mirror in telescopes, leading to the construction of the first liquid mirror telescope in New Zealand more than two centuries ago (1872). In the modern context, the implementation of optics in microfluidic platforms is particularly appropriate considering the precision of the microfluidic fabrication process, the optical smoothness of fluidic interfaces, and the high reconfigurability of liquid media. These three features make optofluidic techniques a unique solution to the challenges in the development of miniaturized imaging systems—challenges that would be extremely difficult to overcome using solid components.

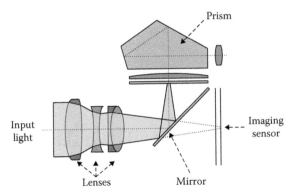

FIGURE 9.1 A simple schematic showing the basic components of a camera including lenses, mirrors, prisms, and imaging sensors. (Adapted from Wikipedia under the GNU Free Documentation License.)

In this chapter, we present a brief introduction to various mechanisms used to create optofluidic components, particularly those that can be integrated in miniaturized imaging systems. We begin by looking at a simple schematic of a classic imaging device—the camera (Figure 9.1). The camera consists of three basic imaging elements: lenses, mirrors/prisms, and imaging sensors. A lens or collection of lenses is located at the front end of the camera, collecting light from the objects to be imaged and ensuring proper focusing. The light is then steered and relayed within the body of the camera via a series of lenses, mirrors, and prisms, to an imaging sensor (e.g., film or CCD). The imaging sensor records the transmitted light and uses the information to reconstruct the image. This kind of "imaging trilogy," or its variant, can be found in almost every type of imaging device. Therefore when reviewing optofluidic imaging elements, it is intuitive to break the discussion into three sections: optofluidic lenses, mirrors/prisms, and imaging sensors.

9.2 Optofluidic Lenses

Our discussion on optofluidic imaging elements begins with the optofluidic microlens. The microlens has always been at the heart of miniaturized optical imaging systems. In traditional macroscale imaging systems, lenses are often made of solid materials (such as glass or plastic); therefore, the tuning and adaptation of a lens is often achieved through the change of the lens' position rather than the change of the physical properties of the lens itself. In solid-based miniaturized imaging systems, the manufacturing of microlenses and the actuation of movable lenses becomes very difficult.

The dynamic properties of fluids give rise to the possibility of directly adjusting the properties of lenses (such as lens curvature and refractive index) to adjust the lens' focusing power. The relatively simple lens mechanism used for variable-focus optofluidic lens focusing is particularly attractive for the design of small camera heads limited by space. Thus, the development of optofluidic microlenses largely centers on improving methods to establish the tunable lens surface through which the light can be bent and brought into focus. This type of lens is called a refractive lens and is used in the vast majority of commercial lenses. In this section, we will mainly focus on the refractive lens, and touch upon another type of lens—the gradient refractive index (GRIN) lens—at the end of the discussion. There are many ways to categorize tunable microlenses based on factors such as the material of the lens or the tuning mechanism. In this chapter, we divide the lens into two groups: out-of-plane microlenses and in-plane microlenses. As indicated by their names, out-of-plane microlenses focus light in the out-of-plane direction while in-plane microlenses focus light along the device plane. Today, the majority of imaging microlenses are out-of-plane microlenses.

9.2.1 Out-of-Plane Optofluidic Microlenses

9.2.1.1 Membrane-Contained Optofluidic Microlenses

A common technique for the fabrication of a tunable lens is to trap liquid within an elastic-membrane-sealed cavity. The cavity often has a cylindrical body, whose circular aperture is sealed with a thin elastic membrane. The membrane can bend outward or inward by applying positive or negative pressure, producing temporary convex (focusing) or concave (divergent) structures. The curvature of the membrane can then be altered through various means, thereby adjusting the focal length. Such a method was introduced several decades ago and has become popular in recent years (Zhang et al., 2003, 2004; Ren et al., 2006) due to the widespread use of soft-lithography (Xia and Whitesides, 1998) and PDMS molding techniques. Today, there exists a wide variety of liquids of refractive indexes equal to or higher than that of PDMS ($n \sim 1.41$), which can be used for the purpose of lens construction.

Figure 9.2 shows the fabrication process for a PDMS elastic membrane used in a liquid tunable lens (Chronis et al., 2003). The mold for the cylindrical fluid chamber and the PDMS membrane were fabricated from an SU-8 photoresist using a two-step SU-8 photolithography technique. The first step forms the mold for the connection channel used for fluid injection and pressure supply. A second, thicker layer of SU-8 was then patterned on top of the first layer to form a circular chamber. A simplified one-step mold fabrication process in which the connection channel and fluidic chamber have identical height is also possible; however, the conjunction of the channel and chamber is a defect to the circular membrane and can result in irregularity in lens shape. The final step involved casting the PDMS onto the mold to create the fluidic chamber sealed with a thin PDMS membrane. The thickness of the membrane can be controlled by adjusting the spin rate and was set to be around 40 μm.

Using the previously discussed method, a tunable microlens array can be fabricated for imaging purposes. An array of elastic membranes that sealed fluidic chambers was integrated with an underlying microfluidic network. Thereby, the focal lengths of all microlenses can be simultaneously controlled by pneumatically regulating the pressure inside the microfluidic network (Figure 9.3a). The pressure control was realized via a pneumatic pump with a pressure regulator. By pumping high-refractive-index oil into the fluidic chamber, the focal length can be tuned from several hundred microns to several millimeters. Figure 9.3b shows the variable focusing results characterized by a photograph imaged through the tunable lens. Increased resolution from none (0 kPa) to 4 μm (10 kPa) to 3 μm (20 kPa) was achieved by changing the pneumatic pressure.

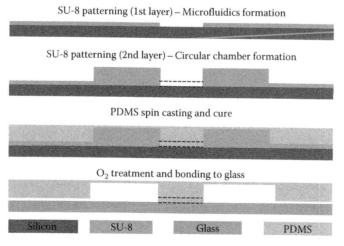

FIGURE 9.2 An example for fabricating a PDMS elastic membrane for a liquid tunable lens. The fabrication used a two-step photolithography and a PDMS mold replica technique. (Reproduced from Chronis, N. et al., *Opt. Express*, 11, 2370, 2003. With permission.)

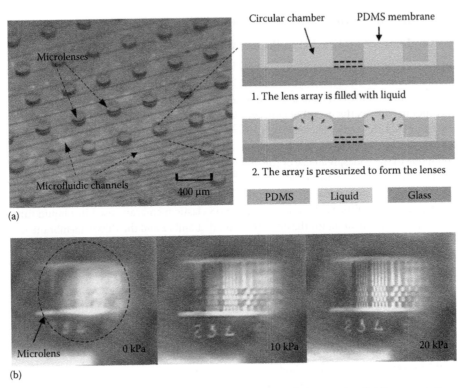

(a)

(b)

FIGURE 9.3 (a) Microscopic image of a microlens array with an underlining microfluidic network for pressure regulation (left) and schematic of the microlens actuation via pressure control. (b) Variable focusing results due to a change in pressure. (Reproduced from Chronis, N. et al., *Opt. Express*, 11, 2370, 2003. With permission.)

The above-mentioned example uses an external pneumatic pump and pressure regulator to control fluidic pressure; however, in many cases, it is highly desirable to eliminate the bulky external pressure control unit and to enable on-chip pressure control. There are several mechanisms that can be adapted for in situ control of membrane curvature. Lee and Lee (2007) introduced a tunable liquid lens with an integrated electromagnetic actuator to control membrane deformation. Figure 9.4 is a schematic view of the device and depicts the operating mechanism of the lens. The device consists of a fluidic chamber with two openings. Both openings are sealed with a thin PDMS slab to form two deformable PDMS membranes, one for the lens surface and one for the driving unit that carries a coiled electrode to generate the Lorentz force for lens actuation. The membrane is placed within a transverse external magnetic field. When current is applied to the electric coil, a Lorentz force acts in the direction normal to the actuating membrane, causing it to deform (Figure 9.4a). Because the volume of the fluidic chamber is fixed and the liquid is incompressible, the PDMS membrane for the lens also deforms to offset the volumetric change induced by the deformation of the actuating PDMS membrane (Figure 9.4b). The deformed membrane and high-refractive-index fluid inside the chamber result in a tunable lens whose curvature can be actuated by the current passing through the coil.

Figure 9.5a illustrates the fabrication process of the lens. The fabrication is a combination of conventional silicon micromachining and soft lithography. Silicon wet etching was first used to create two thin Si/SiO_2 thin films from the backside of the wafer. Two-step Au deposition, together with an SU-8 passivation process, was used to define the electric coil structure for generating the Lorentz force. Similar to the previous example, a PDMS membrane was fabricated on top of the defined electric components by spin coating the uncured PDMS. In this case, however, the thickness of the membrane can be controlled by both spin rate and by reducing the viscosity of the PDMS solution via dilution (Figure 9.5a inset).

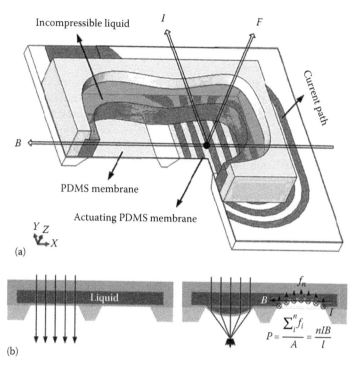

FIGURE 9.4 (a) Schematic of the electromagnetic actuator integrated with an elastic membrane. (b) Mechanisms of tunable lens using an electromagnetic actuator. (Reprinted from Lee, S.W. and Lee, S.S., *Appl. Phys. Lett.*, 90, 121129, 2007. With permission. Copyright 2009, American Institute of Physics.)

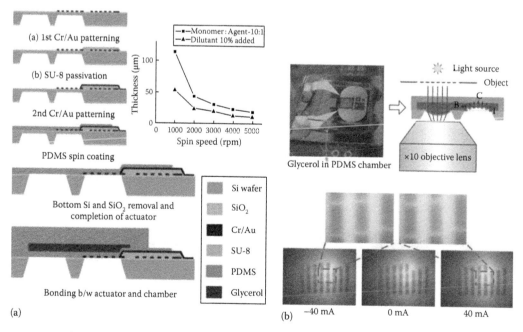

FIGURE 9.5 (See color insert following page 11-20.) (a) Schematic of the electromagnetic actuator and lens fabrication process. (b) The characterization process of variable focusing. (Reprinted from Lee, S.W. and Lee, S.S., *Appl. Phys. Lett.*, 90, 121129, 2007. With permission. Copyright 2009, American Institute of Physics.)

A membrane as thin as 11.0 μm can be fabricated by optimizing the dilution factors and spin coat rate. After the PDMS spin coat, the residual Si/SiO$_2$ thin film was removed from the backside via bulk micromachining, and the PDMS membrane was released. The device is completed by the bonding of a PDMS fluid chamber on top of the membrane.

The completed device including the driving circuit and a pair of magnets (to establish the magnetic field around the coil) is shown in the upper section of Figure 9.5b. The fluidic chamber was filled with 99.0% glycerol, and the variable focusing was characterized via the photograph through the microlens (lower Figure 9.5b). Changes in the image of the micropattern were obvious when the applied current changed from −40 to 40 mA.

In work by Pang et al. (2005), they took a step away from the commonly used spherical lens surface and cylindrical fluid-filled chamber. Instead, they introduced an optofluidic lens that operates as a set of two tunable cylindrical lenses formed by rectangular fluid-filled chambers (Figure 9.6). The device is a four-layer PDMS structure. The fluid chambers in the two top layers (1 and 2) and the two bottom layers (3 and 4) are aligned. Chambers 2 and 3 are interconnected and perpendicular to each other. Between two overlapping top layers and bottom layers, there are flexible membranes of ~ 200 μm thickness that physically separate the two channels. Each chamber contains a rectangular liquid chamber that is filled with fluids of different refractive indices. Chambers 1 and 4 contain a high-refractive-index fluid, and chambers 2 and 3 contain a low-refractive-index fluid. The optical function area is a square window where the top two channels cross the lower two channels. The densities of the two fluids were intentionally matched to each other in order to reduce the influence of gravity and mechanical shocks to the lens. The advantage of having two overlapping adaptive cylindrical lenses is the flexibility to focus light along two orthogonal axes independently. Since the two membranes do not interfere with each other, the focal lengths of both lenses can be adjusted independently and continuously between 40 and 23 mm, and the minimum focused beam width is around 40 μm. Compared to spherical lens, this is of particular interest for correction of aberration and astigmatism in imaging.

9.2.1.2 Immiscible Fluid Optofluidic Microlenses

The aforementioned membrane-contained optofluidic lenses can be effectively tuned for variable light focusing. However, the thin elastic membranes which are integral to these designs can be challenging to fabricate with consistency, and the elasticity of the membrane may vary over time. This necessitates the constant calibration of the focal length and can be inconvenient in long-term practical applications. Another type of tunable liquid lens operates through manipulation of the interface between two immiscible fluids (i.e., oil-water, water-air). In this case, the membrane is no longer needed to contain the fluids. On the microscale, the behavior of the fluid interface between two immiscible fluids

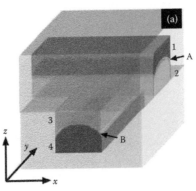
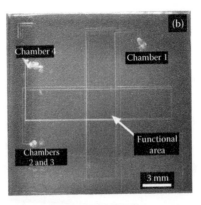

FIGURE 9.6 (a) Schematic for the fabrication of an orthogonal adaptive cylindrical lens set using the multilayer soft-lithography method. (b) Microscopic image of the assembled lens. (Reproduced from Pang, L. et al., *Opt. Express*, 13, 9003, 2005. With permission.)

is dominated by interfacial tension, and therefore its shape is close to a perfectly spherical surface, a profile that is widely used in commercial lenses. When using two immiscible fluids of different refractive indices, light can be bent and brought to focus when passing through the fluid–fluid interface. The optically smooth fluidic interface also makes it possible to achieve high imaging quality in a cost-effective fashion. A number of clever tuning mechanisms have been developed to change the focusing power of the immiscible fluid, optofluidic lens. Among these mechanisms, electrowetting is perhaps the most successful one and is close to commercial application.

Electrowetting (Mugele and Baret 2005; Shamaia et al., 2008) is a method to modify the wetting properties of hydrophobic surfaces by applying an electric field. Figure 9.7 demonstrates a classic configuration for electrowetting on dielectric (EWOD). A drop of water is placed between an electrode and a dielectric insulating surface. Prior to applying the electric field ($U_0 = 0\,$V), the droplet forms a natural contact angle at the liquid–solid interface. When the voltage is applied, the existence of the electric field within the droplet induces the change of the electric charge distribution in the droplet and the reduction of surface energy, thus the contact angle. The contact angle can be modified to varying degrees. The stronger the applied voltage, the more the water seeks to wet the electrode surface and lesser the contact angle.

Figure 9.8 shows a miniaturized tunable microlens developed by Kuiper and Hendriks (2004) of Philips Research Eindhoven (the Netherlands). The miniaturized microlens features a cylindrical housing that contains two immiscible fluids with different refractive indices. The properties of the fluids were carefully selected in order to (1) form stable menisci that center around the optical axis of the lens, and (2) adaptively adjust the curvature of the immiscible fluid interface. First, the densities of the fluids were matched with each other so the meniscus can become insensitive to external vibrations, shocks, or orientation. The buoyancy-neutral environment also eliminates any deformation of the fluidic interface caused by gravity, making the shape of the meniscus perfectly spherical. Second, one of the fluids is set electrically conducting and the other is set insulating in order to satisfy electrowetting conditions. In this case, nonpolar oil (insulating, refractive index ~ 1.58) and salt solution (conductive, refractive index ~ 1.38) were used. The addition of salt can also adjust the density of the fluids. The top and the cylindrical wall were coated with hydrophobic coating so that the conductive aqueous solution wet the bottom surface, maintaining the orientation of the lens.

Figure 9.8a shows that before a voltage is applied, a spherical interface forms between the two fluids. In this case, the lens is divergent. The application of the voltage lowers the contact angle. As the voltage

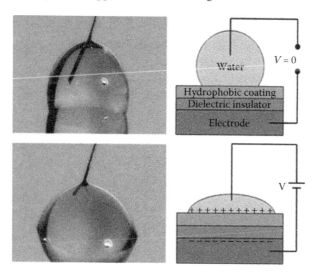

FIGURE 9.7 Principle of electrowetting. The contact angle of a conducting drop on an insulated conducting substrate can be modified by applying an applied voltage. (Reproduced from Shamaia, R. et al., *Soft Matter*, 4, 38, 2008. With permission of The Royal Society of Chemistry.)

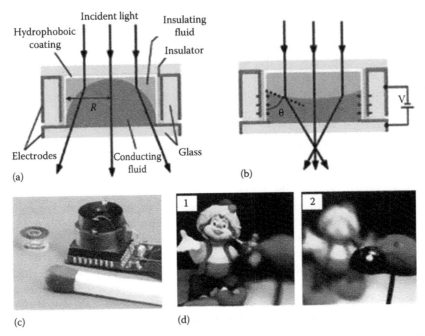

FIGURE 9.8 Schematic of the device configuration for the electrowetting tunable lens and the shape of the meniscus (a) prior to and (b) after the electrowetting actuation. (c) Images of the final assembled camera with the liquid lens integrated with a CMOS chip. (d) 1 and 2: the variable focusing results characterized via imaging through the tunable lens. (Reprinted from Kuiper, S. and Hendriks, B.H.W., *Appl. Phys. Lett.*, 85, 1128, 2004. With permission. Copyright 2009, American Institute of Physics.)

increases, the contact angle reduces and the lens surface transition from divergent to flat to convergent (Figure 9.8b).

The use of a variable-focus optofluidic lens in the microscope allows the user to focus and image different cross sections of the target sample. Figure 9.5c shows an assembled camera with a liquid lens integrated with a CMOS chip. Figure 9.5d and e are two images captured by a camera which was focused at 50 and 2 cm, respectively—variable focusing results were obvious. Electrowetting is now behind several commercially available fluidic lenses from manufacturers such as Philips and Varioptic. A schematic of the Varioptic design is shown in Figure 9.9.

Besides electrowetting, there exist other methods for the manipulation of an immiscible fluid interface. For example, López et al. (2005) introduced a tunable microlens based on redox surfactants. Similar

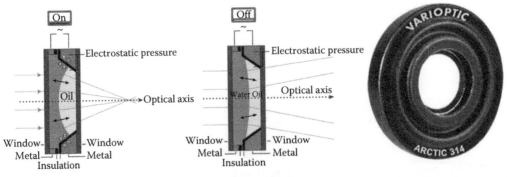

FIGURE 9.9 Schematic (left) and actual image (right) of a commercial electrowetting liquid lens. (Courtesy of Varioptic, Lyon, France.)

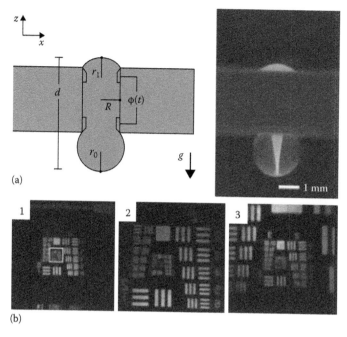

FIGURE 9.10 (a) Schematic of the redox-surfactant tunable liquid lens showing pinned hanging meniscus and the electrode embedded at each contact circle (left) and the image of a green laser source via the liquid lens (right). (b) 1 and 2, images of the standard target imaged through pure water (2 is a magnified view of the square in 1). 3 is the image of the same area after the actuation of the lens. (Reprinted from López, C.A. et al., *Appl. Phys. Lett.*, 87, 134102, 2005. With permission. Copyright 2009, American Institute of Physics.)

to electrowetting, this method takes advantage of the interfacial tension between two immiscible fluids. It utilizes a surfactant that can be electrochemically manipulated to modify surface tension.

Figure 9.10a shows the schematic of the lens tuning mechanism. The lens is formed by overfilling a small through-hole (diameter less than 1 mm) in a roughly 3 mm thick Teflon plate. The non-wetting property of Teflon ensures that the contact lines of the liquid are pinned at the opening of the holes and the optical axis of the droplet are aligned with the axis of the hole. The hanging meniscus has a spherical shape and can be used to focus light. The curvature of the hanging meniscus is determined by the balance of surface tension of the liquid at both the top and bottom of the contacting surface and the hydrostatic pressure due to the height of the droplet. In order to tune the curvature of the lens, a water soluble ferrocenyl surfactant, whose surface activity can be electrochemically controlled, was added to the water. Surfactants can increase (when oxidized) or decrease (when reduced) surface tension over a total range of 16 dyn/cm (more than 1/5 of the surface tension of pure water). The surface tension variation results in a change in the droplets' balance position and hanging meniscus curvature. The process can be controlled electrochemically by varying the potential of the cathode and anode (ring electrode embedded at the top and bottom of the through-holes). An obvious change in the resolution after electrochemical activation of the liquid lens can be seen in Figure 9.10b.

In the previous two examples, the actuation of an immiscible fluid lens was achieved through modification of surface tension electrically or electrochemically. In fact, an immiscible fluid liquid lens can also be activated via mechanical actuation by physically changing the shape of the lens that would otherwise be determined by the surface tension.

Figure 9.11 describes an interesting design of an adaptive optofluidic lens developed by Dong et al. (2006). The lens consists of a water-filled, stimuli-responsive hydrogel ring that is sandwiched between a glass substrate and an aperture slip whose opening is aligned with the hydrogen ring. The hydrogen

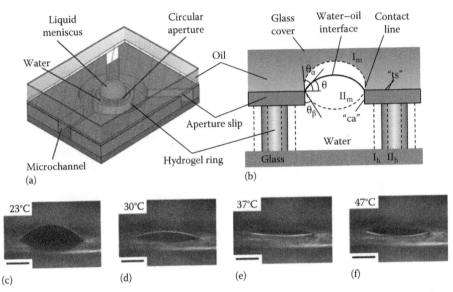

FIGURE 9.11 (a) Schematic of the assembly of the hydrogel ring with the microfluidic components for pinning the water–oil interface. (b) Variable-focusing mechanism via the deformation of stimuli-responsive hydrogel. (c–f) Change of the fluidic lens curvature due to external stimuli (temperature variation). (Reprinted from Dong, L. et al., *Nature*, 442, 551, 2006. With permission. Copyright 2009, American Institute of Physics.)

ring is first filled with water through the nearby microfluidic network. Afterward, oil is added from the top of the water meniscus and capped with an additional glass cover slip. In the design of the aperture, the surfaces of the sidewall and bottom ("ca" in Figure 9.11b) were modified to be hydrophilic while the top surface ("ts" in Figure 9.11b) was treated to be hydrophobic. Such treatment is crucial to establish a stable water–oil interface needed for lens tuning. It ensures that the boundary between the water and oil phases is always pinned along the hydrophobic–hydrophilic contact lines (edge of the apertures opening). When proper stimuli (e.g., pH, temperature, light) are provided, the hydrogel underneath the aperture can expand or shrink, which results in a volumetric change of the space inside the ring. The volume changes cause the over-fill of the water enclosed by the ring and top oil layer, and cause a pressure which acts across the water–oil interface. The curvature of the interface, and hence the focal length of the liquid lens, change as a response to the pressure difference as shown in Figure 9.11c.

9.2.2 In-Plane Optofluidic Microlenses

In the previous section, we examined several optofluidic tunable lens mechanisms that can be used for imaging purpose by focusing light in a plane perpendicular to the device plane. Another group of so-called in-plane optofluidic microlenses, which focus light within the device plane, must also be mentioned. Although these lenses are more preferable for on-chip detection purposes involving the enhancement of light intensity and do not have direct relevance to imaging, they are still worth examining to provide a thorough background. More importantly, due to the limitations of planar device structures, it is difficult to apply the lens tuning mechanisms of out-of-plane microlenses to in-plane ones. Therefore, many unique lens fabrication and tuning mechanisms have been proposed to circumvent the constraints imposed by limited space. Researchers believe that these novel methods may extend into future development of out-of-plane imaging lenses.

For developing in-plane tunable microlenses, the ultimate challenge is to implement a lens tuning mechanism within the constraints imposed by the two-dimensional planar structure. Ideally, an in-plane tunable lens should be fabricated using a single-step soft-lithography process so the lens can be seamlessly integrated with other on-chip components to facilitate on-chip detection. Therefore, it is rather difficult to implement methods that require complex fabrication such as the previously discussed electrowetting technique. There are a handful of tunable lens designs that permit tunable in-plane focusing. Figure 9.12 shows two examples using liquid-filled PDMS chambers (Godin et al., 2006, Figure 9.12a) and pinned menisci for variable lens focusing (Dong and Jiang, 2007, Figure 9.12b). These methods are similar to those of out-of-plane lenses. However, among in-plane tunable optofluidic lenses, one unique property that has rarely been exploited in out-of-plane lenses has been utilized for constructing tunable lenses. At the small scale, the Reynolds number (which characterizes the presence of turbulent flow) is extremely low, and thus the flow always remains laminar. Two fluid flows joining in a channel will not mix via turbulence; thus, diffusion alone is the driving mechanism for mixing. Because diffusion occurs slowly, fluid interfaces in microfluidic devices remain distinct. This distinct interface between flowing fluids can be exploited to form the lens surface and the immiscible fluid interface. Additionally, due to laminar flow, the shape of such a fluid–fluid virtual interface can be manipulated hydrodynamically via adjusting the flow injection conditions. Therefore, no additional components are required for lens tuning. Such a concept has been realized in the in-plane tunable lens designs described in the following paragraphs.

Figure 9.13 depicts an in-plane optofluidic tunable lens developed by Mao et al. (2007). It consists of two fluids (5 M $CaCl_2$ solution, $n = 1.445$ and H_2O, $n = 1.335$) flowing within a microfluidic channel that is brought into a 90° curve. Upon entering the curve, the two fluids experience a centrifugal force. The portion of each fluid flowing in the middle of the channel height (where the flow velocity is the highest) experiences a higher centrifugal force than the portions of the fluid closer to the top and bottom of the channel. As a result, the fluid in the middle of the channel height is pulled toward the outer channel wall, and the fluid at the top and bottom of the channel is swept toward the inner channel wall, to ensure the continuity of the fluid ("Dean flow" effect). The originally flat fluidic interface bows outward, creating a cylindrical lens. The amount of interface bowing, and hence the focal point of the lens, can be altered by changing the flow rates of the two fluids. Higher flow rates generate a lens with greater curvature and hence shorter focal length. As with the previously discussed lenses, the smooth fluid interface of this lens provides an excellent optical surface. The curvature of the fluid interface is not optically ideal, but it is shaped well enough to result in sufficient focusing, and the lens is tunable by changing the flow rates of the liquid. It should be noted that this is the only tunable lens (to the authors' knowledge) that brings light in the z-direction to a focal point in the plane of the device (where the plane of the device is the xy-plane).

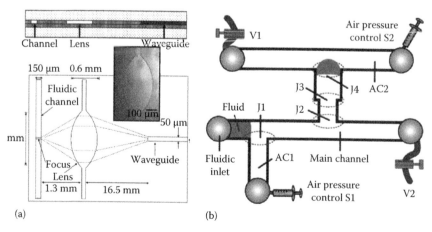

(a)

(b)

FIGURE 9.12 (a) In-plane tunable optofluidic lens using a liquid-filled PDMS chamber. (Reprinted from Godin et al., *Appl. Phys. Lett.*, 89, 061106, 2006. With permission. Copyright 2009, American Institute of Physics.) (b) In-plane tunable optofluidic lens using pinned meniscus. (Reprinted from Dong, L. and Jiang, H., *Appl. Phys. Lett.*, 91, 041109, 2007. With permission. Copyright 2009, American Institute of Physics.)

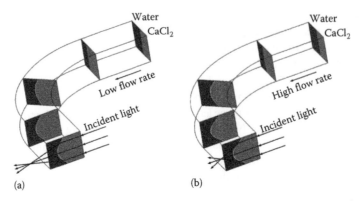

(a)   (b)

FIGURE 9.13 (a) Mechanism of in-plane tunable fluid-fluid lens for focusing of light in the z-direction with respect to the device plane (where the plane of the device is the xy-plane). The tuning of the lens curvature is via the "Dean flow" effect. (b) Change of the lens curvature and focal distance can be observed by changing the flow rate. (Reproduced from Mao, X. et al., *Lab Chip*, 7, 1303, 2007. With permission of The Royal Society of Chemistry.)

The distinct interface between flowing fluids at the micron scale is again employed quite cleverly by the following device developed by Tang et al. (2008). This fluid lens consists of two cladding flows that sandwich a core flow; the refractive index of the cladding liquid is lower than that of the core liquid. The microfluidic channel suddenly increases in width, and then returns to its previous size, thus broadening the core and cladding flows into the shape of a lens. The shape of the lens, and thus its focal point, can be altered by adjusting the flow rates of the core flow and two cladding flows. Meniscus, plano-convex, and biconvex shapes can be achieved. A schematic of the lens (Figure 9.14a), along with a photograph of

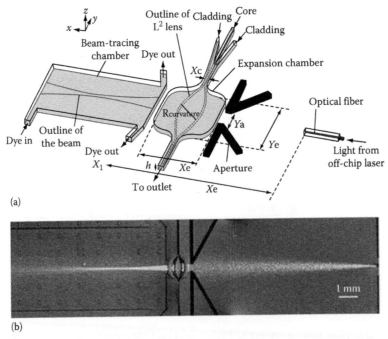

(a)

(b)

FIGURE 9.14 (See color insert following page 11-20.) (a) Schematic of the liquid-core/liquid cladding tunable optofluidic lens. (b) Microscopic image of the lens in action. (Reproduced from Tang, S.K.Y. et al., *Lab Chip*, 8, 395, 2008. With permission of The Royal Society of Chemistry.)

the lens in action (Figure 9.14b), is shown. In the schematic, a biconvex lens is achieved by keeping the flow rates of the two cladding flows equivalent. The smooth fluid interfaces of this lens provide excellent optical surfaces that result in strong focusing. Additionally, the focal point of this lens can be varied across a wide tunable range from 5 to 12 mm, and the width of the focused beam is only 16 μm.

The previously mentioned optofluidic tuning mechanisms provide a cost-effective method to obtain an optically smooth fluidic lens surface; however, the operation of the lenses relies on continual pumping at a relative high flow rate to minimize diffusion across the virtual fluidic interface and to sustain the operation of the lens. This requires significant amounts of supply fluid, presenting a considerable disadvantage for use in imaging applications.

One final tunable lens design that focuses light in-plane must be noted. The lenses researchers have developed thus far are known as refractive lenses. As discussed above, it is very difficult to create and sustain a tunable refractive microlens surface, even with state-of-the-art microfabrication and optofluidic lens manipulation methods. These difficulties have led researchers to search for alternative solutions. Mao et al. (2009) recently reported a novel tunable lens configuration that does not require a refractive lens surface. They named the method liquid gradient refractive index (L-GRIN) lens because the focusing is realized via GRIN (Figure 9.15a) established by controlling transverse diffusion of solution across laminar flows (Figure 9.15b). The L-GRIN lens has a transversely variable refractive index and a flat lens structure, in contrast to the curved lens surface in refractive lenses. The light traveling

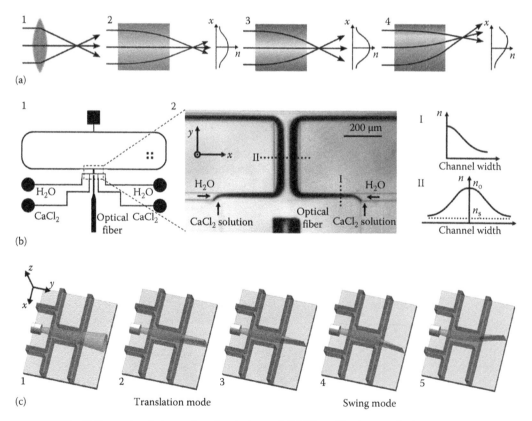

FIGURE 9.15 (a) Comparison between the refractive lens and GRIN lens. The changes of refractive index gradients in GRIN lens can result in variations of focal length and light direction. (b) Schematic of the L-GRIN device that utilizes the diffusion of the $CaCl_2$ in the laminar flow to establish the light-focusing refractive index profile. (c) The focusing power and the light direction of the L-GRIN lens can be conveniently adjusted by changing the flow conditions. (Reproduced from Mao, X. et al., *Lab Chip*, 9, 2050, 2009. With permission of The Royal Society of Chemistry.)

along the GRIN material can be gradually bent toward the optical axis and brought to focus, rather than abruptly changing its direction at the refractive lens surface. The refractive index distribution, and hence the focusing power of the lens, can be conveniently adjusted by changing the fluidic injection rate. More importantly, the lens can swing the focal point from side to side with an unsymmetrical refractive index profile, steering the focused light left or right (Figure 9.15c). The combination of tunable focusing and light steering capabilities allows the lens to direct the light to any given point within the plan of a microfluidic device.

In the L-GRIN lens, diffusion enables (rather than hinders) the operation and tuning of the lens, and thus the working flow rates of the optofluidic lens are much reduced, making the device more sustainable. L-GRIN lenses combine the advantages of in-plane and out-of-plane lenses. On one hand, the L-GRIN lens is easy to fabricate and the lens tuning mechanism is relatively straightforward. On the other hand, it is rather robust and sustainable when compared with other in-plane fluid–fluid lens. Currently, the L-GRIN lens has only been demonstrated in planar structures and can only focus and steer light in one dimension. However, the same concept can be conveniently adapted for 3D structures (such as in a coaxial injection tube) needed for 2D imaging. One of the potential uses for such a 3D L-GRIN lens is in the imaging probe for an endoscope; such a probe would permit a doctor to scan an internal body location without having to move the probe itself.

9.3 Optofluidic Mirrors and Prisms

Mirrors and prisms act in an imaging system to steer and relay the transmitted image. As discussed, electrowetting has been featured as a major mechanism to change the profile of the immiscible fluid interface for optofluidic microlenses. In fact, it is also a major mechanism for actuating liquid-based micromirrors and microprisms.

Wan et al. (2006) introduced a tunable micromirror using a mercury droplet driven by electrowetting. The mirror (Figure 9.16) is constructed by sandwiching a mercury droplet between a transparent indium tin oxide (ITO) glass (top electrode) and a substrate with a bottom electrode. The gap between the top and bottom electrodes is selected such that the droplet is nearly a half-hemisphere at rest state. The incident light from the top was then diverted to a different direction by the curved mercury–air interface. Upon the application of voltage, the reduction of surface energy causes the mercury to form a much larger flat surface to reflect the light.

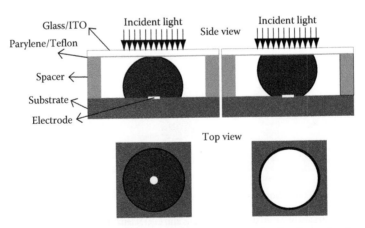

FIGURE 9.16 Schematic of the tunable micromirror using a mercury droplet driven by electrowetting. (Left) "Off" status of the mirror when the meniscus is at the rest state, and (right) "on" status of the mirror when electrowetting changes the shape of the meniscus, which produces a large, flat reflecting area. (Reprinted from Wan, Z. et al., *Appl. Phys. Lett.*, 89, 201107, 2006. With permission. Copyright 2009, American Institute of Physics.)

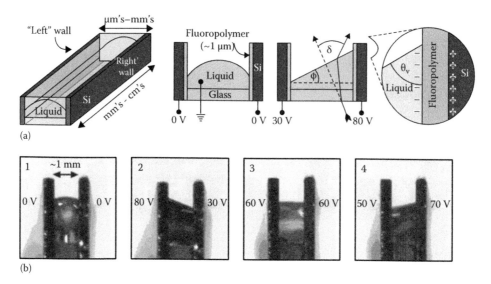

(a)

(b)

FIGURE 9.17 (See color insert following page 11-20.) (a) Schematic of the electrowetting microprism (EMP). (b) Image of the EMP at different operating states: (1) rest state, (2) through (4) tilt of the refractive surface by applying appropriate voltage at different sides of the electrode. (Reproduced from Smith, N.R. et al., *Opt. Express*, 14, 6557, 2006. With permission.)

Another interesting example that uses an electrowetting actuated fluidic boundary to steer the light is the electrowetting microprism (EMP) introduced by Smith et al. (2006). The schematic of the device is shown in Figure 9.17a. A salt solution is filled in a small rectangular cell with a transparent bottom and a pair of parallel electrodes on opposite side walls. The electrode is coated with an oxide layer that serves as an insulator and a hydrophobic coating. The salt solution is electrically grounded. At the rest state (0 V), the liquid forms a curved liquid–gas interface due to surface tension. When an appropriate voltage on the sidewall is applied, the boundary becomes flat due to the changing contact angle along each sidewall, forming a prism. The prism face can be horizontal or tilted (Figure 9.17b), depending on whether the voltages on both sides are equal. The input light from the bottom transparent window can be steered toward different directions.

9.4 Imaging Sensors

The sensor unit (e.g., film, CCD) is the end of an imaging system. Traditionally, relay and magnification of the image is usually carried out through the lens and mirror/prism systems before the light can be projected onto the sensor unit to form the final image. The development of imaging sensors does not necessarily fall into the field of optofluidics. However, researchers believe that through the integration of microfluidic components with image sensors, biological samples such as cells can be brought in close proximity to the image sensor to eliminate the need for lenses, mirrors, and prisms. Such a method would significantly reduce the complexity and cost of the imaging system. However, one of the disadvantages is that the image resolution is limited by the image sensor's pixel size, which is typically 5 μm or larger. Therefore high resolution imaging would be difficult without the magnification provided by a classic lens system. Recently, this problem was addressed by an optofluidic microscope (OFM) developed by Heng et al. (2006) and further advanced by Cui et al. (2008). The device is composed of a PDMS microfluidic channel for sample introduction and an underlying CMOS sensor array for imaging (Figure 9.18a). A layer of metal was coated on the CMOS sensor to block stray light. A line of holes was drilled through the metal layer to permit the transmission of light to the CMOS imaging plane. The holes were 1 μm in diameter and 10 μm apart. The microfluidic channel was bonded on top of the metal

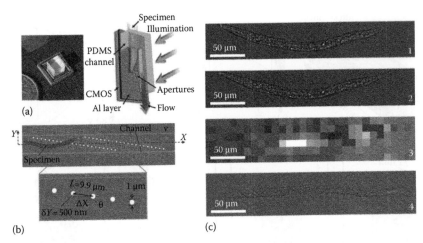

FIGURE 9.18 Principle of the optofludic microscope (OFM). (a) The OFM is composed of a top PDMS device for samples and underlying CMOS sensor array. (b) Lines of holes are aligned with the microfluidic channel. The image of samples can be reconstructed by recording the transmitted light through holes using the underlying CMOS sensor array when samples are passing thorough the sensing area. (c) 1 and 2, Images of *Caenorhabditis elegans* obtained via the OFM. 3, The imaging result when the sample is directly placed on top of the imaging sensor without the nanoholes. 4, The image obtained via a traditional microscope with a 20X objective lens. (Reproduced from Cui, X. et al., *Proc. Natl. Acad. Sci.*, 105, 10670, 2008. With permission. Copyright 2009, National Academy of Science, U.S.A.)

membrane to allow the introduction of a sample. During imaging, the device was illuminated from the top, and a sample was passed through the microfludic channel. As a sample passed over top of a hole, it blocked the holes, and the change in transmitted light was recorded by the sensor array. The key to reconstructing the image at high resolution is to align the line of holes and the microfluidic channel at a small angle so that across the channel the holes are spaced closely enough (500 nm) to achieve high resolution, and yet the distance between holes is still large enough that the light transmitted through each individual hole falls onto only one imaging pixel (10 μm apart). The transmitted signal is recorded and timed as the object passes through each hole (Figure 9.18b). Thus, an image can be reconstructed with information about the transmission signal and time delay for every hole (Figure 9.18c). Such a compact and lens-free OFM can be particularly attractive for developing point-of-care platforms for applications such as blood analysis, cancer diagnosis, and drug screening, where a convenient and affordable cell imaging element is needed.

9.5 Future Perspective

Optics is a well-developed discipline. Microfluidics, though much less mature, has developed at a rapid rate for nearly two decades. There are endless possibilities for emerging optofluidic technologies, which stem from a rich body of knowledge regarding the manipulation of both light and fluids. Many optofluidic imaging devices have been successfully demonstrated based on techniques discussed in this chapter. However, the maturity of most of these devices falls far below the standard for practical applications. The tuning of many optical imaging elements still relies on external pressure sources or fluid injection, which is highly undesirable for applications that require miniaturization and integration. Electrowetting-based devices also face challenges such as complex fabrication processes and high voltage requirements. Many on-going efforts have been focused on addressing these problems. For example, integrated on-chip pumping and valve systems (Unger et al., 2000; Laser and Santiago, 2004) may be used for on-chip pressure control, and optimization of electrowetting electrode coating (Berry et al., 2006) has shown to dramatically reduce the voltage needed for effective contact-angle actuation. The constant stream of technological innovation in microfluidics is also expected to provide simpler, faster, and more cost-effective fluidic manipulation mechanisms for future optofluidic imaging elements.

Acknowledgments

This work was supported by the National Science Foundation (ECCS-0824183 and ECCS-0801922) and the Penn State Center for Nanoscale Science (MRSEC).

References

Berry, S., Kedzierski, J., and Abedian, B. 2006. Low voltage electrowetting using thin fluoroploymer films. *Journal of Colloid and Interface Science* 303: 517–524.

Chronis, N., Liu, G., Jeong, K. H. et al. 2003. Tunable liquid-filled microlens array integrated with microfluidic network. *Optics Express* 11: 2370–2378.

Cui, X., Lee, L. M., Heng X. et al. 2008. Lensless high-resolution on-chip optofluidic microscopes for *Caenorhabditis elegans* and cell imaging. *Proceedings of the National Academy of Sciences* 105: 10670–10675.

Descour, M. R., Kärkkäinen, A. H. O., Rogers, J. D. et al. 2002. Toward the development of miniaturized imaging systems for detection of pre-cancer. *IEEE Journal of Quantum Electronics* 38: 122–130.

Dong, L. and Jiang, H. 2007. Tunable and movable liquid microlens in situ fabricated within microfluidic channels. *Applied Physics Letters* 91: 041109.

Dong, L., Agarwal, A. K., Beebe D. J. et al. 2006. Adaptive liquid microlenses activated by stimuli-responsive hydrogels. *Nature* 442: 551–554.

Godin, J., Lien, V., and Lo, Y. H. 2006. Demonstration of two-dimensional fluidic lens for integration into microfluidic flow cytometers. *Applied Physics Letters* 89: 061106.

Heng, X., Erickson, D., Baugh, L. R. et al. 2006. Optofluidic microscopy—A method for implementing a high resolution optical microscope on a chip. *Lab on a Chip* 6: 1274–1276.

Kuiper, S. and Hendriks, B. H. W. 2004. Variable-focus liquid lens for miniature cameras. *Applied Physics Letters* 85: 1128–1130.

Kwon, S. and Lee, L. P. 2004. Micromachined transmissive scanning confocal microscope. *Optics Letters* 29: 706–708.

Laser, D. J. and Santiago, J. G. 2004. A review of micropumps. *Journal of Micromechanics and Microengineering* 14: 35–64.

Lee, S. W. and Lee, S. S. 2007. Focal tunable liquid lens integrated with an electromagnetic actuator. *Applied Physics Letters* 90: 121129.

López, C. A., Lee, C. C., and Hirsa, A. H. 2005. Electrochemically activated adaptive liquid lens. *Applied Physics Letters* 87: 134102.

Mao, X., Lin, S.-C. S., Lapsley, M. I. et al. 2009. Tunable liquid gradient refractive index (L-GRIN) lens with two degrees of freedom. *Lab on a Chip* 9: 2050–2058.

Mao, X., Waldeisen, J. R., Juluri, B. K. et al. 2007. Hydrodynamically tunable optofluidic cylindrical microlens. *Lab on a Chip* 7: 1303–1308.

Monat, C., Domachuk, P., Eggleton, B. J. et al. 2007. Integrated optofluidics: A new river of light. *Nature Photonics* 1: 106–114.

Mugele, F. and Baret, J. 2005. Electrowetting: From basics to applications. *Journal of Physics Condensed Matter* 17: 705–774.

Pang, L., Levy, U., Campbell, K. et al. 2005. Set of two orthogonal adaptive cylindrical lenses in a monolith elastomer device. *Optics Express* 13: 9003–9013.

Psaltis, D., Quake, S. R., Yang, C. et al. 2006. Developing optofluidic technology through the fusion of microfluidics and optics. *Nature* 442: 381–386.

Ren, H., Fox, D., Anderson, P. A. et al. 2006. Tunable-focus liquid lens controlled using a servo motor. *Optics Express* 14: 8031–8036.

Shamaia, R., Andelman, D., Bergec, B. et al. 2008. Water, electricity, and between on electrowetting and its applications. *Soft Matter* 4: 38–45.

Smith, N. R., Abeysinghe, D. C., Haus, J. W. et al. 2006. Agile wide-angle beam steering with electrowetting microprisms. *Optics Express* 14: 6557–6563.

Tang, S. K. Y., Stan, C. A., and Whitesides, G. M. 2008. Dynamically reconfigurable liquid-core liquid-cladding lens in a microfluidic channel. *Lab on a Chip* 8: 395–401.

Unger, M. A., Chou, H. P., Thorsen, T. et al. 2000. Monolithic microfabricated valves and pumps by multilayer soft lithography. *Science* 288: 113–116.

Wan, Z., Zeng, H., Feinerman, A. et al. 2006. Area-tunable micromirror based on electrowetting actuation of liquid-metal droplets. *Applied Physics Letters* 89: 201107.

Xia, Y. and Whitesides, G. M. 1998. Soft lithography. *Annual Reviews in Materials Science* 28: 153–184.

Zhang, D. Y., Justis, N., and Lo, Y. H. 2004. Fluidic adaptive lens of transformable lens type. *Applied Physics Letters* 84: 4194–4196.

Zhang, D. Y., Lien, V., Berdichevsky, Y. et al. 2003. Fluidic adaptive lens with high focal length tunability. *Applied Physics Letters* 82: 3171–3172.

10

Optofluidic Switches and Sensors

Steve Zamek
Boris Slutsky
Lin Pang
Uriel Levy
Yeshaiahu Fainman

In the beginning of the 1980s, an intersection of physics, chemistry, and nanotechnology laid the foundation for *microfluidics*. Microfluidics allowed fast and controlled manipulation of very small volumes of fluid, and these capabilities opened new avenues in optics. The integration of microfluidics with optoelectronic components became known as *optofluidics* (Psaltis et al. 2006, Monat et al. 2007). This integration throve twofold. First, it allowed the integration of optical components into lab-on-a-chip devices, known also as "micro total analysis systems" (μTAS). Second, it inherited the methods used in microfluidics to build new optical elements and attain new functionalities. In this chapter, we focus on those optical elements and devices that are based on integrated optofluidic components. Throughout this chapter, we use the term *fluid* in its broad sense, meaning liquid or gaseous phases of substances, and pure or mixed liquids including solutions and colloids.

The diverse field of optofluidics has been steadily penetrating application areas of optical communications, data storage, display technologies, bioengineering, medical devices, imaging, metrology, computing, and many others. The ever-growing field of microfluidics enabled fast and easy fabrication, versatile and modular designs, the construction of simulation tools, and a robust integration of fluids into optoelectronic components. In the following sections, we discuss two particular families of optofluidic devices—switches and biochemical sensors.

10.1 Integrated Optofluidic Switching and Tuning

10.1.1 Beam Manipulation: Switching, Deflection, and Scanning

Optical switching technologies were advanced by the fast-developing field of telecommunications. Various physical phenomena employed for optical switching applications include electro-optic (Kaminow 1961, Motoki 1973, Kenan et al. 1974, Alferness 1988), acousto-optic (Dixon 1967, Smith et al. 1996), magneto-optic (Shirasaki et al. 1981), thermo-optic (Haruna and Koyanna 1982, Zatykin et al. 1985), and electrowetting (Hayes and Feenstra 2003, Heikenfeld and Steckl 2005a,b) effects, and micro-mechanical components (Petersen 1977, Wu et al. 2006). One of the first fluid-based switches was based on magneto-optic fluids (Brady et al. 1983, Ginder et al. 1994, Saito et al. 1995, Hong 1999). All-optical switching based on changing the physical properties of black oils was theoretically and experimentally studied (Da Costa 1986). In these switches, the surface of a liquid film was deformed using an optical beam, altering the phase and intensity distribution of the reflected and transmitted laser beams.

Despite numerous development efforts on optofluidic switches, these devices are still in their embryonic stage. The ever-growing field of communications requires fast multi-port switching with short delays, wide bandwidths, and low insertion losses. The very compelling optofluidic technology set a few records trying to address these requirements during the last decade. A broad scope of effects was employed to perform optical switching using fluids.

Since the timescales on which fluids can be displaced (replaced) are commonly on the order of milliseconds, these components promise to benefit optical protection switches, in which fast real-time switching is not crucial. Such switches reconfigure the interconnection of $N \times N$ input–output ports in case of an interruption of service, while one or more faulty optical transmission lines are repaired or replaced. The major advantages of optofluidic switches in such applications are comparatively low insertion losses over a wide bandwidth and of low complexity for multi-port routing and switching. Some of the optofluidic switches are also merely sensitive to polarization. In this section, we review two types of optofluidic switches based on the total internal reflection (TIR) and tunable diffraction gratings.

10.1.1.1 Switches Based on Total Internal Reflection

The TIR of an otherwise transparent surface was achieved by the replacement of fluids with different refractive indices by one another (Campbell et al. 2004), bubble generation (Jackel and Tomlinson 1990) and manipulation (Ware 2000, Hengstler et al. 2003a,b, Uebbing et al. 2006), fluid–fluid interface control using electrowetting (Beni and Hackwood 1981, Heikenfeld and Steckl 2005a,b, Hou et al. 2007), thermocapillary effect (Makihara et al. 1999), and hydrodynamic spreading (Nguyen et al. 2007).

Switching of guided waves using TIR by bubbles was commercially developed by Agilent Technologies, Palo Alto, California (Fouquet et al. 1998, Fouquet 2000). In its suggested design, multiple waveguides are created in planar lightwave circuits, intersecting at several cross points, as shown in Figure 10.1. At these cross points, step discontinuities are introduced. These discontinuities are filled with liquid, whose refractive index is matched to the index of the waveguide, so that the optical mode travels unimpeded through the cross point. When a bubble is inserted into the cross point, the light is reflected into another waveguide. These bubbles can be formed and removed hundreds of times per second, providing a fast and reliable switching function. This technology is similar to that used in ink-jet printers, indicating that such bubble switches should be mass producible.

Another implementation of a TIR switch was based on the oil-latching interfacial tension variation effect (Makihara et al. 1999). The switch exploits the thermocapillary effect to move trapped bubbles in a capillary slit. The light path is switched when the refractive-index-matching oil moves in the slit due to surface tension variation caused by heating (thermocapillarity). A high extinction ratio (>50 dB), a low crosstalk (<–50 dB), and a response time below 10 ms were achieved in a 16×16 switch (Sakata et al. 2001, Venkatesh et al. 2002). The effects of surface properties on time response were also investigated (Sukhanov et al. 2004).

As an example of an optofluidic switch, we describe the free-space TIR switch demonstrated by Campbell et al. (2004). This device, shown schematically in Figure 10.2a through c, was used to switch between the transmission (bypass) and the reflection (exchange) of a beam incident onto a

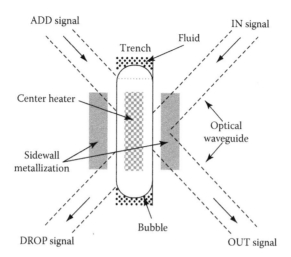

FIGURE 10.1 Top view of the bubble switch element. In this design, the bubbles are generated by a heater, shown in the center. (Reprinted from Hengstler, S. et al., *Proceedings of IEEE/LEOS International Conference on Optical MEMS (OMEMS 2003)*, Santa Clara, CA, pp. 117–118, 2003. With permission.)

channel filled with liquid. The device is implemented in polydimethylsiloxane (PDMS), whose refractive index is 1.41. When the channel is filled with a liquid whose refractive index is matched to the index of the PDMS, an incident beam is transmitted through the channel. When the difference in the refractive indices is large enough, TIR occurs. The fluid in the channel is manipulated by a microfluidic delivery system that consists of two layers. The channels of one layer (the flow layer) are used to deliver the liquids into the mirror channel. The master mold used for the fabrication of this layer is shown in Figure 10.2a. The channels of the second layer (the control layer) are used to actuate the microvalves, enabling fully controlled manipulation of the liquid in the mirror channel. These microvalves are pressure-actuated integrated "push-up" membranes (Studer et al. 2004). The master mold for this layer with the flow layer on top is shown in Figure 10.2b, and the two integrated layers are shown in Figure 10.2c, with the valves formed at the junctions of the control and the flow channels. The channels connecting the inlets and the outlet with the mirror are completely sealed by the flexible membranes when a pressure is applied to the membranes through the control channels. The switch had an insertion loss smaller than 1 dB and an extinction ratio on the order of 20 dB. The device could switch between transmission (bypass) and reflection (exchange) modes within less than 20 ms.

10.1.1.2 Switches Based on Tunable Gratings

Diffraction gratings are widely used in optics for beam splitting and spatiotemporal filtering. Maxima in the far-field diffraction pattern of a grating under plane-wave illumination are given by (Born and Wolf 1999)

$$n_i \sin\alpha_i - n_o \sin\alpha_o = \frac{\lambda}{\Lambda} m \qquad (10.1)$$

where
λ is the free-space wavelength of the incident beam
Λ is the period of the grating

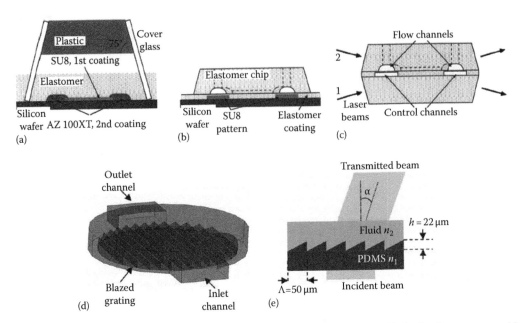

FIGURE 10.2   Two types of optofluidic switches. (a, b, c) 2×2 switch based on TIR: (a) The flow layer master mold with the PDMS cast- and facet-forming structure on top of it. (b) The PDMS chip with engraved flow layer channels aligned on top of the control layer mold coated with a thin layer of the elastomer. (c) The complete three-layer elastomer chip with four optical quality facets. The arrows on the left show directions of incident laser beams 1 and 2, and the arrows on the right show transmitted and reflected beams. (Reprinted from Campbell, K. et al., *Appl. Phys. Lett.*, 85, 6119, 2004. With permission.) (d, e) 1×4 switch based on tunable diffraction grating: (d) The functional area of the device (diffraction channel), consisting of a circular microchannel with a blazed grating imprinted onto its bottom. (e) Concept of operation showing the blazed grating with the incident and transmitted laser beams. (Reprinted from Groisman, A. et al., *Opt. Express*, 16, 13499, 2008. With permission.)

α_i and α_o are the angles of the incident and the output (diffracted) plane waves with respect to the
    normal to the interface

n_i and n_o are the refractive indices of the media on the incident and the output sides

m is an integer, called the diffraction order

When free space is assumed on both sides of a grating, Equation 10.1 gives the deflection angle under paraxial approximation to be $\Delta\alpha \approx m \cdot \lambda/\Lambda$. Optofluidics enables the design and fabrication of tunable gratings, which allow sweeping, scanning, and switching of $\Delta\alpha$. Two major families are tunable geometry and tunable refractive index gratings.

An example of the first family of tunable gratings is self-assembled flowing lattices of bubbles (Hashimoto et al. 2006). A robust control over the size and volume fraction of the bubbles was employed to generate gratings with a pitch tunable from 12 to 51 μm (Garstecki et al. 2004). This allowed continuous angle scanning, stable operation for extended periods of time, and both two-dimensional (2D) and one-dimensional (1D) configurations. The drawbacks of bubble-based gratings are long switching times on the order of seconds and low diffraction efficiency (high insertion losses).

Another family of tunable gratings is based on refractive index tuning. A 1×4 free-space optical switch based on a transmission blazed grating was recently reported (Groisman et al. 2008, Zamek et al. 2008). A blazed grating is submerged into a liquid, as shown in Figure 10.2d and e. Such a grating exhibits highly efficient diffraction of an incident beam when the fluid satisfies the blazed grating condition (Neviere 1991):

$$m = \frac{h\Delta n}{\lambda}$$

(10.2)

where

h is the height of the grating profile

Δn is the difference between the refractive indices of the fluid and the material used to construct the grating

Introducing a liquid with a refractive index that satisfies Equation 10.2 results in beam deflection (diffraction) by an angle given by Equation 10.1. It should be noticed that the integer m in Equation 10.2 determines the dominant diffraction order. A beam normally incident on the grating will be deflected by an angle $\Delta\alpha \approx m \cdot \lambda/\Lambda$, as already mentioned above.

The design of the 1×4 free-space optical switch was based on salt solutions in water, which provided low-viscosity fluid for fast operation and a wide range of refractive indices from 1.33 up to 1.41 depending on the salt concentration.

One of the promises of these suggested switches based on tunable gratings is a low complexity $N \times N$ switching. For example, only two 1×4 switches are required to construct a 4×4 switch, which would otherwise consist of five 2×2 crossbars or sixteen 1×2 switches (Saleh and Teich 1991). These switches are easily scalable to larger number of ports, and can be tailored to a desired wavelength and deflection angles. Both diffraction- and TIR-based switches are fabricated in PDMS, which also suggests easy integration into lab-on-a-chip devices.

10.1.1.3 Deflectors and Beam Scanners

The capability of optical beam scanning is crucial in applications such as optical storage, diagnostics, biomedical imaging, projection displays, optical screening and quality control, spectroscopy, microscopy, sensing, optical communications, and many others. Conventional beam deflectors are based on mechanical or electromechanical reflectors (Schlafer and Fowler 1965, Alsabrook 1966, Dostal 1966), variable refractors (Giarola and Billeter 1963, Aas and Erf 1964, Boer 1964, Chen et al. 1964, Fowler et al. 1964, Haas et al. 1964, Kalibjian et al. 1965, Liu and Walters 1965), and birefringent deflection (with polarization discriminator) (Kulcke et al. 1964, Nelson 1964, Schmidt 1964, Tabor 1964), with more of the early techniques reviewed by Fowler and Schlafer (1966). More recent techniques reported in the literature are based on ferroelastic materials (Tsukamoto et al. 1980, Guilbert and Czapla 2001), liquid crystals (LCs) (Resler et al. 1996, Nashimoto et al. 1998, Wang et al. 2000), and lead lanthanum zirconate titanate (PLZT) (Thomas et al. 1997). Mature microelectromechanical systems (MEMS) paved the way for spatial light modulators (Hornbeck 1997), such as DMD (digital micromirror device) and DLP™ (digital light processing), which were introduced by Texas Instruments (Dudley et al. 2003, Dudley and Dunn 2005). With the advent of optofluidics, new capabilities for beam deflection and scanning became available.

An optical deflector for continuous beam scanning based on an electrowetting microprism with millisecond response time was recently demonstrated (Smith et al. 2006). Beam scanning in the range of $\pm 7°$, with high steering efficiency, polarization-independent operation, and wide steering range, with electrowetting was achieved. An even wider range of deflection angles were achieved using membrane-based micro-mirrors (Werber and Zappe 2006). The micro-mirror is mounted on the membrane and fixed with a thin silicon hinge to the sidewall of the cavity, as shown in Figure 10.3. The application of a differential pressure between the cavity and the ambient pressures deflects the PDMS membrane, leading to a tilting motion. The micro-mirror, fixed to the membrane, is deflected. By varying the pressure, the tilting angles were varied from 0° to 75°, relative to the substrate surface. This application to medical devices with a variable mirror setup, used for in vivo diagnostics was further suggested by the author.

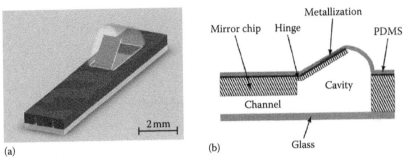

(a)   (b)

FIGURE 10.3 Deflected micro-mirror. (a) The silicon chip (top) is bonded to the Pyrex substrate (bottom), and the mirror is attached to the inside of the PDMS membrane. The pneumatic access channels are seen on the left. (b) Cross-sectional view of the mirror device. This sketch shows the mirror in a deflected position at an arbitrary angle. (Reprinted from Werber, A. and Zappe, H., *J. Opt. A—Pure Appl. Opt.*, 8, S313, 2006. With permission.)

10.1.1.4 Beam Shaping

Optofluidics took a significant step toward miniaturization of tunable lenses. These lenses are discussed in further detail in Section 10.1.2.1. Here, we will only briefly mention their application to beam shaping.

Along with imaging and aberrations control, liquid-filled lenses can be employed for biaxial beam shaping and optical signal processing. A set of two cylindrical orthogonal lenses (Pang et al. 2005) for beam shaping is reviewed here in greater detail. Figure 10.4a shows the device consisting of three

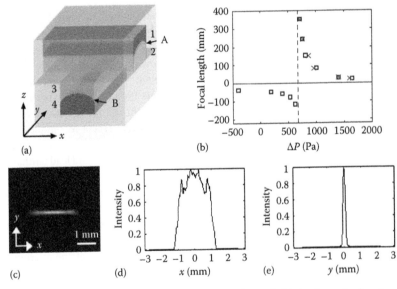

FIGURE 10.4 A set of cylindrical orthogonal lenses (Pang et al. 2005). (a) A schematic drawing of the device, demonstrating its operation as a positive cylindrical lens along the y axis and a negative cylindrical lens along the x axis. Numbers and letters A and B designate chambers (and layers) and membranes, respectively. Chambers 1 and 4 are filled with a low-refractive-index liquid, whereas the high-refractive-index liquid is introduced into chambers 2 and 3. (b) Focal lengths of the cylindrical lenses focusing light along the x axis (squares) and y axis (crosses) as functions of the pressure differences, ΔP_x and ΔP_y, respectively. (c,d,e) Laser beam shaping with a CCD camera placed at $z = 200$ mm behind the device. The driving pressures are $\Delta P_x = 1.2$ kPa and $\Delta P_y = 0.8$ kPa. (c) Patterns of light on the CCD. (d,e) Intensity profiles in the x and y directions (along a line going through the center of the laser spot). (Reprinted from Pang, L. et al., *Opt. Express*, 13, 9003, 2005. With permission.)

chambers separated by two PDMS membranes and filled with liquids of different refractive indices. Differences in the pressures applied to the three chambers result in bending of the membranes. The focal length as a function of pressure is shown in Figure 10.4b. Under a proper design, the shape of the bending membrane is nearly cylindrical. Therefore, the shape of the beam becomes highly elliptical with an aspect ratio of approximately 10, as suggested by Figure 10.4c through e. An experimental comparison between the beam intensity profiles suggested that focusing along one direction has no appreciable effect on the other. Therefore, variations of differential pressures are directly translated into changes in the focal distances with practically no cross talk. Although the optofluidic lenses close our discussion of switching and tuning, we will return to this topic later in Section 10.1.2.1.

10.1.2 Pneumatically Driven Membrane-Based Devices

Numerous optofluidic tunable devices based on soft polymer membranes have recently been reported. In such devices, the geometry of the optical element is altered by the application of pressure to deform soft polymer elements. In this section, we review two types of devices based on polymer membranes. We start with a tunable lens based on a simple polymer membrane, and continue the discussion with a more complicated but a more compelling composite membrane technology. Some review of the mechanical properties of a thin bending membrane is suggested to provide an insight into more complicated mechanics of general structures, such as composite membranes discussed later in this section.

To avoid confusion, we should note first that the term *membrane* used in optics is somewhat inconsistent with its definition in mechanics. We use the term membrane broadly, so its mechanical behavior can follow different mechanical models. Membrane-based devices commonly consist of a chamber with a soft distensible wall, typically made of a soft polymer. When a pressure is applied to the chamber, the wall bends, altering the geometry of the element. The deflection regime under applied pressure is commonly described in terms of the middle line, $u(x, y)$, and the thickness, $t(x, y)$, of the bending wall. Depending on the design and the materials used, the bending wall can be modeled as a *shell* or a *plate*. The major difference is that a plate can take bending stresses in addition to tensile and compressive stresses observed in shells. Some guidelines to the proper choice of the model are found in the thin-plates and shells theory (Mollmann 1981, Reddy 1993, Krauthammer and Ventsel 2001).

Two simplified cases of a circular bending membrane with clamped edges are given by the shell (Poisson 1829, Prandtl 1903) and thin-plate (Reddy 1993, Krauthammer and Ventsel 2001) approximations:

$$\text{Shell:}\quad u(r)=\frac{P}{4S}\left(r_0^2-r^2\right) \tag{10.3}$$

$$\text{Plate:}\quad u(r)=\frac{P}{64D}\left(r_0^2-r^2\right)^2 \tag{10.4}$$

where

  P and S are the uniform pressure and the isotropic tension per unit length
  $r^2 \equiv x^2+y^2$, where x and y are the coordinates in the support plane

Equation 10.3 is called the membrane equilibrium equation. It describes the displacement of an axisymmetric elastic shell with a simply supported contour. Equation 10.4 is given for a circular plate with an axisymmetric uniform load, P, with a clamped edge, where $D=Eh^3 (1 - v^2)^{-1}/12$ is the stiffness (flexural rigidity) of the plate, E is the modulus of elasticity, v is Poisson's ratio, and h is the thickness of the plate.

We shall now clarify the distinctions between the two models. The shell model assumes "large" initial tension and "small" pump pressure, P (i.e., the actuation-induced strain is negligible compared with the initial strain); linear response; and no resistance to bending (Hartog 1987). This model was applied to liquid-filled tunable lenses introduced in the 1970s (Hamilton and Lanckton 1973, Burrows and Hamilton 1974). These lenses were filled with liquid, and therefore had to retain considerable amounts of liquid and overcome distortions caused by gravity. The radius of the curvature of the apex is readily calculated to be $2SP^{-1}$. The effective focal length of the whole aperture is found by applying a best fit of Equation 10.3 to a spherical surface. More recently, this model was applied to small aperture lenses based on elastic films (Rawicz and Mikhailenko 1996).

Fast operation and small form factors require the use of comparatively thick polymer membranes, whose bending stresses cannot be neglected. Therefore, the plate model is widely adopted in polymer-based devices (Campbell et al. 2007). Under small deformations, the membrane thickness can be assumed constant, and the bending profile is explicitly given by Equation 10.4. With larger deformations, numerical techniques might be required to account for a nonconstant flexural rigidity due to thickness variations.

10.1.2.1 Pneumatically Driven Membrane-Based Lenses

The fast development of optoelectronics in the past decades has kept pushing miniaturization of imaging systems to new limits. Traditional imaging systems with variable focal lengths incorporate a pair of lenses with a variable distance between them. This approach is too cumbersome in many application areas, where bulky optics with moving parts is intolerable.

A significant step toward miniaturization of adaptive lenses was made in the 1980s with the advent of LCs, which allowed a successful implementation of LC-based lenses (Sato 1979, Berreman 1980, Kowel 1984, Naumov et al. 1998, Commander et al. 2000). Such lenses are based on the electro-optic effect, which allows control over the refractive index of the constituent medium. LC-based lenses suffer from many limitations: first, strong fields are required to produce a noticeable change in the refractive index with the electro-optic effect; second, LCs produce a birefringent optical medium, and so LC lenses are polarization sensitive. Despite recent works (Ren and Wu 2006, Chiu et al. 2008, Mao et al. 2009, Miccio et al. 2009), fast tunable lenses insensitive to polarization and with a large aperture continue to be challenging.

Optofluidic lenses overcame many of the limitations imposed by the LC lenses. Liquid-filled and polymer-based optofluidic lenses allowed focal tuning from $-\infty$ to $+\infty$; fast response times; easy, low-cost fabrication; and a whole range of optical aperture sizes—from meters down to several millimeters. Two major types of optofluidic lenses evolved—liquid-filled and pneumatic lenses. Liquid-based lenses incorporate liquid encapsulated in a shell (chamber) whose shape is altered mechanically or electrically. These lenses commonly offer very wide focal tuning but very limited aperture. Pneumatic lenses, on the contrary, allow large apertures for high-performance optics, at the expense of the focal tuning range. So these two types represent a trade-off between the aperture size and the focal tuning range, and span a wide range of application areas, including vision devices (von Waldkirch et al. 2005, Varioptic, Campbell et al. 2007, Tsai et al. 2008), microscopy (Gambin et al. 2006, Tsai et al. 2007, 2008), photography (Kuiper and Hendriks 2004), optical data storage (Hendriks et al. 2005), bioengineering and medicine (Divetia et al. 2005, Mao et al. 2007), biochemical and temperature sensing (Dong et al. 2006, Wang and Fang 2007), and lab-on-a-chip devices (Roulet et al. 2001, Seo and Lee 2004, Gambin et al. 2006). These lenses can be driven mechanically (Chronis et al. 2003, Zhang et al. 2003, Chen et al. 2004, Jeong et al. 2004, Kuwano et al. 2005, Ren and Wu 2005, Moran et al. 2006, Mao et al. 2007, Tang et al. 2008), thermally (Dong and Jiang 2006, Lee et al. 2006, 2007, Yang et al. 2006), electrochemically (López et al. 2005, Dorrer et al. 2007), by the electrowetting effect (Berge and Peseux 2000, Kuiper and Hendriks 2004), and by radiation pressure (Casner and Delville 2001). These diverse means of manipulation and control make optofluidic lenses versatile and appealing for a wide scope of tasks.

The very old idea of using liquid-based lenses (Gordon 1918, Wright 1968, Knollman et al. 1971, Hamilton and Lanckton 1973, Burrows and Hamilton 1974) has been recently revisited for portable adaptive optical lenses with mechanical actuators (Kaneko et al. 1997, Ahn and Kim 1998, Chronis et al. 2003, Zhang et al. 2003). A micro-lens array integrated into a microfluidic device was demonstrated (Chronis et al. 2003). Trade-offs between focusing and angle of view were analyzed (Agarwal et al. 2004). Fast actuation with a response time of milliseconds was achieved using piezoelectric actuators (Oku et al. 2004). A compact varifocal micro-lens with an integrated thermal actuator and sensor was implemented (Wang et al. 2005); a set of integrated cylindrical lenses with independent biaxial tuning was constructed (Pang et al. 2005); variable focus was also achieved by replacing liquids with different refractive indices; chemical actuation of a pH-sensitive hydrogel was also employed to design a varifocal lens (Dong and Jiang 2006, Dong et al. 2006); the hydrodynamical interface between two liquids flowing in a curved microchannel was exploited to construct a variable focus lens (Mao et al. 2007). Additional designs and actuation techniques have been continuously reported (Chen et al. 2004, Ren and Wu 2005, 2007, Werber and Zappe 2005, Chandra et al. 2007, Dein and Lin 2007, Lee and Lee 2007).

The lenses mentioned above exhibit aspheric surfaces leading to optical aberrations, which were analyzed for a variety of configurations (Sugiura and Morita 1993, Rawicz and Mikhailenko 1996, Chronis et al. 2003, Werber and Zappe 2005, Wang et al. 2007, Beadie et al. 2008). These aberrations can be diminished by using compound structures (as in acoustic lenses) (Feldman 1969, Tannaka and Koshikaw 1973), the introduction of an aperture (Werber and Zappe 2005), composite membranes (Campbell et al. 2006, 2007), and two different lens curvatures (Jeong et al. 2004).

Liquid-based lenses are based on the curvature of the fluid contained in a polymer micro-chamber. Although such lenses are already commercially available from Varioptic (Berge 2005), scaling them to large apertures is challenging. The presence of liquid makes the device slow and vulnerable to mechanical shocks, and the optical performance is compromised by formation of undesired bubbles. These limitations are lifted by polymer-based devices. In a polymer-based lens for example, it is the curvature of the polymer that creates differences in the optical path length, and therefore determines the optical properties of the lens.

Tunable polymer-based lenses can be actuated thermally (Glebov et al. 2004, Lee et al. 2006, Yang et al. 2006) or pneumatically (Campbell et al. 2007). For thermally tuned lenses, both planar (Glebov et al. 2004) and free-space (Lee et al. 2006) configurations were demonstrated. In the free-space configuration, a thermally actuated polymeric lens is composed of a flexible PDMS lens, silicon conducting ring, and silicon heater (Lee et al. 2006). The mismatch of the coefficients of thermal expansion and stiffness between PDMS and silicon leads to the deformation of the polymer lens during heating, so as to further change its focal length. The difficulty to control the thermal expansion of a large area limits the aperture to hundreds of micrometers for any practical design. Pneumatically actuated lenses do not bear these limitations, and are discussed in further detail below.

Although pneumatically actuated lenses, discussed further, do not contain liquids, they can still be considered optofluidic elements following the definition of fluid. A pneumatically actuated lens is shown on Figure 10.5a and b. The membrane is integrated in compound camera lenses that contain two more elements attached to the same mount: a planoconvex glass lens and a diaphragm between the membrane and the lens. The mount is sealed by the membrane and the lens, and the pressure of air in it is adjusted through a connector on a side. The application of vacuum to the interior of the mount pulls the membrane inward. The shape of the deformed membrane is modeled as a thin circular plate with clamped edges. The measured profile follows the model discussed in the beginning of this section (Equation 10.4), as shown in Figure 10.5c. The diaphragm is integrated with a set of lenses to reduce spherical aberrations. Such lenses allow construction of large apertures not available in other types of fluidic lenses along with millisecond transition times. Pneumatically actuated lenses can be driven at 500 Hz with four diopter variation in the refractive power (Campbell et al. 2007). A considerable focusing power can be obtained at low pressures (see Figure 10.5d). Therefore, these lenses can be used for fast longitudinal scanning in three-dimensional (3D) imaging.

The advantages of pneumatically driven lenses as opposed to liquid-filled lenses become obvious. First, the optical performance of pneumatically actuated lenses is not compromised by the bubbles that form in the liquid. Second, apertures much larger than in the other types of fluidic lenses are readily

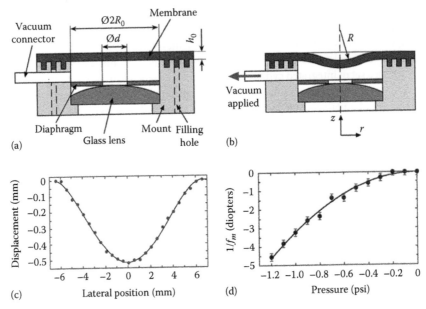

FIGURE 10.5 A set of constant and variable lenses. (a) Schematic of an adaptive compound camera lens with a flexible membrane; (b) the same camera lens with the membrane pulled inward when vacuum is applied; (c) displacement of the membrane in lens 2 at the gauge pressure of vacuum applied to the lens, $P=-0.5\,$psi, as a function of lateral position with respect to the membrane center, fitted by equation $\xi=-A(R_0^2-r^2)^2$, with $R_0=6.6\,$mm; and (d) refractive power, $1/f_m$, of the meniscus lens formed by the membrane in lens 2 as a function of P. (Reprinted from Campbell, K. et al., *Appl. Phys. Lett.*, 91, 171111, 2007. With permission. Copyright 2009, American Institute of Physics.)

available. Finally, the lens carries no additional mass of the liquid, so the response time is faster, and the impact of mechanical shocks and gravitational distortions is diminished. One drawback is a lower focusing power (diopters) than in the other types of fluidic lenses.

10.1.2.2 Composite Membrane Devices

One significant drawback of the above-mentioned polymer-based devices is the very limited geometry of the deflecting membrane. It might be of great interest in optics to allow tailoring of arbitrary bending profiles. As we show here, this goal is easily achieved by using the so-called composite membrane technology.

Composite structures in this context are a combination of two or more materials, each retaining its own elastic properties. Certain combinations of several materials with an appropriate design allow the deformation to be highly adjustable and easily tailored. This feature was successfully employed to build tunable optical devices (Campbell et al. 2006). The pattern of epoxy grafted into the membrane defines its mode of deformation under pressure. A planar architecture allows standard, easy, and precise soft-photolithography fabrication techniques to be used for adaptive optical devices.

Composite membranes were used to construct tunable gratings, which are an alternative to the ones discussed in Section 10.1.1.1. Another application can potentially benefit optical beam scanners based on mirrors mounted on distensible composite membranes (Werber and Zappe 2005). Recently, a new type of device suggested by Campbell et al. (2006), opened a path to full 2D scanning. The proposed technique is based on a binary grating mounted on a PDMS membrane. In these devices, a beam incident onto the grating is split into several diffraction orders. The grating is stretched or rotated to linearly deflect the diffracted beams or cause an in-plane beam rotation. The two devices are called the stretcher and the rotator, respectively, and are reviewed below in greater detail.

Figure 10.6a shows the fabrication concept of a composite membrane. The great advantage of this technique is the leverage of well-developed lithographic tools, which allow high precision along with cheap high-throughput production. In the stretcher, the grafted pieces of epoxy focus the pressure-induced extension of the membrane to a thin strip of PDMS (see Figure 10.6b), and the pressure-induced

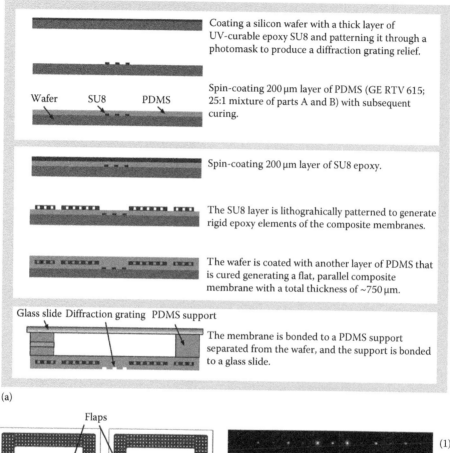

Coating a silicon wafer with a thick layer of UV-curable epoxy SU8 and patterning it through a photomask to produce a diffraction grating relief.

Spin-coating 200 µm layer of PDMS (GE RTV 615; 25:1 mixture of parts A and B) with subsequent curing.

Wafer   SU8   PDMS

Spin-coating 200 µm layer of SU8 epoxy.

The SU8 layer is lithograhically patterned to generate rigid epoxy elements of the composite membranes.

The wafer is coated with another layer of PDMS that is cured generating a flat, parallel composite membrane with a total thickness of ~750 µm.

Glass slide Diffraction grating PDMS support

The membrane is bonded to a PDMS support separated from the wafer, and the support is bonded to a glass slide.

(a)

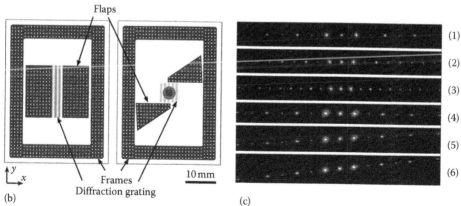

Flaps

(1)
(2)
(3)
(4)
(5)
(6)

y
x
                    Frames   10 mm
          Diffraction grating
(b)   (c)

FIGURE 10.6 Composite membrane devices with stretching and rotating geometries. (a) Fabrication steps. (b) Schematic drawings of the composite membranes in the stretcher (left) and the rotator (right). (c) Photographs of diffraction patterns generated at different pressures by the diffraction gratings engraved in the membranes of the stretcher, (1) through (3), and of the rotator, (4) through (6). For the patterns (1) through (6), the pressures are 0, 0.75, 1.5, 0, −0.75, −1.5 psi. (Reprinted from Campbell, K. et al., *Appl. Phys. Lett.*, 89, 154105, 2006. With permission. Copyright 2009, American Institute of Physics.)

extension provides a tunable grating period. Approximately, 50% elongation of the grating period was achieved at an applied pressure below 1.5 psi. This strong deformation at low pressures in the region of interest is obtained by grafting pieces of epoxy into the membrane. The composite structure transfers most of the deformation onto the thin strip of a soft membrane patterned with a grating profile, causing angular deflection of the diffracted beams, as can be seen from Figure 10.6c (images 1–3).

In the rotator, the composite structure exhibits in-plane rotation of the grating. Again, epoxy grafted into the membrane decreases the membrane's torsion in some areas, while increasing it in the others. This allows achieving angular rotation of the grating and, consequently, in-plane rotation of the diffracted beams by up to 8° at an applied pressure of approximately 2 psi. The rotated far-field intensity profiles are shown in Figure 10.6c (images 4–6).

To summarize, the advantages offered by the polymer-based tunable optics are obvious. First, the optical components can be created by easy and cheap replication techniques, which allow a cost-efficient fabrication of large volumes. Second, easy tailoring of the mechanical properties of composite polymers allows the implementation of an arbitrary optical design. Furthermore, large apertures and fast actuation, difficult to achieve with liquid-based components, are inherent features of polymer-based components.

10.1.3 Tuning of Resonant Structures

The vast majority of optical devices discussed so far requires microfluidic components, which commonly include actuators, mixers, and containers, to retain sufficient amounts of fluids. As a necessary step toward real on-chip integration, these functions are to be implemented in a planar architecture. Such fully integrated devices, recently demonstrated (Erickson et al. 2006, Levy et al. 2006), showed great promise for future integrated chip-scale components. A wide range of tunable fiber-based resonators were recently reported (Mach et al. 2002, Kerbage and Eggleton 2003, Domachuk et al. 2004). Mixing of two liquids with different refractive indices is one of the techniques used in a number of tunable devices, such as tunable waveguides (Wolfe et al. 2004), resonators, and filters (Levy et al. 2006).

The possibility of the integration of microfluidic mixers with photonic elements on the same substrate allowed on-chip tuning of optical resonators. Macroscopic dye lasers traditionally tuned by dye concentration were recently miniaturized with the use of these microfluidic mixers (Bilenberg et al. 2003). A similar approach was recently adopted in ring lasers (Galas et al. 2005), distributed feedback lasers (DFB) (Gersborg-Hansen and Kristensen 2007), and micro-ring resonators (Levy et al. 2006). Tunable gratings written in the core of a microstructured fiber (Kerbage and Eggleton 2003) are another type of optofluidic tuning of resonators. In this case, the resonant transmission through the fiber is caused by periodically spaced microfluidic plugs that are infused into the air holes of a fiber cladding.

Optical micro-ring resonator tuning with a variable-index liquid as an upper cladding was demonstrated (Levy et al. 2006). The fine tuning of an optical micro-ring resonator device (MRD) by a dynamic variation of the refractive index of the surrounding medium is demonstrated in Figure 10.7a and b. The MRD is positioned at the bottom of a flow-through microchannel, which is a part of a microfluidic chip (see Figure 10.7c and d). The liquid injected into the microchannel constitutes the upper cladding of the MRD waveguides. The variation of the refractive index of the liquid is achieved by on-chip mixing of desired proportions of two source liquids with different indices of refraction. The refractive index of the liquid controls the resonance wavelengths and the strength of coupling between the bus waveguide and the resonator. An extinction ratio of 37 dB was achieved by refractive index tuning (see Figure 10.7e).

The constructed optofluidic device combines the optical MRD with the microfluidic channel system, which has two inlets and three outlets. The liquids injected into inlets 2 and 1, a solution and a pure solvent, flow through a square microchannel network of the type introduced by Jeon et al. (Jeon et al. 2000). The network generates repeated splitting and mixing, so that the concentration of the solute varies

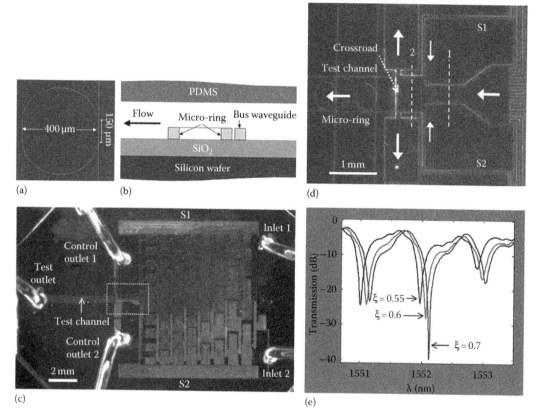

FIGURE 10.7 (See color insert following page 11-20.) MRD integrated with an on-chip mixer. (a) Scanning electron microscope (SEM) micrograph of the MRD. (b) Schematic drawing showing the cross-section of the test microchannel of the MRD. (c) Micrograph of the integrated optofluidic device with water and green dye injected into inlets 1 and 2, respectively. (d) Magnified image of the region inside the dashed rectangle in (c). Arrows show direction of flow. (e) Transmission versus wavelength for three different values of ξ—the dimensionless control parameter of the on-chip mixer. (Reprinted from Levy, U. et al., *Appl. Phys. Lett.*, 88, 111107, 2006. With permission. Copyright 2009, American Institute of Physics.)

linearly across the stream emerging from the network along the dashed line 1 in Figure 10.7d. There are two additional serpentine channels, S1 and S2, in Figure 10.7c and d, directly connecting the two inlets with the channel carrying the linear profile stream, and adding plateaus of zero and the maximal solute concentration to the concentration profile of the stream along the dashed line 2 in Figure 10.7d.

A micro-ring resonator coupled to a bus waveguide is commonly used for filtering and biochemical sensing. Such an MRD shows very narrow margins to fabrication tolerances. Consequently, typical extinction ratios for micro-ring resonator filters are in the range of 10–15 dB (Niehusmann et al. 2004). Higher extinction ratios normally require some post-fabrication trimming, for example, modification of the refractive index of the cladding via photo-oxidation using UV radiation (Chu et al. 1999, Sparacin et al. 2005). Alternatively, high extinction ratios can be reached by using the thermo-optic effect and tuning the effective refractive index by the variation of temperature (Christiaens et al. 2004, Rabiei and Gunter 2005). This latter approach also allows dynamic tuning of the parameters of the MRD. For the tuning to be efficient, the effective refractive index usually has to be varied by 10^{-3} or more, and controlled within $\sim 10^{-5}$. Even for polymeric waveguides with a typically high thermo-optic coefficient, $\partial n/\partial T \approx 10^{-4} °C^{-1}$, these numbers imply a temperature variation of a few $10°C$ and stabilization within $0.1°C$. Thus, tuning of an MRD with the thermo-optic effect requires significant consumption of power for temperature stabilization. In addition, the density of individually adjustable devices on a chip is limited by the thermal conductivity. Therefore, the suggested optofluidic tuning of an optical resonator is advantageous in several applications.

Another approach allowing the tunability of resonant structures is electrowetting. This approach was recently used to tune the resonance frequency and the extinction ratio of an MRD (Shamai and Levy 2009). Tunability was achieved by controlling the wetting angle of a droplet that was partially covering an MRR made of a polymer waveguide. By applying a voltage to the droplet, its wetting angle is modified, and the droplet covers a larger area of the MRD. This results in an increase in the effective refractive index of the MRD waveguide; thus, the resonant wavelength and the transmission through the device can be modified. In addition to the tuning of the resonant wavelength, the authors also demonstrated a significant tuning of the extinction ratio by positioning the droplet on top of the coupling region between the MRD and the bus waveguide, thus allowing control of the coupling coefficient of the device. Figure 10.8 shows (a) the transmission spectrum of the device in the off (dashed curve) and the on (solid curve) states, together with (b and c) microscope images showing the MRD and the droplet in the off and the on states, respectively.

As can be seen, a noticeable shift in the resonant wavelength occurs upon the application of voltage, while variations in the extinction ratio are relatively small. This is because the droplet is located far away from the coupling region. In contrast, Figure 10.9 shows the transmission spectrum of the device for a case where the droplet covers the coupling region in the on state. In such a case, the extinction ratio varies drastically.

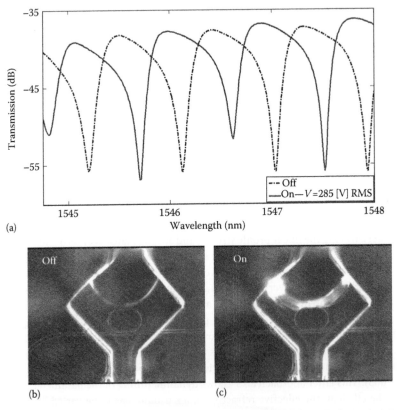

FIGURE 10.8 (a) Transmission spectrum of the device in the off (dashed curve) and the on (solid curve) states; and (b, c) microscope images showing the MRD and the droplet in the off and the on states, respectively. (Reprinted from Shamai, R. and Levy, U., *Opt. Express*, 17, 1116, 2009. With permission.)

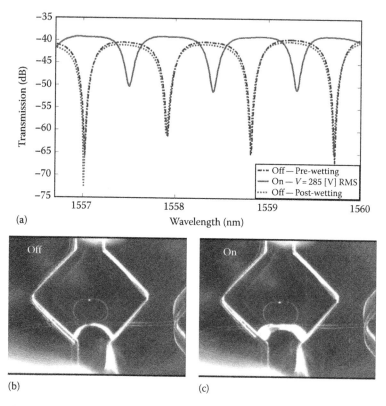

FIGURE 10.9 (a) Transmission spectrum of the device in the off (dashed curve and dotted curve) and the on (solid curve) states; and (b,c) microscope images showing the MRD and the droplet in the off and the on states, respectively. (Reprinted from Shamai, R. and Levy, U., *Opt. Express*, 17, 1116, 2009. With permission.)

10.1.4 Summary

In this section, we reviewed a number of tunable optofluidic devices. In these devices, gases and liquids are used to manipulate and control light in its many forms: planar-guided and free-space, coherent and broadband. These devices are comprised of two major parts—the microfluidic device used as an actuator and the optical element itself that performs the optical adaptation function. We provided a detailed review of adaptive and tunable optical elements, such as lenses, gratings, scanners, and beam shapers. Future integration of these elements with their actuation devices and other optical components allows the construction of miniature multifunctional optical devices, enabling further integration into micro- and macroscale systems.

The advantages of optofluidic components are numerous. Optofluidic switches for instance, use liquids, and thus can operate with an ultrawide spectral bandwidth and, simultaneously, be invariant to the state of the polarization of optical beams; these properties cannot be achieved with other technologies. $N \times 1$ optical switching based on optofluidic components is made possible and allows significant simplification of $N \times N$ optical switching and interconnections. Both the TIR and diffraction phenomena were exploited to construct optofluidic switches.

Pneumatically driven compound lenses overcome the pitfalls of conventional liquid-filled tunable lenses. The absence of liquid allows faster operation and improved performance, which is not compromised by bubbles under mechanical stresses and vibrations. Moreover, apertures much larger than in liquid-filled lenses are readily available. These lenses were shown to produce considerable refractive power under comparatively low applied pressures. We anticipate these lenses to be applicable mostly in the area of vision devices, since their focal tuning range is quite limited compared to liquid-filled lenses.

The composite membrane technology presented in this chapter is based on soft polymer membranes patterned with rigid epoxy inclusions. These inclusions are designed to obtain the desired mechanical properties of the membrane and allow good control of the deformation profile. As the fabrication relies on planar lithographic techniques, the pattern can be easily produced in high volumes with high accuracy. Although composite membranes were so far exploited for tunable gratings (Campbell et al. 2006), they show much potential for any adaptive and tunable optical MEMS, where an arbitrary geometry of deformation is desired.

The on-chip tuning of optical devices became possible with the perfection of fabrication and integration toolsets provided by microfluidics. On-chip mixers allow coarse and fine tuning of the refractive index in a very large range, which is not available using electro-optic, magneto-optic, thermo-optic, and acousto-optic techniques. These mixers can potentially relax the tight fabrication requirements from optical micro-components, dictated by the design.

To summarize, the fast development of optofluidics in the last decade was primarily stimulated by the preceding development of microfluidics. In this early stage of development of optofluidics as a field of research, every new device opens new perspectives and stimulates research showing improved performance and new application areas. The significant progress of micro- and nano-fabrication techniques is striving to provide smoother surfaces, smaller form factors, robustness and better repeatability, faster prototyping, and, therefore, lower costs of optofluidic devices.

10.2 Optofluidic Plasmonic Sensors

Optofluidic plasmonics, consisting of microfluidics integrated with optics and plasmonics, enables the advancement of fundamentals in surface sciences of plasmonic fields with unique implications on numerous potential applications in chemistry, biochemistry, biology, medicine, and engineering. Plasmons enable the localization of optical fields beyond the diffraction limit. These highly confined optical modes enhance light–matter interactions in systems with free electrons in micro- and nanoscale structures. New applications and devices that are expected to directly benefit from these light-confined modes include biochemical sensors, surface-enhanced Raman scattering (SERS), surface-enhanced resonance Raman scattering (SERRS), optical nonlinearities (e.g., SHG), near-field probes and data storage, nanoscale lasers, left handed-materials and "perfect" lenses, enhanced light extraction and detection, detectors and thermo-photovoltaics, subdiffraction-limit lithography, modulators, spectral filters, and interconnects.

Some of these aspects are covered in other chapters of this book, including Chapter 18, Plasmonics, in Part III. This section reviews the advances in biochemical sensors made possible through an integration of plasmonics with microfluidics.

10.2.1 Surface Plasmons

The electron charges on a metal–dielectric boundary can perform coherent fluctuations, called surface-plasma oscillations (Ritchie 1957). The experimental demonstration of this phenomenon dates back to 1960, when Powell and Swan conducted electron-energy-loss experiments. These charge oscillations are accompanied by a mixed transverse and longitudinal electromagnetic field, which is maximal at the boundary and disappears far away from the interface. The field corresponds to a surface wave, which is described by the dispersion relation, $\omega(k)$. Sometimes this phenomenon is called the surface-plasmon polariton (SPP), to emphasize the strong coupling of the electromagnetic field with oscillating dipoles in the metal. Extensive theoretical treatments of surface plasmons (SPs) can be found in several textbooks (Agranovich and Mills 1982, Raether 1998, Maier 2007). For the reader's convenience, we will review the major properties of SPs for the completeness of this chapter.

We describe an SPP on the metal–dielectric interface as an electromagnetic wave, $H = H_0 \exp(ik_x x \pm ik_{j_z} z - i\omega t)$, where "+" is for a solution in the domain $z \geq 0$ (dielectric; $j=2$) and "−" is for $z \leq 0$ (metal; $j=1$). Here, k_z is imaginary, resulting in a solution that is bound for $z \to \infty$ and $z \to -\infty$, as shown in Figure 10.10a. The wavevector, k_x (parallel to the interface), is complex and satisfies the dispersion relation, $\omega(k_x)$:

$$k_x = \frac{\omega}{c}\left(\frac{\varepsilon_1 \varepsilon_2}{\varepsilon_1 + \varepsilon_2}\right)^{1/2} \tag{10.5}$$

with the dielectric functions $\varepsilon_1(\omega)$ and $\varepsilon_2(\omega)$ for the metal and the dielectric, respectively. For the dielectric function of the metal, $\varepsilon_1 = \varepsilon_1' + i_1\varepsilon_1''$, we notice that $|\varepsilon_1'| > |\varepsilon_1''|$ at the optical frequencies, which leads to the following approximations:

$$k_x' \equiv \mathrm{Re}\{k_x\} = \frac{\omega}{c}\left(\frac{\varepsilon_1' \varepsilon_2}{\varepsilon_1' + \varepsilon_2}\right)^{1/2} \tag{10.6a}$$

$$k_x'' \equiv \mathrm{Im}\{k_x\} = \frac{\omega}{c}\left(\frac{\varepsilon_1' \varepsilon_2}{\varepsilon_1' + \varepsilon_2}\right)^{3/2} \frac{\varepsilon_1''}{2(\varepsilon_1')^2} \tag{10.6b}$$

We should notice that for metals, $\varepsilon_1' < 0$, and the condition $|\varepsilon_1'| > \varepsilon_2$ should be satisfied in order for the expression in Equation 10.6a to become real. Both conditions are satisfied for metals (at optical frequencies) and doped semiconductors. From Equations 10.6a and b, the transverse wavevectors in the direction normal to the interface are obtained, and the spatial extensions of the field into the metal and the dielectric are derived. The depth at which the field falls by a factor of e is given by

$$\text{in the metal}: \quad z_1 = \frac{\lambda}{2\pi}\left(\frac{\varepsilon_1' + \varepsilon_2}{\varepsilon_1'^2}\right)^{1/2}$$

$$\text{in the dielectric}: \quad z_2 = \frac{\lambda}{2\pi}\left(\frac{\varepsilon_1' + \varepsilon_2}{\varepsilon_2'^2}\right)^{1/2} \tag{10.7}$$

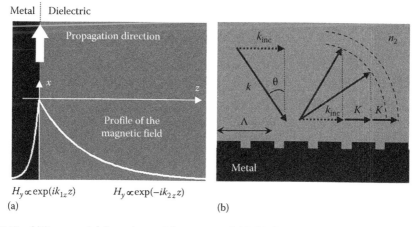

Metal : Dielectric

Propagation direction

Profile of the magnetic field

$H_y \propto \exp(ik_{1z}z)$ $H_y \propto \exp(-ik_{2z}z)$
(a)

(b)

FIGURE 10.10 (a) Exponential dependence of the magnetic field. (b) Illustration of the excitation of the SPP with a grating, according to Equation 10.8b with $p=2$. The two arcs have the radii $n_2 w/c$ (inner arc) and $\mathrm{Re}\{k_{\mathrm{SPP}}\}$ (outer arc). The zero- and the first-order diffracted waves are shown.

For noble metals at optical frequencies, the dielectric function ε_1 is such that the decay length in the dielectric (z_2) is smaller than the wavelength. For example, for a wavelength of 560 nm, the confinement of a plasmon is within ~100 nm from the gold–air interface. This high confinement with a maximum energy flow at the boundary is what makes an SPP so attractive for numerous applications, especially for biochemical sensing, where a thin layer (potentially a monolayer) of molecules, adsorbing to a surface, is to be detected. Upon the formation of the adlayer adsorbed to the surface of the metal, the wavevector, k_x, is perturbed. For shorter wavelengths, the perturbation of the SPP is higher, due to a larger spatial overlap between the mode and the volume, in which the perturbation of the dielectric function occurs (Rothenhäusler 1988, Homola 2006).

The propagation length, defined as the distance at which the intensity decreases to $1/e$th of the original intensity, is given by $L = \left(2k_x''\right)^{-1}$. It typically decreases at higher frequencies (shorter wavelengths) and increases in the near infrared (NIR) and infrared (IR). More on the optical response of metals can be found in Chapter 18 by Sinton et al. As we shall see in the following sections, the reflection spectrum of a nanohole array in a metal film exhibits resonances associated with an excitation of SPs. If material losses are dominant in the SP behavior, the quality factor, Q, of the resonances is directly proportional to the propagation length, L. Therefore, we notice there is a trade-off between the field confinement to the surface (and hence sensitivity to a thin adlayer) and the quality factor of the resonance. This trade-off should be considered in design when choosing the operational wavelength for a plasmon-based sensor.

10.2.2 Excitation of Surface Plasmons

The excitation of SPs can be performed with electron or laser beams. Optical excitation does not suffer from the severe limitations imposed on the electron beam, and allows excitation of SPs at low values of k_x (Raether 1998). Popular optical techniques for the excitation of SPs on the plane of a metal–dielectric interface include grating couplers (Kretschmann 1971), or prism couplers (Otto 1968). These methods are described in more detail in Chapter 18, Plasmonics. A multitude of sensors based on plasmons excited with the two techniques was reported in the literature. In this section we review another technique, which employs subwavelength holes in a metal film for the excitation of SPs.

An array of subwavelength holes in a metal film exhibits all the diffraction anomalies known for 1D and 2D metallic gratings. Diffraction anomalies of metallic gratings were first introduced by Wood in 1902 (Wood 1902), were interpreted by Lord Rayleigh as an emergence of a new diffraction order from evanescence, and were analytically treated by Fano (1941). It was later pointed out by Hessel and O'Liner (1965) that there is an additional type of diffraction anomaly: the excitation of a guided surface wave supported by the structure. The effects of SPs excited on a metal grating at optical frequencies were experimentally studied and analyzed in the 1960s (Hägglund and Sellberd 1966, Ritchie et al. 1968). An extensive overview of the diffraction anomalies of gratings can be found in Petit (1980). The two types of anomalies in the diffraction of a plane wave by a 1D grating are summarized by the following two conditions:

$$\text{Wood–Rayleigh anomalies}: \quad \frac{\omega}{c}n_2 = k_{\text{inc}} + mK \qquad (10.8a)$$

$$\text{surface plasmons}: \quad k_{\text{SPP}} = k_{\text{inc}} + pK \qquad (10.8b)$$

where
    k_{inc} is the projection of the incident k vector onto the plane of the grating
    K is the grating k vector ($K = 2\pi/\Lambda$, where Λ is the period of the grating)
    n_2 is the refractive index of the medium on the incident side

k_{SPP} is the real part of the wavevector of the SP, given by Equation 10.6a

m and p are integers, called the diffraction orders

Equation 10.8b is commonly referred to as the phase-matching condition between the incident and the surface guided waves.

We should further notice that $k_{inc} = (\omega/c)n_2 \sin\theta$, θ being the angle of incidence from a medium with refractive index n_2. The incident wavevector, k_{inc}, is changed by varying the angle or the wavelength of the incident plane wave. While k_{inc} is changing, as it passes the point for which Equation 10.8a is satisfied, violent redistribution of powers between the diffracted orders takes place. As it passes the point for which the second condition of Equation 10.8b is satisfied, the excitation of a guided surface wave on the interface of the metal and the dielectric takes place, causing a dip (a peak) in the reflection (transmission) of the grating. For a 2D grating or a conical diffraction by a 1D grating, an energy exchange between the two polarization states of the diffracted, reflected, or transmitted beams occurs. The later phenomenon is treated in greater detail in the following section.

10.2.3 Sensing through an Array of Subwavelength Holes in Metal

The experiments conducted by Ebbesen et al. in 1998 (Ebbesen et al. 1998) showed that the classical diffraction theory (Bethe 1944) fails to explain the strong transmission of light through subwavelength holes in a metal film. These experiments stimulated a great interest in the role of SPs in the transmission through perforated metallic structures (Porto et al. 1999, Takakura 2001, Cao and Lalanne 2002, Muller et al. 2003, Barnes et al. 2004, Gordon et al. 2004, Pendry et al. 2004, Hibbins et al. 2005, Lomakin and Michielssen 2005, Rokitski et al. 2005, Tetz et al. 2005, Fainman et al. 2006, Hooper et al. 2006, Ozbay 2006, Tetz et al. 2006, Pang et al. 2007a,b). An additional discussion on nanohole arrays in metal films is found in Chapter 18 by Sinton et al.

The first SP-based sensors worked in reflection (Nylander et al. 1982, Homola et al. 1999, Homola 2003). Such sensors are nowadays commercially available from a number of companies (Biacore, McDonnell 2001). Recently these sensors were revisited in the light of new findings in plasmonics. Biochemical sensors based on the transmission through a nanohole array in a metal film were studied by several groups (Brolo et al. 2004, Williams et al. 2004, Rindzevicius et al. 2005, Tetz et al. 2006). One of their main advantages is the transmission architecture, which is well suited for a dense integration of the sensor on a chip.

The general idea of sensing based on nanohole arrays is illustrated in Figure 10.11. A laser beam is incident upon a metal film perforated with subwavelength holes. The transmission spectrum exhibits peaks at the resonance that satisfies the phase-matching condition for the SP:

$$k_{SPP} = k_{inc} + pK_x + qK_y \tag{10.9}$$

where

k_{SPP} is the wavevector of the SP

k_{inc} is the projection of the incident wavevector onto the plane of the metal–dielectric interface

p and q are integers that are called the orders of the grating along x and y axes

Grating vectors K_x and K_y are given by $K_x \equiv 2\pi/\Lambda_x\ \hat{x}$ and $K_y \equiv 2\pi/\Lambda_y\ \hat{y}$, where Λ_x and Λ_y are the periods of the array along x and y axes, respectively. When a nanohole array is functionalized with receptors designed to trap specific biochemical agents from a solution, a thin adlayer of such agents will accumulate on the surface, as shown in Figure 10.11b. The adsorption of the target cells or the molecules will perturb the optical properties of the structure, altering the magnitude of k_{SPP}. This manifests in the variation of the transmission spectrum of the nanohole array, which can be measured using a spectrometer (Brolo

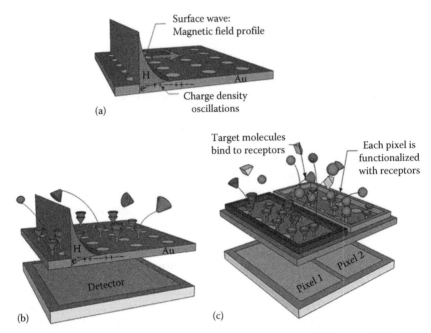

FIGURE 10.11 Concept of high-throughput biochemical sensing with a nanohole array. (a) Excitation of the SP on the metal–dielectric boundary; (b) functionalization of the surface with receptors for the detection of molecular binding; and (c) partitioning the nanohole array into subarrays for high-throughput monitoring of multiple reactions.

et al. 2004, Williams et al. 2004, Rindzevicius et al. 2005), by scanning the wavelength of the source (Pang et al. 2007a,b) or scanning the incident angle of the incident beam (Tetz et al. 2006). Alternative designs, based on the measurement of the intensity of a transmitted (reflected) light, are also possible (Lesuffleur et al. 2008, Yang et al. 2008). Figure 10.11c shows the concept of multi-parallel biochemical sensing using the proposed architecture. Multiple nanohole arrays can be functionalized with different types of receptors to enable a parallel detection of various biochemical agents at the same time. Using the idea of Rothenhäusler and Knoll (1988), the nanohole arrays can be imaged onto a detector to allow fast, highly parallel readout.

A wave incident on one side of the nanohole array excites the SP, which propagates and decouples back into the radiating wave on the other side of the array. Since k_{SPP} does not necessarily lie in the plane of incidence, polarization conversion occurs. The phenomenon of polarization enhancement through the excitation of surface modes was studied for 1D gratings using conical diffraction formulation (Elston et al. 1991). The effect of polarization conversion was employed by Tetz et al. (2006) to construct an SP-based biochemical sensor, illustrated in Figure 10.12a. The spectral sensitivity of the nanohole array to the bulk refractive index of the dielectric was analyzed as a function of the wavelength (Pang et al. 2007a,b), and was shown to be comparable to the sensitivities of grating-based SPR sensors. For an array with periods $\Lambda_x = \Lambda_y = \Lambda$, the sensitivity to the refractive index of the dielectric at normal incidence is given by

$$S_\lambda \equiv \frac{\Delta\lambda}{\Delta n} = \frac{\Lambda}{\sqrt{(p)^2 + (q)^2}} \left(\frac{\varepsilon_m}{n^2 + \varepsilon_m} \right)^{3/2} \tag{10.10}$$

where
n is the refractive index of the dielectric on the side of the SP
p and q are the orders used to excite the SP

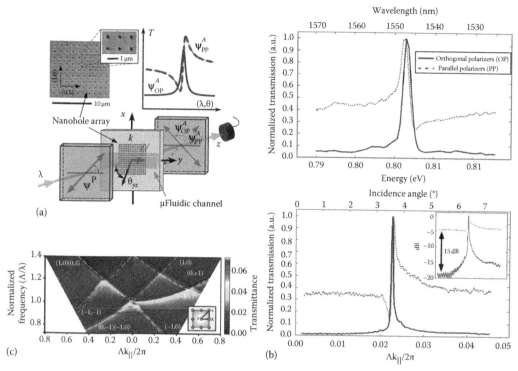

FIGURE 10.12 (See color insert following page 11-20.) Polarization-resolved measurement of a transmittance lineshape of a nanohole array. (a) Experimental setup; also shown is an SEM image of a representative sample. (b) Measured spectral (top) and angular (bottom) transmittance of the nanohole array when the polarizer and the analyzer are aligned (PP) and orthogonal (OP). In the PP condition, a Fano lineshape is produced by the interference of light directly transmitted through the nanoholes and light reradiated via the SPP. In OP, the directly transmitted component is suppressed, resulting in a Lorentzian lineshape. Each curve has been normalized to its maximum to clearly illustrate the respective lineshape functions, except in the inset, where the same data is plotted in a logarithmic scale to show the ~15–20 dB background-level reduction for the Lorentzian versus Fano-type resonances. (c) Transmittance through a 2D array of 350 nm holes in a 100 nm thick aluminum film on GaAs substrate. To cover the domain of normalized frequencies, Λ/λ, shown, data from two samples with hole periods $\Lambda = 1.4\,\mu m$ and $\Lambda = 1.6\,\mu m$ are combined; the horizontal white line at $\Lambda/\lambda \sim 0.95$ indicates the boundary between the two data sets. (Reprinted from Fainman, Y. et al., *Opt. Photon. News*, 17, 24, 2006. With permission.)

Resonant transmission through a nanohole array can be investigated in more detail with the setup in Figure 10.12a. A beam from a tunable laser source (1520–1570 nm)collimated to ~1 mm diameter is incident on the nanohole array at a small angle, θ_{yz}, in the yz plane. The nanohole-array spacing, $\Lambda = 1.4\,\mu m$, is chosen to satisfy the phase-matching condition (Equation 10.8b), for the excited (1, 0) SPP mode propagating in the $-y$ direction. The zero-order diffracted beam from the back side of the sample is projected through the polarizer on the detector in the far field. The directly transmitted background maintains the polarization of the input, which is set at 45° with respect to the incidence plane yz. When the polarizers are aligned to transmit the same polarization state, spectral transmission exhibits a Fano-shape resonance. However, when the output polarizer is orthogonal to the input polarization state, the directly transmitted light is rejected, and the transmission spectrum assumes a Lorentzian shape (see Figure 10.12b). The polarizers' extinction is 60 dB, which assures an almost zero background in spectral transmission measurements.

The measured full widths at half maximum (FWHM) for wavelength interrogation are 1.28 meV (2.47 nm) and 2.86 meV (5.53 nm) in the orthogonal polarizers (OPs) and parallel polarizers (PPs) conditions, respectively, and the PP transmission peak is red-shifted from that of the OP by 0.40 meV (0.77 nm). Similarly, the measured FWHM for angular interrogation (the lower plot in Figure 10.12b) is $0.0012 \Lambda k_{inc}/2\pi$ (0.092°) and $0.011 \Lambda k_{inc}/2\pi$ (0.87°) for the OP and the PP, respectively, and the corresponding red shift is 0.0005 (0.04°).

10.2.4 Integrated Surface-Plasmon Sensors

In this section, we describe an SP sensor integrated with a microfluidic delivery system (Tetz et al. 2006). This setup consists of a sample holder (see Figure 10.13a), placed between an orthogonally crossed polarizer–analyzer pair (see Figure 10.13c) in such a way that the surface wave is excited by a projection of the incident electric field polarization. The reradiated resonant field is projected onto the analyzer, and the greater part of the nonresonant optical transmission is rejected. The transmitted light is simultaneously used to image the sample onto an InGaAs camera for alignment and to measure light transmission using a photodiode. Any change of the in-plane wavevector is achieved by rotating the sample in the xz plane (angle θ in Figure 10.13c) via a mechanical rotation stage with a 0.001° angle resolution.

Before the biorecognition experiment, the sensitivity of the sensor was determined using a series of ethylene glycol (EG) solutions ranging from 0% to 9% by volume. Figure 10.14a depicts normalized transmission spectra of the EG solution series. Figure 10.14b shows the resonant wavelength as a function of time. At $t=0$, the distilled water (DI) was devoid of EG. The EG solutions were added to the channel in increasing concentrations of about 2% until a maximum of 9% EG was reached, as depicted by each jump along the time axis. Following the 9% EG test, the DI water was introduced into the

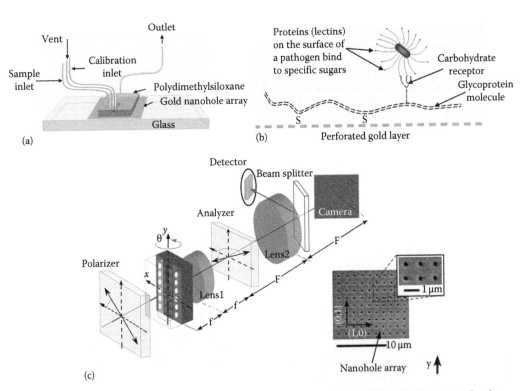

FIGURE 10.13 Integrated optofluidic plasmonic sensor. (a) An SPP sample holder; (b) illustration of pathogen captured via a carbohydrate receptor and protein binding; and (c) schematic of the 2D nanohole-array SPP transmission setup. Inset: scanning electron micrograph of a section of the gold nanohole array. (Reprinted from Hwang, G.M. et al., *Sens. J., IEEE*, 8, 2074, 2008. With permission.)

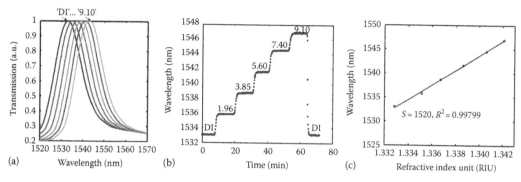

FIGURE 10.14 Sensor calibration. (a) Normalized transmission spectra corresponding to EG solution series (%): 0, 1.96, 3.85, 5.60, 7.40, 9.10. Data were obtained at an incidence angle of 18° from the normal. (b) Representation of resonant wavelength versus time; EG concentrations in percentages are shown. (c) Representation of resonant wavelength versus RIUs. The resonant wavelength and RIUs are linearly related ($R^2 = 0.99$, Pearson's). Open circles represent data based on five-shot averaging. The line represents a linear fit to experimental data. Sensitivity $S = 1520$ nm/RIU. Total shift = 13.9 nm. (Reprinted from Hwang, G.M. et al., *Sens. J., IEEE,* 8, 2074, 2008. With permission.)

channel, returning the SPP resonant wavelength to 1533 nm. (Accurate concentrations of the solution were 1.96%, 3.85%, 5.60%, 7.40%, and 9.10%.)

The wavelength–time trace exhibits stability at each EG concentration, and the increase in the SPP resonant wavelength is proportional to the increase in the EG concentration. Figure 10.14c shows the resonant wavelength as a function of the refractive index of the fluid at the overlayer. The resonant wavelength and refractive index units (RIUs) are linearly related ($R^2 = 0.99$, Pearson's). Open circles represent data based on five-shot averaging. The line represents a least squares linear regression on the experimental data. The sensitivity (S_λ) of the (1, 0)-type SP is computed from Equation 10.6a by taking the derivative of the resonant wavelength with respect to the index of refraction, and is found to be $S_\lambda = 1526$ nm/RIU, which matches well with the experimental value of 1520 nm/RIU. A source-scanning repeatability of 0.1 nm corresponds to a resolution of 6.6×10^{-5} RIU (0.1 nm/(1520 nm/RIU)). In the following two paragraphs, we review the typical methods for studying protein–carbohydrate and protein–protein reactions with this sensor.

Figure 10.15a shows a protein–protein binding using monoclonal anti-bovine serum albumin (mAb) to bovine serum albumin (BSA) in the SPP setup (Pang et al. 2007a,b). The integrated microfluidic SPR sensor device is first cleaned by running 2% sodium dodecyl sulfate (SDS) solution through the surface of the gold nanohole array. The SDS solution is then replaced by a water rinse. The phosphate-buffered saline (PBS) solution is then introduced; this defines the baseline of the sensor for this experiment. 2 mg/mL BSA solution is then introduced into the channel and left overnight. The PBS solution is introduced into the channel to wash out the excess unattached BSA. 26.6 µg/mL purified monoclonal anti-BSA solution is introduced into the channel to allow for specific binding to the BSA attached to the gold surface. After 60 min, PBS is introduced to wash out the unbound anti-BSA, although a disassociation of the weakly bound BSA from the gold surface could also have occurred. The resulting resonant wavelength is 1536.3 nm, which corresponds to a binding shift $\Delta\lambda = 0.7$ nm.

Figure 10.15b shows a protein–carbohydrate binding using Con A and ovomucoid (Hwang et al. 2008). Prepared solutions consist of (1) 0.01 M PBS prepared by dissolving the PBS powder in water with a pH value of 7.4 at 25°C; (2) 1 mg/mL Con A in PBS; and (3) an aqueous solution of ovomucoid (1 mg/mL) heated for 10 min at 95°C, stirred for 30 min, passed through a 0.2 µm pore syringe filter (Fisherbrand, nylon), and allowed to cool on ice.

The integrated microfluidic SPR sensor device is pretreated with DI water, and a PBS solution is introduced to define the baseline of the sensor for the experiment. A solution with 1 mg/mL of ovomucoid is then introduced into the channel. The PBS is added to the channel for flushing away excess

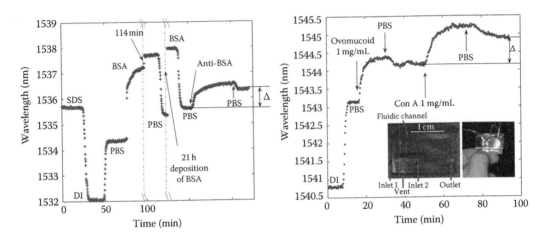

FIGURE 10.15 (See color insert following page 11-20.) Detection and real-time monitoring of (a) protein–protein and (b) protein–carbohydrate bindings on a gold nanohole array observed by monitoring the resonant wavelength of a single SPP mode at the fluid–metal overlayer in a transmission setup. Insets are photographs of an SPP sample holder with (right) and without (left) tubing. (Reprinted from Hwang, G.M. et al., *Sens. J., IEEE,* 8, 2074, 2008. With permission.)

unbound ovomucoid. The purified Con A solution is introduced into the channel to allow for specific binding to the ovomucoid attached to the gold surface. After 25 min, PBS is introduced to wash out the unbound Con A, although a disassociation of the weakly bound ovomucoid from the gold surface could also occur. The resulting resonant wavelength is 1544.9 nm, which corresponds to a binding shift $\Delta\lambda = 0.7$ nm.

The example shown above implemented a label-free method to use glycoproteins as carbohydrate receptor molecules on a gold nanohole array to capture and detect pathogen stimulants in a transmission-based nanohole-array SPP setup. The sensitivity of the nanohole-array-based SP sensor, which is still low compared to the conventional SP sensor, can be improved by better wavelength-scanning repeatability, experiments with different illumination wavelength bands, fabrication techniques to achieve smoother surfaces, and more uniform periodicity. The nanohole array is easily scalable to a larger number of sensing elements for highly parallel biochemical detection.

10.2.5 Summary

SPR sensors are more appealing today than ever before, due to an increasing number of application areas, which include the examination of protein–protein or protein–DNA interactions (Mernagh et al. 1998), the detection of conformational changes in an immobilized protein (Sota et al. 1998), protein–carbohydrate interactions (Hwang et al. 2008), cellular membranes (Kim et al. 1998), the quantification of cell receptors and ligands interactions (Garcia et al. 1997), and antibody–peptide interactions (Sibille and Strosberg 1997). The great promise is not only a label-free detection of analytes, but also the capability of examining kinetic and thermodynamic constants of biomolecular interactions.

Ever since the first application of SPR to biosensing was demonstrated in 1982 (Nylander et al. 1982), there has been a steadily growing interest, and far more advanced systems have been developed. One shortcoming of the existing commercial SPR instruments is their bulkiness due to the optics used to excite the SPs. In this respect, the nanohole arrays presented in this chapter hold a great promise, since they allow effective fabrication, on-chip integration and packaging, scalability to large numbers of detection spots, and easy integration with microfluidic delivery in a complete optofluidic system.

References

Aas H.G. and R.K. Erf. 1964. Application of ultrasonic standing waves to the generation of optical beam scanning. *J. Acoust. Soc. Am.* 36: 1906–1913.

Agarwal M., R.A. Gunasekaran, P. Coane, and K. Varahramyan. 2004. Polymer-based variable focal length microlens system. *J. Micromech. Microeng.* 14: 1665–1673.

Agranovich V.M. and D.L. Mills (eds). 1982. *Surface Polaritons*, North Holland, Amsterdam, the Netherlands.

Ahn S.-H. and Y.-K. Kim. 1998. Design and fabrication of variable focusing lens. *Proc. SPIE* 3515: 270–277.

Alferness R.C. 1988. Waveguide electrooptic switch arrays. *Sel. Areas Commun.* 6(7): 1117–1130.

Alsabrook C.M. 1966. A multicolor laser display. *18th Annual National Aerospace Electronics Conference,* Dayton, OH, pp. 325–331.

Barnes W.L., W.A. Murray, J. Dintinger, E. Deveaux, and T.W. Ebbesen. 2004. Surface plasmon polaritons and their role in the enhanced transmission of light through periodic arrays of subwavelength holes in a metal film. *Phys. Rev. Lett.* 92: 107401.

Beadie G., M.L. Sandrock, M.J. Wiggins, R.S. Lepkowicz, J.S. Shirk, M. Ponting, Y. Yang, T. Kazmierczak, A. Hiltner, and E. Baer. 2008. Tunable polymer lens. *Opt. Express* 16(16): 11847–11857.

Beni G. and S. Hackwood. 1981. Electro-wetting displays. *Appl. Phys. Lett.* 38: 207.

Berge B. 2005. Liquid lens technology: Principle of electrowetting based lenses and applications to imaging. *Proceedings of the MEMS 2005 Conference,* Miami, FL.

Berge B. and J. Peseux. 2000. Variable focal lens controlled by an external voltage: An application of electrowetting. *Eur. Phys. J. E* 3: 159–163.

Berreman D.W. 1980. U.S. Patent 4,190,330.

Bethe H.A. 1944. Theory of diffraction by small holes. *Phys. Rev.* 66(7/8): 163–182.

Bilenberg B., B. Helbo, J.P. Kutter, and A. Kristensen. 2003. Tunable microfluidic dye laser. *12th International Conference on Transducers, Solid-State Sensors, Actuators and Microsystems,* Boston, MA, vol. 1, pp. 206–209.

Boer K.W. 1964. Franz-Keldysh effect for appreciable deflection of light beams. *Phys. Status Solidi* 8: K179–854.

Born M. and E. Wolf. 1999. *Principles of Optics,* 7th edn, Cambridge University Press, Cambridge, U.K.

Brady M.J., L.V. Gregor, and M. Johnson. 1983. U.S. Patent 4,384,761.

Brolo A.G., R. Gordon, B. Leathem, and K.L. Kavanagh. 2004. Surface plasmon sensor based on the enhanced light transmission through arrays of nanoholes in gold films. *Langmuir* 20(12): 4813–4815.

Burrows A.A. and V.E. Hamilton. 1974. Stereopsis using a large aspheric field lens. *Appl. Opt.* 13(4): 739.

Campbell K., A. Groisman, U. Levy, L. Pang, S. Mookherjea, D. Psaltis, and Y. Fainman. 2004. A microfluidic 2×2 optical switch. *Appl. Phys. Lett.* 85: 6119–6121.

Campbell K., U. Levy, Y. Fainman, and A. Groisman. 2006. Pressure-driven devices with lithographically fabricated composite epoxy-elastomer membranes. *Appl. Phys. Lett.* 89: 154105.

Campbell K., Fainman Y., and Groisman A. 2007. Pneumatically actuated adaptive lenses with millisecond response time. *Appl. Phys. Lett.* 91 (17): 171111

Cao Q. and P. Lalanne. 2002. Negative role of surface plasmons in the transmission of metallic gratings with very narrow slits. *Phys. Rev. Lett.* 88: 057403.

Casner A. and J.-P. Delville. 2001. Adaptive lensing driven by the radiation pressure of a continuous-wave laser wave upon a near-critical liquid-liquid interface. *Opt. Lett.* 26: 1418–1420.

Chandra D., S. Yang, and P.-C. Lin. 2007. Strain responsive concave and convex microlens arrays. *Appl. Phys. Lett.* 91: 251912.

Chen F.S., J.E. Geusic, S.K. Kurtz, J.G. Skinner, and S.H. Wemple. 1964. The use of Perovskite paraelectrics in beam deflectors and light modulators. *Proc. IEEE* (Correspondence) 52: 1258–1259.

Chen J., W. Wang, J. Fang, and K. Varahramyan. 2004. Variable-focusing microlens with microfluidic chip. *J. Micromech. Microeng.* 14: 675–680.

Chiu C., Y. Lin, P. Chao, and A. Fuh. 2008. Achieving high focusing power for a largea perture liquid crystal lens with novel hole-and-ring electrodes. *Opt. Express* 16: 19277–19284.

Christiaens I., D.V. Thourhout, and R. Baets. 2004. *International Conference on Indium Phosphide and Related Materials*, Kagoshima, Japan, p. 425.

Chronis N., G. Liu, K.-H. Jeong, and L. Lee. 2003. Tunable liquid-filled microlens array integrated with microfluidic network. *Opt. Express* 11: 2370–2378.

Chu S.T., W. Pan, S. Sato, T. Kaneko, B.E. Little and Y. Kokubun. 1999. An eight-channel add-drop filter using vertically coupled microringresonators over a cross grid. *IEEE Photon. Technol. Lett.* 11: 688.

Commander L.G., S.E. Day, and D.R. Selviah. 2000. Variable focal length microlenses. *Opt. Commun.* 177: 157–170.

Da Costa G. 1986. All-optical light switch using interaction between low-power light beams in a liquid film. *Opt. Eng.* 25: 1058.

Dein S. and C.-W. Lin. 2007. Design and analysis of an asymmetrical liquid-filled lens. *Opt. Eng.* 46: 123002.

Divetia A., T.-H. Hsieh, J. Zhang, Z. Chen, M. Bachman, and G.-P. Li. 2005. Dynamically focused optical coherence tomography for endoscopic applications. *Appl. Phys. Lett.* 86: 103902.

Dixon R.W. 1967. Photoelastic properties of selected materials and their relevance for applications to acoustic light modulators and scanners. *J. Appl. Phys.* 38: 5149.

Domachuk P., H. Nguyen, and B.J. Eggleton. 2004. Transverse probed microfluidic switchable photonic crystal fiber devices. *IEEE Photon. Technol. Lett.* 16: 1900–1902.

Dong L. and H. Jiang. 2006. pH-adaptive microlenses using pinned liquid-liquid interfaces actuated by pH-responsive hydrogel. *Appl. Phys. Lett.* 89: 211120.

Dong L., A.K. Agarwal, D.J. Beebe, and H. Jiang. 2006. Adaptive liquid microlenses activated by stimuli-responsive hydrogels. *Nature* 442: 551–554.

Dorrer C., O. Prucker, and J. Ruhe. 2007. Swellable surface-attached polymer microlenses with tunable focal length. *Adv. Mater.* 19(3): 456.

Dostal F. 1966. The fork as a scanner: A new twist. *Electron. Communicator* 1: 45.

Dudley D. and C. Dunn. 2005. DLP Technologie-nicht nur für Projektoren und Fernsehen. *Photonik* 1: 32–35.

Dudley D., W.M. Duncan, and J. Slaughter. 2003. Emerging digital micromirror device (DMD) applications. *Proc. SPIE* 4985: 14–25.

Ebbesen T.W., H.J. Lezec, H.F. Ghaemi, T. Thio, and P.A. Wolff. 1998. Extraordinary optical transmission through sub-wavelength hole arrays. *Nature (London)* 391: 667–669.

Elston S.J., G.P. Bryan-Brown, and J.R. Sambles. 1991. Polarization conversion from diffraction gratings. *Phys. Rev. B* 44: 6393–6400.

Erickson D., T. Rockwood, T. Emery, A. Scherer, and D. Psaltis. 2006. Nanofluidic tuning of photonic crystal circuits. *Opt. Lett.* 31: 59–61.

Fainman Y., K. Tetz, R. Rokitski, and L. Pang. 2006. Surface plasmonic fields in nanophotonics. *Opt. Photon. News* 17: 24–29.

Fano U. 1941. The theory of anomalous diffraction gratings and of quasi-stationary waves on metallic surfaces (Sommerfeld's waves). *J. Opt. Soc. Am.* 31: 213–222.

Feldman H. 1969. Nearly spherical acoustic lenses. *J. Acoust. Soc. Am.* 45: 868.

Fouquet J.E. 2000. Compact optical cross-connect switch based on total internal reflection in a fluid containing planar lightwave circuit. *Optical Fiber Communication Conference and Exhibit (OFC)*, Baltimore, MD, pp. 7–10.

Fouquet J.E., S. Venkatesh, M. Troll, D. Chen, H.F. Wong, and P.W. Barth. 1998. A compact, scalable crossconnect switch using total internal reflection due to thermally-generated bubbles. *IEEE LEOS Annual Meeting*, Orlando, FL.

Fowler V. and J. Schlafer. 1966. A survey of laser beam deflection techniques. *Appl. Opt.* 5: 1675–1682.

Fowler V.J., C.F. Buhrer, and L.R. Bloom. 1964. Electro-optic light beam deflector. *Proc. IEEE* (Correspondence) 52: 193–194.

Galas J.C., J. Torres, M. Belotti, Q. Kou, and Y. Chen. 2005. Microfluidic tunable dye laser with integrated mixer and ring resonator. *Appl. Phys. Lett.* 86: 264101.

Gambin Y., O. Legrand, and S.R. Quake. 2006. Microfabricated rubber microscope using soft solid immersion lenses. *Appl. Phys. Lett.* 88: 174102.

Garcia C.K., M.D. Tallquist, L.R. Pease et al. 1997. Alpha-beta T cell receptor interactions with syngeneic and allogeneic ligands: Affinity measurements and crystallization. *Proc. Natl. Acad. Sci. USA* 94: 13838–13843.

Garstecki P., I. Gitlin, W. DiLuzio, G.M. Whitesides, E. Kumacheva, and H.A. Stone. 2004. Formation of monodisperse bubbles in a microfluidic flow-focusing device. *Appl. Phys. Lett.* 85: 2649.

Gersborg-Hansen M. and A. Kristensen. 2007. Tunability of optofluidic distributed feedback dye lasers. *Opt. Express* 15: 137–142.

Giarola A.J. and T.R. Billeter. 1963. Electroacoustic deflection of a coherent light beam. *Proc. IEEE* (Correspondence) 51: 1150–1151.

Ginder J.M., J.T. Remillard, and W.H. Weber. 1994. U.S. Patent 5,351,319.

Glebov A.L., L.D. Huang, S. Aoki, M. Lee, and K. Yokouchi. 2004. Planar hybrid polymer-silica microlenses with tunable beamwidth and focal length. *IEEE Photon. Technol. Lett.* 16: 1107–1109.

Gordon R.L. 1918. U.S. Patent 1,269,422.

Gordon R., A.G. Brolo, A. McKinnon, A. Rajora, B. Leathem, and K.L. Kavanagh. 2004. Strong polarization in the optical transmission through elliptical nanohole arrays. *Phys. Rev. Lett.* 92: 037401.

Groisman A., S. Zamek, K. Campbell, L. Pang, U. Levy, and Y. Fainman. 2008. Optofluidic 1×4 switch. *Opt. Express* 16(18): 13499–13508.

Guilbert L. and Z. Czapla. 2001. Interferences in light deflection by ferroelastic domain walls. *Appl. Opt.* 40: 125–128.

Haas W., R. Johannes, and P. Cholet. 1964. Light beam deflection using the Kerr effect in single crystal prisms of BaTiO3. *Appl. Opt.* 3: 988–989.

Hamilton V.E. and A.H. Lanckton. 1973. Stereopsis using a large liquid-filled aspheric field lens stereoscope. *Photogramm. Eng. Rem. Sens.* 39(3): 294.

Hartog D.J.P. 1987. *Advanced Strength of Materials*, Courier Dover Publications, New York, p. 11.

Haruna M. and J. Koyama. 1982. Thermooptic deflection and switching in glass. *Appl. Opt.* 21: 3461–3465.

Hashimoto M., B. Mayers, P. Garstecki, and G.M. Whitesides. 2006. Flowing lattices of bubbles as tunable, self-assembled diffraction gratings. *Small* 2(11): 1292–1298.

Hayes R.A. and B.J. Feenstra. 2003. Video-speed electronic paper based on electrowetting. *Nature* 425: 383–385.

Hägglund J. and F. Sellberd. 1966. Reflection, absorption, and emission of light by opaque optical gratings. *J. Opt. Soc. Am.* 56: 1031–1040.

Heikenfeld J. and A.J. Steckl. 2005a. Intense switchable fluorescence in light wave coupled electrowetting devices. *Appl. Phys. Lett.* 86(1): 011105.

Heikenfeld J. and A.J. Steckl. 2005b. High-transmission electrowetting light valves. *Appl. Phys. Lett.* 86(15): 151121.

Hendriks B.H.W., S. Kuiper, M.A.J. Van As, C.A. Renders, and T.W. Tukker. 2005. Electrowetting-based variable-focus lens for miniature systems. *Opt. Rev.* 12(3): 255–259.

Hengstler S., J.J. Uebbing, and P. McGuire. 2003a. Laser-activated optical bubble switch element. *14th International Conference on Integrated Optics and Optical Fibre Communication (ECOC-IOOC 2003)*, Rimini, Italy.

Hengstler S., J.J. Uebbing, and P. McGuire. 2003b. Laser-activated optical bubble switch element. *Proceedings of IEEE/LEOS International Conference on Optical MEMS (OMEMS 2003)*, Santa Clara, CA, pp. 117–118.

Hessel A. and A.A. Oliner. 1965. A new theory of wood's anomalies on optical gratings. *Appl. Opt.* 4: 1275–1297.

Hibbins A.P., B.R. Evans, and J.R. Sambles. 2005. Experimental verification of designer surface plasmons. *Science* 308(5722): 670.

Homola J. 2003. Present and future of surface plasmon resonance biosensors. *Anal. Bioanal. Chem.* 377(3): 528–539.

Homola J. 2006. *Surface Plasmon Resonance Based Sensors*, Springer, New York.

Homola J., S.S. Yee, and G. Gauglitz. 1999. Surface plasmon resonance sensors: Review. *Sens. Actuator B: Chem.* 54(1–2): 3–15.

Hong C.Y. 1999. Optical switch devices using the magnetic fluid thin films. *J. Magn. Magn. Mater.* 201: 178–181.

Hooper I.R., T.W. Preist, and J.R. Sambles. 2006. Making tunnel barriers (including metals) transparent. *Phys. Rev. Lett.* 97: 053902.

Hornbeck L.J. 1997. Digital light processing for high-brightness, high-resolution applications. *Proc. SPIE* 3013: 27–41.

Hou L., N.R. Smith, and J. Heikenfeld. 2007. Electrowetting manipulation of any optical film. *Appl. Phys. Lett.* 90(25): 251114.

Hwang G.M., L. Pang, E.H. Mullen, and Y. Fainman. 2008. Plasmonic sensing of biological analytes through nanoholes. *Sens. J., IEEE* 8(12): 2074–2079.

Jackel J.L. and W.J. Tomlinson. 1990. Bistable optical switching using electrochemically generated bubbles. *Opt. Lett.* 15(24): 1470.

Jeon N.L., S.K.W. Dertinger, D.T. Chiu, I.S. Choi, A.D. Stroock, and G.M. Whitesides. 2000. Generation of solution and surface gradients using microfluidic systems. *Langmuir* 16(22): 8311–8316.

Jeong K.H., G. Liu, N. Chronis, and L. Lee. 2004. Tunable microdoublet lens array. *Opt. Express* 12(11): 2494–2500.

Kalibjian R., T. Huen, C. Maninger, and J. Yee. 1965. Laser deflection modulation in a CdS prism. *Proc. IEEE* (Correspondence) 53(5): 539.

Kaminow I.P. 1961. Microwave modulation of the electro-optic effect in KH2PO4. *Phys. Rev. Lett.* 6: 528–530.

Kaneko T., T. Ohmi, N. Ohya, N. Kawahara, and T. Hattori. 1997. A new, compact, and quick-response dynamic focusing lens. *Proceedings of the 9th International Conference on Solid-State Sensors and Actuators (Transducers '97)*, Chicago, IL, pp. 63–66.

Kenan R.P., C.M. Verber, and Van E. Wood. 1974. Wide-angle electro-optic switch. *Appl. Phys. Lett.* 24: 428.

Kerbage C and B.J. Eggleton. 2003. Tunable microfluidic optical fiber gratings. *Appl. Phys. Lett.* 82(9): 1338–1340.

Kim E., S.J. DeMarco, S.M. Marfatia, A.H. Chishti, M. Sheng, and E.E. Strehler. 1998. Plasma membrane Ca^{2+} ATPase Isoform 4b binds to membrane-associated guanylate kinase (MAGUK) proteins via their PDZ (PSD-95/Dlg/ZO-1) domains. *J. Biol. Chem.* 273: 1591–1595.

Knollman G.C., J.L.S. Bellin, and J.L. Weaver. 1971. Variable focus liquid filled hydroacoustic lense. *J. Acoust. Soc. Am.* 49(1): 253.

Kowel S., D. Cleverly, and P. Kornreich. 1984. Focusing by electrical modulation of refraction in a liquid crystal cell. *Appl. Opt.* 23: 278–289.

Krauthammer T. and E. Ventsel. 2001. *Thin Plates and Shells—Theory, Analysis and Applications*, Marcel Dekker, New York.

Kretschmann E. 1971. Determination of optical constants of metals by excitation of surface plasmons. *Zeitschrift für Physik* 241(4): 313.

Kuiper S. and B.H.W. Hendriks. 2004. Variable-focus liquid lens for miniature cameras. *Appl. Phys. Lett.* 85: 1128.

Kulcke W., T.J. Harris, K. Kosanke, and E. Max. 1964. A fast, digital-indexed light deflector. *IBM J.* 8: 64–67.

Kuwano R., T. Tokunaga, Y. Otani, and N. Umeda. 2005. Liquid pressure varifocus lens. *Opt. Rev.* 12(5): 405–408.

Lee S.W. and S.S. Lee. 2007. Focal tunable liquid lens integrated with an electromagnetic actuator. *Appl. Phys. Lett.* 90: 121129.

Lee S.N., H.W. Tung, W.C. Chen, and W.L. Fang. 2006. Thermal actuated solid tunable lens. *IEEE Photon. Technol. Lett.* 18: 2191–2193.

Lee S.Y., W.C. Chen, H.W. Tung, and W. Fang. 2007. Microlens with tunable astigmatism. *IEEE Photon. Technol. Lett.* 19: 18.

Lesuffleur A., H. Im, N.C. Lindquist, K.S. Lim, S.-H. Oh. 2008. Laser-illuminated nanohole arrays for multiplex plasmonic microarray sensing. *Opt. Express* 16: 219–224.

Levy U., K. Campbell, A. Groisman, S. Mookherjea, and Y. Fainman. 2006. On-chip microfluidic tuning of an optical microring resonator. *Appl. Phys. Lett.* 88: 111107–111109.

Liu S.G. and W.L. Walters. 1965. Optical beam deflection by pulsed temperature gradients in bulk GaAs. *Proc. IEEE* (Correspondence) 53: 522–523.

Lomakin V. and E. Michielssen. 2005. Enhanced transmission through metallic plates perforated by arrays of subwavelength holes and sandwiched between dielectric slabs. *Phys. Rev. B* 71: 235117.

López C.A., C.C. Lee, and A.H. Hirsa. 2005. Electrochemically activated adaptive liquid lens. *Appl. Phys. Lett.* 87: 134102.

Mach P., M. Dolinski, K.W. Baldwin, J.A. Rogers, C. Kerbage, R.S. Windeler, and B.J. Eggleton. 2002. Tunable microfluidic optical fiber. *Appl. Phys. Lett.* 80: 4294.

Maier Stefan A. 2007. *Plasmonics: Fundamentals and Applications*, Springer, New York.

Makihara M., M. Sato, F. Shimokawa, and Y. Nishida. 1999. Micromechanical optical switches based on thermocapillary integrated in waveguide substrate. *Lightwave Technol.* 17(1): 14.

Mao X., J.R. Waldeisen, B.K. Juluri, and T.J. Huang. 2007. Hydrodynamically tunable optofluidic cylindrical microlens. *Lab Chip* 7: 1303–1308.

Mao, X., S. Lin, M.I. Lapsley, J. Shi, B.K. Juluri, and T.J. Huang. 2009. Tunable liquid gradient refractive index (L-GRIN) lens with two degrees of freedom. *Lab Chip* 9: 2050–2058.

McDonnell J.M. 2001. Surface plasmon resonance: Towards an understanding of the mechanisms of biological molecular recognition. *Curr. Opin. Chem. Biol.* 5(5): 572–577.

Mernagh D.R., P. Janscak, K. Firman and G.G. Kneale. 1998. Protein–protein and protein–DNA interactions in the Type I restriction endonuclease R.EcoR124I. *Biol. Chem.* 379: 497–503.

Miccio L., A. Finizio, S. Grilli, V. Vespini, M. Paturzo, S. De Nicola, and P. Ferraro. 2009. Tunable liquid microlens arrays in electrode-less configuration and their accurate characterization by interference microscopy. *Opt. Express* 17: 2487–2499.

Mollmann H. 1981. *Introduction to the Theory of Thin Shells*, John Wiley & Sons, Chichester, U.K.

Monat C., P. Domachuk, and B.J. Eggleton. 2007. Integrated optofluidics: A new river of light. *Nat. Photon.* 1: 106–114.

Moran P.M., S. Dharmatilleke, A.H. Khaw, K.W. Tan, M.L. Chan, and I. Rodriguez. 2006. Fluidic lenses with variable focal length. *Appl. Phys. Lett.* 88: 041120.

Motoki T. 1973. Low voltage optical modulator using electrooptically induced phase gratings. *Appl. Opt.* 12: 1472–1476.

Muller R., V. Malyarchuk, and C. Lienau. 2003. Three-dimensional theory on light-induced near-field dynamics in a metal film with a periodic array of nanoholes. *Phys. Rev. B* 68(20): 205415.1–205415.9.

Nashimoto K., S. Nakamura, H. Moriyama, M. Watanabe, and E. Osakabe. 1998. Electro-optic beam deflector using epitaxial Pb(Zr,Ti)O3 waveguides on Nb-doped SrTiO3. *Appl. Phys. Lett.* 73: 303.

Naumov A.F., M.Y. Loktev, I.R. Guralnik, and G. Vdovin. 1998. Liquid-crystal adaptive lenses with modal control. *Opt. Lett.* 23: 992–994.

Nelson T.J. 1964. Digital light deflection. *Bell Syst. Tech. J.* 43: 821–845.

Neviere M. 1991. Electromagnetic study of transmission gratings. *Appl. Opt.* 30: 4540–4547.

Nguyen N.-T., T.-F. Kong, J.-H. Goh, and C.L.-N. Low. 2007. A micro optofluidic splitter and switch based on hydrodynamic spreading. *J. Micromech. Microeng.* 17: 2169–2174.

Niehusmann J., A. Vörckel, P. Bolivar, T. Wahlbrink, W. Henschel, and H. Kurz. 2004. Ultrahigh-quality-factor silicon-on-insulator microring resonator. *Opt. Lett.* 29: 2861–2863.

Nylander C., B. Liedberg and T. Lind. 1982. Gas detection by means of surface plasmon resonances. *Sens. Actuators* 3: 79.

Oku H., K. Hashimoto, and M. Ishikawa. 2004. Variable-focus lens with 1-kHz bandwidth. *Opt. Express* 12: 2138–2149.

Otto A. 1968. Excitation of nonradiative surface plasma waves in silver by the method of frustrated total reflection. *Zeitschrift für Physik* 216(4): 398–410.

Ozbay E. 2006. Plasmonics: Merging photonics and electronics at nanoscale dimensions. *Science* 311: 189.

Pang L., U. Levy, K. Campbell, A. Groisman, and Y. Fainman. 2005. Set of two orthogonal adaptive cylindrical lenses in a monolith elastomer device. *Opt. Express* 13: 9003–9013.

Pang L., K.A. Tetz, and Y. Fainman. 2007a. Observation of the splitting of degenerate surface plasmon polariton modes in a two-dimensional metallic nanohole array. *Appl. Phys. Lett.* 90: 111103–111105.

Pang L., G. Hwang, B. Slutsky, and Y. Fainman. 2007b. Spectral sensitivity of two-dimensional nanohole array surface plasmon polariton resonance sensor. *Appl. Phys. Lett.* 91: 123112.

Petit R. 1980. *Electromagnetic Theory of Gratings*, Topics in Applied Physics, Springer, New York.

Pendry J.B., L. Martín-Moreno, and F.J. Garcia-Vidal. 2004. Mimicking surface plasmons with structured surfaces. *Science* 305(5685): 847–848.

Petersen K.E. 1977. Micromechanical light modulator array fabricated on silicon laser display applications. *Appl. Phys. Lett.* 31(8): 521–523.

Porto J.A., F.J. Garcia Vidal, and J.B. Pendry. 1999. Transmission resonances on metallic gratings with very narrow slits. *Phys. Rev. Lett.* 83: 2845.

Prandtl L. 1903. Torsion of Prismatic rods. *Phys. Z.* 4: 758–759.

Psaltis D., S.R. Quake, and C. Yang. 2006. Developing optofluidic technology through the fusion of microfluidics and optics. *Nature* 442: 381–386.

Rabiei P. and P. Gunter. 2005. Dispersion-shifted LiNbO3 waveguides for wide-band parametric amplifiers. *IEEE Photon. Technol. Lett.* 17(1): 133–135.

Raether H. 1998. *Surface Plasmons on Smooth and Rough Surfaces and on Gratings*, Springer, Berlin, Germany.

Rawicz A.H. and I. Mikhailenko. 1996. Modeling a variable-focus liquid-filled optical lens. *Appl. Opt.* 35(10): 1587–1589.

Reddy J.N. 1993. *Theory and Analysis of Elastic Plates*, Taylor & Francis, Irvine, CA.

Ren H. and S.T. Wu. 2005. Variable-focus liquid lens by changing aperture. *Appl. Phys. Lett.* 86: 211107.

Ren H. and S.T. Wu. 2006. Adaptive liquid crystal lens with large focal length tenability. *Opt. Express* 14: 11292–11298.

Ren H. and S.T. Wu. 2007. Variable-focus liquid lens. *Opt. Express* 15(10): 5931–5936.

Resler D.P., D.S. Hobbs, R.C. Sharp, L.J. Friedman, and T.A. Dorschner. 1996. High-efficiency liquid-crystal optical phased-array beam steering. *Opt. Lett.* 21: 689–691.

Rindzevicius T.A.Y., A. Dahlin, F. Hook, D.S. Sutherland, and M. Kall. 2005. Plasmonic sensing characteristics of single nanometric holes. *Nano Lett.* 5(11): 2335–2339.

Ritchie R.H. 1957. Plasma losses by fast electrons in thin films. *Phys. Rev.* 106(5): 874–881.

Ritchie R.H., E.T. Arakawa, J.J. Cowan, and R.N. Hamm. 1968. Surface-plasmon resonance effect in grating diffraction. *Phys. Rev. Lett.* 21: 1530–1533.

Rokitski R., K.A. Tetz, and Y. Fainman. 2005. Propagation of femtosecond surface plasmon polariton pulses on the surface of a nanostructured metallic film: Space-time complex amplitude characterization. *Phys. Rev. Lett.* 95(17): 177401.

Rothenhäusler B. and W. Knoll. 1988. Surface–plasmon microscopy. *Nature* 332: 615–617.

Roulet J.C., R. Völkel, H.P. Herzig, E. Verpoorte, N.F. de Rooij, and R. Dändliker. 2001. Microlens systems for fluorescence detection in chemical Microsystems. *Opt. Eng.* 40: 814–821.

Saito M., M. Takakuwa, and M. Miyagi. 1995. Optical constants of magnetic fluids and their application to optical switches. *IEICE Trans. Electron.* E78-C: 1465.

Sakata T., H. Togo, M. Makihara, F. Shimokawa, and K. Kaneko. 2001. Improvement of switching time in a thermocapillarity optical switch. *J. Lightwave Technol.* 19(7): 1023–1027.

Saleh B.E.A. and M.C. Teich. 1991. Photonic switching and computing. *Fundamentals of Photonics*, John Wiley & Sons, New York.

Sato S. 1979. Liquid-crystal lens-cells with variable focal length. *Jpn. J. Appl. Phys.* 18: 679–1684.

Schlafer J. and V.J. Fowler. 1965. A precision, high speed, optical beam steerer. *International Electron Devices Meeting*, Washington, DC.

Schmidt U.J. 1964. A high speed digital light beam deflector. *Phys. Lett.* 12: 205–206.

Seo J. and L.P. Lee. 2004. Disposable integrated microfluidics with self-aligned planar microlenses. *Sens. Actuators B: Chem.* 99(2–3): 615–622.

Shamai R. and U. Levy. 2009. On chip tunable micro ring resonator actuated by electrowetting. *Opt. Express* 17(2): 1116–1125.

Shirasaki M., N. Takagi, T. Obokata, and K. Shirai. 1981. Bistable magnetooptic switch using YIG crystal with phase matching films. *IEEE J. Quantum Electron.* 17(12): 2498–2499.

Sibille P. and A.D. Strosberg. 1997. A FIV epitope defined by a phage peptide library screened with a monoclonal anti-FIV antibody. *Immunol. Lett.* 59: 133–137.

Smith D.A., R.S. Chakravarthy, Z.Y. Bao, J.E. Baran, J.L. Jackel, A. dAlessandro, D.J. Fritz, S.H. Huang, X.Y. Zou, S.M. Hwang, A.E. Willner, and K.D. Li. 1996. Evolution of the acousto-optic wavelength routing switch. *J. Lightwave Technol.* 14: 1005–1019.

Smith N.R., D.C. Abeysinghe, J.W. Haus, and J. Heikenfeld. 2006. Agile wide-angle beam steering with electrowetting microprisms. *Opt. Express* 14(14): 6557–6563.

Sota H., Y. Hasegawa, and M. Iwakura. 1998. Detection of conformational changes in an immobilized protein using surface plasmon resonance. *Anal. Chem.* 70: 2019–2024.

Sparacin D., C. Hong, L. Kimerling, J. Michel, J. Lock, and K. Gleason. 2005. Trimming of microring resonators by photo-oxidation of a plasma-polymerized organosilane cladding material. *Opt. Lett.* 30: 2251–2253.

Studer G.H., A. Pandolfi, M. Ortiz, W.F. Anderson, and S.R Quake. 2004. Scaling properties of a low-actuation pressure microfluidic valve. *J. Appl. Phys.* 95: 393.

Sugiura N. and S. Morita. 1993. Variable-focus liquid-filled optical lens. *Appl. Opt.* 32(22): 4181–4186.

Sukhanov V.I., F.L. Vladimirov, and D.A. Monakhov. 2004. How the surface properties of the liquid element affect the response time of a liquid optical deflector. *J. Opt. Technol.* 71(3): 183–186.

Tabor W.J. 1964. The use of Wollaston prisms for a high-capacity digital light deflector. *Bell Syst. Tech. J.* 43: 1153–1154.

Takakura Y. 2001. Optical resonance in a narrow slit in a thick metallic screen. *Phys. Rev. Lett.* 86: 5601.

Tang S.K.Y., C.A. Stan and G.M. Whitesides. 2008. Dynamically reconfigurable liquid-core liquid-cladding lens in a microfluidic channel. *Lab Chip* 8: 395–401.

Tannaka Y. and T. Koshikaw. 1973. Solid-liquid compound hydroacoustic lens of low aberration. *J. Acoust. Soc. Am.* 53(2): 590–595.

Tetz K.A., L. Pang, and Y. Fainman. 2006. High-resolution surface plasmon resonance sensor based on linewidth-optimized nanohole array transmittance. *Opt. Lett.* 31(10): 1528–1530.

Tetz K.A., R. Rokitski, M. Nezhad, and Y. Fainman. 2005. Excitation and direct imaging of surface plasmon polariton modes in a two-dimensional grating. *Appl. Phys. Lett.* 86(11): 111110.

Thomas J.A., M. Lasher, Y. Fainman, and P. Soltan. 1997. PLZT-based dynamic diffractive optical element for high-speed, random-access beam steering. *Opt. Scanning Syst.: Des. Appl., Proc. SPIE* 3131: 124–132.

Tsai P.S., B. Migliori, K. Campbell, T.N. Kim, Z. Kam, A. Groisman, and D. Kleinfeld. 2007. Spherical aberration correction in nonlinear microscopy and optical ablation using a transparent deformable membrane. *App. Phys. Let.* 91: 191102.

Tsai F.S., S.H. Cho, Y.-H. Lo, B.Vasko, and J. Vasko. 2008. Miniaturized universal imaging device using fluidic lens. *Opt. Lett.* 33(3): 291–293.

Tsukamoto T., J. Hatano, and H. Futama. 1980. Light deflection from ferroelastic domains. *J. Phys. Soc. Jpn.* 49(Suppl. B): 155–159.

Uebbing J.J., S. Hengstler, D. Schroeder, D. Schroeder, S. Venkatesh, and R. Haven. 2006. Heat and fluid flow in an optical switch bubble. *J. Microelectron. Mech. Syst.* 15(6): 1528–1539.

Varioptic—The Liquid Lens Company, www.varioptic.com.

Venkatesh S., J.-W. Son, J.E. Fouquet, R.E. Haven, D. Schroeder, H. Guo, W. Wang, P. Russell, A. Chow, and P.F. Hoffmann. 2002. Recent advances in bubble actuated photonic cross-connect. *Silicon-Based and Hybrid Optoelectronics IV*, San Jose, CA.

von Waldkirch M., P. Lukowicz, and G. Troster. 2005. Oscillating fluid lens in coherent retinal projection displays for extending depth of focus. *Opt. Commun.* 253(4–6): 407–418.

Wang W.S. and J. Fang. 2007. Variable focusing microlens chip for potential sensing applications. *IEEE Sens.* 7(1–2): 11–17.

Wang X., D. Wilson, R. Muller, P. Maker, and D. Psaltis. 2000. Liquid-crystal blazed-grating beam deflector. *Appl. Opt.* 39: 6545–6555.

Wang W., J. Fang, and K. Varahramyan. 2005. Compact variable-focusing microlens with integrated thermal actuator and sensor. *IEEE Photon. Technol. Lett.* 17(12): 2643–2645.

Wang Z., Y. Xu, and Y. Zhao. 2007. Aberration analyses for liquid zooming lenses without moving parts. *Opt. Commun.* 275(1): 22–26.

Ware A. 2000. New photonic-switching technology for all-optical networks. *Lightwave* 17(3): 92–98.

Werber A. and H. Zappe. 2005. Tunable microfluidic microlenses. *Appl. Opt.* 44(16): 3238–3245.

Werber A. and H. Zappe. 2006. Tunable, membrane-based, pneumatic micro-mirrors. *J. Opt. A—Pure Appl. Opt.* 8(7): S313–S317.

Williams S.M., K.R. Rodriguez, S. Teeters-Kennedy, S. Shah, T.M. Rogers, A.D. Stafford, and J.V. Coe. 2004. Scaffolding for nanotechnology: Extraordinary infrared transmission of metal microarrays for stacked sensors and surface spectroscopy. *Nanotechnology* 15(10): S495–S503.

Wolfe D.B., R.S. Conroy, P. Garstecki, B.T. Mayers, M.A. Fischbach, K.E. Paul, M. Prentiss, and G.M. Whitesides. 2004. Dynamic control of liquid-core/liquid-cladding optical waveguide. *Proc. Natl. Acad. Sci. USA.* 101(34): 12434–12438.

Wood R.W. 1902. On a remarkable case of uneven distribution of light in a diffraction grating spectrum. *Phil. Mag.* 4: 396.

Wright B.M. 1968. Improvements in or relating to variable focus lenses. English patent 1,209,234.

Wu M.C., A. Solgaard, and J.E. Ford. 2006. Optical MEMS for lightwave communication. *J. Lightwave Technol.* 24(12): 4433–4454.

Yang H., Y.-H. Han, X.-W. Zhao, K. Nagai, and Z.-Z. Gu. 2006. Thermal responsive microlens arrays. *Appl. Phys. Lett.* 89: 111121.

Yang J.-C., J. Ji, J.M. Hogle, and D.N. Larson. 2008. Metallic nanohole arrays on fluoropolymer substrates as small label-free real-time bioprobes. *Nano Lett.* 8: 2718–2724.

Zamek S., K. Campbell, L. Pang, A. Groisman, and Y. Fainman. 2008. Optofluidic 1×4 switch. *Conference on Lasers and Electro-Optics*, San Jose, CA, paper CTuR4.

Zatykin A.A., S.K. Morshnev, and A.V. Frantsesson. 1985. A thermooptic fiber switch. *Kvantovaya Elektronika* 12(1): 211–213.

Zhang D.-Y., V. Lien, Y. Berdichevsky, J. Choi, and Y.-H. Lo. 2003. Fluidic adaptive lens with high focal length tenability. *Appl. Phys. Lett.* 82: 3171.

11

Optofluidic Ring Resonators

Jonathan D. Suter
Xudong Fan

11.1 Introduction and Background

11.1.1 Introduction to Ring Resonators

Ring resonators are fascinating both for their geometric simplicity and for their ability to efficiently confine optical energy within small mode volumes. In the past couple of decades, optical ring resonators have been employed in many applications, including cavity quantum electrodynamics (QED) (Lefevre-Seguin and Haroche 1997, Vernooy et al. 1998), solid-state microlasers (Baer 1987, Liu et al. 2004), nonlinear optics (Raman (Lin and Campillo 1997), hyper-Raman (Klug et al. 2005), Brillouin scattering (Culverhouse et al. 1991), and four-wave mixing (Klug et al. 2005)), electro-optic modulators (Cohen et al. 2001), optical signal processing (such as filters (Little et al. 2004), dispersion compensation (Madsen et al. 1999), and slow-light generation (Heebner et al. 2002)), optical frequency combs (Del'Haye et al. 2007), optical gyroscopes (Armenise et al. 2001), opto-mechanics (including strain (Huston and Eversole 1993), radiation pressure (Kippenberg et al. 2005), and ultrasound sensing (Ashkenazi et al. 2004)), and frequency stabilization (Sakai et al. 1991).

Recently, incorporation of ring resonators with fluidics has come under intensive investigation. Such an integration belongs to a newly defined area called optofluidics (Psaltis et al. 2006, Monat et al. 2007, Horowitz et al. 2008), which describes optical systems fabricated with fluids, crafted to enable manipulation of photonics on the microscale and build additional functionality into chip-based microfluidics. Optofluidic ring resonators, broadly speaking, are optical devices in which the ring resonator and fluidic systems are combined to produce synergistic functionality that could not otherwise be achieved. The fluidic handling structures involved with such a system may be externally fashioned or may be inherent to the ring resonator geometry.

Optofluidic ring resonators open the door for a number of very exciting applications, among which the most important and most intensively investigated are sensing and microfluidic lasers. Ring resonator sensors include those based on refractive index (RI) detection (Hanumegowda et al. 2005a,

Xu et al. 2007b, Bernardi et al. 2008, Zamora et al. 2008), light absorption (Nitkowski et al. 2008), fluorescence (Blair and Chen 2001), Raman scattering (Sennaroglu et al. 2007), and cavity-enhanced, surface-enhanced Raman spectroscopy (SERS) (White et al. 2007a). With regard to microfluidic lasers, optofluidics provide the benefit of small sample volumes, easy sample delivery, and large spectral coverage and tunability. Optofluidic ring resonator lasers have been demonstrated using both microdroplets and solid resonators that rely on external fluid as a gain medium. These lasers require the use of dyes (Lin et al. 1986), quantum dots (QDs) (Schafer et al. 2008), or a Raman signal-generating substance (Kwok and Chang 1993, Sennaroglu et al. 2007). A more thorough discussion of optofluidic lasers using other optical cavities can be found in Chapter 12 by Kristensen and Mortensen.

In the following sections, we will discuss some of the important properties of optical modes in optofluidic ring resonators and their applications, with a particular focus on ring-resonator-based sensing and microfluidic lasers. These applications and the highlighted examples will provide a good overview of the breadth of the subject matter defined by optofluidic ring resonators.

11.1.2 Optical Mode in a Ring Resonator

The basic optical phenomenon that allows ring resonators to function is the circular resonant mode, which relies on total internal reflection (TIR) to circulate light at resonant wavelengths. Confinement is accomplished via high RI contrast between the ring material (with high RI) and its surrounding media (with lower RI). Due to Snell's law, photons within the ring may encounter one or more reflections at the interfacial boundary due to TIR. This can lead to a condition where they constructively interfere as they complete a full round-trip inside the ring. Under these conditions, a ring resonator is formed with the majority of the electric field concentrated at the interface. This optical phenomenon is what is known as the whispering-gallery mode (WGM). The wavelength at which the resonant condition is met can be modeled based on Mie theory (Gorodetsky and Ilchenko 1999).

Any optical cavity, including an optical ring resonator, can be characterized in terms of a quality factor, or Q-factor. It is defined as the amount of stored energy within the cavity divided by the amount of energy radiated per resonant cycle (Gastine et al. 1967), which is equivalent to dividing the resonant wavelength by the linewidth of the resonance. In Figure 11.1, the spectra of cavity resonances with different Q-factors are presented for comparison at a wavelength of 1550 nm. As this figure shows, narrower linewidths correspond to higher Q-factors. Generally speaking, the higher the Q-factor, the better

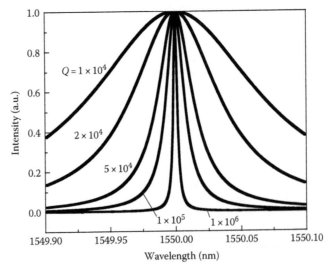

FIGURE 11.1 Lorentzian model showing normalized spectra of cavity resonances centered at 1550 nm.

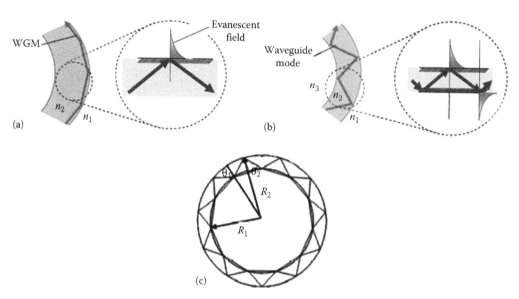

FIGURE 11.2 Schematic of (a) a thick-walled ring resonator with whispering gallery modes, (b) a thin-walled ring resonator with circulating waveguide modes, and (c) high-order nonevanescent modes. (Reprinted from Ling, T. and Guo, L.J., *Opt. Express*, 15, 17424, 2007. With permission.)

the optical cavity in confining the optical energy, which also responds to the more round trips each photon in the cavity makes within it before being absorbed, scattered, or removed via another mechanism.

In a ring resonator, there exist three types of these modes that can be utilized for various applications: the WGM, the circulating waveguide mode, and certain special high-order modes that will be discussed later. The difference between the WGM and the circulating waveguide mode is that the WGM is guided by the TIR at the outer boundary of the ring resonator, whereas the circulating waveguide mode is confined by the TIR at both the outer and inner boundaries (see Figure 11.2a and b). The former case occurs in a microsphere, a microdroplet, a microdisk, and a thick-walled capillary ring resonator, whereas the latter one is represented by a ring-shaped waveguide and thin-walled capillary ring resonators.

Both these types of modes can be solved within the framework of Mie theory or the Mie model (Bohren and Huffman 1983, Gorodetsky and Ilchenko 1999). According to Mie theory, there are a large number of resonant modes that fit within any ring resonator of a given size. They are designated by radial-mode order numbers. Typically, ring resonators make use of relatively low-order modes, because these exhibit generally higher Q-factors. However, higher-order modes tend to become more sensitive, with a greater fraction of light intensity in the evanescent field. It has been shown that ultrahigh-order modes are capable of producing very high sensitivities (Ling and Guo 2007).

The third type of mode present in a ring resonator corresponds to extremely high-order modes in the Mie model (Ling and Guo 2007). This type of mode differs from the first two in terms of the light behavior at the inner boundary, i.e., the TIR at the inner boundary is no longer valid. The resonance of this mode forms due to positive interference that takes place between refracted and reflected rays at the inner boundary (Figure 11.2c). While all three of these mode types can be considered as unique, they will, for convenience, be referred to as WGMs throughout this chapter.

11.1.3 Excitation of WGMs

In the past 20 years, a number of methods have been developed to efficiently excite and couple out the WGMs in a ring resonator. Most of the excitation and outcoupling methods are based on evanescent coupling. Coupling between a ring resonator and a waveguide or other optical structures requires

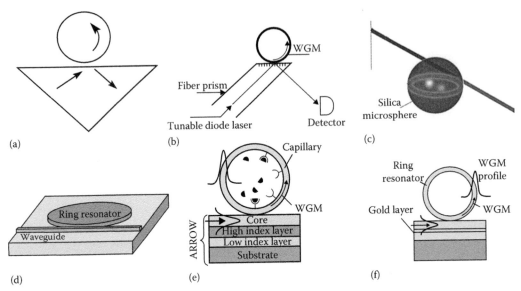

FIGURE 11.3 (a) Ring resonator coupling methods include prism. (Reprinted from Little, B.E. et al., *Opt. Lett.*, 25, 73, 2000. With permission.) (b) Angle-polished fiber. (Reprinted from Hanumegowda, N.M. et al., *Sens. Act. B*, 120, 207, 2006. With permission. Copyright 2009, Elsevier.) (c) Tapered fiber. (Reprinted from Vahala, K.J., *Science*, 424, 839, 2003. With permission. Copyright 2009, Macmillan Publishers Ltd.) (d) Planar waveguide. (From White, I.M. et al., *IEEE Sens. J.*, 7, 28, 2007c. With permission. ©2009 IEEE.) (e) ARROW. (Reprinted from White, I.M. et al., *Appl. Phys. Lett.*, 89, 191106, 2006b. With permission. Copyright 2009, American Institute of Physics.) (f) Gold-cladded waveguides. (Reprinted from White, I.M. et al., *Opt. Express*, 15, 646, 2007b. With permission.)

that the evanescent fields of both structures be overlapping. Additionally, the mode within the guiding structure must be phase matched with the WGM.

Figure 11.3 shows various methods to excite and out-couple the WGM. They include prisms (Gorodetsky and Ilchenko 1999, Mazzei et al. 2005) (Figure 11.3a), angle-polished fibers (Ilchenko et al. 1999, Hanumegowda et al. 2006, Zhu et al. 2006) (Figure 11.3b), tapered optical fibers (Cai et al. 2000, Vollmer et al. 2002, Armani et al. 2003, Vahala 2003, Vollmer et al. 2003, Armani and Vahala 2006, Hossein-Zadeh and Vahala 2006, White et al. 2006a) (Figure 11.3c), planar waveguides (Ksendzov et al. 2004, Ksendzov and Lin 2005, Ramachandran et al. 2008) (Figure 11.3d), anti-resonant reflecting optical waveguides (ARROWs) (Little et al. 2000, White et al. 2006b) (Figure 11.3e), and gold-cladded waveguides (White et al. 2007b) (Figure 11.3f). With planar waveguides, either vertical (Chu et al. 1999, Djordjev et al. 2002, Kim et al. 2008) or side coupling (Boyd et al. 2003, Chao and Guo 2003, Almeida et al. 2004a, Baehr-Jones et al. 2004) may be used. Vertically coupled rings are arranged with the waveguide running underneath or above the ring, coupling through its top or bottom face. Side coupling waveguides are in the same plane as the ring and couple through its side. Figure 11.4 provides an example of a planar ring that has an optical signal delivered to it by an adjacent waveguide. In Figure 11.4a, a photograph of the light intensity scattering off of the ring boundaries shows light confinement. Figure 11.4b shows a simulated field distribution for WGMs in an ideal structure with no surface roughness. The discrete nodes of high intensity represent positive interference from resonance.

Surface plasmon resonance (SPR)–assisted coupling can also be achieved using gold-coated prisms (Hon and Poon 2007) or tapered fibers with gold nanorod-coated microspheres (Shopova et al. 2008a). Furthermore, spiral-shaped output couplers for planar rings have been fabricated that are integrated into the rings, as shown in Figure 11.5 (Chern et al. 2003, Luo and Poon 2007, Wu et al. 2008). These provide a nonevanescent coupling method that is rotationally asymmetric and selective.

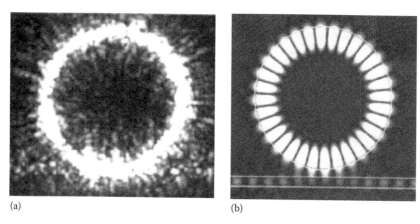

(a)   (b)

FIGURE 11.4 (a) Snapshot of scatter from a 15 μm planar ring resonator with waveguide on resonance. (Reprinted from Krioukov, E. et al., *Opt. Lett.*, 27, 1504, 2002. With permission.) (b) Simulation of field distribution for similar resonator configuration. (Reprinted from Boyd, R.W. et al., *J. Mod. Opt.*, 50, 2543, 2003. With permission of the publisher (Taylor & Francis Group, http://informaworld.com.)

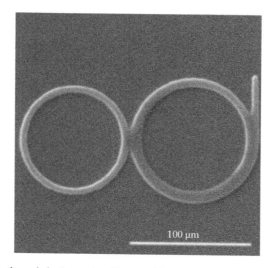

FIGURE 11.5 A ring-spiral coupled microcavity. (Reprinted from Wu, X. et al., *Appl. Phys. Lett.*, 93, 081105, 2008. With permission. Copyright 2009, American Institute of Physics.)

In addition to the method discussed above, WGM excitation can be done through free-space coupling (Tzeng et al. 1984, Nuhiji and Mulvaney 2007). This is a method that is easy to realize, in which unguided light is irradiated upon the ring, and the scattered signal is collected through carefully aligned optics, although the intensity of collected light will be low, and therefore some applications like Raman spectroscopy and detection of low-concentration analytes become challenging.

11.2 Applications of Optofluidic Ring Resonators

11.2.1 Label-Free Sensing

11.2.1.1 Principles of Refractometric and Absorption-Based Sensing

All ring resonator sensing approaches utilize the evanescent field created by the WGM as a transduction mechanism. The evanescent field created by the WGM typically projects about 100 nm from the ring into the adjacent media (Ilchenko and Matsko 2006). This means that the sensors are limited to interrogation of fluid volumes very near to the surface only; however, this can be used to a real advantage to differentiate molecules that bind to the ring resonator surface from those in the bulk solution.

Refractometric and absorption-based sensing belong to the so-called label-free detection methods in which no fluorescent label on analytes is used (Fan et al. 2008), enabling detection of native, unmodified analytes. While some label-free ring resonator sensors use optical absorption (the imaginary part of the complex RI) for sensing, most optical label-free ring resonators are categorized as refractometers, directly detecting changes in the RI (the real part of the complex RI) in the media overlapping the WGM.

Refractometric sensing relies on changing the effective RI encountered by the WGM, n_{eff}. The resonant condition governing WGMs is defined, in general, by the following relationship (Gorodetsky and Ilchenko 1999):

$$2\pi n_{eff} r = m\lambda \tag{11.1}$$

where
  r is the radius of the ring resonator
  λ is the resonant wavelength value
  m is an integer

The WGM spectral position shifts when n_{eff} is altered due to the bulk RI change, in which the RI change occurs in the region longer than the evanescent decay length (which is about 100 nm), or molecules attached to the ring resonator surface, in which the RI change occurs in a distance much shorter than the evanescent-field decay length. Thus, by monitoring the WGM spectral position in real time, it is possible to acquire the kinetic and quantitative information regarding molecular interaction.

In RI sensing (and absorption sensing as well), the fraction of the WGM that interacts with the analyte is important. To take this fractional interaction effect into account, Equation 11.1 can be approximately rewritten as (Zhu et al. 2007a)

$$2\pi r \cdot \left[n_1 \eta_1 + n_2 \eta_2 + n_3 \eta_3 \right] \approx m \cdot \lambda \tag{11.2}$$

where η_1, η_2, and η_3 are the fractions of the WGM present outside the resonator, in the resonator material, and inside the resonator core (if applicable), respectively. n_1, n_2, and n_3 are the corresponding RIs. From Equation 11.2, the bulk RI sensitivity (BRIS), S, can be deduced (Zhu et al. 2007a):

$$S = \left(\frac{\delta\lambda}{\delta n_1} \right) \approx \frac{\lambda}{n_2} \eta_1 \tag{11.3}$$

where we assume that the RI sensing occurs near the outer surface of the ring resonator. For sensing near the inner surface, η_1 can simply be replaced with η_3. BRIS values for several different ring resonators are presented in Table 11.1.

For detection of molecules attached to the ring resonator surface, Equation 11.3 is no longer valid. Vollmer et al. developed an analytical model using the first-order perturbation theory to relate the WGM spectral shift, $\delta\lambda$, i.e., sensing signal, to the molecular density on the microsphere surface, σ_p (Vollmer et al. 2002, Vollmer et al. 2003):

$$\frac{\delta\lambda}{\lambda} = \frac{\sigma_p \alpha_{ex}}{\varepsilon_0 (n_s^2 - n_b^2) r} \tag{11.4}$$

TABLE 11.1 Ring Resonator Characteristics and Performance Benchmarks

Platform	RI LOD	Q-Factor	BRIS (nm/RIU)	Analyte (LOD)	Reference
Microsphere	3×10^{-7}	8×10^{9}	30	Virus InfA (10 fM)	Vollmer et al. (2008)
Microdroplet	NA	1.15×10^{5}	NA	NA	NA
Toroid	NA	2×10^{8}	NA	Heavy water (1 ppm)	Armani and Vahala (2006)
Microcapillary	5×10^{-6}	1×10^{7}	37	DNA (10 pM)	Suter et al. (2008)
Planar ring	1.8×10^{-5}	4.5×10^{5}	141	Streptavidin (3 pg/mm²)	Xu et al. (2008)
Slotted planar ring	1×10^{-4}	2.7×10^{4}	490	Acetylene gas (100%)	Robinson et al. (2008)
Microfiber loop	1×10^{-7}	1×10^{6}	700	Isopropanol (1.5% v/v)	Xu et al. (2008)
LRROS	1×10^{-6}	$\sim 1.5 \times 10^{4}$	800	NA	NA

where

ε_0 is the vacuum permittivity

n_s and n_b are the RI for the sphere and buffer solutions

α_{ex} is excess polarizability for molecules in water

Recently, another more practical relationship has been established between the BRIS, S, and molecular surface density (Zhu et al. 2007a):

$$\frac{\delta\lambda}{\lambda} = \sigma_p \alpha_{ex} \frac{2\pi\sqrt{n_2^2 - n_3^2}}{\varepsilon_0 \lambda^2} \frac{n_2}{n_3^2} S \tag{11.5}$$

in which

$\delta\lambda$ is the WGM wavelength shift

ε_0 is the vacuum permittivity

S is the BRIS

n_2 and n_3 are the RI for the ring resonator and buffer solution

Using Equation 11.5, it is possible to predict the ring resonator sensing performance for biological/chemical molecules using a simple BRIS characterization.

Experimentally, monitoring the WGM can be done by either collecting scattered light or observing the transmission spectrum of a waveguide coupled to the resonator. Two methods are commonly employed to extract the WGM spectral position (Chao and Guo 2006). The first method is to scan across a range of wavelengths that includes the resonant position and track the value of the transmission minimum (Figure 11.6a). The drawback of this method is that tunable light sources can be quite expensive, and the data analysis is complicated. The second way is to use a fixed wavelength set at the steepest point of the slope on a resonance edge. As the WGM shifts, the transmitted intensity at this wavelength varies rapidly (Figure 11.6b). This method can provide a much more sensitive RI detection method than dip tracking, but only when the measured shifts are very small. When the WGM shifts more than a quarter of a linewidth, the intensity response becomes extremely nonlinear and ceases to be useful.

While tracking WGM shifts in real time, the spectral resolution will be limited by noises presented in the sensor system, in particular, thermally induced noise and nonspecific binding. Various methods have been proposed to reduce thermally induced WGM spectral fluctuations (Armani et al. 2007a, Han and Wang 2007, Suter et al. 2007, He et al. 2008). These involve modifying the optical cavity in order to cancel out all thermal effects. When the temperature of the ring changes, the refractive properties

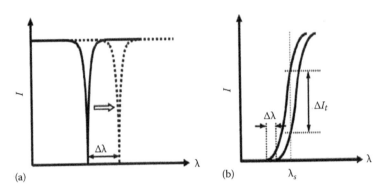

FIGURE 11.6 (a) Transmission spectrum showing resonant wavelength shift ($\Delta\lambda$) tracking by identifying the minimum. (b) Intensity shift measurement method showing half of a resonant linewidth. (Reprinted from Chao, C.-Y. and Guo, L.J., *J. Lightwave Technol.*, 24, 1395, 2006. With permission. ©2009 IEEE.)

of both the resonator and surrounding fluid change, and thermal expansion changes the ring radius. Both thermo-optic and thermal expansion effects lead to WGM noise; however, they can be used to compensate for one another. The thermo-optic effect in fluids tends to cause negative (or blue) WGM shifts in response to a temperature increase, while thermal expansion and thermo-optic effects in silica (the material that forms the body of the resonator) cause positive (or red) WGM shifts. Therefore, if the fractional light intensities in the ring and in the evanescent field are balanced perfectly, then, in theory, the ring should be insensitive to temperature fluctuations. Furthermore, a reference channel can also be implemented along with the sensing channel to cancel out the common-mode noise such as thermally induced noise and nonspecific bindings.

To complement refractometric sensing, absorption can be a useful tool. This method leverages the absorptive properties of certain analytes that can manifest themselves by reducing the Q-factor of the cavity (Nadeau et al. 2002, Armani and Vahala 2006, Nitkowski et al. 2008) or by changing the intensity of the resonant dips observed from the coupling fiber or waveguide (Farca et al. 2007). In the former case, the decrease in Q-factor is simply caused by additional losses to absorption, which are dependent on the absorption coefficients of a given molecular species at a given wavelength. In the latter case, dip depths change in response to changes in the coupling efficiency between the resonator and the waveguide.

11.2.1.2 Solid Dielectric Microspheres

Solid dielectric microspheres serve as excellent optical ring resonators. They can be made out of optically transparent materials such as fused silica (Gorodetsky and Ilchenko 1999, Arnold et al. 2003, Teraoka et al. 2003, Vollmer et al. 2003, Hanumegowda et al. 2005a, Hanumegowda et al. 2005b), chalcogenides (Grillet et al. 2008), or even polymers like polystyrene (Ma et al. 2003, Zijlstra et al. 2007, Lutti et al. 2008). Among different types of microspheres, the fused silica-based microsphere is particularly attractive, as it provides extremely high Q-factors. A Q-factor as high as 10^9 has been reported when the fused silica microsphere is in air (Gorodetsky and Ilchenko 1999). When immersed in water, the microsphere has a Q-factor in excess of 10^7 (Arnold et al. 2003, Hanumegowda et al. 2005a), which results in a high spectral resolution, and hence a low limit of detection (LOD). Additionally, bioconjugation procedures for silica surfaces have been well developed for covalent attachment, which can easily be adapted for fused silica microsphere-based biosensing.

Fabrication of microspheres is usually accomplished by melting the tip of an optical fiber. Using a flame or CO_2 laser, the tip of a fiber is allowed to bead under its own surface tension, typically to around 100 μm in diameter (Vollmer et al. 2002, Hanumegowda et al. 2006). In this way, it is left on its fiber stem

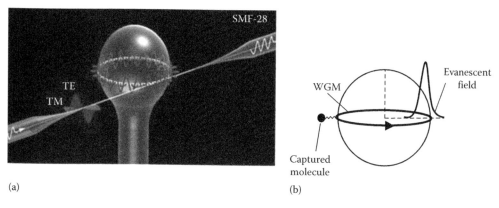

(a)   (b)

FIGURE 11.7 (a) Schematic of a microsphere and fiber taper showing relative mode orientations. (Reprinted from Topolancik, J. and Vollmer, F., *Biophys. J.*, 92, 2223, 2007. With permission. Copyright 2009, Elsevier.) (b) Microsphere schematic showing molecule immobilization and field intensity.

and can be easily manipulated (see Figure 11.7a). Also, commercially available silica microspheres can be annealed to achieve optical quality (Snee et al. 2004).

Fused silica microspheres have been employed in both refractometric sensing and biological sensing applications. Figure 11.7b shows how immobilized molecules interact with the WGM at the equator where evanescent intensity is high. Once fabricated, the spheres can be inserted into fluidic chambers or aligned with the input and collection optics. For refractometric sensing, a BRIS of approximately 30 nm/RIU with a detection limit of 3×10^{-7} RIU was achieved (Hanumegowda et al. 2005a). For biological sensing, the surface of the microsphere is first coated with a layer of biorecognition molecules, such as single-strand DNA (Hanumegowda et al. 2005a, Nuhiji and Mulvaney 2007), antibodies (Francois et al. 2008), aptamers (Zhu et al. 2006), and peptides (Hanumegowda et al. 2005b).

Specific biosensing applications have included several studies on protein detection (Vollmer et al. 2002, Arnold et al. 2003, Noto et al. 2007, Zhu et al. 2007a). In 2002, a 300 µm microsphere was used to investigate bovine serum albumin (BSA) adsorption as well as the interaction between streptavidin and a BSA-biotin conjugate. Figure 11.8 shows the sensorgram resulting from streptavidin binding to biotin

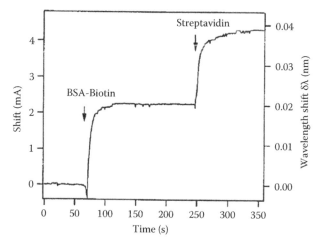

FIGURE 11.8 Sensorgram showing specific binding of BSA-biotin and streptavidin to a microsphere ring resonator. (Reprinted from Vollmer, F. et al., *Appl. Phys. Lett.*, 80, 4057, 2002. With permission. Copyright 2009, American Institute of Physics.)

on the sphere surface (Vollmer et al. 2002). A positive net WGM shift was observed, indicating higher biomass loading on the sphere surface. A further study with BSA using similar adsorption techniques showed that detection of only a few protein molecules was possible (Arnold et al. 2003). In 2007, one further BSA effort was published, focusing on the difference in microsphere sensitivity using different polarizations (Noto et al. 2007). Specifically, the ratio of WGM shift of transverse electric (TE) and transverse magnetic (TM) modes was used to determine the extent of BSA's polarizability anisotropy. Figure 11.7a shows the relative orientation of TE and TM modes within a microsphere. As an extension of this protein work, it is possible to monitor enzymes by monitoring the cleaving of protein molecules. This was demonstrated using a BSA-trypsin model in which trypsin cleaves a part of BSA originally on the WGM (Hanumegowda et al. 2005b). As a result, a negative shift in the WGM spectral position was observed. Clinically relevant biomolecules have also been detected with the microsphere. For example, thrombin plays an important role in the process of blood coagulation. The detection of thrombin levels has many potential applications including therapies for cardiovascular diseases like atherosclerosis and surgical procedures (Becker and Spencer 1998). Using an aptamer-coated microsphere, thrombin in buffer was detected with an LOD of around 1 NIH unit/mL, or roughly 6 nM (Zhu et al. 2006).

One further BSA experiment that is worthy of note was conducted by Francois and Himmelhaus (2009) using dye-doped microsphere lasers. In this particular study, polystyrene microspheres with Nile Red dopant were deposited on a glass slide and lased using a pulsed excitation source. The lasing spectrum from WGMs, which will be talked about extensively in Section 11.2.4, was tracked as reagents were flowed over the spheres. BSA deposition was monitored and analyzed in terms of adsorption kinetics (Francois and Himmelhaus 2009).

DNA detection is also of keen interest for a number of purposes. Typically, detection is accomplished by attracting the target DNA molecule via complementary strand hybridization. Vollmer et al. were able to detect as little as 6 pg/mm^2 11-base target strands and to differentiate single-base mismatched DNA from complementary ones (Vollmer et al. 2003). The results further demonstrated the detection of single-base-mismatched DNA. Another effort used much smaller, 7.5 μm silica microspheres coated with dye-labeled DNA probes (Nuhiji and Mulvaney 2007). The dye, tetramethyl rhodamine, was excited from an external source, and its fluorescence was coupled into the WGMs in the spheres and was then collected through a free-space collection system to indicate the WGM spectral position. Hybridization between the probe and complementary target DNA strands, then, caused a shift in the characteristic WGM peaks within the fluorescence spectrum. The DNA targets were 5 μM, using lengths between 10 and 40 nucleotides. Similar experiments were also carried out with polystyrene beads with diameters between 1.5 and 20 μm impregnated with Nile Red dye to assay solutions of both BSA and the polyelectrolytes poly(allylamine hydrochloride) and poly(sodium 4-styrenesulfonate) (Weller et al. 2008). An LOD of roughly 213 pg/mm^2 was achieved. Despite the relatively low spectral resolution (because of relatively low Q-factors), this method provides an easy method for monitoring the WGM spectral position without any sophisticated in/out coupling systems (such as fiber tapers, prism, etc.), which may open a door for in vivo and remote in vitro WGM sensing (Meissner and Allen 2005, Weller et al. 2008).

Larger analytes have been investigated with microsphere biosensors, specifically with whole cell bacteria and virus particles. *E. coli* was immobilized on a microsphere surface by starting with a poly-L-lysine coating (Ren et al. 2007). In addition to monitoring the WGM shift through a coupled optical fiber, the surface of the sphere was monitored by a microscope in order to ascertain surface densities. It was discovered that the LOD of this approach was 10^2 bacteria/mm^2, corresponding to approximately 44 total cells on the microsphere. Very recently, single viral particle detection has been achieved using 50 μm silica microspheres using 10 fM of influenza A virions (Vollmer et al. 2008). A defining characteristic of single-particle adsorption was presented, showing step-wise shifts from individual viral particles. This is in sharp contrast to typical adsorption curves in lower-sensitivity systems where discrete WGM shifts are impossible to resolve. Interestingly, the sizes of these steps are not uniform, which is caused by the difference in sensitivity as particles bind closer or farther away from the "equator" of the WGM where the optical energy density is the highest.

There have been many theoretical efforts to characterize the microsphere's performance as a sensor for biological analytes (Keng et al. 2007, Ren et al. 2007). These include theoretical models such as Equation 11.4, which relate the WGM shift to surface densities of molecules on a microsphere surface (Vollmer et al. 2003). Such work has also been extended to the case of virus particles and cells whose sizes are larger than the WGM evanescent decay length. The work by Keng et al. explored the ability of the microsphere to study Brownian motion in microparticles (Keng et al. 2007), which is an important phenomenon to understand in order to interpret results from biosensing experiments. From monitoring diffusion processes, particle sizes can also be inferred. Such measurements are applicable to chemical as well as biological sensing.

Some important chemical sensing examples include the detection of mercury ions in water down to 50 ppb (Hanumegowda et al. 2006), polar crystals (pentacene and terrylene) (Norris et al. 1997), and even for the removal of silica monolayers from the sphere using hydrofluoric acid (HF) (White et al. 2005). The removal of 4 pm of silica (0.4 fmol of silica) from the sphere caused a detectable shift in the WGM, indicating that sub-monolayer thickness changes can be detected. This was an important demonstration of the microsphere's ability to detect molecules smaller than those typically targeted in biomolecular assays.

To this point, the microsphere has been discussed as a refractometer used for tracking WGMs. Since microspheres have very high Q-factors, the WGM spectral linewidth (instead of spectral shift) can also be used to generate sensing signal. This is particularly useful when detecting the optical absorption (the imaginary part of the complex RI) and hence the concentration of the analyte present in the WGM field, since molecules, such as dyes, will cause a degradation of the measured Q-factor (Nadeau et al. 2002).

Although the microsphere ring resonator has demonstrated highly sensitive detection, mass production and integration with microfluidics for efficient sample delivery continues to be very challenging. In order for microspheres to be practical beyond the research laboratory, these issues must be addressed more conveniently in future work.

11.2.1.3 Microdroplet

Geometrically, a microdroplet is nearly identical to the solid microsphere discussed in the previous section. However, the microdroplet is unique in that it functions as a microfluidic element and a ring resonator. In contrast to a solid microsphere, the predominant electric field of the WGM that forms along the microdroplet surface lies inside the microdroplet, resulting in a large light–analyte interaction that can be explored for many applications (such as in sensing and microfluidic lasers).

Microdroplets of water (Lin et al. 1992), methanol (Saito et al. 2008), and ethanol (Tzeng et al. 1984) can be formed by an aerosol generator (Lin et al. 1986), or in oil (Hossein-Zadeh and Vahala 2006, Tanyeri et al. 2007) (Figure 11.9a), on an ultrahydrophobic surface (Dorrer and Ruhe 2006, Hossein-Zadeh and Vahala 2006, Kiraz et al. 2007a, Sennaroglu et al. 2007, Kiraz et al. 2008) (Figure 11.9b), or levitated by electrodes (Arnold and Folan 1986, Azzouz et al. 2006) or optical tweezers (Hopkins et al. 2004). Many of these control mechanisms are also useful for microdroplet lasers (discussed in Section 11.2.4.2). The size of the microdroplet can be as small as a few micrometers, which corresponds to a total fluidic volume on the order of picoliters. Such a small volume is highly desirable for the development of biosensors with low sample consumption and in fundamental research such as single-molecule detection (see Chapter 13 by Cipriany and Craighead) (Barnes et al. 1993). The tuning of the droplet size has been demonstrated using the electrowetting method (Mugele and Baret 2005, Hossein-Zadeh and Vahala 2006, Kiraz et al. 2008).

Microdroplets offer the best surface quality of any ring resonators available, resulting in a negligible surface-scattering-induced loss. Consequently, the Q-factor of a microdroplet is mainly determined by the optical absorption of the solvent. Within the visible spectrum, water, methanol, and ethanol absorption is very low ($\sim$0.001 cm^{-1}) (Stone 1972), resulting in a Q-factor of up to 1×10^8.

Despite these merits, microdroplets suffer from difficult integration into practically viable devices. Excitation of the WGM through waveguiding structures such as tapers and prisms is very challenging.

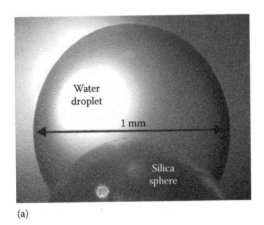

(a)

(b)

FIGURE 11.9 (a) Droplet under oil positioned on a silica sphere. (Reprinted from Hossein-Zadeh, M. and Vahala, K.J., *Opt. Express*, 14, 10800, 2006. With permission.) (b) Droplet under oil positioned on an ultrahydrophobic surface in air. (Reprinted from Dorrer, C. and Ruhe, J., *Langmuir*, 22, 7652, 2006. With permission. Copyright 2009, American Chemical Society.)

Furthermore, the time-dependence of resonance in an evaporating droplet is a significant practical limitation (Hossein-Zadeh and Vahala 2006) (although immersion in oil helps prevent evaporation, the system becomes more complicated). To date, only a very few refractometric sensing or label-free biosensing applications have been demonstrated on microdroplets, leaving this area an attractive field for further exploration. One of these applications involves detection of single *E. coli* cells in a PBS buffer droplet doped with R6G (Tanyeri and Kennedy 2008). WGM peaks from the R6G emission are suppressed upon adding cells due to absorption and scattering effects. This procedure is label-free despite the use of dyes because the cells are not actually tagged with them.

Most of the applications of droplets have been in fluorescence-based sensing (Tanyeri et al. 2005) and microfluidic lasers, which we will discuss in detail in Section 11.2.2 and Section 11.2.4.2, respectively.

11.2.1.4 Cylinder and Microcapillary

Cylindrically shaped ring resonators may be created out of a solid post, such as a segment of an optical fiber, or a hollow capillary. The former configuration can be used in exactly the same manner as the solid microsphere in biochemical sensing that has been discussed in Section 11.2.1.2, where an extrinsic microfluidic chamber or channel is required to deliver samples to the ring resonator. In contrast, the hollow capillary configuration differs from the solid cylinder ring resonator drastically, as it serves both as a resonator and as a microfluidic channel that conducts sample fluid. Furthermore, the capillary-based ring resonator relies on the evanescent field on or near the inner surface (rather than the outer surface) of the capillary as shown in Figure 11.10.

In order to obtain a sufficient electric field of the WGM near the capillary inner wall, the wall thickness of the capillary cannot exceed a few micrometers (White et al. 2006a, Zhu et al. 2007a, Zamora et al. 2008). These thin-walled capillaries are almost exclusively fabricated out of silica glass starting with a large capillary preform (White et al. 2006a, Zamora et al. 2008). Under intense heat from a torch or CO_2 laser, the glass is softened until it can be easily pulled to very small dimensions. This process is similar to the manufacture of microspheres, except that in the microsphere case the glass is heated until it completely liquefies, whereas in capillary fabrication, relatively low heat and high pulling speed are needed to ensure that the capillary does not collapse and that the original aspect ratio (wall thickness vs. outer diameter) remains the same. In the end, the capillary will be on the order of 10–100 μm in diameter with a thickness as thin as 1–2 μm. If needed, the wall is etched thinner with hydrofluoric acid (HF).

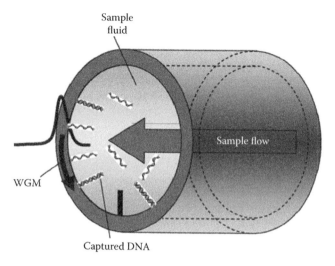

FIGURE 11.10 Thin-walled capillary demonstrating refractometric DNA detection and showing evanescent penetration into core. (Reprinted from Suter, J.D. et al., *Biosens. Bioelectron.*, 23, 1003, 2008b. With permission. Copyright 2009, Elsevier.)

When the WGMs become sensitive to the RI of the interior fluid, the configuration can be utilized as a sensor for biological and chemical analytes (Zhu et al. 2006, White et al. 2007a,c, Zhu et al. 2007b, 2008a,b, Shopova et al. 2008b, Yang et al. 2008). This type of sensor has demonstrated BRIS benchmarks of 37 nm/RIU with thin-walled capillaries (Suter et al. 2008b) (in fact, according to unpublished results over 100 nm/RIU has been achieved) and can provide Q-factors of over 10^7 (White et al. 2007c), leading to an LOD on the order of 10^{-6}–10^{-7} RIU (White et al. 2007c). A dual capillary scheme called the optofluidic coupled ring resonator (OCRR) has been explored for an RI LOD estimated at around 1×10^{-9} (Shopova et al. 2008b) (Figure 11.11). In this case, two parallel capillaries are coupled to each other, and

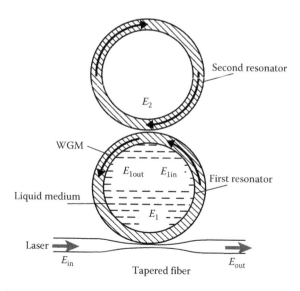

FIGURE 11.11 Coupled capillaries with coresonant WGMs tuned by the first resonator. (Reprinted from Shopova, S.I. et al., *Microfluid. Nanofluid.*, 6, 425, 2008b. With permission from Springer Science + Business Media.)

interference between coresonant WGMs create split modes that shift relative to one another based on the RI of the fluid inside the capillaries (Smith et al. 2003).

The microcapillary ring resonator has been demonstrated for a wide variety of biosensing purposes. Biorecognition molecules are first coated on the interior surface of the capillary, and then the fluid sample is flowed through the capillary using a mechanical pump, capillary force, or electroosmosis. The thin-walled capillary is able to detect single-stranded DNA molecules at concentrations down to 10 pM and discriminate a single mismatched nucleotide (Suter et al. 2008b). It can also detect streptavidin down to 100 pM using a peptide-binding phage (Zhu et al. 2008a). Rapid and sensitive virus detection has also been presented. M13 viral particles down to 2.3×10^3 pfu/mL were detected in only 5–10 min. (Zhu et al. 2008b). Given the small sample volume used in this experiment, only a few tens of viral particles are detected.

In each of these sensing applications it can be quite useful to be able to estimate the amount of material immobilized on the sensor surface. Equation 11.5, furnished in Section 11.2.1.1, shows how surface density of molecules is related to the WGM shift (Zhu et al. 2007a). The relationship is dependent on the excess polarizability (α_{ex}) of the molecule causing the shift. In the DNA detection experiment by Suter et al. (2008b), surface densities were estimated for probe and target DNA immobilization steps, and were used to estimate the hybridization efficiency of the system. The same model has been used to determine the surface density of BSA protein layers in a capillary (Zhu et al. 2007a).

Another microcapillary design uses microtubes that are composed of rolled-up semiconductor materials (Prinz et al. 2000, Schmidt and Eberl 2001, Mendach et al. 2006, Songmuang et al. 2006, Songmuang et al. 2007, Bernardi et al. 2008). By growing very thin alternating layers (submicron) of materials with induced strain at the interface, the material can be caused to roll up into a tube ranging from 2 to 5 μm in outer diameter (Figure 11.12). These layers have been created with InGaAs and GaAs, which have a lattice mismatch that naturally places strain on the interface (Prinz et al. 2000, Kipp et al. 2006, Mendach et al. 2006). However, Si/SiO$_x$ layers have also been recently demonstrated (Songmuang et al. 2007, Bernardi et al. 2008). When the layer underneath them is etched away, this strain causes the rolling effect. After rolling, the tube is annealed at very high temperature (around 850°C for Si/SiO$_x$ (Songmuang et al. 2007)) and can then be used as a cylindrical ring resonator. These resonators have shown BRIS values of up to 62 nm/RIU (Bernardi et al. 2008) and Q-factors around 3×10^4 (Kipp et al. 2006). While these tubes are much better suited for on-chip integration, their performance is limited by this low Q-factor, caused by irregularities in the tube wall.

So far in this section we have discussed evanescent label-free detection using thin-walled silica capillaries and rolled tubes. Next, we will discuss two special cases of capillary ring resonators that are able to push the WGM into the sample fluid inside the capillary core. In the first case, the majority of the electric field may also be pushed into the fluid by simply etching the wall thickness to extremely small

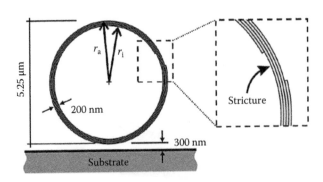

FIGURE 11.12 A rolled-up InGaAs/GaAs microtube. (Reprinted from Kipp, T. et al., *Phys. Rev. Lett.*, 96, 077403, 2006. With permission. Copyright 2009, American Institute of Physics.)

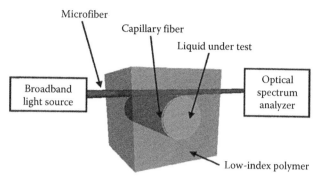

FIGURE 11.13   Liquid ring resonator optical sensor. (Reprinted from Sumetsky, M. et al., *Opt. Express*, 15, 14376, 2007. With permission.)

dimensions. In such a case, the pre-etched capillary must be embedded within a very low RI polymer for support. This was successfully accomplished by embedding the glass capillary in a polymer with RI = 1.384 and etching the wall with HF until the optical resonances in the glass disappeared (Sumetsky et al. 2007). This configuration has been named the liquid ring resonator optical sensor (LRROS) (in Figure 11.13). Running a fluid through the core with RI greater than that of the polymer, ultrahigh-sensitivity refractometric measurements could be made. BRIS was observed to be 800 nm/RIU. While this is one of the best sensitivities available from ring resonator sensors, the need to use solvents with RI greater than the RI of the encasing polymer is one of its limitations. Furthermore, the biosensing applications of this type of microcapillary have yet to be explored.

Extreme thinning of the wall is an interesting approach, but it is not necessarily required in order to push WGM energy into the capillary core. In the second case, with a 32 μm thick wall, low-order WGMs are completely confined within the wall, and only the WGMs of higher order can be utilized for sensing (Ling and Guo 2007). This phenomenon of sensing using thicker walls has been investigated looking at two types of modes, both having very high radial-order numbers. The first of these, a 35th-order radial mode, has an evanescent field decaying into the core, but the majority of the WGM intensity remains within the wall (Figure 11.14a). The BRIS of this configuration is capable of reaching

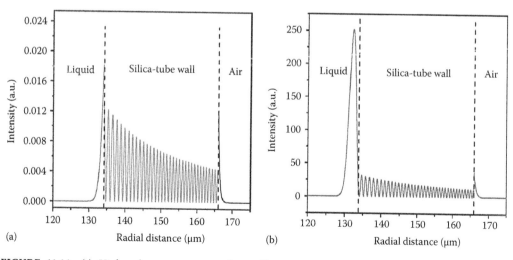

FIGURE 11.14   (a) High-order evanescent mode profile and (b) nonevanescent-interference mode profile for thick-walled capillaries (32 μm). (Reprinted from Ling, T. and Guo, L.J., *Opt. Express*, 15, 17424, 2007. With permission.)

approximately 100 nm/RIU. The second type of mode is of an even higher radial-order number of 37 and arises from positive interference between reflected and refracted beams incident upon the wall/core interface, as discussed earlier (see Section 11.1.2 and Figure 11.2c). The majority of this non-evanescent mode's field intensity energy lies within the core (but not on the interior surface) (Figure 11.14b), enabling BRIS up to 600 nm/RIU with a Q-factor of 2×10^4. This mode has very high BRIS, but it cannot be used for biosensing where an evanescent field is needed. In the case of biosensing, the bulk RI may play too much of a role in the resultant optical signal. As a result, it becomes difficult to differentiate the actual specific binding occurring on the interior surface from the interfering analytes flowing through the capillary that do not bind to the surface, if this type of mode were used for biomolecule binding detection.

All of the capillary-based ring resonators discussed above share the same fluidic integration advantages. As a consequence, they are highly compatible with column-based analytical chemistry technologies and can be employed to develop novel capillary electrophoresis (Zhu et al. 2007b) or micro-gas chromatography (Shopova et al. 2008c, Sun and Fan 2008, Sun et al. 2008) with real-time, sensitive, on-column detection capability. In the electrophoresis study, glycerol was used in order to establish the proof of concept, and flow was controlled by means of electroosmotic flow (Zhu et al. 2007b). In the gas chromatography study, the inside of a capillary was coated with a vapor-sensitive polymer, which interacts with transmitted gas species. Separation and detection of various vapors have been demonstrated with a subnanogram LOD.

11.2.1.5 Planar Ring Resonators

Microfabrication techniques have advanced to the point where top-down construction of micro- and nano-sized structural features is a routine task. Soft-lithography techniques enable fabrication of features below 100 nm using photomask-based replica molding (Xia et al. 1997), and down to 2 nm with imprinting technologies using PDMS on molds made using chemical vapor deposition (Hua et al. 2004). E-beam lithography, focused ion beam, and related technologies have also paved the way for construction on the subwavelength scale. This has enabled chip-based ring resonator structures, which are excellent candidates for biological and chemical sensing.

Planar ring resonators, as they may be called, are a category of optical cavities that are manufactured using any combination of layer deposition and removal phases on top of a semiconductor substrate. A common subset of these devices is SOI, or silicon-on-insulator, ring resonators, in which silicon structures are fabricated on a layer of insulating material, commonly SiO_2 (Almeida et al. 2004a, Baehr-Jones et al. 2004, Niehusmann et al. 2004, Martinez and Lipson 2006, Padigi et al. 2006, Passaro and De Leonardis 2006, De Vos et al. 2007, Passaro et al. 2007, Xu et al. 2007a, Dai et al. 2008, Xu et al. 2008a) (Figure 11.15a and b). Polymer ring structures (Chao and Guo 2002, Chao et al. 2006, Zijlstra et al. 2007), polyimide (Lee et al. 2002), sapphire (Shaforost et al. 2008), and silicone nitride (SiON) (Schweinsberg et al. 2007), have also been used. Notable exceptions also exist, for instance, using high RI fluids in channels fabricated with soft lithography (Li et al. 2007) (Figure 11.15b).

One of the greatest benefits to using planar ring resonator sensors concerns packaging and reproducibility. Once templates and protocols for manufacture have been optimized, rings can be duplicated to very high tolerances. This is not always true of other ring resonator platforms. Once fabricated on-chip, sensors require an effective fluidics-handling system. Often, the solution to this entails building a flow channel on top of the ring and controlling fluid flow via a peristaltic pump, syringe pump, or other method (Yalcin et al. 2006, De Vos et al. 2007). Flow chambers are frequently constructed using PDMS-based soft lithography techniques (such as those proposed in Figure 11.16). These PDMS layers can then be plasma-treated and bonded to the ring substrate.

Planar ring resonators have shown excellent BRIS performance—in the range of 100–200 nm/RIU (Yalcin et al. 2006)—but their Q-factors tend to suffer from surface roughness problems (Rahachou and Zozoulenko 2003, Chao and Guo 2004). An SOI ring 200 μm in diameter with a waveguide width of 450 nm yields a Q-factor of 1.2×10^5 in air (Nitkowski et al. 2008). This is among the best values

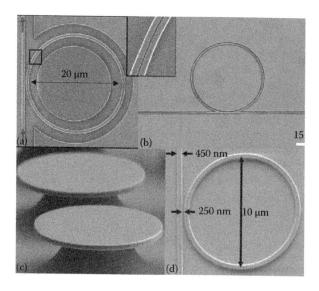

FIGURE 11.15 (a) SOI slot waveguide. (Reprinted from Robinson, J.T. et al., *Opt. Express*, 16, 4296, 2008. With permission.) (b) Liquid waveguide resonator in PDMS. (Reprinted from Li, Z. et al., Optofluidic microring dye laser, *IEEE/LEOS Summer Topical Meetings*, Portland, OR, pp. 70–71, 2007. With permission. ©2009 IEEE.) (c) Silica elevated disks. (Reprinted from Ghulinyan, M. et al., *Opt. Express*, 16, 13218, 2008. With permission.) (d) Thin silicon planar ring resonator. (Reprinted from Almeida, V.R. et al., *Opt. Lett.* 29, 2867, 2004a. With permission.)

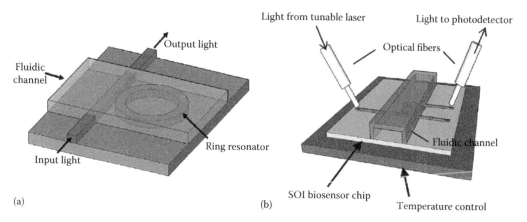

FIGURE 11.16 (a) Proposed flow channels for planar waveguides with straight-through output coupling. (Reprinted from Nitkowski, A. et al., *Opt. Express*, 16, 11930, 2008. With permission.) (b) Add/drop output couplers. (Reproduced from De Vos, K. et al., *Opt. Express*, 15, 7610, 2007. With permission.)

demonstrated for planar ring resonators. The use of different materials can lead to slightly better performance than this, however. For example, sapphire discs have shown Q-factors up to 4.5×10^5 (Shaforost et al. 2008). Despite low Q-factor, planar rings have been a popular ring resonator configuration due to the fact that they can be reproducibly manufactured and easily integrated into chip-based processes.

Both disc (Boyd and Heebner 2001) (Figure 11.15c) and circular ring-shaped planar waveguides (Boyd et al. 2003) (Figure 11.15d) have been proposed for pathogen sensing. In order to decrease the footprint of the ring while maintaining a large interaction area, folded and spiral SOI shapes are also possible (Xu et al. 2008a). Streptavidin–biotin affinity has been used as a proof of concept in a few setups because of their well-known binding affinities for each other (Boyd et al. 2003, Yalcin et al. 2006). An LOD of 250 pg/mm² for streptavidin binding to a polystyrene ring covered with biotin has been demonstrated (Chao et al. 2006). With a folded ring SOI cavity, streptavidin detection down to 3 pg/mm² has been demonstrated (Xu et al. 2008a). This value represents mass loading of protein on the sensor surface. LOD for the bulk concentration of streptavidin has also been reported down to 10 ng/mL, corresponding to roughly 150 pM (De Vos et al. 2007).

A good example of how planar rings can be integrated into functional systems has been published, using an array of 100 µm rings packed into a very small area (Ramachandran et al. 2008). The waveguides collecting signal from each ring were U-shaped, which allows the sensors to be placed side-by-side. Each ring was fabricated out of hydex glass that was demonstrated with several different biomolecules, including antibodies, DNA, bacterial cells, and proteins.

In 2004, a very unique solution to optical confinement in low RI was proposed in which a configuration called the "slot waveguide" was utilized (Almeida et al. 2004b) (Figure 11.15a). The device is composed of two concentric planar rings that are separated by a distance of between 40 and 200 nm (Almeida et al. 2004b, Baehr-Jones et al. 2005, Barrios et al. 2007, Dai and He 2007, Barrios et al. 2008a,b, Dai and He 2008, Robinson et al. 2008, Vivien et al. 2008). The exact spacing required depends entirely upon the wavelength of light being used. This unique mode of confinement relies on solutions to Maxwell's equations in which the electric field of the optical mode undergoes a large discontinuity. If the two concentric waveguides are close enough together to interact, then the majority of the field intensity must exist within the low-index gap (Almeida et al. 2004b). The slot waveguide has been demonstrated for RI detection of different chemicals, including cyclohexane and isopropanol and has been proposed for other biological or chemical sensing purposes (Barrios et al. 2007). It has also been demonstrated for the detection of BSA and anti-BSA antibodies with LODs of 16 and 28 pg/mm², respectively (Barrios et al. 2008a). A polymer ring was recently used for detecting glucose with a BRIS close to 200 nm/RIU (Kim et al. 2008). The LOD for glucose was estimated to be approximately 200 µM.

Annular Bragg resonators and onion-like resonators are unique ring resonators that are non-WGM based (Scheuer and Yariv 2003, Scheuer et al. 2005) (Figure 11.17). Using concentric rings of alternating RI materials, an omnidirectional Bragg grating is created. Introducing a defect, then, creates a ring-shaped Bragg resonator capable of Q-factors on the order of 1×10^3 and BRIS approaching 400 nm/RIU (Scheuer et al. 2005). The majority of optical field intensity is then concentrated in the defect, where light–matter interaction takes place. These structures have been used to detect RI changes down to 5×10^{-4} (Scheuer et al. 2006).

Similar to microspheres, planar ring resonators can also be employed for optical absorption measurements (Nitkowski et al. 2008). A 200 µm diameter planar ring was used to detect N-methylaniline by measuring the Q-factor changes due to absorption. This chemical, which absorbs strongly near 1500 nm,

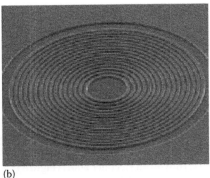

(a)   (b)

FIGURE 11.17   (a) Side view and (b) top view of a PMMA annular Bragg grating on SiO₂. (From Scheuer, J. et al., *IEEE J. Sel. Top. Quantum Electron.*, 11, 476, 2005. With permission. ©2009 IEEE.)

was used to demonstrate that ring resonators can measure absorbance with accuracy similar to that of commercial spectrophotometers.

11.2.1.6 Microtoroid

The first microtoroid was proposed in 2001 and was essentially a bottled segment of a silica fiber (Ilchenko et al. 2001). This effort produced a resonator with a Q-factor of 10^7 and was aimed at reducing the number of supported modes within the resonator compared with the microsphere platform. The contemporary microtoroid on-chip, however, consists of an etched elevated disk that has smoothed edges and an even smaller mode volume (Kippenberg et al. 2004). Typically, microtoroids are fabricated by photolithographically depositing circular SiO_2 pads on a silicon substrate. Then, the silicon is etched out from underneath until the circular pad is suspended in air by a thin pedestal. The surface of the disk is then irradiated with a CO_2 laser, causing the edges to bead into the desired toroid shape (Armani et al. 2003) as shown in Figure 11.18.

The microtoroid on-chip has gained a lot of attention over the last few years, due to the greatly advanced Q-factor it offers over the previously explored elevated disk platform due to the significant reduction in scattering loss resulting from the surface roughness (Armani et al. 2003, Kippenberg et al. 2003, 2004, Martin et al. 2004, Armani et al. 2005, Armani and Vahala 2006, Armani et al. 2007a,b, Del'Haye et al. 2007, He et al. 2008, Kulkarni et al. 2008). In fact, Q-factors up to 2×10^8 in water have been reported (Armani et al. 2007a).

It is also possible to fabricate the toroids out of PDMS molds (Martin et al. 2004, Armani et al. 2007b). The molded polymer toroid resonators, being only as smooth as the original PDMS mold, were only able to achieve Q-factors up to 3×10^5 (Armani et al. 2007b); however, this methodology allows the use of more cost-effective materials and does not require clean-room facilities. By comparison, the Q-factor achieved by these molded toroids is still competitive to, or better than, the Q-factors achievable with normal elevated disk resonators.

These resonators have been demonstrated as sensors for heavy water (Armani and Vahala 2006) and interleukin-2 in serum (Armani et al. 2007a). In the former case, the difference in optical absorption between heavy water and normal water is observed based on its impact on the Q-factor of the cavity (Armani and Vahala 2006). In the latter case, raw WGM mode shifts are tracked in order to observe interleukin binding to immobilized antibodies (Armani et al. 2007a). This work claims single-molecule detection based on an additional WGM shift enhancement from thermo-optics effects. If the Q-factor of the cavity is extraordinarily high (over 10^8, in this case), the power within the cavity is sufficient to actually cause immobilized molecules to heat up. This heating will cause additional changes to the RI of the toroid. The thermo-optic coefficient of silica is positive, which means that increased temperature increases the resonant wavelength of the WGM, thus adding to the shift from the adsorbed molecule.

FIGURE 11.18 Silica microtoroid ~120 μm in diameter with Q-factor of 1×10^8. (Reprinted from Armani, A.M. et al., *Nature*, 421, 925, 2003. With permission. Copyright 2009, Macmillan Publishers Ltd.)

Regarding microfluidics, microtoroids suffer from similar problems as microspheres and planar ring resonators in that the fluid handling components must be individually fabricated and integrated. In the case where a fiber taper is used to couple to the WGMs, alignment is very delicate as optimal coupling will take place only at the exact "equator" of the toroid, where the WGM field energy is concentrated. The movement of fluid around this coupling system risks disturbing the alignment and corrupting the data. Another limitation is that analytes must bind to this same equator region in order to generate a large WGM response. If the binding takes place elsewhere closer to the hub, the WGM will be less sensitive, or even completely insensitive, to it (Armani et al. 2007a). Given the relatively small area of the high-sensitivity region on the toroid compared with its entire surface area, detection of ultralow concentrations may be restricted.

11.2.1.7 Microfiber Loop

Tapered silica optical fibers have often been used as in- and outcoupling waveguides for microspheres (Vollmer et al. 2002, Vollmer et al. 2003), microtoroids (Armani and Vahala 2006), and microcapillaries (White et al. 2006a, Suter et al. 2008b). When pulled under heat to very small diameter, they are increasingly flexible and may be bent into extremely small radii of curvature without breaking. It has been shown, therefore, that they can easily be manipulated into microscale loops that act as ring resonators as shown in Figure 11.19 (Tong et al. 2003, Sumetsky 2004, Sumetsky et al. 2005, Sumetsky et al. 2006, Jiang et al. 2007, Vienne et al. 2007, Xu and Brambilla 2007a,b,c, Xu et al. 2007b, Vienne et al. 2008, Xu and Brambilla 2008, Xu et al. 2008b).

The very first demonstration of an optical-fiber-based ring resonator was in 1982, using a fiber 3 m in length (Stokes et al. 1982). This fiber resonator was relatively weakly guiding and required a coupler of substantial dimensions, necessitating its immense size. Seven years later, an optical fiber was tapered down to 8.5 μm with a flame and arranged in a 2 mm loop (Caspar and Bachus 1989). By bringing the fiber into close enough proximity to itself, the loop was able to approximate a ring resonator. Contemporary microfiber resonators exhibit the benefits of stronger guiding due to high RI contrast and self-coupling (Sumetsky 2004). It is now possible for a fiber loop to support moderately good Q-factors (on the order of 1×10^5) at diameters below 200 μm (Vienne et al. 2007), even down to 15 μm (Tong et al. 2003). At this size scale, the loops are well suited for compact integrated evanescent sensing. One method for this involves wrapping a 2.5 μm fiber around a template rod (as in Figure 11.19a) and embedding the whole assembly in polymer. This system has been employed with BRIS values as large as 40 nm/RIU, which is comparable to that of microsphere (Xu and Brambilla 2008). The use of an embedding polymer reduces somewhat the light–matter interaction that allows the ultrahigh BRIS systems mentioned above.

(a) (b) (c)

FIGURE 11.19 (a) A coiled multiloop resonator. (Reprinted from Sumetsky, M., *Opt. Express*, 12, 2303, 2004. With permission.) (b) Double loop microfiber resonator. (Reprinted from Sumetsky, M., *Opt. Express*, 12, 2303, 2004. With permission. Copyright 2009, Macmillan Publishers Ltd.) (c) Microknot resonator. (Reprinted from Tong, L. et al., *Nature*, 426, 816, 2003. With permission.)

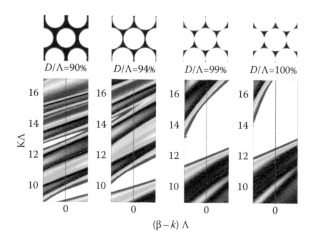

FIGURE 4.11 Index profiles (top) and DOPS plots for four cladding structures (bottom), with ratios of hole diameter to pitch 0.90, 0.94, 0.99, and 1 from left to right. The color scale represents the photonic density-of-states from low (light gray) to high (black). Zero density is highlighted by white color, and shows, for $(\beta - k)\Lambda < 0$, the permissible transmission band location for an HC-PCF possessing such a cladding structure. The vertical line is the air-line, $\beta = k$. (From Benabid, F., *Phil. Tran. R. Soc. A*, 364, 3439, 2006.)

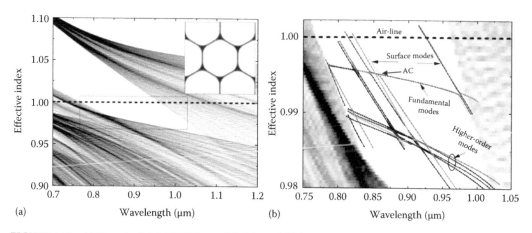

FIGURE 4.13 (a) The calculated DOPS for an HC-PCF and (b) focuses in on the bandgap region near the air light line and shows calculated mode trajectories superimposed. The extension of the guided modes out of the bandgap is due to the DOPS being calculated for the average cladding hole diameter, while the modes are calculated for the actual structure with slight variations in hole diameter across the cladding.

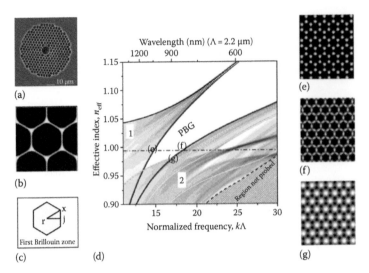

FIGURE 4.15 (a) SEM of an HC-PCF with core guidance at 1064 nm. (b) Details of the cladding structure used for the numerical modeling. (c) Brillouin zone symmetry point nomenclature. (d) Propagation diagram for the HC-PCF cladding lattice. White represents zero DOS and black maximum DOS. The upper x-axis shows the corresponding wavelengths for an HC-PCF guiding at 800 nm (pitch $\Lambda = 2.15\,\mu m$). The solid and dotted lines represent the Γ-point and J-point mode trajectories, respectively. The trajectories of the cladding modes on the edges of the PBG are represented (e) in red for the interstitial "apex" mode with the near field, (f) in blue for the silica "strut" mode, and (g) in green for the "air-hole" mode. The first two modes are shown at an effective index of 0.995 (dash-dotted white line), whereas the air-hole mode is shown at $k\Lambda = 15.5$ and $n_{eff} = 0.973$.

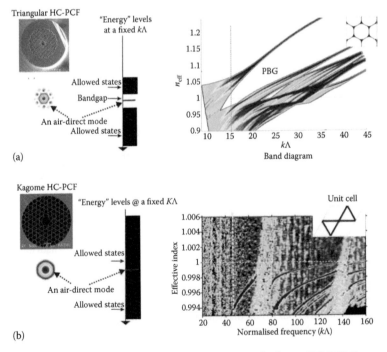

FIGURE 4.17 (a) Top left: Scanning electron micrograph of a triangular-lattice HC-PCF. Bottom left: near-field profile of the fundamental (HE_{11}-like) air-guided core mode lying within a bandgap. Center: schematic representation of the mode spectrum when an air defect mode is guided. Right: DOPS diagram showing the presence of the PBG. (b) Same as (a) but for Kagome-lattice HC-PCF. The fundamental core mode lies within a continuum of cladding modes and the band diagram does not exhibit a PBG.

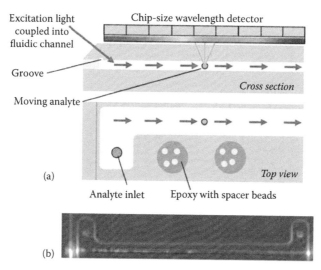

(a)

FIGURE 7.12 (a) (Top) Cross-sectional view of a fluidic chamber. Light is guided within the fluid, thereby continuously exciting the analyte. The chip-sized wavelength detector records the fluorescence spectrum of the analyte while it moves along the detector. (Bottom) Top view of the fluidic chamber. (b) Photograph of the top view of the chamber, demonstrating large-volume fluorescence excitation. (Reproduced from Takagi, J. et al., *Lab Chip*, 5, 778, 2005. With permission of Royal Society of Chemistry.)

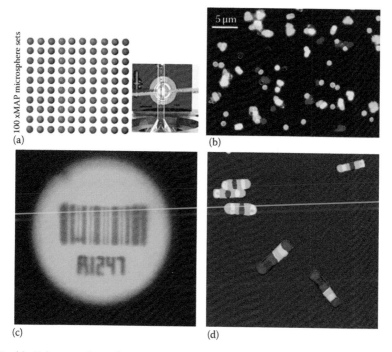

FIGURE 7.18 (a) Color encoding (http://www.luminexcorp.com/index.html). (b) Quantum dots: fluorescence micrograph of a mixture of CdSe/ZnS QD-tagged beads emitting single color. (Reprinted from Macmillan Publishers Ltd: Han, M. et al., *Nat. Biotech.*, 19, 631, 2001. With permission. Copyright 2009.) (c) 2D graphical coding. Polybeads with a barcode patterned by photobleaching. (Reprinted from Macmillan Publishers Ltd: Liu, G. L. et al., *Nat. Mater.*, 5, 27, 2006. With permission. Copyright 2009.) (d) Color barcode using doping with rare-earth ions. (From Dejneka, M.J. et al., *Proc. Natl. Acad. Sci. USA*, 100, 389, 2003. With permission. Copyright 2009, National Academy of Sciences.)

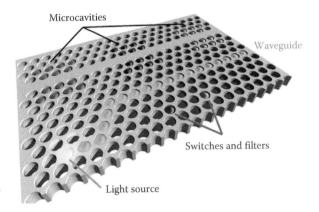

Microcavities

Waveguide

Switches and filters

Light source

FIGURE 8.11 Schematic of a reconfigurable, integrated optofluidic PhC circuit. The basic platform is a planar PhC with air holes. Numerous components can be integrated onto the same chip through selective infiltration of air holes.

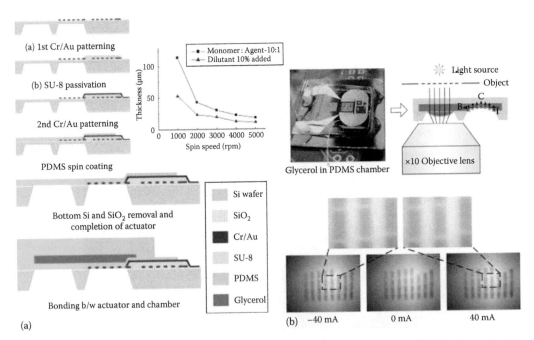

(a) 1st Cr/Au patterning

(b) SU-8 passivation

2nd Cr/Au patterning

PDMS spin coating

Bottom Si and SiO₂ removal and completion of actuator

Bonding b/w actuator and chamber

(a)

- Monomer : Agent-10:1
- Dilutant 10% added

Thickness (μm)

Spin speed (rpm)

Si wafer
SiO₂
Cr/Au
SU-8
PDMS
Glycerol

Light source
Object
C
B
I

Glycerol in PDMS chamber

×10 Objective lens

(b)   −40 mA   0 mA   40 mA

FIGURE 9.5 (a) Schematic of the electromagnetic actuator and lens fabrication process. (b) The characterization process of variable focusing. (Reprinted from Lee, S.W. and Lee, S.S., *Appl. Phys. Lett.*, 90, 121129, 2007. With permission. Copyright 2009, American Institute of Physics.)

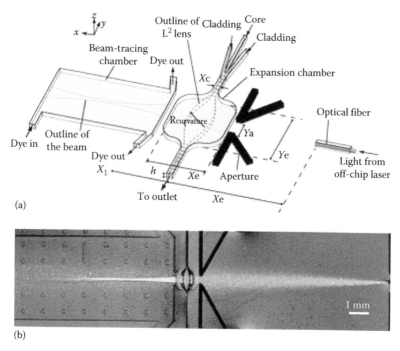

(a)

(b)

FIGURE 9.14 (a) Schematic of the liquid-core/liquid cladding tunable optofluidic lens. (b) Microscopic image of the lens in action. (Reproduced from Tang, S.K.Y. et al., *Lab Chip*, 8, 395, 2008. With permission of The Royal Society of Chemistry.)

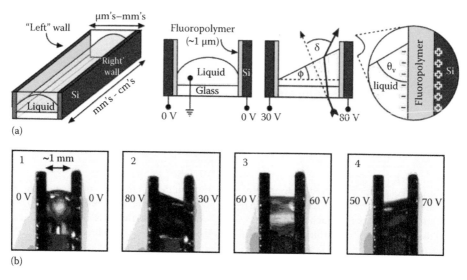

(a)

(b)

FIGURE 9.17 (a) Schematic of the electrowetting microprism (EMP). (b) Image of the EMP at different operating states: (1) rest state, (2) through (4) tilt of the refractive surface by applying appropriate voltage at different sides of the electrode. (Reproduced from Smith, N.R. et al., *Opt. Express*, 14, 6557, 2006. With permission.)

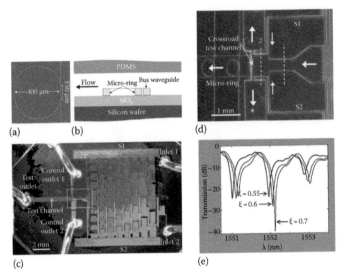

FIGURE 10.7 MRD integrated with an on-chip mixer. (a) Scanning electron microscope (SEM) micrograph of the MRD. (b) Schematic drawing showing the cross-section of the test microchannel of the MRD. (c) Micrograph of the integrated optofluidic device with water and green dye injected into inlets 1 and 2, respectively. (d) Magnified image of the region inside the dashed rectangle in (c). Arrows show direction of flow. (e) Transmission versus wavelength for three different values of ξ—the dimensionless control parameter of the on-chip mixer. (Reprinted from Levy, U. et al., *Appl. Phys. Lett.*, 88, 111107, 2006. With permission. Copyright 2009, American Institute of Physics.)

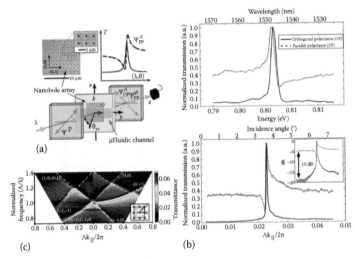

FIGURE 10.12 Polarization-resolved measurement of a transmittance lineshape of a nanohole array. (a) Experimental setup; also shown is an SEM image of a representative sample. (b) Measured spectral (top) and angular (bottom) transmittance of the nanohole array when the polarizer and the analyzer are aligned (PP) and orthogonal (OP). In the PP condition, a Fano lineshape is produced by the interference of light directly transmitted through the nanoholes and light reradiated via the SPP. In OP, the directly transmitted component is suppressed, resulting in a Lorentzian lineshape. Each curve has been normalized to its maximum to clearly illustrate the respective lineshape functions, except in the inset, where the same data is plotted in a logarithmic scale to show the ~15–20 dB background-level reduction for the Lorentzian versus Fano-type resonances. (c) Transmittance through a 2D array of 350 nm holes in a 100 nm thick aluminum film on GaAs substrate. To cover the domain of normalized frequencies, Λ/λ, shown, data from two samples with hole periods Λ = 1.4 μm and Λ = 1.6 μm are combined; the horizontal white line at Λ/λ ~0.95 indicates the boundary between the two data sets. (Reprinted from Fainman, Y. et al., *Opt. Photon. News*, 17, 24, 2006. With permission.)

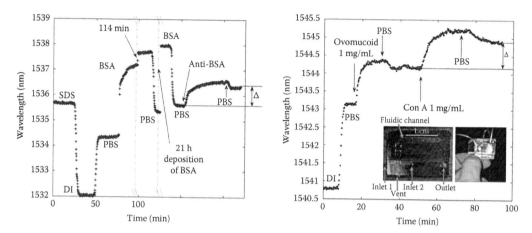

FIGURE 10.15 Detection and real-time monitoring of (a) protein–protein and (b) protein–carbohydrate bindings on a gold nanohole array observed by monitoring the resonant wavelength of a single SPP mode at the fluid–metal overlayer in a transmission setup. Insets are photographs of an SPP sample holder with (right) and without (left) tubing. (Reprinted from Hwang, G.M. et al., *Sens. J., IEEE*, 8, 2074, 2008. With permission.)

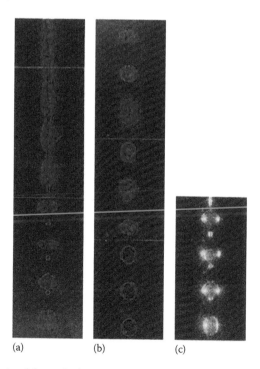

(a)　　　(b)　　　(c)

FIGURE 11.22 Red lasing signal from Rhodamine 590-doped ethanol droplets (a) near the top of the aperture, (b) much farther down, and (c) with the green scattering signal from pump laser. (Reprinted from Qian, S.-X. et al., *Science*, 231, 486, 1986. With permission from AAAS.)

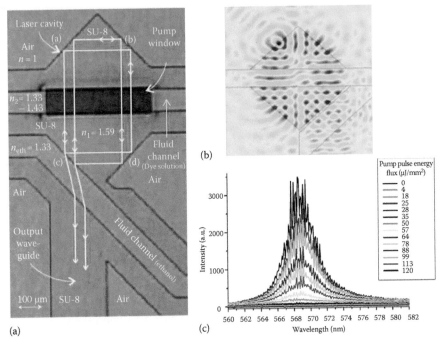

(a)

(b)

(c)

FIGURE 12.13 Optofluidic ring laser. (a) Optical microscope image of the laser resonator structure consisting of two isosceles triangles of the photo definable polymer SU-8 ($n_1 = 1.59$) and a micro-fluidic channel in-between. Two classical trajectories of equal optical path length are drawn. The optical pumping is performed through a window (dark rectangular area in the photo) in a metal mask. (Reproduced from Gersborg-Hansen, M. et al., *Microelectron. Eng.*, 78, 185, 2005. With permission. Copyright 2009 from Elsevier.) (b) Example of simulated standing wave cavity mode. (Reprinted from Gersborg-Hansen, M. and Kristensen, A., *Appl. Phys. Lett.*, 89, 103518, 2006. With permission. Copyright 2009 from American Institute of Physics.) (c) Laser spectra measured at different pump pulse energies when the laser is operated with rhodamine 6 G in ethanol. (Reproduced from Gersborg-Hansen, M. et al., *Microelectron. Eng.*, 78, 185, 2005. With permission. Copyright 2009 from Elsevier.)

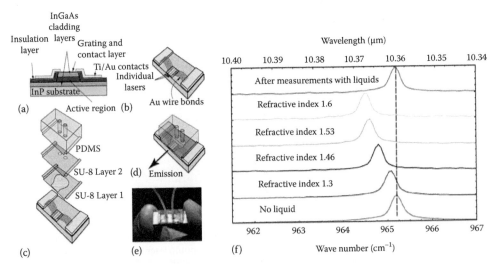

FIGURE 12.16 Panels (a)–(e) show schematics of the microfluidic integration of distributed feedback quantum cascade laser. Panel (f) shows fluidic tuning of lasing wavelength at room temperature. (Reprinted from Diehl, L., *Opt. Express*, 14, 11660, 2006. With permission. Copyright 2009 from Optical Society of America.)

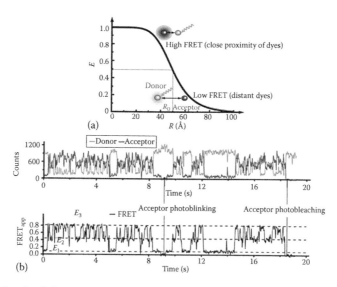

(a)

(b)

FIGURE 13.8 (a) Graphical depiction of FRET energy transfer as a function of separation distance. (b) FRET processes are observed with a mutant hairpin ribosome, where the donor and acceptor separation is dynamic. Time–intensity traces illustrate an anticorrelated behavior for each of the two probe colors and indicate the FRET process. Characteristic photoblinking and photobleaching is observed after prolonged excitation. (Reprinted from Roy, R. et al., *Nat. Methods*, 5, 507, 2008. With permission.)

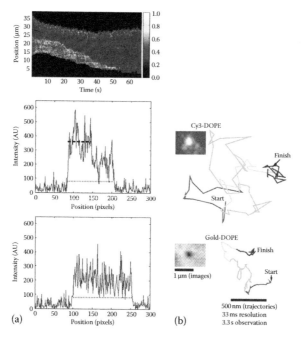

FIGURE 13.12 (a) YOYO-1 labeled T4-bacteriopage DNA is driven into a nanofluidic channel and elongated. The top panel illustrates the normalized fluorescent intensity of an elongated DNA strand as a function of time. The middle and bottom panel are the projected intensities at $t = 15$ s and $t = 55$ s when the molecule is folded and unfolded, respectively. (Reprinted from Levy, S.L. et al., *Nano Lett.*, 8, 3839, 2008. With permission.) (b) The phospholipid DOPE is labeled with the fluorescent dye Cy-3 or a gold nanoparticle, and studied during diffusion through plasma membrane compartments. Anomalous diffusion between compartments, or "hop diffusion," was observed using Gold-DOPE and high-speed imaging. (Reprinted from Murase, K. et al., *Biophys. J.*, 86, 4075, 2004. With permission.)

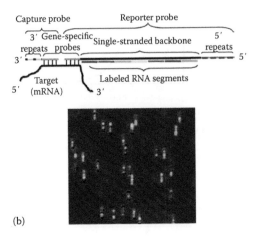

(b)

FIGURE 13.13 (b) Gene expression is studied using mRNA hybridized with a Nanostring reporter probe. The hybrid molecule is bound to a surface using the capture probe and then elongated by electrokinetic forces. Molecular mapping is accomplished with a color-encoded reporter probe that identifies a hybridized gene specific probe–mRNA pair. (Reprinted from Geiss, G.K. et al., *Nat. Biotechnol.*, 26, 317, 2008. With permission.)

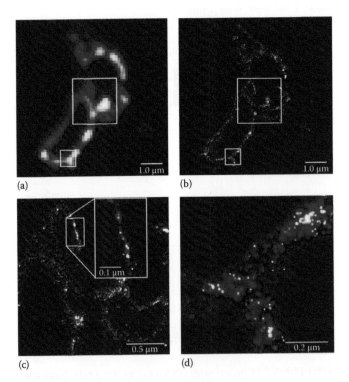

FIGURE 13.17 Comparison of TIRF and FPALM images of a COS-7 cell expressing a transmembrane protein labeled with PA-FP Kaede. (a) TIRF image, (b) FPALM image, (c,d) high-magnification views of FPALM images illustrate single-fluorescent protein molecules, enabling the study of lysosome and endosome interactions. (Reprinted from Betzig, E. et al., *Science*, 313, 1642, 2006. With permission.)

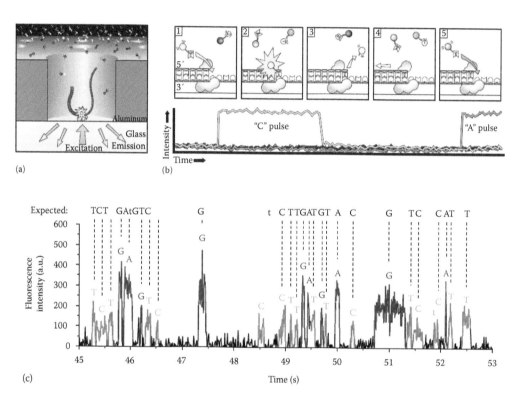

FIGURE 13.20 (a) ZMW filled with a polymerase beneath a sea of fluorescently labeled, free nucleotides. (b) Each type of nucleotide is identified with a different color of fluorophore. Nucleotide pairing during synthesis causes dissociation of the nucleotide from the fluorophore. A momentary increase in fluorescence intensity is observed within the waveguide aperture and is recorded as a function of time. (c) A 9 s portion of an intensity vs. time trace illustrates the individual base readout during the DNA sequencing process. (Reprinted from Eid, J. et al., *Science*, 323, 133, 2009. With permission.)

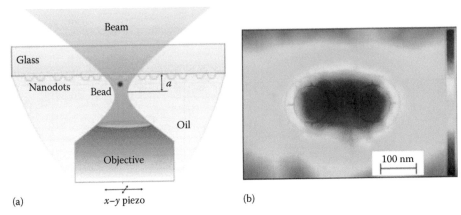

(a)

(b)

FIGURE 14.9 (a) Schematic of metallic, nanostructure-enhanced nanometric optical tweezers. (b) Light power excited by a transverse-magnetic laser light (1064 nm) is shown as a color map and calculated for a plane at a height 200 nm above the nanostructured substrate (the plane is parallel to the glass substrate). (Reproduced from Grigorenko, A.N. et al., *Nat. Photon.*, 2, 365, 2008. With permission.)

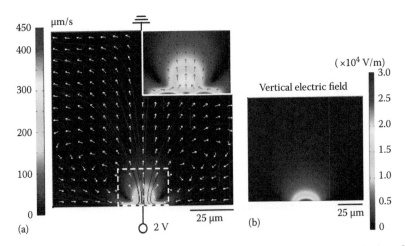

(a)

(b)

FIGURE 14.24 (a) Light-induced ac electroosmosis flow near a virtual electrode. The highest flow velocity exists on the surface at the edge of a virtual electrode. Owing to the symmetrical flow pattern, there exists a stagnant-flow zone near the middle surface of a virtual electrode. This is where the nanoparticles are trapped. (b) Simulated vertical electric field near a virtual electrode during LACE manipulation. (Reprinted from Chiou, P.Y. et al., *J. Microelectromech. Syst.*, 17, 525, 2008. With permission. Copyright 2009 from IEEE.)

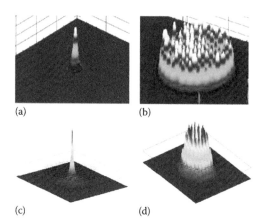

(a)   (b)

(c)   (d)

FIGURE 15.3 Fluid-filled MOF with solid-core at a wavelength of 1200 nm: (a) near field pattern for low refractive index fluid, (b) near field pattern for high refractive index fluid, (c) FEM simulation of field pattern for low refractive index fluid, and (d) FEM simulation of field pattern for high refractive index fluid. (From Wang, Y. et al., 2009. With permission.)

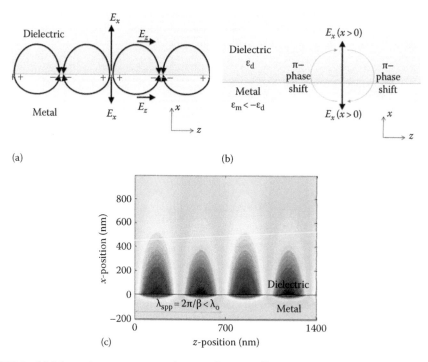

FIGURE 18.3 (a) Schematic representation of SPP as charge oscillations at the interface between a metal and a dielectric. The electric field has a longitudinal (z-direction) component that is $\pi/2$ out-of-phase with the transverse component (x-direction). (b) Schematic of SPP self-consistency relation that is allowed for by a negative relative permittivity in the metal. The electromagnetic wave has a π phase shift upon crossing the boundary, allowing for self-consistency when crossing the boundary twice. This allows the SPP waveguide mode to exist at only a single interface. (c) Calculated transverse magnetic field (y-direction) for an SPP above gold at free-space wavelength of 700 nm. The SPP wavelength is shorter than the free-space wavelength, as described in the text. (Reprinted from Gordon, R., *IEEE Nanotechnol. Mag.*, 2, 12, 2008. With permission.)

FIGURE 18.8 Visual example of absorption characteristics of metallic nanoparticles as a function of nanoparticle properties. This case involves Au shell–silica core particles with differing shell thickness. (Reprinted from Loo, C. et al., *Technol. Cancer Res. Treat.*, 3, 33, 2004. With permission.)

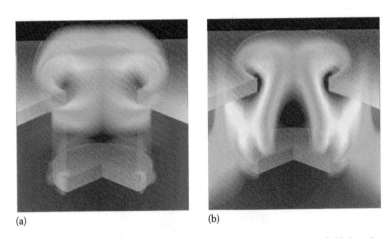

(a)   (b)

FIGURE 18.13 Hole/disk plasmonic structures. Computational results for the EM field distribution for 883 nm separation (a) and 1138 nm (b). The larger-wavelength excitation results in strong coupling between levels. (Reprinted from Stewart, M.E. et al., *Proc. Natl. Acad. Sci. USA*, 103, 17143, 2006. With permission.)

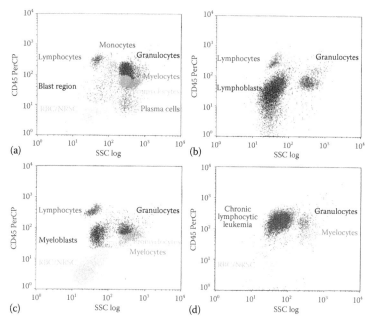

FIGURE 19.4 Analysis of normal and leukemic bone marrow by CD45 side scatter (SSC) analysis. (a) Normal bone marrow showing normal populations of blood, (b) abnormal increase of lymphoblasts and myeloblasts as seen in acute lymphoblastic leukemia (ALL), (c) chronic myelogenous leukemia (CML), and (d) chronic lymphocytic leukemia (CLL). (From Jennings, C.D. and Foon, K.A., *Cancer Invest.*, 15, 384, 1997b. With permission. © American Society of Hematology.)

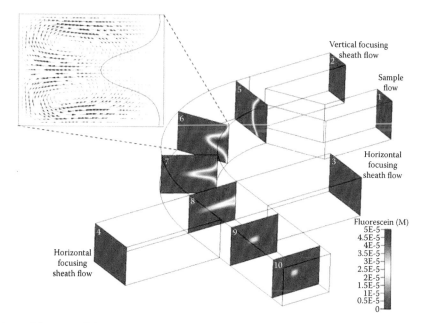

FIGURE 19.7 Schematic of the 3D hydrodynamic focusing process by employing the microfluidic drifting technique. Slices are the cross-sectional profiles of the fluorescein dye concentration in the focusing device. Inset: the simulation of the secondary flow velocity field shows the formation of Dean vortices in the 90° curve. An iso-curve of fluorescein concentration=25 μM is arbitrarily chosen as the boundary of the sample flow. (Reproduced from Mao, X. et al., *Lab Chip*, 7, 1260, 2007. With permission from The Royal Society of Chemistry.)

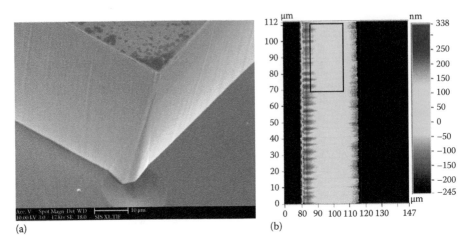

(a)   (b)

FIGURE 19.16 (a) SEM image of a cryogenically etched mold feature. Features are 50 μm deep, etched into a silicon wafer for mold durability. Sidewall roughness is measured by optical profilometry (Veeco NT1100). (b) A sample of the profilometry measurement, along with a sample measurement area (black rectangle). Roughnesses on the order of 30 nm have been measured for cryogenically etched features. (From Godin, J. and Lo, Y.-H., Advances in on-chip polymer optics for optofluidics, *Conference on Lasers and Electro-Optics*, 2009. With permission from Optical Society of America.)

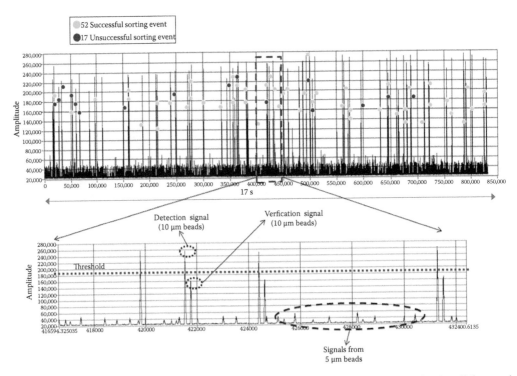

FIGURE 19.26 Sorting of 10 μm fluorescent beads from a mixture containing 10 and 5 μm beads. All the signals shown here have been amplified by the match filter. Firing of the PZT actuator occurs only when the detected signal intensity goes above threshold. For every single sorted particle, the detected signal (upstream) is always followed by a verification signal (downstream). Within the time period (~17 s), 52 out of 69 (~75% sorting efficiency) 10 μm beads have been successfully sorted. Most importantly, no beads were mistakenly sorted, yielding a high enrichment (~1000-fold) ratio for rare events.

Q-factors up to 1×10^6 have been predicted for two-loop structures (Sumetsky 2004) (Figure 11.19b). Due to the very small size of the fibers themselves (submicron in many cases) evanescent exposure to the surrounding media is quite extensive and thus the BRIS can be very high, up to 700 nm/RIU (Xu et al. 2007b) with RI LODs predicted close to 1×10^{-7} RIU (Xu et al. 2008b). This excellent resolution makes them one of the more promising RI sensing platforms available today. Nonetheless, there has been little exploration of applications for RI sensing using microfiber loops. As a relatively new platform, there is much room for future investigation into this area.

In addition to the loop configuration, microfibers have also been arranged into microknots (Tong et al. 2003, Jiang et al. 2006, 2007, Vienne et al. 2007) (Figure 11.19c). The primary advantage of the microknot is that its assembly is much more stable compared to a loop held together by electrostatic and van der Waals forces (Jiang et al. 2006). Once drawn into a knot, the fiber will not accidentally straighten, even if drawn into a very tight radius. The best reported Q-factor for such a structure is 5.7×10^4 in air (Jiang et al. 2006), which is not as good as some other loop configurations because the self-coupling area where the fiber is twisted around itself is not conducive to the highest efficiency coupling.

For microfluidic integration, fiber loops work excellently when embedded, because the alignment of the loops relative to each other is secured, and they can be fashioned into a convenient tube-like geometry. Using a removable rod, the fiber is coiled and coated with a layer of ultralow RI polymer-like Teflon AF (RI = 1.29) (Xu and Brambilla 2007a, b, Xu et al. 2008b). The low RI of the Teflon helps to confine the WGM within the fiber while providing stability for the coil positions. Once the Teflon is cured, the rod can be removed and the evanescent field of the fiber will extend into the resultant hollow core.

However, given the performance of such a configuration, there seems to be very little advantage of using it over the microcapillaries discussed in Section 11.2.1.4. For outcoupling of the optical signal, however, the embedded fiber loop is superior as the delivery fiber is integrated into the ring itself.

As a review, Table 11.1 relates the significant performance parameters for each ring resonator platform discussed in this section as well as the LODs for some important analytes. While BRIS and Q-factor vary significantly between different configurations, the RI LOD (not published for all configurations) can be a good basis for comparison.

11.2.2 Fluorescence Sensing with Ring Resonators

Despite the advantages of label-free detection methods, ring resonators can successfully be used as fluorescence-based sensors as well. This technique can be employed relatively simply by introducing a dye species within the optical field of the WGM, which is used to excite the dye, or the dye may be externally excited and its fluorescence emission may be coupled into the WGM of the resonator. If the dye is conjugated to a molecule of interest, then that target is said to be labeled, and can then be detected via the dye molecule rather than by inducing RI changes at the sensor surface.

The advantage of using a fluorescence method is that they are based on mature protocols that allow fluorescently labeled molecules to be directly imaged with confocal microscopy techniques. Additionally, with these modern confocal methods, limits of detection using fluorescence assays can potentially exceed any other technology available (Nie et al. 1994). Fluorescence can also be leveraged with evanescent optical sensors. This approach was initially explored using straight waveguides (fiber or planar) etched small enough to expose the evanescent field (Sutherland et al. 1984, Lukosz 1995, Abel et al. 1996). Evanescent excitation of labeled fluorophores on the waveguide surface was used to detect the presence of target molecules. This type of sensing technique used fairly long waveguides in order to perform high-sensitivity detection, up to several millimeters in order to detect femtomolar concentrations (Plowman et al. 1996, Krioukov et al. 2002).

In the interest of reducing the size of these sensors, the evanescently excited fluorescence was explored with ring resonator configurations. With a ring resonator cavity the fluorescence emission can be enhanced tremendously because of intracavity light buildup from repeated photon circulation within the cavity (longer photon lifetime) (Blair and Chen 2001). This creates very long effective sensing lengths even though the physical size of the resonator is quite small (Krioukov et al. 2003), which enables not only conservation of sample volume, but also compact sensor integration into a small footprint.

To date, most of the fluorescence-based biochemical sensing is demonstrated with planar ring resonators. Using a silicon nitride microdisk 30 μm in diameter, indocyanine green was detected via fluorescent enhancement (Krioukov et al. 2002). The group projected that their setup was capable of yielding a 40-fold fluorescent enhancement over a straight waveguide configuration. With this enhancement, it was predicted that detection of 20 dye molecules was possible. A similar demonstration with 9.8 μm polystyrene beads covalently immobilized AlexaFluor-488 dyes in order to provide WGM-enhanced fluorescence detection (Schiro and Kwok 2004).

11.2.3 Raman and SERS-Based Sensors

Raman effects are additional nonfluorescent phenomena that can be explored and leveraged with ring resonator platforms (see Chapter 6 by Zhang for a more thorough discussion of the physical principles of Raman scattering). The unique Raman signatures of many chemicals can be effectively used as a fingerprint for detection, however the intensity of these signals are quite weak. The advent of SERS has enabled amplification of weak Raman signals via plasmonic effects as well. Using these plasmon effects in noble metal nanoparticles, Raman enhancements of up to 10^{14} are possible (Michaels et al. 1999). This sort of amplification is significant enough that it has enabled single-molecule detection (Nie and Emory 1997).

When the Raman signal-generating molecules are inside the cavity, like in a microdroplet, the WGM provides another enhancement effect termed "cavity-enhanced Raman scattering" (CERS) due to tremendous light build up in the cavity (Snow et al. 1985, Kwok and Chang 1993, Lin and Campillo 1995, Symes et al. 2004, Symes et al. 2005, Hopkins and Reid 2006, Ausman and Schatz 2008). In this phenomenon, the increased optical path length caused by WGMs allows the build up of intense energy within a droplet, allowing nonlinear processes such as CERS to occur. This approach can help to assay the composition of a microdroplet due to the unique signature of Raman scattering from lower concentrations of sample than would be possible with single-pass optical methods (Symes and Reid 2006).

Cavity enhancement can be even further amplified in conjunction with SERS (Kim et al. 1999, White et al. 2007a, Ausman and Schatz 2008). Using both metal particles (such as gold or silver) and ring resonators, Raman signals that produce plasmonic resonances may be coupled into WGMs that create further enhancement due to optical feedback. In an experiment by White at al. (2007a), the inside of a thin-walled capillary was coated with silver nanoparticles and then R6G was introduced as is shown in Figure 11.20a. The Raman spectrum from the R6G (Figure 11.20b) was amplified by the surface plasmons in the silver, which in turn was amplified by the evanescent field from the WGM. This effect was strong enough that a 400 pM solution was detectable. Such a configuration is capable of 300-fold enhancement over conventional SERS (White et al. 2006c). A SERS-based laser was also demonstrated with a cylindrical ring resonator (Kim et al. 1999). In this study, a quartz tube with a 1 mm outer diameter was filled with R6G and a colloidal silver solution, but the Q-factor of this cavity was measured to be only 1.5×10^4. These experimental results were recently confirmed by a theoretical work based on Mie theory that modeled the electric field intensity within microsphere-based ring resonators (Ausman and Schatz 2008). It is found that non-plasmonic Raman enhancement from WGMs alone was able to achieve nearly the same enhancement factor as from metal particle-based SERS at certain "hot spots" on the microsphere surface near the mode propagation axis.

11.2.4 Microfluidic Lasers

Another important application of the ring resonator is the laser, which was first demonstrated in the 1960s using doped microspheres with the gain medium being in the solid state (Garrett et al. 1961).

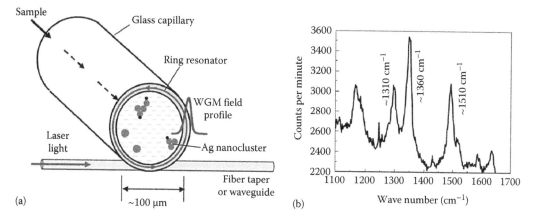

(a)   ~100 µm   (b)

FIGURE 11.20 (a) Setup for SERS detection in a capillary-based ring resonator and (b) the Raman spectrum measured for R6G. (Reprinted from White, I.M. et al., *Opt. Express*, 15, 17433, 2007a. With permission.)

Since then, many solid-state ring resonator lasers have been created using microspheres (Wang et al. 1995, Sandoghdar et al. 1996, Sasagawa et al. 2003, Zijlstra et al. 2007), planar rings (Michler et al. 2000), cylinders (Moon et al. 2004), and even microtoroids (Kippenberg et al. 2006). Cylindrical micro-cavity lasers can also be formed from optical fibers coated with a high RI fluorescent polymer (Frolov et al. 1997) or from polymer ring doped with dye (Wu et al. 2008). Due to the multimode nature of WGMs within ring resonators, the lasing output is typically multipeaked like the spectrum shown in Figure 11.21. The relatively broadband emission of ring resonator lasers can be minimized by etching a grating onto the sphere surface (Sasagawa et al. 2003) or by using the Vernier effect with a second ring of a different size (Wu et al. 2008).

Ring resonators have recently been employed in microfluidic lasers, in which the gain medium is in liquid form. Table 11.2 lists the relevant traits of some interesting microfluidic lasers described in the literature, some of which will also be covered in Chapter 12.

They take advantage of the adaptive nature of liquid, the convenience of changing the liquid gain medium, potential compatibility with microfluidics for easy and safe gain medium, and sample delivery. As a result, these devices are of great interest for the development of integrated, miniaturized tunable laser light sources with broad spectral coverage and for the development of micrototal-analysis system (or lab-on-a-chip) for biological and chemical analysis.

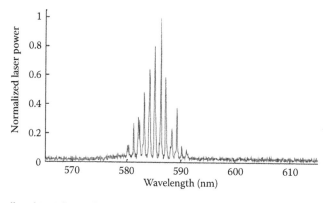

FIGURE 11.21 Broadband R6G lasing from a PDMS fluid planar ring. (Reprinted from Li, Z. et al., Optofluidic microring dye laser, *IEEE/LEOS Summer Topical Meetings*, Portland, OR, pp. 70–71, 2007. With permission.)

TABLE 11.2 Comparison of Lasing Threshold and Lasing Mode for Various Types of Optofluidic Ring Resonators

Platform	Lasing Threshold	Mode (CW/Pulse)	Gain Medium	Reference
Microsphere	$200\,\mu J/pulse$	Pulsed (10 ns)	R6G	An and Moon (2003)
Microdroplet	500 mW	Pulsed (5 ns)	R6G	Azzouz et al. (2006)
Microdroplet	$750\,J/cm^2$	Pulsed (100 ns)	Rhodamine B	Kiraz et al. (2007)
Microdroplet	$35\,W/cm^2$	CW	R6G	Tzeng et al. (1984)
Microdroplet	$0.5\,MW/cm^2$	Pulsed (100 ns)	R6G	Tanyeri et al. (2007)
Microdroplet	$0.4\,J/pulse$	Pulsed (10 ns)	CdSe/ZnS QD	Schafer et al. (2008)
Microdroplet	NA	Pulsed	Glycerol/water (Raman)	Sennaroglu et al. (2007)
Microcapillary	$25\,nJ/mm^2$	Pulsed (10 ns)	R6G	Lacey et al. (2007)
Microfiber loop	$9.2\,\mu J/pulse$	Pulsed (6 ns)	R6G	Jiang et al. (2007)

In ring resonator microfluidic lasers, the gain medium usually consists of fluorescent dyes or QDs (Schafer et al. 2008) dissolved in solvent. For Raman lasers, the gain medium can be the solvent itself. Many solvents, such as carbon disulfide (Cheung et al. 1993), glycerol (Sennaroglu et al. 2007), or ethanol (Snow et al. 1985), have large Raman gain coefficients and therefore can be used as an excellent Raman gain medium. The dye may be carried externally to the resonator cavity using microfluidics (in the case of a solid ring resonator), or it may be embedded or dissolved within it (in the case of a fluid ring resonator). In the first case, the gain region for the dye will be the small area of evanescent overlap. In the second case, the majority of the WGM optical field within the resonator can be used as the gain region.

The optical pump of the gain medium can be from the direct excitation by tuning the pump light into the gain absorption band or from indirect excitation through such a mechanism as energy transfer. It is achieved either by free-space excitation or through an optical coupler such as a fiber taper, waveguide, or prism. The dye emission is coupled into the WGM, which provides positive optical feedback for lasing. Due to the high Q-factor of the ring resonator, low threshold lasing is possible. The laser can be in a continuous-wave (CW) mode or a pulsed mode, depending on how it is pumped. The laser output is typically collected by free-space coupling or through an optical coupler for efficient and directional light delivery.

11.2.4.1 Solid Dielectric Microspheres

Microspheres are among the most widely used formats for microfluidic lasers. Compared to droplet lasers they are much easier to manipulate, and compared to most other geometries they are easier to produce. The world-record Q-factors obtainable with microspheres (Gorodetsky and Ilchenko 1999) also provide extremely high optical feedback for extremely low lasing thresholds.

In order to create a microfluidic laser from a microsphere, the sphere must be immersed in a fluorescent gain medium. Lasing has been observed from microspheres immersed in R6G, obtaining a threshold of $200\,\mu J$ per pulse from 532 nm laser with a 10 ns pulse (An and Moon 2003). This work also showed how the spectral position of the peak lasing emission changed depending on the concentration of R6G used due to self-absorption.

11.2.4.2 Microdroplets

Unlike solid microspheres, the majority of the WGM of a microdroplet is accessible to gain-generating molecules. Microdroplet lasers have been created from ethanol (Tzeng et al. 1984), methanol (Saito et al. 2008), ethylene glycol (Azzouz et al. 2006), and water (Lin et al. 1992, 1998), and are typically cladded with air, providing high RI contrast. The emission of microdroplets is usually collected and excited via free space, due to the difficulty in controlling and positioning the droplets. However, coupling of a tapered optical fiber to a liquid-cladded droplet has been demonstrated as an efficient delivery waveguide for lasing emissions (Hossein-Zadeh and Vahala 2006). Such advancement makes the use of microdroplets much more practical, however positioning of such a fiber, as mentioned in Section 11.2.1.3, is quite difficult.

Manipulation of microdroplet lasers is another significant challenge that has been approached a number of different ways. Often, droplets generated at the end of a pipette or similar orifice can be lased while they hang. With the use of delicate optical alignments, droplets can even lase as they fall through space as shown in Figure 11.22 with Rhodamine 590 in ethanol (Qian et al. 1986). More recently, droplet levitation has been researched (Azzouz et al. 2006, Schafer et al. 2008). Levitating droplet in air prevents distortion and optical loss from contact with a substrate. Ultrahydrophobic substrates, however, can provide an excellent way of controlling droplets while creating good fluid contact angles in order to maintain sphericity (Kiraz et al. 2007b, Sennaroglu et al. 2007). Contact angles over 150° have been measured.

It is also possible to substitute the air cladding for enhanced droplet stability. Saito et al. (2008) demonstrated that it was possible also to suspend a methanol microdroplet within a viscous polysiloxane resin (Figure 11.23a). Doing so provides the benefit of being able to isolate the droplet in a defined position. This manuscript also demonstrated that shape deformations could be induced on the droplet by squeezing the elastomer, which allows tuning of the droplet lasing spectrum as shown in Figure 11.23b. Droplets have also been produced within another liquid by forming water droplets (RI = 1.33) in a lower RI oil (RI = 1.29) (Hossein-Zadeh and Vahala 2006, Tanyeri et al. 2007). While handling and positioning of microdroplets within another liquid can be challenging, it avoids the problem of evaporation which plagues microdroplets in air. Extremely small droplets possess such an extremely large surface-to-volume ratio that evaporation will have an appreciable impact on the volume within seconds. Tanyeri et al. (2007) achieved lasing in R6G-doped water/glycerol microdroplets carried through fluidic chambers filled with oil. This not only presents a good solution

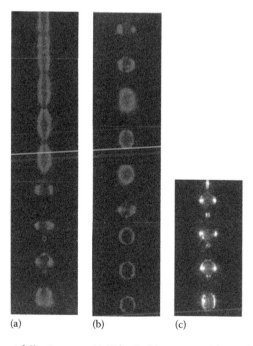

(a)   (b)   (c)

FIGURE 11.22 (See color insert following page 11-20.) Red lasing signal from Rhodamine 590-doped ethanol droplets (a) near the top of the aperture, (b) much farther down, and (c) with the green scattering signal from pump laser. (Reprinted from Qian, S.-X. et al., *Science*, 231, 486, 1986. With permission from AAAS.)

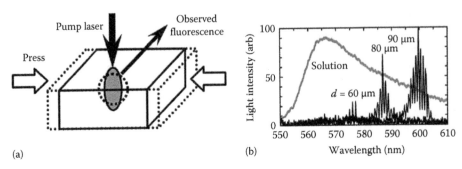

FIGURE 11.23 (a) Schematic of microdroplet laser tuning with mechanical deflection. (b) Laser spectrum dependence on droplet diameter. (Reprinted from Saito, M. et al., *Opt. Express*, 16, 11915, 2008. With permission.)

to prevent droplet evaporation, but also proves that the droplets can be precisely positioned within the fluid given the correct approach. For more discussion on microfluidics and nanofluidics, readers are referred to Chapter 1 by Krishnan and Erickson.

Demonstrations of microdroplet lasers have utilized both conventional dyes and QDs. An example of the latter used 9 μm diameter droplets of water/glycerin with CdSe/ZnS QDs dissolved in them (Schafer et al. 2008). These droplets achieved a threshold of 53 mJ/cm². The most common dye for microdroplet lasing is R6G because of its high quantum efficiency. Early demonstrations produced some relatively high lasing thresholds with falling R6G-doped droplets. For instance, Lin et al. (1986) observed 10⁴ W/cm² using a pulsed pump system and Tzeng et al. (1984) observed 35 W/cm² using a CW pump system. Levitated droplets have shown lasing thresholds of 500 mW using R6G (Azzouz et al. 2006). A similar dye, Rhodamine B, has been used with a pulsed pump laser, exhibiting a lasing threshold of 750 J/cm² in 20 μm droplets (Kiraz et al. 2007b).

Microdroplets have also been explored for Raman lasers. As discussed in Section 11.2.3, weak Raman signals can be amplified by the WGM in a ring resonator. Raman lasing has been demonstrated in microdroplets using strong Raman bands within glycerol (Sennaroglu et al. 2007). Stimulated Raman scattering (SRS), a similar coherent phenomenon, has been demonstrated many times with microdroplets (Qian et al. 1985, Snow et al. 1985, Qian and Chang 1986, Kwok and Chang 1993, Lin and Campillo 1997, Matsko et al. 2003, Lin and Campillo 2004). These experiments are largely interesting for purposes relating to basic research, but also have the potential to eventually provide versatile customizable light sources.

11.2.4.3 Cylinder and Microcapillary

Cylindrical ring resonators bring the same benefits to microfluidic lasers that they do to biological and chemical sensing. With hollow capillaries, dye media can be passed through the center of the capillary continuously, preventing intensity drops due to photobleaching. Additionally, since the dye is flowed through the hollow core, the laser energy can be coupled out via the external capillary surface using a fiber taper (Shopova et al. 2007b) or angle-polished fiber prism (Suter et al. 2008a). This is particularly important because without such outcoupling ability, the laser energy generated by the resonator will be scattered tangentially away and will be much less useful for any practical purposes.

Cylinders can also be used for microfluidic lasers. As with microspheres, a solid cylinder (essentially an optical fiber) must be immersed in the gain medium. An interesting idea involving cylinders requires threading a 125 μm silica fiber inside of a capillary and filling the gap space with R6G in ethanol (RI of ethanol = 1.361) (Moon et al. 2000). This configuration is displayed in Figure 11.24a. The evanescent field from the inner fiber is used to create the evanescent field for gain. The lasing threshold observed for this system was 200 μJ with a 5 ns pulsed laser source, and looks like the photograph in Figure 11.24b.

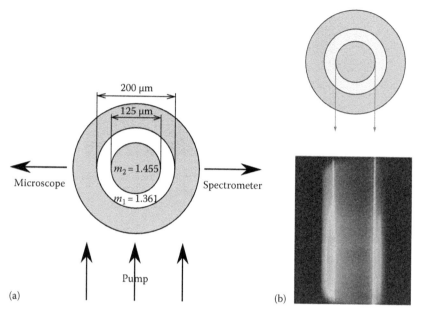

FIGURE 11.24 (a) Capillary with axial silica rod and ethanol-filled fluid cavity with R6G and (b) a photograph of WGM emission. (Reprinted from Moon, H.-J. et al., *Phys. Rev. Lett.*, 85, 3161, 2000. With permission. Copyright 2009, American Institute of Physics.)

Very low lasing thresholds down to $25\,nJ/mm^2$ have been demonstrated with thin-walled capillaries using R6G in ethanol (Lacey et al. 2007). In the same manuscript, it was also demonstrated that capillaries can operate in two different lasing modes depending on the RI of the dye solvent, providing the potential for versatile RI selection. In the case that the RI of the solvent is less than that of the capillary wall (usually around 1.45), the WGM exists almost entirely within the glass wall. In the case that the RI of the solvent is greater, however, such as in the case of solvents like quinoline (RI = 1.626), the WGMs exist almost entirely within the fluid core. Knight et al. (1992) were the first to observe cylindrical WGM lasing with R6G in quinoline using a thick-walled capillary. Several subsequent studies pursued this strategy as well (Knight et al. 1994). A similar study, also using R6G in quinoline, was conducted by Moon et al. (2000), which examined how decreasing the RI of the solvent inside the capillary caused blue shifts in the WGM lasing spectrum. It is possible to obtain the same result using PDMS cavities with a quinoline dye solvent (Suter et al. 2008a).

Thin-walled capillary microlasers can be used as intracavity biosensors. Using an R6G donor and an LDS 722 acceptor, Shopova et al. (2007a) created a FRET laser. A cascaded FRET laser was also demonstrated using Coumarin 480 as the first donor in sequence with R6G and LDS 722. Also, a QD was used as a donor to excite Nile Blue lasing in the same cavity (Shopova et al. 2007a). These demonstrations comprise one of the few efforts to date in the direction of microfluidic laser-based biosensing.

11.2.4.4 Planar Ring Resonators

Planar ring resonators are a relatively uncommon form of microfluidic laser. They are innately well suited for handling dye solution because it can be flowed over the ring surface without disturbing alignments or distorting structures. Microfluidic lasers have been demonstrated with fluid channels fabricated in PDMS polymer (Li et al. 2007). As long as the fluid RI is maintained above that of PDMS, the structure works as a ring resonator. In this case, the channels were filled with a methanol/ethylene glycol mixture, which had an RI just large enough to confine WGMs (RI = 1.409). The channels were pumped with a pulsed laser resulting in a lasing threshold of $9\,nJ/pulse$.

FIGURE 11.25 Microfluidic laser from 450 μm microknot immersed in an R6G solution. (Reprinted from Jiang, X. et al., *Appl. Phys. Lett.*, 90, 233501, 2007. With permission. Copyright 2009, American Institute of Physics.)

11.2.4.5 Microfiber Loops

The microfiber loop is a relatively new form of ring resonator, and as such, is still being developed toward its full potential. However, it has been proven to be quite functional as a microfluidic laser. Because the ring resonator (i.e., the loop) is a part of the fiber, the lasing signal supported by the ring resonator can be coupled out directly into the fiber without additional optical couplers such as fiber tapers, fiber prisms, or waveguides.

Jiang et al. (2007) immersed a microknot into an R6G solution and launched a pulsed pump signal through it. The evanescent field from the fiber excited the surrounding dye, which then coupled into the ring's WGMs. The laser signal generated from the WGM had a threshold of 9.2 μJ/pulse. The resulting system can be seen in Figure 11.25. The arrows in this figure represent the direction taken by the pump light and the resultant laser signal.

To summarize, Table 11.2 gives a quick review of some of the optofluidic ring resonator laser technologies discussed in the preceding section. They can be compared based on the gain medium used for lasing as well as the threshold achieved. Note that CW mode operation is quite difficult to achieve due to its relatively low average power and is thus rather rare in the literature.

11.2.5 Other Applications

There are a few other interesting applications that have not been covered in-depth in this chapter. In one instance, fluid can be used as a cladding for rings, including planar configurations (Levy et al. 2006). This is accomplished by changing the fluid running on top of the ring. As a result, the resonant wavelength and the coupling efficiency between the ring resonator and the adjacent waveguide can be controllably changed, which is important in many reconfigurable photonic devices. Other advanced application topics of study utilizing optofluidic integrated ring resonators include Raman spectroscopy (Symes and Reid 2006) and two-photon phenomena (Krioukov et al. 2005). While the scope of this chapter does not allow an in-depth treatment of these subjects, they are very interesting future directions for optofluidics.

11.3 Conclusion

In this chapter, we have provided an overview of the recent progress in the emerging field of optofluidic ring resonators. By confining light in a very small volume, optofluidic ring resonators serve as extremely versatile instruments, especially for the purposes of sensing and microfluidic lasers. As sensors, optofluidic ring resonators are compact and scalable to an array format, and have competitive LOD (potentially down to single-molecule detection) and extremely low sample volume (picoliter to nanoliter). As microfluidic lasers, optofluidic ring resonators provide excellent optical feedback resulting from their high Q-factors. Furthermore, they offer wide spectral coverage and tunability, and therefore, can be a very promising technology platform for development of micrototal-analysis systems.

In addition to sensors and microfluidic lasers, the optofluidic ring resonator technology has even broader implications in various optofluidic linear/nonlinear and active/passive photonic components/ systems. It can also be used for development of novel analytical chemical instruments in gas/liquid chromatography and electrophoresis. Additionally, the optofluidic ring resonator can serve as an excellent model system for fundamental research in cavity QED, thresholdless lasers, ultrafast lasers, single-photon emitters, single-molecule detection, high-resolution spectroscopy, and nonlinear optics. It can be envisioned at this point that important improvements, additional applications, and even commercialized products will emerge in this field within only the next few years.

Future development of these ideas will require significant improvement in microfluidics and optics design and manufacture. Primarily, this will involve better and more complete integration of optofluidic devices into monolithic structures as some components remain necessarily external, a fact that limits miniaturization and mass production. These improvements will be manifested in three general directions. First, microfluidics will continue to be developed to enhance optical functionality and improve sensing performance. Second, fluidic components such as pumps, valves, and mixers need to be almost entirely integrated to enhance the fluidic handling capability. Finally, whenever possible, light sources and detectors must be integrated on the same platform as the optofluidics device. Such integration will completely eliminate the need for external bulky optical devices, an important step toward the realization of a true lab-on-a-chip or micrototal analysis system.

Acknowledgments

This work is partially supported by the Wallace H. Coulter Foundation Early Career Award, NSF (ECCS-0729903 and CBET-0747398), and the NIH Biodetectives Training Grant.

References

Abel, A.P., Weller, M.G., Duveneck, G.L., Ehrat, M., and Widmer, H.M., 1996. Fiber-optic evanescent wave biosensor for the detection of oligonucleotides. *Analytical Chemistry* 68: 2905–2912.

Almeida, V.R., Barrios, C.A., Panepucci, R.R. et al., 2004a. All-optical switching on a silicon chip. *Optics Letters* 29: 2867–2869.

Almeida, V.R., Xu, Q., Barrios, C.A., and Lipson, M., 2004b. Guiding and confining light in void nano-structure. *Optics Letters* 29: 1209–1211.

An, K. and Moon, H.-J., 2003. Laser oscillations with pumping-independent ultrahigh cavity quality factors in evanescent-wave-coupled-gain microsphere dye lasers. *Journal of the Physical Society of Japan* 72: 773–776.

Armani, A.M. and Vahala, K.J., 2006. Heavy water detection using ultra-high-Q microcavities. *Optics Letters* 31: 1896–1898.

Armani, A.M., Kippenberg, T.J., Spillane, S.M., and Vahala, K.J., 2003. Ultra-high-Q toroid microcavity on a chip. *Nature* 421: 925–928.

Armani, A.M., Armani, D.K., Min, B., Vahala, K.J., and Spillane, S.M., 2005. Ultra-high-Q microcavity operation in H_2O and D_2O. *Applied Physics Letters* 87: 151118.

Armani, A.M., Kulkarni, R.P., Fraser, S.E., Flagan, R.C., and Vahala, K.J., 2007a. Label-free, single-molecule detection with optical microcavities. *Science* 317: 783–787.

Armani, A.M., Srinivasan, A., and Vahala, K.J., 2007b. Soft lithographic fabrication of high Q polymer microcavity arrays. *Nano Letters* 7: 1823–1826.

Armenise, M.N., Passaro, V.M.N., De Leonardis, F., and Armenise, M., 2001. Modeling and design of a novel miniaturized integrated optical sensor for gyroscope systems. *Journal of Lightwave Technology* 19: 1476–1494.

Arnold, S. and Folan, L.M., 1986. Fluorescence spectrometer for a single electrodynamically levitated microparticle. *Review of Scientific Instruments* 57: 2250–2253.

Arnold, S., Khoshsima, M., Teraoka, I., Holler, S., and Vollmer, F., 2003. Shift of whispering-gallery modes in microspheres by protein adsorption. *Optics Letters* 28: 272–274.

Ashkenazi, S., Chao, C.-Y., Guo, L.J., and O'Donnell, M., 2004. Ultrasound detection using polymer microring optical resonator. *Applied Physics Letters* 85: 5418–5420.

Ausman, L.K. and Schatz, G.C., 2008. Whispering-gallery mode resonators: Surface enhanced Raman scattering without plasmons. *The Journal of Chemical Physics* 129: 054704.

Azzouz, H., Alkhafadiji, L., Balslev, S. et al., 2006. Levitated droplet dye laser. *Optics Express* 14: 4374–4379.

Baehr-Jones, T., Hochberg, M., Walker, C., and Scherer, A., 2004. High-Q ring resonators in thin silicon-on-insulator. *Applied Physics Letters* 85: 3346–3347.

Baehr-Jones, T., Hochberg, M., Walker, C., and Scherer, A., 2005. High-Q optical resonators in silicon-on-insulator-based slot waveguides. *Applied Physics Letters* 86: 081101.

Baer, T., 1987. Continuous-wave laser oscillation in a Nd:YAG sphere. *Optics Letters* 12: 392–394.

Barnes, M.D., Ng, K.C., Whitten, W.B., and Ramsey, J.M., 1993. Detection of single Rhodamine 6G molecules in levitated microdroplets. *Analytical Chemistry* 65: 2360–2365.

Barrios, C.A., Gylfason, K.B., Sanchez, B. et al., 2007. Slot-waveguide biochemical sensor. *Optics Letters* 32: 3080–3082.

Barrios, C.A., Banuls, M.J., Gonzalez-Pedro, V. et al., 2008a. Label-free optical biosensing with slot-waveguides. *Optics Letters* 33: 708–710.

Barrios, C.A., Gylfason, K.B., Sanchez, B. et al., 2008b. Slot-waveguide biochemical sensor: Erratum. *Optics Letters* 33: 2554–2555.

Becker, R.C. and Spencer, F.A., 1998. Thrombin: Structure, biochemistry, measurement, and status in clinical medicine. *Journal of Thrombosis and Thrombolysis* 5: 215–229.

Bernardi, A., Kiravittaya, S., Rastelli, A. et al., 2008. On-chip Si/SiO$_x$ microtube refractometer. *Applied Physics Letters* 93: 094106.

Blair, S. and Chen, Y., 2001. Resonant-enhanced evanescent-wave fluorescence biosensing with cylindrical optical cavities. *Applied Optics* 40: 570–582.

Bohren, C.F. and Huffman, D.R. (Eds.), 1983. *Absorption and Scattering of Light by Small Particles*. New York: Wiley.

Boyd, R.W. and Heebner, J.E., 2001. Sensitive disk resonator photonic biosensor. *Applied Optics* 40: 5742–5747.

Boyd, R.W., Heebner, J.E., Lepeshkin, N.N. et al., 2003. Nanofabrication of optical structures and devices for photonics and biophotonics. *Journal of Modern Optics* 50: 2543–2550.

Cai, M., Painter, O., and Vahala, K.J., 2000. Observation of critical coupling in a fiber taper to a silica-microsphere whispering-gallery mode system. *Physical Review Letters* 85: 74–77.

Caspar, C. and Bachus, E.-J., 1989. Fibre-optic micro ring-resonator with 2 mm diameter. *Electronics Letters* 25: 1506–1508.

Chao, C.-Y. and Guo, L.J., 2002. Polymer microring resonators fabricated by nanoimprint technique. *Journal of Vacuum Science and Technology B* 20: 2862–2866.

Chao, C.-Y. and Guo, L.J., 2003. Biochemical sensors based on polymer microrings with sharp asymmetrical resonance. *Applied Physics Letters* 83: 1527–1529.

Chao, C.-Y. and Guo, L.J., 2004. Reduction of surface scattering loss in polymer microrings using thermal-reflow technique. *IEEE Photonics Technology Letters* 16: 1498–1500.

Chao, C.-Y. and Guo, L.J., 2006. Design and optimization of microring resonators in biochemical sensing applications. *Journal of Lightwave Technology* 24: 1395–1402.

Chao, C.-Y., Fung, W., and Guo, L.J., 2006. Polymer microring resonators for biochemical sensing applications. *IEEE Journal of Selected Topics in Quantum Electronics* 12: 134–142.

Chern, G.D., Tureci, H.E., Stone, A.D., and Chang, R.K., 2003. Unidirectional lasing from InGaN multiple-quantum-well spiral-shaped micropillars. *Applied Physics Letters* 83: 1710–1712.

Cheung, J.L., Kwok, A.S., Juvan, K.A., Leach, D.H., and Chang, R.K., 1993. Stimulated low-frequency emission from anisotropic molecules in microdroplets. *Chemical Physics Letters* 213: 309–314.

Chu, S.T., Little, B.E., Pan, W., Kaneko, T., and Kokubun, Y., 1999. Cascaded microring resonators for crosstalk reduction and spectrum cleanup in add-drop filters. *IEEE Photonics Technology Letters* 11: 1423–1425.

Cohen, D.A., Hossein-Zadeh, M., and Levi, A.F.J., 2001. Microphotonic modulator for microwave receiver. *Electronics Letters* 37: 300–301.

Culverhouse, D., Kalli, K., and Jackson, D.A., 1991. Stimulated Brillouin scattering ring resonator laser for SBS gain studies and microwave generation. *Electronics Letters* 27: 2033–2035.

Dai, D. and He, S., 2007. Highly-sensitive sensor with large measurement range realized with two cascaded-microring resonators. *Optics Communications* 279: 89–93.

Dai, D. and He, S., 2008. Low-index-material-based nano-slot waveguide with quasi-Bragg-reflector buffer. *Electronics Letters* 44: 1354–1356.

Dai, D., Yang, L., and He, S., 2008. Ultrasmall thermally tunable microring resonator with a submicrometer heater on Si nanowires. *Journal of Lightwave Technology* 26: 704–709.

De Vos, K., Bartolozzi, I., Schacht, E., Bienstman, P., and Baets, R., 2007. Silicon-on-insulator microring resonator for sensitive and label-free biosensing. *Optics Express* 15: 7610–7615.

Del'Haye, P., Schliesser, A., Arcizet, O. et al., 2007. Optical frequency comb generation from a monolithic microresonator. *Nature* 450: 1214–1217.

Djordjev, K., Choi, S.-J., Choi, S.-J., and Dapkus, P.D., 2002. Microdisk tunable resonant filters and switches. *IEEE Photonics Technology Letters* 14: 828–830.

Dorrer, C. and Ruhe, J., 2006. Advancing and receding motion of droplets on ultrahydrophobic post surfaces. *Langmuir* 22: 7652–7657.

Fan, X., White, I.M., Shopova, S.I. et al., 2008. Sensitive optical biosensors for unlabeled targets: A review. *Analytica Chimica Acta* 620: 8–26.

Farca, G., Shopova, S.I., and Rosenberger, A.T., 2007. Cavity-enhanced laser absorption spectroscopy using microresonator whispering-gallery modes. *Optics Express* 15: 17443–17448.

Francois, A. and Himmelhaus, M., 2009. Whispering gallery mode biosensor operated in the stimulated emission regime. *Applied Physics Letters* 94: 031101.

Francois, A., Krishnamoorthy, S., and Himmelhaus, M., 2008. Advances in label-free optical biosensing: Direct comparison of whispering gallery mode sensors with surface plasmon resonance. *Single Molecule Spectroscopy and Imaging*. SPIE, San Jose, CA.

Frolov, S.V., Shkunov, M., Vardeny, Z.V., and Yoshino, K., 1997. Ring microlasers from conducting polymers. *Physical Review B: Condensed Matter and Materials Physics* 56: R4363–R4366.

Garrett, C.G.B., Kaiser, W., and Bond, W.L., 1961. Stimulated emission into optical whispering modes of spheres. *Physical Review* 124: 1807–1809.

Gastine, M., Courtois, L., and Dormann, J.L., 1967. Electromagnetic resonances of free dielectric spheres. *IEEE Transactions on Microwave Theory and Techniques* 15: 694–700.

Ghulinyan, M., Navarro-Urrios, D., Pitanti, A. et al., 2008. Whispering-gallery modes and light emission from a Si-nanocrystal-based single microdisk resonator. *Optics Express* 16: 13218.

Gorodetsky, M.L. and Ilchenko, V.S., 1999. Optical microsphere resonators: Optimal coupling to high-Q whispering-gallery modes. *Journal of the Optical Society of America B* 16: 147–154.

Grillet, C., Bian, S.N., Magi, E.C., and Eggleton, B.J., 2008. Fiber taper coupling to chalcogenide microsphere modes. *Applied Physics Letters* 92: 171109.

Han, M. and Wang, A., 2007. Temperature compensation of optical microresonators using a surface layer with negative thermo-optic coefficient. *Optics Letters* 32: 1800–1802.

Hanumegowda, N.M., Stica, C.J., Patel, B.C., White, I.M., and Fan, X., 2005a. Refractometric sensors based on microsphere resonators. *Applied Physics Letters* 87: 201107.

Hanumegowda, N.M., White, I.H., Oveys, H., and Fan, X., 2005b. Label-free protease sensors based on optical microsphere resonators. *Sensor Letters* 3: 1–5.

Hanumegowda, N.M., White, I.M., and Fan, X., 2006. Aqueous mercuric ion detection with microsphere optical ring resonator sensors. *Sensors and Actuators B* 120: 207–212.

He, L., Xiao, Y.-F., Dong, C. et al., 2008. Compensation of thermal refraction effect in high-Q toroidal microresonator by polydimethylsiloxane coating. *Applied Physics Letters* 93: 201102.

Heebner, J.E., Boyd, R.W., and Park, Q.-H., 2002. Slow light, induced dispersion, enhanced nonlinearity, and optical solitons in a resonator-array waveguide. *Physical Review E* 65: 036619.

Hon, N.K. and Poon, A.W., 2007. Surface plasmon resonance-assisted coupling to whispering-gallery modes in micropillar resonators. *Journal of the Optical Society of America B* 24: 1981–1986.

Hopkins, R.J. and Reid, J.P., 2006. A comparative study of the mass and heat transfer dynamics of evaporating ethanol/water, methanol/water, and 1-propanol/water aerosol droplets. *Journal of Physical Chemistry B* 110: 3239–3249.

Hopkins, R.J., Mitchem, L., Ward, A.D., and Reid, J.P., 2004. Control and characterization of a single aerosol droplet in a single-beam gradient-force optical trap. *Physical Chemistry Chemical Physics* 6: 4924–4927.

Horowitz, V.R., Awschalom, D.D., and Pennathur, S., 2008. Optofluidics: Field or technique? *Lab on a Chip* 8: 1856–1863.

Hossein-Zadeh, M. and Vahala, K.J., 2006. Fiber-taper coupling to Whispering-Gallery modes of fluidic resonators embedded in a liquid medium. *Optics Express* 14: 10800–10810.

Hua, F., Sun, Y., Guar, A. et al., 2004. Polymer imprint lithography with molecular-scale resolution. *Nano Letters* 4: 2467–2471.

Huston, A.L. and Eversole, J.D., 1993. Strain-sensitive elastic scattering from cylinders. *Optics Letters* 18: 1104–1106.

Ilchenko, V.S. and Matsko, A.B., 2006. Optical resonators with whispering-gallery modes—Part II: Applications. *IEEE Journal of Selected Topics in Quantum Electronics* 12: 15–32.

Ilchenko, V.S., Yao, X.S., and Maleki, L., 1999. Pigtailing the high-Q microsphere cavity: A simple fiber coupler for optical whispering-gallery modes. *Optics Letters* 24: 723–725.

Ilchenko, V.S., Gorodetsky, M.L., Yao, X.S., and Maleki, L., 2001. Microtorus: A high-finesse microcavity with whispering-gallery modes. *Optics Letters* 26: 256–258.

Jiang, X., Tong, L., Vienne, G. et al., 2006. Demonstration of optical microfiber knot resonators. *Applied Physics Letters* 88: 223501.

Jiang, X., Song, Q., Xu, L., Fu, J., and Tong, L., 2007. Microfiber knot dye laser based on the evanescent-wave-coupled gain. *Applied Physics Letters* 90: 233501.

Keng, D., McAnanama, S.R., Teraoka, I., and Arnold, S., 2007. Resonance fluctuations of a whispering gallery mode biosensor by particles undergoing Brownian motion. *Applied Physics Letters* 91: 103902.

Kim, W., Safonov, V.P., Shalaev, V.M., and Armstrong, R.L., 1999. Fractals in microcavities: Giant coupled, multiplicative enhancement of optical responses. *Physical Review Letters* 82: 4811–4814.

Kim, G.-D., Son, G.-S., Lee, H.-S., Kim, K.-D., and Lee, S.-S., 2008. Integrated photonic glucose biosensor using a vertically coupled microring resonator in polymers. *Optics Communications* 281: 4644–4647.

Kipp, T., Welsch, H., Strelow, C., Heyn, C., and Heitmann, D., 2006. Optical modes in semiconductor microtube ring resonators. *Physical Review Letters* 96: 077403.

Kippenberg, T.J., Spillane, S.M., Armani, D.K., and Vahala, K.J., 2003. Fabrication and coupling to planar high-Q silica disk microcavities. *Applied Physics Letters* 83: 797–799.

Kippenberg, T.J., Spillane, S.M., and Vahala, K.J., 2004. Demonstration of ultra-high-Q small mode volume toroid microcavities on a chip. *Applied Physics Letters* 85: 6113–6115.

Kippenberg, T.J., Rokhsari, H., Carmon, T., Scherer, A., and Vahala, K.J., 2005. Analysis of radiation-pressure induced mechanical oscillation of an optical microcavity. *Physical Review Letters* 95: 033901.

Kippenberg, T.J., Kalkman, J., Polman, A., and Vahala, K.J., 2006. Demonstration of an erbium-doped microdisk laser on a silicon chip. *Physical Review A* 74: 051802.

Kiraz, A., Doganay, S., Kurt, A., and Demirel, A.L., 2007a. Enhanced energy transfer in single glycerol/water microdroplets standing on a superhydrophobic surface. *Chemical Physics Letters* 444: 181–185.

Kiraz, A., Sennaroglu, A., Doganay, S. et al., 2007b. Lasing from single, stationary, dye-doped glycerol/water microdroplets located on a superhydrophobic surface. *Optics Communications* 276: 145–148.

Kiraz, A., Karadag, Y., and Coskun, A.F., 2008. Spectral tuning of liquid microdroplets standing on a superhydrophobic surface using electrowetting. *Applied Physics Letters* 92: 191104.

Klug, M., Kablukov, S.I., and Wellegehausen, B., 2005. Cw hyper-Raman laser and four-wave mixing in atomic sodium. *Optics Communications* 245: 415–424.

Knight, J.C., Driver, H.S.T., Hutcheon, R.J., and Robertson, G.N., 1992. Core-resonance capillary-fiber whispering-gallery-mode laser. *Optics Letters* 17: 1280–1282.

Knight, J.C., Driver, H.S.T., and Robertson, G.N., 1994. Morphology-dependent resonances in a cylindrical dye microlaser: Mode assignments, cavity Q values, and critical dye concentrations. *Journal of the Optical Society of America B* 11: 2046–2053.

Krioukov, E., Klunder, D.J.W., Driessen, A., Greve, J., and Otto, C., 2002. Integrated optical microcavities for enhanced evanescent-wave spectroscopy. *Optics Letters* 27: 1504–1506.

Krioukov, E., Greve, J., and Otto, C., 2003. Performance of integrated optical microcavities for refractive index and fluorescence sensing. *Sensors and Actuators B* 90: 58–67.

Krioukov, E., Klunder, D.J.W., Driessen, A., Greve, J., and Otto, C., 2005. Two-photon fluorescence excitation using an integrated optical microcavity: A promising tool for biosensing of natural chromophores. *Talanta* 65: 1086–1090.

Ksendzov, A. and Lin, Y., 2005. Integrated optics ring-resonator sensors for protein detection. *Optics Letters* 30: 3344–3346.

Ksendzov, A., Homer, M.L., and Manfreda, A.M., 2004. Integrated optics ring-resonator chemical sensor with polymer transduction layer. *Electronics Letters* 40: 63–65.

Kulkarni, R.P., Fraser, S.E., and Armani, A.M., 2008. Characterization of high-Q optical microcavities using confocal microscopy. *Optics Letters* 33: 2931–2933.

Kwok, A.S. and Chang, R.K., 1993. Stimulated resonance Raman scattering of Rhodamine 6G. *Optics Letters* 18: 1703–1705.

Lacey, S., White, I.M., Sun, Y. et al., 2007. Versatile opto-fluidic ring resonator lasers with ultra-low threshold. *Optics Express* 15: 15523–15530.

Lee, H.-P., Park, J.-J., Ryoo, H.-H. et al., 2002. Resonance characteristics of waveguide-coupled polyimide microring resonator. *Optical Materials* 21: 535–541.

Lefevre-Seguin, V. and Haroche, S., 1997. Towards cavity-QED experiments with silica microspheres. *Materials Science and Engineering B* 48: 53–58.

Levy, U., Campbell, K., Groisman, A., Mookherjea, S., and Fainman, Y., 2006. On-chip microfluidic tuning of an optical microring resonator. *Applied Physics Letters* 88: 111107.

Li, Z., Zhang, Z., Scherer, A., and Psaltis, D., 2007. Optofluidic microring dye laser. *IEEE/LEOS Summer Topical Meetings*, Portland, OR, pp. 70–71. IEEE, Portland, OR.

Lin, H.-B. and Campillo, A.J., 1995. Radial profiling of microdroplets using cavity-enhanced Raman spectroscopy. *Optics Letters* 20: 1589–1591.

Lin, H.-B. and Campillo, A.J., 1997. Microcavity enhanced Raman gain. *Optics Communications* 133: 287–292.

Lin, H.-B. and Campillo, A.J., 2004. cw Nonlinear optics in droplet microcavities displaying enhanced gain. *Physical Review Letters* 73: 2440–2443.

Lin, H.-B., Huston, A.L., Justus, B.L., and Campillo, A.J., 1986. Some characteristics of a droplet whispering-gallery-mode laser. *Optics Letters* 11: 614–616.

Lin, H.-B., Eversole, J.D., and Campillo, A.J., 1992. Spectral properties of lasing microdroplets. *Journal of the Optical Society of America B* 9: 43–50.

Lin, H.-B., Eversole, J.D., Campillo, A.J., and Barton, J.P., 1998. Excitation localization principle for spherical microcavities. *Optics Letters* 23: 1921–1923.

Ling, T. and Guo, L.J., 2007. A unique resonance mode observed in a prism-coupled micro-tube resonator sensor with superior index sensitivity. *Optics Express* 15: 17424–17432.

Little, B.E., Laine, J.-P., Lim, D.R. et al., 2000. Pedestal antiresonant reflecting waveguides for robust coupling to microsphere resonators and for microphotonic circuits. *Optics Letters* 25: 73–75.

Little, B.E., Chu, S.T., Absil, P.P. et al., 2004. Very high-order microring resonator filters for WDM applications. *IEEE Photonics Technology Letters* 16: 2263–2265.

Liu, X., Fang, W., Huang, Y. et al., 2004. Optically pumped ultraviolet microdisk laser on a silicon substrate. *Applied Physics Letters* 84: 2488–2490.

Lukosz, W., 1995. Integrated optical chemical and direct biochemical sensors. *Sensors and Actuators B* 29: 37–50.

Luo, X. and Poon, A.W., 2007. Coupled spiral-shaped microdisk resonators with non-evanescent asymmetric inter-cavity coupling. *Optics Express* 15: 17313–17322.

Lutti, J., Langbein, W., and Borri, P., 2008. A monolithic optical sensor based on whispering-gallery modes in polystyrene microspheres. *Applied Physics Letters* 93: 151103.

Ma, X., Lu, J.Q., Brock, R.S. et al., 2003. Determination of complex refractive index of polystyrene microspheres from 370 to 1610 nm. *Physics in Medicine and Biology* 48: 4165–4172.

Madsen, C.K., Lenz, G., Bruce, A.J. et al., 1999. Integrated all-pass filters for tunable dispersion and dispersion slope compensation. *IEEE Photonics Technology Letters* 11: 1623–1625.

Martin, A.L., Armani, D.K., Yang, L., and Vahala, K.J., 2004. Replica-molded high-Q polymer microresonators. *Optics Letters* 29: 533–535.

Martinez, L. and Lipson, M., 2006. High confinement suspended micro-ring resonators in silicon-on-insulator. *Optics Express* 14: 6259–6263.

Matsko, A.B., Savchenkov, A.A., Letargat, R.J., Ilchenko, V.S., and Maleki, L., 2003. On cavity modification of stimulated Raman scattering. *Journal of Optics B: Quantum and Semiclassical Optics* 5: 272–278.

Mazzei, A., Gotzinger, S., Menezes, L.d.S., Sandoghdar, V., and Benson, O., 2005. Optimization of prism coupling to high-Q modes in a microsphere resonator using a near-field probe. *Optics Communications* 250: 428–433.

Meissner, K.E. and Allen, A., 2005. Whispering gallery mode biosensors using semiconductor quantum dots. *IEEE Sensors*, Irvine, CA, p. 4.

Mendach, S., Songmuang, R., Kiravittaya, S. et al., 2006. Light emission and wave guiding of quantum dots in a tube. *Applied Physics Letters* 88: 111120.

Michaels, A.M., Nirmal, M., and Brus, L.E., 1999. Surface enhanced Raman spectroscopy of individual rhodamine 6G molecules on large Ag nanocrystals. *Journal of the American Chemical Society* 121: 9932–9939.

Michler, P., Kiraz, A., Zhang, L. et al., 2000. Laser emission from quantum dots in microdisk structures. *Applied Physics Letters* 77: 184–186.

Monat, C., Domachuk, P., and Eggleton, B.J., 2007. Integrated optofluidics: A new river of light. *Nature Photonics* 1: 106–114.

Moon, H.-J., Chough, Y.-T., and An, K., 2000. Cylindrical microcavity laser based on the evanescent-wave-coupled gain. *Physical Review Letters* 85: 3161–3164.

Moon, H.-J., Park, G.-W., Lee, S.-B., An, K., and Lee, J.-H., 2004. Laser oscillations of resonance modes in a thin gain-doped ring-type cylindrical microcavity. *Optics Communications* 235: 401–407.

Mugele, F. and Baret, J.-C., 2005. Electrowetting: From basics to applications. *Journal of Physics: Condensed Matter* 17: R705–R774.

Nadeau, J.L., Ilchenko, V.S., Kossakovski, D., Bearman, G.H., and Maleki, L., 2002. High-Q whispering-gallery mode sensor in liquids. *Proceedings of SPIE* 4629: 172–180.

Nie, S. and Emory, S.R., 1997. Probing single molecules and single nanoparticles by surface-enhanced Raman scattering. *Science* 275: 1102–1106.

Nie, S., Chiu, D.T., and Zare, R.N., 1994. Probing individual molecules with confocal fluorescence microscopy. *Science* 266: 1018–1021.

Niehusmann, J., Vorckel, A., Bolivar, P.H. et al., 2004. Ultrahigh-quality-factor silicon-on-insulator microring resonator. *Optics Letters* 29: 2861–2863.

Nitkowski, A., Chen, L., and Lipson, M., 2008. Cavity-enhanced on-chip absorption spectroscopy using microring resonators. *Optics Express* 16: 11930–11936.

Norris, D.J., Kuwata-Gonokami, M., and Moerner, W.E., 1997. Excitation of a single molecule on the surface of a spherical microcavity. *Applied Physics Letters* 71: 297–299.

Noto, M., Keng, D., Teraoka, I., and Arnold, S., 2007. Detection of protein orientation on the silica microsphere surface using transverse electric/transverse magnetic whispering gallery modes. *Biophysical Journal* 92: 4466–4472.

Nuhiji, E. and Mulvaney, P., 2007. Detection of unlabeled oligonucleotide targets using whispering gallery modes in single, fluorescent microspheres. *Small* 3: 1408–1414.

Padigi, S.K., Asante, K., Kovvuri, V.S.R. et al., 2006. Micro-photonic cylindrical waveguide based protein biosensor. *Nanotechnology* 17: 4384–4390.

Passaro, V.M.N. and De Leonardis, F., 2006. Modeling and design of a novel high-sensitivity electric field silicon-on-insulator sensor based on a whispering-gallery-mode resonator. *IEEE Journal of Selected Topics in Quantum Electronics* 12: 124–133.

Passaro, V.M.N., Dell'Olio, F., and De Leonardis, F., 2007. Ammonia optical sensing by microring resonators. *Sensors* 7: 2741–2749.

Plowman, T.E., Reichert, W.M., Peters, C.R. et al., 1996. Femtomolar sensitivity using a channel-etched thin film waveguide fluoroimmunosensor. *Biosensors and Bioelectronics* 11: 149–160.

Prinz, V.Y., Seleznev, V.A., Gutakovsky, A.K. et al., 2000. Free-standing and overgrown InGaAs/GaAs nanotubes, nanohelices and their arrays. *Physica E* 6: 828–831.

Psaltis, D., Quake, S.R., and Yang, C., 2006. Developing optofluidic technology through the fusion of microfluidics and optics. *Nature* 442: 381–386.

Qian, S.-X. and Chang, R.K., 1986. Multiorder stokes emission from micrometer-size droplets. *Physical Review Letters* 56: 926–929.

Qian, S.-X., Snow, J.B., and Chang, R.K., 1985. Coherent Raman mixing and coherent anti-Stokes Raman scattering from individual micrometer-size droplets. *Optics Letters* 10: 499–501.

Qian, S.-X., Snow, J.B., Tzeng, H.-M., and Chang, R.K., 1986. Lasing droplets: Highlighting the liquid-air interface by laser emission. *Science* 231: 486–488.

Rahachou, A.I. and Zozoulenko, I.V., 2003. Effects of boundary roughness on a Q factor of whispering-gallery-mode lasing microdisk cavities. *Journal of Applied Physics* 94: 7929–7931.

Ramachandran, A., Wang, S., Clarke, J. et al., 2008. A universal biosensing platform based on optical micro-ring resonators. *Biosensors and Bioelectronics* 23: 939–944.

Ren, H.-C., Vollmer, F., Arnold, S., and Libchaber, A., 2007. High-Q microsphere biosensor—Analysis for adsorption of rodlike bacteria. *Optics Express* 15: 17410–17423.

Robinson, J.T., Chen, L., and Lipson, M., 2008. On-chip gas detection in silicon optical microcavities. *Optics Express* 16: 4296–4301.

Saito, M., Shimatani, H., and Naruhashi, H., 2008. Tunable whispering gallery mode emission from a microdroplet in elastomer. *Optics Express* 16: 11915–11919.

Sakai, Y., Yokohama, I., Kominato, T., and Sudo, S., 1991. Frequency stabilization of laser diode using a frequency-locked ring resonator to acetylene gas absorption lines. *IEEE Photonics Technology Letters* 3: 868–870.

Sandoghdar, V., Treussart, F., Hare, J. et al., 1996. Very low threshold whispering-gallery-mode microsphere laser. *Physical Review A* 54: R1777–R1780.

Sasagawa, K., Yonezawa, Z., Ohta, J., and Nunoshita, M., 2003. Control of microsphere lasing wavelength using $\lambda/4$-shifted distributed feedback resonator. *Electronics Letters* 39: 1817–1819.

Schafer, J., Mondia, J.P., Sharma, R. et al., 2008. Quantum dot microdrop laser. *Nano Letters* 8: 1709–1712.

Scheuer, J. and Yariv, A., 2003. Annular Bragg defect mode resonators. *Journal of the Optical Society of America B* 20: 2285–2291.

Scheuer, J., Green, W.M.J., DeRose, G.A., and Yariv, A., 2005. InGaAsP Annular Bragg lasers: Theory, applications, and modal properties. *IEEE Journal of Selected Topics in Quantum Electronics* 11: 476–484.

Scheuer, J., Green, W.M.J., and Yariv, A., 2006. Annular Bragg resonators (ABR)—The ideal tool for biochemical sensing, nonlinear optics and cavity QED. *Proceedings of SPIE* 6123: 234–245. SPIE, San Jose, CA.

Schiro, P.G. and Kwok, A.S., 2004. Cavity-enhanced emission from a dye-coated microsphere. *Optics Express* 12: 2857–2863.

Schmidt, O.G. and Eberl, K., 2001. Thin solid films roll up into nanotubes. *Nature* 410: 168.

Schweinsberg, A., Hocde, S., Lepeshkin, N.N. et al., 2007. An environmental sensor based on an integrated optical whispering gallery mode disk resonator. *Sensors and Actuators B* 123: 727–732.

Sennaroglu, A., Kiraz, A., Dundar, M.A., Kurt, A., and Demirel, A.L., 2007. Raman lasing near 630 nm from stationary glycerol-water microdroplets on a superhydrophobic surface. *Optics Letters* 32: 2197–2199.

Shaforost, E.N., Klein, N., Vitusevich, S.A., Offenhausser, A., and Barannik, A.A., 2008. Nanoliter liquid characterization by open whispering-gallery mode dielectric resonators at millimeter wave frequencies. *Journal of Applied Physics* 104: 074111.

Shopova, S.I., Cupps, J.M., Zhang, P. et al., 2007a. Opto-fluidic ring resonator lasers based on highly efficient resonant energy transfer. *Optics Express* 15: 12735–12742.

Shopova, S.I., Zhu, H., and Fan, X., 2007b. Optofluidic ring resonator based dye laser. *Applied Physics Letters* 90: 221101.

Shopova, S.I., Blackledge, C.W., and Rosenberger, A.T., 2008a. Enhanced evanescent coupling to whispering-gallery modes due to gold nanorods grown on the microresonator surface. *Applied Physics B* 93: 183–187.

Shopova, S.I., Sun, Y., Rosenberger, A.T., and Fan, X., 2008b. Highly sensitive tuning of coupled optical ring resonators by microfluidics. *Microfluidics and Nanofuidics* 6: 425–429.

Shopova, S.I., White, I.H., Sun, Y. et al., 2008c. On-column micro gas chromatography detection with capillary-based optical ring resonators. *Analytical Chemistry* 80: 2232–2238.

Smith, D.D., Chang, H., and Fuller, K.A., 2003. Whispering-gallery mode splitting in coupled microresonators. *Journal of the Optical Society of America B* 20: 1967–1974.

Snee, P.T., Chan, Y., Nocera, D.G., and Bawendi, M.G., 2004. Whispering-gallery-mode lasing from a semiconductor nanocrystal/microsphere resonator composite. *Advanced Materials* 17: 1131–1136.

Snow, J.B., Qian, S.-X., and Chang, R.K., 1985. Stimulated Raman scattering from individual water and ethanol droplets at morphology-dependent resonances. *Optics Letters* 10: 37–39.

Songmuang, R., Deneke, C., and Schmidt, O.G., 2006. Rolled-up micro- and nanotubes from single-material thin films. *Applied Physics Letters* 89: 223109.

Songmuang, R., Rastelli, A., Mendach, S., Deneke, C., and Schmidt, O.G., 2007. From rolled-up Si microtubes to SiO$_x$/Si optical ring resonators. *Microelectronic Engineering* 84: 1427–1430.

Stokes, L.F., Chodorow, M., and Shaw, H.J., 1982. All-single-mode fiber resonator. *Optics Letters* 7: 288–290.

Stone, J., 1972. Measurements of the absorption of light in low-loss liquids. *Journal of the Optical Society of America* 62: 327–333.

Sumetsky, M., 2004. Optical fiber microcoil resonator. *Optics Express* 12: 2303–2316.

Sumetsky, M., Dulashko, Y., Fini, J.M., and Hale, A., 2005. Optical microfiber loop resonator. *Applied Physics Letters* 86: 161108.

Sumetsky, M., Dulashko, Y., Fini, J.M., Hale, A., and DiGiovanni, D.J., 2006. The microfiber loop resonator: Theory, experiment, and application. *Journal of Lightwave Technology* 24: 242–250.

Sumetsky, M., Windeler, R.S., Dulashko, Y., and Fan, X., 2007. Optical liquid ring resonator sensor. *Optics Express* 15: 14376–14381.

Sun, Y. and Fan, X., 2008. Analysis of ring resonators for chemical vapor sensor development. *Optics Express* 16: 10254–10268.

Sun, Y., Shopova, S.I., Frye-Mason, G., and Fan, X., 2008. Rapid chemical-vapor sensing using optofluidic ring resonators. *Optics Letters* 33: 788–790.

Suter, J.D., White, I.M., Zhu, H., and Fan, X., 2007. Thermal characterization of liquid core optical ring resonator sensors. *Applied Optics* 46: 389–396.

Suter, J.D., Sun, Y., Howard, D.J., Viator, J.A., and Fan, X., 2008a. PDMS embedded opto-fluidic microring resonator lasers. *Optics Express* 16: 10248–10253.

Suter, J.D., White, I.M., Zhu, H. et al., 2008b. Label-free quantitative DNA detection using the liquid core optical ring resonator. *Biosensors and Bioelectronics* 23: 1003–1009.

Sutherland, R.M., Dahne, C., Place, J.F., and Ringrose, A.R., 1984. Immunoassays at a quartz-liquid interface: Theory, instrumentation and preliminary application to the fluorescent immunoassay of human immunoglobulin G. *Journal of Immunological Methods* 74: 253–265.

Symes, R. and Reid, J.P., 2006. Determining the composition of aqueous microdroplets with broad-band cavity enhanced Raman scattering. *Physical Chemistry Chemical Physics* 8: 293–302.

Symes, R., Sayer, R.M., and Reid, J.P., 2004. Cavity enhanced droplet spectroscopy: Principles, perspectives, and prospects. *Physical Chemistry Chemical Physics* 6: 474–487.

Symes, R., Gilham, R.J.J., Sayer, R.M., and Reid, J.P., 2005. An investigation of the factors influencing the detection sensitivity of cavity enhanced Raman scattering for probing aqueous binary aerosol droplets. *Physical Chemistry Chemical Physics* 7: 1414–1422.

Tanyeri, M. and Kennedy, I.M., 2008. Detecting single bacterial cells through optical resonances in microdroplets. *Sensor Letters* 6: 326–329.

Tanyeri, M., Dosev, D., and Kennedy, I.M., 2005. Chemical and biological sensing through optical resonances in pendant droplets. *Proceedings of the SPIE*, 6008: 60080Q.

Tanyeri, M., Perron, R., and Kennedy, I.M., 2007. Lasing droplets in a microfabricated channel. *Optics Letters* 32: 2529–2531.

Teraoka, I., Arnold, S., and Vollmer, F., 2003. Perturbation approach to resonance shifts of whispering-gallery modes in a dielectric microsphere as a probe of a surrounding medium. *Journal of the Optical Society of America B* 20: 1937–1946.

Tong, L., Gattass, R.R., Ashcom, J.B. et al., 2003. Subwavelength-diameter silica wires for low-loss optical wave guiding. *Nature* 426: 816–819.

Topolancik, J. and Vollmer, F., 2007. Photoinduced transformations in bacteriorhodopsin membrane monitored with optical microcavities. *Biophysical Journal* 92: 2223.

Tzeng, H.-M., Wall, K.F., Long, M.B., and Chang, R.K., 1984. Laser emission from individual droplets at wavelengths corresponding to morphology-dependent resonances. *Optics Letters* 9: 499–501.

Vahala, K.J., 2003. Optical microcavities. *Science* 424: 839–846.

Vernooy, D.W., Furusawa, A., Georgiades, N.P., Ilchenko, V.S., and Kimble, H.J., 1998. Cavity QED with high-Q whispering gallery modes. *Physical Review A* 57: R2293–R2296.

Vienne, G., Li, Y., and Tong, L., 2007. Effect of host polymer on microfiber resonator. *IEEE Photonics Technology Letters* 19: 1386–1388.

Vienne, G., Grelu, P., Pan, X., Li, Y., and Tong, L., 2008. Theoretical study of microfiber resonator devices exploiting a phase shift. *Journal of Optics A: Pure and Applied Optics* 10: 025303.

Vivien, L., Marris-Morini, D., Griol, A. et al., 2008. Vertical multiple-slot waveguide ring resonators in silicon nitride. *Optics Express* 16: 17237–17242.

Vollmer, F., Braun, D., Libchaber, A. et al., 2002. Protein detection by optical shift of a resonant microcavity. *Applied Physics Letters* 80: 4057–4059.

Vollmer, F., Arnold, S., Braun, D., Teraoka, I., and Libchaber, A., 2003. Multiplexed DNA quantification by spectroscopic shift of two microsphere cavities. *Biophysical Journal* 85: 1974–1979.

Vollmer, F., Arnold, S., and Keng, D., 2008. Single virus detection from the reactive shift of a whispering-gallery mode. *Proceedings of the National Academy of Sciences of the United States of America* 105: 20701–20704.

Wang, Y.Z., Lu, B.L., Li, Y.Q., and Liu, Y.S., 1995. Observation of cavity quantum-electrodynamic effects in a Nd: Glass microsphere. *Optics Letters* 20: 770–772.

Weller, A., Liu, F.C., Dahint, R., and Himmelhaus, M., 2008. Whispering gallery mode biosensors in the low-Q limit. *Applied Physics B* 90: 561–567.

White, I.M., Hanumegowda, N.M., and Fan, X., 2005. Subfemtomole detection of small molecules with microsphere sensors. *Optics Letters* 30: 3189–3191.

White, I.M., Oveys, H., and Fan, X., 2006a. Liquid-core optical ring-resonator sensors. *Optics Letters* 31: 1319–1321.

White, I.M., Oveys, H., and Fan, X., 2006b. Integrated multiplexed biosensors based on liquid core optical ring resonators and antiresonant reflecting optical waveguides. *Applied Physics Letters* 89: 191106.

White, I.M., Oveys, H., and Fan, X., 2006c. Increasing the enhancement of SERS with dielectric microsphere resonators. *Spectroscopy* 21: 36–42.

White, I.M., Gohring, J., and Fan, X., 2007a. SERS-based detection in an optofluidic ring resonator platform. *Optics Express* 15: 17433–17442.

White, I.M., Suter, J.D., Oveys, H., and Fan, X., 2007b. Universal coupling between metal-clad waveguides and optical ring resonators. *Optics Express* 15: 646–651.

White, I.M., Zhu, H., Suter, J.D. et al., 2007c. Refractometric sensors for lab-on-a-chip based on optical ring resonators. *IEEE Sensors Journal* 7: 28–35.

Wu, X., Li, H., Liu, L., and Xu, L., 2008. Unidirectional single-frequency lasing from a ring-spiral coupled microcavity laser. *Applied Physics Letters* 93: 081105.

Xia, Y., McClelland, J.J., Gupta, R. et al., 1997. Replica molding using polymeric materials: A practical step toward nanomanufacturing. *Advanced Materials* 9: 147–149.

Xu, F. and Brambilla, G., 2007a. Embedded optical microfiber coil resonator. *OECC/IOOC*, Yokohama, Japan.

Xu, F. and Brambilla, G., 2007b. Embedding optical microfiber coil resonators in teflon. *Optics Letters* 32: 2164–2166.

Xu, F. and Brambilla, G., 2007c. Manufacture of 3-D microfiber coil resonators. *IEEE Photonics Technology Letters* 19: 1481–1483.

Xu, F. and Brambilla, G., 2008. Demonstration of a refractometric sensor based on optical microfiber coil resonator. *Applied Physics Letters* 92: 101126.

Xu, D.-X., Densmore, A., Waldron, P. et al., 2007a. High bandwidth SOI photonic wire ring resonators using MMI couplers. *Optics Express* 15: 3149–3155.

Xu, F., Horak, P., and Brambilla, G., 2007b. Optical microfiber coil resonator refractometric sensor. *Optics Express* 15: 7888–7893.

Xu, D.-X., Densmore, A., Delage, A. et al., 2008a. Folded cavity SOI microring sensors for high sensitivity and real time measurement of biomolecular binding. *Optics Express* 16: 15137–15148.

Xu, F., Pruneri, V., Finazzi, V., and Brambilla, G., 2008b. An embedded optical nanowire loop resonator refractometric sensor. *Optics Express* 16: 1062–1067.

Yalcin, A., Popat, K.C., Aldridge, J.C. et al., 2006. Optical sensing of biomolecules using microring resonators. *IEEE Journal of Selected Topics in Quantum Electronics* 12: 148–155.

Yang, G., White, I.M., and Fan, X., 2008. An opto-fluidic ring resonator biosensor for the detection of organophosphorus pesticides. *Sensors and Actuators B* 133: 105–112.

Zamora, V., Diez, A., Andres, M.V., and Gimeno, B., 2008. Refractometric sensor based on whispering-gallery modes of thin capillaries. *Optics Express* 15: 12011–12016.

Zhu, H., Suter, J.D., White, I.M., and Fan, X., 2006. Aptamer based microsphere biosensor for thrombin detection. *Sensors* 6: 785–795.

Zhu, H., White, I.H., Suter, J.D., Dale, P.S., and Fan, X., 2007a. Analysis of biomolecule detection with optofluidic ring resonator sensors. *Optics Express* 15: 9139–9146.

Zhu, H., White, I.M., Suter, J.D., Zourob, M., and Fan, X., 2007b. Integrated refractive index optical ring resonator detector for capillary electrophoresis. *Analytical Chemistry* 79: 930–937.

Zhu, H., White, I.H., Suter, J.D., and Fan, X., 2008a. Phage-based label-free biomolecule detection in an opto-fluidic ring resonator. *Biosensors and Bioelectronics* 24: 461–466.

Zhu, H., White, I.H., Suter, J.D., Zourob, M., and Fan, X., 2008b. Opto-fluidic micro-ring resonator for sensitive label-free viral detection. *The Analyst* 133: 356–360.

Zijlstra, P., van der Molen, K.L., and Mosk, A.P., 2007. Spatial refractive index sensor using whispering gallery modes in an optically trapped microsphere. *Applied Physics Letters* 90: 161101.

12

Optofluidic Light Sources

Anders Kristensen
N. Asger Mortensen

12.1 Introduction

Lab-on-a-chip research and applications (Janasek et al. 2006) seek changes of paradigm through the miniaturization of chemical and biochemical analyses on microchips. Optics and optical probes have historically played an important role in analytical chemistry. In this context, optofluidics offers potential solutions through *optofluidic light sources*, where light is generated or manipulated in a microfluidic architecture (Psaltis et al. 2006, Monat et al. 2007).

Optofluidic light sources offer an attractive alternative to the otherwise inherent challenge of coupling light into integrated optics. Light sources, including lasers, can be integrated in the lab-on-a-chip platform by infiltrating dedicated microfluidic components with liquid light emitters such as liquid laser dye solutions. A number of such architectures for optofluidic light sources have been realized, and three examples are outlined in Figure 12.1, including the first demonstration Helbo et al. presented at the IEEE MEMS Conference in Kyoto, Japan, 2003 (Helbo et al. 2003a). As a common feature, the on-chip light source is basically added to the lab-on-a-chip microsystem without adding further steps in the fabrication procedure (Balslev et al. 2006). As an additional advantage, there are, by virtue, no alignment issues with the light sources. All optical components are defined in the same lithography process and benefit from the high placement accuracy of modern microlithography.

This chapter is intended to give an introduction to the optofluidic integration of light sources, particularly lasers, on the lab-on-a-chip platform.

12.2 Brief Review of the Laser Principle

LASER is an acronym for *light amplification by stimulated emission of radiation*. The stimulated photon emission supports a coherent population of photon states, and is thereby the underlying physical mechanism for achieving the characteristics of output emission from a laser: monochromaticity and coherence. This is in contrast to other light sources such as Edison's glow bulb and light-emitting diodes

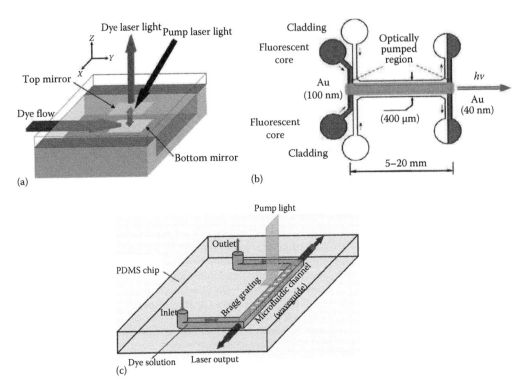

FIGURE 12.1 Examples of architectures for optofluidic light sources. Panel (a) shows the laser reported by Helbo et al. (2003a,b, 2004) (Reprinted from Helbo, B. et al., *Sens. Actuator A Phys.*, 111, 21, 2004. With permission. Copyright 2009 from Elsevier.), panel (b) shows the implementation by Vezenov et al. (2005a) (Reprinted from Vezenov, D.V. et al., *J. Am. Chem. Soc.*, 127, 8952, 2005a. With permission. Copyright 2009 from American Chemical Society.), and panel (c) shows the work by Li et al. (2006a) (Reprinted from Li, Z.Y. et al., *Opt. Express*, 14, 696, 2006a. With permission. Copyright 2009 from Optical Society of America.)

(LED) that emit noncoherent light, which has a phase that varies randomly with time and position, in a wide band of frequencies.

The laser can at the simplest be described as a feedback coupled amplifier for photons, as illustrated in Figure 12.2a, which to first order oscillates at a resonance frequency determined by the spectral properties of the feedback and gain medium. This is quite analogous to feedback in audio systems, for example, when the output from the loudspeakers is picked up in the microphone and fed back to the amplifier. Both the amplitude and the frequency of these oscillations are very sensitive to the feedback conditions, for example, by modifying the amplification or changing the microphone position. In lasers, the input power, I_0, can be provided by different means, but most commonly the energy is supplied by an electrical current or by the absorption of incident photons. These types of lasers are referred to as electrically or optically pumped devices, respectively. The optical feedback in lasers is generally provided by classical optical components like mirrors and gratings, see Figure 12.2b, while the amplification relies on quantum physics and the statistical properties of photons. The optical gain and amplification in the laser medium is obtained through the process of stimulated emission of photons, where existing photons stimulate the emission of further "identical" photons having the same frequency and phase properties. Through the optical feedback in combination with the emission spectrum of the gain medium, a narrowband photon population with a common phase is built up in the optical resonator. A fraction of these photons are extracted from the cavity, for example, through a minute transmittance through one of the mirrors, to form monochromatic and coherent laser emissions. For more details, see Chapter 5 in this text or for a thorough introduction to laser physics, we refer to Svelto's classical textbook in (Svelto 1998).

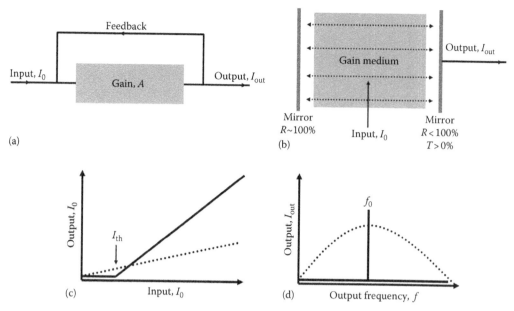

FIGURE 12.2 The laser can be described as a feedback coupled amplifier for photons, see panel (a). The amplification provided by a gain medium inserted in an optical resonator, for example formed by two mirrors, see panel (b). Panel (c) outlines the input–output characteristics of a laser (solid line) with a characteristic threshold, I_{th} for onset of lasing, opposed to the behavior in the absence of lasing, e.g., florescence (dashed line). Panel (d) shows the corresponding characteristic narrow-line emission spectrum from a laser (solid line) compared to the broad emission spectrum of traditional light source.

As illustrated in Figure 12.2c and d, the output from a laser has several profound signatures from the underlying mechanism of stimulated spontaneous emission, which distinguish the laser (solid lines) from classical incoherent light sources (dotted lines). In the spectral response, the laser ultimately manifests itself by emitting one color at a well-defined frequency f_0. This is referred to as monochromaticity, as opposed to multicolor or spectrally broad emission from conventional sources. Also, the output power I_{out} of a laser exhibits a very characteristic, sudden increase in I_{out}, when the input power I_0 reaches a threshold, I_{th}. The obvious interpretation is that the threshold, I_{th}, marks the onset of lasing corresponding to a situation where the round-trip gain experienced by the photons overcomes the round-trip cavity losses, leading to a positive net feedback, and stimulated emission of further photons. Above the lasing threshold, the slope $\partial I_{out}/\partial I_0$ attains a finite value describing the efficiency by which pump energy is converted into laser output energy.

12.3 Wavelength Ranges of Interest for Sensing

Formally, the terminology of *fluids* includes both viscous *liquids* and less viscous and dilute *gases*. However, in the context of lab-on-a-chip applications the majority of applications involve the detection and quantification of chemicals and chemical reactions occurring in the liquid phase. Typically, liquid buffers will have salinities resembling biological liquids. In terms of optical properties, ordinary water is quite representative of many buffers. Figure 12.3 illustrates water's properties for a range of optical wavelengths. The polar nature of the water molecules causes a refractive index close to that of vacuum ($n \sim 1$) at low wavelengths (corresponding to high frequencies), where the water molecules respond so slowly to the rapidly varying electromagnetic field that they only cause a modest change in refractive index compared to vacuum. At longer wavelengths, the refractive index increases, attaining a value of around 1.33 in the visible regime. More importantly, the absorption varies strongly and while water

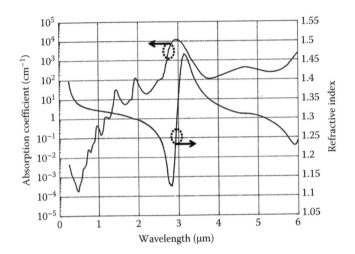

FIGURE 12.3 Spectral dependence of absorption (left vertical axis) and the refractive index (right vertical axis) of water. (Reprinted from Hillman, T.R. and Sampson, D.D., *Opt. Express*, 13, 1860, 2005. With permission. Copyright 2009 from Optical Society of America.)

is transparent in the visible regime, the liquid rapidly turns opaque when approaching wavelengths of 1 μm. In many contexts, this limits optical techniques to a window in the electromagnetic spectrum around the visible wavelengths. Fortunately, there are many examples of biochemicals responding effectively to visible and ultraviolet electromagnetic radiations. Important examples include the *green fluorescent protein* (Ormo et al. 1996) (the discovery being honored by the 2008 Nobel Prize in chemistry) and more generally the techniques of fluorescence microscopy (see Chapter 6 in this text or Ref. (Wouters 2006) for a review) and microfluidic implementations of flow cytometry (see Chapter 19 in this text or Ref. (Chung and Kim 2007)). In the ultraviolet DNA exhibits pronounced molecular absorption (see Ref. Umazano and Bertolotto (2008) and references therein) and proteins may also be studied by means of UV excitation (see Ref. Tiihonen et al. (2007) and references therein). The UV excitation is targeted at the proteins or more explicitly at the aromatic amino acids, particularly tryptophan (Trp) and tyrosine (Tyr) (Lippitz et al. 2002), and also nucleotides such as NADH, which are involved in cell metabolism (Lakowicz 1999). These biomolecules have strong absorption bands in the spectral region ranging from 280 to 340 nm and emitting in the 300–600 nm region. In the blue part of the spectrum, the fluorescence detection of DNA is an important example (Han and Craighead 2000, Tegenfeldt et al. 2004) and likewise Raman spectroscopy provides important information in the visible regime where on-chip surface-enhanced Raman spectroscopy (SERS) may be used to monitor minute concentrations (Chen and Choo 2008)—see Chapter 17 in this text. Finally, the infrared domain is interesting for refractive index–based sensing schemes with applications including cell mortality determination (Liang et al. 2007).

12.4 Sources with Fluids as Active Medium: Classical Dye Lasers, On-Chip Dye Lasers, and Dye-Filled Fiber Lasers

12.4.1 Classical Dye Lasers

The first demonstration in 1966 by Sorokin and Lankard (1966) of lasing at 755 nm from the dye chloro-aluminium phthalocyanine dissolved in ethanol and optically pumped by a ruby laser, initiated the era of dye lasers, using organic dye molecules as a gain medium. When optically pumped, the organic dye molecules exhibit gain over a broad wavelength range, often 50–100 nm or more. This enables continuous wavelength tuning of the laser emission, controlled by the laser resonator feedback. Lasing has been demonstrated with more than 600 different dyes (Maeda 1984). Today, a range of different dye molecules provide optical gain over the full spectrum—the ultraviolet to the near infrared.* Since the 1970s, liquid dyes have been widely used in frequency tunable and high-power lasers in the visible frequency range, as reviewed in the book of Schäfer (1990) and the more recent review by Duarte (2003).

The organic dye Rhodamine 6G was applied in both liquid and solid state dye lasers in 1967 (Soffer and Mcfarlan 1967), and has since been widely used in conventional dye lasers as well as in most of the optofluidic light sources discussed in this chapter. Rhodamine 6G has a strong absorption peak in the green, around a vacuum wavelength of 530 nm, and emits in the yellow (vacuum wavelength 570–600 nm) with a high quantum yield (Kubin and Fletcher 1982). Figure 12.4a shows (part of) the energy level diagram for the Rhodamine 6G molecule. The spectrum contains bands of molecular rotation/vibration states (S_0, S_1, and T_1) separated by energy gaps. The spectrum exhibits a singlet–triplet system (S_i and T_i), with the lasing transition between the first excited S_1 and ground (S_0) singlet bands.

Under optical pumping, molecules are excited from the S_0 rotation/vibration ground state to the excited rotation/vibration states in the S_1 band, followed by fast non-radiative decay (thermalization) to the bottom of the the S_1 band. The transition times τ for different transitions are indicated on the figure.

The triplet state T_1 is metastable with a decay time, τ_{ph}, of microseconds. Decay from the singlet band S_1 into the triplet band T_1 can potentially destroy lasing. If a significant population occupies triplet bands, these molecules will not only be unavailable for lasing, but will also enhance cavity losses through triplet–triplet absorption, T_1 to T_2, which has a spectral overlap with the lasing transition, see also Figure 12.4b. Triplet band population is minimized by lowering the molecule-pump radiation interaction time below the S_1 to T_1 transition time $\tau_{TS} = 160$ ns. In addition to triplet band population, the organic laser dyes also bleach under continuous optical pumping and heating (Schäfer 1990, Nilsson et al. 2004). The dye molecule may decompose ("photobleach") directly through one or two photon processes or indirectly by heating of the molecule during non-radiative decay.

Classical or macroscopic dye lasers operating in pulsed mode have been realized using pulsed lasers as well as flashlamps for optical pumping. High power and continuous wave (CW) lasing is also obtained in macroscopic lasers, where the required cooling and small interaction time between the dye molecules

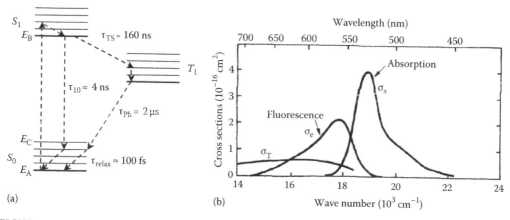

(a)   (b)

FIGURE 12.4 The laser dye-rhodamine 6G emits in the yellow, 570–600 nm when optically excited with green light (532 nm). (a) Simplified energy diagram for rhodamine 6G. (b) shows the cross sections for singlet–singlet absorption, fluorescence, and triple–triplet absorption for rhodamine 6G dissolved in ethanol. (Reproduced from Svelto, O., *Principles of Lasers*, Springer, Heidelberg, Germany, 1998. With permission. Copyright 2009 from Springer Science and Business Media.)

---

* Exciton laser dyes, http://www.exciton.com.

and the pump light is achieved by flushing the laser dye through the dye laser resonator at high speed using a thin sheet jet stream, with fluid velocities typically in the range of 5 m/s. The external handling apparatus for the liquid dye in general makes conventional dye lasers bulky and inconvenient to handle, and the appearance of solid-state device alternatives during the past decade has been welcomed by many users.

12.4.2 On-Chip Dye Lasers

In any application of integrated optics, the coupling of light offers a significant challenge both in terms of reproducibility and in terms of assembly costs. The microfluidic platform offers a simple and cost-efficient way to solve these challenges: on-chip dye lasers. A microfluidic dye laser, see also Figure 12.1, consists of a microfluidic channel with an embedded optical resonator that is filled with a liquid laser dye. The on-chip dye laser is "just another microfluidic channel," which in most cases can be added to the microfluidic chip without adding further process steps. In addition, the alignment of the laser device to other on-chip components is handled with the precision of today's planar microfabrication technology.

The authors' first demonstrations of microfluidic dye lasers in 2003 (Helbo et al. 2003a,b) were based on a pair of opposing flat metallic mirrors, with the gain dye solution located in-between the mirrors, see Figures 12.1a and 12.5. The microfluidic channel was defined photolithographically in a 10 μm thick film of negative resist, SU-8 (Microchem) sandwiched between two glass substrates. These devices were operated with Rhodamine 6 G laser dye dissolved in either ethanol, ethylene glycol, or benzyl alcohol, and emitted at vacuum wavelengths between 560 and 590 nm when optically pumped at 532 nm, using a pulsed, frequency doubled Nd:YAG laser. The 1 mm by 1 mm metallic top mirror consisted of 5 nm chrome (Cr) and 40 nm gold (Au). This layer configuration was designed to give a power reflectance of 0.72, a power transmittance of 0.06, and a power absorbance of 0.22 at a wavelength of 570 nm (Palik 1998). The bottom mirror layer stack of 10 nm Cr and 100 nm Au was correspondingly designed with an expected power reflectance of 0.83, a power transmittance close to 0, and a power absorbance of 0.17 at a wavelength of 570 nm (Palik 1998).

Figure 12.5c shows a series of five emission spectra from the microfluidic dye laser, when operated with a dye solution of 10^{-2} mol/L Rhodamine 6 G in ethanol. In the experiment, an external syringe pump was used to establish a replenishing dye flow of 10 μL/min. The frequency doubled Nd:YAG pumping laser (532 nm) provided 10 ns pulses with a repetition frequency of 10 Hz. Figure 12.5d shows a plot of the peak value (output power) and full width at half maximum (FWHM) as a function of optical pumping power density. The data exhibit the characteristics of the onset of laser oscillations: a change in the slope of output versus input power accompanied by a drop in linewidth as a threshold optical pump power density around 30 mW/cm² is passed. The inset shows a series of data taken with different concentrations of Rhodamine 6 G in ethanol, varying from 10^{-4} to 10^{-1} mol/L. No clear signatures of lasing were observed for concentrations below 10^{-2} mol/L.

The lasing wavelength of the dye laser is determined by a combination of the available modes in the laser resonator, the gain profile of the dye, the concentration (Ali et al. 1991, Schäfer 1990), and refractive index of the dye solution. From classical dye lasers, it is well known that the concentration of dye in the liquid solution must be kept at a certain value to ensure reproducible functionality of the laser system, and this issue was studied in detail for the vertically emitting microfluidic dye laser by Helbo et al. (2004).

The microfluidic platform enables fast and precise mixing of liquids (Janasek et al. 2006). This was applied to realize the first tunable, on-chip dye laser in 2003 (Bilenberg et al. 2003, 2006). In this work, the vertically emitting metal mirror laser was integrated with a microfluidic mixer and the response to varying parameters of the fluid was investigated, see Figure 12.6. The wavelength of the laser output can be tuned by on-chip mixing of liquids controlling the refractive index and the concentration of dye solution. Figure 12.6b illustrates the dynamics of the dye gain under increasing optical pumping of an ensemble of dye molecules. As the population is gradually inverted, the net absorption decreases and the net gain will increase. The Stokes shift between the net gain and the absorption will therefore cause

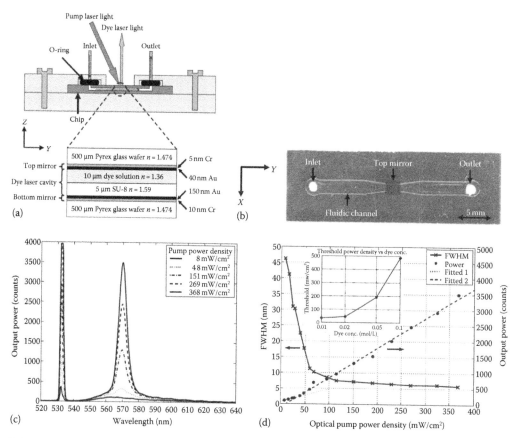

FIGURE 12.5 Vertically emitting microfluidic dye laser. (a) Cross-sectional outline of the microfluidic channel with embedded Fabry–Perot type laser resonator formed by two metallic mirrors in the ceiling and floor of the channel. (b) Photo of the laser chip with the microfluidic channel and the 1 mm by 1 mm area metal top mirror. (c) Emission spectra from the laser at different pump laser intensities. The device is operated with 10^{-2} mol/L of rhodamine 6 G in ethanol and a dye flow rate of 10 μL/min. The laser is optically pumped by a frequency doubled Nd:YAG pumping laser (532 nm), which provides 5 ns pulses with a repetition frequency of 10 Hz. A strong peak appears at 570 nm. For the highest pump laser intensity 368 mJ/cm², or pump pulse energy fluency of 37 mJ/cm², a series of additional peaks appear at wavelengths 582.8, 591.3, 600.5, 610.1, 619.1, 628.5, and 638 nm. These are identified as longitudinal cavity modes of the Fabry-Perot laser resonator. (d) Measured output power and FWHM for the emission peak at 570 nm versus optical pumping power density. The threshold lasing is estimated to be 34 mW/cm², or 3.4 mJ/cm² by linear fits to the high- and low-power parts of the output power data. The measured dependency on dye concentration of the threshold pumping power density for lasing is plotted in the inset. (Reproduced from Helbo, B. et al., Micro-cavity fluidic dye laser, in *IEEE The 16th Annual International Conference on Micro Electro Mechanical systems, MEMS-03*, Kyoto, Japan, pp. 235–238, 2003a. With permission. Copyright 2009 from IEEE; Helbo, B. et al., *J. Micromech. Microeng.*, 13, 307. With permission.)

a wavelength shift of the total gain. In a laser, the total gain will increase with increased pumping power until the lasing threshold, where the total gain will be clamped. The rate at which the gain increases is determined by the amount of dye molecules, whereby the wavelength where the laser reaches the lasing threshold will depend on the dye concentration.

The response time of the tunable laser is measured in Figure 12.6c. Initially the laser resonator has a steady flow of 5×10^{-3} mol/L solution Rhodamine 6 G in ethanol. At $t=0$, the liquid flows at the two input ports are changed to achieve a final dye concentration of 2×10^{-2} mol/L. The lasing spectra are

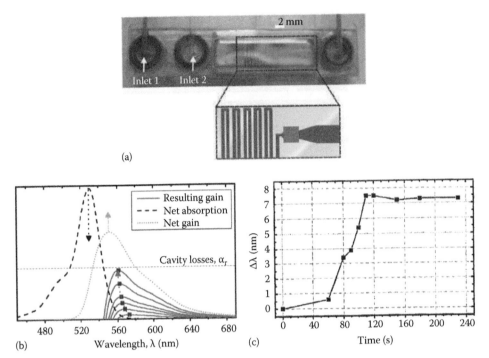

FIGURE 12.6 First tunable microfluidic dye laser. (a) A meander mixing channel is placed in front of the vertically emitting microfluidic laser resonator. In the photograph, the tunable dye laser is mounted in a sample holder fitted with inlet and outlet connections. In the experiments, an ethanolic rhodamine 6 G solution (inlet 1) is mixed with pure ethanol (inlet 2). (b) Illustration of the dye gain dynamics during increasing optical pumping. (c) Shift in lasing wavelength ($\Delta\lambda$) as a function of the elapsed time in seconds in response to a change in relative flow rates at $t=0$. The flow rates in the two inlets are changed from 2.5 μL/h rhodamine 6 G (2×10^{-2} mol/L) and 7.5 μL/h pure ethanol to 5 μL/h rhodamine 6 G and 0 μL/h ethanol. This corresponds to a shift in concentration from 5×10^{-3} mol/L to 2×10^{-2} mol/L. (Reprinted from Bilenberg, B. et al., *J. Appl. Phys.*, 99, 023102, 2006. With permission. Copyright 2009 from American Institute of Physics.)

recorded continuously, and the lasing wavelength is plotted versus time. The plotted data show that the lasing wavelength, and hence the concentration inside the laser resonator begins to increase after 60 s and saturates after 110 s. The onset of change in the lasing wavelength corresponds with the expected time of 76 s from the design of the device (distance divided by flow velocity).

For integration with planar optics and microfluidic networks, devices with in-plane emission would in most cases be preferred. Such laterally emitting devices require an in-plane optical resonator, where the simple metal mirror resonator cannot readily be realized. In addition, the confinement of light to the liquid presents another challenge.

Balslev and Kristensen (2005) presented the first laterally emitting optofluidic dye laser, see Figure 12.7. In this device, an array of polymer walls inside a microfluidic channel provides optical feedback. Similarly to the previously discussed vertically emitting device, the microfluidic channel structure is defined in a single UV lithography process on an 8 μm thick film of SU-8 resist on a glass substrate and sealed with a glass lid using a 4 μm layer of PMMA (Bilenberg et al. 2004). The laser is optically pumped through the glass lid. The 22 polymer walls have a width of 26.1 μm, and the spacing is 17.8 μm. When the dye laser is operated with Rhodamine 6 G dissolved in ethanol (refractive index $n=1.36$), the polymer walls ($n=1.596$) provide a high order (~130) Bragg reflection for the Rhodamine dye emission at vacuum wavelengths around 570 nm. The spacing between the two central polymer walls is doubled, forming a λ/4 phase shift element, which serves to yield a single resonance for each Bragg reflection,

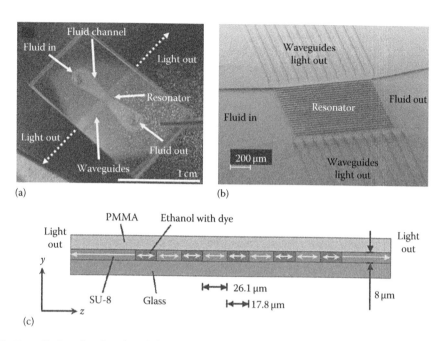

FIGURE 12.7 High order distributed feedback (DFB) microfluidic dye laser, defined by photolithography in an 8 μm thick film of SU-8 resist, and sandwiched between two glass substrates by polymer adhesive wafer bonding. (a) Photograph of the laser chip. The laser is pumped optically at 532 nm by a Nd:YAG laser beam, imping-ing normal to the chip plane. Dye laser emission is coupled into planar SU-8 waveguides, and guided to the edge of the chip. (b) Scanning electron micrograph of the central part of the microfluidic channel laser resonator. The laser resonator consists of 22 SU-8 bars arranged to form a high order (~130) DFB resonator with a central λ/4 phase shift element. (c) Cross-sectional outline of the laser resonator. The periodic modulation of refractive index between ethanol ($n = 1.33$ in the microfluidic channels and SU-8 polymer bars ($n = 1.59$) yields optical feedback through high order Bragg reflection. (Reprinted from Balslev, S. and Kristensen, A., *Opt. Express*, 13, 1860, 2005. With permis-sion. Copyright 2009 Optical Society of America; Balslev, S., Polymer dye lasers, PhD thesis, DTU, 2006, ISBN 87-8993582-9. With permission.)

as commonly used for distributed feedback (DFB) lasers (Hunsperger 2002). In a 1D model, where the fluidic channels and polymer segments are considered as infinite planes, the nominal spacing between the longitudinal resonance modes is estimated to be 2.46 nm (Balslev and Kristensen 2005).

The layered structure of PMMA, SU-8, and glass forms a slab waveguide, which determines the transverse modes of the laser. As the SU-8 core is rather thick (8 μm) and has a high refractive index of ~1.59, yielding a large index step to both the glass buffer and PMMA cladding layers, the slab waveguide supports 16 propagating modes at vacuum wavelengths around 570 nm. As illustrated in Figure 12.7c, the fluidic segments of the resonator do not guide light, since the refractive index of the fluid (ethanol, $n = 1.36$) is lower than the refractive index of the surrounding PMMA ($n = 1.49$) and glass ($n = 1.48$). This implies that light is diffracted when it traverses the fluidic element, and not all energy is coupled from one guiding polymer segment to the next. Finite difference beam propagation method (FDBPM) cal-culations show that the losses experienced by the propagating modes supported by the slab waveguide increase with increasing mode number, $m = 0, 1, 2, \ldots$, thus providing a mode discrimination mecha-nism, which favors lasing in a single propagating slab-waveguide mode.

The resulting round-trip loss for the resonator modes can be estimated by combining the mode-dependent losses calculated from the FDBPM calculations with transmission matrix elements. The differ-ence in phase evolution in the guiding (polymer) and the antiguiding (fluidic) parts of the grating period gives rise to significant variations in strength of the optical feedback in different Bragg reflection orders,

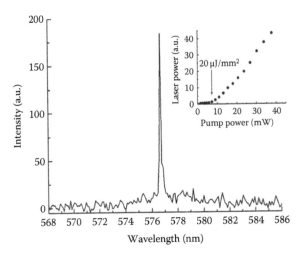

FIGURE 12.8 Output spectrum and pump curve from a fluid high order Bragg grating DFB laser fabricated with UV lithography in SU-8. (Reprinted from Balslev, S. and Kristensen, A., *Opt. Express*, 13, 1860, 2005. With permission. Copyright 2009 from Optical Society of America; Balslev, S., Polymer dye lasers, PhD thesis, DTU, 2006, ISBN 87-8993582-9. With permission.)

or longitudinal modes. Together with the transverse mode-dependent losses in the liquid segments of the resonator, this provides a sufficient mode discrimination for the laser to operate in a single mode.

Figure 12.8 shows a spectrum and a pump curve for a SU-8 defined laser device. The device was operated with solution of 20 mmol/L Rhodamine 6 G in ethanol, and was optically pumped with a pulsed (5 ns pulses, repetition frequency 10 Hz) frequency double Nd:YAG laser ($\lambda = 532$ nm). The dye solution was pumped through the microfluidic channel with a flow rate of 10 μL/h. This provided a sufficient convective flow to exchange the dye solution in the resonator area about every 1 s, eliminating problems with photobleaching. The dye laser pulse energy emerging from one side of the laser was measured to be 1.2 μJ at a pump pulse energy of 105 μJ impinging the laser area. The lasing threshold was 20 μJ mm^{-2}.

A laterally emitting Fabry–Perot cavity laser was realized by Kou et al. (2006), see Figure 12.9. This laser was fabricated by a hybrid packaging scheme, where metal mirrors are deposited on the cleaved surfaces of optical fibers before they are fitted into slots on the polymer chips. The device was operated with a mixture of two dyes, rhodamine 6 G (0.001 mol/L) and sulforhodamine 101 (0.01 mol/L), in ethanol, to demonstrate simultaneous, collinear emission at two wavelengths, 559 nm and 597 nm, when the device was pumped at 532 nm from a pulsed, frequency doubled Nd:YAG laser above a threshold pulse energy density of 2 μJ/mm^2.

The lack of light confinement within the liquid dye leads to losses in the laterally emitting laser resonators and thereby increased threshold for lasing. Within the "lab-on-a-chip" and microfluidics fields, in general it is a key challenge to confine light and liquids in the same volume—to maximize the interaction between light and sample—due to the relatively low refractive index of most liquids. This challenge has stimulated a large research effort into light guiding through liquid filled capillaries — liquid core waveguides. On-chip liquid core waveguides have been realized by construction of anti-resonant reflecting optical waveguides (ARROW) (Duguay et al. 1986, Schmidt and Hawkins 2008) and by applying low-index materials such as Teflon AF (Wang et al. 2003).

Vezenov et al. 2005a,b demonstrated liquid core waveguiding optofluidic lightsources by defining the microfluidic channels in polydimethylsiloxane(PDMS), of refractive index 1.406, in combination with Rhodamine dye dissolved in ethylene glycol (EG), which has a refractive index of 1.432. The device is outlined in Figure 12.10. The devices were fabricated by replica molding (Mcdonald et al. 2002), and a lateral Fabry–Perot laser cavity was realized by depositing metal (Au) mirrors on the

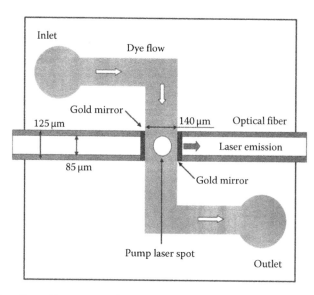

FIGURE 12.9 Outline of laterally emitting Fabry–Perot cavity microfluidic laser. The metal mirrors defining the laser resonator are deposited on optical fibers before they are fitted into the polymer chip. The microfluidic channel, defined in polysdimethylsiloxane (PDMS) by casting, is approximately 50 μm deep. (Reprinted from Kou, Q. et al., *Appl. Phys. lett.*, 88, 091101, 2006. With permission. Copyright 2009 from American Institute of Physics.)

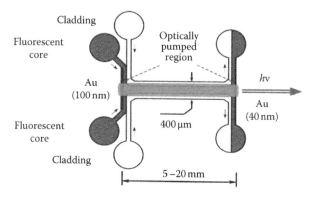

FIGURE 12.10 Liquid core waveguide optofluidic light source. (Reprinted from Vezenov, D.V. et al., *J. Am. Chem. Soc.*, 127, 8952, 2005a. With permission. Copyright 2009 from American Chemical Society.)

vertical side-walls of the microfluidic channel. The central microfluidic channel, which forms the laser resonator is terminated by T-junctions, which allows lateral outcoupling of light emission from the cavity. The T-junction also enables the formation of a lateral laminar flow sheet stack, where the liquid dye stream is sandwiched between two liquid cladding streams of, e.g. methanol, of refractive index 1.326. This *liquid core liquid cladding* (LL) waveguide construction can be used to tune the waveguide properties, such as propagating modes—providing wavelength tunability—and numerical aperture. The light guiding in the dye solution allowed for using a long optical cavity, 5–20 mm, yielding a large gain volume that is favorable for obtaining a low threshold for lasing. The emission from the device exhibits a clear drop in spectral linewidth and change in output efficiency at threshold optical pump pulse energy of 22 μJ. In the experimental demonstration, the long optical path length in the Fabry–Perot resonator prohibits the resolution of longitudinal cavity modes, and it is not resolved whether the observed threshold marks the onset of lasing or amplified spontaneous emission (ASE) (Svelto 1998).

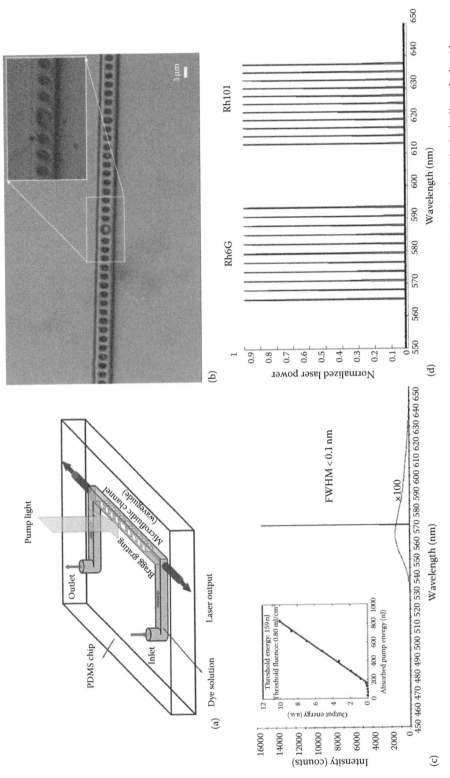

FIGURE 12.11 Single mode liquid core waveguide DFB laser. (a) Schematic of the device, fabricated in PDMS (refractive index of 1.406) which allows for liquid-core wave-guiding with laser dye dissolved in, e.g. ethylene glycol. PDMS posts in the microfluidic channel form a DFB resonator. (b) Optical micrograph of the device. The inset shows the central phase-shift element, which is inserted to obtain lasing in a single, longitudinal mode. (c) Emission spectrum and efficiency curve (inset) for the laser. (d) Tunable single mode lasing is observed using different laser dyes and by controlled straining of the PDMS chip. (Reprinted from Li, Z.Y. and Psaltis, D., *IEEE J. Select. Top. Quant. Electron.*, 13, 185, 2007. With permission. Copyright 2009 from IEEE.)

The first demonstration of single mode lasing from liquid core PDMS devices was obtained by Li et al. (2006a,b) and Li and Psaltis (2007), see Figure 12.11. A single mode liquid core waveguide was realized when the 5 μm wide and 2 μm high microfluidic channel was infiltrated with a liquid of refractive index slightly higher than the refractive index for PDMS, 1.406. Rhodamine laser dye in a mixture of methanol and ethylene glycol was used to to keep the refractive index step between the liquid core and the PDMS cladding below 0.003, required for single mode waveguiding. As a frequency selective element, an array of μm sized PDMS posts are embedded in the microfluidic channel providing a periodic modulation of the effective refractive index for the propagating modes in the waveguide—forming a DFB resonator.

The grating period was 3 μm, whereby the device relies on 15th order Bragg reflections in the periodic modulation of effective refractive index imposed by the PDMS posts. Like the device of Balslev et al. (Balslev and Kristensen 2005), the laser resonator is fitted with a central phase-shift element to lift the mode degeneracy and obtain lasing in a single longitudinal mode.

The laser was operated with a 1 mM Rhodamine 6 G dye solution and optically pumped with a pulsed laser at 532 nm. The pump laser beam was focused with a cylindrical lens to a 100 μm wide stripe covering the entire laser resonator, and the dye laser output was collected from the edge of the chip. Single mode laser output at 570 nm was observed above the pump fluence threshold lasing for lasing of 0.8 mJ/cm², see Figure 12.11c.

The PDMS is a soft elastomer with a Young's modulus of ~750 kPa, which allows for mechanical tuning of the laser resonator, by simply stretching the device (Li et al. 2006a). The Bragg reflection wavelength, λ, changes proportionally to the change in grating period, Λ:

$$\frac{\Delta\lambda}{\lambda} = \frac{\Delta\Lambda}{\Lambda} \tag{12.1}$$

Combining an approximate 5% elongation and two different laser dyes, Rhodamine 6 G and Rhodamine 101, they demonstrated a lasing wavelength tuning range of 60 nm without destroying the single mode liquid core waveguiding, see Figure 12.11d.

In general, the tuning range of a laser resonator is limited by the free spectral range (FSR) of the resonator. To avoid mode hopping when the cavity resonance is modified, the mode spacing should also be larger than the gain width of the amplifying medium. This favors the design of DFB lasers operating in a low Bragg reflection order. In addition, the light confinement improves with low Bragg orders. The mode spacing decreases with increasing Bragg order, corresponding to an increased transverse—or out of plane—momentum, implying cavity losses as the modes are more easily coupled out of the device plane. This is also the reason why semiconductor DFB and DBR lasers employ low (first or third) order Bragg reflections (Hunsperger 2002).

Gersborg-Hansen and Kristensen (2007) demonstrated a third-order optofluidic dye laser, see Figure 12.12, by miniaturizing the device layout of the 30th order DFB laser of Balslev and Kristensen (2005). A planar waveguide structure supports a single propagating TE-TM mode. The basic waveguide structure employs a SiO_2 (refractive index ($n=1.46$) buffer substrate, a 300 nm thick polymer core layer of SU-8 ($n=1.59$), and a top cladding made of polymethylmethacrylate (PMMA) ($n=1.49$)). The device structure takes advantage of a nanofluidic channel architecture defined lithographically in the SU-8 film by combined e-beam and UV lithography (CEUL) (Gersborg-Hansen et al. 2007b). The laser resonator consists of an array of 300 nm high nanochannels of period $\Lambda=601$ nm, which comprises a third-order Bragg grating with a central $\pi/2$ phase shift, embedded in a 500 μm wide shallow nanochannel. This DFB laser resonator operates in a third Bragg mode at the lasing wavelength of the applied Rhodamine 6 G dye emission, 560–610 nm and facilitates a FSR of 290 nm—half the lasing wavelength! Liquid core waveguiding conditions were obtained using a dye solvent mixture of ethylene glycol ($n=1.43$) and benzyl alcohol ($n=1.546$), and a threshold for lasing of ~7 μJ/mm² without correcting for the absorption efficiency being far less than 100%. The large FSR enables wavelength tuning over the full gain spectrum

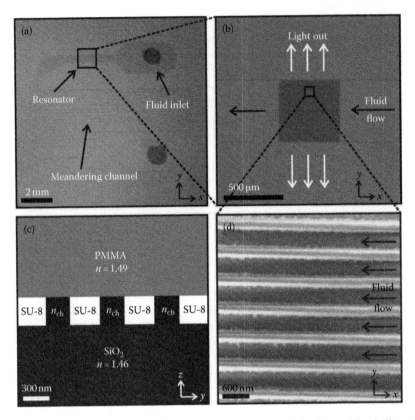

FIGURE 12.12 Third order optofluidic dye laser. (a) Top-view picture of the device. The shallow 300 nm high meandering channel facilitates capillary filling of the embedded laser resonator. (b) Optical micrograph of the DFB laser resonator embedded in the shallow meandering channel. (c) Side-view outline showing the layers of the DFB laser resonator. (d) Scanning electron micrograph showing details of the third order Bragg grating of nanofluidic channels, which constitutes the DFB laser resonator. (Reprinted from Gersborg-Hansen, M. and Kristensen, A., *Appl. Phys. Lett.*, 89, 103518, 2006. With permission. Copyright 2009 from American Institute of Physics.)

of the dye, and 45 nm tunability is achieved using a single laser dye by changing the grating period and dye solution refractive index (Gersborg-Hansen and Kristensen 2007).

In addition to Fabry–Perot and DFB resonators, ring resonators have proven to be a third attractive device architecture for on-chip optofluidic dye lasers. The first demonstrated optofluidic ring resonator dye laser by Gersborg-Hansen et al. (2004, 2005), see Figure 12.13, employing total internal reflection at vertical sidewalls of UV lithography defined triangular polymer "cat-eye" reflectors. The resulting cavity modes have a simple standing wave interpretation (Gersborg-Hansen et al. 2006), as illustrated by the finite element simulation results shown in Figure 12.13b. The laser dye is flushed through a central microfluidic channel, and the cavity output is optofluidically controlled by a second microfluidic channel along one of the reflecting sidewalls. When this is infiltrated with a (low refractive index) liquid, light is evanescently coupled—historically also referred to as *frustrated total internal reflection* (Zhu et al. 1986)—out of the cavity and into a polymer waveguide. This cavity design was later adapted in a PDMS device and integrated with a microfluidic pump and mixer by Galas et al. (2005) to realize a tunable device. Li and Psaltis (2007) used the same type of microfluidic PDMS ring-pump as liquid core fluidic microring laser, see Figure 2.14.

Shopova et al. (2007) exploited the high Q, >10^7, waveguide modes in silica capillaries to realize liquid core optofluidic ring dye lasers, see Figure 12.15. Thin-walled silica capillaries are infiltrated with

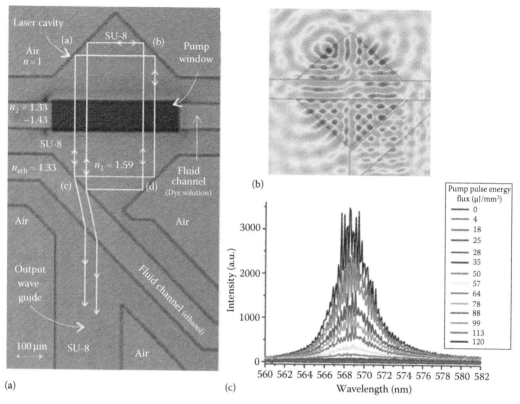

(a)   (b)

(c)

FIGURE 12.13 (See color insert following page 11-20.) Optofluidic ring laser. (a) Optical microscope image of the laser resonator structure consisting of two isosceles triangles of the photo definable polymer SU-8 ($n_1 = 1.59$) and a micro-fluidic channel in-between. Two classical trajectories of equal optical path length are drawn. The optical pumping is performed through a window (dark rectangular area in the photo) in a metal mask. (Reproduced from Gersborg-Hansen, M. et al., *Microelectron. Eng.*, 78, 185, 2005. With permission. Copyright 2009 from Elsevier.) (b) Example of simulated standing wave cavity mode. (Reprinted from Gersborg-Hansen, M. and Kristensen, A., *Appl. Phys. Lett.*, 89, 103518, 2006. With permission. Copyright 2009 from American Institute of Physics.) (c) Laser spectra measured at different pump pulse energies when the laser is operated with rhodamine 6 G in ethanol. (Reproduced from Gersborg-Hansen, M. et al., Microelectron. Eng., 78, 185, 2005. With permission. Copyright 2009 from Elsevier.)

liquid dye. The wall-thickness is adjusted to a few μm, whereby the mode-overlap of waveguide modes of the silica wall with the infiltrated dye solution is increased to provide lasing. This approach gives a versatile opto-fluidic ring resonator (OFRR) dye laser that can be operated regardless of the refractive index (RI) of the liquid (Lacey et al. 2007). These types of resonators are discussed in detail in Chapter 11 in this text. Although presently not integrated on a chip, silica capillary–based devices inherently support high-Q modes, which are attractive for optofluidic light sources. Another promising approach was demonstrated by Vasdekis et al. (2007), who obtained multi narrow-linewidth fiber laser emission from liquid infiltrated pieces of hollow optical fibers.

In addition to the optical design and properties of the on-chip dye lasers discussed earlier, the intrinsic challenge of dye bleaching requires further engineering to realize the vision of miniaturized light sources and integrated optical sensing systems. The majority of the demonstrated devices have relied on a convective dye-replenishment scheme actuated by external syringe pumps—analogous to macroscopic dye laser systems. Alongside the development of the laser resonators, different layouts for integrated pumps (Galas et al. 2005) and system-in-a-package (Balslev et al. 2004) approaches have been explored.

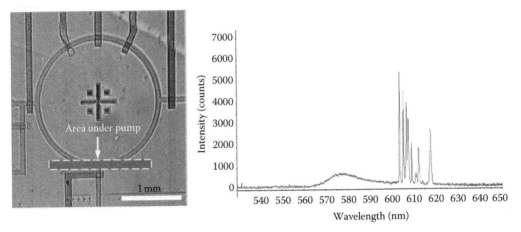

FIGURE 12.14 Optofluidic microring laser, realized by operating a PDMS microring pump with a laser dye. (a) Optical micrograph of the device. The shaded area is optically pumped. (b) Emission spectrum from the device. (Reproduced from Li, Z.Y. and Psaltis, D., *IEEE J. Select. Top. Quant. Electron.*, 13, 185, 2007.)

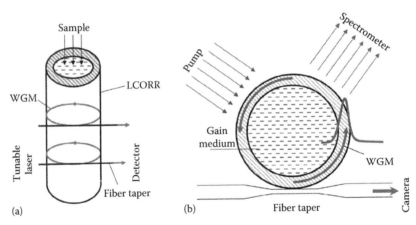

FIGURE 12.15 (a) Outline of an optofluidic ring laser, with optical feedback provided by a liquid core optical ring resonator (LCORR). In this configuration, light is guided in a capillary glass wall and amplification is provided through evanescent coupling to the liquid laser dye inside the capillary tube, as illustrated in panel (b). (Reprinted from Shopava, S.I. et al., *Appl. Phys. Lett.*, 90, 221101, 2007. With permission. Copyright 2009 from American Institute of Physics.)

Gersborg-Hansen et al. (2007a) investigated the diffusion dynamics of the microfluidic dye lasers, and demonstrated the feasibility of dye diffusion alone to provide the required dye replenishment.

12.5 Optofluidic Laser Intra-Cavity Sensors

Tuning and sensing constitute a general theme within the field of optofluidics. The optofluidic platform offers a precise control of placement and volume of samples to be sensed. This is particularly attractive for applying optofluidic lasers for intra-cavity sensing, where small perturbations of the optical resonator leads to changes in lasing wavelength and threshold. Laser intra-cavity sensing schemes can, to the first order, be divided into two categories: (1) refractometry, where a change in refractive index perturbs optical path lengths and thereby the resonance condition, and (2) absorption spectroscopy, where increased absorption contributes to the overall cavity loss and thus changes the lasing threshold.

12.5.1 Laser Intra-Cavity Refractometry

Evanescent wave sensing (Brecht and Gauglitz 1995) is a well-suited and widely used technique in optical devices for biosensing. An active reagent is immobilized at the surface of an optical waveguide. Due to evanescent waves at the surface of the waveguide, the immobilized reagent is detected as a change in the effective refractive index, Δn_{eff}, of the waveguide. If the sensing element is placed inside a laser cavity, the change n_{eff}, will change the modes in the cavity and hence shift the lasing wavelength. It has been predicted (Lindvold and Lading 1998), that the resolution, $\Delta n_{eff}/\langle n_{eff}\rangle$, of such intra-cavity sensors can be improved 4–6 orders of magnitude compared to configurations with the sensing element outside the cavity.

The basic idea is that upon liquid infiltration of a device, its optical response will be perturbed. In general, the shift of a resonance $\Delta\lambda$ due to a shift Δn in the liquid index will be given by (Mortensen et al. 2008)

$$\Delta\lambda = \frac{\partial\lambda}{\partial n}\Delta n = f\frac{\lambda}{n}\Delta n \tag{12.2}$$

where the filling fraction $0<f<1$ is a dimensionless number quantifying the relative optical overlap with the liquid part of space. In the literature, there has been a strong attention to the particular value of $\Delta\lambda/\Delta n$, i.e. the wavelength shift per refractive-index unit change. The above result may be derived with the aid of standard electromagnetic perturbation theory. In addition to a large wavelength shift it is quite often equally important to have a low detection limit (DL), i.e. the lowest possible refractive-index change that can be detected (White and Fan 2008). Of course, this is limited by the measurement setup and the noise level. However, due to noise it is challenging to quantify shifts that are much smaller than the resonance line width. In other words, the line width of the resonance will typically govern the detection limit. Introducing the resonance Q factor, we thus arrive at

$$DL = \min\{\Delta n\} = \frac{n}{f\times Q} \sim \frac{1}{f\times Q} \tag{12.3}$$

so that a low DL calls for a high-Q mode offering also a decent optical overlap with the fluid. This is by no means trivial and often a high Q will jeopardize the filling fraction.

Optofluidic refractometric intra-cavity sensing has been explored with quantum cascade lasers (Loncar et al. 2007) and with polymer photonic crystal band edge lasers (Christiansen et al. 2009).

12.5.2 Fluidic Quantum Cascade Lasers

The quantum cascade laser was first introduced in 1994 by Capasso and coworkers in the paper by Faist et al. (1994). As later emphasized by Capasso, "Band-structure engineering has led to a fundamentally new laser with applications ranging from highly sensitive trace-gas analysis to communications" (Capasso et al. 2002). More recently, the same group has demonstrated optofluidic versions where the quantum cascade laser is tuned by microfluidic means (Diehl et al. 2006, Loncar et al. 2007). In addition to the tuning capabilities, the optofluidic quantum cascade laser may also be applied for intra-cavity absorption spectroscopy (Belkin et al. 2007), as will be discussed later in Section 12.5.3.

The first report of a quantum cascade laser taking advantage of fluidic tuning (Diehl et al. 2006) utilized a distributed feedback quantum cascade laser integrated in a microfluidic chip, see Figure 12.16. The relative shift $\Delta\lambda/\lambda$ of the lasing wavelength is of the order of 10^{-3}, suggesting that the field overlap of

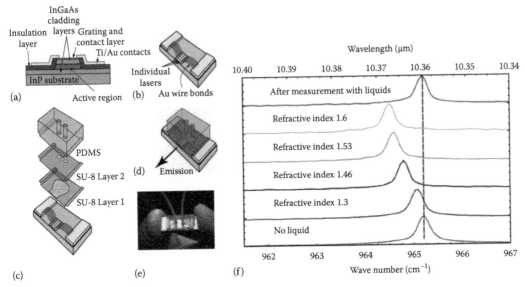

FIGURE 12.16 (See color insert following page 11-20.) Panels (a)–(e) show schematics of the microfluidic integration of distributed feedback quantum cascade laser. Panel (f) shows fluidic tuning of lasing wavelength at room temperature. (Reprinted from Diehl, L., *Opt. Express*, 14, 11660, 2006. With permission. Copyright 2009 from Optical Society of America.)

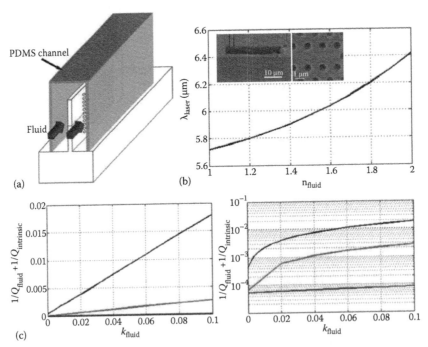

FIGURE 12.17 Quantum cascade laser employing a photonic crystal optical cavity that can be tuned by optofluidic means. (a) Schematic of the device. (b) Optofluidic tuning curve: laser emission wavelength as function of refractive index of the fluid in the microfluidic channel. (c) $1/Q = 1/Q_{intrinsic} + 1/Q_{fluid}$ of laser immersed in fluid as a function of imaginary part of the refractive index of the fluid-plotted on linear and semilog. (Reproduced from Loncar, M. et al., *Opt. Express*, 15, 4499, 2007. With permission. Copyright 2009 from the Optical Society of America.)

the lasing mode with the liquid is modest and is as low as $f \sim 5 \times 10^{-4}$. Nevertheless, the device has been employed for intra-cavity detection in volumes down to 10 pL (Belkin et al. 2007).

Porous structures such as photonic crystal may, if designed properly, in principle allow for higher field overlaps with a perturbing liquid (Mortensen et al. 2008). More recently reported devices take advantage of high-Q cavities in photonic crystals (Loncar et al. 2007), see Figure 12.17. Potential applications in lab-on-a-chip systems for intra-cavity chemical and biological sensing are emphasized.

12.5.3 Polymer Photonic Crystal Bandedge Laser Sensing

Christiansen et al. (2009) have realized an optofluidic cell sensor employing a polymer photonic crystal band edge laser (Arango et al. 2007, Christiansen et al. 2008), see Figure 12.18. The lasers are 300 nm thick slab waveguides of Ormocer hybrid material doped with a laser dye (pyrromethene). A 2D, low-index contrast photonic crystal is defined by periodic indentations in the waveguide upper surface, which create the required periodic modulation of effective refractive index for the guided slab-waveguide modes. Selectivity is provided by surface chemistry on the laser device, which serves to immobilize the cells under study, HeLa cells in these initial studies. The laser emission wavelength is observed to vary linearly with the cell coverage.

This observation indicates that the entire optical mode of the photonic crystal laser is perturbed by the immobilized cells. We emphasize that the cells are much larger than the period of the photonic

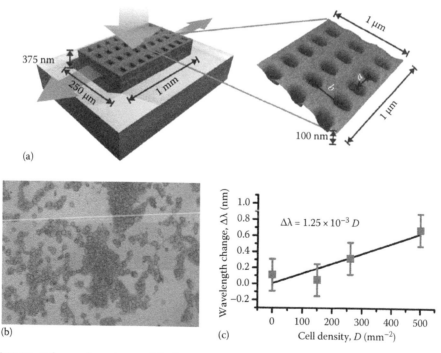

FIGURE 12.18 Polymer photonic crystal dye lasers as optofluidic cell sensors. (a) Outline of polymer dye photonic crystal bandedge laser with an inset AFM image of the surface. The laser is made of Ormocore hybrid polymer doped with the laser dye Pyrromethene 597. It is fabricated on a glass substrate. The lasers are pumped optically from above and emit in the chip plane. The rectangular PhC lattice increases pump absorption and provides laser feedback. (b) Optical micrograph of a laser device covered with HeLa cells. (c) Measured data and linear fit for the wavelength change as a function of cell surface density. (Reprinted from Christiansen, M.B. et al., *Appl. Phys. Lett.*, 93, 231101, 2008. With permission. Copyright 2009 from American Institute of Physics.)

crystal lattice, and the coverage is very irregular, see Figure 2.18. If the index difference becomes too large, the distributed feedback of the laser simply breaks down. It is also important to note that f is quite small for these lasers. This limits the extent to which the mode can be influenced. It also limits the sensitivity, however, so a trade-off must be made when designing lasers for sensing larger objects. This is in contrast to the requirements for measuring sub-wavelength objects or bulk refractive index, where f, and thus the sensitivity is normally maximized, see Equations 12.2 and 12.3.

12.5.4 Laser Intra-Cavity Absorption Spectroscopy

Laser intra-cavity absorption spectroscopy has been widely applied as one of the most sensitive methods for the spectral analysis of gases, chemical reaction kinetics, and in plasma research. The basic idea is that a high intrinsic Q-factor leads to a low lasing threshold, which is perturbed by additional intra-cavity absorption of the analyte.

Since

$$Q_{total}^{-1} = Q^{-1} + Q_{abs}^{-1}$$ (12.4)

where (Mortensen et al. 2008)

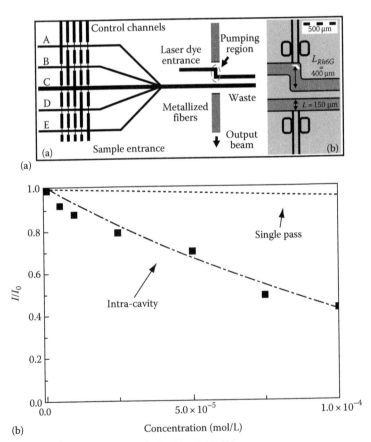

FIGURE 12.19 Optofluidic laser intra-cavity absorption device. (a) Outline and optical micrograph of the device. (b) Device output intensity as a function of methylene blue concentration. (Reprinted from Galas, J.C. et al., *Appl. Phys. Lett.*, 89, 224101, 2006. With permission. Copyright 2009 from American Institute of Physics.)

$$Q_{abs} = f^{-1} \frac{2\pi n}{\alpha \lambda} \tag{12.5}$$

a high intrinsic Q is required in order to observe a low absorption. The detection limit becomes

$$DL = \min\{\alpha\} = \frac{1}{f \times Q} \frac{2\pi n}{\lambda} \sim \frac{1}{f \times Q} \frac{1}{\lambda} \tag{12.6}$$

Thus, similar to the case of intra-cavity refractometry, $f \times Q$ should be maximized, see Equation 12.3.

The requirement of high $f \times Q$ is a challenge to optofluidic lasers and resonators. Galas et al. (2006) reported on optofluidic laser intra-cavity absorption measurements on low concentration analytes. The applied optofluidic laser device was fitted with a lateral Fabry–Perot resonator using the optical fiber packaging approach of Kou (Kou et al. 2006). Two microfluidic channels pass the Fabry–Perot resonator: one for the laser dye and one for the analyte, see Figure 12.19. The resulting width of the Fabry–Perot laser cavity was ~600 µm, and the emitted linewidth (FWHM) was 6 nm. In the reported experiments, methylene blue dissolved in water was applied to the sample channel, and the device emission intensity was measured as a function of the methylene blue concentration.

Optofluidic quantum cascade lasers have also been explored for intra-cavity absorption spectroscopy (Belkin et al. 2007), showing competitive values of the $f \times Q$ product.

12.6 Conclusion

In this chapter, we have discussed the implementation of liquid-based light sources on lab-on-a-chip and their potential applications for sensing. The concept of optofluidic light sources, based on a liquid light-emitting medium, goes beyond a bare miniaturization of the well-established dye laser technology. At first sight the optofluidic light sources offer a way to add new functionality, on-chip light generation, to microsystems without adding further process steps—the optofluidic light source is "just another micro-fluidic channel," and the light source is aligned to the other components on the chip with the precision of modern planar microfabrication technology. However, the microfluidic platform with its capability to handle small amounts of liquid fast and with high precision, has been utilized to demonstrate adaptive components based on fast and precise mixing of liquids to control index of refraction and contents of liquid solutions, and precise control of laminar flow sheets to configure and redirect liquid core liquid cladding waveguides. The microfluidic platform has also facilitated new concepts like diffusion-enabled dye replenishment. The presented research efforts on optofluidic light sources have the character of conceptual studies, although this has relied on development both in the design and fabrication of high Q optical resonators to facilitate single mode lasing with a low threshold for lasing. In addition to the demonstration of new concepts and figures of merit that can compete with existing technologies, the issue of prototyping for large-scale and low-cost manufacturability must also be addressed to transform the research efforts into a technology.

References

Ali, M.A., Moghaddasi, J., and Ahmed, S.A., 1991. Optical-properties of cooled rhodamine-b in ethanol, *J. Opt. Soc. Am. B*, 8 (9), 1807–1810.

Arango, F.B., Christiansen, M.B., Gersborg-Hansen, M., and Kristensen, A., 2007. Optofluidic tuning of photonic crystal band edge lasers, *Appl. Phys. Lett.*, 91 (22), 223503.

Balslev, S. and Kristensen, A., 2005. Microfluidic single-mode laser using high-order bragg grating and antiguiding segments, *Opt. Express*, 13 (1), 344–351.

Balslev, S., Roxhed, N., Griss, P., Stemme, G., and Kristensen, A., 2004. Microfluidic dye laser with compact, low-cost liquid dye dispenser, *Micro Total Anal. Syst.*, 2, 375–377.

Balslev, S., Jørgensen, A.M., Bilenberg, B., Mogensen, K.B., Snakenborg, D., Geschke, O., Kutter, J.P., and Kristensen, A., 2006. Lab-on-a-chip with integrated optical transducers, *Lab Chip*, 6 (2), 213–217.

Belkin, M.A., Loncar, M., Lee, B.G., Pflugl, C., Audet, R., Diehl, L., Capasso, F., Bour, D., Corzine, S., and Hofler, G., 2007. Intra-cavity absorption spectroscopy with narrow-ridge microfluidic quantum cascade lasers, *Opt. Express*, 15 (18), 11262–11271.

Bilenberg, B., Helbo, B., Kutter, J.P., and Kristensen, A., 2003. Tunable microfluidic dye laser, in: *Proceedings of the 12th International Conference on Solid-State Sensors, Actuators and Microsystems, Transducers 03*, Boston, MA, vols. 1 and 2, pp. 206–209.

Bilenberg, B., Nielsen, T., Clausen, B., and Kristensen, A., 2004. Pmma to su-8 bonding for polymer based lab-on-a-chip systems with integrated optics, *J. Micromech. Microeng.*, 14 (6), 814–818.

Bilenberg, B., Rasmussen, T., Balslev, S., and Kristensen, A., 2006. Real-time tunability of chip-based light source enabled by microfluidic mixing, *J. Appl. Phys.*, 99 (2), 023102.

Brecht, A. and Gauglitz, G., 1995. Optical probes and tranducers, *Biosens. Bioelectron.*, 10 (9–10), 923–936.

Capasso, F., Gmachl, C., Sivco, D.L., and Cho, A.Y., 2002. Quantum cascade lasers, *Phys. Today*, 55 (5), 34–40.

Chen, L.X. and Choo, J.B., 2008. Recent advances in surface-enhanced raman scattering detection technology for microfluidic chips, *Electrophoresis*, 29 (9), 1815–1828.

Christiansen, M.B., Kristensen, A., Xiao, S., and Mortensen, N.A., 2008. Photonic integration in k-space: Enhancing the performance of photonic crystal dye lasers, *Appl. Phys. Lett.*, 93 (23), 231101.

Christiansen, M.B., Lopancinska, J.M., Jakobsen, M.H., Mortensen, N.A., Dufva, M., and Kristensen, A., 2009. Polymer photonic crystal dye lasers as optofluidic cell sensors, *Opt. Express*, 17 (4), 2722–2730.

Chung, T.D. and Kim, H.C., 2007. Recent advances in miniaturized microfluidic flow cytometry for clinical use, *Electrophoresis*, 28 (24), 4511–4520.

Diehl, L., Lee, B.G., Behroozi, P., Loncar, M., Belkin, M.A., Capasso, F., Aellen, T., Hofstetter, D., Beck, M., and Faist, J., 2006. Microfluidic tuning of distributed feedback quantum cascade lasers, *Opt. Express*, 14 (24), 11660–11667.

Duarte, F.J., 2003. Organic dye lasers: Brief history and recent developments, *Opt. Photon. News*, 14 (10), 20–25.

Duguay, M., Kokubun, Y., Koch, T., and Pfeiffer, L., 1986. Antiresonant reflecting optical wave-guides in SiO_2-Si multilayer structures, *Appl. Phys. Lett.*, 49 (1), 13–15.

Faist, J., Capasso, F., Sivco, D.L., Sirtori, C., Hutchinson, A.L., and Cho, A.Y., 1994. Quantum cascade laser, *Science*, 264 (5158), 553–556.

Galas, J.C., Torres, J., Belotti, M., Kou, Q., and Chen, Y., 2005. Microfluidic tunable dye laser with integrated mixer and ring resonator, *Appl. Phys. Lett.*, 86 (26), 264101.

Galas, J.C., Peroz, C., Kou, Q., and Chen, Y., 2006. Microfluidic dye laser intracavity absorption, *Appl. Phys. Lett.*, 89 (22), 224101.

Gersborg-Hansen, M. and Kristensen, A., 2006. Optofluidic third order distributed feedback dye laser, *Appl. Phys. Lett.*, 89 (10), 103518.

Gersborg-Hansen, M. and Kristensen, A., 2007. Tunability of optofluidic distributed feedback dye lasers, *Opt. Express*, 15 (1), 137–142.

Gersborg-Hansen, M., Balslev, S., Mortensen, N.A., and Kristensen, A., 2004. A coupled cavity micro fluidic dye ring laser, in: *Proceedings of Micro- and Nano-Engineering (MNE) Conference*, Rotterdam, The Netherlands: MNE.

Gersborg-Hansen, M., Balslev, S., Mortensen, N.A., and Kristensen, A., 2005. A coupled cavity microfluidic dye ring laser, *Microelectron. Eng.*, 78–79, 185–189.

Gersborg-Hansen, M., Balslev, S., and Mortensen, N.A., 2006. Finite-element simulation of cavity modes in a microfluidic dye ring laser, *J. Opt. A: Pure Appl. Opt.*, 8 (1), 17–20.

Gersborg-Hansen, M., Balslev, S., Mortensen, N.A., and Kristensen, A., 2007a. Bleaching and diffusion dynamics in optofluidic dye lasers, *Appl. Phys. Lett.*, 90 (14), 143501.

Gersborg-Hansen, M., Thamdrup, L., Mironov, A., and Kristensen, A., 2007b. Combined electron beam and UV lithography in SU-8, *Microelectron. Eng.*, 84 (5–8), 1058–1061.

Han, J. and Craighead, H.G., 2000. Separation of long DNA molecules in a microfabricated entropic trap array, *Science*, 288, 1026–1029.

Helbo, B., Kristensen, A., and Menon, A., 2003a. Micro-cavity fluidic dye laser, in: *IEEE The 16th Annual International Conference on Micro Electro Mechanical Systems, MEMS-03*, Kyoto, Japan, pp. 235–238.

Helbo, B., Kristensen, A., and Menon, A., 2003b. A micro-cavity fluidic dye laser, *J. Micromech. Microeng.*, 13 (2), 307–311.

Helbo, B., Kragh, S., Kjeldsen, B.G., Reimers, J.L., and Kristensen, A., 2004. Investigation of the dye concentration influence on the lasing wavelength and threshold for a micro-fluidic dye laser, *Sens. Actuator A-Phys.*, 111 (1), 21–25.

Hillman, T.R. and Sampson, D.D., 2005. The effect of water dispersion and absorption on axial resolution in ultrahigh-resolution optical coherence tomography, *Opt. Express*, 13 (6), 1860–1874.

Hunsperger, R.G., 2002. *Integrated Optics, Springer Series in Optical Sciences*, vol. 33, 5th ed., Berlin, Germany: Springer, Verlag.

Janasek, D., Franzke, J., and Manz, A., 2006. Scaling and the design of miniaturized chemical-analysis systems, *Nature*, 442 (7101), 374–380.

Kou, Q., Yesilyurt, I., and Chen, Y., 2006. Collinear dual-color laser emission from a microfluidic dye laser, *Appl. Phys. Lett.*, 88 (9), 091101.

Kubin, R.F. and Fletcher, A.N., 1982. Fluorescence quantum yields of some rhodamine dyes, *J. Lumines.*, 27 (4), 455–462.

Lacey, S., White, I.M., Sun, Y., Shopova, S.I., Cupps, J.M., Zhang, P., and Fan, X.D., 2007. Versatile optofluidic ring resonator lasers with ultra-low threshold, *Opt. Express*, 15, 15523–15530.

Lakowicz, J.R., 1999. *Principles of Fluorescence Spectroscopy*, 2nd ed., New York: KA/PP.

Li, Z.Y. and Psaltis, D., 2007. Optofluidic distributed feedback dye lasers, *IEEE J. Select. Top. Quant. Electron.*, 13 (2), 185–193.

Li, Z.Y., Zhang, Z.Y., Emery, T., Scherer, A., and Psaltis, D., 2006a. Single mode optofluidic distributed feedback dye laser, *Opt. Express*, 14 (2), 696–701.

Li, Z.Y., Zhang, Z.Y., Scherer, A., and Psaltis, D., 2006b. Mechanically tunable optofluidic distributed feedback dye laser, *Opt. Express*, 14 (22), 10494–10499.

Liang, X.J., Liu, A.Q., Lim, C.S., Ayi, T.C., and Yap, P.H., 2007. Determining refractive index of single living cell using an integrated microchip, *Sens. Actuator A-Phys.*, 133 (2), 349–354.

Lindvold, L.R. and Lading, L., 1998. Evanescent field sensing: Cavity-coupled refractive index sensor (cris), *Proc. SPIE*, 3276 (1), 220–227.

Lippitz, M., Erker, W., Decker, H., van Holde, K.E., and Basche, T., 2002. Two-photon excitation microscopy of tryptophan-containing proteins, *Proc. Natl. Acad. Sci. U. S. A.*, 99 (5), 2772–2777.

Loncar, M., Lee, B.G., Diehl, L., Belkin, M., Capasso, F., Giovannini, M., Faist, J., and Gini, E., 2007. Design and fabrication of photonic crystal quantum cascade lasers for optofluidics, *Opt. Express*, 15 (8), 4499–4514.

Maeda, M., 1984. *Laser Dyes. Properties of Organic Compounds for Dye Lasers*, Tokyo, Japan: Academic Press.

Mcdonald, J., Chabinyc, M., Metallo, S., Anderson, J., Stroock, A., and Whitesides, G., 2002. Prototyping of microfluidic devices in poly(dimethylsiloxane) using solid-object printing, *Anal. Chem.*, 74 (7), 1537–1545.

Monat, C., Domachuk, P., and Eggleton, B.J., 2007. Integrated optofluidics: A new river of light, *Nature Photonics*, 1 (2), 106–114.

Mortensen, N.A., Xiao, S., and Pedersen, J., 2008. Liquid-infiltrated photonic crystals: Enhanced light-matter interactions for lab-on-a-chip applications, *Microfluidics Nanofluidics*, 4 (1–2), 117–127.

Nilsson, D., Nielsen, T., and Kristensen, A., 2004. Solid state microcavity dye lasers fabricated by nanoimprint lithography, *Rev. Sci. Instrum.*, 75 (11), 4481–4486.

Ormo, M., Cubitt, A.B., Kallio, K., Gross, L.A., Tsien, R.Y., and Remington, S.J., 1996. Crystal structure of the aequorea victoria green fluorescent protein, *Science*, 273 (5280), 1392–1395.

Palik, E.D., 1998. *Handbook of Optical Constants of Solids*, New York: Academic Press.

Psaltis, D., Quake, S.R., and Yang, C.H., 2006. Developing optofluidic technology through the fusion of microfluidics and optics, *Nature*, 442 (7101), 381–386.

Schäfer, F.P., Ed., 1990. *Dye Lasers*, 3rd ed., Berlin, Germany: Springer.

Schmidt, H. and Hawkins, A.R., 2008. Optofluidic waveguides: I. Concepts and implementations, *Microfluidics Nanofluidics*, 4 (1–2), 3–16.

Shopova, S.I., Zhou, H., Fan, X., and Zhang, P., 2007. Optofluidic ring resonator based dye laser, *Appl. Phys. Lett.*, 90 (22), 221101.

Soffer, B.H. and Mcfarlan, B.B., 1967. Continuously tunable narrow-band organic dye lasers, *Appl. Phys. Lett.*, 10 (10), 266–267.

Sorokin, P.P. and Lankard, J.R., 1966. Stimulated emission observed from an organic dye chloro-aluminum phthalocyanine, *IBM J. Res. Dev.*, 10 (2), 162.

Svelto, O., 1998. *Principles of Lasers*, Heidelberg, Germany: Springer.

Tegenfeldt, J.O., Prinz, C., Cao, H., Chou, S., Reisner, W.W., Riehn, R., Wang, Y.M., Cox, E.C., Sturm, J.C., Silberzan, P., and Austin, R.H., 2004. The dynamics of genomic-length DNA molecules in 100-nm channels, *Proc. Natl. Acad. Sci. U. S. A.*, 101 (30), 10979–10983.

Tiihonen, M., Pasiskevicius, V., and Laurell, F., 2007. Tailored UV-laser source for flourescence spectroscopy of biomolecules, *Op. Lasers Eng.*, 45 (4), 444–449.

Umazano, J. and Bertolotto, J., 2008. Optical properties of DNA in aqueous solution, *J. Biol. Phys.*, 34 (1–2), 163–177.

Vasdekis, A.E., Town, G.E., Turnbull, G.A., and Samuel, I.D.W., 2007. Fluidic fibre dye lasers, *Opt. Express*, 15 (7), 3962–3967.

Vezenov, D.V., Mayers, B.T., Conroy, R.S., Whitesides, G.M., Snee, P.T., Chan, Y., Nocera, D.G., and Bawendi, M.G., 2005a. A low-threshold, high-efficiency microfluidic waveguide laser, *J. Am. Chem. Soc.*, 127 (25), 8952–8953.

Vezenov, D.V., Mayers, B.T., Wolfe, D.B., and Whitesides, G.M., 2005b. Integrated fluorescent light source for optofluidic applications, *Appl. Phys. Lett.*, 86 (4), 041104.

Wang, Z., Cai, W., Wang, Y., and Upchurch, B., 2003. A long pathlength liquid-core waveguide sensor for real-time pco(2) measurements at sea, *Ma. Chem.*, 84 (1–2), 73–84.

White, I.M. and Fan, X., 2008. On the performance quantification of resonant refractive index sensors, *Opt. Express*, 16 (2), 1020–1028.

Wouters, F.S., 2006. The physics and biology of fluorescence microscopy in the life sciences, *Contemp. Phys.*, 47 (5), 239–255.

Zhu, S., Yu, A.W., Hawley, D., and Roy, R., 1986. Frustrated total internal-reflection - a demonstration and review, *Am. J. Phys.*, 54 (7), 601–606.

III

Bioanalysis

13

Single-Molecule Detection

Benjamin Cipriany
Harold Craighead

13.1 A Motivation to Study Single Molecules

Developing technologies and methods are allowing practical access to single-molecule detection (SMD), manipulation, and analysis. New optical detectors and imaging devices permit rapid high-sensitivity and low-noise capture of optical signatures from individual molecules. Chemically selective optical labels such as fluorophores or nanoscale scattering objects enhance the optical signature of selected molecules and enable their recognition in a complex mixture. Micro- and nanofluidic systems are advancing technological capabilities that can couple optical analysis to small volumes of fluid and also select and separate molecules of interest. Nanoscale structuring and optical methods combined with fluid handling are being integrated into lab-on-a-chip tools for radically new approaches to analytical chemistry or biomolecular study. These tools are being directed toward ultrasensitive detection and identification of molecules, even if rare in number or rate of occurrence. Future generations of DNA sequencing and gene expression may utilize single-molecule analysis (Eid et al. 2009). SMD methods provide direct visualization of mRNA dynamics within live *Escherichia coli* cells (Golding and Cox 2004; Golding et al. 2005), allowing access to study the rich biodiversity present within a given species of an animal or organism. This attention to biomolecule individuality and uniqueness enables studies and discoveries previously obscured in ensemble measurements.

Single-molecule measurement techniques have advanced substantially since their origin in fluorescence correlation spectroscopy (FCS) (Magde et al. 1972, 1974, 1978). The FCS technique assembled many single-molecule observations into an ensemble mathematical description of a system useful for studying diffusion and chemical kinetics. Subsequently developed techniques were based upon fluorescence, optical cavity resonance (OCR), and surface-enhanced Raman scattering (SERS). Recent experimental work has turned to the study of single-molecule dynamics within cells, to more accurately understand biological processes (Steyer and Almers 2001; Toomre and Manstein 2001). These experiments have stimulated commensurate development in nanoparticle and fluorophore labeling technologies, allowing gene expression (Fortina and Surrey 2008; Geiss et al. 2008) and membrane transport (Steyer and Almers 2001) to be observed directly and recorded in real time. Single-molecule techniques now challenge the conventional limitations of optical imaging put forth by Abbe's law, which impose a diffraction limit on the spatial resolution of conventional imaging. Image processing techniques combined with stochastic actuation of fluorophore emitters, in techniques such as fluorescence photoactivation localization microscopy (FPALM) (Betzig et al. 2006; Hess et al. 2006) now provide localization of fluorophore emitters with single-nanometer accuracy. Optofluidic devices now strive to implement these powerful and mathematically rigorous single-molecule techniques in compact systems.

In this chapter, we present a comprehensive view of SMD techniques. We introduce Poisson statistical descriptions as the basis for single-molecule isolation and statistical threshold to achieve signal-to-noise separation and single-molecule identification or counting. While a variety of optical methods for sensing and detection are available, we provide a brief survey focused on those that have demonstrated SMD, with fluorescence being the most mature and commonly used. Our continued presentation of fluorescence methods surveys fluorescent probes and three different imaging modalities, each with a discussion of the microscopy equipment, physical processes, and fluidic devices that enable SMD. These detailed discussions are complemented with short, practical illustrations of each imaging mode to elucidate its capabilities and state of the art. The applications of SMD are sampled in a discussion of two miniaturized analytical devices that achieve unprecedented analytical rigor and throughput in the study of genome sequencing, DNA structure and polymer mechanics, and biochemical reactions. While these and most SMD capable devices presently require microscopy-based study, we emphasize the opportunities for convergence of single-molecule techniques with highly integrated optofluidic components in the development of complete, lab-on-a-chip systems. We conclude with comments that emphasize the interdisciplinary culmination of physics, chemistry, biology, and engineering inherent to the field of optofluidics and the importance of device integration to make these advanced tools accessible to society.

13.2 A Mathematical Foundation for Single-Molecule Detection

The observation and identification of a single molecule requires distinct, unique identifier or a physical volume not crowded by other interfering molecules. Nanofluidic systems can enhance single-molecule observation by restricting the volume of material being observed and thereby significantly reduce optical scattering and fluorescence that leads to background noise in optical detection. A single-molecule observation implies a detection event that occurs rarely when compared to the total number of possible observations, a characteristic suitably modeled with Poisson statistics. An in-depth theoretical basis for SMD is left to the reader (Enderlein et al. 1998; Berezhkovskii et al. 1999; Barkai et al. 2004). In application, the Poisson probability-density function (PDF) describes the condition for single-molecule observation or occupancy in a given volume of solution as determined by the concentration of the species and the volume under investigation, modified from (Feller 1968)

$$P_c(x) = \frac{c^x e^{-c}}{x!} \tag{13.1}$$

where

 c is the average number of molecules per unit volume
 x is the number of molecules that instantaneously occupy that volume

Since $P_c(x > 1)$, the probability of observing more than one molecule at a time is nonzero in most cases, there is a finite probability of detecting more than one molecule. Multiple molecule occupancy can introduce error in experiments (Nie et al. 1995; Enderlein et al. 1998); however, in highly dilute systems this error becomes virtually negligible. To minimize the probability of detecting more than one molecule at an instant, either the volume or the concentration of the solution, or both, must be reduced to a $P_c(x > 1)$ approaching zero.

Most single-molecule techniques utilize a threshold to separate single-molecule signaling events from a background-level observation, or noise. One method for definition of the threshold measures the Poisson PDF's mean and then establishes a threshold that is 3–5 times the mean. This is analogous to establishing a minimum signal-to-noise ratio (SNR) of 3–5 for an experiment (Figure 13.1).

Observation of single molecules has been achieved using this Poisson PDF-based method for thresholding in high-throughput counting experiments within highly confined attoliter volumes (Foquet et al. 2002). By contrast, full-field images of single molecules, typical of molecular mapping experiments (Herrick and Bensimon 1999; Geiss et al. 2008), utilize highly dilute solutions (sub-picomolar) and are often accompanied by image processing techniques. These techniques correct for image background intensity fluctuations and examine nearest-neighbor image pixels to establish regions containing single molecules. Full-field imaging of single molecules offers highly parallel observation of biomolecules in a fluid environment (Geiss et al. 2008).

The SNR in single-molecule experiments can be broadly attributed to tradeoffs in detection sensitivity and time, which together affect the experiment's observation statistics. In imaging-based experiments, the SNR is maximized by increasing image integration time, but this reduces the number of molecules that can be individually observed in a given time (Michalet et al. 2003, 2007; Moerner and Fromm 2003). In flow-driven experiments with microfluidic channels, the rate of flow controls

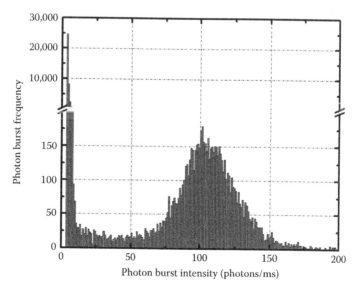

FIGURE 13.1 Photon counting histogram (PCH) illustrating the Poisson-distributed photon noise and a Gaussian-distributed population of single molecules, each with a single fluorophore label. A threshold of 6 photons/ms would provide distinct separation between these distributions and a signal-to-noise ratio greater than 15.

the number of molecules that can be observed in a given time, but limits the SNR if flow occurs too quickly. Therefore, given a system where all things are equal, the SNR is inversely proportional to the number of molecules observed per unit time. This limitation has motivated the development of new fluorophores, high-numerical-aperture optical components, and innovative methods for perturbation and observation for advancing single-molecule research techniques, which will be further discussed in Section 13.6.

13.3 Methods and Materials to Study Single Molecules in Fluidic Environments

Optical methods achieve SMD with a high SNR using a variety of readily availabe measurement tools. We will focus on three optical methods that have all demonstrated SMD with biological systems: fluorescence microscopy, optical cavity resonance, and surface-enhanced Raman spectroscopy. Our emphasis on fluorescence-based techniques is intentional, as it is currently considered by many to be the "gold standard" of detection and has found widespread application within the biomedical and analytical biochemistry fields (Michalet et al. 2003; Giepmans et al. 2006).

13.3.1 Fluorescence

Fluorescence single-molecule spectroscopy (SMS) methods rely on the specific, chemical attachment of spectrally distinct fluorescent dyes, proteins, and cellular components. Fluorescence-based SMS is possible due to the binding chemistries of organic fluorescent dyes—each dye molecule has a single binding site that is highly selective (Kapanidis and Weiss 2002) and while it is possible to have a single biomolecule with many bound organic dyes, generally, the opposite is not possible. This chemical binding relationship is the basis for conferring single-molecule identity during detection. The single-nanometer size and strong covalent bonds formed during attachment of most fluorophores are important to prevent adverse effects of dye-exclusion or dissociation during the labeling process or detection experiment. By contrast, the smallest of nanoparticles can still support multiple binding sites, and the attachment of multiple molecules, evidenced by their ability to self-aggregate and support multiple oligonucleotide binding sites (Alivisatos 2004).

The photophysical process of fluorescence in most fluorophores is observed in the near-ultraviolet and visible wavelength portions of the electromagnetic spectrum (see Chapter 6). The Jablonski energy diagram is a useful tool to describe the fluorescence (Bohmer and Enderlein 2003; Michalet et al. 2003). S0 identifies the ground state where no photon energy has been added to the system and S1 is the first excited singlet state. A short-wavelength or high-energy photon excites the fluorophore from the ground state to an adjacent, small collection of energy levels slightly higher than S1, called the vibration states. Dissipation of excess energy occurs as a molecular vibration, until the remaining energy can be dissipated by fluorescence. Fluorescence emitted during a radiative transition produces a photon of energy equal to S1–S0. The triplet state or forbidden state, T1, provides a nonradiative transition between S1 and T1. Statistically, a portion of the total energy absorbed from the excitation photons is always dissipated into the triplet state until there is either sufficient energy to be promoted back to the singlet state, or a transition back to the ground state can occur through nonradiative processes or phosphorescence. There are additional mechanisms responsible for a short-lived "dark state," and for photobleaching (Bohmer and Enderlein 2003; Michalet et al. 2003) (Figure 13.2).

Fluorescence single-molecule experiments are commonly performed using visible wavelength microscopy. Epi-illumination, typical of inverted microscopes, is often preferred in SMS experiments since excitation light is introduced into the underside of a sample and dissipated into free space, reducing the collection of scattered light and making fluorescence emission filtering more efficient in sensitive experiments. A broadband illumination source, typically a mercury or xenon arc lamp, or a monochromatic

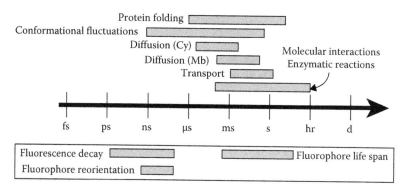

FIGURE 13.2 Fluorescence decay is the characteristic timescale for fluorescence photoemission, which occurs much faster than biophysical processes and diffusion. The fluorophore life span is influenced most by oxidation and photobleaching. The fluorophore is a useful reporter in the study of biological systems over a wide range of timescales. (Reprinted from Michalet, X. et al., *J. Mod. Opt.*, 54, 239, 2007. With permission.)

illumination source, typically a laser, is used to provide fluorescence excitation at wavelengths from the near-ultraviolet to near-infrared. Unlike a laser source, the lambertian, or light bulb-like, property of a lamp provides no control over optical coherence, directionality, or polarization, making it an excellent illumination source for wide-field imaging (Moerner and Fromm 2003; Petty 2007). Laser induced fluorescence (LIF) is advantageous in experiments where demanding SNR limitations require maximum fluorescence collection within the visible spectrum or independent control of the excitation power at each wavelength (Moerner et al. 2003).

Water and oil immersion microscope objectives have become the centerpiece of many fluorescence SMS experimental setups. These objectives maximize the photon collection efficiency by enhancing the effective numerical aperture to greater than 1, making it easier to perform single-molecule experiments where fluorophores typically emit thousands, or fewer, of photons per second. The immersion objective reduces reflection losses through a near-perfect refractive index match at the objective and fluidic sample interface. The solid angle of collected light approaches an entire 2π sterradians, and these objectives are often corrected for observation through coverslip thickness (~170 μm) glass.

Fluorescence filters are used to separate laser/lamp excitation and fluorescence emission wavelengths. Multilayer stacks of evaporated or sputtered thin films are deposited with angstrom-level thickness control to form these filters. There are three main filter types: excitation or cleanup, dichroic mirrors, and emission. Excitation filters restrict the light of an arc lamp to a narrow spectral width (50–100 nm), and cleanup filters remove all spectral content (such as amplified spontaneous emission (ASE) or other laser wavelengths) from a laser, except the desired lasing wavelength. These filters are oriented at an angle normal to the optical path to achieve maximum filtering. Dichroic mirrors are oriented at a 45° angle, relative to the optic axis, and typically reflect the incident excitation light and pass the emitted fluorescence. A passband emission filter is placed normal to the optic axis and provides efficient transmission of fluorescence within a desired spectral bandwidth, nominally 30–80 nm, and rejects all other wavelengths. In general, most single-color fluorescence filter systems achieve over 60 dB of optical isolation for excitation and emission filters and 20 dB for most dichroic mirrors. Advances in filter design and fabrication allow current optical elements to achieve sub-10 nm transitions between pass and stop bands, and over 90% transmission. Optical isolation is decreased when these systems are designed to examine multiple colors simultaneously.

A highly sensitive photon detector is the final component of a fluorescence SMS setup. Single-photon counting avalanche photodiodes (SPC-APDs) and photomultiplier tubes (PMTs) provide a single photodetector "pixel" for detecting and quantifying fluorescence with nanosecond time resolution (Michalet et al. 2007). Electron-multiplied charge-coupled device (EMCCD) and intensified charge-coupled

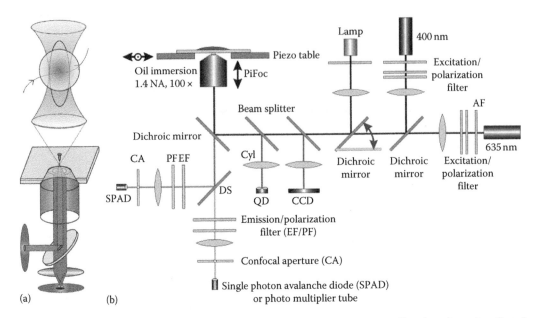

FIGURE 13.3 (a) Conceptual drawings of fluorescence microscopy optics. An incident laser beam is reflected by a dichroic mirror and focused onto a fluid sample exciting fluorescence from a single molecule. Fluorescence is collected and then transmitted through the dichroic mirror, a bandpass emission filter, and a pinhole aperture. (Reprinted from Walter, N.G. et al., *Nat. Methods*, 5, 475, 2008. With permission.) (b) A detailed drawing of a confocal fluorescence microscope configured for SMD using either LIF or lamp-based wide-field fluorescence imaging. (Reproduced from Bohmer, M. and Enderlein, J., *Chemphyschem*, 4, 793, 2003. With permission.)

device (ICCD) imagers provide dense "pixel" arrays for studying dim fluorescent signals within an entire image field (Ambrose et al. 1999; Moerner et al. 2003). We will discuss these technologies later in context with their application. However, suffice it to say that when these detectors are incorporated into a fluorescence confocal microscope, single molecules have been observed with SNRs of 10–20 (Michalet et al. 2003; Moerner et al. 2003) (Figure 13.3).

13.3.2 Optical Cavity Resonance

OCR is a sensitive method for the detection of biological samples. Labeling, either by fluorophore or nanoparticle, is not required to achieve detection, which simplifies sample preparation and makes this method attractive for "label-free" lab-on-a-chip applications. OCR occurs when the lifetime of photons captured inside a cavity exceeds the photon's round-trip transit time within the cavity. Changes in resonant frequency can be induced by small perturbations in the effective refractive index of the cavity mode using biomolecules selectively bound or flowed over the cavity surface. Frequency perturbations are directly proportional, often linearly, to the amount of biomolecule present. Cavities can be designed with resonant frequencies in the visible and infrared portions of the spectrum and with physical dimensions suitable for confining or binding single molecules.

Integrated photonic components and fabrication technology are essential to SMD using OCR. These components provide light guiding and confinement at dimension scales comparable to the wavelength. Modern lithographic techniques routinely make these components with dimensions ranging from sub-100 nm to 10 cm on the same optical device and with nanometer-scale control over surface roughness. These structures can be formed of low-loss optical materials such as silica, silicon, silicon nitride, silicon oxynitride, and alumina, for guiding the infrared or visible wavelengths or both. Silica exhibits the lowest material propagation losses, particularly at near-infrared wavelengths, and can be functionalized

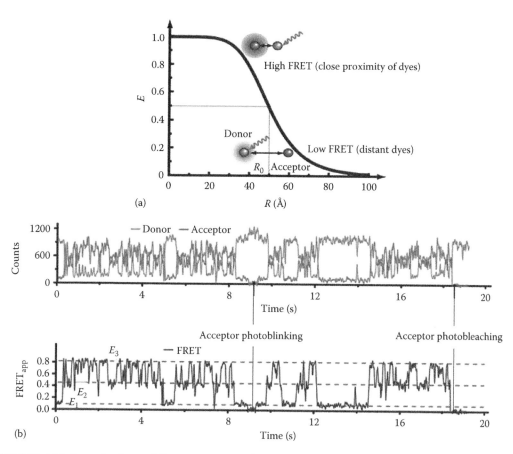

FIGURE 13.8 (See color insert following page 11-20.) (a) Graphical depiction of FRET energy transfer as a function of separation distance. (b) FRET processes are observed with a mutant hairpin ribosome, where the donor and acceptor separation is dynamic. Time–intensity traces illustrate an anticorrelated behavior for each of the two probe colors and indicate the FRET process. Characteristic photoblinking and photobleaching is observed after prolonged excitation. (Reprinted from Roy, R. et al., *Nat. Methods*, 5, 507, 2008. With permission.)

presence of its complementary nucleic acid target and becomes bound. These probes have been proven useful for signaling gene transcription and cell regulatory events (Manganelli et al. 1999; Li et al. 2002), plant virus detection (Eun and Wong 2000), and in SMD studies and quantification of DNA (Zhang et al. 2005).

Quantum dots (QDs) are nanoparticles of semiconductors that can be surface functionalized to support biomolecular attachment (Bruchez et al. 1998). The QD is spherically shaped with a semiconductor core of CdSe or CdS and a shell of silica or other oxide. This construction protects the semiconductor core, reduces the toxicity to biomolecules, and provides a convenient surface to functionalize with a streptavidin or silane-based self-assembled monolayer (Michalet et al. 2005). The fluorescence emission spectrum is a function of the core diameter, which becomes more red-shifted as the diameter of the core increases (Alivisatos 2004). This is due to increased spacing within the energy levels of the semiconductor as the size of the particle decreases, giving rise to increased quantum confinement. Almost all QDs are excited with a near-UV wavelength, and produce fluorescence in a narrow (<40 nm) and symmetric spectrum. The UV can cause damage to biological specimens, and results in lower optical transmission efficiency through some grades of glass. A common feature of QDs is "blinking" that occurs during fluorescence—periods when the particle becomes dark despite being illuminated. Blinking is excitation intensity dependent and related to the shell thickness of the quantum dot, causing "off" states that can range from microseconds to seconds (Nirmal et al. 1996). Some of these disadvantageous properties have incited the development of C-dots, a QD-like particle that utilizes a core of many organic fluorophores imbedded in a silica shell. Though not as bright as quantum dot fluorescence, these particles overcome the toxicity inherent in the use of semiconductor components (Ow et al. 2005; Burns et al. 2006). UV excitation is not required with C-dots, and their size is also slightly smaller (3.3–6 nm), an important characteristic for clinical and nanomedicine applications (Burns et al. 2009).

The labor-intensive process of labeling a biomolecule with a fluorescent dye is one of the principle drawbacks of fluorescence microscopy. Chemical attachment of dyes can require significant sample preparation and purification steps following the labeling procedure to remove excess dye not bound during the reaction. Furthermore, careful consideration must be given to the role of dye specificity, cytotoxicity, and fluorescence excitation conditions on a sample-by-sample basis.

13.5 Single-Molecule Detection: Fluorescence Spectroscopy Methods

13.5.1 Laser-Induced Fluorescence Techniques

LIF experiments provide the highest level of control over fluorescence SNR parameters in single-molecule experiments. Focusing confinement and polarization, due to the laser beam's highly directional, intense (MW/cm^2), coherent, and polarized properties, allows measurements otherwise not accessible with broadband illumination sources. Argon, krypton, and argon–krypton gas ion lasers are the mainstay lasers of fluorescence spectroscopy. Diode and diode-pumped solid-state lasers are also available at visible wavelengths, with cost-efficient, stable, and low-maintenance configurations.

Spatial filtering using a confocal aperture, or pinhole, is an essential component in fluorescence SMD to maximize the light collection SNR. The term "confocal" comes from the "conjugate focal" plane, the location of this spatial filter. An appropriately sized confocal aperture, typically 25–100 μm in diameter, blocks the collection of fluorescence from adjacent focal planes during imaging. Fluorescence observed is uniquely due to the focal plane of the microscope objective. A pinhole in an opaque material is the most straightforward implementation of the confocal aperture. Coupling the fluorescence to the core of a multimode optical fiber (normally 50–250 μm in diameter) also serves the same purpose, but with both the added benefit of flexible light routing and the added detriment of increased propagation loss due to the optical fiber.

With light levels on the order of 1000 photons or fewer emitted by a single-fluorophore avalanche photodiodes (APDs) or PMTs are essential detectors. Both photodetectors amplify or multiply the collected photons to a level that can be transduced electronically (Michalet et al. 2007). The glass PMT housing encapsulates charged metal plates in a vacuum. Photons incident on these charged metal plates produce free electrons under the photoelectric effect. These electrons are cascaded through a series of anodes and dynodes, and at each impact produce a secondary electron emission multiplying the incident electron number. It is very sensitive to damage by static electricity discharge or light saturation. APDs utilize a semiconductor with sharply doped profiles that are under a high, reverse electrical bias. This reverse bias creates a high electric field within the semiconductor that energizes carriers and induces impact ionization events. Photons incident on an APD produce an electron and a hole, which undergo a rapid succession or avalanche of these impact ionization events, each producing more electrons and subsequently gain in the current output from the detector. APDs operating at high gains are typically cooled to reduce noise and dark current levels so that detection of single-photon events is possible. Dark current is a stochastic occurrence of impact ionization events that produce electrical signals when the photodiode is not exposed to light. The dark current is the effective noise floor of the detector. The avalanching process inherent to APD operation causes an after-pulsing condition, whereby impact ionization traps electrons and holes and causes additional photocurrent or photon counts to be observed at a delayed time (after pulsing does not occur with PMTs). The APD can be incorporated with various

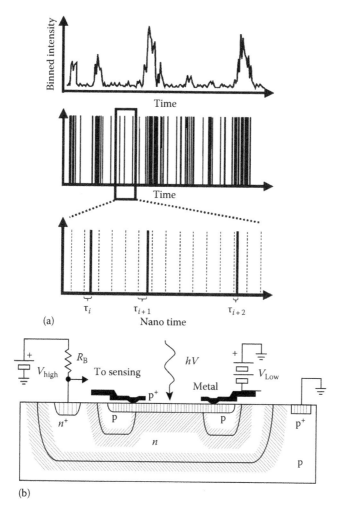

(a)

(b)

FIGURE 13.9 (a) SPC-APDs (Geiger mode) output pulses representing three different photon detection events, within a series of periodic time windows. Pulse groupings are formed, or binned, to a relevant timescale to represent the number of photons associated with a single-molecule fluorescence event. (b) An APD junction under high reverse voltage bias. (Reprinted from Michalet, X. et al., *J. Mod. Opt.*, 54, 239, 2007. With permission.)

protection circuits to prevent detection saturation. The output signal can be an analog current or, if operating in SPC or Geiger mode, a digital waveform that encodes the number of photons observed (see Figure 13.9).

The high quantum efficiency (QE) of silicon APDs at the visible wavelengths, as compared to the PMT, make them the popular choice for most fluorescence applications. The electrical signal produced by an APD or PMT is delivered to photon counting hardware and/or hardware correlators that are connected to a computer. Counting and correlator hardware components are customized for each application and corresponding experiment.

13.5.1.1 Fluorescence Correlation Spectroscopy

Single-molecule measurement techniques in fluidic environments are often traced to the development of fluorescence correlation spectroscopy (FCS), more generally known as fluctuation correlation spectroscopy. Fluorescence fluctuations are created by a fluorophore-labeled molecule during transit through a laser beam and then autocorrelated. The focal volume of the focused laser beam can be approximated with an ellipsoid shape that contains a small number of molecules at any instant. Mathematical fitting of the FCS autocorrelation curve provides indirect measurement of the translational diffusion constant of a molecule, the average number of molecules contained within the focal volume, and the size of the focal volume (Magde et al. 1972, 1974; Elson and Magde 1974). FCS is more commonly used to measure a small collection of molecules (<10) at a time; however, SMD can be achieved with extremely dilute solutions (Figure 13.10).

Timescale correlations made using FCS measurements are characteristic of photophysical and molecular phenomena. The fluorescence radiative recombination lifetime of a fluorophore is about 10 ns, near the temporal resolution of most APDs. Triplet-state lifetimes occur at the 1–10 μs scale and alter the curvature of the correlation function significantly when the fluorophore is being overexcited. Translational diffusion of the molecule and its residency time within the laser focal volume comprise the 0.1–10 ms timescale. Abrupt changes in the correlation function curvature at the diffusion timescales are often attributed to samples with multiple subpopulations of molecules (Webb 2001; Samiee et al. 2005). Study of these subpopulations is difficult and attempts to deconvolve superpositioned correlation functions have been published using the maximum entropy method of FCS (Sengupta et al. 2003a,b).

Correlation functions represent an ensemble average over time, so there is no record or time stamp associated with the observation of each single molecule. Time correlation is made using a multiple-Tau algorithm, and is often implemented using dedicated hardware, such as field-programmable gate arrays (FPGA), to achieve high-speed correlation from the 100 ns to 100 s timescale. Correlation at the sub-100 ns time range is generally not of interest, except when studying fluorophore radiative lifetimes or rotational diffusion.

Fluctuation cross-correlation spectroscopy (FCCS) extends correlation measurements to two or more independent fluorophores, sharing time-coincident fluorescence fluctuations. Binding reactions involving differently color-coded biomolecules are studied by this method and in tandem with FCS to provide a measure of both unbound and bound subpopulations and their respective diffusion coefficients (Schwille et al. 1997). Binding reactions must incur significant changes in molecule size to adequately measure changes in the diffusion coefficient. The laser focal volume limits the molecule

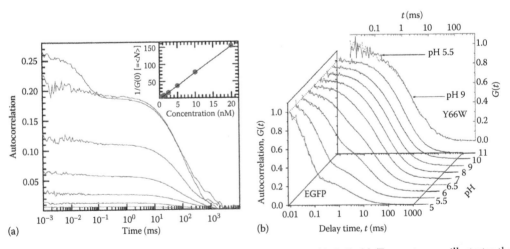

FIGURE 13.10　(a) FCS studies of rhodamine 6G at concentrations 20–0.62 nM. Topmost curve illustrates the triplet excited state for the 0.62 nM sample under conditions of 70× higher excitation power. (From Maiti, S. et al., *Proc. Natl. Acad. Sci. U.S.A.*, 94, 11753, 1997. With permission.) (b) The effect of protonation on enhanced green fluorescent protein (eGFP) is studied by varying the solution pH values. FCS autocorrelation is dominated by diffusion-driven relaxation at high pH and chemical relaxation (flickering) at low pH. A mutant GFP, GFP-Y66W (shown in the inset), is not affected by deprotonation and does not flicker as a function of pH. (From Haupts, U. et al., *Proc. Natl. Acad. Sci. U.S.A.*, 95, 13573, 1998. With permission.)

size that can be studied using FCS, which in some instances has encouraged the adaptation of FCS techniques to wide-field imaging methods to monitor fluorescent processes throughout an entire cell (Schwille et al. 1999).

The theme of molecule confinement established by FCS has prompted the development of other SMD techniques. A diffraction-limited focal volume of laser light illuminates a volume on the order of a picoliter, while containing a few molecules at a time. Attoliter and zeptoliter focal volume confinement has been achieved by the projection of a highly focused laser beam onto fluidic channels (Foquet et al. 2004) and subwavelength apertures (Levene et al. 2003). The SMD condition is achieved at higher sample concentrations as a result of these decreased volume containers and the rate of detection can be dramatically improved through a pressure-driven or electrokinetic controlled flow. FCS measurements conducted within these structures require a modified correlation function that accounts for the superposition of both an optical geometry and a fluidic geometry providing confinement. The correlation function becomes flow dominated, instead of diffusion dominated, in fluidic channels as given by (Foquet et al. 2002, 2004)

$$G(\tau) = G(\tau)_{\text{Diff}} \exp\left(-\frac{\tau^2}{(w_x/v_x)^2(1 + 4D/w_x^2\tau)}\right);$$

$$G(\tau)_{\text{Diff}} = \frac{1}{N} \frac{1}{\sqrt{(1 + 4D/w_x^2\tau)}} \frac{1}{\sqrt{(1 + 4D/w_y^2\tau)}} \frac{1}{\sqrt{(1 + 4D/w_z^2\tau)}} \qquad (13.2)$$

The unrestricted diffusion describes

a focal volume of width w_i (the e^{-2} radius along a given axis)

D the diffusion coefficient

N the average number of molecules within the focal volume

τ is the correlation time during a molecule's transit time across the focal volume

v_i is the flow velocity along a given axis (as applicable)

The effective volume can be extracted from a ratio of N/C, where C is the sample concentration. The correlation function is further modified by the geometry of the subwavelength aperture in zero-mode waveguides (ZMWs), and its mathematical derivation can be found in Levene et al. 2003; Samiee et al. 2005.

13.5.1.2 Time-Correlated Single-Photon Counting

Time-correlated single-photon counting (TCSPC) is one of the most information-rich methods for SMD. Photons emitted during single-molecule fluorescence are counted as a function of time using an APD or PMT. Mathematical models can be applied to the measured quantities single-molecule position, observation time, and photon emission intensity. These models allow derivation of biomolecule flow velocity and electrophoretic mobility (Stavis et al. 2005a,b,c), binding configuration (Stavis et al. 2005a,b,c, 2007), and size (Foquet et al. 2002; Reccius et al. 2008) within a microfluidic channel. The TCSPC technique can be modified to monitor light polarization and derive molecular orientation, using single-molecule fluorescence polarization techniques (Harms et al. 1999).

The photon-counting histogram (PCH) is a compact illustration of single-molecule observations, which both preserves information about individual SMD events and illustrates trends of the ensemble population (Ambrose et al. 1999; Chen et al. 1999; Hillesheim and Muller 2003). Recorded TCSPC events are grouped or binned as photon bursts, a collection of many photons arriving at the detector within a relevant timescale, such as a diffusion, flow, or reaction time. Photon bursts are then evaluated with a threshold parameter established from the Poisson PDF to separate SMD events from the noise. A PCH histogram is constructed to summarize the number of detection events observed and their respective photon intensity. The PCH is an important tool when discussing SMD results in relation to ensemble measurements made with other techniques, while still elucidating the presence of rare single-molecule events.

Experiments performing LIF within microfluidic channels are well suited to TCSPC methods and PCH analysis. Electrokinetic and pressure-driven flows within microfluidic channels establish directed passage of molecules through the laser focal volume and ensure a single measurement of each molecule. Approximately uniform illumination of each molecule is achieved through a reduction of channel dimensions below the Gaussian width of the laser spot size (Stavis et al. 2005a,b,c). This analysis method has been used to correlate single-molecule photon-burst intensities to DNA molecule size using a ladder of HindIII-digested lambda DNA labeled with YOYO-1 (see Figure 13.11) (Foquet et al. 2002). This method demonstrates the ability to use nanofluidics, instead of conventional agarose gels, to size DNA without the need for a separation medium.

Multicolor TCSPC, where each fluorophore encodes the identity of a different molecule, examines the spatial and temporal coincidence of biological processes. Two-color DNA sandwich assays have detected anthrax-protective antigen within a flow cell (Castro and Okinaka 2000) using this

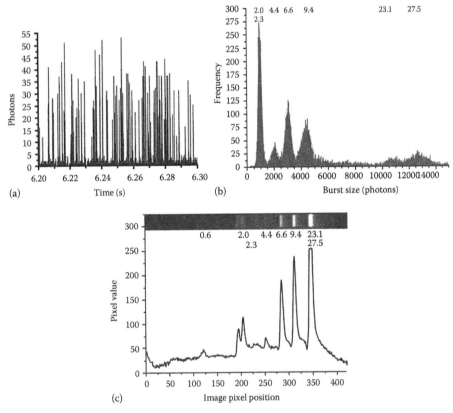

(a)

(b)

(c)

FIGURE 13.11   (a) Single molecules are visualized as bursts of photons as a function of time, measured using TCSPC. (Reprinted from Foquet, M. et al., *Anal. Chem.*, 76, 1618, 2004. With permission.) (b) Binned photon bursts are complied into a PCH to illustrate DNA sizing within a sub-micrometer fluidic channel. Sizing of HindIII-digested lambda DNA is performed on the basis of single-molecule fluorescence intensity. Labels above each peak indicate the measured DNA size in kbp. The 600 bp fragment was also detected, but not shown. (Reprinted from Foquet, M. et al., *Anal. Chem.*, 74, 1415, 2002. With permission.) (c) DNA sizing as performed with an agarose gel electrophoresis measurement. Labels above each peak indicate the measured DNA size in kbp. (Reprinted from Foquet, M. et al., *Anal. Chem.*, 74, 1415, 2002. With permission.)

technique. Polymerase chain reaction (PCR) amplification has been studied at the single-molecule level using pairs of differently colored primers to produce two-color amplicon products (Stavis et al. 2007). This method, as compared to the standard PCR technique, provides *in situ* SMD and identification of both amplicon products and primers without post-amplification separation. Time-coincidence histograms (TCH) illustrate the coincidence frequency, based upon the Poisson analysis and an SMD algorithm conducted for each fluorescence color. Single molecules with two-color labels are counted and binned as a function of their temporal coincidence. The resulting TCH is similar to the cross-correlation curve for the two-color labels, but directly illustrates the number of time-coincident SMD events. This two-color method is very powerful, particularly in situations intolerant to SMD errors. The probability of false-positive detection is given by the joint probability of $P_c(x > 1)$ for the independent populations, which can be on the order of 3×10^{-5}, depending on sample concentration (Stavis et al. 2007).

13.5.1.3 Practical Considerations: Maximizing the Signal-to-Noise Ratio

Filter design—Optical system design with few, high-transmission elements is essential to observe the scarce photons available in fluorescence SMD. Filters designed with the fewest number of passband transitions tend to have the best stopband rejection, increasing the overall system SNR. In some cases, it is possible to excite two colors of fluorophore with the same laser source, reducing the equipment needed. Spectral cross talk is an important consideration when observing two or more fluorophore colors. The red-shifted emission spectrum of organic dyes can overlap causing fluorescence from one dye to be collected as the fluorescence of another. Filter selections that minimize the overlap of the filtered fluorescence will reduce spectral cross talk in experiments where each molecule contains a similar number of fluorophores of each color. Fluorophore and filter selections must both be considered in experiments where each molecule is labeled with drastically different quantities of each fluorophore color, as this configuration can suffer the greatest cross talk. While a reasonable estimate of cross talk is possible, it ultimately requires empirical study.

Material autofluorescence—Materials used in the construction of fluidic channels and flow cells have a significant role in the SNR of fluorescence observations. Most materials contain trapped impurities that autofluoresce when illuminated, particularly at short wavelengths and with high-intensity light (Piruska et al. 2005). Polymer-based materials are notoriously autofluorescent. Micro- and nanofabrication techniques do have a limited set of compatible materials. Polydimethylsiloxane (PDMS), xenor, parylene, SU-8, and photocurable silicone have all become popular materials for use in micro- and nanofabricated devices for "lab-on-a-chip" applications. However, each of these polymers exhibit autofluorescence when illuminated with visible laser radiation that can be comparable or overwhelm single-molecule fluorescence (Desai et al. 2008). These versatile materials can still be used in SMD experiments, but may reduce the color palette of fluorophore options and the fluorescence SNR possible. Boron and phosphorous doped glasses, such as borofloat, also exhibit significant autofluorescence. Fused silica is one of the few materials that has a very low autofluorescence throughout the entire visible spectrum, and is demonstrated for SMD experiments involving a single-fluorophore label (Stavis et al. 2005a,b,c).

Quantum efficiency and yield—QY provides a measure of the total fluorescence process efficiency at a given excitation wavelength. The QY is a useful parameter to compare the excitation requirements for fluorescent particles and, for organic dyes, typically ranges from 0.30 to 0.70. While differences in QY can be compensated by increased excitation intensity, this adjustment will increase the photobleaching, material autofluorescence, and photodamage to the biological sample. QE is used to characterize the APD and PMT detectors, describing the efficiency with which photons converted into a representative electrical current. QE varies as a function of the photon wavelength, the semiconductor material (for APDs), and the physical process used for conversion.

Photobleaching—The photobleaching process essentially causes oxidation of the fluorophore. A variety of reducing agents do exist that can be used to reduce photobleaching effects, such as 2-mercaptoethanol (Nie et al. 1995). Care must be used since these chemicals do scavenge oxygen from the atmosphere and also produce strong odors.

13.5.2 Wide-Field Fluorescence Imaging Techniques

13.5.2.1 Experimental Setup and Equipment

Wide-field fluorescence imaging (WFFI) extends SMD techniques to a length scale sufficiently large to simultaneously image many single molecules. Distinguishing characteristics of a WFFI setup often include a charge-coupled device (CCD)-based imager for sensitive fluorescence detection throughout the entire image plane, and an arc lamp excitation source for broadband, incoherent illumination of the image plane. Hybrid imaging configurations of LIF and WFFI exist, with the most prominent examples being laser-scanning confocal imaging and total internal reflection fluorescence (TIRF).

CCD detectors comprised of large pixel arrays, up to 1024 by 1024 pixels, are available for fluorescence imaging in the visible spectrum. Image readout is performed sequentially, row by row, through a shared analog-to-digital (A/D) converter, producing a full-frame image capture on the order of 1–100 ms. Complementary metal-oxide-semiconductor (CMOS) imagers utilize an independent A/D element for each pixel row to acquire images more quickly, but introduce more row-level variations in image sensitivity and the recorded image. Therefore, CCD images are preferred in many scientific applications. CCD noise is the principle limitation in single-photon detection for these imagers. The relevant noise sources (shot, thermal, and imager readout) limit the detector sensitivity (Michalet et al. 2007). Readout noise is often the practical limitation in fluorescence SMD. It is a measure of the aggregate noise produced in the charge transfer and A/D process (Andor 2009).

ICCD and EMCCD are often used in single-molecule fluorescence studies. The intensifier in an ICCD is analogous to a PMT. The intensifier can be "gated" or turned on/off, with microsecond duration, for rapid image collection. However, the millisecond CCD readout limits the repetition rate of gated image collection. Like the PMT, the intensifier can be easily damaged by light saturation. The EMCCD achieves image gain in a method analogous to the APD. Charge from each row of the CCD is readout through a gain register, where impact ionization occurs, resulting in electron multiplication of the signal (Andor 2009). Since photon collection is performed before amplification, light saturation protection circuits can be implemented. Fewer image distortions and optical losses occur in EMCCDs, which cause the overall image quality and detection QE to be better than ICCDs in most cases (Michalet et al. 2007; Andor 2009). Both ICCDs and EMCCDs are monochromatic imagers, so additional optical design or multiple imagers are required for color image acquisition.

13.5.2.2 Single-Molecule Tracking

Digital video fluorescence (DVF) of single molecules creates a time-resolved record of single-molecule position and fluorescence intensity. Image-processing algorithms can be applied to DVF to perform single-molecule tracking (SMT), identifying the centroid of fluorescence intensity for an arbitrary shape or point particle, and tracking the centroid over many video frames. Much like TCSPC, the information captured in DVF can be studied with physical models applied to this centroid. EMCCD and ICCD technology greatly enhance the sensitivity of DVF, but do not provide an absolute measure of the fluorescence observed unlike APDs and PMTs in TCSPC. Image integration time, camera gain settings, and background intensity compensation to null arc lamp illumination fluctuations, must be kept constant to perform comparative analysis of DVF images. A minimal collection of equipment is needed in most SMT experiments, namely a microscope, fluorescence separation optics, an arc lamp, and imager.

Single-molecule transport within confined environments, imposed by both artificially and naturally occurring geometries, is a common theme of investigation with SMT. The square displacement of single lipid molecules in both a supported phospholipid membrane and a polymer-stabilized phospholipid monolayer have been studied to understand the lateral mobility within the membrane (Schutz et al. 1997), and similar experiments have even introduced the notion of compartmentalized membranes that cause a diffusive "hop" for lipids between compartments (Figure 13.12b) (Murase et al. 2004). Single proteins have been tracked during their diffusion within a cellular environment (Goulian and Simon 2000).

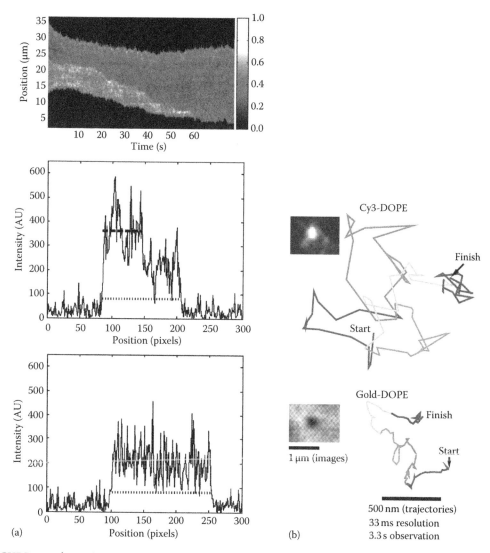

FIGURE 13.12 (See color insert following page 11-20.)   (a) YOYO-1 labeled T4-bacteriopage DNA is driven into a nanofluidic channel and elongated. The top panel illustrates the normalized fluorescent intensity of an elongated DNA strand as a function of time. The middle and bottom panel are the projected intensities at $t=15\,s$ and $t=55\,s$ when the molecule is folded and unfolded, respectively. (Reprinted from Levy, S.L. et al., *Nano Lett.*, 8, 3839, 2008. With permission.) (b) The phospholipid DOPE is labeled with the fluorescent dye Cy-3 or a gold nanoparticle, and studied during diffusion through plasma membrane compartments. Anomalous diffusion between compartments, or "hop diffusion," was observed using Gold-DOPE and high-speed imaging. (Reprinted from Murase, K. et al., *Biophys. J.*, 86, 4075, 2004. With permission.)

Biopolymer mechanics have been studied within silica nanochannel geometries to elucidate the compression and free expansion of DNA (Tegenfeldt et al. 2004; Reccius et al. 2005; Mannion et al. 2006). Entropic unfolding of a T4-bacteriophage dsDNA molecule has also been observed within a nanofluidic channel, allowing measurement of the entropic repulsive forces involved in DNA self-avoidance (see Figure 13.12a) (Levy et al. 2008).

13.5.2.3 Molecular Combing and Mapping

Molecular combing and mapping is similar to the process often used to solve a giant jigsaw puzzle—scatter the pieces onto a large table so you can see each one and then pattern match them together. Using a solution with DNA segments the experimenter can "comb" the DNA onto a surface so that each segment is separate and easily imaged using WFFI techniques. Various color-encoded sequences are identified and pattern matched using a database, allowing a digital reconstruction of the once continuous DNA strand.

Shear gradient caused by the no-slip condition at a solid-surface is frequently used in molecular combing experiments. A solid substrate with a chemically-treated surface is dipped into a solution containing fluorophore-labeled DNA. Surfaces formed by a coating of poly-methylmethacrylate (PMMA) (Allemand et al. 1997) or functionalized with a monolayer of silane (Bensimon et al. 1994; Michalet et al. 1997) have been reported in literature. Constant-rate substrate removal from the solution causes a meniscus to form and DNA within the sheer gradient to elongate, binding to the surface. The removal rate influences the magnitude of the shear gradient and the resulting stretched length of DNA. The substrate with elongated DNA coating is then observed with DVF. Dilute DNA solutions combed with this technique have allowed demonstrated imaging on elongated collections of hundreds of single molecules, using volumes of 5 μL, and chromosomal *E. coli* DNA fragment has been examined with a 420 μm contour length using molecular combing (see Figure 13.13a) (Bensimon et al. 1994).

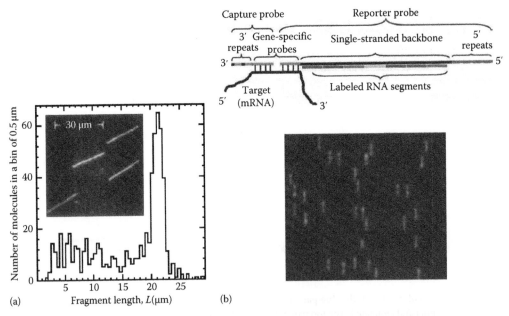

(a)   Fragment length, *L*(μm)   (b)

FIGURE 13.13 (a) Lambda DNA labeled with YOYO-1 is stretched on a silanized surface. Molecular combing elongates the DNA to an average length of 21.5 μm. (Reprinted from Bensimon, A. et al., *Science*, 265, 2096, 1994. With permission.) (b) **(See color insert following page 11-20.)** Gene expression is studied using mRNA hybridized with a Nanostring reporter probe. The hybrid molecule is bound to a surface using the capture probe and then elongated by electrokinetic forces. Molecular mapping is accomplished with a color-encoded reporter probe that identifies a hybridized gene specific probe–mRNA pair. (Reprinted from Geiss, G.K. et al., *Nat. Biotechnol.*, 26, 317, 2008. With permission.)

The second molecular combing technique has origins in the field of fluorescence *in situ* hybridization (FISH), whereby a molecule is tethered to a surface using a specific binding reaction and then elongated. Fiber FISH, a technique that is used to study genomic DNA for single-nucleotide polymorphisms (SNPs) and nucleotide translocations (Femino et al. 1998), has been adapted to implementation within a microfluidic channel. The surface of the fluidic channel can be functionalized to tether the DNA and a flow is established to elongate the DNA, stretching it uniformly in the same direction. Parallel, multiplex microfluidic channels further scale this technique to examine large quantities of genetic material. This method has recently been implemented in mRNA profiling using a collection of color barcodes, which are hybridized to specific sequences and then flowed in a collection of microfluidic channels (Figure 13.13b) (Geiss et al. 2008). This is a powerful technique for massively parallel study of gene expression.

Molecular mapping enhances the functionality of molecular combing to identify specific fluorescence signatures. Fluorescence images are analyzed using digital signal processing techniques to localize fluorescent sources, calculate spacing and sequence information, and, in some cases, perform combinatorial matching to digitally "reassemble" the original nucleotide sequence. Fluorophore color barcodes indicate specific nucleotide sequences using staining methods similar to those in FISH (Michalet et al. 1997; Herrick and Bensimon 1999) or by hybridization with an oligonucleotide probe decorated with multiple fluorophore colors (Herrick and Bensimon 1999).

13.5.2.4 Total Internal Reflection Fluorescence

Confined fluorescence excitation near a liquid–solid interface is challenging with conventional fluorescence microscopy. TIRF overcomes this challenge, using total internal reflection (TIR) to couple light into the solid substrate (typically silica or sapphire), allowing evanescent excitation of fluorescent molecules within approximately 100 nm of the liquid–solid interface (Axelrod et al. 1984). The reduced excitation profile enables TIRF imaging to study transport across cell membranes (Steyer and Almers 2001), cellular signaling (Han et al. 1999; Steyer and Almers 2001), kinetics at surfaces (Axelrod et al. 1984), and other processes occurring at solid surfaces. Fluorescence from adjacent focal planes is not excited when using TIRF, therefore image quality is often better than with most laser-scanning confocal microscopes (Figure 13.14).

The critical angle of incidence necessary to achieve the TIR condition can be achieved with both laser and arc lamp sources, with the former being the most common. Light coupling at the critical angle is strongly wavelength dependent and is often performed by prism coupling, thin-film interference, or by off-optical-axis transmission through a microscope objective. The evanescent excitation profile can be used to study single fluorophores near a surface when TIRF is imaged with an ICCD or EMCCD camera.

Applications of TIRF microscopy have been particularly numerous within the field of cellular biology. Studies of exocytosis, a process by which proteins are secreted across the cell membrane through vesicles, used TIRF to reveal vesicle tethering and fusion steps with direct, real-time observation (Steyer and Almers 2001; Toomre and Manstein 2001). Kinesin molecular motors have been observed traversing along microtubules (Vale et al. 1996). This has provided a fascinating insight to single-molecule kinetics and activity. RNA polymerase binding to DNA, a crucial step in gene transcription and expression has been studied using TIRF combined with a laser trap to elongate the DNA and observe polymerase movement (Harada et al. 1999). Enzyme-substrate interactions have been studied with a FRET pair that becomes separated upon uptake by a ribozyme, causing a fluorescence color change (Zhuang et al. 2000; Ha 2001a,b). TIRF and FRET are quite complementary, and with a growing collection of fluorescent proteins, we anticipate there will be many opportunities to advance cellular and molecular biology through their simultaneous implementation.

13.5.2.5 Practical Considerations

Limitations of Single-Molecule Detection—Conditions of extreme dilution are common to most SMD experiments using WFFI techniques; however, this can prove disadvantageous when dealing with

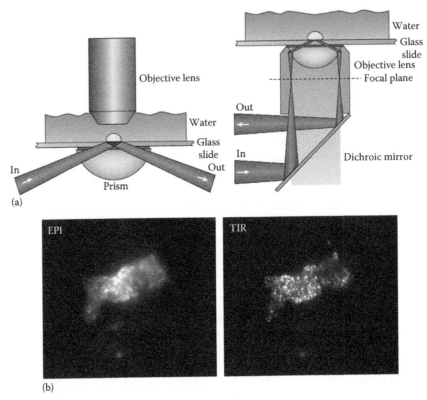

(a)

(b)

FIGURE 13.14   (a) Prism-coupled and objective-coupled TIRF illumination modes. (Reprinted from Steyer, J.A. and Almers, W., *Nat. Rev. Mol. Cell Biol.*, 2, 268, 2001. With permission.) (b) A bovine chromaffin cell with secretory granules marked with GFP–atrial naturetic protein is compared in conventional epi-illumination and TIRF imaging modes. Fluorescence is excited by an argon laser and imaged with a cooled CCD camera. (Reprinted from Axelrod, D., *Traffic*, 2, 764, 2001. With permission.)

chromatin, cells, proteins, which have shown dissociation, cooperation, and aggregation behaviors, respectively, dependent upon concentration (Claudet et al. 2005). Photobleaching is a central limitation to WFFI experiments, restricting continuously excited observations to tens of seconds and requiring short "snapshots" in time-lapsed experiments. As with LIF techniques, materials with low autofluorescence are necessary to maximize the SNR, especially in imaging experiments with only one or a few fluorophores attached to each molecule. TIRF experiments continue to challenge the single-label regime of SMD, and require synergy of EMCCD cameras with extremely high quantum efficiencies, low autofluorescence materials, and high numerial aperture oil immersion microscope objectives to maximize photon capture.

13.5.3 Sub-Diffraction-Limit Fluorescence Imaging

13.5.3.1 Experimental Setup and Equipment

The diffraction limit restricts spatial imaging resolution to the order of a half-wavelength. Sub-diffraction-limit fluorescence imaging or "super-resolution" fluorescence imaging (SRFI) techniques surpass the diffraction limit to achieve imaging at the nanometer scale. Ensemble imaging with nano-scale resolution was demonstrated using stimulated emission depletion (STED) and reversible saturable optical fluorescence transitions (RESOLFT) techniques, which leverage photophysical and photochemical process, respectively (Hell 2007).

Recently developed single-molecule SRFI techniques combine the stochastic photoactivation of fluorophores with image-processing techniques to create dramatic composite images that localize each emitter. Conventional, diffraction-limited WFFI with densely packed collections of fluorophores produce blurred images because all dye molecules are excited simultaneously. In SRFI, photoactivation causes a spatially sparse collection of fluorophores to be stochastically selected for activation. Subsequent fluorescence excitation is therefore able to collect an image of well-separated fluorophores during each imaging cycle, which can then be overlaid in a super-resolution composite to image each fluorophore regardless of packing density on the biological sample. Photoactivation causes breaking and formation of *cis* and *trans* bonds, controlling the fluorescent state of the dye, and can be reversible or irreversible depending on the fluorescent dye. These single-molecule SRFI methods are performed by FPALM and stochastic optical reconstruction microscopy (STORM) (Hell 2007) (see Figure 13.15).

Single-molecule SRFIs are collected during a multistep process. The first steps of photoactivation with a short-wavelength laser (typically UV or deep blue), fluorescence excitation by a visible wavelength laser, and fluorescence image collection with an EMCCD form the basic imaging cycle. One cycle produces an image of fluorophore emitters separated more than half a wavelength apart, whose spatial position is localized with nanometer accuracy using a mathematical fitting to a point-spread function (PSF). This PSF fitting was first incorporated into fluorescence imaging with one nanometer accuracy (FIONA), reporting the highest resolution localization to date during particle tracking with the molecular motor myosin (Yildiz et al. 2003) (Figure 13.16).

The precision of localization is limited by the number of photons collected and, to second order, by the stability of the apparatus used during observation and image collection (Yildiz et al. 2003; Hess et al. 2006):

$$\sigma_x^2 = \frac{r_0^2 + q^2/12}{N} + \frac{8\pi r_0^4 b^2}{q^2 N^2} \tag{13.3}$$

The localization precision σ_x is described by
   N is the total photons observed
   b is the background photon noise of each image pixel
   q is the size of an image pixel
   r_0 is the standard deviation of the PSF

The imaging cycle is repeated many times to photoactivate the complete set of fluorophore emitters in an image field. The fluorophore positions in each imaging cycle are compiled and combined in a super-resolution composite.

The burgeoning field of single-molecule SRFI is currently devoted to the study of fixed tissues and cells suspended in liquid media. It has produced dramatic imaging results, which have been overlaid onto various scanning electron micrograph (SEM) and transmission electron micrograph (TEM) images to provide an unprecedented record of structure in biological samples with nanometer spatial resolution. However, SRFI's ultimate implementation and incorporation into the broader collection of WFFI techniques, specifically molecular mapping and TIRF, seems apparent.

13.5.3.2 Fluorescence Photoactivation Localization Microscopy

The ability to localize fluorophores in densely packed configurations is the distinguishing characteristic of FPALM over its predecessor FIONA (Betzig et al. 2006; Hess et al. 2006). Photoactivation is achieved with photoactivatable fluorophores (PA-F) and photoactivatable fluorescent proteins (PA-FPs) illuminated with near-UV light: a 405 nm wavelength diode laser is a popular option. FPALM utilizes an

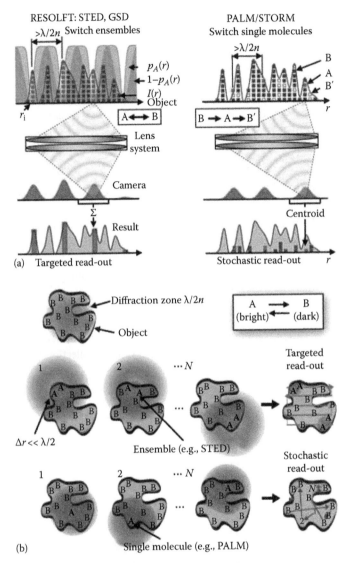

FIGURE 13.15 (a) Photoactivation techniques of STED and GSD activate ensembles of fluorophores within the photoactivation spot $I(r)$, with a probability of activation into the bright state $p_A(r)$. Single-molecule photoactivation with PALM or STORM is achieved by a stochastic photoactivation, not limited by the size of the photoactivation beam size. (b) Fluorescence readout for ensemble and single-molecule super-resolution imaging. (Reprinted from Hell, S.W., *Science*, 316, 1153, 2007. With permission.)

imaging cycle, similar to the one previously described, involving a two- or three-step sequence, depending on the use of irreversible or reversible photoactivatable dyes. The FPALM cycle deviates during the fluorescence imaging step, wherein the molecule is either photobleached permanently or photodeactivation reverses the molecule into a dark state by a second illumination with the UV laser. Imaging time varies as a function of photoactivation rate, the fluorophore lifetime, and localization precision, with total acquisition times currently reported between 2 and 12 h (Betzig et al. 2006). Findings reported using FPALM already include impressive images of fixed sections of lysosomes and mitochondria (Betzig et al. 2006) (see Figure 13.17) and three-dimensional image reconstruction of caged fluorescein beads (Juette et al. 2008).

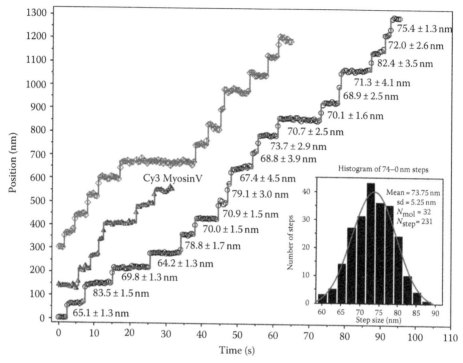

FIGURE 13.16 Molecular motor myosin V is observed "walking" or "stepping" down a microtubule using single-molecule TIRF and FIONA to track progression. The hand-over-hand model of movement is verified, shown by the average 74 nm step made by myosin V during 231 stepping events. The PSF fitting used in FIONA allows each step to be localized with single-nanometer precision. (Reprinted from Yildiz, A. et al., *Science*, 300, 2061, 2003. With permission.)

FPALM is a versatile technique because it can be performed with a wide variety of PA-Fs or PA-FPs. It has been demonstrated in combination with other imaging methods, such as SEM and TEM, to elucidate cell machinery with nanometer resolution (Betzig et al. 2006).

13.5.3.3 Stochastic Optical Reconstruction Microscopy

STORM pairs PSF localization with the photoactivatable fluorophore switch pairs (PA-FSP), to achieve single-molecule imaging in dense-packed fluorophore samples. The experimental method is similar to FPALM, involving a three-step process of photoactivation, imaging, and photodeactivation (Rust et al. 2006). The PA-FSP is composed of cyanine dyes, such as Cy3 and Cy5, which are bound together to share intersystem crossing behaviors during photoactivation and deactivation. It has been shown that these PA-FSPs do not function well without B-mercaptoethanol or glucose oxidase to scavenge oxygen in buffer solutions (Bates et al. 2005).

In STORM, all PA-FSPs are deactivated or "switched" into a dark state at the start of an experiment using a long-wavelength laser. Unlike FPALM, the blue-shifted fluorophore in the PA-FSP determines the photoactivation wavelength, typically between 450 and 500 nm, which is both longer and less photo-toxic than the near-UV wavelengths of FPALM. Excitation and imaging is performed on this molecule, until it is in a dark state or is switched off. This switch to the dark state is not a permanent photobleach. This process occurs for many thousands of imaging cycles before permanent photobleaching occurs. STORM has been used to image Rec-A coated circular-plasmid DNA (Rust et al. 2006) and to understand the network formation of mitochondria–microtubules contacts within cells in both 2-dimensional and 3-dimensional imaging formats (Huang et al. 2008a,b) (Figure 13.18).

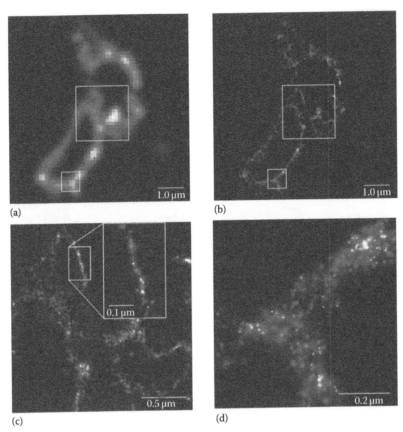

(a)

(b)

(c)

(d)

FIGURE 13.17 (See color insert following page 11-20.) Comparison of TIRF and FPALM images of a COS-7 cell expressing a transmembrane protein labeled with PA-FP Kaede. (a) TIRF image, (b) FPALM image, (c,d) high-magnification views of FPALM images illustrate single-fluorescent protein molecules, enabling the study of lysosome and endosome interactions. (Reprinted from Betzig, E. et al., *Science*, 313, 1642, 2006. With permission.)

13.5.3.4 Practical Considerations

Ultraviolet Photoactivation—Long term exposure to near-UV wavelengths does pose concerns with phototoxicity. These wavelengths are also prone to induce significant autofluorescence and experience higher absorptive losses in all but the purest of materials. UV-grade fused silica is necessary for efficient transmission, and chromatic-aberration-compensated optics are essential.

Imaging Rate—The FPALM and STORM techniques require minimal upgrade from the equipment commonly used in LIF and wide-field fluorescence imaging, bringing nanometer imaging within reach of many laboratories. Image acquisition times on the order of 2–10 h continue to be a bottle-neck; however, the continued improvement of ICCD and EMCCD technologies, with regard to the image field size, acquisition rates, and sensitivity will improve the overall experiment speed.

13.6 Single-Molecule Detection: Applications within Fluidic Environments

Single-molecule observations at concentrations greater than a nanomolar require a spatial or volume confinement in the sub-attoliter regime. Lithographically defined, nanofluidic structures have demonstrated sample volume confinement to physical dimensions at and below the diffraction limit for rapid

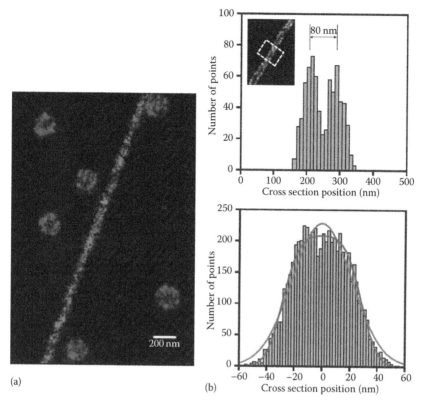

FIGURE 13.18 (a) Microtubules (fibril) and clathrin pits (clusters) in a BS-C-1 cell imaged with STORM. (b) The super-resolution image provides distinct and separable localization of two crossing microtubules with a 80 nm gap (top) and measurement of the microtubule diameter of 80 nm (bottom). (Reprinted from Bates, M. et al., *Science*, 317, 1749, 2007. With permission.)

study of many single molecules or samples at physiologically relevant concentrations. Examples of these fluidic structures are discussed in the following sections.

13.6.1 Zero-Mode Waveguides

ZMWs extend the concept of single-molecule confinement to the physical limits of direct lithographic patterning. Subwavelength dimension holes are formed in a thin (~100 nm) aluminum layer atop a fused silica substrate. Zeptoliter (10^{-21} L) volume confinement is achieved with diameters ranging from 20 to 100 nm. This ultrasmall optical excitation volume enables the study of single molecules at concentrations of physiological relevance, 10 nM to 10 μM, which are inaccessible using most micro- and nanofluidic structures (Levene et al. 2003) (Figure 13.19). Diffusive transport introduces fluorescently labeled molecules into the ZMW aperture and defines the rate of single-molecule observation. The ZMW is easily formed in arrays with over 100 elements that can be studied simultaneously.

The ZMWs were first demonstrated in a single-molecule study of DNA polymerase activity, synthesizing double stranded M13 DNA from a coumarin-dCTP-rich olution (Levene et al. 2003). The polymerization reaction was observed in real time using TCSPC and FCS to study the entire enzymatic DNA synthesis process from start to finish, temporally resolving individual base-pairing events. This ZMW technology is being developed by Pacific Biosciences into a massively parallel, single-molecule DNA-sequencing technology. Sequencing is performed using phospholinked nucleotide-fluorophore

FIGURE 13.19 A photomicrograph and a electron micrograph sequence illustrating the extreme packing densities possible with ZMWs and suggesting a highly parallel SMD architecture. (Reprinted from Levene, M.J. et al., *Science*, 299, 682, 2003. With permission.)

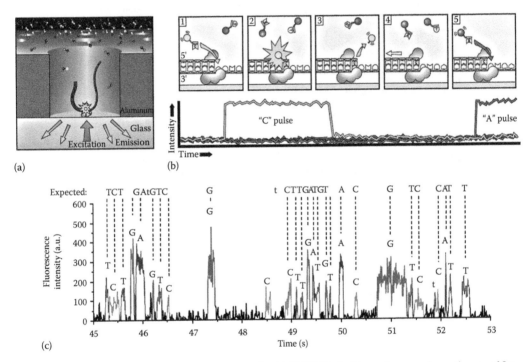

FIGURE 13.20 (See color insert following page 11-20.)   (a) ZMW filled with a polymerase beneath a sea of fluorescently labeled, free nucleotides. (b) Each type of nucleotide is identified with a different color of fluorophore. Nucleotide pairing during synthesis causes dissociation of the nucleotide from the fluorophore. A momentary increase in fluorescence intensity is observed within the waveguide aperture and is recorded as a function of time. (c) A 9 s portion of an intensity vs. time trace illustrates the individual base readout during the DNA sequencing process. (Reprinted from Eid, J. et al., *Science*, 323, 133, 2009. With permission.)

molecules that break apart during DNA synthesis (Foquet et al. 2008; Eid et al. 2009) (Figure 13.20). At the time of dissociation, the fluorescence intensity increases and is recorded using four-color TCSPC, where each fluorophore encodes a different type of nucleotide.

High throughput, simultaneous SMD on thousands of ZMWs has been achieved using holographic phase-shift masks to create a matching number of laser spots, each to form the ultrasmall ZMW focal volume (Lundquist et al. 2008). Improvements in the ZMW process have been made (Foquet et al. 2008) and EMCCD technology is now being used to perform fluorescence readout of the base incorporation at rates faster than 10 ms, while maintaining single-fluorophore sensitivity in this application. For more details on ZMWs, see Chapter 16.

13.6.2 Rapid Single-Molecule Detection in Nanofluidics

Fluidic structures with sub-micrometer and nanometer cross-sectional dimensions provide single-molecule confinement at 10 nM concentrations and less. We have demonstrated a variety of methods and materials for creating channels on this size scale including direct lithographic patterning of fused silica (Stavis et al. 2005a,b,c; Reccius et al. 2008), sacrificial layer patterning with chemical vapor deposition (CVD) polysilicon and silica (Foquet et al. 2002), and sacrificial layer patterning with electrospun fibers (Verbridge et al. 2005); see Figure 13.21. Fused silica is our choice fluidic channel material for its ultralow autofluorescence and repeatability in manufacture by microfabrication processes.

We have implemented these nanofluidics for a variety of applications, similar to existing bioanalytical techniques, using SMD to demonstrate the unparalleled quantization and sensitivity of a single-molecule experiment. In contrast to the ZMW, nanofluidics used for SMD can analyze each and every molecule, making this technology viable to study single molecules with a rare rate of occurrence, even during rapid counting. Electrokinetically driven flow produces millimeter/second flow speeds and plug-like profile to achieve SMD at a rate of over 1000 molecules/min (Stavis et al. 2005a,b,c). TCSPC with photon-counting APDs and a high-speed hardware correlator are necessary at these SMD observation rates.

Our original demonstration of this technology was used to size HindIII-digested lambda DNA, clearly resolving each of seven different lengths in solution. Unlike conventional gel electrophoresis, these fluidics require no size-dependent gel media, and place almost no limitation on the DNA length (Foquet et al. 2002). We have further developed this technique to investigate the conformation of individual DNA during passage through a fluidic channel. Conformation was identified using the fluorescence intensity profile emitted by folded DNA during transit through a laser beam and an algorithm to fit this profile to a collection of physical models (Reccius et al. 2008) (Figure 13.22).

This technique demonstrated molecule length resolution of 114 nm and an impressive 20 ms analysis time per molecule. The ability to study conformation *in situ* suggests opportunities to evaluate the questions of structure–function biology at the single-molecule level.

We have also demonstrated SMD detection of dual-color fluorophore-labeled PCR amplification using nanofluidics. Two-color time coincidence measurements allow singly labeled PCR primers to be distinguished from dual-color labeled amplicon product, measured *in situ* without post-amplification purification (Stavis et al. 2007); see Figure 13.23.

This experiment encourages application of nanofluidics to a variety of binding assays that are currently limited by ensemble averaging and labor-intensive post-purification. It also underscores this as a

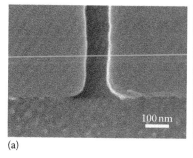 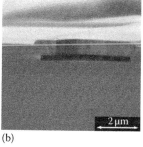 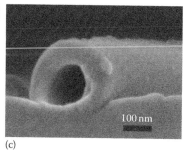

(a)   (b)   (c)

FIGURE 13.21 Nanofluidic structures are formed using a variety of fabrication techniques. (a) Direct lithographic patterning and reactive ion etching are used to form a 100 nm-wide fluidic channel in fused silica for studies of DNA elongation.(Reprinted from Reccius, C.H. et al., *Biophys. J.*, 95, 273, 2008. With permission.) (b) Sacrificial layer processing removes the solid-core of polysilicon from within a silica sub-micrometer fluidic channel, used for DNA sizing. (Reprinted from Foquet, M. et al., *Anal. Chem.*, 74, 1415, 2002. With permission.) (c) An electrospun fiber is used as a sacrificial template to form nanofluidic capillaries. (Reprinted from Verbridge, S.S. et al., *J. Appl. Phys.*, 97, 124317, 2005. With permission.)

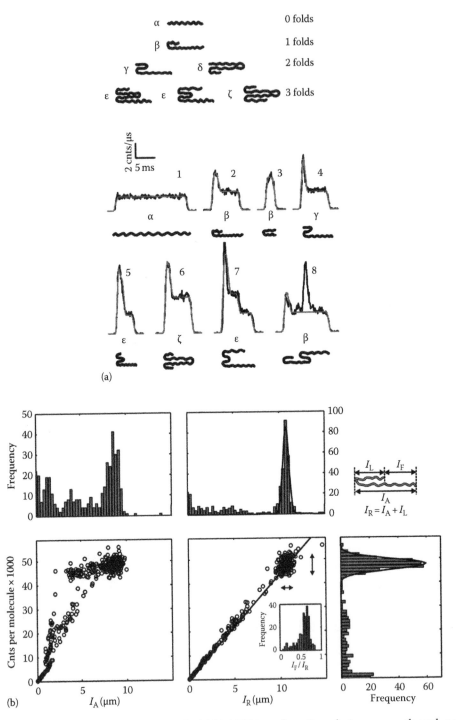

FIGURE 13.22 High-speed analysis of individual folded DNA conformations during passage through a nanofluidic channel. (a) Possible folded DNA configurations and the modeled intensity vs. time corresponding to those configurations. (b) The distribution of photon counts vs. apparent length (I_A) for 416 molecules is noticeably different than their measured real length (I_R) of 10.7 µm, indicating that a folded configuration is observed. The looped length of molecule, (I_L), and the free length of molecule, (I_F), provide further agreement with these results. (Reprinted from Reccius, C.H. et al., *Biophys. J.*, 95, 273, 2008. With permission.)

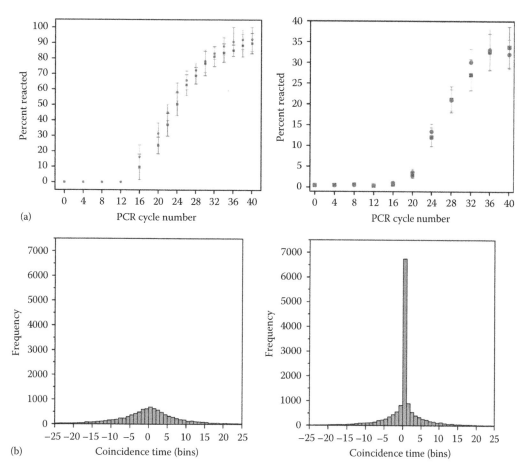

FIGURE 13.23 A PCR amplification is performed on 325 bp segments of 16S rDNA from *Thermobifida fusca*. Primers used in the reaction each have one fluorophore, either green or red, and amplification produces dual-colored amplicon products. (a) TCH of a PCR reaction at Cycle 0 (left) and Cycle 40 (right) illustrate the presence of a dual-color labeled amplicon product. (b) Comparison of amplified product as measured with gel electrophoresis (left) and by single-molecule techniques in a sub-micrometer fluidic channel (right). (Reprinted from Stavis, S.M. et al., *Biomicrofluidics*, 1, 034105, 2007. With permission.)

powerful detection technique, since the probability of false-coincident detection can be further reduced with increased color combinations.

13.6.3 Opportunities and Trends for Optofluidic Integration

Many fluidic devices that have reported SMD require a considerable infrastructure of supporting equipment often found in a research laboratory environment. Equipment such as excitation sources (arc lamps and gas lasers), highly sensitive detectors (EMCCDs and APDs), pumps and macroscale fluidic distribution networks, biological sample preparation tools (cell lysis and DNA amplification), and microscopy equipment are commonly found crowding the small, microfabricated devices used to study biomolecules. We believe that some of these technologies are amenable to on-chip miniaturization, while others continue to be more suitable for off-chip implementation.

A variety of on-chip integrated technologies that would aid in the SMD process are reported in literature. Cell lysis and sample filtering as preparation methods have found application in microfluidic architectures using chemical (Li and Harrison 1997; El-Ali et al. 2005) and electrical (McClain et al.

2003; Lu et al. 2005) lysis, and integrated polymer monoliths (Svec et al. 2000; Yu et al. 2001). These have even been reported in combination with microfluidic chambers for PCR-based amplification on-chip (Krishnan et al. 2004; Easley et al. 2006). These integrated devices would enable investigation of biological materials derived from low-count cell populations, offering exciting opportunities to showcase the exceptional sensitivity and low quantity sample consumption of SMD techniques. Increasingly complicated fluidic mixing and separation tasks have been achieved with herringbone pattern chaotic mixers (Stroock et al. 2002), seperation by entropic trapping of DNA (Han and Craighead 2000; Bakajin et al. 2001), and integrated-waveguide-based near-field capture of DNA (Yang et al. 2009). Fluidic distribution networks have also been shown with valves buried within multilayer-PDMS microfluidics to achieve high-count combinatorial mixing of solutions (Thorsen et al. 2002; Hong and Quake 2003) and reduction of sample dead volumes. However, many of these valve-based implementations still require considerable off-chip hardware. We foresee that application-driven requirements will encourage the merger and integration of these various existing technologies.

The gradual replacement of microscopy components using integrated optical elements is a growing trend in research, and is applicable to each of the methods we have discussed in this chapter—fluorescence, OCR, and SERS. Recent examples of fluorescence-based methods utilize integrated parabolic reflectors affixed to the fluidic chip (Krogmeier et al. 2007) and ARROW-based optofluidic waveguides (Yin et al. 2005, 2006) to aid in the delivery of excitation light and capture of emitted fluorescence. Performing both directed excitation and fluorescence capture in an efficient manner is the fundamental challenge to these devices, establishing a demanding standard of SNR performance to be comparable with confocal microscopy. Deformable optical elements, such as liquid lenses (Tang et al. 2008) and light-guiding sheath flows (Wolfe et al. 2004, 2005), offer creative new approaches to the miniaturized manipulation of light within these microfluidic environments, and may prove helpful to bridge the size and geometry "gaps" between conventional and integrated optics. In contrast, SMD performed using OCR and SERS in fluidic devices do not necessarily require microscopy elements, but imposes the challenges of reduced SNR, as compared to fluorescence methods, and require precise thermal control to achieve stable cavity resonance.

The integration of highly stable and controllable light sources and single-photon-counting detectors present a formidable challenge for monolithic integration with microfluidic systems. In recent years, solid-state lasers have become available at a variety of visible wavelengths, offering more compact and stable alternatives to their gas lasers. New compound semiconductor lasers, using the wide bandgap material gallium nitride, are available as laser pointers, and complete the spectrum of blue, green, and red wavelengths. We believe that these solid-state, visible laser sources and light-emitting diodes (LEDs) provide solutions for increasingly compact systems (Burns et al. 1998). Detector technology integration is currently limited by packaging and thermal management components necessary to achieve single-photon observation. While some authors have demonstrated a "sandwich" of fluidics with a detector array, or form their fluidics on top of active electronic substrates (Burns et al. 1998; Webster et al. 2001), these configurations currently lack sufficient sensitivity and speed required for rapid SMD.

13.7 Summary

Single-molecule studies continue to challenge the *status quo* ensemble measurement techniques currently available in many molecular biology and biology laboratories. SMD offers unique capabilities for studying the identity and characteristics of each molecule within a population. One of the most elegant aspects of SMD is the breadth of information accessible from measurements of time, position, and intensity. Poisson statistics and mathematical models of physical and chemical phenomena provide a powerful toolbox to derive measurements of diffusion, concentration, charge, size, and conformational state.

We have provided a focused review of demonstrated technologies for SMD. Fluorescence-based methods have received the greatest attention and development within the scientific community, but techniques based upon SERS and optical resonance offer attractive options in systems integration and label-free detection of single molecules. The stringent premium currently placed on analyte processing and purification in single-molecule techniques will continue to be a central theme, driving research and development of methods requiring less pre-processing and purification of biological samples.

Systems integration is the ultimate embodiment of many single-molecule techniques we have discussed (Craighead 2006). Indicative of the collective desire to achieve "lab-on-a-chip" miniaturization with full functionality, there has been a considerable increase in the number of publications discussing highly multiplexed analysis, simplified materials and fabrication methods for fluidics, and more complete integration of fluorescence excitation sources and emission detectors on a common substrate. We believe this trend will continue as these technologies strive toward a mature status, entering the mainstream marketplace for diagnostic tools at the hospital bedside, the clinical research lab, and even at home.

References

Alivisatos, P. (2004). The use of nanocrystals in biological detection. *Nature Biotechnology* 22(1): 47–52.

Allemand, J. F., D. Bensimon, L. Jullien, A. Bensimon, and V. Croquette (1997). pH-dependent specific binding and combing of DNA. *Biophysical Journal* 73(4): 2064–2070.

Ambrose, W. P., P. M. Goodwin, J. H. Jett et al. (1999). Single molecule fluorescence spectroscopy at ambient temperature. *Chemical Reviews* 99(10): 2929–2956.

Andor (2009). *Digital Camera Fundamentals*. Retrieved May 1, 2009, from http://www.andor.com/pdfs/Digital%20Camera%20Fundamentals.pdf.

Armani, A. M., R. P. Kulkarni, S. E. Fraser, R. C. Flagan, and K. J. Vahala (2007). Label-free, single-molecule detection with optical microcavities. *Science* 317(5839): 783–787.

Axelrod, D. (2001). Total internal reflection fluorescence microscopy in cell biology. *Traffic* 2(11): 764–774.

Axelrod, D., T. P. Burghardt, and N. L. Thompson (1984). Total internal-reflection fluorescence. *Annual Review of Biophysics and Bioengineering* 13: 247–268.

Baehr-Jones, T., M. Hochberg, C. Walker, and A. Scherer (2004). High-Q ring resonators in thin silicon-on-insulator. *Applied Physics Letters* 85(16): 3346–3347.

Bakajin, O., T. A. J. Duke, J. Tegenfeldt et al. (2001). Separation of 100-kilobase DNA molecules in 10 seconds. *Analytical Chemistry* 73(24): 6053–6056.

Barkai, E., Y. J. Jung, and R. Silbey (2004). Theory of single-molecule spectroscopy: Beyond the ensemble average. *Annual Review of Physical Chemistry* 55: 457–507.

Barrios, C. A., V. R. Almeida, R. R. Panepucci, B. S. Schmidt, and M. Lipson (2004). Compact silicon tunable Fabry-Perot resonator with low power consumption. *IEEE Photonics Technology Letters* 16(2): 506–508.

Bates, M., T. R. Blosser, and X. W. Zhuang (2005). Short-range spectroscopic ruler based on a single-molecule optical switch. *Physical Review Letters* 94(10): 108101.

Bates, M., B. Huang, G. T. Dempsey, and X. W. Zhuang (2007). Multicolor super-resolution imaging with photo-switchable fluorescent probes. *Science* 317(5845): 1749–1753.

Bensimon, A., A. Simon, A. Chiffaudel et al. (1994). Alignment and sensitive detection of DNA by a moving interface. *Science* 265(5181): 2096–2098.

Berezhkovskii, A. M., A. Szabo, and G. H. Weiss (1999). Theory of single-molecule fluorescence spectroscopy of two-state systems. *Journal of Chemical Physics* 110(18): 9145–9150.

Betzig, E., G. H. Patterson, R. Sougrat et al. (2006). Imaging intracellular fluorescent proteins at nanometer resolution. *Science* 313(5793): 1642–1645.

Bohmer, M. and J. Enderlein (2003). Fluorescence spectroscopy of single molecules under ambient conditions: Methodology and technology. *Chemphyschem* 4(8): 793–808.

Bruchez, M., M. Moronne, P. Gin, S. Weiss, and A. P. Alivisatos (1998). Semiconductor nanocrystals as fluorescent biological labels. *Science* 281(5385): 2013–2016.

Burns, A., H. Ow, and U. Wiesner (2006). Fluorescent core-shell silica nanoparticles: Towards Lab on a Particle architectures for nanobiotechnology. *Chemical Society Reviews* 35(11): 1028–1042.

Burns, M. A., B. N. Johnson, S. N. Brahmasandra et al. (1998). An integrated nanoliter DNA analysis device. *Science* 282(5388): 484–487.

Burns, A. A., J. Vider, H. Ow et al. (2009). Fluorescent silica nanoparticles with efficient urinary excretion for nanomedicine. *Nano Letters* 9(1): 442–448.

Castro, A. and R. T. Okinaka (2000). Ultrasensitive, direct detection of a specific DNA sequence of *Bacillus anthracis* in solution. *Analyst* 125(1): 9–11.

Chen, Y., J. D. Muller, P. T. C. So, and E. Gratton (1999). The photon counting histogram in fluorescence fluctuation spectroscopy. *Biophysical Journal* 77(1): 553–567.

Claudet, C., D. Angelov, P. Bouvet, S. Dimitrov, and J. Bednar (2005). Histone octamer instability under single molecule experiment conditions. *Journal of Biological Chemistry* 280(20): 19958–19965.

Craighead, H. (2006). Future lab-on-a-chip technologies for interrogating individual molecules. *Nature* 442(7101): 387–393.

Desai, S. P., B. A. Taff, and J. Voldman (2008). A photopatternable silicone for biological applications. *Langmuir* 24(2): 575–581.

Easley, C. J., J. M. Karlinsey, J. M. Bienvenue et al. (2006). A fully integrated microfluidic genetic analysis system with sample-in-answer-out capability. *Proceedings of the National Academy of Sciences of the United States of America* 103(51): 19272–19277.

Eid, J., A. Fehr, J. Gray et al. (2009). Real-time DNA sequencing from single polymerase molecules. *Science* 323(5910): 133–138.

El-Ali, J., S. Gaudet, A. Gunther, P. K. Sorger, and K. F. Jensen (2005). Cell stimulus and lysis in a microfluidic device with segmented gas-liquid flow. *Analytical Chemistry* 77(11): 3629–3636.

Elson, E. L. and D. Magde (1974). Fluorescence correlation spectroscopy. 1. Conceptual basis and theory. *Biopolymers* 13(1): 1–27.

Enderlein, J., D. L. Robbins, W. P. Ambrose, and R. A. Keller (1998). Molecular shot noise, burst size distribution, and single-molecule detection in fluid flow: Effects of multiple occupancy. *Journal of Physical Chemistry A* 102(30): 6089–6094.

Eun, A. J. C. and S. M. Wong (2000). Molecular beacons: A new approach to plant virus detection. *Phytopathology* 90(3): 269–275.

Feller, W. (1968). *An Introduction to Probability Theory and Its Applications*. New York, John Wiley & Sons.

Femino, A., F. S. Fay, K. Fogarty, and R. H. Singer (1998). Visualization of single RNA transcripts *in situ*. *Science* 280(5363): 585–590.

Foquet, M., J. Korlach, W. Zipfel, W. W. Webb, and H. G. Craighead (2002). DNA fragment sizing by single molecule detection in submicrometer-sized closed fluidic channels. *Analytical Chemistry* 74(6): 1415–1422.

Foquet, M., J. Korlach, W. R. Zipfel, W. W. Webb, and H. G. Craighead (2004). Focal volume confinement by submicrometer-sized fluidic channels. *Analytical Chemistry* 76(6): 1618–1626.

Foquet, M., K. T. Samiee, X. X. Kong et al. (2008). Improved fabrication of zero-mode waveguides for single-molecule detection. *Journal of Applied Physics* 103(3): 034301.

Fortina, P. and S. Surrey (2008). Digital mRNA profiling. *Nature Biotechnology* 26(3): 293–294.

Geiss, G. K., R. E. Bumgarner, B. Birditt et al. (2008). Direct multiplexed measurement of gene expression with color-coded probe pairs. *Nature Biotechnology* 26(3): 317–325.

Giepmans, B. N. G., S. R. Adams, M. H. Ellisman, and R. Y. Tsien (2006). Review—The fluorescent toolbox for assessing protein location and function. *Science* 312(5771): 217–224.

Golding, I. and E. C. Cox (2004). RNA dynamics in live *Escherichia coli* cells. *Proceedings of the National Academy of Sciences of the United States of America* 101(31): 11310–11315.

Golding, I., J. Paulsson, S. M. Zawilski, and E. C. Cox (2005). Real-time kinetics of gene activity in individual bacteria. *Cell* 123(6): 1025–1036.

Goulian, M. and S. M. Simon (2000). Tracking single proteins within cells. *Biophysical Journal* 79(4): 2188–2198.

Gurrieri, S., K. S. Wells, I. D. Johnson, and C. Bustamante (1997). Direct visualization of individual DNA molecules by fluorescence microscopy: Characterization of the factors affecting signal/background and optimization of imaging conditions using YOYO. *Analytical Biochemistry* 249(1): 44–53.

Ha, T. (2001a). Single-molecule fluorescence methods for the study of nucleic acids. *Current Opinion in Structural Biology* 11(3): 287–292.

Ha, T. (2001b). Single-molecule fluorescence resonance energy transfer. *Methods* 25(1): 78–86.

Han, J. and H. G. Craighead (2000). Separation of long DNA molecules in a microfabricated entropic trap array. *Science* 288(5468): 1026–1029.

Han, W. P., Y. K. Ng, D. Axelrod, and E. S. Levitan (1999). Neuropeptide release by efficient recruitment of diffusing cytoplasmic secretory vesicles. *Proceedings of the National Academy of Sciences of the United States of America* 96(25): 14577–14582.

Harada, Y., T. Funatsu, K. Murakami et al. (1999). Single-molecule imaging of RNA polymerase-DNA interactions in real time. *Biophysical Journal* 76(2): 709–715.

Harms, G. S., M. Sonnleitner, G. J. Schutz, H. J. Gruber, and T. Schmidt (1999). Single-molecule anisotropy imaging. *Biophysical Journal* 77(5): 2864–2870.

Haupts, U., S. Maiti, P. Schwille, and W. W. Webb (1998). Dynamics of fluorescence fluctuations in green fluorescent protein observed by fluorescence correlation spectroscopy. *Proceedings of the National Academy of Sciences of the United States of America* 95(23): 13573–13578.

Hell, S. W. (2007). Far-field optical nanoscopy. *Science* 316(5828): 1153–1158.

Herrick, J. and A. Bensimon (1999). Imaging of single DNA molecule: Applications to high-resolution genomic studies. *Chromosome Research* 7(6): 409–423.

Hess, S. T., T. P. K. Girirajan, and M. D. Mason (2006). Ultra-high resolution imaging by fluorescence photoactivation localization microscopy. *Biophysical Journal* 91(11): 4258–4272.

Hillesheim, L. N. and J. D. Muller (2003). The photon counting histogram in fluorescence fluctuation spectroscopy with non-ideal photodetectors. *Biophysical Journal* 85(3): 1948–1958.

Hong, J. W. and S. R. Quake (2003). Integrated nanoliter systems. *Nature Biotechnology* 21(10): 1179–1183.

Huang, B., S. A. Jones, B. Brandenburg, and X. W. Zhuang (2008a). Whole-cell 3D STORM reveals interactions between cellular structures with nanometer-scale resolution. *Nature Methods* 5(12): 1047–1052.

Huang, B., W. Q. Wang, M. Bates, and X. W. Zhuang (2008b). Three-dimensional super-resolution imaging by stochastic optical reconstruction microscopy. *Science* 319(5864): 810–813.

Invitrogen (2009). *The Handbook—A Guide to Fluorescent Probes and Labeling Technologies*. Retrieved May 2, 2009, from http://www.invitrogen.com/site/us/en/home/References/Molecular-Probes-The-Handbook.html.

Juette, M. F., T. J. Gould, M. D. Lessard et al. (2008). Three-dimensional sub-100 nm resolution fluorescence microscopy of thick samples. *Nature Methods* 5(6): 527–529.

Kanda, T., K. F. Sullivan, and G. M. Wahl (1998). Histone-GFP fusion protein enables sensitive analysis of chromosome dynamics in living mammalian cells. *Current Biology* 8(7): 377–385.

Kapanidis, A. N. and S. Weiss (2002). Fluorescent probes and bioconjugation chemistries for single-molecule fluorescence analysis of biomolecules. *Journal of Chemical Physics* 117(24): 10953–10964.

Kneipp, K., Y. Wang, H. Kneipp et al. (1997). Single molecule detection using surface-enhanced Raman scattering (SERS). *Physical Review Letters* 78(9): 1667–1670.

Kneipp, K., H. Kneipp, V. B. Kartha et al. (1998). Detection and identification of a single DNA base molecule using surface-enhanced Raman scattering (SERS). *Physical Review E* 57(6): R6281–R6284.

Kneipp, K., H. Kneipp, I. Itzkan, R. R. Dasari, and M. S. Feld (1999). Ultrasensitive chemical analysis by Raman spectroscopy. *Chemical Reviews* 99(10): 2957–2976.

Krishnan, M., D. T. Burke, and M. A. Burns (2004). Polymerase chain reaction in high surface-to-volume ratio SiO_2 microstructures. *Analytical Chemistry* 76(22): 6588–6593.

Krogmeier, J. R., I. Schaefer, G. Seward, G. R. Yantz, and J. W. Larson (2007). An integrated optics microfluidic device for detecting single DNA molecules. *Lab on a Chip* 7(12): 1767–1774.

Levene, M. J., J. Korlach, S. W. Turner et al. (2003). Zero-mode waveguides for single-molecule analysis at high concentrations. *Science* 299(5607): 682–686.

Levy, S. L., J. T. Mannion, J. Cheng, C. H. Reccius, and H. G. Craighead (2008). Entropic unfolding of DNA molecules in nanofluidic channels. *Nano Letters* 8(11): 3839–3844.

Li, P. C. H. and D. J. Harrison (1997). Transport, manipulation, and reaction of biological cells on-chip using electrokinetic effects. *Analytical Chemistry* 69(8): 1564–1568.

Li, J. W. J., X. H. Fang, and W. H. Tan (2002). Molecular aptamer beacons for real-time protein recognition. *Biochemical and Biophysical Research Communications* 292(1): 31–40.

Lu, H., M. A. Schmidt, and K. F. Jensen (2005). A microfluidic electroporation device for cell lysis. *Lab on a Chip* 5(1): 23–29.

Lundquist, P. M., C. F. Zhong, P. Q. Zhao et al. (2008). Parallel confocal detection of single molecules in real time. *Optics Letters* 33(9): 1026–1028.

Magde, D., W. W. Webb, and E. Elson (1972). Thermodynamic fluctuations in a reacting system—Measurement by fluorescence correlation spectroscopy. *Physical Review Letters* 29(11): 705–708.

Magde, D., E. L. Elson, and W. W. Webb (1974). Fluorescence correlation spectroscopy. 2. Experimental realization. *Biopolymers* 13(1): 29–61.

Magde, D., W. W. Webb, and E. L. Elson (1978). Fluorescence correlation spectroscopy. 3. Uniform translation and laminar-flow. *Biopolymers* 17(2): 361–376.

Maiti, S., U. Haupts, and W. W. Webb (1997). Fluorescence correlation spectroscopy: Diagnostics for sparse molecules. *Proceedings of the National Academy of Sciences of the United States of America* 94(22): 11753–11757.

Manganelli, R., E. Dubnau, S. Tyagi, F. R. Kramer, and I. Smith (1999). Differential expression of 10 sigma factor genes in *Mycobacterium tuberculosis*. *Molecular Microbiology* 31(2): 715–724.

Mannion, J. T., C. H. Reccius, J. D. Cross, and H. G. Craighead (2006). Conformational analysis of single DNA molecules undergoing entropically induced motion in nanochannels. *Biophysical Journal* 90(12): 4538–4545.

McClain, M. A., C. T. Culbertson, S. C. Jacobson et al. (2003). Microfluidic devices for the high-throughput chemical analysis of cells. *Analytical Chemistry* 75(21): 5646–5655.

Measor, P., L. Seballos, D. L. Yin et al. (2007). On-chip surface-enhanced Raman scattering detection using integrated liquid-core waveguides. *Applied Physics Letters* 90(21): 211107.

Michalet, X., R. Ekong, F. Fougerousse et al. (1997). Dynamic molecular combing: Stretching the whole human genome for high-resolution studies. *Science* 277(5331): 1518–1523.

Michalet, X., A. N. Kapanidis, T. Laurence et al. (2003). The power and prospects of fluorescence microscopies and spectroscopies. *Annual Review of Biophysics and Biomolecular Structure* 32: 161–182.

Michalet, X., F. F. Pinaud, L. A. Bentolila et al. (2005). Quantum dots for live cells, *in vivo* imaging, and diagnostics. *Science* 307(5709): 538–544.

Michalet, X., O. H. W. Siegmund, J. V. Vallerga et al. (2007). Detectors for single-molecule fluorescence imaging and spectroscopy. *Journal of Modern Optics* 54(2–3): 239–281.

Moerner, W. E. and D. P. Fromm (2003). Methods of single-molecule fluorescence spectroscopy and microscopy. *Review of Scientific Instruments* 74(8): 3597–3619.

Murase, K., T. Fujiwara, Y. Umemura et al. (2004). Ultrafine membrane compartments for molecular diffusion as revealed by single molecule techniques. *Biophysical Journal* 86(6): 4075–4093.

Nie, S. M. and S. R. Emery (1997). Probing single molecules and single nanoparticles by surface-enhanced Raman scattering. *Science* 275(5303): 1102–1106.

Nie, S. M., D. T. Chiu, and R. N. Zare (1995). Real-time detection of single-molecules in solution by confocal fluorescence microscopy. *Analytical Chemistry* 67(17): 2849–2857.

Nirmal, M., B. O. Dabbousi, M. G. Bawendi et al. (1996). Fluorescence intermittency in single cadmium selenide nanocrystals. *Nature* 383(6603): 802–804.

Ow, H., D. R. Larson, M. Srivastava et al. (2005). Bright and stable core-shell fluorescent silica nanoparticles. *Nano Letters* 5(1): 113–117.

Panchuk-Voloshina, N., R. P. Haugland, J. Bishop-Stewart et al. (1999). Alexa dyes, a series of new fluorescent dyes that yield exceptionally bright, photostable conjugates. *Journal of Histochemistry & Cytochemistry* 47(9): 1179–1188.

Petty, H. R. (2007). Fluorescence microscopy: Established and emerging methods, experimental strategies, and applications in immunology. *Microscopy Research and Technique* 70(8): 687–709.

Piruska, A., I. Nikcevic, S. H. Lee et al. (2005). The autofluorescence of plastic materials and chips measured under laser irradiation. *Lab on a Chip* 5(12): 1348–1354.

Reccius, C. H., J. T. Mannion, J. D. Cross, and H. G. Craighead (2005). Compression and free expansion of single DNA molecules in nanochannels. *Physical Review Letters* 95(26): 268101.

Reccius, C. H., S. M. Stavis, J. T. Mannion, L. P. Walker, and H. G. Craighead (2008). Conformation, length, and speed measurements of electrodynamically stretched DNA in nanochannels. *Biophysical Journal* 95(1): 273–286.

Robinson, J. T., L. Chen, and M. Lipson (2008). On-chip gas detection in silicon optical microcavities. *Optics Express* 16(6): 4296–4301.

Roy, R., S. Hohng, and T. Ha (2008). A practical guide to single-molecule FRET. *Nature Methods* 5(6): 507–516.

Rust, M. J., M. Bates, and X. W. Zhuang (2006). Sub-diffraction-limit imaging by stochastic optical reconstruction microscopy (STORM). *Nature Methods* 3(10): 793–795.

Rye, H. S., S. Yue, D. E. Wemmer et al. (1992). Stable fluorescent complexes of double-stranded DNA with bis-intercalating asymmetric cyanine dyes—Properties and applications. *Nucleic Acids Research* 20(11): 2803–2812.

Samiee, K. T., M. Foquet, L. Guo, E. C. Cox, and H. G. Craighead (2005). Lambda-repressor oligomerization kinetics at high concentrations using fluorescence correlation spectroscopy in zero-mode waveguides. *Biophysical Journal* 88(3): 2145–2153.

Schmidt, H. and A. R. Hawkins (2008). Optofluidic waveguides: I. Concepts and implementations. *Microfluidics and Nanofluidics* 4(1–2): 3–16.

Schutz, G. J., H. Schindler, and T. Schmidt (1997). Single-molecule microscopy on model membranes reveals anomalous diffusion. *Biophysical Journal* 73(2): 1073–1080.

Schwille, P., F. J. Meyer-Almes, and R. Rigler (1997). Dual-color fluorescence cross-correlation spectroscopy for multicomponent diffusional analysis in solution. *Biophysical Journal* 72(4): 1878–1886.

Schwille, P., J. Korlach, and W. W. Webb (1999). Fluorescence correlation spectroscopy with single-molecule sensitivity on cell and model membranes. *Cytometry* 36(3): 176–182.

Sengupta, P., K. Garai, J. Balaji, N. Periasamy, and S. Maiti (2003a). Measuring size distribution in highly heterogeneous systems with fluorescence correlation spectroscopy. *Biophysical Journal* 84(3): 1977–1984.

Sengupta, P., K. Garai, B. Sahoo et al. (2003b). The amyloid beta peptide (A beta[1–40]) is thermodynamically soluble at physiological concentrations. *Biochemistry* 42(35): 10506–10513.

Stavis, S. M., J. B. Edel, Y. G. Li et al. (2005a). Detection and identification of nucleic acid engineered fluorescent labels in submicrometre fluidic channels. *Nanotechnology* 16(7): S314–S323.

Stavis, S. M., J. B. Edel, Y. G. Li et al. (2005b). Single-molecule mobility and spectral measurements in submicrometer fluidic channels. *Journal of Applied Physics* 98(4): 044903.

Stavis, S. M., J. B. Edel, K. T. Samiee, and H. G. Craighead (2005c). Single molecule studies of quantum dot conjugates in a submicrometer fluidic channel. *Lab on a Chip* 5(3): 337–343.

Stavis, S. M., S. C. Corgie, B. R. Cipriany, and H. G. Craighead (2007). Single molecule analysis of bacterial polymerase chain reaction products in submicrometer fluidic channels. *Biomicrofluidics* 1(3): 034105.

Steyer, J. A. and W. Almers (2001). A real-time view of life within 100 nm of the plasma membrane. *Nature Reviews Molecular Cell Biology* 2(4): 268–275.

Stroock, A. D., S. K. W. Dertinger, A. Ajdari et al. (2002). Chaotic mixer for microchannels. *Science* 295(5555): 647–651.

Svec, F., E. C. Peters, D. Sykora, and J. M. J. Frechet (2000). Design of the monolithic polymers used in capillary electrochromatography columns. *Journal of Chromatography A* 887(1–2): 3–29.

Tan, W. H., K. M. Wang, and T. J. Drake (2004). Molecular beacons. *Current Opinion in Chemical Biology* 8(5): 547–553.

Tang, S. K. Y., C. A. Stan, and G. M. Whitesides (2008). Dynamically reconfigurable liquid-core liquid-cladding lens in a microfluidic channel. *Lab on a Chip* 8(3): 395–401.

Tegenfeldt, J. O., C. Prinz, H. Cao et al. (2004). The dynamics of genomic-length DNA molecules in 100-nm channels. *Proceedings of the National Academy of Sciences of the United States of America* 101(30): 10979–10983.

Thorsen, T., S. J. Maerkl, and S. R. Quake (2002). Microfluidic large-scale integration. *Science* 298(5593): 580–584.

Toomre, D. and D. J. Manstein (2001). Lighting up the cell surface with evanescent wave microscopy. *Trends in Cell Biology* 11(7): 298–303.

Tsien, R. Y. (1998). The green fluorescent protein. *Annual Review of Biochemistry* 67: 509–544.

Tyagi, S. and F. R. Kramer (1996). Molecular beacons: Probes that fluoresce upon hybridization. *Nature Biotechnology* 14(3): 303–308.

Vale, R. D., T. Funatsu, D. W. Pierce et al. (1996). Direct observation of single kinesin molecules moving along microtubules. *Nature* 380(6573): 451–453.

Verbridge, S. S., J. B. Edel, S. M. Stavis et al. (2005). Suspended glass nanochannels coupled with micro-structures for single molecule detection. *Journal of Applied Physics* 97(12): 124317.

Vollmer, F. and S. Arnold (2008). Whispering-gallery-mode biosensing: Label-free detection down to single molecules. *Nature Methods* 5(7): 591–596.

Walter, N. G., C. Y. Huang, A. J. Manzo, and M. A. Sobhy (2008). Do-it-yourself guide: How to use the modern single-molecule toolkit. *Nature Methods* 5(6): 475–489.

Webb, W. W. (2001). Fluorescence correlation spectroscopy: Inception, biophysical experimentations, and prospectus. *Applied Optics* 40(24): 3969–3983.

Webster, J. R., M. A. Burns, D. T. Burke, and C. H. Mastrangelo (2001). Monolithic capillary electrophoresis device with integrated fluorescence detector. *Analytical Chemistry* 73(7): 1622–1626.

Wolfe, D. B., R. S. Conroy, P. Garstecki et al. (2004). Dynamic control of liquid-core/liquid-cladding optical waveguides. *Proceedings of the National Academy of Sciences of the United States of America* 101(34): 12434–12438.

Wolfe, D. B., D. V. Vezenov, B. T. Mayers et al. (2005). Diffusion-controlled optical elements for optofluidics. *Applied Physics Letters* 87(18): 181105.

Xu, H. X., E. J. Bjerneld, M. Kall, and L. Borjesson (1999). Spectroscopy of single hemoglobin molecules by surface enhanced Raman scattering. *Physical Review Letters* 83(21): 4357–4360.

Xu, Q. F., V. R. Almeida, R. R. Panepucci, and M. Lipson (2004). Experimental demonstration of guiding and confining light in nanometer-size low-refractive-index material. *Optics Letters* 29(14): 1626–1628.

Yang, A. H. J., S. D. Moore, B. S. Schmidt et al. (2009). Optical manipulation of nanoparticles and biomolecules in sub-wavelength slot waveguides. *Nature* 457(7225): 71–75.

Yildiz, A., J. N. Forkey, S. A. McKinney et al. (2003). Myosin V walks hand-over-hand: Single fluorophore imaging with 1.5-nm localization. *Science* 300(5628): 2061–2065.

Yin, D. L., J. P. Barber, A. R. Hawkins, and H. Schmidt (2005). Waveguide loss optimization in hollow-core ARROW waveguides. *Optics Express* 13(23): 9331–9336.

Yin, D. L., D. W. Deamer, H. Schmidt, J. P. Barber, and A. R. Hawkins (2006). Single-molecule detection sensitivity using planar integrated optics on a chip. *Optics Letters* 31(14): 2136–2138.

Yu, C., M. H. Davey, F. Svec, and J. M. J. Frechet (2001). Monolithic porous polymer for on-chip solid-phase extraction and preconcentration prepared by photoinitiated in situ polymerization within a microfluidic device. *Analytical Chemistry* 73(21): 5088–5096.

Zhang, C. Y., S. Y. Chao, and T. H. Wang (2005). Comparative quantification of nucleic acids using single-molecule detection and molecular beacons. *Analyst* 130(4): 483–488.

Zhuang, X. W., L. E. Bartley, H. P. Babcock et al. (2000). A single-molecule study of RNA catalysis and folding. *Science* 288(5473): 2048–2051.

14

Optical Trapping and Manipulation

Eric Pei-Yu Chiou

14.1 Optical Manipulation with Direct Optical Forces

14.1.1 Introduction

Optical forces for particle manipulation have been widely applied in various fields since the technique was first introduced by Ashkin (1970). Through sculpturing the spatial distribution of light fields, versatile functional manipulations including optical trapping (Ashkin et al., 1986), rotation (Curtis and Grier, 2003), sorting of micro- and nanoscale particles (MacDonald et al., 2003; Wang et al., 2005), and large-scale trapping (Garces-Chavez et al., 2006) have been demonstrated. Accurate control of optical forces also makes optical tweezers a powerful tool for studying fundamental biological sciences, including molecular forces in molecular motors (Svoboda et al., 1993; Svoboda and Block, 1994; Mehta et al., 1999), DNAs (Purohit et al., 2003; Chemla et al., 2005) and cell mechanics (Samuel and Berg, 1996; Jass et al., 2004). Recent progress in near-field trapping (Yang et al., 2009) and plasmonic trapping (Yannopapas, 2008) also promise optical manipulation of nanoscale particles with near-field intensity enhanced by nanostructures. Integrating optical manipulation with microfluidic systems is also another vibrant field since optical forces are biocompatible and provide unique features of trapping, transport, and sorting functions (MacDonald et al., 2003; Wang et al., 2005). Several semi-optical trapping and manipulation mechanisms have also been proposed in recent years. Optoelectronic tweezers (OET) utilize a light-patterned electric field to achieve particle manipulation functions (Chiou et al., 2005). Since the light field in OETs is used to trigger an electric field, the optical power required to sculpt a potential profile is orders of magnitude lower than conventional optical tweezers. This promises optical manipulation on a large area without facing the trade-off between trap forces and optical power. Optical manipulation with other indirect optical forces such as light-actuated ac electroosmosis (LACE) (Chiou et al., 2008), optomagnetic tweezers

(Mehta et al., 2008), and anti-Brownian electrokinetic (ABEL) trap (Cohen and Moerner, 2005, 2006) will also be discussed.

14.1.2 Principles of Optical Manipulation

14.1.2.1 Mechanical Momentum of Light

Light is an electromagnetic wave and carries both energy and momentum. Light illumination on a light-absorbing object not only creates heat but also imparts mechanical momentum to the illuminated object. The energy and momentum carried by a light beam is discrete and has a minimum unit of a photon. From the classical mechanics point of view, a photon is a very unique type of particle that carries energy and momentum but has no mass. The energy of a photon (E_{photon}) is given by the product of Planck's constant ($h=6.626\times10^{-34}\,J\cdot s$) and the frequency of the light (ν): $E_{photon}=h\nu$. The momentum of a photon is expressed as $p=h/\lambda$, where λ is the wavelength of light. Since the energy of a single photon is small, a light beam usually carries large numbers of photons during propagation. For example, a 1.0 mW He–Ne laser beam with a wavelength of 632.8 nm delivers 3.2×10^{15} photons/s. Assuming the beam is fully absorbed by a particle, this particle experiences a force of 3.2 pN calculated by Newton's second law: $F=dp/dt$. If the illuminated particles are transparent, the momentum transfer comes from the optical reflection and refraction by the particles, which change the propagation (or momentum) direction of the illumination light beam.

14.1.2.2 Optical Radiation Pressure and Early Optical Manipulation

Since the radiation pressure of a photon is small, it was not clear how to use it for practical applications until the invention of the laser in 1960. In 1970, Ashkin demonstrated that with a focused laser beam, one can use radiation pressure to significantly change the dynamics of small, transparent, micrometer-sized particles (Ashkin, 1970). He identified two types of forces: one is the scattering force in the beam-propagation direction and the other one is the gradient force that tries to pull particles into high-intensity regions. In his experiment, it was observed that latex particles were pulled toward the beam axis and pushed forward in the light-propagation direction. Both forces originate from radiation pressure as shown in Figure 14.1. Assume a transparent sphere with a diameter larger than the laser and an optical refractive index larger than its surrounding medium is positioned slightly off the axis of a mildly focused Gaussian beam. Neglect all reflection and absorption effects, and consider a pair of light beams "a" and "b" striking the sphere. The amplitude of these two rays after refraction remains the same but the direction is changed. Figure 14.1b and c illustrates how the momentum of ray "a" and "b" is changed before and after refraction by the sphere. To fulfill the conservation of momentum, ray "a" and "b" must exert momentum $P_{a,particle}$ and $P_{b,particle}$, respectively,

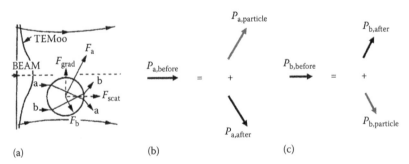

FIGURE 14.1 (a) Ray-optics analysis of optical forces exerted on a transparent dielectric sphere illuminated by a mildly focused Gaussian beam. (Reprinted from Ashkin, A., *Phys. Rev. Lett.*, 24, 156, 1970. With permission. Copyright 2009 from American Physical Society.) (b) Momentum change of ray "a" before and after refraction. (c) Momentum analysis of ray "b." The net optical force on the sphere is the sum of the momentum transferred to the particle from all refracted rays.

on the sphere with a direction and a magnitude as shown. The summation of all the rays incident on the sphere gives a net optical force that can be decomposed into two components: the scattering force, F_{scat}, pushing the sphere along the propagation direction and the gradient force, F_{grad}, pulling the sphere toward the high-intensity region. The optical force calculated using ray optics has been shown to give good agreement with experimental results for particles with sizes much larger than the wavelength (Mie regime, particle diameter $d \gg \lambda$).

For a particle with a size much smaller than the wavelength, optical forces can be obtained by calculating the dipole force, $F_D = 1/2\ \alpha\nabla E^2$, induced on the particle in the optical field, where α is the optically induced polarization of the particle and E is the electric field (Rayleigh regime, particle diameter $d \ll \lambda$). This dipole force pushes particles with a refractive index higher than the surrounding medium into regions with a high-optical intensity as shown in Figure 14.2. Since the induced polarization of a particle is linearly proportional to the volume of a particle, optical forces decay fast as the size decreases ($F \sim d^3$). Forming stable optical traps to confine nanoscale particles needs a higher laser power to overcome the strong Brownian motion.

For particles with sizes close to the wavelength, the calculation of optical forces requires more rigorous methods. One way is to numerically simulate the electromagnetic field around the particle and integrate the Maxwell stress tensor over the entire object surface to get the optical force (Grigorenko et al., 2008).

14.1.2.3 Achieving a Stable Three-Dimensional Optical Trap

The key to achieving a stable three-dimensional (3D) optical trap is to overcome the scattering force that pushes particles down the optical axis. This can be accomplished by illuminating two counter-propagating Gaussian beams with the same laser power on a particle as shown in Figure 14.3. The scattering

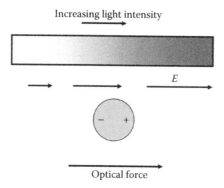

FIGURE 14.2 Optical forces on particles much smaller than the wavelength of light can be calculated by assuming an electric dipole is induced on the particle. This interaction between this induced dipole and the optical field generates a gradient force pushing the particle into strong intensity regions.

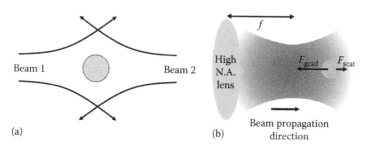

FIGURE 14.3 (a) A stable three-dimensional optical trap created by two counter-propagating, divergent laser beams. (b) A stable 3D trap created by a single laser beam focused by a high-N.A. objective lens.

forces from these two beams cancel with each other, and the gradient forces add up together to trap particles in the center of these two beams. If one of the beams is blocked, the trapped particles will be pushed away from the center due to the scattering force from the unblocked beam.

A 3D stable optical trap can also be accomplished by focusing a laser beam using a high-numerical aperture (N.A. typically >1) objective lens. The first 3D single-beam optical trap, know as optical tweezers, was demonstrated in 1986 (Ashkin et al., 1986). A laser beam focused by a high-N.A. objective lens can create strong gradient forces near the focal point to overcome the axial scattering forces as shown in Figure 14.3b.

The single-beam optical trap has wide applications in biological fields, especially for cell manipulation and for studying forces generated by biomolecules. In 1987, it was demonstrated that tobacco mosaic viruses (rod shaped, 20 nm in diameter, and 300 nm long) and bacteria can be trapped by optical tweezers constructed with a 100 mW, 514 nm argon laser (Ashkin and Dziedzic, 1987; Ashkin et al., 1987). In this early experiment of optical trapping of biological samples, the trapped bacteria were killed almost instantly by the strong laser intensity ($\sim 10^7$ W/cm^2) at the focal point since certain molecules in bacteria absorbed the illumination light and heated up the trapped bacteria. This is a phenomenon called "opticution," that is, death by light. This severe photodamage effect can be greatly reduced by changing the laser wavelength to near infrared (780–1064 nm), the transparent window to biological materials. Ashkin observed that a bacterium trapped by a 50 mW, 1.06 µm laser beam can grow in size and divide into two. Since then, optical tweezers have been widely applied to trap various biological samples including bacteria, mammalian cells, subcellular organelles, and even DNA molecules, and the applications of optical tweezers in biological fields have boomed. Optical tweezers have been used to probe the viscoelastic properties of DNAs, cell membranes, proteins, and characterize the forces exerted by molecular motors such as myosin, kinesin, processive enzymes, and ribosomes (Svoboda and Block, 1994; Mehta et al., 1999; Neuman et al., 1999; Purohit et al., 2003; Jass et al., 2004; Zhang and Liu, 2008; Moffitt et al., 2009).

Despite the success of near-infrared optical tweezers, some photodamage effects on biological cells still exist, especially for high-power optical tweezers and having a long trapping time. It has been observed that the viability of human sperm is decreased when exposed to 300 mW, 1060 nm laser tweezers for 120 s (Liu et al., 1996). The motility of *Escherichia coli* also decreases after being trapped in 100 mW, 1064 nm laser tweezers for 600 s (Neuman et al., 1999).

14.1.3 Holographic Optical Tweezers

To expand optical tweezers' functionality for more complex optical manipulations, recent advances in physical optics enable crafting versatile optical field landscapes. Multiple optical traps can be achieved by scanned optical tweezers, in which a laser beam is steered by a programmable scanning mirror before being focused by a high-N.A. objective lens (Sasaki et al., 1991, 1997). A rapidly scanning laser beam can achieve multiple optical traps by dwelling on each particle briefly and moving to the next one as shown in Figure 14.4a. Scanned optical tweezers can achieve dynamic optical traps on a 2D focal plane, but not in the optical axis direction. The complexity of optical manipulation using scanned optical tweezers is also limited by the time required to reposition multiple wandering particles.

Multiple optical traps can also be achieved by inserting a diffractive optical element (DOE) in the laser beam path. As shown in Figure 14.4b, a laser beam is split into multiple beams by a diffractive beamsplitter and imaged by a set of telescope optics to the pupil of the objective lens to form optical traps near the focal plane. This technique is called holographic optical tweezers (HOT). The pattern produced by the DOE controls the number, the strength, and the location of these optical traps. To understand how it works, imagine multiple laser beams that pass through the objective lens to form optical traps. The superposition of these beams form an interference-field pattern on the DOE. The function of the DOE is to convert the wavefront of an incident laser beam into this interference pattern. Using a

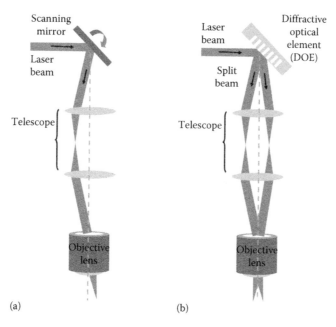

FIGURE 14.4 (a) Scanned optical tweezers and (b) holographic optical tweezers.

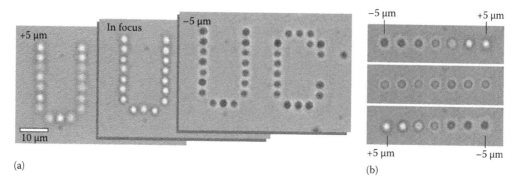

FIGURE 14.5 (a) Example of multiparticle holographic optical traps. (b) 3D holographic optical traps. (Reprinted from Curtis, J.E. et al., *Opt. Commun.*, 207, 169, 2002. With permission. Copyright 2009 from Elsevier.)

programmable spatial light modulator (SLM), such as a liquid crystal array, the interference-field pattern can be reconfigured in real time to create dynamic optical traps. Optical traps created by HOT are not limited to the 2D focal plane as in scanned optical tweezers. HOT allows generating multiple 3D traps for more complex 3D optical manipulation. Figure 14.5a demonstrates HOT for the dynamic trapping of 34 silica beads. The location of trapped particles can be shifted by $\pm 5\,\mu m$ out of the focal plane as shown in Figure 14.5b.

14.1.4 Optical Manipulation Using Evanescent Waves

Optical tweezers require a high-N.A. objective lens to form stable optical traps. The small field of view of a high-N.A. lens, typically smaller than $100\,\mu m \times 100\,\mu m$, also limits the effective optical manipulation area for parallel processing. In order to obtain a large manipulation area, optical manipulations using evanescent waves near structures with high refractive indexes under the condition of total internal reflection (TIR) have been proposed (Kawata and Sugiura, 1992; Kawata and Tani, 1996). The rapid decay of electromagnetic fields into the low-refractive-index material provides strong gradient forces to trap particles near the interface. The penetration depth of evanescent waves is typically in the range of 100–200 nm in typical materials such as water (n_2) and glass (n_1). Figure 14.6 depicts the working principle of a particle attracted and propelled by optical forces near an interface where TIR occurs. Since the gradient force provided by evanescent waves only provides confinement in the direction perpendicular to the interface, particles are free to move on the plane parallel to the interface if other types of forces, such as the optical scattering force, are applied on the particle as shown in Figure 14.7. In the TIR condition, the scattering force on the particle is in the light-propagation direction. The light beam used for near-field optical manipulation is often s-polarized since it produces stronger near fields than those associated with p-polarized light (Kawata and Sugiura, 1992).

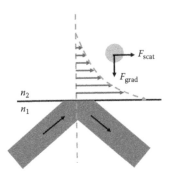

FIGURE 14.6 Optical manipulation using evanescent waves generated by total internal reflection (TIR) near the interface. The rapidly decaying evanescent field provides gradient forces to confined particles near the interface. The scattering force pushes the particle to move on the surface.

Near-field optical manipulation using strip optical waveguides has also attracted great interests recently (Kawata and Tani, 1996; Takuo and Sadahiko, 2000; Grujic and Hellesø, 2007; Schmidt et al., 2007). An evanescent wave near an optical waveguide structure provides 2D confinement to trapped particles and allows the optical scattering forces to transport them along a waveguide whose shape and length can be arbitrarily patterned by microfabrication. Particles trapped by a waveguide can be transported along a complex trajectory for a long distance, limited only by the optical power loss of the waveguide. One drawback of optical manipulation using waveguides is that the majority part of the optical energy is confined in the solid core of the waveguide and only a small portion of the optical energy is used for evanescent wave trapping. This feature makes trapping small nanoparticles inefficient, requiring high laser power.

Almeida et al. proposed a slot waveguide nanophotonic structure that can overcome this challenge (Almeida et al., 2004). Figure 14.8a illustrates the transverse electric field distribution in individual slab waveguides that the exponentially decaying evanescent waves do not overlap. When these two individual waveguides are positioned close enough, the evanescent waves from them start to interact and

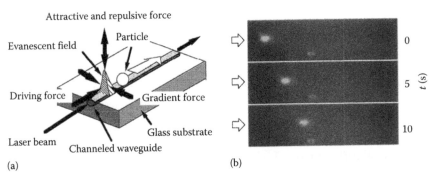

FIGURE 14.7 (a) Schematic of microparticle transport along a channeled waveguide. (b) A 5.1 µm latex sphere is trapped and transported along the waveguide. (Reproduced from Kawata, S. and Tani, T., *Opt. Lett.*, 21, 1768, 1996. With permission.)

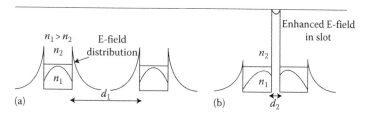

FIGURE 14.8 Schematic E-field distribution of a TM mode in (a) two independent optical waveguides and in (b) two closely positioned waveguides where evanescent waves from them overlap to form a new propagation mode having a strong E-field in the slot.

couple to form a new fundamental eigenmode, in which the electric field strength in the slot is greatly enhanced and is higher than the field of an individual waveguide, providing a tightly confined optical field in the low-refractive-index material. Such an enhancement is only true for the TM polarization direction in which there is a discontinuity in the E-field that results in an enhancement in the low-index channel. For TE polarization, the E-field is continuous across the interfaces and there is no enhancement, just the sum of two decaying exponentials.

Such a slot waveguide structure can produce a highly localized optical field that is confined in a volume smaller than the optical diffraction limit. Since optical forces on nanoparticles are proportional to the gradient of the optical intensity, such a tight confinement can provide strong optical trapping for confining nanoparticles or macromolecules in the slot. It has been demonstrated that λ-phage DNA molecules and 75 nm dielectric spheres can be trapped and transported in a subwavelength slot (120–60 nm) by an optical power less than 300 mW (Yang et al., 2009).

14.1.5 Surface-Plasmon–Enhanced Optical Trap

Tools providing strong optical trapping on nanometer particles could be useful for many studies and analyses in biology. However, the smallest spot size of a free-space laser beam is limited by the optical diffraction of the lenses used for forming optical traps. A 1064 nm Gaussian beam focused by an oil-immersed objective lens (N.A. 1.3) gives a diffraction-limited spot diameter of $(2\lambda/\pi NA) \sim 500$ nm. Nanoparticles trapped within this optical trap experience strong Brownian motion, and the precise position control of trapped particles is difficult. To enhance the strength of the optical trap, one can increase the power of the laser beam or use particles with large polarizability. However, these refinements are not appropriate for many systems, especially for biological objects whose polarizability is low and high laser-power irradiation could cause photodamage. Slot waveguides introduced in the previous section are one example of providing subdiffraction-limited confinement to nanoparticles through cleverly designed waveguide structures.

Subwavelength trapping volumes can also be accomplished by exciting surface plasmonic waves on metallic nanostructures. Recent progress in plasmonics has enabled several important breakthroughs in control, confinement, and enhancement of optical fields including optical imaging with a subdiffraction limit resolution using metallic superlenses (Fang et al., 2005), high-sensitivity biomolecule detections using surface-plasmon-enhanced Raman spectroscopy (SERS) (Kneipp et al., 1997), and plasmon-enhanced fluorescent detection (Gryczynski et al., 2002; Song et al., 2005). For more information, see Chapter 18.

Surface plasmons are collective oscillations of free electrons on metal surfaces. The electromagnetic field distribution near metal surfaces and the electron oscillation frequencies can be engineered by patterning the metal nanostructures (Yannopapas, 2008). This permits the creation of tightly confined optical fields with a small trapping volume governed by the surrounding metal structures to overcome the optical diffraction limit. Furthermore, choosing a laser beam with an appropriate wavelength that matches one of the surface-plasmon resonance frequencies of the excited metal structures creates an enhanced electromagnetic field stronger than the background excitation light. For biological applications, the resonance frequencies can be engineered to the NIR window to prevent photodamage to biological samples.

Optical manipulation benefits from surface-plasmon effects mainly from two aspects. First, highly localized optical fields provide deep and sharp potential wells for accurate positioning of trapped nanometer-sized particles. Second, the enhanced electromagnetic field and the large field gradient due to tight confinement near the metal structure enables plasmonic trapping of nanoparticles with lower optical power than free-space optical tweezers.

Figure 14.9a depicts the schematic of subwavelength plasmonic optical traps produced near the surface of an array of tightly patterned gold nanodot pairs on a glass substrate (Grigorenko et al., 2008). Such geometry allows control over the critical gap between the pair and the frequencies of the localized plasmon resonance. A normal incidence light beam is used to excite the surface-plasmon resonances. The trapping position is produced by the superposition of the fields from the incident Gaussian beam and the near-field patterns produced by surface-plasmon resonances. Shown in Figure 14.9b is the simulated light-power distribution at a height 200 nm above a nanodot pair excited by a transverse-magnetic (TM) 1064 nm laser beam. Such a nanodot pair structure can provide a trap size smaller than the optical diffraction limit and an enhanced trap force by two orders of magnitude when trapping 200 nm polystyrene beads.

Plasmonic optical tweezers often couple with thermal heating effects to some extent. Excitation of surface-plasmon resonances relies on the absorption of the electromagnetic energy from the illumination light. The kinetic energy of the oscillating electrons converts into lattice heat in a few picoseconds. When the illumination power is high, significant heating could induce temperature increases near the metal structures. A temperature gradient could induce fluidic convection and other physical phenomena such as thermophoresis of particles (Garces-Chavez et al., 2006). It was demonstrated that enhanced optical forces and optically induced thermophoretic and convective forces produced by surface-plasmon-polariton excitation can be applied to form large-scale ordering and trapping of colloidal aggregations. For plasmonic optical manipulation of biological samples, heating effects might be a concern and have to be carefully handled to prevent perturbation of the manipulated biological systems such as protein denaturing or cell killing.

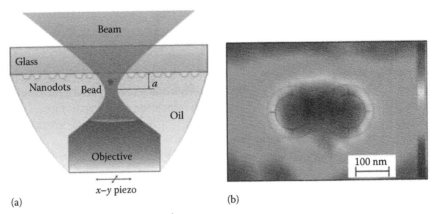

FIGURE 14.9 (See color insert following page 11-20.)   (a) Schematic of metallic, nanostructure-enhanced nanometric optical tweezers. (b) Light power excited by a transverse-magnetic laser light (1064 nm) is shown as a color map and calculated for a plane at a height 200 nm above the nanostructured substrate (the plane is parallel to the glass substrate). (Reproduced from Grigorenko, A.N. et al., *Nat. Photon.*, 2, 365, 2008. With permission.)

14.1.6 Optical Manipulation and Trapping in a Liquid-Core Waveguide

Lab-on-a-chip (LOC) systems provide numerous advantages such as small sample volume, increased sensitivity, and rapid reaction over their macroscale analogues in biomedical applications. In LOCs, microfluidic channels are usually used to transport biological samples such as cells, DNAs, and proteins to detection regions for analysis. Optical detection methods such as fluorescence analysis, fluorescence resonance energy transfer (FRET), and surface-enhanced Raman scattering (SERS) are popular in many LOC systems.

The function of a microfluidic channel is to provide mechanical confinement to guide the flow of fluids, which is similar to an optical waveguide guiding the flow of light. Recent advances in opto-fluidics have shown that liquid flow and light flow can be guided in the same mechanical structure such as a liquid-core optical waveguide. This promises integrated microfluidic and optical systems for on-chip biomedical analysis that could potentially provide higher sensitivity, higher efficiency, and lower cost.

One potential challenge of developing liquid-core optical waveguides for biomedical applications is the low refractive index of water (n=1.33), compared to many widely used dielectric cladding materials such as SiO_2 (1.46). As a result, TIR, the principle commonly used for guiding light in waveguides, cannot be applied for guiding light in a liquid-core waveguide. To solve this issue, several methods have been proposed including the use of low-refractive index materials such as fluorinated polymers or nanoporous structures as the cladding layers, or using interference-based techniques in the cladding layers such as Bragg-interference structures (Yeh et al., 1977, 1978), photonic-crystal fibers (PCFs) (Cregan et al., 1999; Russell, 2003), and antiresonant reflecting optical waveguides (ARROWs) (Hawkins and Schmidt, 2008; Schmidt and Hawkins, 2008).

Figure 14.10a shows the structure of a Bragg waveguide where the dielectric cladding layers are repeated periodically (theoretically to infinity). The electric field propagating inside the periodic medium can be described by

$$E_K(x,z) = E_K(x)e^{iKx}e^{i\beta z}$$

where $E_K(x)$ is a periodic function with period K. The Bloch wave vector, K, reflects the periodicity of the structure, the wave frequency, ω, and the indices of the dielectric materials. Real values of K correspond to propagating waves, and lead to allowed zones or so-called bands if K is plotted in an ω–β diagram, while imaginary values of K represent evanescent waves and lead to forbidden bands. Such a periodic Bragg lattice is also called a 1D photonic crystal (PC). To utilize a Bragg lattice for liquid-core waveguides, two conditions have to be fulfilled for the wavelength of interest. First, since the Bragg lattice needs to be highly reflective to reduce optical losses, one has to operate in a forbidden

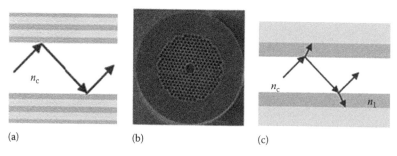

(a) (b) (c)

FIGURE 14.10 (a) Bragg waveguide structure, (b) cross-section view of a hollow-core photonic crystal fiber, and (c) ARROW waveguide principle. (Reprinted from Schmidt, H. and Hawkins, A., *Microfluid. Nanofluid.*, 4, 3, 2008. With permission. Copyright 2009 from Springer Science + Business Media.)

band of K. Second, the transverse component of the wave vector has to satisfy the phase resonance condition as in index-guiding waveguides to allow light propagation in the z-direction. In Bragg waveguides, the core index can be smaller than any of the refractive indices in the periodic dielectric cladding layers, a property that makes Bragg-interference structure appropriate for liquid-core waveguides for biomedical applications.

Photonic crystal fibers (PCFs) guide light based on a photonic band-gap effect (a Bragg-interference structure can be considered a 1D PC). In these devices, a periodic lattice of air capillaries surrounds the core, creating a photonic band gap in the cladding layer to confine light of wavelengths within the low-index core as shown in Figure 14.10b.

ARROWs are also based on an interference effect, but do not require the periodicity of a multilayer structure as in Bragg structures, which could present fabrication challenges in many applications. In ARROWs, a single dielectric layer is sufficient to provide low-loss propagation of light as illustrated in Figure 14.10c. A light ray impinging from the low-index core on the high-index ARROW layer (index n_1) is refracted into the ARROW layer due to the absence of TIR. The thickness of ARROW layers is chosen such that the round trip phase shift, φ_{RT}, of the transverse wave component in medium n_1 fulfills an antiresonance condition $\varphi_{RT} = m\pi$ (m odd). Light is reflected back into the core with high efficiency and enables low-loss light propagation in the low-index core. The guiding efficiency of ARROW waveguides can be improved by adding more layers that fulfill the antiresonance condition. Unlike Bragg structures, these layers do not have to form a periodic stack that provides additional design flexibility. These and other liquid-core waveguides and their use in optofluidics are discussed in more detail in Chapter 16 Integrated Optofluidic Waveguides.

Figure 14.11 presents an example of transporting microparticles in a liquid-core PCF. Particles being transported in the core experience three different optical forces, an axial scattering force, F_{scat}, pushing particles to move along the fiber, an axial gradient force, F_{grad}, due to optical losses along the length of the fiber, and the gradient forces in the radial directions due to variations of the mode intensity profile. It has been shown that by launching a 120 mW, 488 nm light beam into a 20 μm diameter hollow-core PCF, 3 μm polystyrene beads can be transported at an average speed of 70 μm/s over a distance greater than 2 cm. The particle propulsive velocity, v, is extremely sensitive to the particle size D (v α D^5 in the scattering-dominated limit). Such a long interaction distance between light and particles, and the high-sensitivity relationship between particle size and transport speed, make PCFs a promising high-efficiency microparticle sorting tool when utilizing optical forces.

Figure 14.12a demonstrates an integrated optofluidic platform for single-particle detection, manipulation, and analysis (Measor et al., 2008). ARROW waveguides are used as a microfluidic channel to delivery fluid and particles from the reservoir to the fluorescent-detection zones. In this device, a laser beam is delivered through a solid-core waveguide into the liquid-core ARROW waveguide for propelling particles in the channel as shown in Figure 14.12b. Since this particular ARROW waveguide is

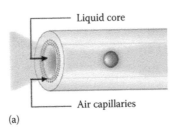

(a)

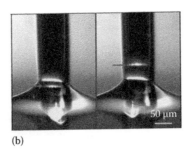

(b)

FIGURE 14.11 (a) Microparticles transported in a liquid-core photonic crystal optical fiber. (b) Example of optically induced transport of 3 μm polystyrene beads in this liquid core. The arrow indicates the position of the band of beads. (Reprinted from Mandal, S. and Erickson, D., *Appl. Phys. Lett.*, 90, 184103, 2007. With permission. Copyright 2009 from American Institute of Physics.)

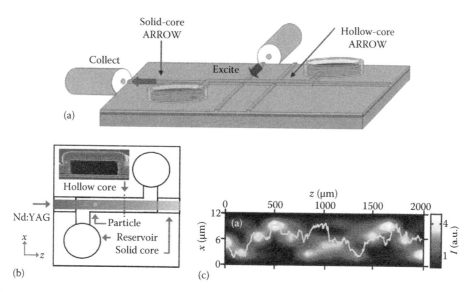

FIGURE 14.12 (a) An integrated optofluidic platform for single-particle detection, optical manipulation, and analysis. (b) Cross section of the ARROW waveguide structure. The light beam from a solid-core waveguide is coupled into the liquid-core ARROW waveguide where microparticles are transported. (c) Microparticle trajectories in the ARROW waveguide. (Reproduced from Measor, P. et al., *Opt. Lett.*, 33, 672, 2008. With permission.)

multimode, the particle trajectories in the channel depend on the superposition of the mode profiles, and the beating interference effects can be observed in the axial direction as shown in Figure 14.12c. Stable trapping of particles inside the ARROW waveguide can also be achieved by launching two counter-propagating beams in the waveguide with certain loss (Kuehn et al., 2008). The concept is similar to the case of using two counter-propagating and divergent beams to form a stable 3D trap as shown in Figure 14.3a, except the reduction of axial scattering forces in the ARROW waveguide case is controlled by the loss of the waveguide. A traditional dual-beam trap with diverging beams across a microfluidic channel has also been demonstrated (Cran-McGreehin et al., 2006).

14.1.7 Optical Sorting of Microscale Particles

Microfluidic devices have been widely applied for biological and chemical analysis in the past decade. Advances in microfluidic control have enabled large-scale microfluidic devices capable of conducting parallel analysis in thousands of microfluidic chambers on a single microfluidic chip (Thorsen et al., 2002). Integration of optical manipulation with microfluidic devices can provide many unique microfluidic functions such as cell manipulation and cell sorting.

Integrating free-space single-beam optical tweezers with microfluidic devices for single-cell manipulation can be easily achieved as long as the material used for fabricating the device is transparent and the substrate thickness is less than $150\,\mu m$ to allow focusing laser beams coming out from a high-N.A. objective lens to be focused at the microfluidic layer for cell and particle manipulation.

In addition to single-beam laser tweezers, more complex optical potential landscapes can be patterned for sorting microscopic particles and biological cells. Shown in Figure 14.13a is a 3D optical lattice formed by a five-beam interference pattern and projected onto a microfluidic channel. When a mixture of particles is pushed through this optical lattice, different types of particles are guided to flow in different directions, depending on particle size, refractive index, and particle shape. Shown in Figure 14.13b are trajectories of two $2\,\mu m$ and one $4\,\mu m$ diameter protein microcapsules flowing across the optical lattice at a flow speed of $20\,\mu m/s$.

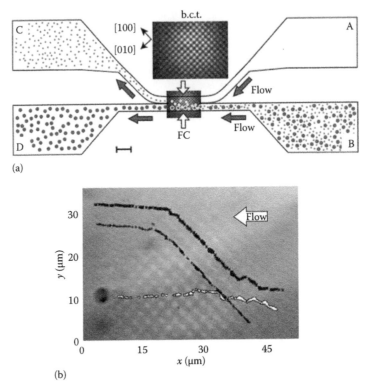

(a)

(b)

FIGURE 14.13 (a) Microparticle sorting in an optical lattice. A five-beam interference pattern is projected to the pathway of microparticles in a microfluidic channel. Microparticles with different optical properties are sorted into different channels and collected. (b) Trajectories of 2 μm protein microcapsules (black dots) and 4 μm ones (white dots) in the optical lattice. (Reprinted from MacDonald, M.P. et al., *Nature*, 426, 421, 2003. With permission. Copyright 2009 from Macmillan Publishers Ltd.)

Optical forces have also been utilized as an optical switch and integrated on a microscale fluorescent-activated cell sorter (FACS) for sorting small number of cells (Wang et al., 2005), which is difficult to do with conventional FACS that needs at least 100,000 cells to start to obtain reliable sorting efficiency. In Figure 14.14, cells in the sample input channel are focused to the middle of the fluorescent-detection channel by two sheath flows. Identified target cells are steered into a collection channel by a 20 W optical switch laser focused by a 0.2 N.A. objective lens. This N.A. is relatively low compared to conventional optical tweezers (usually N.A. > 1). However, optical trapping is not required here. The lower divergence and long depth-of-focus of the beam allows moving cells in lateral directions regardless of their initial depth in the microfluidic channel. Switching a cell into the target channel takes about 2–4 ms under this laser power. This optical switch-based microscale FACS can sort 105 cells/s with a 73.6% recovery rate.

14.1.8 Commercial Optical Tweezers Systems

In the past, optical tweezers were usually custom built in research labs. Now, there are several commercial optical tweezers systems available for users with little laser background. Elliot Scientific provides optical tweezers systems to biochemistry laboratories, chemistry laboratories, and classrooms with a turnkey system. MMI Molecular Machines & Industries also sells a single-beam

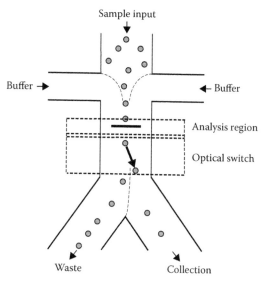

FIGURE 14.14 Optical-switch-based microscale FACS. (Reproduced from Wang, M.M. et al., *Nat. Biotechnol.*, 23, 83, 2005. With permission.)

optical tweezers systems for manipulating single cells or subcellular particles in a size range of 0.1–200 μm. JPK Instruments Inc. sells an optical tweezers systems with a 3D particle-tracking capability. Arryx® Technology provides HOT that use a holographic device (SLM) to sculpt the light from a laser into hundreds of independently controllable optical traps that can be positioned and moved in three dimensions. Carl Zeiss's PALM MicroLaser Systems further combines optical tweezers with laser microdissection technology for noncontact microsurgery and micromanipulation applications.

14.2 Optical Manipulation with Indirect Optical Forces

One of the major advantages of optical manipulation is the ability for reconfiguring optical fields for dynamic manipulation. Optically trapped particles can be transported and positioned with high accuracy to any location in three dimensions by simply changing the projected optical fields. However, optical manipulation using direct optical forces is commonly considered as a low-throughput progress. For free-space optical tweezers, it takes about 1 mW of optical power to form a stable trap to overcome thermal forces. This optical power increases linearly with the number of traps, and the maximum number of optical traps is currently limited to a few thousands, utilizing a high-power laser with an output of a few Watts. Higher laser power may result in potential damage to the optical components used for constructing optical traps. To increase the throughput, several indirect optical-manipulation methods using light-addressed electrical or fluidic forces have been proposed to lower the optical power requirement of conventional optical tweezers.

Figure 14.15 illustrates how a group of microscopic particles are trapped in potential minima sculptured by optical fields. Reconfiguring the optical fields creates dynamic motion of trapped particles. In addition to optical fields, such dynamic potential landscapes can also be formed using light-patterned electric, magnetic, and fluidic fields for indirect optical manipulation. Several examples of indirect optical manipulation will be discussed in the following sections, including OET, LACE, optomagnetic tweezers, and ABEL traps (Enderlein, 2000; Cohen and Moerner, 2006).

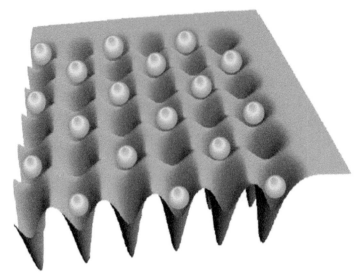

FIGURE 14.15 Potential landscape sculptured by optical fields for parallel trapping of microscope particles. Such a potential landscape can also be formed using light-patterned electric, magnetic, or fluidic fields for indirect optical manipulation. (Reproduced from Dholakia, K. et al., *Nat. Materials*, 4, 579, 2005.)

14.2.1 Optoelectronic Tweezers

The operation principle of OETs is based on light-induced dielectrophoretic (electrical based) forces for micro- and nanoparticle manipulation. Dielectrophoresis is a phenomenon in which a force is exerted on a dielectric particle when it is subjected to a nonuniform electric field. The time-averaged DEP force can be expressed as

$$F_{DEP} = 2\pi a^3 \varepsilon_m Re[K^*(\omega)]\nabla(E_{rms}{}^2)$$

$$K^*(\omega) = \frac{\varepsilon_p^* - \varepsilon_m^*}{\varepsilon_p^* - 2\varepsilon_m^*}, \quad \varepsilon_p^* = \varepsilon_p - j\frac{\sigma_p}{\omega}, \quad \varepsilon_m^* = \varepsilon_m - j\frac{\sigma_m}{\omega}$$

where
  E_{rms} is the root-mean-square magnitude of the imposed ac electric field
  a is the particle radius
  ε_m and ε_p are the permittivities of the surrounding medium and the particle, respectively
  σ_m and σ_p are the conductivities of the medium and the particle, respectively
  ω is the angular frequency of the applied electric field
  $K^*(\omega)$ is the Clausius–Mossotti (CM) factor (Jones, 1995)

CM factor is a complex number representing the particle's dielectric signature (similar to the refractive index of particles in optical fields), and its real part, $Re[K^*(\omega)]$, can have a value ranging from 1 to −0.5, depending on the particle size, material composition, structure, surface charges, ac frequency, and the medium's dielectric properties. If $Re[K^*(\omega)] > 0$, the particle moves toward the strongest electric field region as shown in Figure 14.16a, a phenomenon known as positive DEP; if $Re[K^*(\omega)] < 0$, a negative DEP force is induced and particles are pushed toward the weakest electric field region as shown in Figure 14.16b.

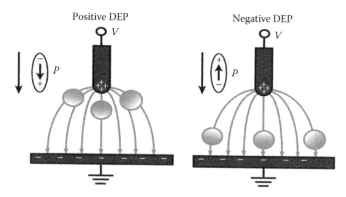

FIGURE 14.16 Illustration of (a) positive DEP and (b) negative DEP forces.

Figure 14.17 illustrates a typical OET device structure as proposed by Chiou et al. (2005). It is a sandwich structure including two electrodes: a transparent conductive ITO electrode, and a photoconductive electrode consisting of multiple featureless layers including an ITO layer, a 100 nm film of $n+$ hydrogenated amorphous silicon a-Si:H, a 1 μm intrinsic film of a-Si:H, and a 20 nm thick silicon nitride film. An aqueous medium containing cells or microparticles is sandwiched between these two electrodes. An ac voltage is applied between the top and the bottom electrodes. In the absence of light, the majority of the applied voltage drops across the a-Si:H layer due to its large electrical impedance. In the area under light illumination, the photoconductivity of the a-Si:H layer increases so that its electrical impedance becomes smaller than that of the aqueous layer. This optical illumination results in voltage switching from the a-Si:H layer to the aqueous layer, like a light-patterned virtual electrode turned on. The electric

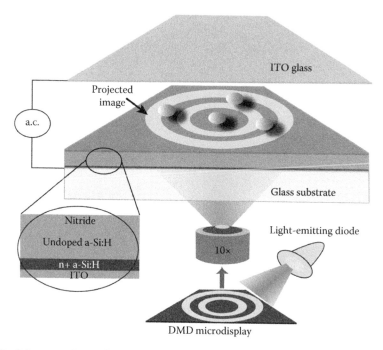

FIGURE 14.17 Schematic of OET devices. (Reproduced from Chiou, P.Y. et al., *Nature*, 436, 370, 2005. With permission.)

field near the virtual electrode is highly nonuniform. This results in a DEP force near the a-Si:H surface. For particles experiencing positive DEP forces, they are attracted to the illuminated spot; while for particles experiencing negative DEP forces, they are repelled away. The featureless photoconductive layer in OET allows continuous addressing of light-induced DEP forces on the 2D OET surface.

A comparison of optical tweezers and OET can be better understood with Figure 14.18. Optical tweezers uses direct optical forces for particle manipulation. The optical energy is directly converted to mechanical forces, and the trapping force is proportional to the optical power. While in OET, the optical energy is used for triggering electrical force powered by externally applied ac bias. As a result, the required optical energy in OET can be very low as long as it is strong enough to turn on virtual electrodes. This allows optical manipulation using OET for massively parallel processing of a large number of particles or cells. Another advantage of OET is that it allows optical manipulation using incoherent light sources such as LEDs, which are easier to obtain and implement than coherent laser sources.

Figure 14.19 shows the simulated electric field distribution in the liquid layer excited by a 17 μm diameter laser spot with an ac bias of 10 V. The conductivity of the liquid is 1 mS/m and the thickness is 15 μm. The photoconductivity of the amorphous silicon layer is assumed to have a Gaussian distribution, following the profile of the incident light with a peak conductivity of 10 mS/m at the center. Since the DEP force is proportional to the gradient of E^2, this electric field distribution shows that the OET can generate DEP forces in both the lateral and vertical directions near the photoconductive surface. One of the key reasons for choosing a-Si:H as the photoconductive layer is its short carrier diffusion length (~100 nm) (Schwarz et al., 1993). This allows patterning high-resolution virtual electrodes with a resolution limited by the optical diffraction limit. If a photoconductive material with higher carrier mobility is used, for example, single-crystalline silicon, the carrier diffusion length is longer than the dimension of optical patterns, which reduces the resolution of virtual electrodes and OET's ability for single-cell manipulation.

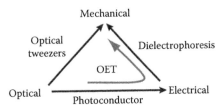

FIGURE 14.18 Comparison of optical tweezers and optoelectronic tweezers.

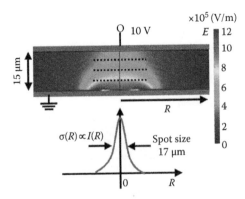

FIGURE 14.19 Electric field distribution in the OET device excited by a 17 μm diameter laser beam.

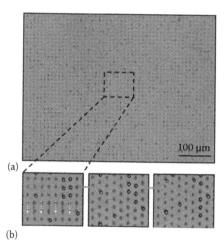

FIGURE 14.20 (a) 15,000 OET traps are created over a 1.3 mm × 1.0 mm area. The 4.5 μm diameter polystyrene beads experiencing negative DEP forces are trapped in the dark area. (b) Snapshots showing parallel transportation of single particles. (Reproduced from Chiou, P.Y. et al., *Nature*, 436, 370, 2005. With permission.)

Figure 14.20a demonstrates 15,000 dynamic microparticle traps been patterned using a digital micromirror device (DMD) and a 10 mW LED over an area of 1.3 × 1.0 mm² on an OET device. The particles are trapped in the dark area by the negative DEP forces. The size of each trap is optimized to capture a single 4.5 μm diameter polystyrene bead. The pitch between particles is 12 μm. By reconfiguring the projected images, the trapped particles can be individually moved in parallel. Shown in Figure 14.20b are snapshots of the captured video images showing the particles in adjacent columns moved in opposite directions.

Since the real part of the CM factor of manipulated particles is dependent upon the ac frequency, the particle size, and the structure of the particles, it is possible to tune the ac frequency to a point where one type of particles experience positive DEP forces and the other type experience negative DEP for sorting applications. Figure 14.21 shows a mixture of live and dead cells being sorted out using dynamic shrinking optical ring patterns. A 100 kHz ac bias is applied to the OET device and the electrical conductivity of the aqueous medium is adjusted to 0.01 S/m in this experiment. At this frequency and medium conductivity, live cells experience positive DEP forces while negative DEP forces are induced on dead cells. The dashed optical rings selectively pick up live cells and transport them to the center while leaving the dead cells in the dark area.

14.2.1.1 Phototransistor OET: Manipulation of Biological Samples in Physiological Buffers

One of the most important applications of OET is its ability to manipulate cells and other biological samples in liquid media with significantly lower optical intensities than optical tweezers. Although OET has been previously used for the manipulation of red and white blood cells (Ohta et al., 2006) and the separation of live and dead human B cells (Chiou et al., 2005), a-Si:H-based OET devices, however, have been limited to the manipulation of particles in liquids whose conductivities are lower than 100 mS/m. This limitation arises from the fact that an a-Si:H-based OET is incapable of effectively switching the ac voltage from the photoconductive layer to the liquid layer due to the relatively small photoconductivity of the amorphous silicon layer.

The conductivity of regular physiological media is usually in the range of 0.5–1.5 S/m, a value higher than an a-Si:H-based OET can operate without dilution. However, in many biological applications, it is essential to manipulate cells in regular physiological media to maintain cell viability. To overcome this limitation of the a-Si:H-based OET device, a possible modification would utilize an

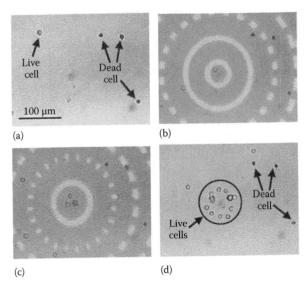

FIGURE 14.21 Selective concentration of live human white blood cells from a mixture of live and dead (appear black in the figure) cells with dynamic shrinking concentric ring patterns. A 100 kHz ac electrical bias is applied to the OET device. (Reproduced from Chiou, P.Y. et al., *Nature*, 436, 370, 2005. With permission.)

N⁺PN phototransistor structure to replace the photoconductive a-Si:H layer (Hsu et al., 2007). This gives two orders of magnitude larger photoconductivity due to the higher carrier mobility in single-crystalline silicon and the current gain in the phototransistor structure. This novel OET device is called the phototransistor OET (phOET). Figure 14.22a shows a schematic of the phOET structure. It consists of a top transparent ITO electrode and a bottom N⁺PN phototransistor structure with an ac bias applied between them. The liquid layer containing the particles of interest is sandwiched between these two surfaces.

The photoconductivity of the phototransistor is 100 times higher than that of a-Si:H, as shown in Figure 14.22b. This allows the phOET to operate in regular physiological aqueous media such as phosphate-buffered saline (PBS) with optical intensity as low as 1 W/cm². A phOET typically operates at frequencies in the MHz range and traps cells with negative DEP forces.

14.2.2 Light-Actuated AC Electroosmosis

OETs trap particles based on dielectrophoretic forces induced by light-patterned virtual electrodes. One of the factors that limits the maximum trap force of OET is the magnitude of the voltage allowed across the top and the bottom electrodes. This limit comes from the maximum voltage the photoconductive layer can hold without too much leakage current in the dark state. For a 1 μm thick a-Si:H layer, it can hold an ac voltage up to $20 V_{pp}$. As a result, OET is an effective tool for manipulating particles with sizes larger than 1 μm. Strong confinement of nanoscopic particles using OET is not easy.

To allow OET for manipulating nanoscale particles, one can reduce the frequency of ac bias to allow light-patterned virtual electrodes to activate another electrokinetic mechanism called ac electroosmosis, which has been demonstrated to be capable of manipulating nanoscale particles with low electric field strength. Wong has utilized ac electroosmosis to concentrate a variety of nanoscale particles and biomolecules including *E. coli*, λ-phage DNA, and 20-base single-strand DNA fragments with electric field strength on the order of 10^4 V/m (Wong et al., 2004). Such an electric field strength is in the range that a regular a-Si:H-based OET can generate.

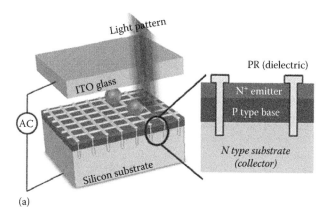

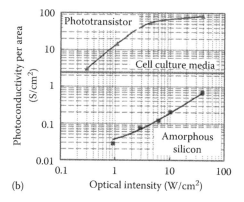

FIGURE 14.22 (a) Schematics of a phototransistor-based OET (phOET). (b) The measured photoconductivity of a phototransistor and an a-Si:H photoconductor. The photoconductivity in phototransistor is more than 100 times larger than in a-Si:H. It requires less than 1 W/cm² to turn on the virtual electrodes in cell culture media. (Reprinted from Hsu, H.Y. et al., Phototransistor-based optoelectronic tweezers for cell manipulation in highly conductive solution, in *Proceedings of the International Conference on Transducer*, Lyon, France, pp. 477–480, 2007. With permission. Copyright 2009 from IEEE.)

Electroosmosis is a widely used mechanism for microfluid pumping (Schasfoort et al., 1999; Mpholo et al., 2003; Daiguji et al., 2004). The tangential electric field in a fluidic channel interacts with ions in the double layer and generates electrostatic forces to drive the boundary layer of liquid to flow at a slip velocity that can be calculated using the Helmholtz–Smoluchowski equation (Lyklema, 1991):

$$v_{slip} = -\frac{\varepsilon \zeta E_t}{\eta}$$

where
v_{slip} is the slip velocity
ε is the permittivity of the liquid medium
ζ is the zeta potential at the interface between the liquid and the channel wall
E_t is the tangential component of electric field, and η is the fluidic viscosity

Electroosmosis is usually observed in the dc mode, in which the zeta potential is determined by the material of the channel wall, the type of electrolytes, and the ionic strength. Recent progresses show

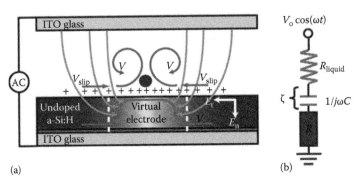

FIGURE 14.23 Illustration of the LACE mechanism. (a) A microfluidic vortex generated by a light-patterned virtual electrode through driving the double-layer charges via the tangential component of an electric field. (b) The equivalent circuit model along an electric field line. LACE operates at frequencies close to the double-layer relaxation frequency. (Reproduced from Chiou, P.Y. et al., *J. Microelectromech. Syst.*, 17, 525, 2008. With permission. Copyright 2009 from IEEE.)

that ac electroosmosis can also generate continuous dc flow by coupling the tangential electric field with the field-induced double-layer charges (Ramos et al., 1999; Green et al., 2000; Hayward et al., 2000; Urbanski et al., 2006).

Figure 14.23a illustrates the device structure and the working principle of LACE (Chiou et al., 2008). It utilizes exactly the same device structure as regular a-Si:H-based OET. An aqueous medium containing nanoparticles and molecules of interest is sandwiched between a transparent ITO and photoconductive electrodes. What is different between OET and LACE is the frequency of the applied ac voltage. In regular OET, the ac frequency needs to be high enough such that there is almost no voltage drop across the interfacial electric double-layer capacitance. Voltage is switched between the a-Si:H layer and the bulk liquid layer, and the DEP force is the dominating force for particle manipulation. In LACE, on the other hand, the frequency is reduced to a level so that part of the voltage is switched to the double-layer capacitance to allow ion accumulation at the interface. The accumulated ions interact with the tangential electric field components at the interface and generate electrostatic forces to drive the interface layer to flow, similar to electroosmosis. Figure 14.23a illustrates the charge polarity, the electric field direction, and the fluidic flow been induced by illuminating a circular virtual electrode. When the polarity of the electric field reverses, the polarity of the accumulated ions also changes. This generates electrostatic forces always pointing to the center of the virtual electrode even though an ac voltage is applied, and induces a continuous microfluidic vortex circulating around the virtual electrode as shown.

Figure 14.24a shows the flow pattern of the microfluidic vortex being induced around a circular virtual electrode. Nanoscale particles near the virtual electrode are driven to flow with the vortex and brought to the surface of the photoconductive electrode. Since the flow pattern of the microfluidic vortex is symmetric to the center axis of the virtual electrode, there exists a stagnant point right on the middle surface of the virtual electrode, where the strongest vertical electric field gradient occurs as shown in Figure 14.24b. Nanoscale particles swept by the vortex flow into the stagnant zone are being held by the strong vertical electric field possibly by positive DEP forces (rigorous theory is still under development).

Figure 14.25 shows the snapshots of 200 nm polystyrene beads being trapped and transported by a virtual electrode created by a laser spot of 17.6 μm using the LACE mechanism with a 1.6 kHz, 6 V_{pp} ac voltage. LACE has also been demonstrated to be able to concentrate and transport other types of nanoscale particles such as λ-phase DNA, 60 nm polystyrene beads, and quantum dots (Chiou et al., 2008).

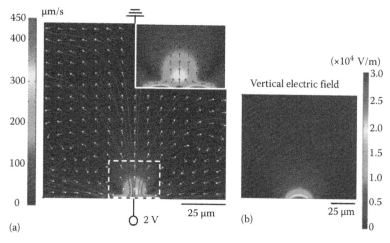

FIGURE 14.24 (See color insert following page 11-20.) (a) Light-induced ac electroosmosis flow near a virtual electrode. The highest flow velocity exists on the surface at the edge of a virtual electrode. Owing to the symmetrical flow pattern, there exists a stagnant-flow zone near the middle surface of a virtual electrode. This is where the nanoparticles are trapped. (b) Simulated vertical electric field near a virtual electrode during LACE manipulation. (Reprinted from Chiou, P.Y. et al., *J. Microelectromech. Syst.*, 17, 525, 2008. With permission. Copyright 2009 from IEEE.)

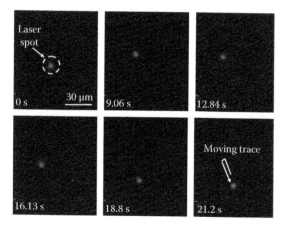

FIGURE 14.25 LACE concentration of 200 nm fluorescent polystyrene particles using a 5 mW, 632 nm laser beam. The diameter of the illuminating laser spot is 17.6 μm. The maximum transport speed is 7.6 μm/s. (Reprinted from Chiou, P.Y. et al., *J. Microelectromech. Syst.*, 17, 525, 2008. With permission. Copyright 2009 from IEEE.)

14.2.3 Magnetic Nanowire-Enhanced Optomagnetic Tweezers

Single-beam optical tweezers have been proven to be valuable tools for manipulation of microscale objects, notably in biological studies. However, the exertion of highly localized forces on nanoscale particles remains challenging since optical tweezers are limited by a combination of the minimum focal width governed by optical diffraction, as well as high optical powers required to trap nanoscale objects. Enhanced optical near fields around metallic structures or in slot waveguides have been used to confine subwavelength particles to small trapping volumes and provide large optical forces on nanoparticles.

However, optical manipulation using these nanostructure-enhanced optical fields will sacrifice the advantage of 3D manipulation of free-space optical tweezers.

Magnetic forces are also popularly used in many biological studies since most biological samples contain no magnetic materials, and applying magnetic fields gives minimum perturbation to the manipulated biological systems(Gosse and Croquette, 2002; de Vries et al., 2005). Furthermore, magnetic forces could provide larger manipulation forces than optical tweezers. A 30 nm magnetic particle could generate pN level forces for activating certain mechanotransduction pathways (Dobson, 2008). The trapping mechanism of magnetic tweezers is similar to optical tweezers to a certain extent. One has to sculpture a potential profile using magnetic fields, instead of optical fields, in space to trap magnetic particles in potential minima. Magnetic forces generated on magnetic particles are proportial to the local gradient of the magnetic field and the magnetic dipole (can be permanent or induced) of the manipulated particles. To exert strong magnetic forces on nanomagnetic particles, sharpened tips of magnetic materials have been used to exert forces over small volumes in electromagnetic tweezers, but such approaches require complicated pole fabrication processes and high electrical currents to generate large magnetic forces, potentially heating samples and necessitating cooling components.

Nanowire-enhanced optomagnetic tweezers is a hybrid tool combining optical tweezers and magnetic tweezers to enable a 3D high-resolution magnetic manipulation (Mehta et al., 2008). Figure 14.26a shows how a nanowire-enhanced optomagnetic tweezers is constructed. Optical tweezers are used to

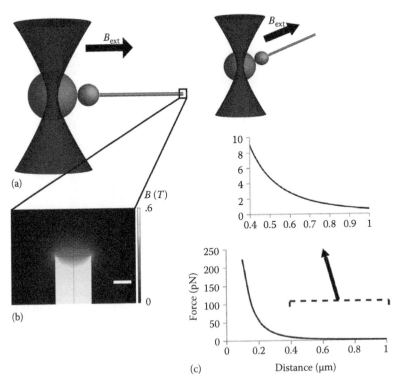

FIGURE 14.26 (a) Schematic of optomagnetic tweezers. An optically trapped polystyrene bead is used to attach to a magnetic bead that attracts a magnetic nanowire under the application of a magnetic field. Translational motion is achieved by translating the optical trap, and rotational by rotating the constant field. (b) Simulation results illustrating the tightly confined magnetic field, B, and distribution around a 200 nm diameter nanowire tip (scale bar is 100 nm). (c) Calculated force on a uniformly magnetized 200 nm diameter sphere as a function of its center's distance from the nanowire tip along the nanowire's long axis. (Reprinted from Mehta, K.K. et al., *Appl. Phys. Lett.*, 93, 254102, 2008. With permission. Copyright 2009 from American Institute of Physics.)

trap a biotinylated polystyrene bead, which is conjugated to a streptavidin-coated paramagnetic bead. Under the application of a magnetic field, the magnetic bead exerts an attractive force on magnetic nanowires in a solution. Attaching the magnetic bead to the optically trapped polystyrene bead is necessary since optical trapping of the magnetic bead itself is difficult, due to the strong scattering and absorption resulting from metallic elements within the magnetic beads. Therefore, by using the optically trapped polystyrene bead as a liaison between the optical and magnetic components of the system, full 3D manipulation of the nanowire is possible—the optical trap trivially allows for translational motion in all three directions, and rotations can be induced by controlling the orientation of the external magnetic field. Due to their morphologies, nanowires composed of magnetic materials naturally generate strong magnetic field gradients over small volumes near their tips that allow for controllable and localized manipulation of magnetic nanoparticles.

Figure 14.27a shows an example of trapping a single 200 nm magnetic nanoparticle near the tip of a magnetic Ni nanowire in a viscous solution of 95% glycerol by volume, with a measured viscosity $\eta = 0.6\,\text{Pa\dot{c}s}$; an average force of 0.33 pN over a range of approximately 0.63–0.8 μm (Mehta et al., 2008). Figure 14.27b demonstrates a fluorescent magnetic particle that is trapped by a nanowire and moved in a circular pattern by rotating the magnetic field. On the other end of the nanowire is the optically trapped polystyrene bead that behaves as an anchor point of the rotating nanowire.

14.2.4 Anti-Brownian Electrokinetic Trap

ABEL trap is an idea of trapping single nanoparticles or molecules by providing active electrokinetic forces on trapped particles through optical feedback control to cancel out particles' displacement due to Brownian motion (Enderlein, 2000; Cohen and Moerner, 2006; Kühn et al., in preparation). An ABEL trap using fluorescent microscopy and electrokinetic forces has been shown to be able to suppress Brownian motion for 20–100 nm fluorescent polystyrene beads, individual protein molecules, single virus particles, lipid vesicles, and fluorescent semiconductor nanocrystals (Cohen and Moerner, 2005, 2006).

Figure 14.28a and b shows the device for an ABEL trap proposed by Cohen in 2005. It was constructed on a Au-electrodes-patterned glass substrate and a PDMS microfluidic channel. During operation, this device is positioned on a fluorescent microscopy to detect the fluorescent signals from a nanoparticle and

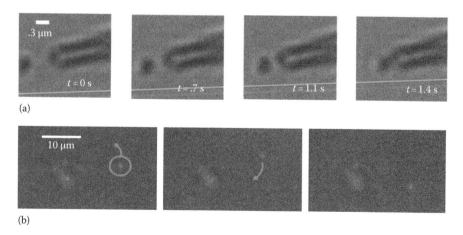

(a)

(b)

FIGURE 14.27 (a) Snapshots showing a single 200 nm magnetic nanoparticle attracted to the tip of a magnetic nanowire. (b) Fluorescence images demonstrating a magnetic nanoparticle trapped at the end of a nanowire and moved in a circular pattern by rotating the nanowire. (Reprinted from Mehta, K.K. et al., *Appl. Phys. Lett.*, 93, 254102, 2008. With permission. Copyright 2009 from American Institute of Physics.)

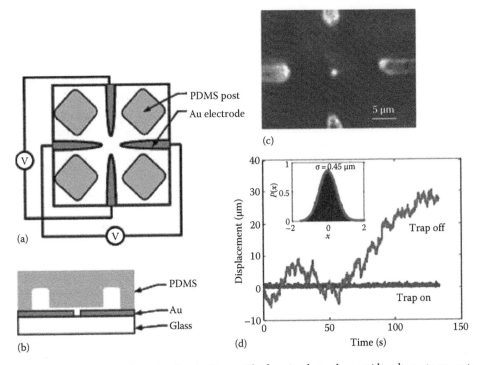

FIGURE 14.28 (a,b) Device schematic of an ABEL trap. The four Au electrodes provide voltages to generate electrokinetic forces on trapped particles in the middle area. (c) Average of 30 video frames showing a single 200 nm particle trapped in the middle. (d) Comparison of trajectories of a 200 nm diameter particle with the ABEL trap on and off. (Reprinted from Cohen, A.E. and Moerner, W.E., *Appl. Phys. Lett.*, 86, 093109, 2005. With permission. Copyright 2009 from American Institute of Physics.)

determine its locations. Without an ABEL trap, a nanoparticle observed in the middle area soon diffuses away due to strong Brownian motion. With the ABEL trap on, the displacement of a nanoparticle due to Brownian motion can be canceled out by a feedback-controlled electrokinetic force on the particle to induce a displacement in the opposite direction. Figure 14.28c shows an overlap of 30 video frames of a 200 nm particle trapped by ABEL. A comparison of the diffusion displacement of a 200 nm particle with the ABEL trap on and off is shown in Figure 14.28d. A significant suppression of the Brownian motion is observed when ABEL trap is on. The trapped location is not necessary in the middle of these four Au electrodes. The trapped particle can be programmed to follow a predetermined two-dimensional trajectory. The concept of ABEL trap has also recently been demonstrated on an integrated optofluidic chip, in which the optical detection part is achieved by fluorescent excitation through two offset waveguides. This eliminates the need of using bulky microscopy and camera, and realizes a compact ABEL trap (Kühn et al., in preparation).

14.3 Summary

Optical manipulation has shown great utilities for biological researches after more than 30 years development in this field. To further expand its applications to other fields, current trends of optical manipulation include (1) expanding the effective optical manipulation such that more objects can be handled in parallel to increase the throughput, (2) reducing optical power for trapping nanoscale particles such that an high throughput or a less photodamage effect is induced in living cell experiments, (3) reducing the device-footprint size of optical manipulation such that a compact and portable optical-manipulation system can be used by people who do not have the expertise in optics. The future developments of optical manipulation are not limited to conventional manipulation using direct optical forces, since several indirect optical force manipulation have shown great promises to provide complementary functions to conventional direct optical force manipulation. A hybrid optofluidic manipulation system combining direct and indirect optical forces may provide broader applications in the future, especially in the area of biomedical diagnostics.

References

Almeida, V. R., Xu, Q. F., Barrios, C. A., and Lipson, M. (2004) Guiding and confining light in void nanostructure. *Optics Letters*, 29, 1209–1211.

Ashkin, A. (1970) Acceleration and trapping of particles by radiation pressure. *Physical Review Letters*, 24, 156–159.

Ashkin, A. and Dziedzic, J. M. (1987) Optical trapping and manipulation of viruses and bacteria. *Science*, 235, 1517–1520.

Ashkin, A., Dziedzic, J. M., Bjorkholm, J. E., and Chu, S. (1986) Observation of a single-beam gradient force optical trap for dielectric particles. *Optics Letters*, 11, 288–290.

Ashkin, A., Dziedzic, J. M., and Yamane, T. (1987) Optical trapping and manipulation of single cells using infrared-laser beams. *Nature*, 330, 769–771.

Chemla, Y. R., Aathavan, K., Michaelis, J. et al. (2005) Mechanism of force generation of a viral DNA packaging motor. *Cell*, 122, 683–692.

Chiou, P. Y., Ohta, A. T., Jamshidi, A., Hsu, H. Y., and Wu, M. C. (2008) Light-actuated ac electroosmosis for nanoparticle manipulation. *Journal of Microelectromechanical Systems*, 17, 525–531.

Chiou, P. Y., Ohta, A. T., and Wu, M. C. (2005) Massively parallel manipulation of single cells and microparticles using optical images. *Nature*, 436, 370–372.

Cohen, A. E. and Moerner, W. E. (2005) Method for trapping and manipulating nanoscale objects in solution. *Applied Physics Letters*, 86, 093109.

Cohen, A. E. and Moerner, W. E. (2006) Suppressing Brownian motion of individual biomolecules in solution. *Proceedings of the National Academy of Sciences of the United States of America*, 103, 4362–4365.

Cran-McGreehin, S., Krauss, T. F., and Dholakia, K. (2006) Integrated monolithic optical manipulation. *Lab on a Chip*, 6, 1122–1124.

Cregan, R. F., Mangan, B. J., Knight, J. C. et al. (1999) Single-mode photonic band gap guidance of light in air. *Science*, 285, 1537–1539.

Curtis, J. E. and Grier, D. G. (2003) Structure of optical vortices. *Physical Review Letters*, 90, 133901.

Curtis, J. E., Koss, B. A., and Grier, D. G. (2002) Dynamic holographic optical tweezers. *Optics Communications*, 207, 169–175.

Daiguji, H., Yang, P. D., and Majumdar, A. (2004) Ion transport in nanofluidic channels. *Nano Letters*, 4, 137–142.

de Vries, A. H. B., Krenn, B. E., van Driel, R., and Kanger, J. S. (2005) Micro magnetic tweezers for nanomanipulation inside live cells. *Biophysical Journal*, 88, 2137–2144.

Dobson, J. (2008) Remote control of cellular behaviour with magnetic nanoparticles. *Nature Nanotechnology*, 3, 139–143.

Enderlein, J. (2000) Tracking of fluorescent molecules diffusing within membranes. *Applied Physics B: Lasers and Optics*, 71, 773–777.

Fang, N., Lee, H., Sun, C., and Zhang, X. (2005) Sub-diffraction-limited optical imaging with a silver superlens. *Science*, 308, 534–537.

Garces-Chavez, V., Quidant, R., Reece, P. J. et al. (2006) Extended organization of colloidal microparticles by surface plasmon polariton excitation. *Physical Review B*, 73, 085417.

Gosse, C. and Croquette, V. (2002) Magnetic tweezers: Micromanipulation and force measurement at the molecular level. *Biophysical Journal*, 82, 3314–3329.

Green, N. G., Ramos, A., Gonzalez, A., Morgan, H., and Castellanos, A. (2000) Fluid flow induced by nonuniform ac electric fields in electrolytes on microelectrodes. I. Experimental measurements. *Physical Review E*, 61, 4011–4018.

Grigorenko, A. N., Roberts, N. W., Dickinson, M. R., and Zhang, Y. (2008) Nanometric optical tweezers based on nanostructured substrates. *Nature Photonics*, 2, 365–370.

Grujic, K. and Hellesø, O. G. (2007) Dielectric microsphere manipulation and chain assembly by counter-propagating waves in a channel waveguide. *Optics Express*, 15, 6470–6477.

Gryczynski, I., Malicka, J., Shen, Y. B., Gryczynski, Z., and Lakowicz, J. R. (2002) Multiphoton excitation of fluorescence near metallic particles: Enhanced and localized excitation. *Journal of Physical Chemistry B*, 106, 2191–2195.

Hawkins, A. and Schmidt, H. (2008) Optofluidic waveguides: II. Fabrication and structures. *Microfluidics and Nanofluidics*, 4, 17–32.

Hayward, R. C., Saville, D. A., and Aksay, I. A. (2000) Electrophoretic assembly of colloidal crystals with optically tunable micropatterns. *Nature*, 404, 56–59.

Hsu, H. Y., Ohta, A. T., Chiou, P. Y., Jamshidi, A., and Wu, M. C. (2007) Phototransistor-based optoelectronic tweezers for cell manipulation in highly conductive solution. In *Proceedings of the International Conference on Transducer*, Lyon, France, pp. 477–480.

Jass, J., Schedin, S., Fallman, E. et al. (2004) Physical properties of *Escherichia coli* P pili measured by optical tweezers. *Biophysical Journal*, 87, 4271–4283.

Jones, T. B. (1995) *Electromechanics of Particles*. New York: Cambridge University Press.

Kühn, S., Lunt, E. J., Phillips, B. S., Hawkins, A. R., and Schmidt, H. (2009) Ultralow power trapping and fluorescence detection of single particles on an optofluidic chip (In press), *Lab on a Chip*.

Kawata, S. and Sugiura, T. (1992) Movement of micrometer-sized particles in the evanescent field of a laser-beam. *Optics Letters*, 17, 772–774.

Kawata, S. and Tani, T. (1996) Optically driven Mie particles in an evanescent field along a channeled waveguide. *Optics Letters*, 21, 1768–1770.

Kneipp, K., Wang, Y., Kneipp, H. et al. (1997) Single molecule detection using surface-enhanced Raman scattering (SERS). *Physical Review Letters*, 78, 1667–1670.

Kuehn, S., Measor, P., Lunt, E. J., Hawkins, A. R., and Schmidt, H. (2008) Particle manipulation with integrated optofluidic traps. In *IEEE/LEOS Summer Topical Meetings*, Acapulco, Mexico.

Liu, Y., Sonek, G. J., Berns, M. W., and Tromberg, B. J. (1996) Physiological monitoring of optically trapped cells: Assessing the effects of confinement by 1064-nm laser tweezers using microfluorometry. *Biophysical Journal*, 71, 2158–2167.

Lyklema, J. (1991) *Fundamentals of Interface and Colloid Science*. London, U.K.: Academic Press.

MacDonald, M. P., Spalding, G. C., and Dholakia, K. (2003) Microfluidic sorting in an optical lattice. *Nature*, 426, 421–424.

Mandal, S. and Erickson, D. (2007) Optofluidic transport in liquid core waveguiding structures. *Applied Physics Letters*, 90, 184103.

Measor, P., Kuhn, S., Lunt, E. J. et al. (2008) Hollow-core waveguide characterization by optically induced particle transport. *Optics Letters*, 33, 672–674.

Mehta, A. D., Rief, M., Spudich, J. A., Smith, D. A., and Simmons, R. M. (1999) Single-molecule biomechanics with optical methods. *Science*, 283, 1689–1695.

Mehta, K. K., Wu, T.-H., and Chiou, E. P. Y. (2008) Magnetic nanowire-enhanced optomagnetic tweezers. *Applied Physics Letters*, 93, 254102–254103.

Moffitt, J. R., Chemla, Y. R., Aathavan, K. et al. (2009) Intersubunit coordination in a homomeric ring ATPase. *Nature*, 457, 446–450.

Mpholo, M., Smith, C. G., and Brown, A. B. D. (2003) Low voltage plug flow pumping using anisotropic electrode arrays. *Sensors and Actuators B-Chemical*, 92, 262–268.

Neuman, K. C., Chadd, E. H., Liou, G. F., Bergman, K., and Block, S. M. (1999) Characterization of photodamage to Escherichia coli in optical traps. *Biophysical Journal*, 77, 2856–2863.

Ohta, A. T., Chiou, P. Y., Jamshidi, A. et al. (2006) Spatial cell discrimination using optoelectronic tweezers. In *LEOS Summer Topical Meetings IEEE*, Piscataway, NJ.

Park, S., Pan, C., Wu, T. H. et al. (2008) Floating electrode optoelectronic tweezers (FEOET): Light driven droplet manipulation in electrically insulating oil. *Applied Physics Letters*, 92, 151101.

Purohit, P. K., Kondev, J., and Phillips, R. (2003) Mechanics of DNA packaging in viruses. *Proceedings of the National Academy of Sciences of the United States of America*, 100, 3173–3178.

Ramos, A., Morgan, H., Green, N. G., and Castellanos, A. (1999) AC electric-field-induced fluid flow in microelectrodes. *Journal of Colloid and Interface Science*, 217, 420–422.

Russell, P. (2003) Photonic crystal fibers. *Science*, 299, 358–362.

Samuel, A. D. T. and Berg, H. C. (1996) Torque-generating units of the bacterial flagellar motor step independently. *Biophysical Journal*, 71, 918–923.

Sasaki, K., Fujiwara, H., and Masuhara, H. (1997) Optical manipulation of a lasing microparticle and its application to near-field microspectroscopy. *Journal of Vacuum Science and Technology B*, 15, 2786–2790.

Sasaki, K., Koshioka, M., Misawa, H., Kitamura, N., and Masuhara, H. (1991) Pattern-formation and flow-control of fine particles by laser-scanning micromanipulation. *Optics Letters*, 16, 1463–1465.

Schasfoort, R. B. M., Schlautmann, S., Hendrikse, L., and van den Berg, A. (1999) Field-effect flow control for microfabricated fluidic networks. *Science*, 286, 942–945.

Schmidt, B. S., Yang, A. H., Erickson, D., and Lipson, M. (2007) Optofluidic trapping and transport on solid core waveguides within a microfluidic device. *Optics Express*, 15, 14322–14334.

Schmidt, H. and Hawkins, A. (2008) Optofluidic waveguides: I. Concepts and implementations. *Microfluidics and Nanofluidics*, 4, 3–16.

Schwarz, R., Wang, F., and Reissner, M. (1993) Fermi-level dependence of the ambipolar diffusion length in amorphous-silicon thin-film transistors. *Applied Physics Letters*, 63, 1083–1085.

Song, J. H., Atay, T., Shi, S. F., Urabe, H., and Nurmikko, A. V. (2005) Large enhancement of fluorescence efficiency from CdSe/ZnS quantum dots induced by resonant coupling to spatially controlled surface plasmons. *Nano Letters*, 5, 1557–1561.

Svoboda, K. and Block, S. M. (1994) Force and velocity measured for single kinesin molecules. *Cell*, 77, 773–784.

Svoboda, K., Schmidt, C. F., Schnapp, B. J., and Block, S. M. (1993) Direct observation of kinesin stepping by optical trapping interferometry. *Nature*, 365, 721–727.

Takuo, T. and Sadahiko, Y. (2000) Optically induced propulsion of small particles in an evanescent field of higher propagation mode in a multimode, channeled waveguide. *Applied Physics Letters*, 77, 3131–3133.

Thorsen, T., Maerkl, S. J., and Quake, S. R. (2002) Microfluidic large-scale integration. *Science*, 298, 580–584.

Urbanski, J. P., Thorsen, T., Levitan, J. A., and Bazant, M. Z. (2006) Fast AC electro-osmotic micropumps with nonplanar electrodes. *Applied Physics Letters*, 89, 143508.

Wang, M. M., Tu, E., Raymond, D. E. et al. (2005) Microfluidic sorting of mammalian cells by optical force switching. *Nature Biotechnology*, 23, 83–87.

Wong, P. K., Chen, C. Y., Wang, T. H., and Ho, C. M. (2004) Electrokinetic bioprocessor for concentrating cells and molecules. *Analytical Chemistry*, 76, 6908–6914.

Yang, A. H. J., Moore, S. D., Schmidt, B. S. et al. (2009) Optical manipulation of nanoparticles and biomolecules in sub-wavelength slot waveguides. *Nature*, 457, 71–75.

Yannopapas, V. (2008) Optical forces near a plasmonic nanostructure. *Physical Review B*, 78, 045412.

Yeh, P., Yariv, A., and Hong, C.-S. (1977) Electromagnetic propagation in periodic stratified media. I. General theory. *Journal of the Optical Society of America*, 67, 423–437.

Yeh, P., Yariv, A., and Marom, E. (1978) Theory of Bragg fiber. *Journal of the Optical Society of America*, 68, 1196–1201.

Zhang, H. and Liu, K. K. (2008) Optical tweezers for single cells. *Journal of the Royal Society Interface*, 5, 671–690.

15

Fluid-Filled Optical Fibers

Michael Barth
Hartmut Bartelt
Oliver Benson

15.1 Introduction

The central idea of optofluidics is to control the flow of liquids and light in a combined system at the micron scale. This enables the realization of highly versatile devices for biochemical sensing and adaptive optical elements (Monat et al. 2007). Optical fibers are natural candidates for these kinds of applications, since they provide the means to confine light in microscopic dimensions while guiding it over macroscopic distances. This provides the opportunity to establish strong and well-controlled interactions with fluidic systems, in particular if the fiber itself contains microfluidic channels to facilitate and control the liquid flow. This chapter is dedicated to these so-called microstructured optical fibers and their application as biochemical sensors.

We will start with a general introduction to the field of optical fibers in Section 15.2, providing an overview on the different light guiding mechanisms and properties as well as on common fabrication techniques. In Section 15.3 the basic concepts and criteria for optofluidic sensing with microstructured fibers are presented, while Section 15.4 covers important issues concerning the injection and flow of liquids in these fibers. The current state of the art of fiber-based sensing is then reviewed in Sections 15.5 through 15.7, discussing the advantages and drawbacks of various fiber designs and sensing methods. As a slightly different application, the concept of light-induced particle transport inside microstructured fibers is introduced in Section 15.8, reviewing also some of the recent experiments. Finally, a brief summary and discussion on the current challenges and prospects of biochemical sensing with optical fibers is given in Section 15.9.

15.2 Microstructured Optical Fibers: General Properties and Technology

The concept of guided light in optical fibers has been successfully applied for many years in optical telecommunication due to a low attenuation and a large optical bandwidth for optical pulse transmission. However, the specific properties and advantages of guided light in optical fibers find an increasing amount of applications in other fields such as optical fiber sensors. The specific guiding principles and the available technologies are the basis of such current and future applications.

15.2.1 Light Guiding in Optical Fibers

The guiding concept in an optical fiber requires a reflection at the core–cladding interface to keep the light coupled into a fiber within the core region during propagation. In principle, several different reflection mechanisms can be used:

- Reflection at a metallic mirror
- Reflection at a dielectric interface (total internal reflection)
- Reflection at a dielectric mirror/dielectric multilayer
- Reflection at a photonic crystal structure

Depending on the principle of reflection, light guiding may require a core with a specific refractive index or light guiding can be achieved with a hollow core and without specific requirements concerning the refractive index of the core (Figure 15.1). Although fibers with a hollow core may be also filled with fluid or gaseous media, such fibers are distinguished as solid and as hollow-core fibers. Holey regions might also be used around a fiber core or in the cladding region of a fiber. The different types of such holey fibers are of special interest for sensing applications such as fiber-based fluid sensing.

The reflection at metallic mirrors is associated with relatively high losses and is, therefore, only applicable for short transmission lengths (metal tube waveguides) (Harrington 2000, Matsuura et al. 2002).

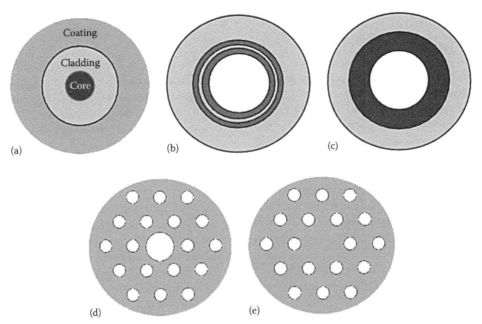

FIGURE 15.1 Different fiber types: (a) step-index fiber, (b) Bragg fiber, (c) ring-core fiber, (d) microstructured fiber (bandgap guiding), and (e) microstructured fiber (effective index guiding).

For such a fiber, there are no specific conditions for the refractive index of the core material. This means that guiding is also possible with a hollow core or with a core which is filled by liquids or gases.

Total internal reflection provides a very efficient reflection principle and is mostly used in today's optical fibers. Extremely low attenuation values with less than 1 dB/km are achievable with this concept of light guiding. Total internal reflection requires a dielectric boundary between materials of different refractive index. A conventional step index optical fiber consists of a fiber core with refractive index n_{core} and a fiber cladding with a refractive index n_{clad}, where the condition $n_{core} > n_{clad}$ is fulfilled Figure 15.1a. This type of fiber requires typically a filled fiber core, and a hollow-core fiber is usually not possible with this guiding principle. In special cases, anomalous dispersion properties of materials (frustrated total internal reflection) that achieve a refractive index $n_{clad} < 1$ can be used allowing a hollow core (Harrington 2000). This effect is, however, in practice combined with high attenuation (typically a few dB/m with best values down to 0.1 dB/m depending on wavelength) and is applicable only for limited wavelength ranges. Therefore, this concept has found application only in specific wavelength ranges such as the infrared region.

An important parameter of fibers based on total internal reflection is the numerical aperture NA which describes the maximum acceptance angle of the light which can be coupled into the core and which is guided along the fiber as long as the fiber is not bent:

$$NA = \sqrt{n_{core}^2 - n_{clad}^2}. \tag{15.1}$$

Another important parameter connected with the numerical aperture is the *V*-parameter

$$V = \frac{2\pi r}{\lambda} \sqrt{n_{core}^2 - n_{clad}^2}, \tag{15.2}$$

where
  λ is the wavelength
  r is the radius of the core of the step index fiber

In an optical fiber, different propagation states (modes) are possible. In a circular symmetric optical fiber structure, a fundamental mode is possible for any value of the *V*-parameter. Higher order modes are only allowed for sufficiently large values of the *V*-parameter. The condition for guiding of the first higher order mode is $V > 2.405$. With this value, a so-called cutoff wavelength is connected which describes the wavelength limit for single-mode propagation:

$$\lambda_{cutoff} = \frac{2\pi r}{2.405} \sqrt{n_{core}^2 - n_{clad}^2}. \tag{15.3}$$

With increasing index contrast in the fiber and increasing diameter of the fiber core, more modes are allowed, where the number of modes *M* can be approximated in a step index fiber for large values of *V* by

$$M \approx \left(\frac{V}{\pi}\right)^2. \tag{15.4}$$

The guided light in an optical fiber propagates not only in the fiber core itself but also partly in the cladding region due to the extension of the evanescent wave. Therefore, an overlap of the electromagnetic field of the propagating light with an analyte close to the fiber core is possible (see Section 15.3.2).

This effect is the basis of many sensor applications with fluid-filled fibers. The penetration depth of the evanescent light depends on the refractive index parameters at the boundary, the angle of the light at the interface, the wavelength, and the polarization state. Typically, the penetration depth is on the order of the wavelength. Therefore, the overlapping region (and the overlap integral of the field in an analyte) is usually small and requires a reasonable interaction length to become usable for a sensor. The overlap integral can be optimized by a small fiber core or by designing specific layers at the boundary. For higher order modes the overlap integral is larger compared to lower order modes in the same fiber. The evanescent wave interaction is also sensible to the bending of a fiber. The bending of a fiber may increase the penetration depth of evanescent waves (increasing the interaction with an analyte) and may also result in leaky waves (additional loss). The additional loss coefficient caused by bending can be described as

$$\alpha_{bend} = c_1 \exp(-c_2 R), \tag{15.5}$$

where
  c_1 and c_2 are constants
  R is the bending radius

15.2.2 Design Possibilities for Optical Fibers

Additional fiber properties such as dispersion (material dispersion, waveguide dispersion, mode dispersion, polarization mode dispersion) are important, particularly when transmitting pulses. There are several modifications to the above described step index fibers having a more complex refractive index structure such as graded index fibers with a continuously varying index distribution (typically with a parabolic index profile) or multilayered structures with several different refractive index values. Another modification concerns the shape of the core. Instead of a circular shape of the light guiding core, the core may be shaped asymmetrically (e.g., with an elliptical core or an unsymmetrical structural birefringence). Then, the fiber propagation is polarization dependent. These effects can be used, for example, for polarization-preserving fibers or for polarizing fibers. Typical examples of birefringent fibers with structural elements close to the core are so called panda-type or bow-tie fibers.

Another specific design possibility, which is of particular interest for filling and sensing, is related to ring-shaped fiber cores (Figure 15.1c), where propagating modes have a ring-like structure with an evanescent wave component penetrating into the central fiber area. In this case the inner part of a fiber can be hollow and allows improved interaction via evanescent waves with filling media.

Efficient optical reflection as a guiding concept can also be achieved with (periodic) dielectric multilayers consisting of materials with different refractive index values similar to optical interference filters (Bragg fiber, OmniGuide fiber, Figure 15.1b) (Yeh et al. 1978, Xu et al. 2000, Johnson et al. 2001, Ibanescu et al. 2003). The reflectivity from such a multilayered film can be described as Bragg reflection depending on the number of layers and their optical properties. Such mirrors are typically limited in their spectral range and have to be designed accordingly. Light guiding in fibers using this principle does not require a solid fiber core and allows the use of hollow cores, which can be filled with gases or liquids.

During recent years the development of microstructured optical fibers (MOFs) or photonic crystal fibers has further extended the design possibilities of optical fibers (Bjarklev et al. 2003, Knight 2003, Russell 2003, 2006, Mortensen and Nielsen 2004). Such fibers represent a generalization of earlier developed and simpler fibers with a structured core and/or cladding region. In such MOFs, holey structures are introduced in the fiber cladding area and also possibly in the fiber core area (Figure 15.1d and e). Its underlying principles were reviewed in the chapter by Benabid and Roberts. This type of fiber offers several advantages compared to conventional fibers:

- Light guiding can be achieved from a single material (no need for materials with different refractive index properties).
- The optical guiding properties can be modified precisely over a wide parameter range.
- Asymmetrical fiber properties can be introduced in a simple way.
- Light guiding can be achieved with a solid core (effective index guiding) or with a hollow core (bandgap guiding) depending on the structural parameters.
- Single-mode propagation is possible independent of the applied wavelength.

There is a great design flexibility concerning the variation of hole diameter d, the hole-to-hole distance (pitch length) Λ, and the number of hole layers. The core of the fiber is realized as a defect from the regular holey structure (e.g., one or many missing holes, hole of different diameter). Additional design options include asymmetrical distribution or size of the holes and combination of different materials or material layers with different indices of refraction (Kirchhof et al. 2004, Bartelt et al. 2008).

The light guiding can be achieved in photonic or microstructured fibers by different principles (see also Section 15.3). In the case of effective index guiding, the effective index of the core and the cladding region (i.e., the averaged index of the solid material and the holey fraction) has to fulfill the condition of total internal reflection. Therefore, either a solid core region or a small air fraction in the core area is required. Since only averaged refractive index values are relevant, a strict periodicity of holes is not necessary. The guiding is efficiently possible over a wide wavelength range. A V-parameter value can be introduced in analogy to conventional step index fibers (Birks et al. 1997, Mortensen et al. 2003):

$$V = \left[\left(\frac{2\pi\Lambda}{\lambda} \right)^2 (n_{\text{fiber}}^2 - 1) f(\lambda) \right]^{1/2}. \tag{15.6}$$

Here, $f(\lambda)$ is the air filling fraction of the MOF.

As an example for the wide design flexibility, in Figure 15.2 several MOFs are shown covering a range in numerical aperture from less than 0.1 to about 0.7. If the holey region in such an MOF is filled with a fluid, the propagation properties depend strongly on the relative refractive index values of silica and the fluid medium (Figure 15.3). For a fluid medium with lower refractive index than glass, light is guided in the central solid core. In the case of a refractive index higher than glass, the light is guided mainly in the liquid medium resulting in a near field pattern consisting of an array of peaks. In an asymmetrical MOF, high values of birefringence can be achieved (Ortigosa-Blanch et al. 2000, Steel and Osgood 2001, Geernaert et al. 2008).

In the case of bandgap guiding the reflection is based on the periodic hole structure resulting in a periodic index modulation in the cladding region (Bragg reflection). Similar to guiding in a fiber with a cylindrically layered structure (Bragg fiber), fibers with a hollow core (or with a core of lower effective refractive index than the core) are possible. Due to the high refractive index contrast between glass and air in typical photonic crystal fibers, a lower number of periods or layers is sufficient compared to solid multilayered fibers. The pitch length Λ should be smaller than the bandgap wavelength, which requires typically sub-micrometer structures. The d/Λ ratio must satisfy the condition $d/\Lambda > 0.5$ for fiber bandgap guiding, which means that a high air content is required. Despite these demanding conditions, microstructured bandgap fibers have been produced successfully with good attenuation values (Russell 2003, Hansen et al. 2004, Fini 2004). However, due to the use of the Bragg reflection concept, the bandgap (i.e., the range of wavelengths with efficient guiding) is limited. The best attenuation values achieved are approximately 1.7 dB/km, while typical values lie around 20 dB/km with bandgap widths of about 100 nm. Several orders of the Bragg reflection effect may be observed offering different wavelength windows for propagation.

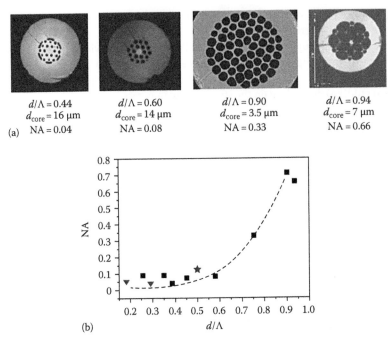

FIGURE 15.2 (a) Examples of MOFs with different lattice parameters and (b) the corresponding variation of the numerical aperture.

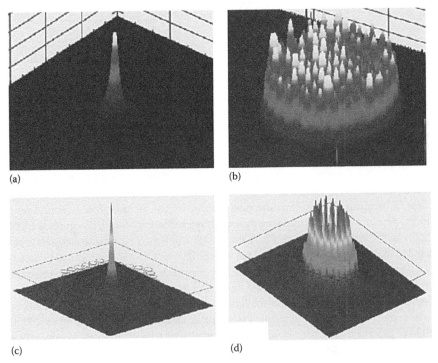

FIGURE 15.3 (See color insert following page 11-20.) Fluid-filled MOF with solid-core at a wavelength of 1200 nm: (a) near field pattern for low refractive index fluid, (b) near field pattern for high refractive index fluid, (c) FEM simulation of field pattern for low refractive index fluid, and (d) FEM simulation of field pattern for high refractive index fluid. (From Wang, Y. et al., 2009. With permission.)

15.2.3 Fiber Materials and Fabrication Techniques

Besides the general flexibility in designing the structure of an optical fiber, another option for modifying the propagation properties is given by the use of specific materials. Most commonly used optical fibers are made from silica glass, which may additionally be doped to modify the refractive index values. Silica glass is available in high quality and allows preparation of optical fibers with extremely low attenuation (down to less than 0.2 dB/km with total internal reflection guiding). At the same time it can be used over a wide spectral range from visible wavelengths to near infrared wavelengths. Silica glass can also be reproducibly manufactured to thin optical fibers without degradation of its optical properties and provides good mechanical properties in combination with long-term stability. For specific applications, especially if light guiding is necessary in wavelength regions beyond the transmission windows of silica glass (e.g., in the infrared region), other glass types may be used. Such fibers typically have much higher attenuation due to absorption or scattering, but values on the order of several dB/m are often acceptable for sensor applications, where no long transmission lengths are required. Besides glasses, polymer materials are also used for preparation of optical fibers. The main advantage of such polymer fibers compared to glass fibers is a simpler handling concerning cutting or splicing. On the other hand, polymer fibers are limited in attenuation (typical values lie around 100–200 dB/km at a wavelength of 650 nm), spectral range, and especially in the applicable temperature range (the glass transition temperature of PMMA is around 100°C, alternative polymers reach up to 200°C) (Kuzyk 2007).

The typical method for fiber fabrication is based on fiber preform preparation with a subsequent fiber drawing process in a fiber draw tower. For the preform fabrication different processes can be used such as rod-in-tube (combination of a glass rod and a glass tube with different refractive index) or the MCVD technology (modified chemical vapor deposition: layer deposition from the gas phase inside a glass tube, collapsing of this structure to a solid preform) or the OVD technology (layer deposition on the outside of a tube). In the case of photonic crystal or microstructured fibers the stack-and-draw method is typically used. The preform is prepared by arranging glass rods and glass tubes (glass capillaries) according to the desired fiber structure. Instead of using circular glass tubes or rods, rectangular or hexagonal shapes have also been applied for fabrication of such preforms. Typical sizes of rods or tubes are in the millimeter or sub-millimeter range. This arrangement can be fixed or glued in order to assure stability for the subsequent drawing process (Figure 15.4). In the final fiber drawing process the preform is heated and stretched to the final fiber shape in a fiber drawing tower. The final geometrical fiber properties can be adjusted by proper use of the preform feed speed and the fiber drawing speed. The glass fiber is then coated by a protective coating (typically acrylate or polyimide), which gives good bending strength and assures that humidity cannot penetrate into the silica matrix. For MOFs with holey regions, additional pressurizing technology is necessary during the fiber drawing process in order to avoid collapsing of the holes. By modification of the applied pressure values and considering the temperature, the fiber parameters of the MOF (air hole size, air fraction) can be varied. It is, therefore, possible to obtain different types of MOFs from the same preform. As an alternative, fibers can also be produced directly from melting glasses or by an extrusion technique. For such methods, usually materials with low melting temperatures compared to silica glass are used. The extrusion method is typically applied for MOFs in polymer materials.

In Figure 15.5 different optical fibers are shown as practical examples of fibers with holey portions which are especially applicable for liquid fiber sensing: a ring-core fiber, an MOF with solid core and holey cladding, a photonic crystal fiber with hollow core, an MOF with strong asymmetry, an MOF with holey core and cladding region and additional high index layers in the core region, and an MOF with small solid core and large air holes (suspended-core fiber).

15.3 Principles of Sensing with Fluid-Filled Optical Fibers

The idea of using optical fibers for optofluidic sensing applications has been pursued for several decades, as fiber-based sensors promise significant advantages compared to conventional sensing devices which

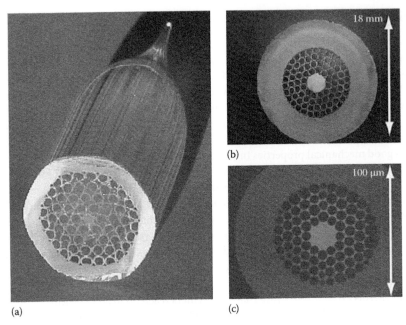

FIGURE 15.4 Fabrication of microstructured or photonic crystal fibers: (a) fiber preform, (b) cross section of the preform, and (c) cross section of the microstructured fiber.

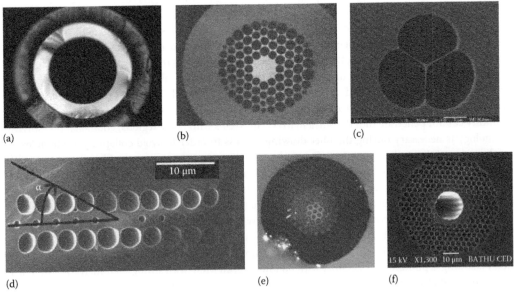

FIGURE 15.5 Examples of MOFs (cross sections): (a) ring-core structure, (b) microstructured fiber with solid core, (c) suspended-core fiber (steering-wheel fiber), (d) asymmetrical microstructured fiber. (Reprinted from Geernaert, T. et al., *IEEE Photon. Technol. Lett.*, 20, 554, 2008. With permission.), (e) microstructured fiber with holey core region and additional high index layers in the core region, and (f) hollow-core bandgap fiber. (Reprinted from Knight, J.C., *Nature*, 424, 847, 2003. With permission.)

rely on free-space optics. First and foremost, fibers offer a much longer interaction length between the guided light and the sample material (e.g., gases or liquids), which in principle is only limited by the length of the fiber. This, in turn, leads to an enhancement of the corresponding detection signal (e.g., in an absorption or fluorescence measurement) and thus increases the overall detection efficiency. For comparison, in standard absorption/fluorescence spectrometry the size of the glass cuvette (containing the sample liquid) is usually restricted to a few centimeters, while fiber sensors can easily reach lengths of several meters.

In order to make a fiber suitable for sensing, the guided optical field has to have a certain overlap with the sample material such that an efficient interaction can take place. In the first practical implementations with standard step-index fibers this was achieved by completely removing the coating and cladding of the fiber, exposing the bare fiber core (Glass et al. 1987, Kao et al. 1998). The latter could then by immersed into the sample liquid and the guided light interacted via the evanescent part of the field (Carniglia et al. 1972). The drawback of this method, however, is that the bare core is usually very fragile and only a very small fraction (<1%) of the light intensity is actually located within the liquid. On the other hand, the solid fiber core avoids complete blocking of transmitted light by dirt or dust (which can happen in a standard spectrometer cell) and assures a certain level of light transmission even for contaminated fluids.

Since the advent of MOFs (Eggleton et al. 2001) it has become possible to deliver sample material directly to the light-guiding region of the fiber (via the holey structure) without the need to strip its cladding. This considerably improves the robustness of the device and provides a vast range of design possibilities to optimize the light–sample interaction. Furthermore, the sample volume needed for sensing is reduced to the sub-microliter regime due to the small dimensions (on the order of a few micrometers) of the incorporated holes. On the other hand, this immediately raises issues concerning the efficient filling of such small capillaries, which is discussed in detail in Section 15.4. We will begin with a general description of the different operation schemes possible with fluid-filled MOFs (Section 15.3.1) as well as a derivation of some figures of merit needed to evaluate and compare the sensing performance of various fiber designs (Section 15.3.2).

15.3.1 Operation Principles of Fiber Sensors

Depending on the structure of the MOF and the method used for fluid infiltration different light-guiding regimes can be distinguished. These regimes shall be illustrated exemplarily for two specific fiber designs, namely a solid-core MOF (Figure 15.6a) as well as a hollow-core MOF (Figure 15.6b). When filled with air, light is predominantly guided in the core of the solid-core MOF via total internal reflection (we will neglect any cladding modes in this simple analysis), while the hollow-core MOF supports a mode which is mainly confined to the central hole via the photonic bandgap effect (Cregan et al. 1999). Both fibers may be readily employed as gas sensors, with the key difference that in the case of the solid-core MOF the interaction of the guided light with the gas only takes place via the evanescent part of the field (Hoo et al. 2002), while in the case of the hollow-core MOF nearly the entire field participates in the interaction (Ritari et al. 2004, Benabid et al. 2005).

When all holes are filled with a low-index liquid, the principal behavior stays the same for both types of fibers as long as the refractive index of the liquid (n_{liquid}) is smaller than that of the fiber material (n_{fiber}). The light confinement will, in general, be reduced due to the smaller refractive index contrast in the holey structure. This weaker confinement can be advantageous in the case of the solid-core MOF, since the evanescent field penetrates further into the liquid, increasing the light–sample interaction. Therefore, this is the regime in which evanescent fiber devices are typically operated. In the case of the hollow-core MOF, a complete filling of the holey structure usually degrades its sensing performance considerably, as the photonic bandgap becomes much narrower (due to the weaker refractive index contrast), severely limiting the wavelength range in which the device can be operated. This drawback can be circumvented by injecting liquid into the core hole only, thus creating a liquid-core/air-cladding hybrid

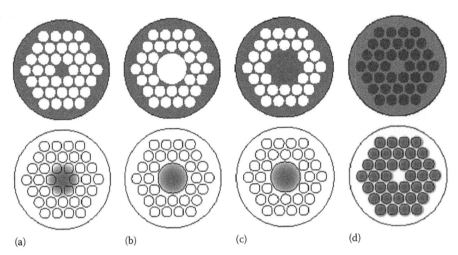

(a)   (b)   (c)   (d)

FIGURE 15.6 Schematic representation of different fiber geometries (upper row) and the corresponding light-guiding regions (lower row): (a) solid-core fiber, (b) hollow-core fiber, (c) selectively filled hollow-core fiber, and (d) completely filled solid-core fiber ($n_{liquid} > n_{fiber}$).

fiber (Figure 15.6c). Such a selective infiltration changes the light-guiding mechanism from bandgap to index-guiding if the effective refractive index of the core is larger than that of the cladding structure (Fini 2004). The key difference to solid-core MOFs, however, is that in this case the light is still propagating mainly within the liquid filament itself, providing a very intense light–sample interaction. Consequently, such a scheme offers the most promising sensing performance, at the expense of a more difficult preparation and infiltration procedure. These issues will be discussed in detail in Sections 15.4 and 15.6.

Finally, the (complete) infiltration of a high-index liquid ($n_{liquid} > n_{fiber}$) shall be considered. In this case, light can be guided in the liquid filaments via total internal reflection, regardless of the specific structure of the fiber (Figure 15.6d). Such a scheme resembles a bundle of parallel capillaries, each guiding light individually (Konorov et al. 2005). However, it is usually restricted in its applicability as not many liquids exhibit refractive indices larger than that of common fiber materials (e.g., silica with $n_{silica} = 1.45$). In particular, it is not applicable for sensing with aqueous solutions ($n_{water} = 1.33$), which is often desired in biological and medical applications.

15.3.2 Figures of Merit for Fiber Sensors

The most suitable choice among the different operation schemes described above for a specific application depends on several factors, such as the quantity which is detected (e.g., intensity, phase, or polarization), the way in which light is launched and extracted into/from the fiber (forward, backward, or sideways), and restrictions imposed by the sample material itself (refractive index, viscosity, etc.). However, there are some general criteria which can help to evaluate and compare the sensing performance of various MOFs.

The most important figure of merit is the fraction of optical field energy propagating within the sample material (often referred to as "overlap"), which is defined as (Cordeiro et al. 2006a)

$$\gamma = \frac{\int_{sample} \mathrm{Re}\{\mathbf{E} \times \mathbf{H}^*\}\hat{\mathbf{z}}\,dA}{\int_{total} \mathrm{Re}\{\mathbf{E} \times \mathbf{H}^*\}\hat{\mathbf{z}}\,dA} \tag{15.7}$$

where

    E and **H** are the electric and magnetic field vectors, respectively

    $\hat{z}$ is the unit vector pointing in the direction of propagation, and the integration is performed either over the total cross section of the fiber or only over the regions containing sample material

γ is the dominant factor for all sensing schemes relying on the detection of transmission signals such as absorption or phase measurements. It directly determines how much attenuation or phase shift the guided light accumulates when propagating through a specific length L of the fiber. Consequently, the Beer–Lambert law, which relates the outgoing light intensity, I, to the incoming intensity, I_0, in an absorption measurement has to be modified such that (Jensen et al. 2004)

$$I(\lambda) = I_0(\lambda) \cdot \exp[-\varepsilon(\lambda)c\gamma L] = I_0(\lambda) \cdot \exp[-\varepsilon(\lambda)cL_{\text{eff}}]. \tag{15.8}$$

Here, $\varepsilon(\lambda)$ and c are the wavelength-dependent molar extinction coefficient and the concentration of the absorbing species (e.g., molecules in a solution), respectively. L_{eff} can be interpreted as an effective interaction length that a free-space setup (for which $\gamma = 1$) with a comparable absorption sensitivity would have. From Equation 15.8 it becomes apparent that the main advantage of fiber sensors, namely, the longer interaction length, can completely diminish if the overlap γ is very small. Instead of γ sometimes the so-called sensitivity coefficient $r = (n_{\text{sample}}/n_{\text{eff}})\gamma$ is used to quantify the sensing efficiency of an MOF, where n_{sample} is the refractive index of the sample material and n_{eff} is the effective index of the guided mode (Cordeiro et al. 2006a).

Another important property is the propagation loss α of the fiber, which includes leakage of the guided modes, scattering due to surface roughness and fabrication imperfections, as well as parasitic absorption in the fiber material itself. Relating to the expression in Equation 15.8 it leads to an additional reduction of the detected signal according to

$$I(\lambda) = I_0(\lambda) \cdot \exp[-\varepsilon(\lambda)c\gamma L] \cdot \exp[-\alpha(\lambda)L] \tag{15.9}$$

and may thus limit the maximum fiber length L which can be employed in the sensing device. To achieve a good sensing performance, $\alpha(\lambda) \ll \varepsilon(\lambda)c\gamma$ should be fulfilled.

While in pure transmission measurements γ and α are usually sufficient to characterize the sensing efficiency of an MOF, additional criteria need to be taken into account if the detected signal is generated inside the fiber, e.g., in fluorescence or Raman measurements. In this case γ not only determines the excitation efficiency of these signals but also their coupling strength to the guided modes of the fiber. The fraction of light which is actually radiated into these guided modes and finally channeled to the detector can then be expressed through the collection efficiency, ξ. In general, ξ depends on the specific optical mode structure of the fiber, as the spatial radiation pattern of the emitting (or scattering) species is altered compared to the free-space behavior due to the modified local optical density of states. However, in the simple case where the emission (or scattering) can be approximated as isotropic and light is guided via total internal reflection in the fiber core, the collection efficiency can be expressed as (Smolka et al. 2007b)

$$\xi = \frac{1}{2}\left[1 - \frac{n_{\text{eff}}^{\text{cladding}}}{n_{\text{eff}}^{\text{core}}} \right]. \tag{15.10}$$

Here, $n_{\text{eff}}^{\text{core}}$ and $n_{\text{eff}}^{\text{cladding}}$ are the effective refractive indices of the guided core mode and the lowest (lossy) cladding mode, respectively. The factor 1/2 accounts for the detection at only one of the two fiber ends as it is the case in most practical implementations. Equation 15.10 is an expression for the internal acceptance angle for light launched into the fiber from an intrinsic source and does not take into account the coupling strength between the emitter (scatterer) and the guided modes. A better figure of merit is, therefore, the total capture coefficient $\phi = \xi \gamma^2$, which also includes the excitation efficiency. Specific examples of different fiber designs and the corresponding figures of merit will be discussed in Sections 15.5 and 15.6.

15.4 Fluid Injection into Microstructured Optical Fibers

After introducing the basic concepts and criteria for optofluidic sensing with MOFs in the previous section, we now turn to the practical problem of injecting liquids into the holey structure of the fiber. Two issues are of particular interest. First, the speed with which the fiber can be infiltrated, as the filling time directly determines the practicality of fiber sensor devices in real-life applications. A corresponding filling model, adopted from Nielsen et al. (2005), is derived in Section 15.4.1. Second, the problem of selectively injecting liquid into specific holes of the MOF needs to be addressed, as this represents the basis for the fluid-core/air-cladding sensing concept introduced in Section 15.3.1. A number of filling approaches is described and discussed in Section 15.4.2. Finally, beyond the field of optical sensing, we also give a brief introduction to optofluidic tuning of MOFs in Section 15.4.3.

15.4.1 Model of Capillary Filling

The driving mechanism (in the absence of external pressure) which causes a liquid to fill a capillary tube is the surface tension of the liquid at the interface with the capillary wall. Depending on the characteristic contact angle θ between the rim of the liquid and the wall, the surface tension σ leads to the capillary force

$$F_c = 2\pi\sigma r \cos\theta, \tag{15.11}$$

where r is the radius of the (circular) capillary tube. If external pressure is applied, an additional force

$$F_p = \pi r^2 \Delta p \tag{15.12}$$

has to be taken into account, with Δp being the pressure difference between the liquid column and the open end of the capillary. As the typical hole radii encountered in MOFs are on the order of a few micrometers, the flow inside the holes can be assumed as laminar. In this case, the motion of the (incompressible) fluid is governed by Poiseuille's law and a liquid column of length L inside the capillary will encounter a friction force

$$F_r = -8\pi\eta L \frac{dL}{dt}, \tag{15.13}$$

where η is the dynamic viscosity of the liquid. The equation of motion for the filling of a capillary tube with a liquid of density ρ is then given by

$$\frac{d}{dt}\left(\pi\rho r^2 L \frac{dL}{dt}\right) = 2\pi\sigma r \cos\theta + \pi r^2 \Delta p - 8\pi\eta L \frac{dL}{dt}. \tag{15.14}$$

Note that we have neglected the action of gravity on the liquid, which is only valid for horizontally oriented fibers and/or small hole diameters ($r < 5\,\mu m$). For vertically oriented fibers and larger holes one would have to add the gravitational force $F_g = -\pi g \rho r^2 L$ (with g being the gravitational constant) on the right-hand side of Equation 15.14. In this case Equation 15.14 becomes nonlinear and has to be solved numerically. In the absence of gravity, however, it has the analytical solution (Nielsen et al. 2005)

$$L(t) = \left[\frac{A}{B^2} \exp(-Bt) + \frac{At}{B} - \frac{A}{B^2} \right]^{1/2},$$

(15.15)

with the coefficients

$$A = \frac{4\sigma\cos\theta + 2r\Delta p}{\rho r} \quad \text{and} \quad B = \frac{8\eta}{\rho r^2}.$$

(15.16)

Based on this solution, the expected filling behavior for water in silica tubes of different radii is shown in Figure 15.7a for the case when no external pressure is applied. It is apparent that small fiber lengths $L \approx 10\,\text{cm}$ can be infiltrated on a relatively fast timescale of a few minutes, while for longer fibers the filling time soon reaches rather impractical values of hours or more. In this case, the application of overhead pressure to increase the filling speed is inevitable (Figure 15.7b). However, even for pressures of several bars the infiltration length is practically limited to a few meters unless very large holes are used.

This simple analysis sets one of the main advantages of fiber-based sensors, namely, the long interaction length between light and sample material, into a new perspective. Due to the unavoidably large impact of friction in such micrometer-sized fluid channels (Yang et al. 2004) the actual sample length

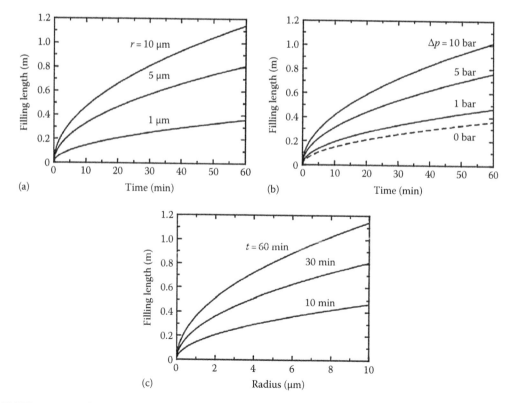

FIGURE 15.7 Calculated filling length versus time of a silica capillary infused with water for (a) different hole radii *r* and no external pressure and (b) constant radius $r = 1\,\mu\text{m}$ and different external pressures Δp. (c) Filling length versus hole radius *r* for different filling times *t* (without external pressure).

(although still much larger than in corresponding free-space devices) will be limited in most real-life applications. This also underlines the necessity to optimize the sensing efficiencies as stated in Section 15.3.2, since a low overlap γ cannot simply be compensated by increasing L arbitrarily.

The dependence of the filling speed on the hole radius r (Figure 15.7c) provides the opportunity to achieve a selective filling effect, as larger holes will be infiltrated much faster than smaller ones. This can be exploited to selectively inject liquids into the core hole of a hollow-core MOF, as will be described in the next section.

15.4.2 Methods for Selective Injection

When preparing a solid-core MOF for optofluidic sensing, it is usually sufficient to infuse the liquid into the entire holey structure, as the basic light-guiding mechanism will not be affected (see Section 15.3.1). However, additional efforts have to be made when using fluid-core MOFs, where the core has to be filled selectively in order to achieve index-guiding within the liquid filament. In the following, two methods, that have been developed and successfully demonstrated for this purpose recently shall be introduced.

The first approach (which we will denote the differential filling speed method) exploits the dependence of the infusion speed of the liquid on the different radii of the holes in the MOF. For example, in a hollow-core MOF the central core hole is usually much larger than the surrounding cladding holes. Consequently, the liquid will have traveled farther in the core than in the cladding holes after a certain period of time. The basic idea then works as follows (Huang et al. 2004) (Figure 15.8a): (1) An optical adhesive is injected into the holey structure of the MOF and cured as soon as there is a significant difference in the infiltration length between the core and cladding holes. Subsequently, the fiber is cleaved at the position indicated in Figure 15.8a such that the core hole is capped with the polymer, while the cladding holes remain open. (2) The filling procedure is then repeated, with the key difference that only the cladding holes are filled, since the core hole is still blocked. After curing the adhesive, the fiber is

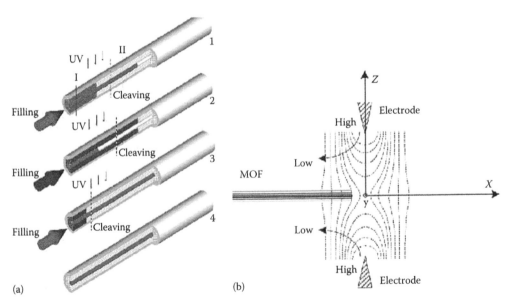

(a)   (b)

FIGURE 15.8 (a) Schematic illustration of the differential filling speed method for selective filling of an MOF. (Reprinted from Huang, Y. et al., *Appl. Phys. Lett.*, 85, 5182, 2004. With permission.) (b) Current and energy density distribution in an arc fusion splicer used to selectively melt the outer cladding holes of a hollow-core MOF. (Reprinted from Xiao, L. et al., *Opt. Express*, 13, 9014, 2005. With permission.)

cleaved again such that all the cladding holes are capped while the core hole remains open. (3) Now the core can selectively be infiltrated with a liquid since the cladding holes are blocked. (4) If necessary, the fiber end can be cleaved once more to remove the remaining polymer. Although providing the desired filling configuration, the differential filling speed method also has some severe drawbacks. First of all, it is relatively time consuming due to the multistep preparation procedure. Furthermore, it is hard to accurately determine the infusion length in the different holes, making it necessary to infiltrate quite large portions of the fiber in order to reliably find the proper position for cleaving. This, in turn, leads to the removal of rather large parts of the fiber before the final configuration is reached.

The second approach (which we will denote the fusion method) relies on the selective melting of the cladding holes when the MOF is exposed to the arc discharge in a conventional fusion splicer (Figure 15.8b). While fusion splicers are usually employed to permanently connect two standard fibers by locally melting and fusing the silica at the corresponding fiber ends, the fusion parameters (such as the current through the electrodes, the duration of the are discharge, and the position of the fiber) can be adjusted such that only the outer part of the MOF melts while the central core hole remains open. The reason for this selective melting is found in the temperature distribution inside the arc discharge as well as the different sizes of core and cladding holes. As the temperature is lowest on the central axis of the arc discharge (where the fiber core is located) (see Figure 15.8b) and the heat transfer from the outer to the inner region of the fiber is hindered due to the holey cladding structure, a temperature gradient builds up across the fiber profile. This is accompanied by a corresponding gradient of the viscosity of the silica, which in combination with the smaller hole radii leads to the selective collapse of the cladding holes. A sequence of different stages of this fusion process is shown in Figure 15.9a through c, while in Figure 15.9d the final result after infiltrating and cleaving the MOF is depicted (Xiao et al. 2005).

The fusion method circumvents several of the drawbacks of the differential filling speed method, as it is fast, reliable, and does not necessitate the removal of a part of the fiber. It has, therefore, developed as the main method employed for applications which require the selective filling of an MOF. However, an essential drawback regarding practical implementations is the insertion of the liquid at the (fusion spliced) end of the fiber, where also the in- or out-coupling of light may be necessary. It is thus often difficult to achieve optical and fluidic access simultaneously. Therefore, some recent work is dedicated to approaches which aim at a lateral access to the holes of the MOF, making the infiltration procedure independent from the optical alignment. Preliminary results using a fusion splicer for this purpose have been obtained by Cordeiro et al. (2006b). Their method relies on the exertion of pressure to the core hole from one end of the fiber while all other holes are sealed. An arc discharge is used to locally melt the silica at the side of the MOF causing the cladding to burst at this point (due to the internal pressure) and thus creating a side hole to the core (Figure 15.10a). The cladding holes are sealed during this process, enabling the selective access to the core hole. However, strong light scattering is induced by this side hole, degrading the performance of the MOF, especially in transmission measurements. A more promising approach, demonstrated by the same group (Cordeiro et al. 2007), uses the nanofabrication capabilities of a focussed ion beam to locally mill holes into the side of an MOF (Figure 15.10b). This not only provides a much better control on the size, shape, and position of the side access, but also helps to reduce losses due to scattering. However, so far this technique has only been applied to solid-core MOFs, where no selective infiltration is needed. In the case of a fluid-core MOF, a combination with a fusion splicing method may be needed (as the cladding holes are not sealed at the position of the side hole) to gain selective access to the core hole.

15.4.3 Tunable Liquid-Filled Fibers

The great variation range in optical material properties of liquids can also be used to achieve tuning and switching functionality in optical fibers (Eggleton et al. 2001). Typical external parameters applicable for such active change or modulation of the optical propagation properties are temperature, electric

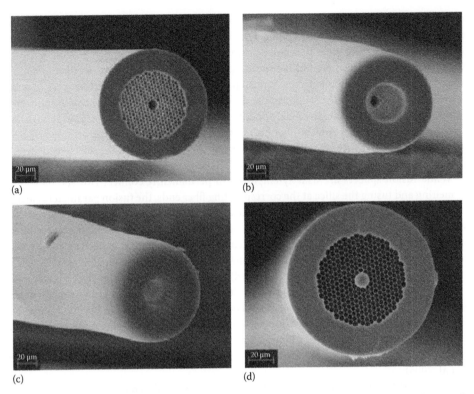

(a)   (b)

(c)   (d)

FIGURE 15.9   End facet of an MOF (a) before and (b,c) after treatment in the arc discharge of a fusion splicer at two different arc currents. (d) Cleaved end facet of the selectively filled MOF. (Reprinted from Xiao, L. et al., *Opt. Express*, 13, 9014, 2005. With permission.)

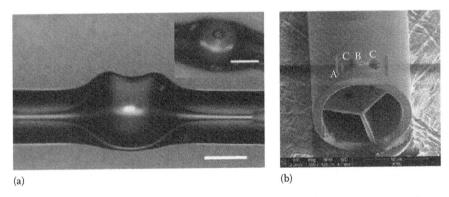

(a)   (b)

FIGURE 15.10   (a) MOF with a side hole opened by applying a local arc discharge and high internal pressure. The white bars are 100 μm wide. (Reprinted from Cordeiro, C.M.B. et al., *Opt. Express*, 14, 8403, 2006b. With permission.). (b) Side holes inscribed into an MOF via focussed ion beam milling. (Reprinted from Cordeiro, C.M.B. et al., *Meas. Sci. Technol.*, 18, 3075, 2007. With permission.)

fields, magnetic fields, or light intensity. Liquids change their refractive index with temperature, which is limited in time response but efficiently applicable for defining guiding properties in a fiber (Larsen et al. 2003, Wang et al. 2009). As pointed out, liquid filling with a refractive index larger than the surrounding (glass) material results in guided modes, whereas a smaller refractive index value will produce losses due to leaky modes. In this way a switching operation in intensity can be achieved. In the case of the periodic cladding structure in a photonic crystal bandgap fiber, the position and range of the band-gaps can be modified by filling with tunable liquids. Liquid crystals are especially applicable to functionalize fibers which are sensitive to electric fields (Du et al. 2003, Lorenz et al. 2008). If the orientation of the liquid crystal can be controlled then it is possible to modify the polarization dependent properties in order to switch or modulate light transmission. Liquid-crystals have also been used for temperature modulation as well as for all-optical modulation of the bandgap of photonic crystal fibers. A change of the temperature effect can be optically induced by controlled absorption of adapted dyes in the liquid (Alkeskjold et al. 2004, Li et al. 2004). Finally, the strong optical nonlinear properties of specific liquids offer the option to modify the propagation properties directly by optical signals. This concept has been proposed for highly efficient supercontinuum generation (Zhang et al. 2006).

15.5 Sensing with Evanescent Field Fibers

While in Section 15.3 some of the basic principles of fiber-based optofluidic sensing were introduced, a review of the recent progress and current state of the art in sensing with solid-core and hollow-core MOFs is presented in this and the next section. Here, we will start with fiber sensors that rely on evanescent field interactions, as is usually the case for solid-core MOFs. The latter offer a number of practical advantages compared to fluid-core fibers, first and foremost the expendability of a selective filling technique, considerably simplifying preparation and handling. Furthermore, as the solid core is usually very small, single-mode operation is easily maintained, being important for applications where phase information is crucial. Also, as light is usually guided via total internal reflection, operation over a broad range of wavelengths is ensured, being only limited by losses of the fiber material itself. However, the main drawback of solid-core fibers is the relatively weak interaction of the propagating light with the sample material, which can be quantified by the overlap γ (see Section 15.3.2) and has long been limited to values of $\gamma < 1\%$. Only in recent years there has been considerable progress in optimizing the light–sample interaction in MOFs, propelled by advances in the fabrication of complex MOF geometries.

The basic aim is to expel as much of the energy of the propagating mode from the fiber core into the evanescent part of the field which interacts with the infiltrated gas or liquid. This can be achieved by reducing the diameter d_{core} of the core relative to the wavelength λ of the injected light. For a simple MOF geometry where the core is basically formed by omitting one hole (see Figure 15.11a), a rule of thumb for

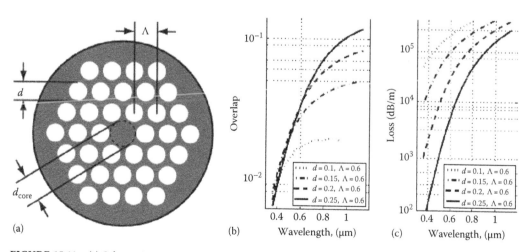

FIGURE 15.11 (a) Schematic representation of a solid-core MOF. (b) Calculated overlap and (c) corresponding loss for different design parameters of the fiber. (Reprinted from Fini, J.M., *Meas. Sci. Technol.*, 15, 1120, 2004. With permission.)

achieving a significant overlap ($\gamma > 1\%$) is $\lambda > 2\Lambda$ (Monro et al. 2001), where Λ is the hole-to-hole distance (also called the fiber pitch). Furthermore, the overlap increases with increasing d/Λ (Monro 2001, Fini 2004), as the air-filling fraction of the cladding structure becomes larger and enables a better interaction between the light and an infiltrated liquid. In Figure 15.11b the estimated overlap (in air) is displayed as a function of the wavelength for a number of different d/Λ combinations. Obviously, values of $\gamma > 10\%$ are achievable using the right choice of design parameters. The performance is even further improved when the liquid is infiltrated into the fiber, since the penetration depth of the evanescent field will increase due to the weaker refractive index contrast. Monro et al. (2001) reported on the fabrication of a solid-core MOF ($d/\Lambda = 0.8$, $\Lambda = 0.75\,\mu m$) with an overlap of $\gamma \approx 35\%$ (in air) at a wavelength of $\lambda = 1.67\,\mu m$. However, as shown in Figure 15.11c, larger ratios of λ/Λ as well as smaller core diameters lead to higher optical losses, thus limiting the achievable sensing performance (see Section 15.3.2). Furthermore, as d_{core} is reduced, the maximum coupling efficiency for light launched into the fiber decreases and the coupling itself becomes more critical and susceptible to external influences. This might severely degrade the handiness of sensor devices in real-life applications.

An alternative way to improve γ without reducing the core diameter d_{core} too much is the incorporation of additional holes into the solid core, thus creating a microstructured-core MOF. An example of such a fiber is shown in Figure 15.12a (Cordeiro et al. 2006a). The holey structure of the core forces the modal field (Figure 15.12b) to penetrate further into the sample material and, consequently,

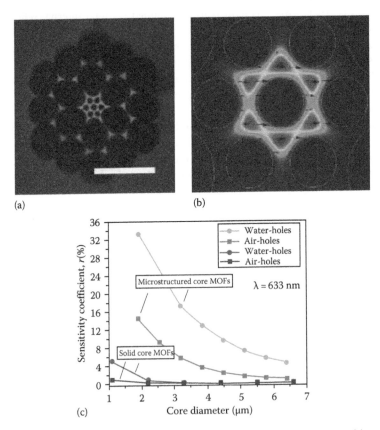

FIGURE 15.12 (a) Polymer MOF with a microstructured core (the white bar is 20 μm wide) and (b) the corresponding calculated mode power and field distribution. (c) Calculated sensitivity coefficient r of a microstructured core as well as a corresponding regular solid core for different diameters and filling conditions. (Reprinted from Cordeiro, C.M.B. et al., *Opt. Express*, 14, 13056, 2006a. With permission.)

increases the overlap compared to an all-solid core of the same dimension. This is quantified in terms of the sensitivity coefficient r (see Section 15.3.2) in Figure 15.12c for a similar fiber design. Apparently, values of $r > 10\%$ (in water at $\lambda = 633\,nm$) can be achieved with core diameters as large as $4\,\mu m$, a performance not possible with simple solid cores of the same size. In the corresponding experimental realization the achieved r was approximately 4% (Cordeiro et al. 2006a) but with further potential for improvement by adjusting the design and the fabrication procedure. The larger core diameter d_{core} considerably helps to enhance the coupling efficiency of light into the fiber, making it much more suitable for practical applications. However, care must be taken as such microstructured-core MOFs usually exhibit higher propagation losses (Fini 2004). Therefore, a trade-off between high overlap and low loss has to be found when designing an MOF for a specific application. Furthermore, the diameter of the holes inside the core should be sufficiently large ($\geq 1\,\mu m$) to still allow for an efficient filling with liquids.

At this point we want to highlight an interesting effect that occurs if the hole diameter inside the microstructured core is reduced to sub-wavelength dimensions. In this case, the electric field can get "trapped" inside the hole (Wiederhecker et al. 2007) due to the boundary conditions which have to be fulfilled at the interface and which lead to a strong increase of the field strength in the hole. This behavior is completely analogous to the field enhancement observed in silicon-on-insulator slot waveguides (Almeida et al. 2004, Xu et al. 2004). The highly localized electric field can, in principle, significantly increase the light–sample interaction in an MOF. However, as this effect requires very small hole sizes on the order of $\lambda/4$, the flow rate/filling time reaches rather impractical values, rendering the usefulness of such an approach for optofluidic sensing applications questionable.

Apart from the MOF geometries considered so far, where the cladding structure typically consists of a large number of holes, another fiber design for evanescent sensing has evolved recently which relies on few very large cladding holes to isolate the core from the outer part of the fiber. In these so-called steering-wheel or suspended-core fibers the core is thus almost freely suspended in air, held only by (usually three) thin struts. An example of such a fiber is shown in Figure 15.13. Steering-wheel MOFs are particularly appealing due to the large hole diameter ($>10\,\mu m$), which is favorable for fast fluid injection and transport. They further promise high overlap values ranging from $\gamma = 7\%$ to $\gamma = 29\%$ (for wavelengths between 850 and 1500 nm) while keeping the propagation loss small ($<1\,dB/m$), as predicted by Zhu et al. (2006). Indeed, overlaps of $\gamma = 29\%$ at a wavelength of $\lambda = 1550\,nm$ were demonstrated experimentally for this type of fiber (Webb et al. 2007), although the authors did not report on the corresponding propagation losses. These losses can be further reduced by using higher index materials (such as soft glasses or polymers) for the fabrication of the MOF. The higher refractive index leads to a tighter confinement of the light to the core, therefore, at first glance, also reducing the sensing performance as the penetration length of the evanescent field is decreased. However, due to the larger index contrast between the fiber and the sample material, regions of high local electric field intensity are formed at

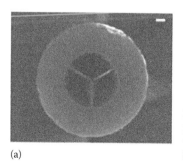
(a)

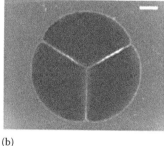
(b)

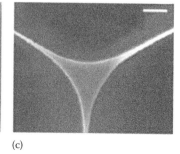
(c)

FIGURE 15.13 Cross sections of a steering-wheel MOF at different magnifications. The white bars have a width of (a) $20\,\mu m$, (b) $10\,\mu m$, and (c) $1\,\mu m$. (Reprinted from Hautakorpi, M. et al., *Opt. Express*, 16, 8427, 2008. With permission.)

the interface between both materials, effectively increasing the light–sample interaction. This effect is very similar to the one discussed above for sub-wavelength holes and has its origin in the discontinuity of the electric field at the boundary. As was demonstrated by Afshar et al. (2007, 2008) the local field enhancement can be exploited to boost the total capture coefficient, ϕ, (see Section 15.3.2) for the intrinsic fluorescence emission from an MOF due to the stronger coupling of the emitters to the modal field. An experimental implementation of soft-glass steering-wheel MOFs for fluorescence sensing was reported by Ruan et al. (2007). The authors recorded emission spectra from quantum dot–labeled proteins infiltrated into the MOF, whereby the fluorescence was excited and collected through the guided modes of the fiber. Although the overlap was only $\gamma = 3\%$ in this case, protein concentrations down to 10^{-9} M could be reliably detected.

Although solid-core MOFs are natural candidates for evanescent-sensing applications, other fiber designs can also be employed for this purpose. In fact, one of the first demonstrations of efficient fiber-based biosensing (with a minimum detectable concentration of $\approx 10^{-7}$ M) was reported for a hollow-core MOF completely filled with an aqueous solution of dye-labeled DNA (Jensen et al. 2004). An image of such a fiber is shown in Figure 15.14c. In this case, the mostly index-guided cladding modes of the fiber were used instead of the photonic bandgap core mode. This situation is equivalent to guiding light in a large number of steering-wheel-like cores, distributed over the whole cladding structure and linked via small silica bridges. The overlap in this fiber was moderate with reported values of $\gamma = 5.2\%$ (Jensen et al. 2004) and $\gamma = 6.5\%$ (Rindorf et al. 2006a), but the main advantage for this type of highly parallel sensing is the light–sample interaction along the entire perimeter of all holes, drastically increasing the surface/volume ratio. For comparison, in solid core MOFs only sample material in the direct vicinity of the core is efficiently sensed. Cladding-guiding hollow-core MOFs may thus be employed in all applications where the interaction of species with a sensor layer at the inner surface of the fiber holes is important. This is typically the case for selective capture experiments, where a specific biomolecule is captured and sensed through binding to a complementary molecule immobilized on the surface. A corresponding scheme was successfully implemented by Rindorf et al. (2006a), but the definition of the sensor layer inside the holes required a multitude of infiltration/washing steps due to the specific surface chemistry of the silica. This procedure can be simplified by using polymer

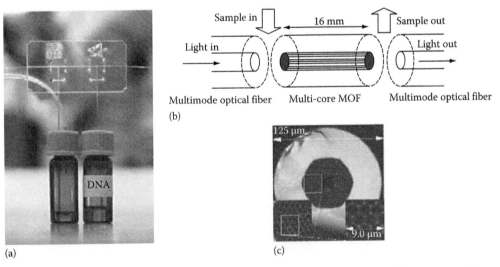

(a)   (c)

FIGURE 15.14 (a) Lab-on-a-chip component with an integrated MOF and external fluid reservoirs. The total inner chip volume is 300 nL. (b) Schematics of the MOF integration. Light is coupled in and out via multimode fibers. A gap between the fibers allows for sample in/outflow. (c) Cross section of the employed hollow-core MOF. (From Rindorf, L. et al., *Anal. Bioanal. Chem.*, 385, 1370, 2006a. With permission.)

fibers instead, as was shown by Jensen et al. (2005). However, we want to elaborate on the work of Rindorf et al. (2006a), since it presents the first proof-of-principle that such a fiber-based sensor can be integrated into a more complex optofluidic system, including capillary tubes for fluid input/output and multimode fibers for light input/output, as depicted in Figure 15.14. Incorporating MOF sensors into on-chip structures is one of the crucial issues for developing devices for real-life applications and future work will have to tackle the subtleties of this important problem. This will be discussed in some more detail in Section 15.9.

So far, in the cited experimental work either absorption or fluorescence measurements were used to probe the infiltrated sample material. However, there are also sensing methods which do not need any extinction or emission signal but directly detect the change of refractive index induced, for example, by a specific kind of molecule. One way to implement such a detection scheme is the use of long-period gratings inscribed in the fiber material via a CO_2 laser, as demonstrated by Rindorf et al. (2006b). The operation principle of this method is based on the beat signal between the core and cladding modes of a solid-core MOF, which arises from the different effective refractive indices, $n_{\text{eff}}^{\text{core}}$ and $n_{\text{eff}}^{\text{cladding}}$ (and thus different propagation constants β), of these modes. The beat signal is resonant with the grating of period, L_{grating}, at the wavelength $\lambda = L_{\text{grating}} \left(n_{\text{eff}}^{\text{core}} - n_{\text{eff}}^{\text{cladding}} \right)$, which can be determined experimentally from transmission measurements. This resonance wavelength shifts when material is infiltrated or deposited in the holes of the MOF, as $n_{\text{eff}}^{\text{core}}$ and $n_{\text{eff}}^{\text{cladding}}$ are affected differently due to the different spatial field distribution of the core and cladding modes. Consequently, changes of the refractive index can be measured quantitatively by monitoring the shift in the resonance wavelength. This approach was successfully employed to measure the presence of a biofilm of DNA molecules with a sensitivity of approximately 1.4 nm wavelength shift per 1 nm film thickness (Rindorf et al. 2006b). It is thus well suited to quantitatively determine the thickness t of a biofilm (down to monolayers of molecules) inside the fiber, as the linearity of this method was ensured for $t < 100$ nm.

15.6 Sensing with Fluid-Core Fibers

Despite the recent advances in the design and fabrication of solid-core MOFs, the overlap between light and sample material is intrinsically limited to a few 10% due to the evanescent nature of the interaction. Higher overlaps and thus better sensing performances can only be achieved by guiding the light directly inside the infiltrated liquid. One way to achieve this without the need for a selective filling technique is by designing the fiber such that it exhibits a photonic bandgap core mode even when completely filled with liquid. For this purpose, fibers made from high-index glasses or polymers are much better suited than silica, as the width of the photonic bandgap scales with the index contrast between sample and fiber material. Cox et al. (2006) demonstrated optical rotation in such a polymer hollow-core MOF filled with an aqueous solution of fructose. Thereby, the central spectral position λ of the photonic bandgap shifts due to the infiltrated liquid with refractive index n_{liquid} according to

$$\lambda = \lambda_0 \left[\frac{1 - \left(n_{\text{liquid}}^2 / n_{\text{fiber}}^2 \right)}{1 - \left(1 / n_{\text{fiber}}^2 \right)} \right]^{1/2}, \tag{15.17}$$

where λ_0 is the original bandgap position of the air-filled fiber. This shift, however, is accompanied by a strong narrowing of the spectral width of the photonic bandgap, increasing the propagation losses and putting severe limitations to the wavelength window in which the sensing device can operate. The approach is thus restricted to applications where light of a single frequency or narrow frequency band is sent through the MOF.

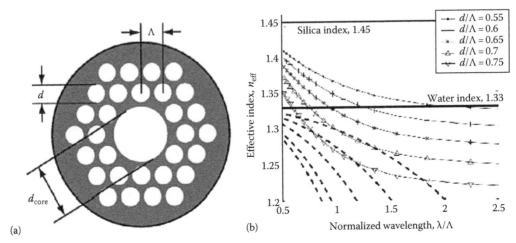

FIGURE 15.15 (a) Schematic representation of a hollow-core MOF. (b) Calculated effective refractive index of the cladding structure (symbols) for different design parameters of the fiber. Also shown are analytical estimates of the core modes (dashed lines). (Reprinted from Fini, J.M., *Meas. Sci. Technol.*, 15, 1120, 2004. With permission.)

To enable broad-band operation selective filling of the core hole is required, using one of the techniques discussed in Section 15.4. This changes the propagation mechanism from bandgap to index-guiding as long as the effective refractive index of the cladding region is smaller than that of the core material (see Section 15.3.1). To fulfill this condition, both the diameter d of the cladding holes as well as the wavelength λ of the guided light have to exceed a certain value, depending on the fiber pitch Λ (Figure 15.15a). Larger holes simply give a larger air-filling fraction of the cladding, while larger wavelengths lead to a weaker confinement of the light in the fiber material, thus effectively decreasing the effective refractive index. As seen in Figure 15.15b, for a simple hollow-core silica MOF the effective index of the cladding drops below that of water for $d/\Lambda > 0.54$ (above a certain wavelength). However, to ensure a strong confinement of the guided mode to the core (with a diameter $d_{core}=2\Lambda$) values of $d/\Lambda > 0.7$ are necessary (Fini 2004). If d/Λ or the core size is further increased, multimode guiding will set in. Consequently, if single-mode operation is desired, the fiber parameters have to be chosen from a narrow parameter window $0.7 < d/\Lambda < 0.8$. The advantages of such a fluid-filled core MOF are clearly the large overlap (which can easily reach values $\gamma > 90\%$), low losses (<1 dB/m) and the good coupling efficiency of light into the fiber due to the relatively large core size.

The first experimental realization of sensing with a fluid-filled core MOF was reported by Yiou et al. (2005). The cladding structure of the hollow-core fiber used in this work exhibited an effective refractive index $n_{eff}^{cladding}=1.25$, clearly lying below that of most common liquids. The core (with a diameter $d_{core}=11\,\mu m$) was selectively filled with ethanol ($n_{liquid}=1.36$) using the fusion method (see Section 15.4.2) and measurements of stimulated Raman scattering demonstrated the suitability of this approach for sensing applications.

A thorough analysis of the sensing efficiency of fluid-core MOFs in absorption as well as fluorescence measurements was performed by Smolka et al. (2007b) and the main results shall be reviewed here. Numerical calculations based on the finite element method were utilized to study the influence of the refractive index n_{liquid} on the overlap γ and on the fluorescence collection efficiency ξ. A schematic illustration of the employed fiber (with parameters $\Lambda=1.6\,\mu m$, $d=1.45\,\mu m$, and $d_{core}=5.3\,\mu m$) is shown in Figure 15.16a together with the corresponding simulated electric field intensity distribution of the fundamental core mode (Figure 15.16b). From the numerical calculations the overlap γ as well as the effective refractive index n_{eff}^{core} of the core mode can be determined. Together with the effective refractive index $n_{eff}^{cladding}=1.15$ of the lowest (lossy) cladding mode, n_{eff}^{core} can be used to estimate the

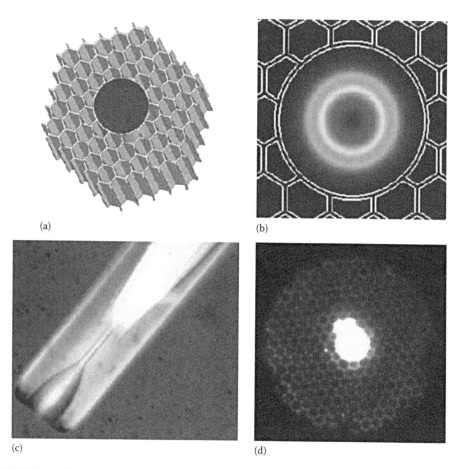

(a)

(b)

(c)

(d)

FIGURE 15.16 (a) Schematic illustration of a fluid-core MOF and (b) the corresponding calculated mode intensity distribution. (c) Side view of a hollow-core MOF after applying the fusion method to seal the cladding holes. (d) White light transmission through the same fiber after infiltration with a liquid. Two adjacent cladding holes are filled in addition to the core, clearly demonstrating index guiding. (Reprinted from Smolka, S. et al., *Opt. Express,* 15, 12783, 2007b. With permission.)

collection efficiency ξ from Equation 15.10. In Table 15.1 the calculated values for γ and ξ are shown for three different injected liquids, namely, water, ethanol, and ethylene glycol. As is evident, the overlap reaches extremely high values (close to unity) due to the almost complete guiding of light within the liquid filament. If higher order modes are taken into account, the overlap reduces to $\gamma \approx 92\%$, but is still much higher than in any evanescent sensing device reported so far. Also, the collection efficiency of the fluid-core MOF is remarkably high, as it corresponds to that of a microscope objective with a numerical aperture of 0.5 (for water), 0.53 (for ethanol), and 0.59 (for ethylene glycol), respectively.

The excellent sensing properties of the fiber were confirmed in spectroscopic measurements of rhodamine 6 G, solved in ethylene glycol and selectively injected into the core of the MOF via the fusion method. A side view of one of the sealed fiber ends is shown in Figure 15.16c, accompanied by a white light transmission image of the cleaved end facet in Figure 15.16d, clearly demonstrating the guiding of light inside the liquid filament via total internal reflection. Absorption and fluorescence spectra obtained from a 10 cm piece of fluid-core MOF are displayed in Figure 15.17a, showing excellent agreement with corresponding free-space measurements. This is not self-evident, as a number of publications report on strong modifications of the spectroscopic features due to the interaction between the detected species

TABLE 15.1 Overlap γ and Collection
Efficiency ξ of a Fluid-Core MOF Filled with
Different Liquids

Sample Liquid	n_{liquid}	γ (%)	ξ (%)
Water	1.33	98.3	6.7
Ethanol	1.36	99.3	7.6
Ethylene glycol	1.43	99.8	9.6

Source: Smolka, S. et al., *Opt. Express*, 15, 12783, 2007b.

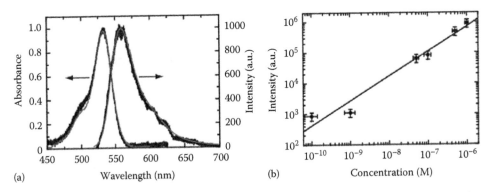

(a)   Wavelength (nm)   (b)   Concentration (M)

FIGURE 15.17   (a) Absorption and fluorescence spectrum of rhodamine 6G solved in ethylene glycol obtained from a fluid-core MOF (black lines) as well as a free-space measurement (gray lines). (b) Peak intensity of the fluorescence signal as a function of the molecular concentration. (Reprinted from Smolka, S. et al., *Opt. Express*, 15, 12783, 2007b. With permission.)

and the inner walls of the fiber (Cordeiro et al. 2006a, Smolka et al. 2007a). Here, these interactions seem to play a minor role since most of the molecules participating in the sensing process are not directly located near the wall of the core and therefore the situation is more comparable to that in a macroscopic cuvette. However, the key difference to a macroscopic measurement is that only 2.2 nL of sample volume are actually needed to record these spectra. By varying the concentration c of the dye molecules and monitoring the optical attenuation through the fiber the overlap γ can be determined from Equation 15.8, yielding a value of $γ=(86\pm10)\%$, in excellent agreement with the theoretical prediction of $γ\approx92\%$. The lowest dye concentration that could reliably be detected in absorption was $c=5\times10^{-8}$ M. In this case, the sensitivity was limited by the accuracy of the reference spectra (acquired by infiltrating pure solvent), since small changes in the coupling conditions of the injected light can lead to significant differences in the transmission spectra. Contrary to this, in fluorescence measurements there is no need for any reference, which enabled an increase of the sensitivity down to concentrations of $c=1\times10^{-10}$ M, corresponding to about 10^5 molecules inside the fiber. As is seen in Figure 15.17b, the detected fluorescence intensity (which in principle can be increased by increasing the integration time) was not the limiting factor in these measurements. Instead, a parasitic background signal originating from the fiber material itself prevented the detection of lower dye concentrations. A similar problem also occurred in other reports (Ruan et al. 2007) and seems to be the major obstacle for further improvements of the performance of many fiber-based sensing devices, including solid-core as well as fluid-core MOFs. Consequently, routes have to be explored how to decrease the impact of this unwanted background fluorescence, some of them being discussed in Section 15.9.

15.7 Surface-Enhanced Raman Scattering in Optical Fibers

We want to conclude the review on the current state of the art in sensing with optical fibers (started in Sections 15.5 and 15.6) with an overview on a particularly promising field of application for these devices, namely, surface-enhanced Raman scattering. Raman scattering is an inelastic scattering process in which an incident photon interacts with a molecule such that photon energy is transferred to or from molecular vibrations causing a spectral shift of the photon to lower or higher frequencies (Stokes or anti-Stokes process), respectively. For a more detailed treatment of background and applications of Raman scattering in optofluidics, see Chapters 6 and 17 by Zhang and Benford, respectively. As the vibrational-level structure is unique for each type of molecule, Raman spectra contain a wealth of information allowing the label-free detection and identification of all sorts of molecular species. For an example of a typical Raman spectrum see Figure 15.18. The cross section for Raman scattering is typically on the order of 10^{-29} cm^2 per molecule, but can be dramatically enhanced by about 10–15 orders of magnitude by attaching the molecule to a metallic nanostructure (Moskovits 1985). This effect has become known as surface-enhanced Raman scattering (SERS) (Kneipp et al. 1997, Nie and Emory 1997). Its origin is yet not fully understood, but one major contribution stems from local field enhancements near the surface of the metal due to the excitation of surface plasmon polaritons. The field enhancement is strongest at the interstice between two metal nanostructures (Xu 1999), leading to so-called "hot spots" in which the SERS signal is extremely pronounced. Therefore, aggregates of metal nanoparticles, deposited on a substrate or directly injected into living cells (Kneipp et al. 2002), are widely used in SERS measurements to take advantage of these hot spots, enabling detection sensitivities down to the level of a single molecule. However, as pointed out by Cox et al. (2007), this sensing environment is highly irregular, leading to strong variations of the SERS signal between different nanoparticle aggregates and thus limiting the potential for quantitative SERS measurements. This drawback can be circumvented by combining the excellent sensitivity of SERS with the advantages of fiber-based sensing. Due to the long interaction length of the light with the sample material inside an MOF, all nanoparticle aggregates (when infiltrated into the fiber) are excited and detected at the same time, instantly giving an average signal without the need to search

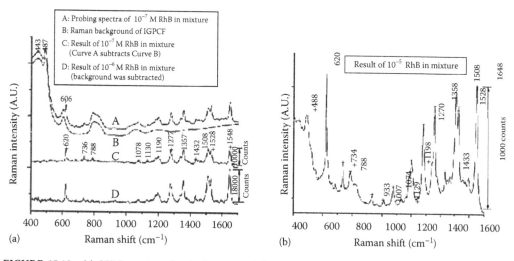

FIGURE 15.18 (a) SERS spectra of a rhodamine B (RhB) solution mixed with gold nanoparticles and infiltrated into a solid-core MOF (denoted here as IGPCF). The pure background signal from the fiber is also shown. (b) Corresponding SERS spectrum at a higher molecular concentration. (Reprinted from Yan, H. et al., *Opt. Express*, 16, 8300, 2008. With permission.)

for specific "active" sites on the sample. This should not only help to improve the reproducibility of SERS measurements but would also allow for a much better quantitative analysis even when analyte concentrations are small.

In recent years, several groups have implemented fiber-based SERS sensors using various kinds of MOFs (hollow-core as well as solid-core) and different types of infiltration schemes. One of the first realizations was reported by Yan et al. (2006), who used a hollow-core MOF with a layer of gold nano-particles (100–200 nm diameter) coated on the inner walls of the fiber. The sample liquid (consisting of a 10^{-5} M aqueous solution of rhodamine B molecules) was injected via capillary forces by dipping one end of the fiber into the solution and letting it dry, thus attaching the molecules to the gold nanopar-ticles. Both, the excitation laser as well as the scattered Raman light were coupled in/out at the opposite end of the fiber. Although a corresponding Raman signal could be detected, the overall performance was poor (compared to a reference sample on silicon), as all holes of the MOF were filled simultane-ously, resulting in weak light confinement and high propagation loss. The first implementation using a selective injection into the fiber core was reported by Zhang et al. (2007), who infiltrated a mixture of silver nanoparticles (40–60 nm diameter) and rhodamine 6G (concentration $c \approx 10^{-5}$ M) into the MOF. Although the general detection scheme was similar to that employed by Yan et al. (2006) (fluid infiltration at one fiber end, light input/output at opposite fiber end), the sensitivity of this device was approximately 100 times higher compared to a reference sample of the same mixture dried on a planar substrate. This improvement very likely results from the much better light–sample interaction and collection efficiency in the fluid-core MOF. However, the ultimate detection limit of this SERS sensor has not fully been explored. A comparison with results from other laboratories is difficult due to differ-ent metal nanoparticles and analyte molecules, different laser powers and wavelengths, and different Raman detectors employed.

A selective infiltration scheme is not necessarily needed to detect Raman signals from lower con-centrations of molecules. This was demonstrated by Cox et al. (2007), who used a polymer MOF exhib-iting a true photonic bandgap guided core mode even when completely filled with water. The detection scheme was again similar to those described above, with a mixture of silver nanoparticles (30–80 nm diameter) and rhodamine 6G as analyte. With this configuration, Raman signals from molecular con-centrations down to 2×10^{-7} M could be reliably measured, a regime not accessible with a correspond-ing free-space setup employed as a reference. Recently, SERS sensing has also been demonstrated in solid-core MOFs with a steering-wheel-like structure (Yan et al. 2008). When a mixture of gold nanoparticles (80 nm diameter) and rhodamine B molecules was infiltrated into the MOF, molecular concentration down to 10^{-7} M could be detected (see Figure 15.18), despite the relatively small overlap value $\gamma \approx 2\%$ in this fiber. Consequently, an optimization of the fiber geometry to maximize γ (which is doubtlessly possible as described in Section 15.5) should allow for even smaller concentrations to be measured.

Up to this point, it is not clear which of the introduced types of MOF is the most suitable for fiber-based SERS sensing. The best results (in terms of analyte concentrations) were in fact obtained for a solid-core MOF with quite modest sensing parameters, which seems to contradict somehow the figures of merit derived in Section 15.3.2. However, as pointed out above, the overall performance of a SERS sensor is determined by a large number of factors, making a direct comparison between results from different laboratories very difficult. It is thus believed that there is still an enormous potential for further improvements of fiber-based SERS sensors, which are already comparable in their sensitivity to other micron-scale devices such as integrated ARROW waveguides (Measor et al. 2007).

15.8 Light-Induced Particle Transport in Fluid-Filled Fibers

After reviewing the development and prospects of optofluidic sensing with MOFs in the previous sec-tions, we will now turn to another interesting application of such systems, namely, the light-induced transport of particles through liquid-filled optical fibers. The idea of using the forces of light to manipulate microscopic objects goes back several decades (Ashkin 1970) and has meanwhile evolved to one of the most versatile fields of application in optical science. Especially the invention of optical tweezing (Ashkin et al. 1986), i.e., the trapping of particles within a strong optical gradient field (such as a tightly focussed laser beam) has found numerous applications particularly in life science, where biological entities are trapped, manipulated, sorted, and probed without any physical contact, using only optical forces (Grier 2003). The implementation of particle trapping in optofluidics is discussed in more detail in Chapters 14 and 16. However, if the focus lies not on trapping an object but on transporting it (via light pressure), conventional free-space optics soon reaches its limitations, since the interaction path length is usually limited to the length of the laser focus (typically a few micrometer). A solution to this problem is provided by optical waveguide structures (Gaugiran et al. 2005, Schmidt et al. 2007, Measor et al. 2008), in which the laser beam is focussed to a small cross section along the entire length of the waveguide, and can thus interact with the particle over very long distances. MOFs are natural candidates for such a particle transportation scheme due to their ability to guide light within air or liquid filaments. Therefore, after giving a short introduction to optical forces in Section 15.8.1, we will review some of the experimental efforts and prospects in this field of application in Section 15.8.2.

15.8.1 Forces in the Rayleigh Particle Limit

Let us consider a dielectric particle of refractive index n_{part} and radius r which has been injected into the fluid core (filled with a liquid of refractive index $n_{\text{liquid}} < n_{\text{part}}$) of an MOF as is shown schematically in Figure 15.19. If light is sent through the fiber, two kinds of optical forces are acting on the particle, namely, a scattering force, F_{scat}, and a gradient force, F_{grad}. Although both are caused by the momentum transfer from the incident photons to the particle, their net effect is quite different. While the scattering force (often also referred to as "light pressure") is proportional to the field intensity I and propels the particle into the propagation direction of the guided laser beam, the gradient force is proportional to the gradient ∇I and pushes the particle toward the region of highest field intensity. Consequently, if we assume single-mode operation with a Gaussian intensity profile, the particle will be drawn to the center of the core (due to F_{grad}) and pushed forward along the fiber (due to F_{scat}) at the same time. In general, the calculation of these forces is nontrivial and governed by the Lorentz–Mie theory. However, in the limiting case of very small particles (compared to the wavelength λ of the light) Rayleigh theory can be applied and the two forces take the simple expressions (Ashkin et al. 1986, Harada and Asakura 1996)

$$F_{\text{scat}} = \frac{128\pi^5 r^6 n_{\text{liquid}}}{3c\lambda^4} \left[\frac{(n_{\text{part}}/n_{\text{liquid}})^2 - 1}{(n_{\text{part}}/n_{\text{liquid}})^2 + 2} \right]^2 I(\rho,\theta,z) \tag{15.18}$$

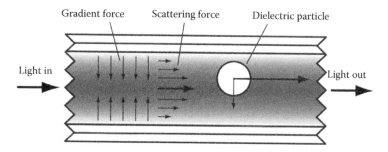

FIGURE 15.19 Schematic illustration of the light-induced forces on a dielectric particle inside an MOF.

and

$$F_{\text{grad}} = \frac{2\pi r^3 n_{\text{liquid}}}{c} \left[\frac{(n_{\text{part}}/n_{\text{liquid}})^2 - 1}{(n_{\text{part}}/n_{\text{liquid}})^2 + 2} \right] \nabla I(\rho, \theta, z), \tag{15.19}$$

where
  c is the speed of light in vacuum
  intensity I has been written as a function of the cylindrical coordinates (ρ, θ, z)

Apart from these optical forces, the particle will encounter a viscous drag force

$$F_{\text{drag}} = -6\pi\eta r v, \tag{15.20}$$

where
  η is the viscosity of the surrounding liquid
  v is the velocity of the particle

In the steady-state case the light forces and the drag force will balance each other and the particle will be transported in forward (+z) direction with the velocity

$$v(z) = \frac{64\pi^4 r^5 n_{\text{liquid}}}{9c\eta\lambda^4} \left[\frac{(n_{\text{part}}/n_{\text{liquid}})^2 - 1}{(n_{\text{part}}/n_{\text{liquid}})^2 + 2} \right]^2 I(\rho, \theta, z)$$

$$+ \frac{r^2 n_{\text{liquid}}}{3c\eta} \left[\frac{(n_{\text{part}}/n_{\text{liquid}})^2 - 1}{(n_{\text{part}}/n_{\text{liquid}})^2 + 2} \right] \frac{\partial I(\rho, \theta, z)}{\partial z}. \tag{15.21}$$

Here, we have taken into account that optical propagation losses will lead to a decaying intensity along the fiber and thus to a gradient $\partial I/\partial z < 0$ which counteracts the scattering force (Mandal and Erickson 2007). Furthermore, we have neglected any influence of gravitational forces, which is valid for a horizontally oriented fiber. Otherwise, a corresponding term $-mg/6\pi\eta r$ has to be added to the right-hand side of Equation 15.21, where m is the effective mass of the particle within the liquid and g is the gravitational constant.

As stated above, Equation 15.21 is strictly valid only as long as $r \ll \lambda$, which is mostly not fulfilled for the experimental implementations discussed in Section 15.8.2, where typically micrometer-sized particles were employed. However, Equation 15.21 can help to gain a qualitative understanding of the underlying physics and its implications. For example, if the propagation losses of the MOF are small enough, the second term in Equation 15.21 can be neglected and the velocity v becomes nearly constant over the entire length of the fiber. Furthermore, as seen from the first term, the velocity strongly depends on the radius of the particle, which opens interesting possibilities to separate and sort particle conglomerates of different sizes.

15.8.2 Experimental Realization of Fiber-Based Particle Transport

The first experimental demonstration of particle transport in a hollow-core fiber was actually realized for Rb atoms, which were guided through a glass capillary of 40 μm diameter in vacuum (Renn et al. 1995). In this case, the laser light was confined to the core due to the so-called grazing-incidence guidance, which exploits the fact that light impinging nearly parallel on the glass interface is almost completely reflected. The fundamental core mode is then a Bessel function, which means that the atoms could be trapped at the central axis of the core, significantly reducing interactions with the glass wall. The same method was later extended to demonstrate the guidance of mesoscopic particles (dielectric, metal, semiconductor, and biological particles) in the size range between 50 nm and 10 μm in air or water-filled capillaries (Renn et al. 1999). For example, a 7 μm polystyrene sphere was propelled through water with a velocity $\upsilon = 300$ μm/s using a 200 mW near-infrared laser beam. During this transport, the particle had virtually no contact with the glass wall due to the radial gradient force. However, this particle transport could only be achieved over distances of a few centimeters at best, since the weak grazing-incidence guidance comes along with high propagation losses.

Hollow-core MOFs can provide light and particle guidance with much lower losses and thus over longer distances, as was demonstrated by Benabid et al. (2002). In these experiments, a 5 μm polystyrene sphere was propelled through the 20 μm core of an air-filled MOF over a total distance of 15 cm. The measured particle velocity was $\upsilon = 1$ cm/s at a laser power of 80 mW. A sequence of snapshots of the particle inside the fiber is shown in Figure 15.20. Also in this case the radial gradient force prevented the particle from colliding with the inner wall, thus ensuring an efficient transport process even through bends of the fiber. However, extending these promising results to fluid-filled MOFs has not been achieved so far, mainly because of the onset of multimode operation when the core is filled with a liquid. Mandal and Erickson (2007) investigated the transport behavior of 3 μm polystyrene spheres in such a fluid-core MOF (with 20 μm core diameter) and observed a maximal guiding distance of only 2 cm (from the entrance of the fiber) at a laser power of 210 mW. This limited transport range may be attributed to relatively high propagation losses, caused by the multimode guidance and the large number of particles introduced into the fiber, which act as scatterers and gradient traps for other particles, leading to the formation of so-called "floating bands."

Particle transport over distances of more than 10 cm is indeed possible in a fluid-core MOF. This was recently demonstrated for 2.56 μm polystyrene spheres in ethanol selectively infiltrated into a fiber with a 8 μm core. Care was taken that only a few particles were distributed over the entire length $L \approx 25$ cm of the MOF. A near-infrared laser beam ($\lambda = 790$ nm) could be launched alternately into both ends of the fiber (which were kept in fluid cells), allowing the transport of one and the same particle into both directions by choosing the corresponding input path, even when the distance to the fiber entrance was more than 15 cm. Snapshots of such a particle during its motion are shown in Figure 15.21. The laser power, measured at the output end of the fiber, was kept constant at 13 mW. However, the highest measured particle velocity $\upsilon = 11$ μm/s was still four times lower than theoretically predicted by numerical simulations. In some cases, the observed velocity even decreased to values of $\upsilon \approx 1$ μm/s. This extremely slow movement very likely resulted from the interaction of the polystyrene spheres with the inner wall

FIGURE 15.20 Sequence of images showing a polystyrene particle (marked by an arrow) which is guided inside an air-filled hollow-core MOF. (Reprinted from Benabid, F. et al., *Opt. Express*, 10, 1195, 2002. With permission.)

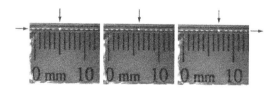

FIGURE 15.21 Sequence of images showing a polystyrene particle (marked by an arrow) which is propelled through a fluid-core MOF filled with ethanol.

of the MOF, as the fiber was still in the multimode regime of light guidance and consequently no radial gradient force prevented the particles from colliding with the wall. This again highlights the importance of single-mode operation in such a transport scheme, which is one of the main issues to be tackled in future work if fluid-core MOFs are indeed to become practical devices for particle transport, separation, and/or investigation.

One way to achieve single-mode guiding (without further decreasing the size of the core) is by increasing the effective refractive index of the cladding structure such that it lies within a narrow parameter window, as described in Section 15.6. This requires the fiber to be designed specifically for a certain liquid and wavelength. A more versatile technique was proposed by Cordeiro et al. (2007) and successfully demonstrated by de Matos et al. (2007). It relies on the (additional) infiltration of the cladding holes with a liquid exhibiting a slightly lower refractive index than the liquid injected into the core. Such a structure, although very demanding in its preparation, can be tuned by changing the fluid in the cladding holes and thus allows to optimize the guiding properties. Finally, we want to note that the light-induced motion of particles inside an MOF may be used to map the intensity distribution of the propagating mode *in situ* as was recently demonstrated for hollow-core ARROW waveguides by Measor et al. (2008).

15.9 Current Challenges and Prospects

The design and application of MOFs has developed rapidly in recent years due to novel fabrication techniques and extensive theoretical and experimental studies. This has led to a multitude of different fiber geometries, all with their specific advantages and drawbacks. Although a large number of experiments have been conducted to test the performance of these structures for chemical and biological sensing applications, it is not yet clear what the optimum design for an optofluidic MOF sensor has to look like. Solid-core MOFs, especially those of the steering-wheel type, are appealing due to their single-mode guiding properties and the ease of infiltration. Fluid-core MOFs, on the other hand, promise higher sensing efficiencies, at the cost of a more complex preparation and handling. Most probably, each specific application will require a slightly different device design, balancing the need for sensitivity, robustness, and cost-effectiveness.

The detection sensitivity currently achieved with MOF-based sensors (solid core as well as hollow core) is limited to molecular concentrations on the order of 1 nM, often due to unwanted background signals (either fluorescence or Raman light) from the fiber itself. Some improvements may be gained by moving to different excitation/detection wavelengths or by applying better background subtraction schemes. However, we want to point out that it is not the purpose of MOF sensors to measure on a single molecule level, since their main advantage, namely, the long interaction length between light and sample material, would be rendered useless if only a single molecules was present in the entire fiber. It is rather the ability of MOFs to probe the injected sample volume and detect nearly *all* present molecules in one single measurement which makes them appealing.

A major issue for real-life applications is the reusability of MOFs once they have been infiltrated. Removing residual molecules (or particles) from previous measurements by flushing the holes with pure solvents or chemicals can, in general, be very time consuming or may be impossible at all. A solution to this problem is the functionalization of the inner walls of the fiber such that the unwanted binding of molecules to the fiber material is prevented. This is crucial for all applications where the MOF is permanently integrated into a more complex optofluidic network and cannot simply be replaced. In other applications a single-use design might be more appropriate, especially as production costs will drop once fabrication methods have evolved to the state of mass production. A typical device would then consist of an optical detection unit which is directly linked to a removable piece of MOF via fiber interconnects, allowing an easy plug-and-play handling and compact product design, without any complicated optical alignment.

References

Afshar, V. S., Warren-Smith, S. C., and Monro, T. M. 2007. Enhancement of fluorescence-based sensing using microstructured optical fibres. *Opt. Express* 15: 17891–17901.

Afshar, V. S., Ruan, Y., Warren-Smith, S. C., and Monro, T. M. 2008. Enhanced fluorescence sensing using microstructured optical fibers: A comparison of forward and backward collection modes. *Opt. Lett.* 33: 1473–1475.

Alkeskjold, T. T., Laegsgaard, J., Bjarklev, A. et al. 2004. All-optical modulation in dye-doped nematic liquid crystal photonic bandgap fibers. *Opt. Express* 12: 5857–5871.

Almeida, V. R., Xu, Q., Barrios, C. A., and Lipson, M. 2004. Guiding and confining light in void nanostructure. *Opt. Lett.* 29: 1209–1211.

Ashkin, A. 1970. Acceleration and trapping of particles by radiation pressure. *Phys. Rev. Lett.* 24: 156–159.

Ashkin, A., Dziedzic, J. M., Bjorkholm, J. E., and Chu, S. 1986. Observation of a single-beam gradient force optical trap for dielectric particles. *Opt. Lett.* 11: 288–290.

Bartelt, H., Kirchhof, J., Kobelke, J. et al. 2008. Preparation and application of functionalized photonic crystal fibres. In *Nanophotonic Materials*, eds. Wehrspohn, R. B., Kitzerow, H.-S., and Busch, K., pp. 289–311. Wiley-VCH, Weinheim, Germany.

Benabid, F., Knight, J. C., and Russell, P. S. J. 2002. Particle levitation and guidance in hollow-core photonic crystal fiber. *Opt. Express* 10: 1195–1203.

Benabid, F., Couny, F., Knight, J. C., Birks, T. A., and Russell, P. St. J. 2005. Compact, stable and efficient all-fibre gas cells using hollow-core photonic crystal fibres. *Nature* 434: 488–491.

Birks, T. A., Knight, J. C., and Russell, P. S. J. 1997. Endlessly single-mode photonic crystal fiber. *Opt. Lett.* 22: 961–963.

Bjarklev, A., Broeng, J., and Bjarklev, A. S. 2003. *Photonic Crystal Fibres.* Kluwer Academic Publishers, Boston, MA.

Carniglia, C. K., Mandel, L., and Drexhage, K. H. 1972. Absorption and emission of evanescent photons. *J. Opt. Soc. Am.* 62: 479–486.

Cordeiro, C. M. B., Franco, M. A. R., Chesini, G. et al. 2006a. Microstructured-core optical fibre for evanescent sensing applications. *Opt. Express* 14: 13056–13066.

Cordeiro, C. M. B., dos Santos, E. M., Brito Cruz, C. H., de Matos, C. J. S., and Ferreira, D. S. 2006b. Lateral access to the holes of photonic crystal fibers—selective filling and sensing applications. *Opt. Express* 14: 8403–8412.

Cordeiro, C. M. B., de Matos, C. J. S., dos Santos, E. M. et al. 2007. Towards practical liquid and gas sensing with photonic crystal fibres: Side access to fibre microstructure and single-mode liquid-core fibre. *Meas. Sci. Technol.* 18: 3075–3081.

Cox, F. M., Argyros, A., and Large, M. C. J. 2006. Liquid-filled hollow core microstructured polymer optical fiber. *Opt. Express* 14: 4135–4140.

Cox, F. M., Argyros, A., Large, M. C. J., and Kalluri, S. 2007. Surface enhanced Raman scattering in a hollow core microstructured optical fiber. *Opt. Express* 15: 13675–13681.

Cregan, R. F., Mangan, B. J., Knight, J. C., Birks, T. A., Russell, P. St. J., Roberts, P. J., and Allan, D. C. 1999. Single-mode photonic band gap guidance of light in air. *Science* 285: 1537–1539.

de Matos, C. J. S., Cordeiro, C. M. B., dos Santos, E. M., Ong, J. S. K., Bozolan, A., and Brito Cruz, C. H. 2007. Liquid-core, liquid-cladding photonic crystal fibers. *Opt. Express* 15: 11207–11212.

Du, F., Lu, Y. Q., and Wu, S. T. 2004. Electrically tunable liquid-crystal photonic crystal fiber. *Appl. Phys. Lett.* 85: 2181–2183.

Eggleton, B. J., Kerbage, C., Westbrook, P. S., Windeler, R. S., and Hale, A. 2001. Microstructured optical fiber devices. *Opt. Express* 13: 698–713.

Fini, J. M. 2004. Microstructure fibres for optical sensing in gases and liquids. *Meas. Sci. Technol.* 15: 1120–1128.

Gaugiran, S., Gétin, S., Fedeli, J. M., Colas, G., Fuchs, A., Chatelain, F., and Dérouard, J. 2005. Optical manipulation of microparticles and cells on silicon nitride waveguides. *Opt. Express* 13: 6956–6963.

Geernaert, T., Nasilowski, T., Chah, K. et al. 2008. Fiber Bragg gratings in germanium-doped highly-birefringent microstructured optical fibers. *IEEE Photon. Technol. Lett.* 20: 554–556.

Glass, T. R., Lackie, S., and Hirschfeld, T. 1987. Effect of numerical aperture on signal level in cylindrical wave-guide evanescent fluorosensors. *Appl. Opt.* 26: 2181–2187.

Grier, D. G. 2003. A revolution in optical manipulation. *Nature* 424: 810–816.

Hansen, T. P., Broeng, J., Jakobsen, C. et al. 2004. Air-guiding photonic bandgap fibers: Spectral properties, macrobending loss and practical handling. *J. Lightwave Technol.* 22: 11–15.

Harada, Y. and Asakura, T. 1996. Radiation forces on a dielectric sphere in the Rayleigh scattering regime. *Opt. Commun.* 124: 529–541.

Harrington, J. A. 2000. A review of IR transmitting, hollow waveguide. *Fiber Integr. Opt.* 19: 211–227.

Hautakorpi, M., Mattinen, M., and Ludvigsen, H. 2008. Surface-plasmon-resonance sensor based on three-hole microstructured optical fiber. *Opt. Express* 16: 8427–8432.

Hoo, Y. L., Jin, W., Ho, H. L., Wang, D. N., and Windeler, R. S. 2002. Evanescent-wave gas sensing using microstructure fiber. *Opt. Eng.* 41: 8–9.

Huang, Y., Xu, Y., and Yariv, A. 2004. Fabrication of functional microstructured optical fibers through a selective-filling technique. *Appl. Phys. Lett.* 85: 5182–5184.

Ibanescu, M., Johnson, S. G., Soljačić, M. et al. 2003. Analysis of mode structure in hollow dielectric waveguide fibers. *Phys. Rev. E* 67: 046608.

Jensen, J. B., Pedersen, L. H., Hoiby, P. E. et al. 2004. Photonic crystal fiber based evanescent-wave sensor for detection of biomolecules in aqueous solutions. *Opt. Lett.* 29: 1974–1976.

Jensen, J. B., Hoiby, P. E., Pedersen, L. H., and Bjarklev, A. 2005. Selective detection of antibodies in microstructured polymer optical fibers. *Opt. Express* 13: 5883–5889.

Johnson, S. G., Ibanescu, M., Skorobogatiy, M. et al. 2001. Low-loss asymptotically single mode propagation in large-core OmniGuide fibers. *Opt. Express* 9: 748–779.

Kao, H. P., Yang, N., and Schoeniger, J. S. 1998. Enhancement of evanescent fluorescence from fiber-optic sensors by thin-film sol-gel coatings. *J. Opt. Soc. Am. A* 15: 2163–2171.

Kirchhof, J., Kobelke, J., Schuster, K. et al. 2004. Photonic crystal fibres. In *Photonic Crystals: Advances in Design, Fabrication, and Characterization*, eds. Busch, K., Lölkes, St., Wehrspohn, R. B., and Föll, H., pp. 266–288. Wiley-VCH, Weinheim, Germany.

Kneipp, K., Wang, Y., Kneipp, H. et al. 1997. Single molecule detection using surface-enhanced Raman scattering (SERS). *Phys. Rev. Lett.* 78: 1667–1670.

Kneipp, K., Haka, A. S., Kneipp, H. et al. 2002. Surface-enhanced Raman spectroscopy in single living cells using gold nanoparticles. *Appl. Spectrosc.* 56: 150–154.

Knight, J. C. 2003. Photonic crystal fibres. *Nature* 424: 847–851.

Konorov, S. O., Zheltikov, A. M., and Scalora, M. 2005. Photonic-crystal fiber as a multifunctional optical sensor and sample collector. *Opt. Express* 13: 3454–3459.

Kuzyk, M. K. 2007. *Polymer Fiber Optics.* Taylor & Francis, Boca Raton, FL.

Larsen, T. T., Bjarklev, A., Hermann, D. S., and Broeng, J. 2003. Optical devices based on liquid crystal photonic bandgap fibres. *Opt. Express* 11: 2589–2596.

Li, J., Gauza, S., and Wu, S. T. 2004. Temperature effect on liquid crystal refractive indices. *J. Appl. Phys.* 96: 19–24.

Lorenz, A., Kitzerow, H.-S., Schwuchow, A., Kobelke, J., and Bartelt, H. 2008. Photonic crystal fiber with a dual-frequency addressable liquid crystal: Behavior in the visible wavelength range. *Opt. Express* 16: 19375–19381.

Mandal, S. and Erickson, D. 2007. Optofluidic transport in liquid core waveguiding structures. *Appl. Phys. Lett.* 90: 184103.

Matsuura, Y., Kasahara, R., Katagiri, T., and Miyagi, M. 2002. Hollow infrared fibers fabricated by glass-drawing technique. *Opt. Express* 10: 488–492.

Measor, P., Seballos, L., Yin, D. et al. 2007. On-chip surface-enhanced Raman scattering detection using integrated liquid-core waveguides. *Appl. Phys. Lett.* 90: 211107.

Measor, P., Kühn, S., Lunt, E. J., Phillips, B. S., Hawkins, A. R., and Schmidt, H. 2008. Hollow-core waveguide characterization by optically induced particle transport. *Opt. Lett.* 33: 672–674.

Monat, C., Domachuk, P., and Eggleton, B. J. 2007. Integrated optofluidics: A new river of light. *Nature Photon.* 1: 106–114.

Monro, T. M., Belardi, W., Furusawa, K., Baggett, J. C., Broderick, N. G. R., and Richardson, D. J. 2001. Sensing with microstructured optical fibres. *Meas. Sci. Technol.* 12: 854–858.

Mortensen, N. A., Folkenberg, J. R., Nielsen, M. D., and Hansen, K. P. 2003. Modal cutoff and the V parameter in photonic crystal fibers. *Opt. Lett.* 28: 1879–1881.

Mortensen, N. A. and Nielsen, M. D. 2004. Modeling of realistic cladding structures for air-core photonic band-gap fibers. *Opt. Lett.* 29: 349–351.

Moskovits, M. 1985. Surface-enhanced spectroscopy. *Rev. Mod. Phys.* 57: 783–826.

Nie, S. and Emory, S. R. 1997. Probing single molecules and single nanoparticles by surface-enhanced Raman scattering. *Science* 275: 1102–1106.

Nielsen, K., Noordegraaf, D., Sørensen, T., Bjarklev, A., and Hansen, T. P. 2005. Selective filling of photonic crystal fibres. *J. Opt. A: Pure Appl. Opt.* 7: L13–L20.

Ortigosa-Blanch, A., Knight, J. C., Wadsworth, W. J. et al. 2000. Highly birefringent photonic crystal fibers. *Opt. Lett.* 25: 1325–1327.

Renn, M. J., Montgomery, D., Vdovin, O., Anderson, D. Z., Wieman, C. E., and Cornell, E. A. 1995. Laser-guided atoms in hollow-core optical fibers. *Phys. Rev. Lett.* 75: 3253–3256.

Renn, M. J., Pastel, R., and Lewandowski, H. J. 1999. Laser guidance and trapping of mesoscale particles in hollow-core optical fibers. *Phys. Rev. Lett.* 82: 1574–1577.

Rindorf, L., Hoiby, P. E., Jensen, J. B., Pedersen, L. H., Bang, O., and Geschke, O. 2006a. Towards biochips using microstructured optical fiber sensors. *Anal. Bioanal. Chem.* 385: 1370–1375.

Rindorf, L., Jensen, J. B., Dufva, M., Pedersen, L. H., Hoiby, P. E., and Bang, O. 2006b. Photonic crystal fiber long-period gratings for biochemical sensing. *Opt. Express* 14: 8224–8231.

Ritari, T., Tuominen, J., Ludvigsen, H., Petersen, J. C., Sørensen, T., Hansen, T. P., and Simonsen, H. R. 2004. Gas sensing using air-guiding photonic bandgap fibers. *Opt. Express* 12: 4080–4087.

Ruan, Y., Schartner, E. P., Ebendorff-Heidepriem, H., Hoffmann, P., and Monro, T. M. 2007. Detection of quantum-dot labeled proteins using soft glass microstructured optical fibers. *Opt. Express* 15: 17819–17826.

Russell, P. St. J. 2003. Photonic crystal fibers. *Science* 299: 358–362.

Russell, P. St. J. 2006. Photonic-crystal fibers. *J. Lightwave Technol.* 24: 4729–4749.

Schmidt, B. S., Yang, A. H. J., Erickson, D., and Lipson, M. 2007. Optofluidic trapping and transport on solid core waveguides within a microfluidic device. *Opt. Express* 15: 14322–14334.

Smolka, S., Barth, M., and Benson, O. 2007a. Selectively coated photonic crystal fiber for highly sensitive fluorescence detection. *Appl. Phys. Lett.* 90: 111101.

Smolka, S., Barth, M., and Benson, O. 2007b. Highly efficient fluorescence sensing with hollow core photonic crystal fibers. *Opt. Express* 15: 12783–12791.

Steel, M. J. and Osgood, R. M. 2001. Elliptical-hole photonic crystal fibers. *Opt. Lett.* 26: 229–231.

Wang, Y., Bartelt, H., Ecke, W. et al. 2009. Thermo-optic fiber switch based on fluid-filled photonic crystal fibers. To be published.

Webb, A. S., Poletti, F., Richardson, D. J., and Sahu, J. K. 2007. Suspended-core holey fiber for evanescent-field sensing. *Opt. Eng.* 46: 010503.

Wiederhecker, G. S., Cordeiro, C. M. B., Couny, F. et al. 2006. Field enhancement within an optical fibre with a subwavelength air core. *Nature Photon.* 1: 115–118.

Xiao, L., Jin, W., Demokan, M. S., Ho, H. L., Hoo, Y. L., and Zhao, C. 2005. Fabrication of selective injection microstructured optical fibers with a conventional fusion splicer. *Opt. Express* 13: 9014–9022.

Xu, H., Bjerneld, E. J., Käll, M., and Börjesson, L. 1999. Spectroscopy of single hemoglobin molecules by surface enhanced Raman scattering. *Phys. Rev. Lett.* 83: 4357–4360.

Xu, Y., Lee, R. K., and Yariv, A. 2000. Asymptotic analysis of Bragg fiber. *Opt. Lett.* 25: 1756–1758.

Xu, Q., Almeida, V. R., Panepucci, R. R., and Lipson, M. 2004. Experimental demonstration of guiding and confining light in nanometer-size low-refractive-index material. *Opt. Lett.* 29: 1626–1628.

Yan, H., Gu, C., Yang, C. et al. 2006. Hollow core photonic crystal fiber surface-enhanced Raman probe. *Appl. Phys. Lett.* 89: 204101.

Yan, H., Liu, J., Yang, C., Jin, G., Gu, C., and Hou, L. 2008. Novel index-guided photonic crystal fiber surface-enhanced Raman scattering probe. *Opt. Express* 16: 8300–8305.

Yang, L.-J., Yao, T.-J., and Tai, Y.-C. 2004. The marching velocity of the capillary meniscus in a microchannel. *J. Micromech. Microeng.* 14: 220–225.

Yeh, P., Yariv, A., and Maron, E. 1978. Theory of Bragg fiber. *J. Opt. Soc. Am.* 68: 1196–1201.

Yiou, S., Delaye, P., Rouvie, A. et al. 2005. Stimulated Raman scattering in an ethanol core microstructured optical fiber. *Opt. Express* 13: 4786–4791.

Zhang, R., Teipel, J., and Giessen, H. 2006. Theoretical design of a liquid-core photonic crystal fiber for supercontinuum generation. *Opt. Express* 14: 6800–6812.

Zhang, Y., Shi, C., Gu, C., Seballos, L., and Zhang, J. Z. 2007. Liquid core photonic crystal fiber sensor based on surface enhanced Raman scattering. *Appl. Phys. Lett.* 90: 193504.

Zhu, Y., Du, H., and Bise, R. 2006. Design of solid-core microstructured optical fiber with steering-wheel air cladding for optimal evanescent-field sensing. *Opt. Express* 14: 3541–3546.

16

Integrated Optofluidic Waveguides

Holger Schmidt

16.1 Introduction

Current developments in optofluidics are concentrated around three major themes: (1) optical devices with functionality defined and controlled by fluids, (2) optical detection of particles in chip-scale fluidic environments for chemical or biological sensing, and (3) optical manipulation of micro- and nanoscale particles. These trends are not only apparent from a look at the scientific literature and topical conferences (IEEE summer Topical Meetings 2006 and 2008, CLEO conference 2009), they are also reflected throughout the various chapters of this book. The main benefits of these efforts are to create novel optical devices with highly tunable properties and to translate well-established microscopy-based methods to a lab-on-chip environment with all its advantages for fluidic analysis.

To this end, the incorporation of integrated optical structures and techniques is indispensable, and comes with the big advantage of potential full planarization of the optofluidic device. This can occur by stacking discrete fluidic and optical layers, as originally proposed by Psaltis (Psaltis et al. 2006), or in hybrid configurations in which fluidics and optics are handled in the same physical plane. Optical waveguides are the cornerstone of either approach, as they enable the transport of light to and from various places on the optofluidic chip. Moreover, waveguides can themselves be part of an optical device, as is the case in Mach–Zehnder interferometers (MZIs), directional couplers, and ring resonators. As we will see, optofluidics has enabled new waveguide functionalities, such as dynamically reconfigurable switches and optical traps.

The basic properties of an optical waveguide have already been covered in Chapter 3, Passive integrated optics, by Janz. One way of categorizing waveguides in the context of optofluidics is by the type of core material, solid or fluid (liquid). Solid-core waveguides are the basis of conventional integrated optics that developed in fiber communications (Hunsperger 2002) and optoelectronics (Coldren and Corzine 1995). They are also often used in optofluidics to provide an interface to fluidic channels or fluidic waveguides. Liquid-core waveguides (LCWs), on the other hand, have experienced a dramatic resurgence in recent years. While initially being considered as the first viable candidates for enabling long-haul optical communication in the 1970s (Payne and Gambling 1972, Stone 1972), it was the advent of optofluidics that has generated a renewed and an increasing interest in these structures. LCWs offer numerous advantages for dealing with liquid materials, including dynamic reconfiguration by changing fluids or flow parameters, small analyte volumes, and near-perfect overlap between the optical energy and particles contained in the core fluid. Figure 16.1 illustrates the decisive and innovative role that LCWs have played in translating established paradigms in all three main optofluidic thrusts to the chip scale.

On the top left, a MEMS-based 1×2 optical switch is shown, in which light is distributed between optical fibers using a movable micromirror. On the bottom left is its counterpart, a 1×3 optofluidic switch implemented by flowing two different liquids through a microchannel. An adjustment of their relative flow speed leads to the redirection of the core liquid, as shown by the dark lines. The center column shows the transfer of single-molecule fluorescence spectroscopy for sensitive particle analysis from high-end microscopy apparatus to the chip scale, using a combination of liquid- and solid-core waveguides. The right part of the figure, finally, depicts the implementation of optical particle trapping and manipulation in a liquid-core slot waveguide that may replace the objective-based optical tweezers shown on the top. All of these examples illustrate the potential of waveguide-driven innovation in optofluidics, and will be discussed in more detail in the following sections.

The remainder of this chapter is organized as follows. After a brief review of the challenges associated with on-chip guiding of light through fluid media, we will introduce the waveguide types most commonly found in contemporary optofluidic devices, and their uses in sensing applications. While solid-core waveguides for evanescent sensing will be covered, the main emphasis of this chapter is on planar, integrated LCWs that provide the most natural fit to biosensing applications. In Section 16.3, we will discuss a couple of representative examples for the use of optofluidic waveguides for biodetection and sensing in more detail. We will conclude in Section 16.4 with a summary and outlook on existing challenges and possible future developments.

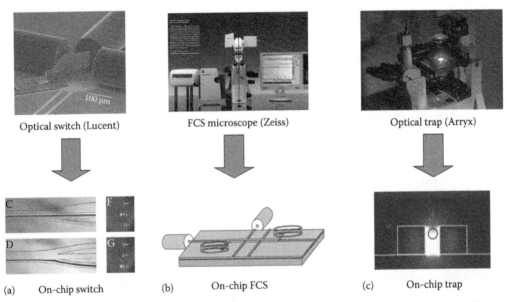

FIGURE 16.1 Examples for the use of LCWs in optofluidics. (a) Optical elements. (Reprinted from Wolfe, D.B. et al., *PNAS*, 101, 12434, 2004.) (b) Particle detection. (After Yin et al., *Opt. Lett.*, 31, 2136, 2006.) (c) Particle manipulation. (Reprinted from Yang et al., *Nature*, 457, 71, 2009a. With permission.)

16.2 Liquid-Core Waveguides

16.2.1 The Waveguiding Challenge

Optical waveguides direct light along a well-defined path in space by using an assembly of dielectric materials around a core material to confine the photon energy to a cross section of finite size. The obvious core material choice for fluidic applications is the sample analyte itself, which contains the molecules of interest or provides the device function. Indeed, glass fibers filled with high-index liquids, such as hexachlorobutadiene, were serious contenders for long-haul telecommunications in the early 1970s (Payne and Gambling 1972, Stone 1972), only to be abandoned in favor of all-solid silica fibers after the material properties of the silica cores had improved sufficiently to provide ultralow loss. It is, thus, surprising that it took several decades for LCWs to start playing a more dominant role in the analysis and manipulation of biological and chemical substances. Table 16.1, which displays the refractive indices of a number of relevant liquids and solids, shows the main reason for this delay.

Low refractive indices of liquids, in particular, aqueous solutions, compared to potential encapsulating materials, make use of conventional waveguiding based on total internal reflection (TIR) a highly nontrivial task (see chapter by Janz). A second reason for the relatively slow pace of LCW developments is the scaling law for the loss of a leaky waveguide, e.g., a fluid-filled capillary surrounded by a high-index cladding, such as glass. Such a structure does not exhibit TIR, but still guides a finite amount of power along the liquid core due to a finite reflection at the liquid–cladding interface. The propagation loss, α, of such a simple lossy waveguide in one dimension is given by (Marcatili and Schmeltzer 1964).

$$\alpha = \frac{\lambda^2}{n_c d_c^3 \sqrt{n_{cl}^2 - n_c^2}} \tag{16.1}$$

where
  λ is the vacuum wavelength
  n_c and d_c are the core index and diameter, respectively
  n_{cl} is the refractive index of the cladding

TABLE 16.1 Refractive Indices of Selected Solid and Liquid Substances

Phase	Materials	Refractive Indices
Liquid	Water	1.33
	Methanol	1.326 (589 nm)
	Acetone	1.357
	Ethanol	1.359 (589 nm)
	Ethylene glycol	1.43
	CaCl$_2$ (40% conc.)	1.44
	Trichloroethylene (TCE)	1.48
Solid	Teflon AF	1.29–1.31
	Polydimethylsiloxane (PDMS)	1.43
	Fused silica	1.45
	Quartz	1.458
	Borofloat glass	1.465
	Silicon nitride (Si$_3$N$_4$)	2.02
	Silicon	3.92
	GaAs	3.84

In order to get a feeling for the ramifications of the strong dependence on the core thickness in Equation 16.1, we can consider the case of a water core (n_c=1.33) surrounded by glass (n_{cl}=1.46) at λ=500 nm for two characteristic core dimensions. For a large channel width of d_c=100 μm, the waveguide loss is only α=3.1×10⁻³/cm, and is typically dwarfed by other loss mechanisms, such as scattering at waveguide imperfections. For a small channel of d_c=5 μm, however, the loss increases dramatically to α=25/cm, too large to be useful even on the short length scales of an optofluidic chip. As interest began to grow in micro- and nanofluidic channels due to their reduced analyte volumes and their potential for more sophisticated optofluidic devices based on single mode waveguide behavior, a need for alternative approaches with smaller waveguide losses arose.

Fortunately, several clever approaches to liquid-core waveguiding have been developed and translated successfully into sensing platforms. In the following sections, we will describe the different types of LCWs, their optical characteristics, and some applications in biological sensing.

16.2.2 Solid-Core Waveguides

Solid-core waveguides are used in optofluidics and biosensing in two ways: (1) by using the evanescent part of the optical mode in a liquid cladding layer that is flown over the waveguide core and (2) to deliver light to fluidic channels. Evanescent sensing has been described in some detail in Chapter 3 by Janz, and has been used successfully for refractive index sensors (Mandal and Erickson 2008) and for surface plasmon resonance (SPR) sensors. The reader can find more details on the latter subject in dedicated reviews (e.g., Homola et al. 1999, Sharma et al. 2007). Here, we briefly describe reverse-symmetry waveguides that are designed to maximize the penetration depth of the evanescent field into the liquid cladding layer. The penetration depth of the evanescent field into the cladding layer is given by

$$\delta = \frac{\lambda}{2\pi\sqrt{n_{\text{eff}}^2 - n_{\text{cl}}^2}} \tag{16.2}$$

where n_{eff} is the effective mode index. Thus, the mode extends predominantly into the higher-index cladding, which is usually the solid cladding on the substrate side. This conventional situation is illustrated in Figure 16.2a. In order to reverse this situation, the index of the lower cladding region needs to be pushed below the index of the liquid, as shown in Figure 16.2b (Horvath et al. 2002). In this way, δ can be increased from a few hundred nanometers to several microns with possible applications to the study of cells and cell membranes bound to the waveguide surface. The desired index profile can be generated by using porous cladding material (Rabus et al. 2007). Fully symmetric profiles using a polymer cladding with index 1.34 have also been realized (Agnarsson et al. 2009). A symmetric profile strikes a good balance between large penetration depth and efficient fiber coupling into the core mode.

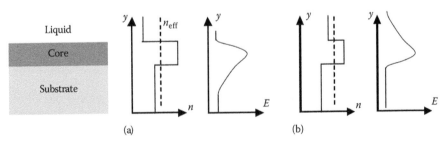

FIGURE 16.2 Conventional waveguide structure. (a) Index and field profiles for evanescent waveguide sensing in conventional geometry and (b) reverse-symmetry waveguide structure.

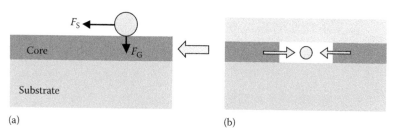

FIGURE 16.3 (a) Evanescent-field trapping and transport of a microbead on top of a waveguide and (b) delivery of optical beams through solid-core waveguides to a particle inside the fluidic channel (particle motion out of the plane).

Evanescent fields from solid-core waveguides have also been used to transport and manipulate particles. Figure 16.3a shows how a microscale particle near the core surface is subjected to two types of forces: a gradient force, F_G, pulling it toward the core–liquid interface and holding it there, and a scattering force, F_S, along the propagation direction of the waveguide mode that can push the particle along the top of the waveguide. This phenomenon was first demonstrated on 4 μm diameter latex beads by Tanaka and Yamamoto (2000). Later, this concept was extended to counterpropagating beams (Grujic and Hellesø 2007), and particle velocities on the order of 28 μm/s were demonstrated for 3 μm polystyrene beads (Schmidt et al. 2007a). The main drawback of using evanescent fields in a device is the small interaction volume, i.e., particles of interest need to be within δ of the interface in order to be detected. Given typical fluidic channel heights of tens to hundreds of μm and additional hydrodynamic focusing effects in the presence of flow, the collection efficiency can be rather small unless an additional force is added that facilitates movement toward the waveguide core.

The use of solid-core waveguides in optofluidics as a means of light delivery is schematically shown in the cross-sectional view of Figure 16.3b, and has been applied to both particle detection and manipulation. A particle moving perpendicular to the depicted plane along the fluidic channel can be optically excited with a single beam that enters the core from the solid waveguide core (dark gray). The resulting signal can be used for fluorescence detection (Vazquez et al. 2009) or flow cytometry (Cho et al. 2009). If two counterpropagating beams are simultaneously present as shown in the image, the equivalent of the original dual-beam trap (Ashkin 1970) is created, as both beams diverge in the liquid channel, creating a trapping potential using the asymmetry in the scattering forces (Cran-McGeehin et al. 2006). The concept of planar optical delivery to a liquid channel becomes even more powerful if the channel itself is an optical waveguide. We will revisit this approach in more detail in the following sections.

16.2.3 Index-Guided Liquid-Core Waveguides

A big advantage of conventional index guiding is that propagation based on TIR is lossless, i.e., energy is only transported along the desired waveguide axis. Any energy loss observed in an actual device is due to imperfections such as absorption or roughness scattering (Hunsperger 2002). Several variations of TIR-based waveguides have been developed over the past years that aim at using this established waveguiding principle with liquid core materials.

16.2.3.1 Liquid-Core Waveguides

Index guiding in aqueous solutions can be accomplished by using low-index claddings made from fluorinated polymers, such as Teflon AF. Teflon AF has a refractive index of 1.29, and is applied as a relatively thin layer on the inside of a higher refractive index material, as illustrated in the side view in Figure 16.4a. LCWs have been built in this way both in cylindrical tubes and with materials that were wafer-bonded to form a rectangular-shaped hollow cross section (Schelle et al. 1999, Datta et al. 2003).

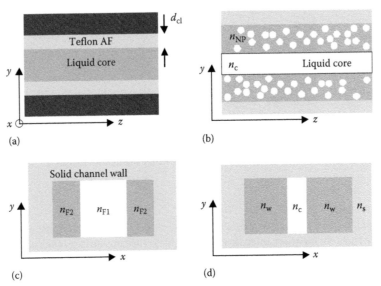

FIGURE 16.4 LCWs with index guiding. (a) LCW with Teflon AF cladding; (b) NP-cladding waveguide; (c) L^2 waveguide; and (d) slot waveguide.

Initially, LCWs were fabricated with large core diameters of 200–500 μm and Teflon AF cladding thicknesses, d_{cl}, on the order of a few microns with relatively large fluctuations. In addition, the softness and moderate optical properties of these fluoropolymers limited their utility as materials for sensor chips (Grewe et al. 2000). Nevertheless, fluoropolymer claddings made from PTFE, PFA, and FEP were used for chemical absorption sensing of ammonia by color changes in bromothymol blue (BTB), and for detection of trichloroethylene (TCE) (Hong and Burgess 1994). More recently, novel fabrication methods for selectively coating PDMS channels with Teflon AF to make optofluidic waveguides were demonstrated (Cho et al. 2009). This technique is reviewed in more detail in Chapter 19 by Chen, where promising applications of these structures for fluorescence detection and flow cytometry are discussed.

16.2.3.2 Nanoporous Cladding Waveguides

Another way to form a low-refractive-index cladding material is to start with a high-refractive-index solid material and add air pores until the average refractive index is low enough to facilitate waveguiding. This has the advantage that the refractive index of the cladding can be tuned by varying the air fraction and that some of the adhesion problems that Teflon AF exhibits can potentially be overcome. Figure 16.4b shows a schematic side view of a slab waveguide that has a nanoporous (NP) cladding with refractive index $n_{NP} < n_c$ embedded within a high-refractive-index substrate. The first NP-cladding waveguides were constructed using a "sacrificial porogen" approach, in which an organic phase is removed from a phase-separated polymer hybrid (Risk et al. 2004). Only one-dimensional (1D) confinement in the transverse (y) direction with a total propagation loss of ~6 dB/cm has been demonstrated using this method. Fluorescence from nanobeads in the liquid core was successfully detected in this proof-of-principle demonstration. If a way to create a lateral NP cladding for complete mode confinement can be developed, NP-cladding waveguides can be a very promising technology due to the existence of low-loss modes and the wide tuning range of the cladding refractive index ($1.15 < n_{NP} < 1.37$).

16.2.3.3 Fluid–Fluid Waveguides

One way to overcome the cladding limitation posed by solid materials is to use two fluids with different refractive indices to define the waveguide core and the cladding, respectively. This concept was first demonstrated by Takiguchi et al. (2003) using liquids inside a glass capillary tube. A cross-sectional view of such a fluid–fluid waveguide is shown in Figure 16.4c. The first optofluidic implementation of this concept was the liquid–liquid or L^2 waveguide, in which two different liquids are flown side by side through a larger fluidic channel to create an optical waveguide. As long as the refractive index of the cladding liquid (n_{cl}) is smaller than that of the core liquid (n_c) and the cladding layer is thicker than a few µm, index guiding in the x direction can be achieved. This concept was first demonstrated using $CaCl_2$ ($n_c = 1.445$) and water as core/cladding liquids, respectively, embedded in PDMS ($n = 1.4$) (Wolfe et al. 2004). These liquids were introduced into the channel through separate fluidic inlets, and exhibited relatively slow mixing along the channel due to the laminar flow conditions. The L^2 waveguide concept is attractive for sensor and device applications, because a control over fluidic properties allows a dynamic modification of the optical performance. We will illustrate this capability in Section 16.3, using the example of a flow-controlled optical switch. Another inherent advantage of L^2 waveguides is that the propagation loss should be independent of the channel roughness (at least along the x direction), although actual loss values of these waveguides have not been published yet. One limitation of L^2 waveguides aside from intermixing of core and cladding liquids is the fact that up to now, L^2 guiding has only been implemented along one dimension (x). In the other dimention (y), the same considerations with respect to the choice of core and cladding materials that were mentioned above need to be made. This issue, however, can be addressed by using hybrid approaches ([Bernini et al. 2008a], see also Section 16.2.5.4).

Another more recent example for the two-fluid paradigm is the case of LA (liquid-core, air-cladding) waveguides, in which the core material is a liquid and the cladding is a gas (air). This structure can be defined in the same way as an L^2 waveguide. The advantages of the LA waveguide lie in the higher index contrast between the core and the cladding material, and the elimination of the problem of analyte molecule diffusion into the cladding. In a specific example, LA waveguides were built on PDMS chips using air as the cladding and ethylene glycol as the core (Lim et al. 2008). By suspending a fluorescent dye in the glycol core and pumping the chip from above, a simple light source for in-plane emission was created.

16.2.3.4 Slot Waveguides

Slot waveguides represent perhaps the most promising approach to realizing index guiding with nanoscale cross sections of the fluidic channel, thereby providing an opportunity for extending waveguide-based sensing into the nanofluidic regime with picoliter or sub-picoliter sample volumes (Almeida et al. 2004, Xu et al. 2004). The cross section of a slot waveguide, as shown in Figure 16.4d, appears quite similar to that of an L^2 waveguide, but operates under a different physical principle. Here, the core medium has the lowest refractive index (n_c) in the structure, while the immediately adjacent medium has the highest refractive index (n_W). Thus, a slab waveguide with five sections is formed in the x direction, and guided mode solutions exist for this configuration. A substantial portion of the optical power can be confined in the low-refractive-index core if the width of the core is smaller than the penetration depth, δ, of the evanescent wave in the core medium, and if x-polarized light is used. The latter choice results in a discontinuity of the electric field (quasi-TE [transverse electric] mode) at the core–cladding interface, which increases the electric field strength in the core by a factor of n_W^2/n_c^2. Typically, on the order of 30% of the optical power can be confined in a 100 nm narrow core for a large refractive-index contrast, a system such as Si/air with a large enhancement factor of 12 (Almeida et al. 2004, Xu et al. 2004). This field enhancement is somewhat reduced for aqueous cores (7 for a Si cladding and 1.2 for PDMS). Additional challenges could arise from molecular interactions with the walls due to the large surface-to-volume ratios of nanofluidic channels. Slot waveguides with horizontal slots and losses of 7 dB/cm have also been constructed, and could be potential candidates for tunneling-based optoelectronic devices (Preston and Lipson 2009). In the field of optofluidics, slot waveguides have recently been used for biomolecule trapping and manipulation (Yang et al. 2009b), and we will revisit this application in more detail in Section 16.3.

16.2.4 Non-TIR-Based Waveguides

16.2.4.1 Metal-Clad Waveguides

An intuitively simple way to eliminate light refraction into the cladding material is to coat the inside of the cladding with a highly reflective metal, as shown in Figure 16.5a. In such a metal waveguide, the outer dielectric material does not affect the guiding properties as long as the metal layer thickness exceeds the skin depth (Jackson 1998). The resulting waveguide loss for a parallel-plate waveguide in the ray optics picture can be calculated from the power loss due to the finite reflection from the metal at the appropriate mode angle (Yeh 2005). The expression for the loss is then

$$\alpha = \frac{(1-R)\lambda}{2n_c d_c^2} \tag{16.3}$$

where R is the water/metal reflectivity at the mode propagation angle. The resulting loss values are approximately two orders of magnitude lower than those for the leaky capillary waveguide discussed above, while exhibiting the same characteristic d_c^{-3} behavior (note that the reflectivity is d_c dependent). Metal-clad waveguides have been realized in practice by coating glass tubes ($d_c \sim 250\,\mu m$) with silver layers of several hundred nanometers thickness. Optical measurements on 50 cm long air-filled fibers were made, and showed an attenuation of $10^{-3}\,cm^{-1}$ at a wavelength of 800 nm (Mohebbi et al. 2002). Silver-coated capillary waveguides were used as multipass absorption cells for capillary electrophoresis (Wang et al. 1991), exhibiting a 40-fold increase in the absorption path length. The major limitations of this approach are difficulties in developing a suitable coating process for on-chip waveguides, and substantially higher experimental losses due to imperfections of the metal coating due to surface roughness (Saleh and Teich 2007). These nonidealities tend to dominate the waveguide performance, resulting in a measurable improvement over leaky capillary waveguides only for very small diameters ($\lesssim 20\,\mu m$) (Grewe et al. 2000).

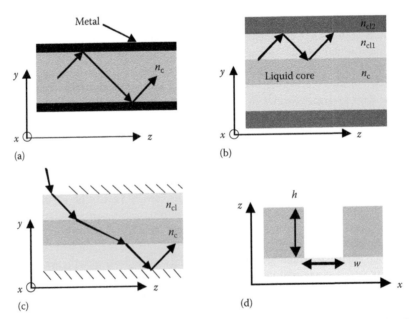

FIGURE 16.5 Non-TIR waveguides. (a) Metal cladding; (b) cladding guiding; (c) refractive guiding; and (d) ZMWG.

16.2.4.2 Waveguides with Cladding Guiding

Another nonconventional approach to liquid-core waveguiding is shown in Figure 16.5b and c. Here, a dielectric cladding is sandwiched between the liquid core and another cladding layer. In Figure 16.5b, the wave is guided by TIR in the first cladding layer, and sensing in the liquid core relies on evanescent-field coupling. This approach was implemented in "fiber-optic capillary" (FOCap) waveguides (Keller et al. 2007) (50–150 μm core diameter) using methanol as the core liquid and fused silica and doped fused silica ($n=1.441$) as the first and second dielectric claddings, respectively. Cladding guiding combined with evanescent sensing is also at work in optofluidic ring resonators (OFRR), which are discussed in more detail in the chapter by Suter and Fan (White et al. 2006a,b, Sumetsky et al. 2007). FOCap waveguides of up to 50 m length were used as chemical sensors by deducing pH changes from the absorption spectrum of fluorescein (Keller et al. 2007). In this case, however, it turned out that evanescent detection, i.e., light propagation in the solid cladding, was more sensitive than the LCW mode of the FOCaps. Figure 16.5c, on the other hand, shows how under an appropriate propagation angle the light can traverse the liquid core repeatedly. This results in effective interaction and good optical overlap with the liquid sample analyte, despite the light not being actually guided in it. Highly efficient reflection from the outer-cladding interface ensures good overall guiding. This can be achieved either by using an outside metal layer (Wang et al. 1991) or a dielectric material (Schmidt et al. 2006, Keller et al. 2007). Silver-coated glass capillaries (inner and outer diameters of 75 and 364 μm, respectively) were made with a 40-fold increase of the path length compared to a single-pass liquid cell. ARGOWs (anti-resonant guided optical waves) used glass slides as the cladding, and air as the outer cladding. Here, the relative path length and, therefore, the interaction within the liquid core are maximized if the ray hits the glass/liquid interface at a near-critical angle and if the core thickness is increased. Relative confinement factors near 60% in a 160 μm thick water layer, and advanced analytical sensing devices were demonstrated (Schmidt et al. 2006, Kiesel et al. 2009). We will revisit these in Section 16.3.2.1.

16.2.4.3 Other Waveguide Types

We conclude this section by briefly discussing two additional, nonconventional waveguide types. Fresnel waveguides appear to be similar to the photonic crystal fibers discussed in the following section in that they incorporate a distribution of holes over the cross section of a solid fiber (Canning et al. 2003, Martelli et al. 2005). Here, however, the hole distribution acts as a Fresnel zone plate to confine the light. Fresnel fibers made of silica with holes of ~5 μm diameter were fabricated, and showed good waveguiding properties when the central hole was filled with water (Martelli et al. 2005). The potential utility of such an approach for optofluidic sensor applications was pointed out, but has not been realized yet.

The final example of atypical LCWs is zero-mode waveguides (ZMWG) (Levene et al. 2003). The side view of a ZMWG in Figure 16.5d shows that its defining elements are nanoscale holes within a metal layer whose width is so small that even optical frequencies are below the cutoff frequency of this cylindrical metal waveguide (Jackson 1998). In a sense, the goal is to create a non-waveguiding structure where only the evanescent tail of an electric field will penetrate into the holes. This leads to extremely small optical excitation volumes on the order of zeptoliters (10^{-21} L) inside the hole. Arrays of ZMWGs were fabricated in an aluminum film (89 nm height) on fused silica substrates using standard nanofabrication methods, resulting in millions of holes with diameters ranging from 30 to 80 nm. ZMWGs do not quite fall within the category of integrated optofluidics since the excitation of the mode volume and fluorescence collection are carried out using traditional microscopy (Levene et al. 2003). Nevertheless, very impressive analytical capabilities have been demonstrated. These include monitoring the DNA polymerase activity with the fluorescence correlation spectroscopy (FCS) at micromolar concentrations (Levene et al. 2003), the detection of oligomerization of a λ-bacteriophage repressor protein (Samiee et al. 2005), and real-time DNA sequencing from single polymerase molecules (Eid et al. 2009). The latter thrust is already being pursued commercially, and has the potential to usher in an era of large-scale genome sequencing at a low cost.

16.2.5 Interference-Guided Liquid-Core Waveguides

A conceptually different approach to solving the refractive index problem for LCWs is to keep surrounding the low-index liquid core with higher-index solid-cladding materials, but to increase the reflection at the liquid–solid interface. This will reduce the waveguide loss and can lead to large improvements over the simple leaky capillary waveguide discussed in Section 16.2.1. Creating the required multilayer or nanostructured devices used to be difficult and prone to fabrication imperfections (Grewe et al. 2000), but recent advances in fabrication technology have helped eliminate or ameliorate many of these problems. The search for alternative waveguide structures was further driven by the growing need for LCWs with small cross sections of a few microns or even nanometer dimensions for single-molecule sensing in sub-picoliter excitation volumes and the emergence of photonic crystals that rely on the spatial structure of the waveguide cladding. The combination of these factors has turned LCWs based on wave interference into a very active field of research, both from a conceptual standpoint and with a view for sensor applications. They represent perhaps the most promising approach toward efficient liquid-core waveguiding due to their favorable balance of low loss, fabrication complexity, and integration potential.

The common principle underlying these waveguides is the use of interference to localize the electromagnetic wave. The refractive index profile of the cladding is structured in the *x*–*y* cross section of the waveguide to create multiple reflections of the electric field that can interfere constructively or destructively. The key idea in the present context is that near-perfect reflection into the original medium can be achieved even if that medium has a lower refractive index than all of the cladding layer materials.

16.2.5.1 Bragg Fibers

A very important implementation of interference-based waveguiding is the special case where dielectric cladding layers are repeated periodically and extend to infinity. Partial reflections at each interface add up in the same way as the Bragg reflections that are well known from x-ray analyses of crystalline materials. In complete analogy, the electric field propagating inside the periodic medium can be described by a Bloch wave vector, K, that reflects the periodicity of the structure (period Λ) and depends on Λ, the wave frequency, ω, and the indices of the dielectric materials. Real (imaginary) values of K correspond to propagating (evanescent) solutions and lead to allowed (forbidden) regions, so-called bands, if K is plotted in an ω-β diagram where β is the wave vector along the waveguide axis.

For a liquid-core Bragg waveguide to form, the Bragg lattice has to be highly reflective, i.e., one has to operate within a forbidden region of K. In addition, the transverse component of the wave vector, β, has to fulfill the same phase resonance ("mode") condition as in the case of index guiding, to allow for the transport of energy along the propagation (z) direction.

Bragg slab waveguides were first proposed by Yeh and Yariv in 1976 (Yeh and Yariv 1976, Yeh et al. 1977). The cross section of a cylindrical Bragg fiber is shown in Figure 16.6a, and illustrates how the periodic layers surround a circular low-refractive-index core, resulting in a refractive index profile that depends on the radial distance from the core center. Cylindrical Bragg fibers were first analyzed

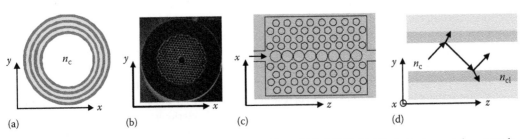

FIGURE 16.6 Interference-based waveguides. (a) Bragg fiber; (b) HC-PCF; (c) 2D photonic crystal waveguide; and (d) ARROW waveguide.

by Yeh et al. (1978) and later refined by Xu et al. (2000). The first experimental demonstrations of light guiding in an air-core Bragg fiber were given by Fink et al. (1999), using a large ~2 mm air core surrounded by alternating layers of tellurium and polymer, followed by 275 μm core As_2Se_3/PES fibers for infrared guiding (Temelkuran et al. 2002). If the Bragg layers are designed to be reflective for all angles of incidence ("omnidirectional" guiding [Winn et al. 1998]), light can even propagate around bends, with a very low loss. This concept has been successfully developed into medical instruments for CO_2 laser delivery for otolaryngology and pulmonology (see www.omni-guide.com).

16.2.5.2 Hollow-Core Photonic Crystal Fibers

The photonic crystal concept is not restricted to 1D refractive index periodicity. In fact, dielectric structures with two-dimensional (2D) periodicity currently play a bigger role in liquid-core waveguiding. Figure 16.6b shows the cross section of a hollow-core photonic crystal fiber (HC-PCF) (Cregan et al. 1999, Russell 2003). In this case, a hollow core with typical diameters between 5 and 20 μm is surrounded by a periodic arrangement of holes inside a silica network. As in the 1D case, the spatial structure of the air/silica refractive index variation determines the propagation properties of the hollow fiber along the z direction. Waveguiding in liquid-core HC-PCFs has successfully been demonstrated by several groups (Mach et al. 2002, Yan et al. 2006, Mandal and Erickson 2007). The main areas of concern for using HC-PCFs in commercial sensors are side access to the fibers and the multimode nature for propagation (Cordeiro et al. 2007). In addition, commercial HC-PCFs are highly multimode in nature, which eliminates their use for interferometric purposes. The number of modes can be reduced by filling the cladding holes with a suitable liquid (Cordeiro et al. 2007). Both the theoretical concepts and the optofluidic applications of these cylindrical fiber structures are reviewed in more detail in Chapter 4 by Benabid and Roberts and Chapter 15 by Barth et al.

16.2.5.3 2D Photonic Crystal Waveguides

The third approach to using a photonic bandgap structure for low-refractive-index waveguiding is schematically depicted in Figure 16.6c. Light propagation in the direction of the arrow is achieved by the Bragg reflection (2D photonic crystal) in the x–z plane (Joannopoulos et al. 1995) and conventional index guiding in the third dimension. This concept was first realized using a central solid-core silicon waveguide surrounded by periodic air holes to facilitate light guiding around tight bends (Loncar et al. 2000). A compelling feature of these PC waveguides compared to their fiber counterparts is their planarity, which is highly attractive for a planar optofluidic integration. Nanofluidic tuning of the optical properties of a photonic crystal waveguide with this geometry was also demonstrated (Erickson et al. 2006). By using a multilayer integration of the optical waveguide layer and a fluidic delivery layer, it was possible to address (fill) individual holes in the 2D PC waveguide, in particular, the central guiding row with the larger holes. The nanoscale spatial control and the level of integration in this platform are encouraging for future developments of fluidically controlled optics (McNab et al. 2003). 2D PC waveguides have not yet been used with liquid guiding channels. Conceptually, this should also be possible and would open up this approach to the optical analysis of sample analytes in an LCW.

16.2.5.4 ARROW Waveguides

The last type of interference-based LCWs is the antiresonant reflecting optical waveguide (ARROW). ARROWs are also based on an interference effect, but do not require the periodicity of a photonic crystal that results in the description of light propagation by means of Bloch wave vectors and allowed/forbidden bands. Thus, a single dielectric layer is sufficient to provide low-loss propagation, as illustrated in Figure 16.6d. A light ray is refracted from the low-refractive-index core into the high-refractive-index ARROW layer (n_{cl}). Low-loss guiding in the core occurs if the ARROW layer thickness fulfills an antiresonance condition for the round trip phase shift of the transverse (y) wave component, $\Phi_{RT} = m\pi$ (m odd), in the cladding layer, and the usual mode condition in the core. In the simple case of multiple ARROW cladding layers made from two alternating cladding materials, analytical expressions for the loss in TE polarization as a function of the core thickness and the number of cladding layers, N, are given by (Archambault et al. 1993)

$$\alpha = \frac{\lambda^2}{n_c d_c^3 \sqrt{n_{cl1}^2 - n_c^2}} \sqrt{\frac{n_{cl2}^2 - n_c^2}{n_{cl1}^2 - n_c^2}}^{-N} \tag{16.4}$$

We see that Equation 16.4 shows the same dependence on the wavelength and the core thickness as Equation 16.1. However, the loss decreases rapidly with the increasing number, N, of ARROW layers. A few layers are sufficient to achieve better performance than with metal claddings (Equation 16.3), resulting in a sufficiently low loss to make cores of a few microns width technologically interesting. Figure 16.7 shows a comparison of the waveguide loss for three different waveguide types: leaky glass capillary waveguide, metal-cladding waveguide, and an ARROW ($N=5$, $n_{cl1}=2.1$, $n_{cl2}=1.45$). In this figure, the loss of a TE mode for a 1D slab waveguide according to Equations 16.1, 16.3, and 16.4, a core index of 1.33 (water), and a wavelength of 632 nm are plotted versus core diameter. All waveguides show the characteristic $1/d^3$ dependence, with the ARROW loss being approximately one thousand times lower than that of the leaky capillary waveguide.

The first ARROWs were silicon dioxide waveguides on top of a silicon substrate (Duguay et al. 1986). ARROWs were later used for semiconductor laser applications (Mawst et al. 1992, Patterson et al. 1999). Hollow-core ARROWs were first demonstrated by Delonge and Fouckhardt using TiO_2 and SiO_2 ARROW layers to confine light in capillaries with a $20\,\mu m \times 20\,\mu m$ cross section (Delonge and Fouckhardt 1995). More recently, LCWs with an ARROW confinement using silicon nitride and silicon dioxide were built using the silicon microfabrication technology. Both large multimode ($d > 100\,\mu m$) (Bernini et al. 2004) and (quasi) single-mode ($d < 10\,\mu m$) (Yin et al. 2004a,b) ARROWs have been demonstrated, and their fabrication methods and loss properties have been analyzed in detail (Barber et al. 2005, 2006, Hubbard et al. 2005, Yin et al. 2005a,b). Figure 16.8 shows cross-sectional views of the two approaches. Both result in rectangular cross sections, typically using the same materials. The main difference is that the hollow channel in Figure 16.8a is produced by etching the substrate followed by wafer bonding, while the ARROW channel in Figure 16.8b is the result of patterning a sacrificial layer that is removed after the top waveguide layers have been deposited (Barber et al. 2005, 2006). Figure 16.8c

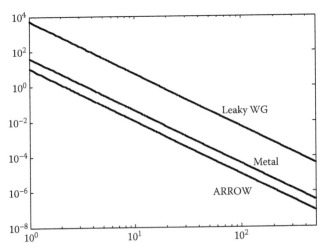

FIGURE 16.7   Waveguide loss (1D slab) versus core diameter for different waveguide types: leaky capillary waveguide, metal-clad waveguide, and five-layer ARROW waveguide.

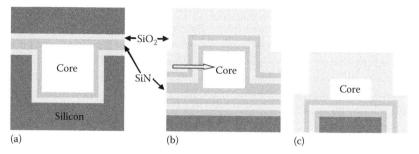

FIGURE 16.8 Optofluidic ARROW structures. (a) Fabrication by wafer bonding; (b) fabrication by sacrificial layer patterning; and (c) sacrificial layer patterning with a single top layer.

shows a variation of the latter waveguide type in which a single overcoating layer of SiO_2 is used. In this case, waveguiding to the top and the sides is provided by the TIR since the terminating layer is air, while the bottom ARROW layers prevent leakage into the high-index silicon substrate. This simplifies processing, eliminates concerns about thickness uniformity for the top layers, and results in improved coupling to solid-core waveguides (Lunt et al. 2008).

An advantage of liquid-core ARROWs built with the sacrificial layer process is the ability to naturally interface them with solid-core ARROWs to facilitate light transport to and from the liquid-core region. The arrow in Figure 16.8b illustrates how light can be guided along the topmost, thick cladding layer (typically SiO_2). In a ray picture, a ray propagating along this layer hits the thinner ARROW cladding layers surrounding the liquid core almost perpendicularly ($\Theta \sim 0°$). When guided in the liquid, on the other hand, the angle of incidence for the fundamental mode is almost grazing (typically $\Theta \sim 3°$). Hence, the layer thickness can be optimized to facilitate effective coupling to and from the solid-core waveguide while maintaining low-loss guiding in the liquid core (Schmidt et al. 2005). This concept is essential for creating interconnected networks of solid- and liquid-core waveguides to enable different optofluidic functionalities.

Quasi-single-mode ARROWs with core dimensions of a few microns have additional characteristics that can be exploited for optofluidic devices. Equation 16.4 suggests a possibly significant wavelength dependence of ARROW guiding for a given cladding structure, i.e., fixed $d_{cl,i}$. This is indeed the case and presents additional design possibilities for liquid-core ARROWs in the context of biodetection and sensing applications. A typical experiment using any of the established optical methods, such as fluorescence detection, Raman scattering, or fluorescence resonance energy transfer (FRET), requires the discrimination between signals at different wavelengths using optical elements such as spectrometers, bandpass, or edge filters (see also Chapter 6 on spectroscopy by Zhang). The requirements for these devices can be rather stringent, often requiring $\geq 50\,dB$ discrimination in the case of separating weak signals from strong excitations. Interference-based waveguides, such as liquid-core ARROWs, are ideally suited for this task, as the interference phenomenon that enables waveguiding is intrinsically dispersive. This is also evident from the equation that governs the thickness, t_i, of the ith ARROW cladding layer, normalized by the vacuum wavelength, λ:

$$\frac{t_i}{\lambda} = \frac{N}{4\sqrt{n_i^2 - n_c^2 + \dfrac{\lambda^2}{4d_c^2}}} \qquad (16.5)$$

where $N = 1, 2, \ldots$ is an integer that results in a highly transmissive (N even) or reflective (N odd) layer. Hence, for a given thickness the structure will either become lossy or guiding, depending on the value of λ. Through careful optimization, the dispersion of the waveguide can be tailored to a particular application. This is even possible for applications that involve multiple wavelengths, such as FRET, where one excitation wavelength needs to be filtered out while two distinct fluorescence wavelengths need to be guided (Hakanson et al. 2007, Schmidt and Hawkins 2008). Emerging developments using selective lithography during the fabrication process will even allow for varying the dispersion across the chip. Another opportunity that arises from the quasi-single-mode operation is to build interferometric structures. We will discuss one of these in more detail in Section 16.3.1.2.

ARROWs with large cross sections can be interfaced with multimode fibers, and have been used to demonstrate refractometric sensors (Campopiano et al. 2004) and flow cytometry (Bernini et al. 2007). In addition, a hybrid structure using liquid–liquid (L^2) waveguiding in the lateral direction and transverse ARROW guiding has been proposed and demonstrated to overcome the limitations of the L^2 waveguide approach (Bernini et al. 2008a). Liquid-core ARROWs with small cross sections have been used for a variety of spectroscopic applications, including the fluorescence detection of dye molecules (Yin et al. 2005c), single dye molecule detection (Yin et al. 2006), surface-enhanced Raman scattering at nanomolar concentration (Measor et al. 2007), and fluorescence correlation spectroscopy of single dye molecules, liposomes, and virus particles (Yin et al. 2007a,b, Rudenko et al. 2009). We will review their biosensing capabilities and the implementations of new approaches for optical particle manipulation in the following section.

16.3 Applications of Liquid-Core Waveguides in Optofluidics

In this section, we will take a closer look at a few representative examples that show the use of LCWs in all three main optofluidic thrusts. The intention is to highlight how diverse the function of the seemingly simple and most basic integrated optical element can be when filled with a liquid core.

16.3.1 Optical Elements

16.3.1.1 1×3 Optofluidic Switch

The first example for an optofluidic device that uses fluids to define and control the optical function showcases the dynamic control afforded by liquid-core liquid-cladding (L^2) waveguides (Takiguchi et al. 2003). As discussed in Section 16.2.3.3, L^2 waveguides work by flowing liquids with different refractive indices through a microfluidic channel such that the liquid with the highest index is sandwiched by a lower-index liquid, thus creating a conventional TIR-based waveguide. Figure 16.9a shows how this concept was physically implemented on a microfluidic chip by Wolfe et al. to create a 1×3 optical switch (Wolfe et al. 2004). Fluidic channels of 75–300 μm width and 100 μm height were fabricated in PDMS,

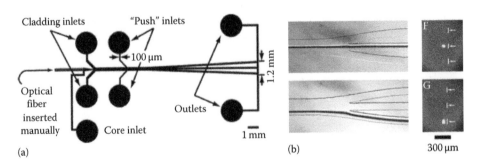

FIGURE 16.9 Optofluidic switch. (a) Fluidic layout of an L^2 waveguide architecture and (b) displacement of core liquid (dark line) depending on cladding flow rates. (Reprinted from Wolfe, D.B. et al., *PNAS*, 101, 12434, 2004. With permission.)

using standard PDMS molding methods (see Chapter 2 on fabrication by Hawkins et al.). The front end of the chip was designed in such a way that fluids from three inlets and light from a fiber could enter the main channel. $CaCl_2$ (index 1.445) and H_2O (index 1.335) were chosen as the core and cladding liquids, respectively. This resulted in a liquid–liquid TIR guiding in the lateral direction and a liquid–solid TIR guiding between $CaCl_2$ and PDMS (index 1.40) in the transverse direction. The liquids were flown through the channels at rates on the order of 1–100 µL/min, to define the core and cladding regions.

The fluidically controlled optical switch was then formed by manipulating the relative flow rates of the two cladding liquids. This effect is illustrated in Figure 16.9b. The core liquid contains a fluorescent dye so that it can be seen as dark lines. As the cladding flow rates are adjusted via the push inlets to differ by as much as 50%, the core liquid and hence the optical signal are displaced from their original path in the center of the channel (top) to one side (bottom). The addition of a three-way splitting in the channel allows directing the optical signal to three different locations at the chip output, thus defining a fluidically controlled 1×3 optical switch. While this device is too slow for communications (switching time ~2 s), it works fast enough for many biosensing applications.

16.3.1.2 Mach–Zehnder Interferometer

An MZI is one of several canonical interferometer types that is particularly well suited for implementation in planar integrated optics because it works with propagating waves and does not require mirrors (Michelson interferometer) or loop structures (Sagnac interferometer) (Saleh and Teich 2007). A top-down view of the MZI layout is shown in Figure 16.10a. A propagating beam is split into two parts that pass through separate interferometer arms before recombining at a second Y junction. While separated,

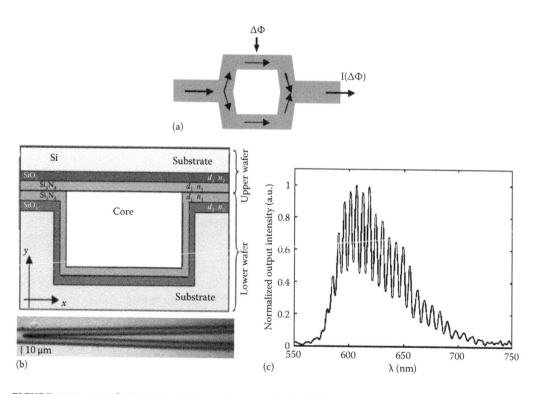

FIGURE 16.10 Optofluidic MZI. (a) Schematic view of MZI. (b) Top: cross section of a liquid-core ARROW; bottom: top view of fabricated Y junction. (Reprinted from Bernini, R. et al., *Appl. Phys. Lett.*, 93, 011106, 2008b. With permission.) (c) Experimentally observed interference pattern of light transmitted through the optofluidic MZI. (Reprinted from Bernini, R. et al., *Appl. Phys. Lett.*, 93, 011106, 2008b. With permission.)

the two beams can acquire a relative phase shift, $\Delta\Phi$, that determines the intensity of the recombined beam due to an interference of the electric fields of the split waves. If the phase shift of one of the arms can be controlled, say using an applied voltage on an electro-optically active waveguide, the MZI can be used as an intensity modulator, operating at frequencies up to tens of gigahertz.

An optofluidic MZI has recently been realized by Bernini et al. using liquid-core ARROW waveguides (Bernini et al. 2008b). Figure 16.10b (top) shows the rectangular cross section of the ARROW waveguides that were fabricated using wafer bonding of two coated silicon wafers, as discussed in Section 16.2.5.4 (Bernini et al. 2006). Waveguiding in the hollow core is achieved by plasma depositions of alternating layers of silicon dioxide ($n = 1.457$, 266 nm) and silicon nitride ($n = 2.227$, 266 nm), designed for low loss at a laser wavelength of 633 nm. Small core dimensions of $5 \times 10\,\mu$m result in high losses for higher-order leaky modes. The resulting quasi-single-mode behavior is essential for creating an interferometric device.

Figure 16.10b (bottom) shows a top-down-view photograph of one of the Y junctions that comprise the interferometer. An asymmetric MZI was created by varying both the separation between the two arms and the parameters of the junction itself. The structure shown here had a total length of 1.5 cm, an arm spacing of 510 μm, and bend radii of 1.71 and 85.6 cm for the two arms, respectively, resulting in an optical path difference of $\Delta L = 47.3\,\mu$m. The functionality of the interferometer was tested using an unpolarized white light source that was coupled into the liquid waveguide core using a multimode fiber (50 μm diameter). Figure 16.10c shows the observed transmission spectrum, which exhibits two main characteristics. On a large wavelength scale, the structure has a high transmission around the ARROW design wavelength of 633 nm, as expected. On a smaller scale, the transmission shows pronounced oscillations due to the dependence of the phase difference on the wavelength. This interference pattern shows a free spectral range (FSR) of 5.8 nm and a peak visibility of $V = I_{max}/I_{min} = 0.375$ at 633 nm. The visibility is limited by the different bend losses in the two interferometer arms, and can be improved in future generations.

This liquid-core waveguide MZI can find applications in biosensing because its transmission at a fixed wavelength is sensitive to changes in the refractive index of the liquid that could result from different (concentrations of) molecules inside the channel.

16.3.2 Particle Detection

16.3.2.1 Optofluidic Spectrometer

The first representative example for particle detection facilitated by LCWs is a distributed spectrometer to detect fluorescence from moving micron-scale objects. In particular, this example illustrates how the waveguides can be designed as parts of a larger system based on the philosophy of a planar architecture.

The idea behind the optofluidic chip introduced by Schmidt et al. is to continuously excite fluorescent particles along the length of a fluidic channel, and collect the fluorescence on a detector array whose spectral response varies with the position along the channel. This creates a distributed spectrometer in which the fluorescence detected from one pixel of the detector array can be assigned to a specific wavelength (Schmidt et al. 2007b). This concept is illustrated in Figure 16.11a, which shows both a side view and a top-down perspective of the chip.

This chip is built by defining 5 mm wide and 50 μm deep grooves in an acrylic sheet using laser cutting. The top of the chip was a glass substrate and both parts were joined using epoxy. Waveguiding in this device occurs as shown in Figure 16.11a. The beam is guided by TIR between the air, and the top and bottom solids, but spends a large fraction of time inside the liquid channel when refracted into it. This has been shown to result in an efficient light–liquid interaction with a confinement up to 90% (Schmidt et al. 2006).

In order to create a device for continuous large-scale spectroscopy, the channel was imaged onto a chip-sized spectrometer consisting of a 12 bit CMOS camera (pixel size $10.6 \times 10.6\,\mu$m) and a linear

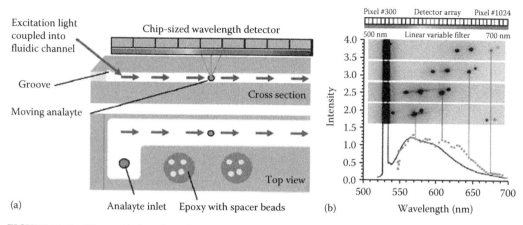

FIGURE 16.11 Waveguide-based on-chip spectrometer. (a) Side and top-down views of chip architecture and (b) cumulative microbead fluorescence spectrum obtained on-chip (dots) compared with a conventional spectrometer (line). (Reproduced from Schmidt, O. et al., *Lab Chip*, 7, 626, 2007b. With permission.)

variable band-pass filter (LVF) with a spectral gradient of 30 nm/mm and covering the range from 380 to 720 nm. In principle, the LVF and the detector array can be attached directly to the chip to create a complete optofluidic system. Fluorescence spectroscopy was then carried out by pumping a suspension of 31 μm diameter polystyrene beads with a syringe pump through the channel (speed ~1.5 μm/s). The beads were continuously excited in the waveguide core with a frequency-doubled Nd:YAG laser at 532 nm. The linear variable filter then picks out the spectral components as the particle moves through the channel. The intensity from each camera pixel corresponds to the intensity at a given wavelength, and can be plotted to reconstruct the fluorescence emission spectrum. This concept of spatially distributed spectrometry is illustrated in Figure 16.11b. In principle, lower fluorescence signals from labeled molecules or cells can be detected by using more sensitive photodetector arrays and by improving the light collection efficiency through a direct integration of the detector components on the chip.

16.3.2.2 Ultrasensitive Virus Detection

The importance of fluorescence analysis and single-molecule detection for biosensing applications has already been described in earlier chapters. The small sample volumes flowing through microfluidic channels should be ideally suited for the fluorescence detection of single particles, and indeed, several examples of confocal microscopy applied to microfluidic structures have been reported in the literature (Foquet et al. 2002, 2004, Lenne et al. 2002, Levene et al. 2003). Fully planar optofluidic-device architectures using LCWs for light delivery and collection, however, have only started to be applied to highly sensitive particle detection.

Here, we review the recently reported demonstration of virus detection on a single-virus level, using liquid-core ARROW waveguides (Rudenko et al. 2009). Figure 16.12 shows the basic chip layout and the detection principle. Solid- and liquid-core ARROW waveguides, as described in Section 16.2.5.4, are lithographically arranged on a silicon chip. This figure also shows scanning electron images of the two waveguide types, indicating the typical core dimensions of 5 μm×12 μm (liquid core) and 3 μm×12 μm (solid core). The liquid core is surrounded by three periods of alternating silicon nitride and oxide layers, whose thicknesses are optimized for low-loss propagation at the emission wavelength of the Alexa 647 dye (670 nm). Reservoirs (10 μL typical volume) can be affixed on top of the chip to facilitate loading of the liquid sample analyte containing the particles of interest. Also shown is a photograph of a fabricated chip with a total footprint of ~1 cm².

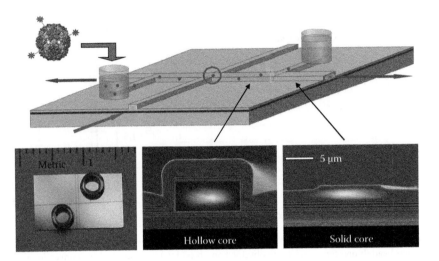

FIGURE 16.12   Optofluidic chip for single-particle fluorescence analysis.

Early experiments showed that fluorescence can be excited and detected collinearly along the liquid waveguide, but this geometry had a sensitivity limit of a few hundred molecules (Yin et al. 2005c). In order to reach single-molecule sensitivity, the intersection geometry shown in Figure 16.12 had to be implemented. Here, the pump signal is coupled via a single-mode fiber into a solid-core waveguide that intersects the liquid channel as shown by the circle. In this way, an excitation/collection region with a volume below 100 fL depending on the waveguide dimensions is created—sufficient to reach the single-molecule detection limit (Yin et al. 2006).

In this recent demonstration, Q-β bacteriophage at picomolar concentrations was prepared in a bicarbonate buffer solution and pipetted into one of the reservoirs, as shown in Figure 16.12. The Q-β phage is part of the Leviviridae RNA virus family and served as a model organism that is harmless to humans. The phage capsid was covalently labeled with the Alexa 647 dye, resulting in multiple attached dyes per virus nanoparticle. Figure 16.13a shows the fluorescence background signal obtained from a pure buffer solution when the waveguide was excited with an HeNe laser at 633 nm and detected with an avalanche photodiode (APD). Figure 16.13b depicts the signal after the virus capsids had been added to the reservoir and moved electrokinetically along the channel by applying a voltage across two electrodes in the reservoir. The characteristic fluorescence bursts from single capsids, and capsid parts are clearly visible above the background. Figure 16.13c shows the autocorrelation trace when the signal was analyzed using FCS, as described in Chapter 6 by Zhang. The data points show correlation values exceeding 1 for short correlation times, confirming that statistically less than one virus was present in the excitation volume at any given time. In addition, a fit of the data to a theoretical model (Yin et al. 2007a) allows for the extraction of the average diffusion coefficient of the virus particles. Further analysis of the data can be found in Rudenko et al. (2009).

This experiment demonstrated that integrated optofluidic waveguides can be used for highly sensitive bioparticle analyses without resorting to high-end microscopy apparatuses. In addition, it opened the possibility for novel devices that detect viruses (or their genetic material) rapidly, specifically, and without the DNA amplification step that is the time-limiting step in current gold standard detection methods based on real-time polymerase chain reactions (RT-PCRs) (Watzinger et al. 2006).

16.3.3 Particle Manipulation

The final two examples highlight recent developments in optofluidic particle trapping and manipulation. Much of the physics behind optical trapping and a variety of examples in optofluidics are discussed

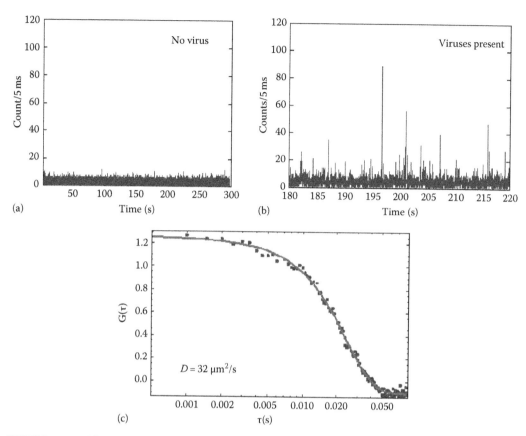

FIGURE 16.13 Ultrasensitive fluorescence detection of bacteriophages. (a) Background detector signal with buffer solution in channel; (b) signal in the presence of labeled virus capsids; and (c) corresponding FCS autocorrelation curve (symbols: data; line: fit) with the extracted average particle diffusion coefficient.

in Chapter 14 by Chiou. Here, we take a closer look at optofluidic particle manipulation that relies on LCWs, specifically slot waveguides and ARROWs. Both types of traps share some essential characteristics that take advantage of guiding light in a liquid-core region: Particles are pushed along a light beam effectively because they move in the region of the highest mode intensity. In addition, scattering forces act along the entire waveguide channel, which creates long-range traps and regions of light–matter interaction that extend across the entire chip.

16.3.3.1 Slot Waveguide Trap

As discussed in Section 16.2.3.4, slot waveguides are formed by defining a nanometer-scale groove (slot) in a high-index region that would ordinarily carry the optical mode. Used with the correct quasi-TE polarization, a substantial fraction of that energy can be concentrated in the slot, despite its small width (Almeida et al. 2004). Since this principle only works effectively with nanoscale slots, slot waveguides are well suited for nanoparticle manipulation, but do not work well with microscale particles for which trapping would essentially revert to evanescent-field trapping.

This potential was explored in a recent set of experiments by Yang et al. (2009a) on silicon slot waveguides on an SiO_2 substrate. The dimensions of the waveguide that is schematically illustrated in Figure 16.14a were $d = 450\,nm$ and $d_S = 60$–$120\,nm$ for the widths of the silicon region and the slot, respectively, and a $200\,nm$ slot height. The waveguides were fabricated using electron beam lithography to achieve the

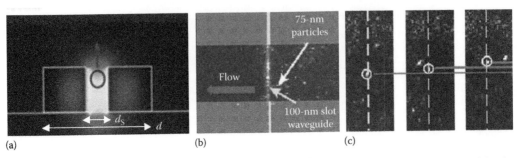

FIGURE 16.14 Optofluidic trapping and particle manipulation in a slot waveguide. (a) Cross section of the slot waveguide showing the nanoparticle and the concentration of optical intensity in the slot. (Reprinted from Yang, A.H.J. et al., *Nano Lett.*, 9, 1182, 2009b. With permission.) (b) Top-down view of fluorescent nanoparticles being trapped in the slot as they move through the larger fluidic channel. (c) Images of the same particle taken at three different times, illustrating optically induced particle movement along the slot waveguide. (Reprinted from Yang, A.H.J. et al., *Nature*, 457, 71, 2009a. With permission.)

desired nanoscale dimensions. The slot waveguides were covered on the top by a PDMS structure that created fluidic channels of 100 μm width and 5 μm height, which were filled with particles suspended in a suitable buffer solution. Optical power on the order of 300 mW was coupled from a laser at 1550 nm into the slot waveguides, which were filled with particles suspended in a suitable buffer solution.

Figure 16.14b is a top-down view of the structure that shows polystyrene beads (75 nm diameter) accumulated in the waveguide slot. The trapping mechanism is the gradient force of the evanescent field protruding from the slot waveguide into the channel. Due to the shallow penetration depth of the evanescent wave, only a small fraction of the beads flowing through the channel are trapped. In addition to this downward trapping force, however, there is an additional scattering force component that pushes the particles along the slot in the direction of the guided optical wave. This effect is illustrated in Figure 16.14c, where snapshots of the same particle at different times are shown. Velocities on the order of 1.5 μm/s were observed for 100 nm particles. In addition, λ-DNA molecules were trapped, demonstrating the ability to trap and transport elongated bioparticles with this long-range trap. More detailed simulations of the particle transport show that the stiffness of this slot waveguide trap is on the order of 1 pN/nm/W and polystyrene bead velocities on the order of 10 μm/s/W are feasible (Yang et al. 2009b).

16.3.3.2 ARROW Waveguide Trap

The ability to optically connect liquid-core ARROWs seamlessly with solid-core waveguides is not only advantageous for sensitive fluorescence detection, it also enables novel approaches and geometries for optical particle trapping and manipulation. This flexibility was first exploited by Measor et al. who recognized that the optical scattering force acting on a microbead in the ARROW channel is position dependent due to the inherent waveguide loss. This, in turn, allowed for extraction of both the waveguide loss and the lateral optical mode profiles by simply recording and analyzing the trajectory of a particle moving along the liquid channel (Measor et al. 2008). This experiment also alluded to the capability of moving particles rapidly between different regions of an optofluidic chip.

In order to hold a particle at a desired position for an extended period of time, however, a trapping mechanism is needed. A new type of dual-beam trap was recently implemented by Kühn et al. (2009a). A dual-beam trap that relies on counterpropagating beams requires a spatial asymmetry between the longitudinal scattering forces to create a longitudinal trapping potential, while using gradient forces in the other two directions. This asymmetry usually results from the two beams focused to different spots (asymmetric beam *area*) (Ashkin 1970). Figure 16.15a (top), on the other hand, shows that this asymmetry can be achieved in a waveguide by taking advantage of the waveguide loss, α (asymmetric beam

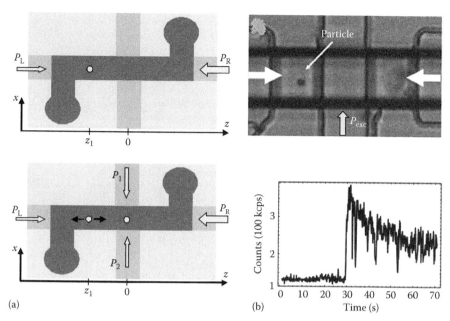

FIGURE 16.15 Optofluidic dual-beam traps. (a) Top: loss-based dual-beam trap controlling the particle position in the liquid-core channel (dark gray); bottom: dual dual-beam trap combining loss-based and divergence-based traps along z and x directions, respectively. (b) Top: micrograph of a polystyrene microbead trapped next to ARROW intersection with trapping and fluorescence excitation beams; bottom: fluorescence signal detected along the LCW as the particle is moved into waveguide intersection at $t=30$ s. (Reproduced from Kühn, S. et al., *Lab Chip*, 9, 2212, 2009a. With permission.)

power). By balancing the input powers of the two beams as shown in this figure ($P_L(z_1) = P_R(z_1)$), the trapping location (vanishing scattering force) can be dynamically defined at any point along the entire channel ($-L/2 < z_1 < L/2$). This concept leaves tremendous room for using an optical waveguide design to determine the properties of a trap, going far beyond what can be done with lens-based dual-beam traps. For instance, the optimum value for the waveguide loss to achieve the tightest trapping is related to the LCW length, L, by $\alpha_{opt} = 2/L$.

A natural choice for the trapping location is, of course, the solid–liquid waveguide intersection described earlier. Figure 16.15b (top) shows the top view of the intersection region with a microbead trapped by two trapping beams ($\lambda = 820$ nm, 300 mW input power at chip facet). At the same time, a third beam for fluorescence excitation with power P_{exc} enters the liquid channel as shown. Figure 16.15b (bottom) shows the fluorescence signal detected in the plane of the chip with an APD detector, as the trapping power balance is adjusted to move the bead in a controlled manner into the excitation region at $t \sim 30$ s. The same type of experiment was also carried out with *E. coli* bacteria. This establishes the ability to perform extended fluorescence studies on the same particle in a completely planar optical geometry. This technique can further be combined with a more conventional divergence-based dual-beam trap (Cran-McGeehin et al. 2006) to create a "dual dual-beam trap," as shown in Figure 16.15a (bottom). Here, each particle is held in place by a different set of optical beams, and their relative distances can be changed by adjusting the beam power of the loss-based dual-beam trap, as shown in this figure.

Another capability afforded by a long-range trap is to act as a particle accumulator or concentrator. This concept is illustrated in Figure 16.16a, where now multiple particles are subjected to scattering forces that drive them to a common trapping point (Kühn et al. 2009b). This effect was demonstrated on a collection of polystyrene microbeads using 125 mW of input power at $\lambda = 820$ nm. Figure 16.16b (top) shows fluorescence from a single dye-labeled microbead excited with an HeNe laser in the usual

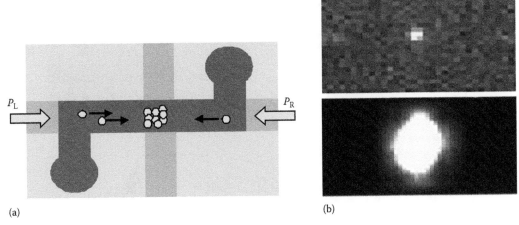

(a)   (b)

FIGURE 16.16 Optical particle concentrator. (a) Schematic view of two trapping beams acting on a particle ensemble and (b) fluorescence detected from a single microbead (top) versus a cloud of ~120 accumulated beads (bottom).

excitation geometry, but observed through a CCD camera from the top. The much larger signal from ~120 microbeads that were optically collected in the excitation is shown in Figure 16.16b (bottom). Here, optical particle collection resulted in an over 60-fold signal increase due to a substantial local enhancement of the particle concentration. This method has found uses in a variety of biosensing applications that utilize optically active particles at low concentrations.

The final example for the versatility of waveguide-based trapping is the implementation of an active feedback trap that operates at a much lower power than conventional all-optical traps (Kühn et al. 2009c). Its operating principle is depicted in Figure 16.17a. Since the particle movement is already restricted in two dimensions by the waveguide channels, the key is to compensate for Brownian motion along the channel direction. Various bulk optical methods for such a trap exist (Enderlein 2000, Decca et al. 2002, Berglund and Mabuchi 2005, Cohen and Moerner 2005). In an integrated optofluidic setting, this task can be accomplished in an elegant way by slightly offsetting the two solid-core excitation waveguides, as shown. If a particle were to diffuse or drift from the channel center to the left, it would fluoresce more strongly when excited with P_1 compared to P_2. By temporally modulating the choice of the input waveguide, one can track the particle's motion at any given time within the limits of the feedback latency and the diffusion coefficient. The detected motion is then compensated electrically by applying a feedback voltage, V, that drives the particle electrokinetically back to the region between the two solid-core waveguides. This type of electro-optical trap is fundamentally different from all-optical versions in that optical power is only needed to elicit a detectable fluorescence signal. It is also a true single-particle trap that can only compensate for the random Brownian motion of one particle at a time (Cohen and Moerner 2005).

In the experiments by Kühn et al., micro- and nanoscale latex beads as well as fluorescently stained *E. coli* bacteria were trapped using optical trapping powers of a few microwatts and feedback voltages of 1–100 V across the fluidic reservoirs. Figure 16.17b shows the position statistics of an *E. coli* bacterium trapped in the electro-optical trap, indicating longitudinal confinement within ±200 nm and a corresponding trap stiffness of $k=91$ nN/m. The inset shows the fluorescing *E. coli* as it is held in the electro-optical trap. Figure 16.17c, finally, depicts the time-dependent fluorescence signal from the *E. coli* (dark line). The labeling dye is photobleached within ~20 s, and as it becomes optically inactive the confinement along the channel is lost (light-gray line). This measurement shows the ability to take fluorescence measurements on single bioparticles using ultralow optical powers in a lithographically

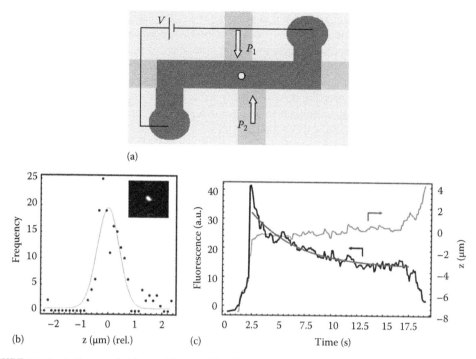

FIGURE 16.17 Active optofluidic particle trap. (a) Schematic view of a waveguide and the excitation geometry along with a feedback voltage, V, applied across fluidic reservoirs; (b) longitudinal particle position histogram showing submicron confinement of trapped *E. coli* (inset: fluorescing *E. coli* bacterium); and (c) time-dependent fluorescence of trapped *E. coli* (dark line) showing exponential decay due to photobleaching (fitted line), and the corresponding longitudinal position in the channel (light-gray line). (From Kühn, S. et al., *Lab Chip.*, (in press) 2009c.)

defined waveguide geometry. The well-known favorable size scaling properties of the anti-Brownian trap (Cohen and Moerner 2005, 2008) give reason to expect that single nanoparticles, such as labeled DNA or viruses, can be trapped and analyzed on this optofluidic chip.

16.4 Summary and Outlook

In this chapter, we have reviewed the current state of integrated optofluidic waveguides. A surprising variety of waveguiding approaches are being pursued to achieve an equally diverse set of tasks on optofluidic chips. All waveguide types need to deal with the problem of creating sufficiently large overlaps between optical fields and sample fluids in light of the inherently low refractive indices of aqueous liquids and gases. In addition to more conventional approaches using evanescent fields from solid-core waveguides, innovative approaches for both index guiding and wave interference guiding have been realized in practice. Despite the breadth of physical implementations and target applications, it is essential to note that LCWs in optofluidics are not merely a means of transporting light from one place on a chip to another. They, themselves become the optical device, defining the function, the optical excitation volume, particle trap potentials, etc. The possibility of dynamically changing waveguide fluids to completely alter the device characteristics, in particular, raises the importance of the waveguide beyond the level seen in conventional solid-state photonics, leaving room for many novel and creative uses of this seemingly simple element. We also saw that the use of LCWs mirrors the broad trends that are generally seen in optofluidics. A few examples were reviewed in more detail in each case to provide a glimpse into the possibilities.

As the field evolves, we can expect several developments to take place. Once basic device function-alities have been established more firmly in proof-of-principle experiments, an increasing amount of efforts will be devoted to incorporating these into larger-scale devices and systems, likely to eventually include integrated electronic circuitry for on-chip signal generation and analysis. A natural evolution would point toward the combination of more complex fluidic and integrated optical functions on the same chip. From the optics perspective, these will likely include on-chip light sources, detectors, optical signal processing (e.g., filtering), or multiple interrogation regions. A proper choice of the optical wave-guide for a particular task at hand will be essential, and we will likely see more hybrid approaches that combine different waveguide types for different purposes in the same device.

Finally, we expect nanotechnology to play a substantial role in integrated optofluidics. Slot waveguides are the first example of this philosophy. Other nanoscale elements can be added to achieve desired func-tionalities on a chip and enable the analysis of biological nanoparticles whose length scales provide a natural match for the feature dimensions. Examples of such nano-inspired optofluidics could include locally defined nanofluidic channels, nanopatterned gratings for optical filtering, and nanoscale gates (nanopores) for controlled particle entry into an optofluidic chip (Rudenko et al. 2007).

It is safe to conclude that waveguide-based optofluidics is still in its infancy, providing countless opportunities for creative research and the future commercialization of a new class of instruments for particle detection and analysis across a wide range of fields.

Acknowledgments

This work was enabled by support from the National Institutes of Health (grants R21EB003430, R01EB006097 and R21ER008802), the National Science Foundation (grant ECS-0528730), and the W.M. Keck Foundation through the W.M. Keck Center for Nanoscale Optofluidics at the University of California, Santa Cruz.

References

Agnarsson, B., Ingthorsson, S., Gudjonsson, T., and Leosson, K. 2009. Evanescent-wave fluorescence microscopy using symmetric planar waveguides. *Opt. Express* 17:5075–5082.

Almeida, V.R., Xu, Q., Barrios, C.A., and Lipson, M. 2004. Guiding and confining light in void nanostruc-tures. *Opt. Lett.* 29:1209–1211.

Archambault, J.L., Black, R.J., Lacroix, S., and Bures, J. 1993. Loss calculations for antiresonant wave-guides. *J. Lightwave Technol.* 11:416–423.

Ashkin, A. 1970. Acceleration and trapping of particles by radiation pressure. *Phys. Rev. Lett.* 24:156–159.

Barber, J.P., Conkey, D.B., Lee, J.R., Hubbard, N.B., Howell, L.L., Yin, D., Schmidt, H., and Hawkins, A.R. 2005. Fabrication of hollow waveguides with sacrificial aluminum cores. *IEEE Photon. Technol. Lett.* 17:363–365.

Barber, J.P., Lunt, E.J., George, Z., Yin, D., Schmidt, H., and Hawkins, A.R. 2006. Integrated hollow wave-guides with arch-shaped cores. *IEEE Photon. Technol. Lett.* 18:28–30.

Berglund, A. and Mabuchi, H. 2005. Tracking-FCS: Fluorescence correlation spectroscopy of individual particles. *Opt. Express* 13:8069–8082.

Bernini, R., Campopiano, S., Zeni, L., and Sarro, P.M. 2004. ARROW optical waveguides based sensors. *Sens. Actuators B* 100:143–146.

Bernini, R., De Nuccio, E., Brescia, F., Minardo, A., Zeni, L., Sarro, P.M., Palumbo, R., and Scarfi, M.R. 2006. Development and characterization of an integrated silicon micro flow cytometer. *Anal. Bioanal. Chem.* 386:1267–1272.

Bernini, R., DeNuccio, E., Minardo, A., Zeni, L., and Sarro, P.M. 2007. Integrated silicon optical sensors based on hollow core waveguide. *Proc. SPIE* 6477: 647714.

Bernini, R., DeNuccio, E., Minardo, A., Zeni, L., and Sarro, P.M. 2008a. Liquid-core/liquid-cladding integrated silicon ARROW waveguides. *Opt. Commun.* 281:2062–2066.

Bernini, R., Testa, G., Zeni, L., and Sarro, P.M. 2008b. Integrated optofluidic Mach-Zehnder interferometer based on liquid core waveguides. *Appl. Phys. Lett.* 93:011106.

Campopiano, S., Bernini, R., Zeni, L., and Sarro, P.M. 2004. Microfluidic sensor based on integrated optical hollow waveguides. *Opt. Lett.* 29:1894–1896.

Canning, J., Buckley, E., and Lyytikainen, K. 2003. Propagation in air by field superposition of scattered light within a Fresnel fiber. *Opt. Lett.* 28:230–232.

Cho, S.H., Godin, J., and Lo, Y.-H. 2009. Optofluidic waveguides in teflon AF-coated PDMS microfluidic channels. *IEEE Photon. Technol. Lett.* 21:1057–1059.

Cohen, A.E. and Moerner, W.E. 2005. Suppressing Brownian motion of individual biomolecules in solution. *PNAS* 103:4362–4365.

Cohen, A.E. and Moerner, W.E. 2008. Controlling Brownian motion of single protein molecules and single fluorophores in aqueous buffer. *Opt. Express* 16:6941–6956.

Coldren, L.A. and Corzine, S.W. 1995. *Diode Lasers and Photonic Integrated Circuits*, 1st edn, Wiley Interscience, New York.

Cordeiro, C.M.B., de Matos, C.J.S., dos Santos, E.M., Bozolan, A., Ong, J.S.K., Facincani, T., Chesini, G., Vaz, A.R., and Brito Cruz, C.H. 2007. Towards practical liquid and gas sensing with photonic crystal fibres: Side access to the fibre microstructure and single-mode liquid-core fiber. *Meas. Sci. Technol.* 18:3075–3081.

Cran-McGreehin, S., Krauss, T.F., and Dholakia, K. 2006. Integrated monolithic optical manipulation. *Lab Chip* 6:1122–1124.

Cregan, R.F., Mangan, B.J., Knight, J.C., Birks, T.A., Russell, P.S.J., Roberts, P.J., and Allan, D.C. 1999. Single-mode photonic band gap guidance of light in air. *Science* 285:1537–1539.

Datta, A., Eom, I., Dhar, A., Kuban, P., Manor, R., Ahmad, I., Gangopadhyay, S., Dallas, T., Holtz, M., Temkin, H., and Dasgupta, P. 2003. Microfabrication and characterization of teflon AF-coated liquid core waveguide channels in silicon. *IEEE Sens. J.* 3:788–795.

Decca, R.S., Lee, C.-W., Lall, S., and Wassall, S.R. 2002. Single molecule tracking scheme using a near-field scanning optical microscope. *Rev. Sci. Instrum.* 73:2675–2679.

Delonge, T. and Fouckhardt, H. 1995. Integrated optical detection cell based on Bragg reflecting waveguides. *J. Chromatogr. A* 716:135–139.

Duguay, M.A., Kokubun, Y., Koch, T., and Pfeiffer, L. 1986. Antiresonant reflecting optical waveguides in SiO_2-Si multilayer structures. *Appl. Phys. Lett.* 49:13–15.

Eid, J. et al. 2009. Real-time DNA sequencing from single polymerase molecules. *Science* 323:133–138.

Enderlein, J. 2000. Tracking of fluorescent molecules diffusing within membranes. *Appl. Phys. B* 71:773–777.

Erickson, D., Rockwood, T., Emery, T., Scherer, A., and Psaltis, D. 2006. Nanofluidic tuning of photonic crystal circuits. *Opt. Lett.* 31:59–61.

Fink, Y., Ripin, D.J., Fan, S., Chen, C., Joannopoulos, J.D., and Thomas, E.L. 1999. Guiding optical light in air using an all-dielectric structure. *IEEE J. Lightwave Technol.* 17:2039–2041.

Foquet, M., Korlach, J., Zipfel, W.R., Webb, W.W., and Craighead, H.G. 2002. DNA fragment sizing by single molecule detection in submicrometer-sized closed fluidic channels. *Anal. Chem.* 74:1415–1422.

Foquet, M., Korlach, J., Zipfel, W.R., Webb, W.W., and Craighead, H.G. 2004. Focal volume confinement by submicrometer-sized fluidic channels. *Anal. Chem.* 76:1618–1626.

Grewe, M., Grosse, A., and Fouckhardt, H. 2000. Theoretical and experimental investigations of the optical waveguiding properties of on-chip microfabricated capillaries. *Appl. Phys. B* 70:S839–S847.

Grujic, K. and Hellesø, O.G. 2007. Dielectric microsphere manipulation and chain assembly by counter-propagating waves in a channel waveguide. *Opt. Express* 15:6470–6477.

Hakanson, U., Measor, P., Yin, D., Lunt, E., Hawkins, A.R., Sandoghdar, V., and Schmidt, H. 2007. Tailoring the transmission of liquid-core waveguides for wavelength filtering on a chip. *Proc. SPIE* 6477:647715.

Homola, J., Yee, S.S., and Gauglitz, G. 1999. Surface plasmon resonance sensors: Review. *Sens. Actuators B* 54:3–15.

Hong, K. and Burgess, L.W. 1994. Liquid-core waveguides for chemical sensing. *Proc. SPIE* 2293:71–79.

Horvath, R., Lindvold, L.R., and Larsen, N.B. 2002. Reverse-symmetry waveguides: Theory and fabrication. *Appl. Phys. B* 74:383–393.

Hubbard, N.B., Howell, L.L., Barber, J.P., Conkey, D.B., Hawkins, A.R., and Schmidt, H. 2005. Mechanical models and design rules for on-chip micro-channels with sacrificial cores. *J. Micromech. Microeng.* 15:720.

Hunsperger, R.G. 2002. *Integrated Optics*, 5th edn, Springer, Berlin, Germany.

Jackson, J.D. 1998. *Classical Electrodynamics*, 3rd edn, Wiley, New York.

Joannopoulos, J.D., Meade, R.D., and Winn, J.N. 1995. *Photonic Crystals: Molding the Flow of Light*, Princeton University Press, Princeton, NJ.

Keller, B.K., DeGrandpre, M.D., and Palmer, C.P. 2007. Waveguiding properties of fiber-optic capillaries for chemical sensing applications. *Sens. Actuators B* 125:360–371.

Kiesel, P., Bassler, M., Beck, M., and Johnson, N.M. 2009. Spatially modulated fluorescence emission from moving particles. *Appl. Phys. Lett.* 94:041107.

Kühn, S., Measor, P., Lunt, E.J., Phillips, B.S., Deamer, D.W., Hawkins, A.R., and Schmidt, H. 2009a. Loss-based optical trap for on-chip particle analysis. *Lab Chip* 9:2212–2216.

Kühn, S., Lunt, E.J., Phillips, B.S., Hawkins, A.R., and Schmidt, H. 2009b. Optofluidic particle concentration by a long-range dual-beam trap. *Opt. Lett.* 34:2306–2308.

Kühn, S., Lunt, E.J., Phillips, B.S., Hawkins, A.R., and Schmidt, H. 2009c. Ultralow power trapping and fluorescence detection of single particles on an optofluidic chip. *Lab Chip*, DOI: 10.1039/B915750F.

Lenne, P.F., Etienne, E., and Rigneault, H. 2002. Subwavelength patterns and high detection efficiency in fluorescence correlation spectroscopy using photonic structures. *Appl. Phys. Lett.* 80:4106–4108.

Levene, M.J., Korlach, J., Turner, S.W., Fouquet, M., Craighead, H.G., and Webb, W.W. 2003. Zero-mode waveguides for single-molecule analysis at high concentrations. *Science* 299:682–686.

Lim, J., Kim, S., Choi, J., and Yang, S. 2008. Fluorescent liquid-core/air-cladding waveguides towards integrated optofluidic light sources. *Lab Chip* 8:1580–1585.

Loncar, M., Nedeljkovic, D., Doll, T., Vuckovic, J., Scherer, A., and Pearsall, T.P. 2000. Waveguiding in planar photonic crystals. *Appl. Phys. Lett.* 77:1937–1939.

Lunt, E.J., Measor, P., Phillips, B.S., Kühn, S., Schmidt, H., and Hawkins, A.R. 2008. Improving solid to hollow core transmission for integrated ARROW waveguides. *Opt. Express* 16:20981–20986.

Mach, P., Dolinski, M., Baldwin, K.W., Rogers, J.A., Kerbage, C., Windeler, R.S., and Eggleton, B.J. 2002. Tunable microfluidic optical fiber. *Appl. Phys. Lett.* 80:4294–4296.

Mandal, S. and Erickson, D. 2007. Optofluidic transport in liquid core waveguiding structures. *Appl. Phys. Lett.* 90:184103.

Mandal, S. and Erickson, D. 2008. Nanoscale optofluidic sensor arrays. *Opt. Express* 16:1623–1631.

Marcatili, E.A.J. and Schmeltzer, R.A. 1964. Hallow metallic and dielectric waveguides for long distance optical transmission and lasers. *Bell Syst. Tech. J.* 43:1783–1809.

Martelli, C., Canning, J., Lyytikainen, K., and Groothoff, N. 2005. Water-core Fresnel fiber. *Opt. Express* 13:3890–3895.

Mawst, L.J., Botez, D., Zmudzinski, C., and Tu, C. 1992. Design optimization of ARROW-type diode lasers. *IEEE Photon. Technol. Lett.* 4:1204–1206.

McNab, S., Moll, N., and Vlasov, Y. 2003. Ultra-low loss photonic integrated circuit with membrane-type photonic crystal waveguides. *Opt. Express* 11:2927–2939.

Measor, P., Lunt, E.J., Seballos, L., Yin, D., Zhang, J.Z., Hawkins, A.R., and Schmidt, H. 2007. On-chip Surface-enhanced Raman scattering (SERS) detection using integrated liquid-core waveguides. *Appl. Phys. Lett.* 90:211107.

Measor, P., Kühn, S., Lunt, E.J., Phillips, B.S., Hawkins, A.R., and Schmidt, H. 2008. Hollow-core waveguide characterization by optically induced particle transport. *Opt. Lett.* 33:672–674.

Mohebbi, M., Fedosejevs, R., Gopal, V., and Harrington, J.A. 2002. Silver-coated hollow-glass waveguide for applications at 800 nm. *Appl. Opt.* 41:7031–7035.

Patterson, S.G., Petrich, G.S., Ram, R.J., and Kolodiejski, L. 1999. Continuous-wave room temperature operation of bipolar cascade laser. *Electron. Lett.* 35:397–397.

Payne, D.N. and Gambling, W.A. 1972. New low-loss liquid-core fibre waveguide. *Electron. Lett.* 8:374–376.

Preston, K. and Lipson, M. 2009. Slot waveguides with polycrystalline silicon for electrical injection. *Opt. Express* 17:1527–1532.

Psaltis, D., Quake, S.R., and Yang, C. 2006. Developing optofluidic technology through the fusion of microfluidics and optics. *Nature* 442:381–386.

Rabus, D.G., DeLouise, L.A., and Ichihashi, Y. 2007. Enhancement of the evanescent field using polymer waveguides fabricated by deep UV exposure on mesoporous silicon. *Opt. Lett.* 32:2843–2845.

Risk, W.P., Kim, H.C., Miller, R.D., Temkin, H., and Gangopadhyay, S. 2004. Optical waveguides with an aqueous core and a low-index nanoporous cladding. *Opt. Express* 12:6446–6455.

Rudenko, M.I., Yin, D., Holmes, M., Hawkins, A.R., and Schmidt, H. 2007. Integration and characterization of SiN nanopores for single-molecule detection in liquid-core ARROW waveguides. *Proc. SPIE* 6444:64440L.

Rudenko, M.I., Kühn, S., Lunt, E.J., Deamer, D.W., Hawkins, A.R., and Schmidt, H. 2009. Ultrasensitive Qβ phage analysis using fluorescence correlation spectroscopy on an optofluidic chip. *Biosens. Bioelectron.*, 24:3258–3263.

Russell, P. 2003. Photonic crystal fiber. *Science* 299:358–362.

Saleh, B.E.A. and Teich, M.C. 2007. *Fundamentals of Photonics*, Wiley, Hoboken, NJ.

Samiee, K.T., Foquet, M., Guo, L., Cox, E.C., and Craighead, H.C. 2005. 1-Repressor oligomerization kinetics at high concentrations using fluorescence correlation spectroscopy in zero-mode waveguides. *Biophys. J.* 88:2145–2153.

Schelle, B., Dreß, P., Franke, H., Klein, K.F., and Slupek, J. 1999. Physical characterization of lightguide capillary cells. *J. Phys. D: Appl. Phys.* 32:3157–3163.

Schmidt, H. and Hawkins, A.R. 2008. Optofluidic waveguides: I. Concepts and implementations. *Microfluid. Nanofluid.* 4:3–16.

Schmidt, H., Yin, D., Barber, J.P., and Hawkins, A.R. 2005. Hollow-core waveguides and 2D waveguide arrays for integrated optics of gases and liquids. *IEEE J. Sel. Top. Quant. Electron.* 11:519–527.

Schmidt, O., Bassler, M., Kiesel, P., Johnson, N.M., and Doehler, G.H. 2006. Guiding light in fluids. *Appl. Phys. Lett.* 88:151109.

Schmidt, B.S., Yang, A.H., Erickson, D., and Lipson, M. 2007a. Optofluidic trapping and transport on solid core waveguides within a microfluidic device. *Opt. Express* 15:14322–14334.

Schmidt, O., Bassler, M., Kiesel, P., Knollenberg, C., and Johnson, N. 2007b. Fluorescence spectrometer-on-a-fluidic-chip. *Lab Chip* 7:626–629.

Sharma, A.K., Jha, R., and Gupta, B.D. 2007. Fiber-optic sensors based on surface plasmon resonance: A comprehensive review. *IEEE Sens. J.* 7:1118–1129.

Stone, J. 1972. Optical transmission in liquid-core quartz fibers. *Appl. Phys. Lett.* 20:239–240.

Sumetsky, M., Windeler, R.S., Dulashko, Y., and Fan, X. 2007. Optical liquid ring resonator sensor. *Opt. Express* 15:14376–14381.

Takiguchi, H., Odake, T., Ozaki, M., Umemura, T., and Tsunoda, K. 2003. Liquid/liquid optical waveguides using sheath flow as a new tool for liquid/liquid interfacial measurements. *Appl. Spectrosc.* 57:1039–1041.

Tanaka, T. and Yamamoto, S. 2000. Optically induced propulsion of small particles in an evanescent field of higher propagation mode in a multimode, channeled waveguide. *Appl. Phys. Lett.* 77:3131–3133.

Temelkuran, B., Hart, S.D., Benoit, G., Joannopoulos, J.D., and Fink, Y. 2002. Wavelength-scalable hollow optical fibers with large photonic bandgaps for CO_2 laser transmission. *Nature* 420:650–653.

Vazquez, R.M., Osellame, R., Nolli, D., Dongre, C., van den Vlekkert, H., Ramponi, R., Pollnau, M., and Cerullo, G. 2009. Integration of femtosecond laser written optical waveguides in a lab-on-chip. *Lab Chip* 9:91–96.

Wang, T., Aiken, J.H., Huie, C.W., and Hartwick, R.A. 1991. Nanoliter-scale multireflection cell for absorption detection in capillary electrophoresis. *Anal. Chem.* 63:1372–1376.

Watzinger, F., Ebner, K., and Lion, T. 2006. Detection and monitoring of virus infections by real-time PCR. *Mol. Aspects Med.* 27:254–298.

White, I.M., Oveys, H., and Fan, X. 2006a. Liquid-core optical ring-resonator sensors. *Opt. Lett.* 31:1319–1321.

White, I.M., Oveys, H., and Fan, X. 2006b. Integrated multiplexed biosensors based on liquid core optical ring resonators and antiresonant reflecting optical waveguides. *Appl. Phys. Lett.* 89:191106.

Winn, J.N., Fink, J., Fan, S., and Joannopoulos, J.D. 1998. Omnidirectional reflection from a one-dimensional photonic crystal. *Opt. Lett.* 23:1573–1575.

Wolfe, D.B., Conroy, R.S., Garstecki, P., Mayers, B.T., Fischbach, M.A., Paul, K.E., Prentiss, M., and Whitesides, G.M. 2004. Dynamic control of liquid-core/liquid-cladding optical waveguides. *PNAS* 101:12434–12438.

Xu, Y., Lee, R.K., and Yariv, A. 2000. Asymptotic analysis of Bragg fiber. *Opt. Lett.* 25:1756–1758.

Xu, Q., Almeida, V.R., Panepucci, R.R., and Lipson, M. 2004. Experimental demonstration of guiding and confining light in nanometer-size low-refractive-index material. *Opt. Lett.* 29:1626–1628.

Yan, H., Gu, C., Yang, C., Liu, J., Jin,G., Zhang, J., Hou, L., and Yao, Y. 2006. Hollow core photonic crystal fiber surface-enhanced Raman probe. *Appl. Phys. Lett.* 89:204101.

Yang, A.H.J., Moore, S.D., Schmidt, B.S., Klug, M., Lipson, M., and Erickson, D. 2009a. Optical manipulation of nanoparticles and biomolecules in sub-wavelength slot waveguides. *Nature* 457:71–75.

Yang, A.H.J., Lerdsuchatawanich, T., and Erickson, D. 2009b. Forces and transport velocities for a particle in a slot waveguide. *Nano Lett.* 9:1182–1188.

Yeh, P. 2005. *Optical Waves in Layered Media*, 2nd edn, Wiley-Interscience, New York.

Yeh, P. and Yariv, A. 1976. Bragg reflection waveguides. *Opt. Commun.* 19:427–430.

Yeh, P., Yariv, A., and Hong, C. 1977. Electromagnetic propagation in periodic stratified media. I. General theory. *J. Opt. Soc. Am.* 67:423–438.

Yeh, P., Yariv, A., and Maron, E. 1978. Theory of Bragg fiber. *J. Opt. Soc. Am.* 68:1196–1201.

Yin, D., Barber, J.P., Hawkins, A.R., and Schmidt, H. 2004a. Integrated ARROW waveguides with hollow cores. *Opt. Express* 12:2710–2715.

Yin, D., Barber, J.P., Hawkins, A.R., Deamer, D.W., and Schmidt, H. 2004b. Integrated optical waveguides with liquid cores. *Appl. Phys. Lett.* 85:3477–3479.

Yin, D., Barber, J.P., Lunt, E.J., Hawkins, A.R., and Schmidt, H. 2005a. Optical characterization of arch-shaped ARROW waveguides with liquid cores. *Opt. Express* 13:10564–10569.

Yin, D., Barber, J.P., Hawkins, A.R., and Schmidt, H. 2005b. Waveguide loss optimization in hollow-core ARROW waveguides. *Opt. Express* 13:9331–9336.

Yin, D., Barber, J.P., Hawkins, A.R., and Schmidt, H. 2005c. Highly efficient fluorescence detection in picoliter volume liquid-core waveguides. *Appl. Phys. Lett.* 87:211111.

Yin, D., Barber, J.P., Deamer, D.W., Hawkins, A.R., and Schmidt, H. 2006. Single-molecule detection using planar integrated optics on a chip. *Opt. Lett.* 31:2136–2138.

Yin, D., Lunt, E.J., Barman, A., Hawkins, A.R., and Schmidt, H. 2007a. Microphotonic control of single molecule fluorescence correlation spectroscopy using planar optofluidics. *Opt. Express* 15:7290–7295.

Yin, D., Lunt, E.J., Rudenko, M.I., Deamer, D.W., Hawkins, A.R., and Schmidt, H. 2007b. Planar optofluidic chip for single particle detection, manipulation, and analysis. *Lab Chip* 7:1171–1175.

17

Raman Detection in Microchips and Microchannels

Melodie Benford
Gerard L. Coté
Jun Kameoka
Miao Wang

17.1 Introduction

In this chapter, we describe the background of Raman spectroscopy, surface-enhanced Raman spectroscopy (SERS), and the development of optofluidic systems toward producing more robust, Raman spectra. Further, applications for the device are described such as the detection of small molecules, the detection of beta amyloid for Alzheimer's disease, the ability to detect the denaturation of proteins, and for use as a point-of-care (POC) diagnostic for acute coronary syndrome (ACS) through the detection of cardiac biomarkers, including, for example, cardiac troponin (cTn). In addition, as one specific example, a new optofluidic platform and methodology is described for early detection and characterization of analytes that could be useful in sensing and diagnosis. This extremely sensitive SERS nanochannel-based optofluidic system is being developed to provide enhancements equal to or exceeding 10^{10} for small molecules at the entrance to a nanochannel, enabling the rapid quantification of femtomolar or smaller levels of targets in fluids.

17.2 Background

17.2.1 Raman Spectroscopy

17.2.1.1 Limitations of Classical Raman Spectroscopy

Raman spectroscopy has been discussed in detail in Chapter 6. Here, the main characteristics of Raman spectroscopy related to the optofluidics approach are summarized. Raman spectroscopy, in general, is the inelastic scattering of a photon from a molecule. The corollary, elastic scattering of a photon from a molecule (i.e., Rayleigh scattering), merely depresses the electron cloud surrounding the molecule, and no energy is lost. In other words, the Rayleigh-scattered light is scattered around at the same color whereas Raman-scattered light, because of energy exchange, is scattered at a different color. If you have ever seen a laser point scatter through your finger you would know that the bulk of the light scattered is elastic at the same color. This is because Raman scattering is an inefficient process and only occurs for 1 in 10^7 scattering events. It perturbs the electric cloud surrounding the molecule, resulting in an energy loss (Figure 17.1). Although Raman scattering is weak in comparison to other processes, such as Rayleigh scattering and fluorescence, the resulting Raman signal is unique to the molecule of interest and exhibits sharp characteristic peaks, whereas fluorescence produces broad peaks. Furthermore, unlike fluorescence, the wavelength of Raman-scattered light is independent of the wavelength of incident light. These sharp bands have applications in not only producing a unique signature spectra that can potentially describe the prominent bonds or chemical makeup, but can also be used for multiplexing signals for detecting multiple, individually labeled analytes (Carey 1982, Tu 1982, Nie and Emory 1997, Kneipp et al. 2002, Beier et al. 2007).

Raman scattering can occur with a change in vibrational, rotational, or electronic energy of a molecule. The energy of a vibrational mode depends on the molecular structure, environment, bond order, molecular substituents, molecular geometry, and hydrogen bonding. Raman scattering is reported in wavenumbers (cm^{-1}) as the difference in energy between incident photon (i.e., incident laser light used to excite the sample, $\lambda_{\text{incident}}$) and the Raman-scattered photon ($\lambda_{\text{scattered}}$). This Raman shift is the energy, or mode, of vibration of the molecule (υ):

$$\bar{\nu} = \frac{1}{\lambda_{\text{incident}}} - \frac{1}{\lambda_{\text{scattered}}}.$$

Raman spectra consist of an exciting line that correlates to the incident laser light, together with weaker lines on either side, symmetrical about the exciting lines. The frequently reported lines are the Stokes lines that are on the low-frequency side of the exciting line. Stokes lines have a higher intensity than anti-Stokes lines (the lines that occur on the other side). The intensity of the Raman band depends on the molecule's polarizability (α), or the ease with which the electron cloud around a molecule can be

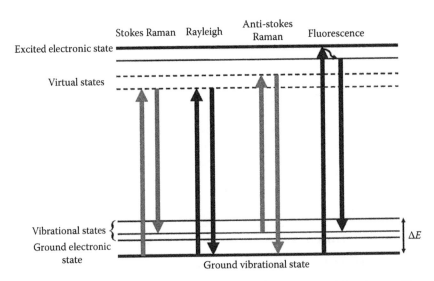

FIGURE 17.1 Jablonksi diagram comparing Raman and Rayleigh scattering and fluorescence. The molecule is excited to a virtual state lower in energy than a real electronic transition.

distorted. The location of the bands depends on the Raman selection rules, and the intensity depends on the strength-induced dipole moment (P) by the electric field (E) of the incident light, with $P=\alpha E$. As a result, Raman signals are partially polarized, which can be seen best with a plane-polarized exciting source in an isotropic medium because the induced electric dipole has components that vary spatially with respect to the coordinates of the molecule. One can calculate the energy levels and selection rules using principles of quantum mechanics, such as Schrodinger's equation and wave mechanics, along with matrix mechanics and operator calculus. The induced dipole emits or scatters light at the frequency of the incident light wave, with Raman scattering occurring due to a change in the polarizability by a molecular vibration. The scattering intensity is proportional to the square of the induced dipole moment. Therefore, if a vibration does not greatly change the polarizability, then the intensity of the Raman band will be low (Wilson et al. 1955, Tu 1982).

Further, Raman spectroscopy and infrared (IR) absorption spectroscopy provide complementary information, although the mechanism of the processes differs (Raman is a scattering process, whereas IR relies on absorption). Raman spectroscopy provides the following advantages over IR spectroscopy: (1) Raman bands are sharper than IR bands for biomolecules in the near infrared, (2) Raman spectra can be obtained for molecules in any form, whereas with IR large interference from water-bending modes occurs and IR measurements require complex preparation techniques, (3) due to the difference in mechanism, Raman has more potential to elucidate molecular properties than IR spectroscopy (Tu 1982).

17.2.1.2 Surface-Enhanced Raman Spectroscopy

The progress of Raman spectroscopy was slow due to the expensive and complex equipment needed to detect such a low signal at trace levels. The discovery of surface-enhanced Raman scattering (SERS) by Fleischmann et al. in 1974 helped Raman spectroscopy gain ground (Fleischmann et al. 1974). At first, Fleischmann postulated that the intense Raman scattering observed from pyridine adsorbed onto a roughened silver electrode surface was due to the greater available surface area for molecules to adsorb. After more investigation and experiments, researchers concluded that increasing the number of scatterers could not account for these enhancements and proposed that these enhancements occurred due to the molecule's adsorbed state. SERS processes can enhance Raman signals by 10^{6}–10^{14}. The qualifications simply require the molecule to be adsorbed to a roughened metal surface. This simple requirement has sparked a myriad of research on viable metal substrates in various shapes and sizes, exhibiting different dielectric properties (Fleischmann et al. 1974, Campion and Kambhampati 1998, Kneipp et al. 2002).

There are two mechanisms thought to be responsible for SERS processes: (1) electromagnetic and (2) chemical. The first mechanism, the best understood and most likely the dominant mechanism, proposes that due to the interaction of light with the metal, enhanced electromagnetic (EM) fields are induced (Figure 17.2). Surface plasmons are collective, dipole electron oscillations that occur in a metal due to the delocalized "sea of electrons." The resonance of surface plasmons amplifies electric fields upon interaction with photons (i.e., incident light). The properties of the metal surface affect plasmon resonance frequency. Therefore, the electromagnetic SERS enhancements are due to the interaction between plasmons of colloidal metal structures, analyte molecules, and incident laser light (Campion and Kambhampati 1998, Stewart and Fredericks 1999, Kneipp et al. 2002, Kelly et al. 2003, Zhang et al. 2006, Aaron et al. 2007, Campbell and Xia 2007).

The proposed chemical mechanism involves the formation of a charge-transfer complex between the surface and analyte molecule. The induced EM field broadens the molecular electronic states by the molecule's interaction with the metal surface. The electronic transitions of many charge-transfer complexes are in the visible range so that resonance enhancement can occur. Researchers are developing controlled nanoscale metal topologies that can delineate the role of electromagnetic and chemical enhancements in the SERS response of adsorbed molecules (Campion and Kambhampati 1998, Kneipp et al. 1998, Kneipp et al. 2002).

SERS differs from the classical Raman scattering in the following aspects: the intensities of bands observed generally fall off with increasing vibrational frequency. Certain classical Raman frequencies

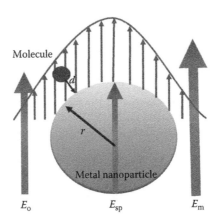

FIGURE 17.2 Induced oscillations of plasmons enhance the local electromagnetic field felt by the molecule, resulting in Raman modes with enhanced intensity. The radius of the metal particle, r, must be less than the wavelength of the incident light. The molecule experiences an enhanced field, E_m, which is a combination of the field of the incident light, E_o, and the induced dipole in the metal nanoparticle, E_{sp}. (Adapted from Campbell, D.J. and Xia, Y., *J. Chem. Educ.*, 84, 91, 2007.)

(such as overtones and combination bands) are not common. The selection rules are relaxed, resulting in the appearance of normally forbidden Raman modes in the spectra. In addition, the SERS spectra tend to be completely depolarized. The bands may be characteristic of the substrate, the adsorbed molecule, or the combined system. Furthermore, the enhancement may be remarkably long ranged, extending tens of nanometers from the surface, depending on substrate morphology. Therefore, the SERS modes have a dependence on the substrate, molecule of interest, and the interactions between the two, along with incident light, whereas traditional Raman signals depend solely on the molecule and the incident light (Campion and Kambhampati 1998).

Noble metals, such as silver, gold, and copper, are the dominant SERS substrates, but work has been reported on some alkali metals. Historically, techniques such as roughening the metal surface or depositing metal nanoparticles on aggregate films have been successfully used as substrates. Further, Raman enhancements observed in nanoparticle aggregates have been attributed to the intense fields of highly localized surface plasmons, or "hot spots," which appear to provide enhancements of the Raman emission from molecules located specifically in those regions of nanoparticle aggregates. Advances in theory and computational techniques have shown that a strong EM field can be invoked between solid nanoparticles, such as colloids, and at sharp boundaries of nanoscale geometries, such as on gratings and island films. Hence, researchers have also developed well-ordered arrays of geometric, nanoscale features on a substrate that can provide amplified EM fields on these surfaces. The largest enhancements occur on roughened surfaces on the scale of 10–100 nm. These include electrode surfaces roughened by one or more oxidation–reduction cycles, island films deposited on glass surfaces at elevated temperatures, films deposited by evaporation or sputtering in a vacuum onto cold substrates, colloids (especially aggregated colloids), single ellipsoidal nanoparticles, and arrays of such particles produced by lithographic techniques. Various designs such as stars, cubes, spheres, triangles, rods, and nanoshells with tunable geometries that affect the magnitude of the local EM field at the nanoparticle surface are also in use. Due to the distance dependency of SERS, attracting and adsorbing the molecule is an area of interest. Much work has been performed to functionalize the metal surfaces with specific chemistries, such as self-assembled monolayers on the roughened metal surface or attaching chemistries to the various nanoparticle designs (Westcott et al. 1998, Michaels et al. 2000, Cao et al. 2002a, Lu et al. 2002, Chiñas-Castillo and Spikes 2003, Culha et al. 2003, Jackson et al. 2003, Kelly et al. 2003, Haes et al. 2004, Mandal et al. 2004a,b, Akarca-Biyikli et al. 2005, Cao et al. 2006, Zhang et al. 2006, Zhixun and Yan 2006, Aaron et al. 2007, Beier et al. 2007, Bonham et al. 2007, Xiaohua Huang et al. 2007, Chou et al. 2008).

Overall, the strength of the induced electromagnetic field has a strong spatial dependence. The Raman enhancement is strongest when the surface plasmon frequency is in resonance with both the excitation and Raman-scattered fields. Plus, the enhancement is dependent on the local nanostructure field of the metal. These strong spatial and local field dependencies are likely responsible for the small, random, localized "hot spots" of EM enhancement known to occur with SERS. Although the potential for trace analyte detection is elevated at these "hot spots," the production of reproducible SERS surfaces with known "hot spot" locations is nearly impossible, especially on a planar surface. The SERS signal intensity varies, depending on size, shape, and aggregation behavior of the colloid or the roughened metal surface. Thus, SERS has problems with reproducibility of the signal enhancement within and between Raman techniques. Many substrate materials and designs have been implemented, as described earlier, to combat these shortcomings but the ability to reproducibly create SERS "hot spots" remains a challenge (Campion and Kambhampati 1998, Kneipp et al. 2002, Beier et al. 2007, Chou et al. 2008).

17.2.2 Optofluidics and Raman Spectroscopy

As mentioned previously, because of the nonuniform distribution of hot spots, it is especially challenging to obtain controllable and consistent enhancement. This inconsistent enhancement is one of the reasons that SERS could not provide reliable and reproducible results in the past. In addition, target molecules are randomly adsorbed on the nanoparticle clusters. This means that the probability of confining target molecules in a hot spot is low. These problems have limited the expansion of SERS applications. To improve the SERS technique, there are some new investigations underway, as described earlier. However, in recent years, there are some reports of integrating optofluidic devices with SERS detection systems. White et al. have developed an optofluidic ring resonator (OFRR) platform (White et al. 2007). By resonating with whispering gallery modes (WGMs), a high-intensity evanescent field is obtained and acts as an enhanced SERS excitation source. The schematic diagram of this setup is shown in Figure 17.3a. With this approach, a detection limit of 400 pM for rhodamine 6G (R6G) has been demonstrated. Measor et al. reported an interconnected solid- and liquid-core antiresonant reflecting optical waveguide (ARROW) for SERS detection (Measor et al. 2007). The schematic diagram of this setup is shown in Figure 17.3d. It also utilizes high excitation intensities in the waveguide to improve the detection limit, with nanomolar sensitivity. In addition to enhanced excitation fields, microfluidic systems were also employed for integrating the preparation, mixing, and detection of the sample on a single chip (Keir et al. 2002, Docherty et al. 2004, Park et al. 2005). Park et al. demonstrated using a PDMS microfluidic device with an alligator-teeth structure to mix silver colloids with analyte solution (Park et al. 2005). The schematic diagram of the alligator-teeth device is shown in Figure 17.3e. This microfluidic system has been applied to analyze duplex dye-labeled DNA oligonucleotides with confocal SERS. The limit of detection was claimed to be 10 pM for a duplex oligomer mixture. Keir et al. reported a microfluidic chip on glass for surface-enhanced resonance Raman scattering (SERRS) (Keir et al. 2002). A stream of silver colloids was formed in the channel by mixing an aqueous solution of silver nitrate and a solution of sodium borohydride in sodium hydroxide solution on the chip. A dye derived from trinitrotoluene (TNT) was used as the analyte and introduced at a downstream junction for SERS detection. A detection limit of 10 fM was obtained. Also integrated with SERRS, Docherty et al. developed a PDMS microfluidic device and successfully detected three dye-labeled oligonucleotides simultaneously (Docherty et al. 2004). Researchers also investigated integrating SERS-active metal nanostructures with microfluidics. For example, nanowells were created on PDMS by soft lithography and deposited with Ag thin film to form a "hot spot" (Liu and Lee 2005). It was then bonded with a glass-based microfluidic chip. The microfluidic chip enabled easy sample delivery and a transparent detection site for SERS. Compared to the smooth Ag layer on PDMS, the detected SERS signal from this microfluidic-nanowell platform was 10^7 times higher. Hunt and Wilkinson and Chen and Choo provide reviews of optofluidic devices for SERS (Hunt and Wilkinson 2008, Chen and Choo 2008).

A novel optofluidics that utilizes the SERS technique to detect small concentrations of molecules will be described. This device has a pinched micro-nanochannel step structure where metallic nanoparticles

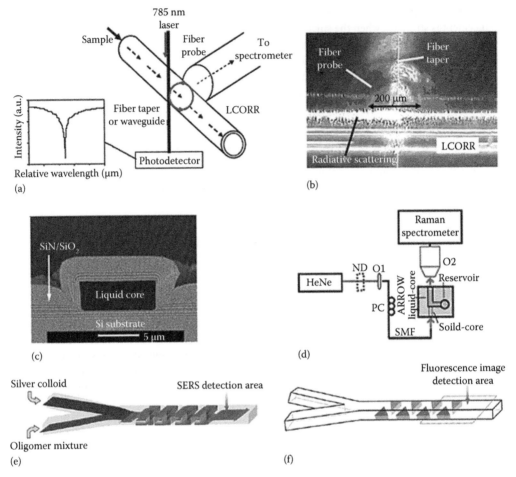

FIGURE 17.3 (a) The schematic diagram of experiment setup for measuring the Raman signal from Rhodamine 6G using Whisper Gallery mode. (b) Snapshot of the real system with fiber probe and capillary tube. (From White, I.M. et al., *Opt. Express*, 15, 17433, 2007. With permission from Optical Society of America.) (c) Cross-sectional SEM image of ARROW waveguide used for measuring Raman signals of Rhodamine 6G. (Reprinted from Measor, P. et al., *Appl. Phys. Lett.*, 90, 211107, 2007. With permission from American Institute of Physics.) (d) Schematic diagram of ARROW experimental setup. (Reprinted from Measor, P. et al., *Appl. Phys. Lett.*, 90, 211107, 2007. With permission from American Institute of Physics.) (e,f) Schematic diagram of alligator teeth-shaped micromixer. Silver colloid and oligomer mixtures are blended because alligator teeth-shaped mixed and SERS signals were detected at the end of channel. (From Park, T. et al., *Lab Chip*, 5, 437, 2005. With permission from The Royal Society of Chemistry.)

and target molecules in aqueous solution can be consistently trapped to form nanoparticle-molecule SERS clusters using capillary force in the channel, as described next.

17.3 Experimental Approach

A schematic diagram of one optofluidic device is shown in Figure 17.4. The microchannel has a depth of 3 μm and a width of 50 μm. The nanochannel has a depth of 40 nm and a width of 2 μm.

In our experiments, gold nanoparticles with a diameter of 60 nm suspended in water were used as the SERS-active metal nanostructures. The gold nanoparticles were mixed with an aqueous sample of interest and then dispensed into the inlet reservoir. Within seconds, the sample solution was drawn into the nanochannel device by capillary force. Since the size of gold nanoparticles is larger than the

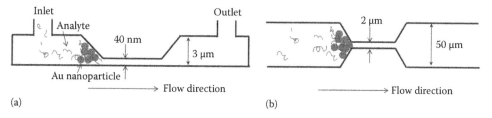

FIGURE 17.4 Schematic diagram of a nanochannel device (a) side view and (b) top view. (From Wang, M. et al., *Microfluid. Nanofluid.*, Figure 2, 6, 411, 2009. With permission from Springer.)

depth of nanochannel, gold nanoparticles were trapped and formed into clusters at the entrance to the nanochannel. Furthermore, capillary force continued to bring analytes through the interstices of the gold clusters and created an area with locally high concentration of analytes around the gold clusters. The highly condensed gold clusters and high concentration of analytes contributed to especially high-detection sensitivity. There are several advantages of this type of optofluidic device:

1. It can improve the detection sensitivity because of an increased local density of nanoparticles/target molecules. This will provide improvement by minimizing the required target molecular concentration. Thus, low volume and low concentration of molecules can be detected.
2. It can provide efficient detection. With a condensed hot spot produced at the nanochannel entrance, the detection area can be visually identified quickly without searching for random hot spots. This will improve the reliability of SERS.
3. It does not require chemical agents or salts to initiate the aggregation of nanoparticles, which can preserve the original form of samples that are sensitive to chemical agents or salts.
4. It can be fabricated on fused silica or borosilicate wafers at low cost with high throughput using standard photolithography. The on-chip property also enables the capability of parallel analysis. This means that the cost of one analysis will be reduced.
5. The consumption of sample of this device is extremely low, in the microliter range, which is preferred for expensive samples.

A prerequisite of using this device for SERS detection is the ability to fabricate it in a reasonable manner. Standard photolithography was used to fabricate optofluidic device on fused silica or borosilicate wafers.

17.3.1 Design and Fabrication of Optofluidic Devices

17.3.1.1 Raman Microscopy Setup

Raman spectroscopy is conventionally performed with green (532 nm), red (633 nm), or near infrared (NIR) (785 nm) lasers, with wavelengths below the first electronic transitions of most molecules, as assumed by scattering theory. Higher wavelength sources (such as NIR) reduce interference from inherently fluorescent molecules. However, this also decreases the scattering efficiency, making longer integration time or higher power necessary (Horiba Jobin Yvon http://www.jobinyvon.com/Raman/Tutorial-Intro).

Raman spectrometers coupled to a microscope are commercially available. A block diagram of one such system is shown in Figure 17.5. The incident laser first passes through a beam expander (in order to fully illuminate the microscope objective lens for delivering the smallest spot size to the sample). The beam is reflected by a series of mirrors to the first notch filter, which rejects the Rayleigh-scattered light in the back-reflected beam from reaching the detector. The notch filter also serves to reflect the incident beam to the microscope, where the beam travels to the sample. The back-reflected light scattered from the sample travels via the same path back to the spectrometer through the notch filter again. The rest of

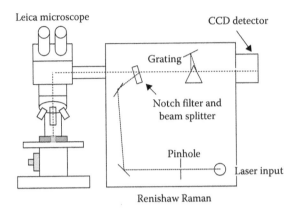

Renishaw Raman

FIGURE 17.5 Block diagram of the microscope and Raman spectrometer.

the back-reflected beam then passes through a second notch filter, which further rejects any Rayleigh-scattered light that might have bled through the first filter. The remaining light then passes through a series of lenses and mirrors to strike a diffraction grating, which spatially separates the light into respective wavelengths. The diffracted light is detected by a thermoelectrically cooled charge-coupled device (CCD) camera (Beier et al. 2007, Chou et al. 2008).

The laser, notch filter, and diffraction grating characteristics are the controllable aspects of the Raman system. The choice of notch filter depends on the wavelength of the incident laser. As the density of the grating (e.g., 600 grooves per mm versus 1200 grooves per mm) is increased, the spectral resolution of the system improves. On the other hand, this improvement comes with a reduction in the viable range of the spectrometer. Increasing the focal length of the spectrometer also adds to an increased resolution by increasing the spectral dispersion of the spectrometer. Miniaturized Raman spectrometers are manufactured for handheld devices, however, with a significant loss in resolution. Further, Raman systems are also available with Fourier transform infrared (FTIR) capabilities, which can provide complementary information to Raman samples. Raman confocal microscopes are also commercially available that provide automated use of three laser sources (green, red, and NIR) and multiple gratings to increase the ease, efficiency, spectral resolution, and mapping capabilities to provide the user with the maximum flexibility in their Raman system (Horiba Jobin Yvon http://www.jobinyvon.com/Raman/Tutorial-Intro, Kaiser Optical Systems www.kosi.com/raman/resources/tutorial/).

17.3.2 Interpreting Raman Spectra

Although Raman modes can arise from vibrational or rotational changes in the molecule, we will focus on vibrational Raman scattering. A simple vibrating molecule can be modeled as a harmonic oscillator whose behavior is described by the wave equation giving rise to selection rules and forbidden energy modes. However, these rules tend to be relaxed when SERS processes are involved. Moreover, for a complex molecule, such as a protein, predicting the Raman modes involves intense, complex calculations involving group theory (Wilson et al. 1955, Carey 1982, Tu 1982).

The total number of vibrations, and therefore Raman modes, for a nonlinear molecule is $3N-6$, where N is the number of atoms. Vibrations include bond stretching (asymmetrical, symmetrical), bending (out-of-plane and in-plane), also known as rocking, twisting, and wagging. Further, cyclic compounds can possess these vibrations plus breathing modes. To further demonstrate the complexity of Raman spectra, a simple aromatic ring can have up to 30 modes of vibration. Fortunately, vibrations of particular groups of atoms tend to appear with characteristic group frequencies that can also depend on the functional group (e.g., C=O stretching vibration band due to a carboxylic acid or a ketone lie in a different range) (Wilson et al. 1955, Carey 1982, Tu 1982).

One method to definitively identify and assign Raman modes is by isotopic substitution, usually with deuterated water (^2H_2O). The isotope changes the mass, resulting in a shift in the band labeled with the isotope. Otherwise, SERS bands are tentatively assigned based on those reported in the literature. Most Raman active molecules are those exhibiting delocalized electrons that can be found in conjugated systems of double bonds and aromatic rings (Carey 1982, Tu 1982).

Interpreting Raman spectra for complex biomolecules, such as proteins, can prove to be a daunting task due to their size and complexity. However, due to the nature of SERS, it can provide localized information pertaining to subgroups or functional groups of the molecule, and can give insight to the molecule's environment. Although Raman spectroscopy can be used for any biomolecule, as an example, we will focus on the complex and growing field of proteomics.

Proteins are composed of chains of amino acids linked by peptide bonds (making up the peptide backbone) and distinguished by the functional groups that give proteins their distinct properties and folding characteristics. Fluorescence spectroscopy is limited to providing information pertaining to the environment of tyrosine and tryptophan amino acid side chains whereas circular dichroism (CD) can provide information on the protein structure, specifically the peptide backbone. CD does not give insight into the side chains of the protein or disulfide bonds, a unique bond that can indicate the integrity of the protein. Plus, both of the previous methods require aqueous solutions. X-ray diffraction is the most powerful method to determine protein structure; however, the molecule must be crystallized, which is not always viable and can potentially change the structure of the protein. Raman spectroscopy can be carried out with a molecule in any form (crystal, powder, gel, and aqueous solution). Plus, as mentioned, Raman spectroscopy can reveal a full spectrum of information such as the peptide backbone, geometry of disulfide bonds (trans versus gauche configuration), environment of some side chains (e.g., buried versus exposed amino acid residues), along with proposed secondary structure (evaluation of the presence of α-helices, β-sheets, and random coil) of the protein by evaluation of amide (specifically I and III) bands. For more information on specific band assignments for aromatic amino acids, disulfide bonds, and so forth, refer to Lippert et al. 1976, Carey 1982, Tu 1982.

In terms of SERS, due to charge transfer effects, Raman bands in close proximity to the metal surface will be enhanced with a strong spatial dependence. This lends itself to the potential for labeling and probing specific sites of complex biomolecules to gain more information than traditional Raman techniques (Greenfield et al. 1967, Lippert et al. 1976, Hu et al. 2001, Li et al. 2003b, Zhao et al. 2004, Brylinski et al. 2005, Choi and Ma 2006, Zhang et al. 2006, Beier et al. 2007, Lefèvre et al. 2007, Chou et al. 2008).

Although β-sheets are a fundamental structure component of native proteins, the β-sheet conformation is also more prevalent in nonnative and misfolded proteins. Thus, if measurable with the optofluidic SERS platform, this can be used as a sensor as an example of prion proteins, common in mad cow disease, or β-amyloid, the protein implicated in the neurotoxicity found in Alzheimer's disease (Koudinova et al. 1999, Huang et al. 2000, Varadarajan et al. 2000, Selkoe 2001, Bacskai et al. 2002, Dahlgren et al. 2002, Haes et al. 2005, Bibl et al. 2006, Beier et al. 2007, Lennon et al. 2007).

17.3.2.1 Design of an Optofluidic System for SERS Detection

An optofluidic device for our SERS applications was fabricated on a 500-μm-thick double-side-polished borosilicate or fused silica wafer using photolithography, dry/wet etching, and bonding. The device has a step structure consisting of a deep microchannel and a shallow nanochannel (Figure 17.6). The deep microchannel has a 20–50 μm width and 2–6 μm depth. The shallow channel has a 40 nm depth, 2–5 μm width, and 40 μm length, which is used for trapping metallic nanoparticles with a size larger than 40 nm at the microchannel–nanochannel junction. The width and depth of the microchannel as well as the width of the nanochannel can be controlled in the fabrication process. Two masks were made—one for the microchannel and another for the nanochannel. In addition, since the nanochannel was too shallow to be seen under the contact aligner, alignment markers were prefabricated for easy alignment of the microchannel and nanochannel during fabrication. With a device fabricated on fused silica as an

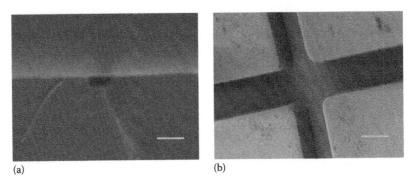

(a) (b)

FIGURE 17.6 (a) The cross-sectional SEM image of nanochannel made by electron beam lithography. The scale bar is 100 nm. (b) Top view SEM image of cross trench for microfluidic device. The scale bar is 50 µm.

example, the wafer was thoroughly cleaned, P20 primer and photoresist S1818 were spin-coated onto the substrate, and alignment markers were patterned. After developing and stripping the photoresist in the selective area, the markers were etched to a depth of ~1 µm by CHF_3/O_2 reactive ion etching. The left photoresist was removed using oxygen plasma. A brief fabrication process after the creation of the alignment markers is listed next, and a flowchart can be found in Figure 17.7:

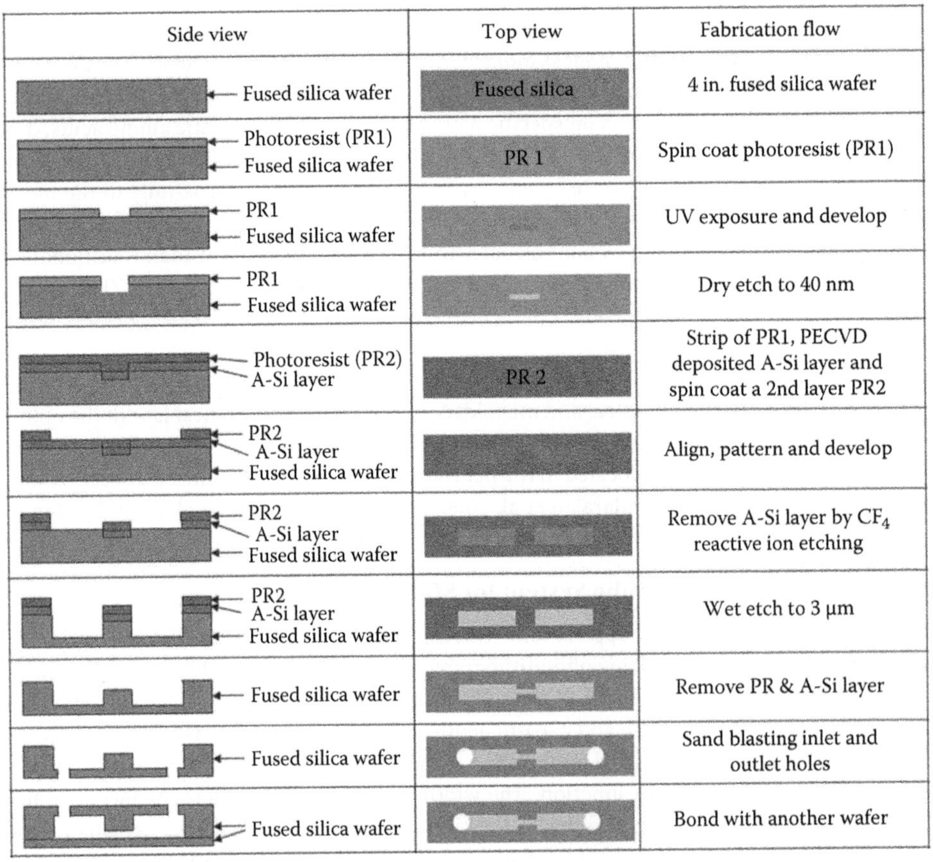

Side view	Top view	Fabrication flow
Fused silica wafer	Fused silica	4 in. fused silica wafer
Photoresist (PR1) / Fused silica wafer	PR 1	Spin coat photoresist (PR1)
PR1 / Fused silica wafer		UV exposure and develop
PR1 / Fused silica wafer		Dry etch to 40 nm
Photoresist (PR2) / A-Si layer	PR 2	Strip of PR1, PECVD deposited A-Si layer and spin coat a 2nd layer PR2
PR2 / A-Si layer / Fused silica wafer		Align, pattern and develop
PR2 / A-Si layer / Fused silica wafer		Remove A-Si layer by CF_4 reactive ion etching
PR2 / A-Si layer / Fused silica wafer		Wet etch to 3 µm
Fused silica wafer		Remove PR & A-Si layer
Fused silica wafer		Sand blasting inlet and outlet holes
Fused silica wafer		Bond with another wafer

FIGURE 17.7 Process flowchart of the fabrication process of nanochannel device. (From Wang, M. et al., *Microfluid. Nanofluid.*, Figure 1, 6, 411, 2009. With permission from Springer.)

1. Generation of nanochannels: P20 primer and photoresist S1813 were spin-coated onto the processed wafer. The channels with a width of 2 μm were patterned onto the substrate. After developing and stripping the photoresist in a selective area, the channels were etched to a depth of 40 nm with reactive ion etching. The left photoresist was removed using oxygen plasma.

2. Generation of microchannels: A layer of amorphous silicon (A-Si) was deposited onto the wafer using plasma-enhanced chemical vapor deposition (PECVD). This layer was used to protect the nanochannel structure from the wet etching of the microchannels. Then P20 primer and photoresist S1818 were spin-coated onto the A-Si-layer-covered wafer. With the microchannel mask and alignment marker, the microchannels were aligned with the nanochannels, patterned, and developed. In the selectively open area for creating microchannels, the A-Si layer was removed by carbon tetrafluoride (CF_4) reactive ion etching. The whole wafer was immersed in concentrated hydrofluoric (HF) acid (49%) to create 3-μm-depth microchannels. The left A-Si layer protected the selective area from the wet etching. After the wet etching, the photoresist and the A-Si layer was removed by oxygen plasma and SF_6/O_2 by reactive ion etching.

3. Creating inlet and outlet holes: The inlet and outlet holes were sand-blasted on the wafer with fine alumina powders.

4. Bonding: The processed wafer was thoroughly cleaned and clung with an intact fused silica wafer using deionized water and annealed at 1050°C for 6 h for permanent bonding. Plastic reservoirs were then attached around the inlets and outlets, respectively, using 5-min epoxy.

A schematic diagram of optofluidic device fabricated in wafer is shown in Figure 17.8 (top view).

17.3.3 Experimental Microscopic Raman System

17.3.3.1 Testing the Nanofluidic Trapping Capability

To test the trapping capability of the optofluidic device, fluorescent beads were used as a sample. A solution of fluorescent polystyrene (PS) nanoparticles (Spherotech Inc., Lake Forest, IL) with a size ranging from ~40 to 90 nm were diluted to 5 mg/L using deionized (DI) water and then introduced into the device from a reservoir. Due to capillary force, the solution was transported into the device within a few seconds. Since the diameter of the fluorescent nanoparticles is larger than the depth of the

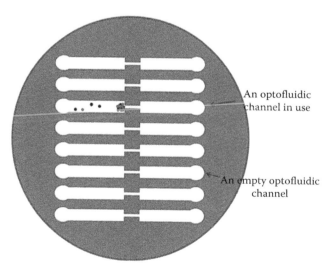

FIGURE 17.8 Schematic diagram of an optofluidic chip (top view). There are 8–12 channel devices on one wafer. Currently, we are working on the new device with more channels on wafer.

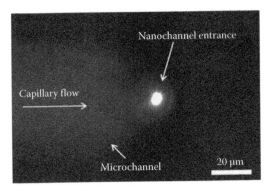

FIGURE 17.9 Fluorescent image of polystyrene nanoparticles trapped at the step boundary of the nanochannel device.

shallow nanochannel, they were trapped at the nanochannel's entrance. The fluorescent image of the PS particles trapped at the entrance of the nanochannel is shown in Figure 17.9. The PS particles emitted extremely high fluorescent signals around the entrance compared to other locations in the microchannel region.

In our SERS experiment, 60-nm gold nanoparticles were mixed with the aqueous sample of interest and dispensed into the device. Figure 17.10a and b shows the optical images of aggregation of gold nanoparticles over time. The trapping process occurred within minutes after the sample solution was dispensed into the inlet reservoir. As the nanoparticles were trapped at the junction of microchannel and nanochannel, they formed clusters, as depicted by the black spot shown in the figure. A scanning electron micrograph (SEM) image around the nanochannel entrance is shown in Figure 17.10c. It is obvious that the majority of the gold nanoparticles are aggregated and located at the entrance to nanochannel. A few isolated gold nanoparticles were found in the microchannel; however, these particles are out of the detection area for the Raman microscope system, and thus they do not contribute to SERS signal.

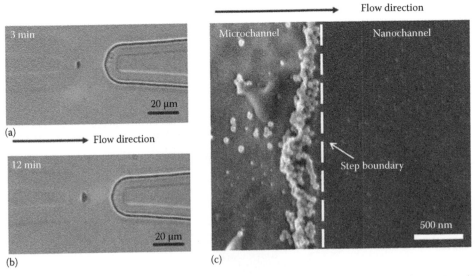

FIGURE 17.10 (a,b) Optical microscopic images: aggregation of Au nanoparticles monitored over time. (From Wang, M. et al., *Microfluid. Nanofluid.*, Figure 3, 6, 411, 2009. With permission from Springer.) (c) SEM image of Au aggregation at the boundary of microchannel and nanochannel formed only at the entrance to the nanochannel.

17.4 Optofluidic SERS Results and Discussion

Before introducing an analyte into the system, it was necessary to ascertain that the gold nanoparticles were indeed aggregating, thereby causing a shift in the ultraviolet-visible (UV-Vis) absorption and transmission spectra. Therefore, in addition to the aggregation tests described previously, the extinction spectra were obtained of the gold colloid alone in a cuvette, the fluid in the microchannel, and the gold nanoparticle clusters at the entrance to the nanochannel (Wang et al. 2007). As shown in Figure 17.11b, the extinction spectrum of gold colloid in solution exhibits a relatively sharp absorption band at 540 nm, with a long tail trailing out toward longer wavelengths. Upon aggregation (Figure 17.11a), the 540-nm band broadens and a broad band appears centered around 800 nm. The changes of the spectra, to high extinction at 800 nm, indicate that gold nanoparticles are clustering at the nanochannel entrance with up to 1500 times the optical density, according to Beer-Lambert's law. This correlates with the microchannel having a depth of 6 μm, whereas the path length of the cuvette used to measure the extinction spectrum of the gold colloid solution was 1 cm.

17.4.1 Small Molecule Detection

To assess the optofluidic device for SERS detection of small molecules, adenine, a small molecule with a known large Raman enhancement, was used. The device was fabricated, as shown in Figure 17.7, using conventional top-down microfabrication techniques that included photolithographic and etching processes on a 500-μm-thick double-side-polished borosilicate wafer (Wang et al. 2007).

The nanoparticles, along with adenine, were introduced into the microchannel port and the enhancement factor of the nanoparticle clusters were estimated and compared to other SERS techniques for adenine. The excitation laser was focused at the nanochannel entrance to obtain the surface-enhanced Raman spectra of adenine molecules. SERS detection was accomplished using a Renishaw System

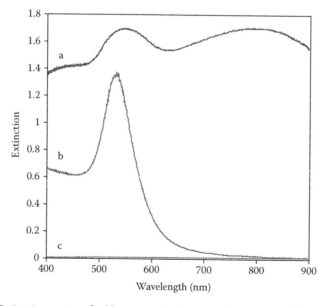

FIGURE 17.11 (a) Extinction spectra of gold nanoparticle clusters at the entrance of the nanochannel, (b) extinction spectra of gold colloid in a cuvette, and (c) extinction spectra collected in the middle of the microchannel channel. Note the broadening and shift in the 540 nm gold colloid peak upon aggregation at the nanochannel entrance. No extinction was detected in the microchannel away from the nanochannel entrance. (Reprinted from Chou, I.-H. et al., *Nano Lett.*, 8, 1729, 2008. With permission from American Chemical Society.)

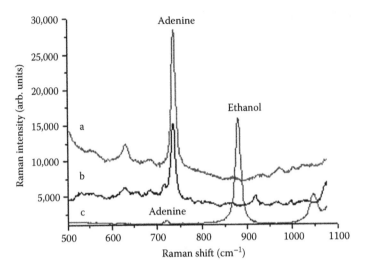

FIGURE 17.12   SERS signals of adenine molecule obtained by (a) Raman signal without SERS active clusters, (b) SERS active clusters made by the conventional colloidal method, (c) SERS active clusters created in an optofluidic device. (From Wang, M. et al., *Lab Chip*, 7, 630, 2007. With permission from The Royal Society of Chemistry.)

1000 Raman Spectrometer coupled to a Leica DMLM microscope (Schaumburg, IL), as depicted in Figure 17.5. The excitation laser source had a wavelength at 785 nm and a power of 8 mW at the sample. The integration time was set to be 2 min and the wavenumber range was from 504 to 1076 cm^{-1}.

There are three signals of adenine molecules shown in Figure 17.12a. As a reference graph, A is the regular Raman signal from a 22-mM solution of adenine on a glass surface without any nanoparticles. The solution was a blend of 10.4 M ethanol and DI water. As depicted, both the adenine peak at 735 cm^{-1} and the ethanol peak at about 882 cm^{-1} are visible. In Figure 17.12b, the signal from a solution of 3.33-μM adenine using a conventional colloidal gold SERS technique is depicted. The sample was prepared by mixing an activation agent of 0.5-M sodium chloride to force the gold nanoparticles to aggregate (Fleischmann et al. 1974, Jeanmarie and VanDuyne 1977, Kneipp et al. 1997, Etchegoin et al. 2003, Félidj et al. 2004). The reading was taken 15 min after the mixing process to allow for the gold nanoparticles to aggregate into clusters. The SERS signal from 3.33-μM adenine obtained with the use of the nanofluidic trapping device is depicted in Figure 17.12c. The signal was obtained immediately after it was dispensed and drawn into the channel via capillary force. As depicted, the SERS signal from the nanofluidic trapping device is the highest of the three approaches. Using Figure 17.12a as a reference, the enhancement factor of the nanofluidic trapping device was calculated to be 10^8, as compared to 10^6 for the colloid in solution (Zhu et al. 2004). Further results with adenine were reported, showing the ability to measure even lower picomolar concentrations (Wang et al. 2007).

17.4.2 Detection of Protein Denaturation

To assess the ability of the SERS optofluidic device to detect protein denaturation due to heat, the spectra of two proteins, insulin from bovine pancreas and bovine serum albumin (BSA), were detected at room temperature (RT) and elevated temperatures (Wang et al. 2009). During heat denaturation, the secondary structure of proteins changes and the proteins unfold from their original form.

For this study, insulin and BSA were both purchased from Sigma Aldrich. Six samples, three BSA and three insulin at different temperatures, with a volume of 5 μL each were dispensed into the inlet reservoirs of six independent optofluidic channels. SERS spectra were collected within 3 min, after visible aggregation of gold nanoparticles under the microscope. Five spectra were collected from three samples for each protein, then normalized and averaged to get the final spectra profile. The spectra shown in Figures 17.13 and 17.14 for different thermal conditions were offset for visual clarity. The peaks are identified in Tables 17.1 and 17.2, as determined from the literature (Yu et al. 1972, Tu 1982, Miura and Thomas 1995, Stewart and Fredericks 1999, Miura et al. 2000, Podstawka et al. 2004).

The first sample, BSA, is a globular protein composed mainly of α-helices and random coils at room temperature (Wetzel et al. 1980, Shanmugam and Polavarapu 2004) and with our optofluidic device, detection at 0.9 ng/L was easily achieved. As temperatures are elevated (>60°C), β-sheet formation with temperature is expected (Wetzel et al. 1980, Shanmugam and Polavarapu 2004). From the three spectra depicted in Figure 17.13, we can tell that as the temperatures increase, the structure of BSA is denatured and β-sheet structure is in fact formed. In particular, there is a marked decrease in the initially intense

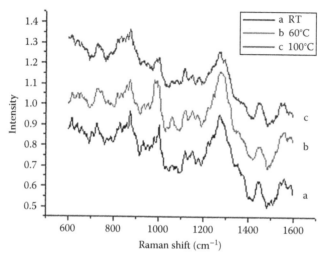

FIGURE 17.13 SERS spectra of BSA at different temperatures (RT, 60°C, 100°C). (From Wang, M. et al., *Microfluid. Nanofluid.*, Figure 4, 6, 411, 2009. With permission from Springer.)

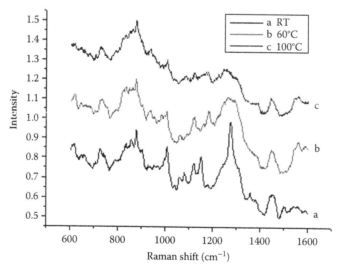

FIGURE 17.14 SERS spectra of insulin at different temperatures (RT, 60°C, 100°C). (From Wang, M. et al., *Microfluid. Nanofluid.*, Figure 5, 6, 2009. With permission from Springer.)

TABLE 17.1 Assignment of Peaks in BSA Spectrum (RT)

Wavenumber	Assignment	
620	COO⁻ wag or phenylalanine	
643	Ring deformation	Tyr
662	C–S stretching vibration	Cys
730	COO⁻ def	Backbone
832	Stretching and ring breathing mode	Tyr
876	Indole ring	Trp
938	α-Helical C–C stretch	
957	C–C stretch	
993	Indole asymmetric breath	Trp
1007	Strong, "breathing" vibration of the benzene ring	Phe
1184		Phe, Tyr
1233	β-Sheet (amide III region)	
1254	Random coil (amide III region)	
1277	α-Helix (amide III region)	
1356	Indole vibration	Trp
1453	CH_2 bending/scissoring mode	
1564	Amide II band	
1593	COO⁻ asymmetric stretch	Backbone

Source: Wang, M. et al., *Microfluid. Nanofluid.*, Table 1, 6, 2009. With permission from Springer.

TABLE 17.2 Assignment of Peaks in Insulin Spectrum (RT)

Wavenumber	Assignment	
620	COO⁻ wag or phenylalanine	
662	C–S stretching vibration	Cys
835	Stretching and ring breathing mode	Tyr doublet
855	Stretching and ring breathing mode	Tyr doublet
938	α-Helical C–C stretching vibration	
963	C–COO⁻ stretch	
1007	Stretching and benzyl ring breathing mode	Phe
1050	β-Sheet (amide III region)	
1066	C–N stretch	
1081	C–N stretch	
1121	NH_3^+ deformation	
1151	NH_3^+ deformation	
1276	α-Helix (amide III region)	
1311	CH_2 wag	
1457	CH_2 scissoring or bending mode	
1552	Amide II band	
1564	Amide II band	
1594	COO⁻ asymmetric stretch	Backbone

Source: Wang, M. et al., *Microfluid. Nanofluid.*, Table 2, 6, 2009. With permission from Springer.

band at 1277 cm^{-1} and amide III band (indicating strong α-helical secondary structure), plus the medium band at 938 cm^{-1} (tentatively assigned to skeletal C-C stretching vibration characteristic of α-helical conformation) decreases by about 25% in the 60°C spectra, and then by 50% in the 100°C, indicating an incomplete loss in α-helical content and supports unfolding of the protein (Tu 1982). However, although diminished, the band still exists, suggesting that the ordered structure is not completely destroyed. For the sample heated to 60°C, the 1277-cm^{-1} band, although still intense, broadens and shifts to the right. For the 100°C sample, a small shoulder develops at 1296 cm^{-1}, indicative of β-sheet structure (Tu 1982), which supports the hypothesis that some β-sheet structure may have been formed upon thermal denaturation (Wetzel et al. 1980, Shanmugam and Polavarapu 2004) and shows our approach can be used to monitor this change. Other noteworthy bands include 1453 cm^{-1} (tentatively assigned to methylene-bending vibration), which remains relatively the same throughout, indicating that this band may not be sensitive to structure change. Lastly, due to the rearrangement of the conformation of the disulfide bonds in BSA, activity in the 655–666-cm^{-1} region is observed upon heating (Tu 1982).

Insulin, a globular protein composed of two polypeptide α- and β-chains joined by disulfide bonds, also exhibits α-helical and random-coil conformations (Jiang and Chang 2005). Figure 17.14 and Table 17.2 represent the SERS spectra of insulin taken at different temperatures and the assignment of peaks for spectrum A, respectively. The secondary structure was changed but not completely destroyed by the heat, as seen previously with BSA. First of all, unfolding of the protein backbone is seen by the dramatic change in the 1050–1200-cm^{-1} region. Plus, the sharp band at 1276 cm^{-1} (the amide III region) decreases considerably and broadens into a shoulder with denaturation (Yu et al. 1972, Lippert et al. 1976, Podstawka et al. 2004). As with BSA, β-sheet conformation is induced upon thermal denaturation, supported by the broadening of this band to include the 1245-cm^{-1} region (tentatively assigned to β-sheet and/or random coil conformation) at 100°C (Yu et al. 1972, Lippert et al. 1976, Podstawka et al. 2004). In general, in the spectra taken at 100°C, the bands that were enhanced have decreased in intensity, with the 1585-cm^{-1} band disappearing altogether, as described more thoroughly in a previous paper that can indicate the peptide groups in close proximity to the gold surface (Wang et al. 2009).

17.4.3 Amyloid β–Detection

A second example of the potential of surface-enhanced Raman spectroscopy and optofluidic devices is the detection of proteins such as β-amyloid peptide (Aβ). This may prove important since one of the primary pathological hallmarks of diseases such as Alzheimer's disease (AD) is the presence of insoluble neuritic plaques, which are composed primarily of β-amyloid peptide (Aβ). The most prevalent species of Aβ present in people with AD are Aβ (1–40) and Aβ (1–42). Both forms of Aβ are found in cerebrospinal fluid (CSF) and blood plasma of all people, regardless of health (Mattson et al. 1997, Varadarajan et al. 2000, Ariga et al. 2001, Selkoe 2001, Wang et al. 2001, Hardy and Selkoe 2002, Gong et al. 2003, Lee et al. 2007). Aβ (1–40), the dominant peptide species, has a concentration of 5 nM in CSF (Selkoe 2001). These species have a strong propensity to aggregate. Many hypothesize that aggregation of Aβ triggers a cascade of events that brings about neuritic dystrophy and neuronal death. To date, Alzheimer's disease can only be diagnosed definitively by postmortem identification of neuritic plaques and neurofibrillary tangles in central nervous system tissue. While methods exist for probable premortem diagnosis of AD, including in vivo imaging of the brain with magnetic resonance imaging or functional positron emission tomography, along with tests of cognitive and psychological function (Duara et al. 1986, Davis et al. 1992), reliable methods of premortem diagnosis are needed and may involve the use of biomarkers of AD such as Aβ. Thus, the feasibility of using a SERS optofluidic device (Figure 17.4) to detect Aβ (1–40), one of the two most prevalent Aβ species in CSF, is described.

Aβ oligomer samples were prepared at concentrations ranging from 11.5 nM to 11.5 pM. The sample was observed at room temperature. Soluble Aβ monomers (1.15 nM) were prepared and stored at 6°C throughout observation period. All Aβ SERS samples were prepared by mixing the protein with gold nanoparticles in solution (volume ratio 1:10). After mixing, 120 µL of the gold colloid-protein mixture was then loaded immediately into the optofluidic reservoir. All SERS spectra were collected using a Renishaw System 1000 Raman Spectrometer coupled to a Leica DMLM microscope (Schaumburg, IL), as depicted in Figure 17.5 and described earlier.

The SERS spectra of Aβ at three different concentrations is shown in Figure 17.15 with the SERS spectral assignments in Table 17.3 based on existing literature pertaining to the spectra of amino acids and proteins (Ortiz et al. 2004, Beier et al. 2007, Wang et al. 2007). Aromatic rings, amides, and carboxylic group vibrations dominate the SERS spectrum of Aβ. The presence of these bands indicates that the previous side chains are in a favorable position with respect to the gold nanoparticle surface due to surface charge interactions and steric constraints of the protein adsorbed to the gold surface. Furthermore,

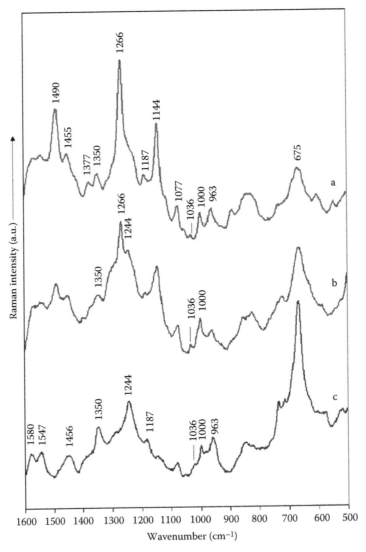

FIGURE 17.15   SERS spectra of Aβ at (a) 11.5 pM, (b) 1.15 nM, (c) 11.5 nM after 24 h in the nanofluidic device. Aβ samples at three different concentrations were prepared in the monomer form and loaded into the nanofluidic device immediately. Seven consecutive scans were taken from the nanoparticle clusters at the entrance to the nano-channel using 50×(NA = 0.75) air objective with a 785 nm excitation laser. (Reprinted from Chou, I.-H. et al., *Nano Lett.*, 8, 1729, 2008. With permission from American Chemical Society.)

TABLE 17.3 Assignment of Bands in SERS Spectra of Aβ

Wavenumber	Assignment	
675		Tyr
823	Stretching and ring breathing mode	Tyr doublet
856	Stretching and ring breathing mode	
C–C stretching vibration	Tyr doublet	
963		
1000	Stretching and benzyl ring breathing mode	Phe
1036	Stretching and benzyl ring breathing mode	Phe
1077	α-Helical C–C stretch	
1144	C–N stretch	
1187	Stretching and benzyl ring breathing mode	Phe & Tyr
1244	β-Sheet (amide III region)	
1266	α-Helix (amide III region)	
1350	C–N stretch	
1455	CH_2 deformation	His
1490	CH_2 wag	
1547	Benzyl ring breathing mode and amide II region	Phe
1580	Benzyl ring breathing mode or COO^- stretch	Tyr or Phe

shifts in the amide III region, 1200–1300 cm^{-1}, reflect the most compelling changes in proteins and are widely used to quantitatively explore secondary structure, which, in this case, reflects the aggregation of β-amyloid (Frushour and Koenig 1974, Kitagawa et al. 1979, Drachev et al. 2004, Takano et al. 2006). At a low concentration (11.5 pM), shown in Figure 17.15a, we observed Raman bands associated with the aromatic side chains at 1000, 1187, and 1488 cm^{-1} (phenylalanine, tyrosine and phenylalanine, and histidine residues, respectively). The phenylalanines, at amino acid 19 and 20 in the Aβ sequence should only be available to interact with the metal surface in unaggregated forms of Aβ (Cavalu et al. 2001). Furthermore, the shifted histidine band at 1488 cm^{-1} (Figure 17.15a) may have been altered due to protein-metal interactions (Asher et al. 2001, Ortiz et al. 2004). These spectra show concentration-dependent changes in the secondary structure (i.e., changes in the amide III region). At low concentrations (e.g., 11.5 pM), the histidine band and a band at 1266 cm^{-1} (in the amide III region) suggest that Aβ has an α-helix structure. However, at a higher concentration (1.15 nM), as seen in Figure 17.15b, the band at 1266 cm^{-1} diminishes and red-shifts to 1244 cm^{-1} (tentatively assigned to the β-sheet conformation), signifies that the proteins are in two different conformational states. With a concentration of 11.5 nM (Figure 17.15c), the 1266-cm^{-1} band is hidden in the shoulder of the new broad band, indicating that the polypeptide backbone of Aβ has taken on a different conformation. Plus, the increase of the 961-cm^{-1} mode (assigned to the C-C stretching mode in the hydrophobic segment of Aβ polypeptide backbone) and the decrease in intensity of aromatic side chain signals suggest that in the more aggregated β-sheet structure of Aβ, the aromatic residues are no longer available to interact with the gold surface and confirms a refolding of the protein (Asher et al. 2001). While most investigators would suggest that increasing the Aβ concentration would result in conformational changes associated with aggregation, few have examined Aβ aggregation at the concentrations used in these studies because of the limits of analytical tools used for structure determination, as described earlier.

17.4.4 Cardiovascular Marker Detection

In order to determine if the nanochannel geometry would provide the needed sensitivity to obtain structure-specific SERS spectra of a cardiac marker, we did a preliminary experiment by placing troponin (TnI) in a nanochannel without surface functionalization. Troponin is currently the cardiac biomarker of choice for acute myocardial infarction and is released upon cardiac necrosis (i.e., accidental death of cells that release toxic byproducts) due to a cardiac insult. It is currently the most sensitive and specific marker. Any increase (currently defined as >99% of a reference healthy population) corresponds to cardiac toxicity, whether due to myocardial infarction or other effects (Wu 2003).

We obtained the SERS spectra using a Renishaw Raman system connected to a Leica microscope (Schaumburg, IL), as depicted in Figure 17.5. SERS spectra of TnI (20 ng/L) in the nanofluidic biosensor are shown in Figure 17.16. The ranges used were on the lower end of those given in the literature for current POC devices yet identifiable peaks were recognized. A small, resolved band at 703 cm^{-1} (assigned to C-S-C stretch) indicates there is some interaction of sulfur with the gold surface, found in cysteine and methione residues. The activity in the 815–885-cm^{-1} region is due to tryptophan and tyrosine residues, indicating that TnI interacts with the surface through these aromatic acid residues. Weak modes attributed to CH$_2$ scissoring, symmetric bend, and wagging modes (1465, 1377, 1281 cm^{-1}, respectively) are also evident in the spectra. Although native TnI is composed of α-helices, the activity in the amide III region suggests a random coil configuration, indicating that the TnI is unfolded in the vicinity of the gold surface, and most likely adsorbs on the surface via the aromatic residues or through the backbone of the peptide (NH$_3$ deformation, 1140 and 1170 cm^{-1}, and the presence of the amide II band). These results indicate that we can get very detailed spectral information from even low concentrations of analyte; however, the spectral information may not be sufficient to distinguish TnI or other cardiac biomarkers from other proteins that might be present in a complex medium such as blood or plasma.

This SERS-based optofluidic approach can be used to measure cardiac biomarkers, such as troponin, which can potentially provide early identification and diagnosis of ACS, ultimately leading to information on the prognosis of the patient. However, there will be an inclination toward the random orientation of the proteins adsorbed to the surface of the gold since ionic bonding is not strong, implying that the molecules easily adsorb and desorb to the surface. Therefore, future work will need to focus on development and optimization of an assay to provide quantitative information.

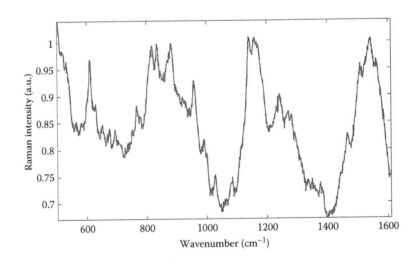

FIGURE 17.16 Normalized SERS spectra of TnI (20 ng/L) in the nanofluidic biosensor.

17.5 Summary and Future Work

As described in this chapter, recent advances in nanotechnology and biotechnology have created an immense opportunity for the use of noble metal nanoparticles as surface-enhanced Raman spectroscopy (SERS) substrates for biological sensing and diagnostics. This is because SERS enhances the intensity of the Raman-scattered signal from an analyte by six orders or more. Many initial studies have been performed using traditional chemically reduced metal colloidal nanoparticles for the SERS detection of a myriad of proteins and nucleic acids. Further, other methods have been examined to create robust SERS substrates, including the creation of thermally evaporated silver island films on microscope glass slides, using the technique of nanosphere lithography (NSL) to create hexagonally close-packed periodic particle arrays of silver nanoparticles on glass substrates, the use of optically tunable gold nanoshell films on glass substrates, and the development of gold nanoshell films. However, the need to induce aggregation for providing the required plasmonic conditions for optimal SERS enhancements compromises the robustness of these nanoparticles for producing reproducible SERS signals. This makes them undesirable options for implementation in any SERS-based biosensor. The need to circumvent the aggregation problem lead to the search for fabricating and/or obtaining robust SERS substrates with plasmonic properties that are not based on aggregation.

An optofluidic device was then described in this chapter as an example of a means to improve the robustness of SERS detection and provide fingerprint information of analytes with a concentration in the nanogram per liter range within minutes. With a pinched micro-nanochannel structure, a "hot spot" can be formed at a predictable location efficiently and consistently. With a well-characterized and repeatable "hot spot," molecular and metal particle enrichment effects enable significant improvement in detection sensitivity. Moreover, the device can detect multiple analytes in aqueous solution simultaneously.

The chapter has shown that optofluidic SERS devices such as these have the potential to be used for multiple applications such as for the diagnosis of Alzheimer's or cardiovascular diseases, for detecting molecular conformational changes, or to identify the denaturation of proteins. The detection volume needed for these approaches was shown to be less than $5\,\mu L$. With conventional photolithography fabrication processes, these device have the potential to be produced at a low cost. Furthermore, multiple optofluidic devices can be fabricated on a single wafer, which enables parallel analysis for higher throughput. These types of optofluidic devices also have the potential applications in the detection of water contaminants and other environmental or security hazards. Future work in this area would need to focus on the optimization of the optofluidic device parameters and functionalization of the nanoparticles to enhance specificity as the solutions to be tested become more complicated.

Acknowledgments

The authors acknowledge the support of the National Institutes of Health (grant no. R21-NS050346-01). Melodie Benford acknowledges the support of the National Science Foundation Graduate Research Fellowship.

References

Aaron, J., N. Nitin et al. 2007. Plasmon resonance coupling of metal nanoparticles for molecular imaging of carcinogenesis *in vivo*. *Journal of Biomedical Optics* 12(3): 034007-1–034007-11.

Akarca-Biyikli, S. S., I. Bulu et al. 2005. Resonant excitation of surface plasmons in one-dimensional metallic grating structures at microwave frequencies. *Journal of Optics A: Pure and Applied Optics* 7: S159–S164.

Ariga, T., K. Kobayashi et al. 2001. Characterization of high-affinity binding between gangliosides and amyloid β-protein. *Archives of Biochemistry and Physics* 388(2): 225–230.

Asher, S. A., A. Ianoul et al. 2001. Dihedral angle dependence of the amide III vibration: A uniquely sensitive UV resonance Raman secondary structural probe. *Journal of the American Chemical Society* 123: 11775–11781.

Bacskai, B. J., W. E. Klunk et al. 2002. Imaging amyloid-β deposits *in vivo. Journal of Cerebral Blood Flow & Metabolism* 22: 1035–1041.

Beier, H. T., C. B. Cowan et al. 2007. Application of surface-enhanced Raman spectroscopy for detection of beta amyloid using nanoshells. *Plasmonics* 3: 55–64.

Bibl, M., B. Mollenhauer et al. 2006. CSF amyloid-ß-peptides in Alzheimer's disease, dementia with Lewy bodies and Parkinson's disease dementia. *Brain* 129: 1177–1187.

Bonham, A. J., G. Braun et al. 2007. Detection of sequence-specific protein-DNA interactions via surface enhanced resonance Raman scattering. *Journal of the American Chemical Society* 129: 14572–14573.

Brylinski, M., L. Konieczny et al. 2005. Early-stage folding in proteins (in silico) sequence-to-structure relation. *Journal of Biomedicine and Biotechnology* 2005(2): 65–79.

Campbell, D. J. and Y. Xia. January 2007. Plasmons: Why should we care? *Journal of Chemical Education* 84(1): 91–96.

Campion, A. and P. Kambhampati. 1998. Surface-enhanced Raman scattering. *Chemical Society Reviews* 27: 241–250.

Cao, Y. C., R. Jin et al. 2002a. Nanoparticles with Raman spectroscopic fingerprints. *Science* 297: 1536–1540.

Cao, H., Z. Yu et al. 2002b. Fabrication of 10 nm enclosed nanofluidic channels. *Applied Physics Letters* 81: 174–176.

Cao, Y.-C., X.-F. Hua et al. 2006. Preparation of Au coated polystyrene beads and their application in immunoassay. *Journal of Immunological Methods* 317: 163–170.

Carey, P. R. 1982. *Biochemical Applications of Raman and Resonance Raman Spectroscopies.* New York: Academic Press, Inc.

Cavalu, S., S. Cinata-Pinzaru et al. 2001. Raman and surface enhanced Raman spectroscopy of 2,2,5,5-Tetramethyl-3-pyrrolin-1-yloxy-3-carboxamide labeled proteins: Bovine serum albumin and cytochrome c. *Biopolymers* 62: 341–348.

Chen, L. and J. Choo. 2008. Recent advances in surface-enhanced Raman scattering detection technology for microfluidic chips. *Electrophoresis* 29: 1815–1828.

Chiñas-Castillo, F. and H. A. Spikes. 2003. Mechanism of action of colloidal solid dispersions. *Journal of Tribology* 125: 552–557.

Choi, S.-M. and C.-Y. Ma. 2006. Structural characterization of globulin from common buckwheat (*Fagopyrum esculentum* Moench) using circular dichroism and Raman spectroscopy. *Food Chemistry* 2006: 150–160.

Chou, I.-H., M. Benford et al. 2008. Nanofluidic biosensing for β-amyloid detection using surface enhanced Raman spectroscopy. *Nano Letters* 8(6): 1729–1735.

Culha, M., D. Stokes et al. 2003. Surface-enhanced Raman scattering substrate based on a self-assembled monolayer for use in gene diagnostics. *Analytical Chemistry* 75: 6196–6201.

Dahlgren, K. N., A. M. Manelli et al. 2002. Oligomeric and fibrillar species of amyloid-beta peptides differentially affect neuronal viability. *Journal of Biological Chemistry* 277: 32046–32053.

Davis, P. C., L. Gray et al. 1992. The consortium to establish a registry for Alzheimer's Disease (CERAD). 3. Reliability of a standardized MRI evaluation of Alzheimer's disease. *Neurology* 42(9): 1676–1680.

Docherty, F. T., P. B. Monaghan, R. Keir, D. Graham, W. E. Smith, and J. M. Cooper. 2004. The first SERRS multiplexing from labeled oligonucleotides in a microfluidics lab-on-a-chip. *Chemical Communications* 1: 118–119.

Drachev, V. P., M. D. Thoreson et al. 2004. Surface-enhanced Raman difference between human insulin and insulin lispro detected with adaptive nanostructures. *Journal of Physical Chemistry B* 108: 18046–18052.

Duara, R., C. Grady et al. 1986. Positron emission tomography in Alzheimer's-disease. *Neurology* 36(7): 879–887.

Etchegoin, P., R. C. Maher et al. 2003. New limits in ultrasensitive trace detection by surface enhanced Raman scattering (SERS). *Chemical Physics Letters* 375: 84–90.

Félidj, N., S. L. Truong et al. 2004. Gold particle interaction in regular arrays probed by surface enhanced Raman scattering. *Journal of Chemical Physics* 120(15): 7141–7146.

Fleischmann, M., P. J. Hendra et al. 1974. Raman spectra of pyridine adsorbed at a silver electrode. *Chemical Physics Letters* 26(2): 163–166.

Frushour, B. G. and J. L. Koenig. 1974. Raman spectroscopic study of tropomyosin denaturation. *Biopolymers* 13: 1809–1819.

Gong, Y., L. Chang et al. 2003. Alzheimer's disease-affected brain: Presence of oligomeric Aβ ligands (ADDLs) suggests a molecular basis for reversible memory loss. *Proceedings of the National Academy of Sciences of the United States of America* 100(18): 10417–10422.

Greenfield, N., B. Davidson et al. 1967. The use of computed optical rotary dispersion curves for the evaluation of protein conformation. *Biochemistry* 6(6): 1630–1637.

Haes, A. J., W. P. Hall et al. 2004. A localized surface plasmon resonance biosensor: First steps toward an assay for Alzheimer's disease. *Nano Letters* 4(6): 1029–1034.

Haes, A. J., L. Chang et al. 2005. Detection of a biomarker for Alzheimer's disease from synthetic and clinical samples using a nanoscale optical biosensor. *Journal of American Chemical Society* 127(7): 2264–2271.

Hardy, J. and D. J. Selkoe. 2002. The amyloid hypothesis of Alzheimer's disease: Progress and problems on the road to therapeutics. *Science* 297: 353–356.

Horiba, J. Y. http://www.jobinyvon.com/Raman/Tutorial-Intro. Raman Tutorial. Retrieved 2008, from http://www.jobinyvon.com/Raman/Tutorial-Intro.

Hu, H.-Y., Q. Li et al. 2001. β-sheet structure formation of proteins in solid state as revealed by circular dichroism spectroscopy. *Biopolymers (Biospectroscopy)* 62: 15–21.

Huang, T. H. J., D.-S. Yang et al. 2000. Structural studies of soluble oligomers of the Alzheimer β-amyloid peptide. *Journal of Molecular Biology* 297: 73–87.

Huang, X., I. H. El-Sayed, W. Qian, and M. A. El-Sayed. 2007. Cancer cells assemble and align gold nanorods conjugated to antibodies to produce highly enhanced, sharp, and polarized surface Raman spectra: A potential cancer diagnostic marker. *Nano Letters* 7(6): 1591–1597.

Hunt, H. C. and J. S. Wilkinson. 2008. Optofluidic integration for microanalysis. *Microfluidics and Nanofluidics* 4: 53–79.

Jackson, J. B., S. L. Westcott et al. 2003. Controlling the surface enhanced Raman effect via the nanoshell geometry. *Applied Physics Letters* 82(2): 257–259.

Jeanmarie, D. L. and R. P. VanDuyne. 1977. Surface Raman spectroelectrochemistry, Part 1: Heterocyclic, aromatic, and aliphatic amines adsorbed on the anodized silver electrode. *Journal Electroanalytical Chemistry* 84: 1–20.

Jiang, C. and J. Y. Chang. 2005. Unfolding and breakdown of insulin in the presence of endogenous thiols. *FEBS Letters* 579: 3927–3931.

Kaiser Optical Systems, I. www.kosi.com/raman/resources/tutorial/. Raman Tutorial. Retrieved August 24, 2007.

Keir, R., E. Igata et al. 2002. SERRS. In situ substrate formation and improved detection using microfluidics. *Analytical Chemistry* 74: 1503–1508.

Kelly, K. L., E. Coronado et al. 2003. The optical properties of metal nanoparticles: The influence of size, shape, and dielectric environment. *Journal of Physical Chemistry* 107: 668–677.

Kitagawa, T., T. Azuma et al. 1979. The Raman spectra of Bence-Jones proteins. Disulfide stretching frequencies and dependence of Raman intensity of tryptophan residues on their environments. *Biopolymers* 18: 451–465.

Kneipp, K., Y. Wang et al. 1997. Single molecule detection using surface-enhanced Raman scattering (SERS). *Physical Review Letters* 78(9): 1667–1670.

Kneipp, K., H. Kneipp et al. 1998. Extremely large enhancement factors in surface-enhanced Raman scattering for molecules on colloidal gold clusters. *Applied Spectroscopy* 52(12): 1493–1497.

Kneipp, K., H. Kneipp et al. 2002. Surface-enhanced Raman scattering and biophysics. *Journal of Physics: Condensed Matter* 14: R597–R624.

Koudinova, N. V., T. T. Berezov et al. 1999. β-amyloid: Alzheimer's disease and brain β-amyloidoses. *Biochemistry (Moscow)* 64(7): 7520757.

Lee, S., E. J. Fernandez et al. 2007. Role of aggregation conditions in structure, stability, and toxicity of intermediates in the Aβ fibril formation pathway. *Protein Science* 16: 723–732.

Lefèvre, T., M.-E. Rousseau et al. 2007. Protein secondary structure and orientation in silk as revealed by Raman spectromicroscopy. *Biophysical Journal* 92: 2885–2895.

Lennon, C. W., H. D. Cox et al. 2007. Probing structural differences in prion protein isoforms by tyrosine nitration. *Biochemistry* 46: 4850–4860.

Li, W., J. O. Tegenfeldt et al. 2003a. Sacrificial polymers for nanofluidic channels in biological applications. *Nanotechnology* 14: 578–583.

Li, G., T. Wang et al. 2003b. Completely reversible aggregation of nanoparticles by varying the pH. *Colloid and Polymer Science* 281: 1099–1103.

Liu, G. L. and L. P. Lee. 2005. Nanowell surface enhanced Raman scattering arrays fabricated by soft-lithography for label-free biomolecular detections in integrated microfluidics. *Applied Physics Letters* 87: 074101.

Lippert, J. L., D. Tyminski et al. 1976. Determination of the secondary structure of proteins by laser Raman spectroscopy. *Journal of the American Chemical Society* 98(22): 7075–7080.

Lu, L., H. Wang et al. 2002. Seed-mediated growth of large, monodisperse core-shell gold-silver nanoparticles with Ag-like optical properties. *Chemical Communications* 144–145.

Mandal, M., N. R. Jana et al. 2004a. Synthesis of Au_{core}-Ag_{shell} type bimetallic nanoparticles for single molecule detection in solution by SERS method. *Journal of Nanoparticle Research* 6: 53–61.

Mandal, M., S. Kundu et al. 2004b. Sniffing a single molecule through SERS using Au_{core}-Ag_{shell} bimetallic nanoparticles. *Current Science* 86(4): 556–559.

Mattson, M. P., R. J. Mark et al. 1997. Disruption of brain cell ion homeostasis in Alzheimer's disease by oxy radicals, and signaling pathways that protect therefrom. *Chemical Research in Toxicology* 10: 507–517.

Measor, P., L. Seballos et al. 2007. On-chip surface-enhanced Raman scattering detection using integrated liquid core waveguides. *Applied Physics Letters* 90: 211107.

Michaels, A. M., J. Jiang et al. 2000. Ag nanocrystal junctions as the site for surface-enhanced Raman scattering of single rhodamine 6G molecules. *Journal of Physical Chemistry* 104: 11965–11971.

Miura, T. and G. J. Thomas. 1995. Raman spectroscopy of proteins and their assemblies. *Subcellular Biochemistry* 24: 55–99.

Miura, T., K. Suzuki et al. 2000. Metal binding modes of Alzheimer's amyloid β-peptide in insoluble aggregates and soluble complexes. *Biochemistry* 39: 7024–7031.

Nie, S. and S. R. Emory. 1997. Probing single molecules and single nanoparticles by surface-enhanced Raman scattering. *Science* 275: 1102–1106.

Ortiz, C., D. Zhang et al. 2004. Identification of insulin variants using Raman spectroscopy. *Analytical Biochemistry* 332(2): 245–252.

Park, T., S. Lee et al. 2005. Highly sensitive signal detection of duplex dye-labelled DNA oligonucleotides in a PDMS microfluidic chip: Confocal surface-enhanced Raman spectroscopic study. *Lab on a Chip* 5: 437–442.

Podstawka, E., Y. Ozaki, and L. M. Proniewicz. 2004. Adsorption of S–S containing proteins on a colloidal silver surface studied by surface-enhanced Raman spectroscopy. *Applied Spectroscopy* 58: 1147–1156.

Selkoe, D. J. 2001. Alzheimer's disease: Genes, proteins, and therapy. *Physiological Reviews* 81(2): 741–766.

Shanmugam, G. and P. L. Polavarapu. 2004. Vibrational circular dichroism spectra of protein films: Thermal denaturation of bovine serum albumin. *Biophysical Chemistry* 111: 73–77.

Stewart, S. and P. M. Fredericks. 1999. Surface-enhanced Raman spectroscopy of peptides and proteins adsorbed on an electrochemically prepared silver surface. *Spectrochimica Acta Part A* 55: 1615–1640.

Takano, K., S. Endo et al. 2006. Structure of amyloid beta fragments in aqueous environments. *Febs J* 273(1): 150–158.

Tu, A. T. 1982. *Raman Spectroscopy in Biology*. New York: John Wiley & Sons, Inc.

Varadarajan, S., S. Yatin et al. 2000. Review: Alzheimer's amyloid β-peptide-associated free radical oxidative stress and neurotoxicity. *Journal of Structural Biology* 130: 184–208.

Wang, L., A. Roitberg et al. 2001. Raman and FTIR spectroscopies of fluorescein in solutions. *Spectrochimica Acta Part A* 57: 1781–1791.

Wang, M., N. Jing et al. 2007. An optofluidic device for surface enhanced Raman spectroscopy. *Lab on a Chip* 7: 630–632.

Wang, M., M. Benford et al. 2009. Optofluidic device for ultra-sensitive detection of proteins using surface-enhanced Raman spectroscopy. *Microfluidics and Nanofluidics* 6: 411–417.

Westcott, S. L., S. J. Oldenburg et al. 1998. Formation and adsorption of clusters of gold nanoparticles onto functionalized silica nanoparticle surfaces. *Langmuir* 14: 5396–5401.

Wetzel, R., M. Becker et al. 1980. Temperature behavior of human serum albumin. *European Journal of Biochemistry* 104: 469–478.

White, I. M., J. Gohring, and X. Fan. 2007. SERS-based detection in an optofluidic ring resonator platform. *Optics Express* 15: 17433–17442.

Wilson, E. B., J. C. Decius et al. 1955. *Molecular Vibrations: The Theory of Infrared and Raman Vibrational Spectra*. New York: Dover Publications, Inc.

Wu, A. H. B. 2003. *Cardiac Markers*. Totowa, NJ: Humana Press Inc.

Yu, N. T., C. S. Liu, and D. C. O'Shea. 1972. Laser Raman spectroscopy and the conformation of insulin and proinsulin. *Journal of Molecular Biology* 70: 117–132.

Zhang, X., N. C. Shah et al. 2006. Sensitive and selective chem/bio sensing based on surface-enhanced Raman spectroscopy (SERS). *Vibrational Spectroscopy* 42: 2–8.

Zhao, H., B. Yuan et al. 2004. The effects of electrostatic interaction between biological molecules and nano-metal colloid on near-infrared surface-enhanced Raman scattering. *Journal of Optics A: Pure and Applied Optics* 6: 900–905.

Zhixun, L. and F. Yan. 2006. SERS of Gold/C_{60} (/C_{70}) nano-clusters deposited on floppy disk and hard disk. *Chemical Physics* 321: 86–90.

Zhu, Z., T. Zhu et al. 2004. Raman scattering enhancement contributed from individual gold nanoparticles and interparticle coupling. *Nanotechnology* 15: 357–364.

18

Plasmonics

David Sinton
Alexandre G. Brolo
Reuven Gordon

18.1 Introduction

Plasmonics is the study and application of surface-bound electromagnetic waves at the interface of a metal and a dielectric called surface plasmons (SPs). Surface plasmon resonance (SPR) results from the coupling of electromagnetic radiation into electron oscillations at the metal–dielectric interface (Ozbay 2006, Stewart et al. 2008). A variety of applications have emerged that exploit the properties of SPs. The confinement of the electromagnetic wave at the surface is a key feature in the application areas developed to date.

The simplest form of SPs is the surface plasmon polariton (SPP) that can be guided at a single flat interface between a metal and dielectric. Shown in Figure 18.1 are the three characteristic length scales associated with SPPs (Barnes et al. 2003). First, the extent of the penetration of the SP electromagnetic field into the metal surface (δ_m in Figure 18.1) is on the order of 10 nm. With respect to SP-based devices, this length provides a minimum feature size. Second, the extent of the penetration of the SP field into the dielectric medium is typically on the order of hundreds of nanometers (δ_d in Figure 18.1). This length provides an upper limit on the size of a structure/device that may be employed to influence SPs, for instance, to focus or direct an SP wave. Third, the characteristic length of propagation along the metal–dielectric interface with excitation in the visible range can vary from microns to tens of microns depending on the loss in the metal (δ_{sp} in Figure 18.1). This length may extend toward the millimeter scale for low-loss metals excited at longer (>1 μm) wavelengths. The loss-dictated propagation length provides a limit, and a challenge, in SP waveguiding applications. Nonetheless, the variance in the length scales associated with SPs presents opportunities (Barnes et al. 2003). For instance, typical propagation lengths could allow for interaction of many individual devices given the minimum device size as dictated by penetration depth. As will be shown in this review, further length scale reductions of SPs can result from nanostructuring the metal.

The confinement of the field enables the SP-based detection of surface-adsorption events, and thus facilitates the application of SP-based detection of chemical and biological species (Homola 2008). SP-based sensing, particularly for characterizing and quantifying biomolecular interactions, has become an increasingly popular research tool dating back to the early 1980s (Gordon and Ernst 1980).

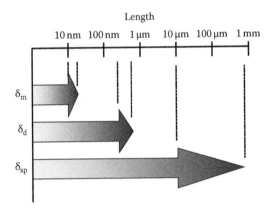

FIGURE 18.1 The various length scales associated with surface plasmons. Representative values for aluminum–dielectric and silver–dielectric interfaces are shown.

An important advantage of sensing using SPs is that surface-adsorption events may be detected in the absence of a label such as a fluorescent tag, and the detection and interpretation of surface binding is thus simplified. Such label-free detection schemes rely on both the sensitivity of the optical system to absorption, but also the specificity of the binding. Several companies produce commercial SPR instrumentation (Biacore by General Electric Health Care Life Sciences and others). The utility of SPP-based sensing and the recent availability of SPR systems has lead to a greatly increased role of SPR in science. A recent review catalogs the many applications of SPR sensors in chemical and biological detection (Homola 2008). In terms of the numbers of published studies, food quality and safety analysis has received the most focus, followed by medical diagnostics and environmental monitoring (Homola 2008).

SPs cannot be excited directly from light in free space due to a momentum mismatch; the wavenumber of light in a dielectric is less than the surface plasmon propagation constant at a metal–dielectric interface (Homola 2008). The bulk of the SPR-sensing applications demonstrated to date employed a Kretschmann-style SPP-excitation method in which the SPP is excited at a noble metal coating on the surface of a prism by the reflection of incident light (as will be discussed in detail in Section 18.2.1). This form of excitation is also referred to as attenuated total internal reflection, or ATR, the Kretschmann–Raether geometry, prism coupling and leaky-wave coupling. With proper alignment, the incident light undergoes total internal reflection on the inner surface of the prism adjacent to the gold surface. An evanescent light field travels through the thin metallic film and generates SPPs at the metal–dielectric interface. Due to optical absorption by the metal, SPP coupling results in a reduced intensity of the reflected light exiting the prism. The resulting SPR angle is sensitive to several factors, including the thickness of the metal film, the wavelength of light, the optical properties of the prism, and the index of refraction of the medium in contact with the metallic surface (Homola 2008). The latter dependency is exploited for sensing of molecular adsorption at the surface. Coupling in this manner can achieve a very broad area of SPP excitation, with dimensions several orders of magnitude larger than the penetration depth of the electromagnetic wave into the dielectric. For sensing, the large area can provide an averaging effect that increases sensitivity to bulk molecular binding at the gold surface. This surface averaging, however, comes at the expense of sensor localization (important for multiplexing and miniaturization) and quantity of sample to be detected (important for low limit of detection applications).

Light may be also coupled into SP modes via interactions with nanostructures. In contrast to propagating SPs, SPs generated with nanostructures are localized to the structure, and are called localized surface plasmons (LSPs). The most common nanostructures employed for exciting LSPs are metal nanoparticles and nanoholes in metal films (Stewart et al. 2008). The excitation wavelength and magnitude of the LSP resonance is a function of nanostructure geometry, size, metal properties, and the properties of the dielectric. Analogous to propagating-SP-based sensing employing prism coupling, nanostructure-based LSP sensing commonly exploits resonance dependencies associated with the refractive index at the metal–dielectric surface. For sensing, LSPs can exhibit several advantages over more traditional propagating SPs in that the sensing surface is highly localized, and thus decreasing the amount of analyte required to produce a signal and increasing the potential for multiplexing. In addition, use of nanostructures allows increased surface-to-volume ratios and the potential to harness rapid diffusion characteristic of nanofluidic transport, in conjunction with the benefits of SP confinement for sensing. For instance, nanoparticles may be packed into small volumes, and nanoholes may be employed as nanochannels.

The confinement of the electromagnetic field at the surface also enables waveguiding applications, and a range of subwavelength optical manipulations (Barnes et al. 2003, Ebbesen et al. 2008). Specifically, light may be guided by SPs with penetration-depth dimensions below the diffraction limit and thus communication using SPs, or SP circuitry, has promise for further downscaling of, and communication between, optical/electrical devices. As will be discussed later in this chapter, the challenges associated with SP waveguiding are in the trade-off between propagation and confinement. Specifically, increased confinement, which is desirable from the perspective of downsizing, leads to increased proximity to the metal and associated increased losses. The result is that a compromise must be made between the degree of confinement and the extent of propagation.

There is potential for SP waveguiding in cases were the required propagation distances are very small, as in the case of communication between plasmonic devices on the scale of tens to hundreds of nanometers. There is also real benefit in incorporating plasmonic waveguiding and plasmonic shaping optics to concentrate and direct SP fields in the vicinity of sensors. An example of this type is plasmonic Bragg gratings surrounding nanohole arrays (Marthandam and Gordon 2007). Examples of these applications of plasmonic waveguiding are provided herein (see Section 18.2.6).

The controlled manipulation of particles and molecules using light, first demonstrated in 1970 (Ashkin 1970) is an important and growing field (see also Chapter 14). External optical control provides a valuable handle with which to study a wide variety of physics at small scales, such as the physical properties of large biomolecules, and for the analysis of single cells. Traditional optical trapping makes use of a combination of optical scattering and gradient forces to manipulate particles exhibiting refractive indexes distinct from their surrounding environment, and has evolved to include multiple traps in structured light fields (Dholakia and Lee 2008). Employing evanescent light for trapping particles close to surfaces has received much recent interest (Righini et al. 2008, Yang et al. 2009). Most notably, the evanescent field from subwavelength structures may be utilized for trapping, circumventing the minimum particle size imposed by the diffraction limit in traditional optical trapping. SPs are particularly attractive in the context of optical trapping for two reasons (Righini et al. 2008, Yang et al. 2009): the intensity enhancement of SP fields leads to decreased intensity requirements, and the subwavelength confinement of the SP field presents an opportunity to trapping on the nanometer scale.

In the context of the field of optofluidics, the marriage of fluidics and optics (see detailed definition and overview in Chapter 1), plasmonics present a unique array of optical tools that may be coupled with fluidics in a variety of ways. The most common application of plasmonics is currently SPR-based sensing in the traditional Kretschmann configuration. Optofluidic integration in that case takes the familiar form of fluidics, enabling by way of sensor functionalization, upstream sample treatment/preconcentration, transport of analytes to the sensor, and subsequent cleaning/reconditioning. In the case of a single large-area SPR detection, the fluidic portion is just an equivalently sized fluid layer. In the case where SPR imaging and several detection spots are used, fluidics have been used to define, functionalize and address the individual sensing areas (Kanda et al. 2004), analogous to microarray technologies. SP applications involving nanostructure-enabled LSPs are very well suited to integration with fluidics. Particularly with respect to sensing, the inherent benefits associated with LSPs can only be realized in conjunction with similarly localized fluid flow and control. Similarly, many synergies exist between emerging plasmonic waveguiding, plasmonic trapping, and microfluidic and nanofluidic transport.

In this chapter, an overview of plasmonics as relevant to the emerging field of optofluidics is presented. The objective is to introduce the field and highlight relevance to optofluidic systems. Reference will be made to some individual contributions as well as many detailed reviews in each sub-area discussed. In the next section, an overview of the physics of different plasmonic phenomena relevant to optofluidic applications is presented. Next, applications involving propagating, nonlocalized SP modes and associated sensing applications are considered. Nanoparticle-based SPs are discussed next, along with sensing applications as well as other emerging applications. Nanohole-based SPs and applications involving these structures are discussed. Enhancement of Raman scattering using plasmonics (surface-enhanced Raman scattering) and its application to sensing is discussed next. Plasmonic waveguiding is discussed as well as its applications in optofluidics. An overview is provided on the application of plasmonics to trapping and recent demonstrations in this area. Lastly, an overview of optofluidic integration and multiplexing with plasmonics is provided with some recent examples.

There are many synergies between this and other chapters in this book. References are made throughout this chapter to other chapters in this book that provide further background, or a more detailed discussion of the given topic.

18.2 Background and Developments

18.2.1 Overview of Relevant Plasmonics Theory

Sommerfeld's study of radiating dipoles at the earth's surface showed the existence of surface wave solutions (Sommerfeld 1909). The practical importance of these solutions led to the understanding that antennae should be placed away from the ground to allow for free-space radiation, instead of exciting the surface waves (a similar phenomenon occurs for radiating molecules close to a metal plane). The study of electron beams by Ritchie showed that such surface waves played an important role in electron loss at metal surfaces (Ritchie 1957). As described in the introduction, there is a momentum mismatch between the surface waves at a metal surface, SPPs, and plane waves in the dielectric. Therefore, some phase-matching element is required to couple into the surface waves. All-optical evanescent coupling was demonstrated by ATR by Kretschmann (1971). This Kretschmann geometry plays an important role in contemporary surface-binding optical sensors since it requires only a prism and a thin metal film. Surface-sensing technologies make use of the fact that the fields associated with SPPs are decaying exponentially away from the surface, and so they are most sensitive to refractive index changes at the surface. It is also possible to couple to SPPs via gratings and waveguides, as will be discussed later. In the context of coupling to SPPs, it is worthwhile to note that SPP loss, which comes from the lossy metal, enables the detection of coupling. A lossless system would reemit all of the energy that was coupled into the SPP, and thereby the signature of reduced out-coupling would be lost.

18.2.1.1 Optical Response of Metals

The optical response of metals in the visible-IR regime is often well approximated by the Drude–Sommerfeld model (Drude 1900). The Drude model basically considers the electrons to be freely conducting, except for scattering events that occur with a time-averaged delay of τ. As a result, the conductivity of the metal can be written as

$$\sigma(\omega) = \frac{Ne^2\tau}{m} \frac{1}{1 + i\omega\tau} \tag{18.1}$$

where

$i = \sqrt{-1}$

$\omega = 2\pi f$ is the angular frequency of the electric field

m is the electron mass
e is the electron charge
N is the free-electron density

Considering the Maxwell–Ampère differential equation

$$\vec{\nabla} \times \vec{B} = \mu_0 \vec{J} + \mu_0 \varepsilon_0 \varepsilon_b \frac{\partial \vec{E}}{\partial t} \tag{18.2}$$

with μ_0, ε_0 being the free-space permeability and permittivity and $\vec{J} = \sigma(\omega)\vec{E}$, then it is possible to formulate an effective relative permittivity for this Drude response given by

$$\varepsilon_r(\omega) = \varepsilon_b - \frac{Ne^2}{\varepsilon_0 m} \frac{1}{\left(\omega^2 + i\omega/\tau\right)} \tag{18.3}$$

where ε_b is the background dielectric permittivity from different polarization mechanisms, which is typically frequency dependent. The plasma frequency is typically defined as

$$\omega_p = \sqrt{\frac{Ne^2}{\varepsilon_0 m}} \tag{18.4}$$

and reasonable values are obtained for noble metals using the Drude theory.

For frequencies less than the plasma frequency, the real part of the relative permittivity is negative. For even smaller frequencies, where the negative real part of the relative permittivity is less than the real part of the dielectric layer above the metal, SPPs can exist. Of course, the Drude model is not a good approximation when interband transitions are also allowed within the metal. For example, gold and copper have interband transitions within the visible region that modify their response considerably; in some cases in the scientific literature, the interband contribution of gold that gives anomalous transmission near 500 nm wavelength has been incorrectly assigned to the bulk plasmon. Figure 18.2 shows the real and imaginary parts of the relative permittivities in the visible–near-IR region for gold, copper, silver, and aluminum. The Drude parameters, as fit to the near-IR part of the spectrum, are also shown for gold, copper, and silver.

18.2.1.2 SPPs as a Waveguide Mode

SPPs are commonly described as electromagnetic waves formed by charge oscillations at the surface of a metal. Figure 18.3a shows a popular schematic, which is useful for visualizing the SPP. We see many interesting features from this figure. First, the electrons in the metal have moved to create positive and negative charge distributions and electric field polarization. The term "polariton" refers to a wave of polarizations, positive and negative charges in a row. Second, the electric field normal to the surface is out-of-phase by $\pi/2$ with respect to the electric field parallel to the surface. Third, the wave must have a TM component.

In a dielectric slab waveguide, light is kept within a high-refractive-index core by total internal reflection. Using the usual reflection picture for a dielectric slab, it is not possible to visualize waveguiding from a single interface. The SPP waveguide can exist because the self-consistency relation can be fulfilled with a phase change across the boundary when the relative permittivities of the materials have opposite sign.

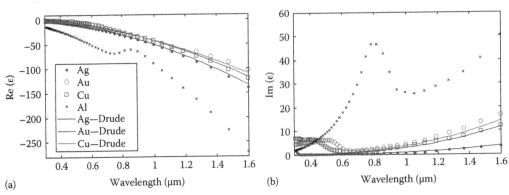

(a)   (b)

FIGURE 18.2 Real (a) and imaginary (b) parts of relative permittivity for silver, gold, copper, and aluminum in the visible and near-IR part of the optical spectrum. Drude parameters used are Ag: $\omega_p = 14 \times 10^{15}$/s, $\tau = 31 \times 10^{-15}$ s; Au: $\omega_p = 13.8 \times 10^{15}$/s, $\tau = 9.3 \times 10^{-15}$ s; Cu: $\omega_p = 13.4 \times 10^{15}$/s, $\tau = 6.9 \times 10^{-15}$ s. (From Johnson, P.B. and Christy, R.W., *Phys. Rev. B*, 6, 4370, 1972. With permission; Okamoto, T., Near-field spectral analysis of metallic beads, in Kawata, S. (ed.), *Near-Field Optics and Surface Plasmon Polaritons*, Springer-Verlag, Berlin, Germany, pp. 97–122, 2001. With permission.)

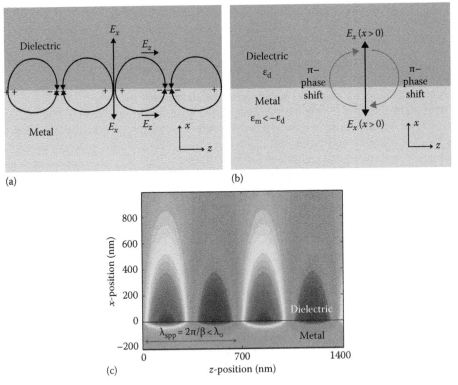

FIGURE 18.3 (See color insert following page 11-20.) (a) Schematic representation of SPP as charge oscillations at the interface between a metal and a dielectric. The electric field has a longitudinal (z-direction) component that is $\pi/2$ out-of-phase with the transverse component (x-direction). (b) Schematic of SPP self-consistency relation that is allowed for by a negative relative permittivity in the metal. The electromagnetic wave has a π phase shift upon crossing the boundary, allowing for self-consistency when crossing the boundary twice. This allows the SPP waveguide mode to exist at only a single interface. (c) Calculated transverse magnetic field (y-direction) for an SPP above gold at free-space wavelength of 700 nm. The SPP wavelength is shorter than the free-space wavelength, as described in the text. (Reprinted from Gordon, R., *IEEE Nanotechnol. Mag.*, 2, 12, 2008. With permission.)

The self-consistency condition for the waveguide modes is that the round-trip phase of the mode is a multiple of 2π, including the phase-of-reflection and propagation. This is required so that the transverse field distribution does not change with propagation. Metals can satisfy the self-consistency condition with only a single interface because some metals have a negative relative permittivity in the visible regime. This means that the normal component of the electric field changes sign when crossing the interface—the normal component of the electric displacement is continuous at the boundary. This sign change is equivalent to a phase shift of π; when crossing the interface twice, the total phase shift is 2π and the self-consistency condition is naturally satisfied. Thereby, a metal–dielectric interface can sustain a waveguide mode. Figure 18.3b shows this phase shift schematically.

Solving Maxwell's equations with exponential solutions decaying away from the interface and propagation in the plane of the interface gives the form of the y-component of the SPP magnetic field:

$$H_y(x,z,t) = \begin{cases} \exp(-\gamma_d x + i\beta z - i\omega t) & x > 0 \\ \exp(\gamma_m x + i\beta z - i\omega t) & x < 0 \end{cases} \tag{18.5}$$

with

$$\gamma_{m,d} = \frac{\omega}{c}\sqrt{\frac{-\varepsilon_{m,d}^2}{\varepsilon_m + \varepsilon_d}}$$

and

$$\beta = \frac{\omega}{c}\sqrt{\frac{\varepsilon_m \varepsilon_d}{\varepsilon_m + \varepsilon_d}}$$

where
$i = \sqrt{-1}$
$\omega = 2\pi f$ is the angular frequency of light
c is the speed of light in vacuum
ε_m is the relative permittivity of the metal
ε_d is the relative permittivity of the dielectric

This equation describes a waveguide mode for $\varepsilon_m < -\varepsilon_d$, ignoring the typically smaller imaginary contributions that lead to loss (loss can be captured by the imaginary part of the propagation constant in Equation 18.5). The remaining electric field components can be found by using the Ampère–Maxwell equation. The optical properties of metals have been considered in detail in past works (e.g., (Johnson and Christy 1972, Rakic et al. 1998, Okamoto 2001)). Figure 18.3c shows the TM field of the SPP. The SPP is a TM (zero magnetic field component in the direction of propagation) waveguide mode.

The propagation constant of the SPP gives an effective index

$$n_{eff} = c\beta/\omega = \sqrt{\frac{\varepsilon_m \varepsilon_d}{\varepsilon_m + \varepsilon_d}} \tag{18.6}$$

that is greater than the refractive index of the dielectric. In other words, the SPP has a shorter wavelength than plane waves in the dielectric. This means that the SPP waveguide mode is not coupled to

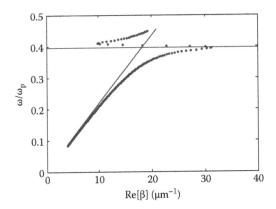

FIGURE 18.4   Dispersion of an SPP at the interface between Ag and vacuum. The solid diagonal line corresponds to the dispersion of a plane wave in vacuum, called the light line. The solid horizontal black line corresponds to the condition that $\mathrm{Re}\{\varepsilon_m\}=-1$.

free-space plane waves; just as light inside a fiberoptic cable stays inside the fiber. In both cases, we can add a prism or a grating to couple plane waves to the SPP or to the waveguide mode inside an optical fiber by a process of phase matching (simultaneous momentum and energy conservation).

It is interesting to note that the effective index becomes large as $\varepsilon_m \to -\varepsilon_d$, which results in a shrinking of the optical wavelength of the SP (Figure 18.4). Around this region, it is possible to image small features at the surface or across a metal film, which is promising for applications in subwavelength imaging, data-storage, and lithography (Pendry 2000, Fang et al. 2005).

18.2.1.3 Nanostructured SPP Waveguiding

Metal–insulator–metal (MIM) and insulator–metal–insulator (IMI) waveguides are variants of SPP at a single interface. MIM waveguides have the advantage of allowing for a much more localized field in the insulator gap region. Figure 18.5a shows the MIM gap waveguide. The solution to the gap is equivalent to the TM dielectric waveguide, except that the relative permittivities of the cladding layers are negative and the TM field has a hyperbolic cosine dependence inside the gap. The equation to find the solution for the effective index of the MIM waveguide mode is

$$\tanh\left(\frac{d}{2}\sqrt{\beta^2 - \left(\frac{\omega}{c}\right)^2 \varepsilon_d}\right) = -\frac{\varepsilon_d}{\varepsilon_m}\sqrt{\frac{\beta^2 - \left(\frac{\omega}{c}\right)^2 \varepsilon_m}{\beta^2 - \left(\frac{\omega}{c}\right)^2 \varepsilon_d}} \tag{18.7}$$

Figure 18.5b plots the effective index of the MIM waveguide mode as a function of gap width, d, for two different wavelengths. Two important results can be seen from this figure:

1. As d is made small, the effective index becomes large. In other words, the phase velocity slows down and the wavelength becomes smaller: we can confine the light more tightly by making the gap smaller. This has many interesting effects. For example, if you want to increase cutoff wavelength of a rectangular hole in a metal, it is better to reduce the size of the hole in the minimum dimension, rather than making it larger (Gordon and Brolo 2005).
2. As d is made smaller, the group index increases; light is slowing down as we make the MIM gap narrower. Therefore, if we gradually taper the gap to a point, light will slow down as it

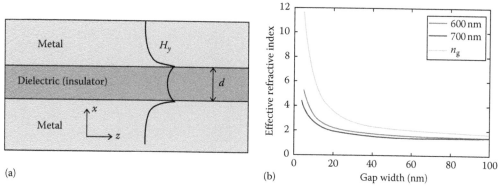

FIGURE 18.5 (a) Schematic of MIM structure with gap d. TM field is plotted vertically, which has a hyperbolic cosine dependence in the gap, an exponential decay into the metal. (b) Effective refractive index of waveguide mode within MIM gap, which increases as the gap is made smaller. Calculations for two free-space wavelengths are shown using the relative permittivity of gold and the difference is used to estimate the group index, which also increases as the gap is made smaller. Therefore, the light slows down as the gap is made smaller. (Reprinted from Gordon, R., *IEEE Nanotechnol. Mag.*, 2, 12, 2008. With permission.)

progresses, and so the energy will build up (provided that the loss is sufficiently small and the taper is adiabatic to minimize reflection). Strong field enhancement due to the slowing down of the SP has been demonstrated at a metal tip where the dielectric surrounds the metal (Stockman 2004). A disadvantage of the MIM structure is that the mode experiences higher loss, so that the propagation distances are greatly reduced. As a rough guide, a 1/10th wavelength gap can exhibit a propagation length on the order of the wavelength (see Section 18.2.6 for more specific dependencies).

The IMI structure has the same solution as the MIM structure, except that the metal and the dielectric relative permittivities are switched. These structures also allow for much greater propagation lengths, which has advantages in terms of interferometric sensors.

18.2.1.4 Coupling to Surface Plasmons: Kretschmann Phase Matching and Waveguide Phase Matching

Since the SPP is a bound waveguide mode, it does not couple directly to free-space radiation. In order to achieve coupling, one should use evanescent coupling, from a high-dielectric medium or a co-propagating waveguide. The Kretschmann coupling geometry is shown in Figure 18.6a. The angle of the plane wave for which the phase matching occurs is given by

$$\theta = \arcsin\left(\mathrm{Re}\left\{ \sqrt{\frac{\varepsilon_m \varepsilon_d}{\varepsilon_m + \varepsilon_d}} \right\} \Big/ n_p \right) \tag{18.8}$$

where n_p is the refractive index of the prism. For example, for gold at a free-space wavelength of 632.8 nm, if $\varepsilon_m = -11.8 + 1.2i$, $\varepsilon_d = 1.7689$ (water), $n_p = 1.56$, the SPR angle is calculated to be 67.48°, which agrees well with the dip in reflection seen in Figure 18.6b. The polarization should also be TM, so the magnetic field is directed out of the plane of incidence to match the polarization of the SPP. The film thickness should be optimized since the light from the plane wave has to tunnel through the metal film. If the thickness of the metal is too great, light will be reflected before it has a chance to couple into the SPP. If the film thickness is too small, the light will couple back rapidly and radiate away through the

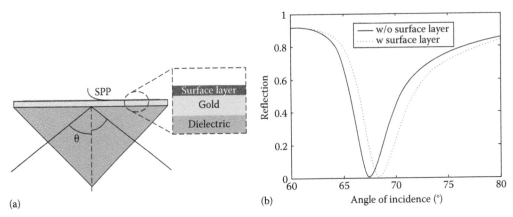

FIGURE 18.6 (a) Schematic of ATR configuration for SPR sensing. The ATR configuration couples plane waves inside the prism into SPP waves when the angle and wavelength are tuned to the phase-matching condition. Since the SPP has loss, the reflection is reduced when there is coupling to the SPP wave. (b) Calculated SPR reflection for a 50 nm gold layer at free-space wavelength of 632.8 nm, with and without a 5 nm surface layer of refractive index 1.5. (It is assumed that above the surface layer is water, with refractive index 1.33, and the prism has a refractive index of 1.56). The shift in the reflection dip is used to detect the surface layer. (Adapted from Gordon, R., *IEEE Nanotechnol. Mag.*, 2, 12, 2008. With permission.)

prism. Since this problem is one dimensional, it is easy to implement a transmission matrix to get a quantitative evaluation of the coupling as a function of angle (Gordon 2008). This transmission matrix approach also allows for calculating the electric field intensity enhancement at the surface for the optimum metal thickness, with respect to the incident plane wave, which is given by (Raether 1988)

$$F = \frac{2\left(\mathrm{Re}\{\varepsilon_m\}\right)^2 \sqrt{-\mathrm{Re}\{\varepsilon_m\}(\varepsilon_d - 1) - \varepsilon_d}}{n_p^2 \, \mathrm{Im}\{\varepsilon_m\}\left(1 - \mathrm{Re}\{\varepsilon_m\}\right)}. \tag{18.9}$$

This enhancement factor can be over 100 for silver films. The transmission matrix formulation can also be used to determine the coupling if a surface layer is added, where a change in the coupling provides surface-sensing capability. This technique is called surface plasmon resonance (SPR) sensing. ATR is the most common method for SPR. It uses phase-matched evanescent coupling between propagating light in a high refractive index medium and the SPP. The phase matching is controlled by tuning the angle and/or wavelength of the incident beam. For the phase-matched condition, light is transferred to the SPP and so there is increased loss (from the metal) and reduced reflection.

As shown in Figure 18.6, the ATR geometry can also be used for an SPR sensor. Figure 18.6b shows the reflection as a function of incidence angle when the refractive index of a 5 nm surface layer is modified to 1.5 (from 1.33 of the aqueous surrounding medium). The shift in the reflection minimum (in this case 0.7°) is used to sense the absorption of molecules at the surface. It is important to include the loss of the metal in this calculation, otherwise all the light that is coupled into the SPP is coupled out of it once again and so the reflection is always 100%. As will be discussed in the next section, SPR has applications in drug development, medical diagnostics, environmental monitoring, and food and water safety (Homola 2008).

18.2.1.5 Waveguide Phase Matching

The condition for waveguide phase matching is similar to Kretschmann phase matching, except instead of using a prism and a plane wave, a waveguide and a waveguide mode are used. In this case, the SPR condition becomes that the propagation constant of the waveguide mode should equal the propagation constant of the SPP.

18.2.1.6 Grating Phase Matching

Gratings can be used to achieve phase matching to couple light from radiation modes, or from a waveguide, into the SPP (as described in Chapter 10). In this case, the grating period provides the wave-vector component to make up for the mismatch between the SPP and the excitation wave. For coupling from a plane wave, the Bragg resonance occurs for the phase-matching angle:

$$\theta = \arcsin\left(\pm\frac{\lambda_0}{2\pi\sqrt{\varepsilon_d}}K_{u,v} + \sqrt{\frac{\varepsilon_m\varepsilon_d}{\varepsilon_m + \varepsilon_d}}\right) \tag{18.10}$$

where

λ_0 is the free-space wavelength

$K_{u,v}$ are the grating resonances in reciprocal space

If we consider a square grating with periodicity d,

$$K_{u,v} = 2\pi\sqrt{\varepsilon_d(u^2 + v^2)}\big/d \tag{18.11}$$

where u,v are integers corresponding to the different order grating resonances. For normal incidence $(\theta=0)$, the free-space wavelength that excites the resonances in a square lattice is given by

$$\lambda_0 = \frac{d}{\sqrt{(u^2 + v^2)}}\sqrt{\frac{\varepsilon_m\varepsilon_d}{\varepsilon_m + \varepsilon_d}} \tag{18.12}$$

It is also possible to sense refractive index changes at the surface by noting changes in the resonant coupling wavelength.

The above descriptions correspond to the most common SPP coupling methods. Variants of these phase-matching configurations are possible. For example, it is possible to have a grating-assisted waveguide coupling configuration.

18.2.1.7 LSPs in Metal Nanoparticles

LSPs are charge oscillations that are bound to a small metal particle or nanostructure. These oscillations can be represented by the displacement of charge in the sphere. For small particles, they are simply Rayleigh scattering, which can have a resonance when the relative permittivity of the scattering particle is negative.

For small particles, where the spatial variation of the electromagnetic field is confined to lengths much smaller than the optical wavelength, the time-dependent contributions to Maxwell's equations may be neglected, so that the solution to the field distribution becomes the same as the electrostatic solution (Jackson 1998). For example, for a metal sphere in a dielectric, the field inside the metal is given by

$$\vec{E}_{in} = \frac{3\varepsilon_d\vec{E}_0}{\varepsilon_m + 2\varepsilon_d} \tag{18.13}$$

where $\vec{E}_0$ is the electric field away from the sphere, which contains the time dependence in the quasistatic approximation. It is clear that the field inside the sphere is maximized when $\mathrm{Re}\{\varepsilon_m\}=-2\varepsilon_d$ (ignoring dielectric losses), where $\mathrm{Re}\{\}$ is the real part. Considering continuity of the parallel component of the electric field at the boundary, this leads to a strong enhancement of the field on the outer surface of the sphere, which is limited in practice by the imaginary part of ε_m. It is clear that the field at the surface of a nanoparticle is enhanced with respect to the incident field; for example, for small spherical silver particles at resonance 350 nm, the electric field intensity is enhanced by close to 500 (Raether 1988).

For a more accurate theory, it should be kept in mind that the bulk permittivities of metals are less accurate for small particles because of electron scattering at the boundaries of the particle (Okamoto 2001); this has been treated phenomenologically by adding (using Matthiessen's rule) a surface-scattering rate to the free-electron scattering, typically given by the Fermi velocity divided by the particle size.

The LSP-enhanced field goes hand-in-hand with an enhanced local dipole, and so the scattering will be most intense near resonance. The equivalent dipole moment of a Rayleigh scatterer is given by

$$\vec{p} = V\varepsilon_0\varepsilon_d \left(1+\kappa\right)\left(\frac{\varepsilon_m-\varepsilon_d}{\varepsilon_m+\kappa\varepsilon_d}\right)\vec{E}_0 \tag{18.14}$$

where
V is the particle volume
$\kappa = 2$ for a sphere

Changing the particle shape influences the resonances, and κ typically increases with aspect ratio for the orientation of the electric field along the long axis. From the point-of-view of sensing and interacting with materials, this shape effect offers possibilities of shifting the resonance wavelength, enhancing scattering and enhancing the local field (Kreibig and Volmer 1995). For example, a silver spheroid with an aspect ratio of 3 has its resonance shifted to 500 nm, and the electric field intensity enhancement is 10^5 (Raether 1988). The far-field scattering cross section of a sphere is

$$\sigma_{scat} = \frac{128\pi^5 a^6 \varepsilon_d^2}{3\lambda_0^4}\left|\frac{\varepsilon_m-\varepsilon_d}{\varepsilon_m+2\varepsilon_d}\right|^2 \tag{18.15}$$

and the absorption cross section is

$$\sigma_{abs} = \frac{8\pi^2 a^3 \sqrt{\varepsilon_d}}{\lambda_0}\mathrm{Im}\left\{\frac{\varepsilon_m-\varepsilon_d}{\varepsilon_m+2\varepsilon_d}\right\} \tag{18.16}$$

where λ_0 is the free-space wavelength and Im{} is the imaginary part.

(Scattering/absorption cross sections can be thought of as the equivalent area over which the incident light field is totally absorbed/scattered. The extinction is the sum of the two and is proportional to the measured change in the transmitted field intensity.) It is clear from these expressions that the scattering and absorption resonances also occur around the region of greatest field enhancement (Figure 18.7). For a single particle, the quality of the resonance is limited by the dispersion of the metal and the dielectric, as is clear from the field-enhancement denominator in Equation 18.13.

In keeping with the quasistatic approximation, we see that the resonance is independent of particle size, but this is not the case for larger spherical particles where Mie theory, including retardation effects, should be used. The retardation effects can broaden and shift the resonances significantly, even for particles ~100 nm (Okamoto 2001).

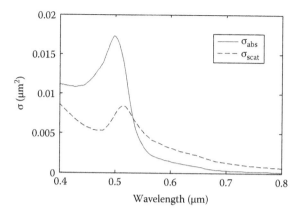

FIGURE 18.7 Scattering cross sections for a 50 nm gold particle in vacuum using the bulk relative permittivity values.

Several opportunities arise in proximity effects from various plasmonic nanoparticles. If particles are brought close together, their coupling modifies the polarizability, which results in substantial shifts in the scattering resonance wavelength. This gives greater sensitivity to the sensing of adsorbed particles if they are tethered with a plasmonic particle (as discussed in Section 18.2.3). Interparticle coupling can also enhance the local field intensities in a way that is useful for enhanced interaction with materials. If we cascade this effect, with smaller and smaller particles in close proximity, the total increase in the local electromagnetic field can be made very large. For example, electric field enhancements of over 1000 have been calculated for self-similar spherical particles in a row (Li et al. 2003).

It is clear from the brief analysis above that there are many potential sensing modalities for LSPR sensing around nanoparticles. Clearly the resonances are dependent upon the local refractive index around the particle. Therefore, shifts in the scattering or extinction spectra can provide information about the local refractive index environment around the particles. One advantage of LSP is that the sensing volume is reduced to the local environment, and so a smaller limit of detection and greater multiplexing are possible. In this case, sensitivity should be considered in terms of the amount of analyte, rather than the bulk refractive index shift measured.

18.2.1.8 LSPs in Nanoholes in Metal Films

It is also possible to get LSPs from holes or slits in metals since the light inside the hole will be reflected at each end (due to impedance and geometric mismatch), leading to standing waves that are localized in the hole, commonly referred to as Fabry–Perot resonances (Garcia-Vidal et al. 2006). For thinner metal films, there is only the lowest-order Fabry–Perot resonance that is peculiar compared with conventional Fabry–Perot resonances, because all of the phase change comes from the phase of reflection at the entrance and the exit of the hole—there is only a small phase change from propagation inside the hole when the mode is close to cutoff. Overall, the total phase change should be close to a multiple of 2π to fulfill the resonance condition. The shape of the hole also has a strong influence on the resonance properties (Gordon et al. 2004), which has been observed in experiments (Degiron et al. 2002).

The LSP inside the hole can have a strong influence on the transmission properties of a single hole, but also plays a role in the extraordinary transmission-hole arrays. Because of the strong localization of the field in single holes (near resonance, or above cutoff), single holes in metals have been used to observe single fluorescent molecules (Levene et al. 2003) and for optical trapping (Kwak et al. 2004). Periodic arrays of nanoholes exhibit additional optical phenomena relevant to optofluidic systems, and this will be covered in detail in Section 18.2.4.

18.2.2 Propagating (Nonlocalized) Surface Plasmons

Propagating SPs are the basis for the most common SPR-detection methodologies, or traditional SPR. As described in Section 18.2.1, traditional SPR detection involves plasmonic excitation at a metal film via prism coupling (see also Figure 18.6a). This Kretschmann-style plasmonic excitation is the basis of most commercial devices currently employing SPR detection (Homola 2008). The relatively recent availability of these commercial devices has resulted in a rapid increase in the applications of SPR in science, and has increased the awareness of the potential of plasmonics in sensing using propagating as well as localized SPs.

In general, traditional SPR using prism coupling methods exploit one-dimensional plasmonics. In other words, the confinement offered by SPs is one characteristic length, and the area over which they act is much larger. This configuration has the advantage of a surface-averaging effect that can by harnessed to provide increased sensor response from many small molecular adsorption events, and the disadvantage of increased size, and increased absolute limit of detection. With respect to the detection area, it is possible to subdivide the surface into much smaller discrete areas, commonly on the order of tens of microns wide and thus achieve a degree of planar multiplexing (Kanda et al. 2004). The result is a pixelated SPR output that can be interpreted as an image, or SPR imaging. The planar multiplexing achieved in SPR imaging is similar to microarrays and is thus well suited to integration with planar microfluidics (see also optofluidic integration in Section 18.2.8).

As plasmonics is a surface effect, the main application of SP sensor technology is in affinity sensing, that is, in the detection of the adsorption of an analyte to the active sensor surface. Detection is facilitated by the change in near-surface refractive index upon binding, and the subsequent sensitivity of the system to that change. Thus the sensitivity of SP-based sensors involves two considerations: how much the analyte adsorption alters the refractive index, and how well that change can be detected. The first consideration may be expressed as (Homola 2008)

$$\Delta n = \frac{\Gamma}{h}\left(\frac{dn}{dc}\right) \tag{18.17}$$

where
  Δn is the change in refractive index
  Γ is the surface concentration (mass/area)
  h is the thickness of the surface layer in which the binding takes place
  dn/dc is the refractive index change per concentration (generally on the order of 0.1 mL/g)
    (Ho et al. 2005)

The response of the SPR device to Δn is the second consideration. The refractive index resolution is commonly quantified in terms of the refractive index unit (RIU). Propagating SPR sensors are typically in the range of 10^{-5} to 10^{-6} RIU, with the best on the order of 10^{-7} RIU. It is noteworthy that under typical conditions, the above parameters lead to a surface-coverage resolution on the order of picograms per square millimeter.

Applications of propagating SPR-based sensors have focused on three main areas (in the order of prevalence) (Homola 2008): food quality/safety, medical diagnostics, and environmental monitoring (Ho et al. 2005). In the area of food quality/safety, the most frequently targeted analytes were bacterial pathogens with limits of detection down to 10^2 cells/mL (Oh et al. 2003). Detection of cancer markers has dominated SPR sensor applications in medical diagnostics with limits of detection on the order of 1 ng/mL achieved in some cases (Wu et al. 2006). Pesticide detection has been most prevalent in the area of environmental monitoring, with limits of detection down to the pg/mL level (Farre et al. 2007).

It is noteworthy that developments in SPR sensor technology and optofluidic integration must be complimented by the development of biorecognition elements and associated immobilization schemes. The benefit of label-free detection achieved using SPR technology relies on biorecognition elements that bind effectively, and highly selectively to their intended target. When working with complex (real) samples, nonspecific adsorption of proteins and general biofouling are major problems. For example, serum, the most commonly employed fluid for SPR-based medical diagnostics, must be highly diluted in order to avoid nonspecific adsorption. While the dilution of analyte is undesirable from a sensing perspective, to date it has been a requirement for serum-based testing, and other complex fluids. These challenges associated with biorecognition are not unique to SPR-based detection; however, they are important when considering ultimate application of SP sensing technologies.

18.2.3 Localized Surface Plasmons: Metal Nanoparticles

Due to a combination of factors, nanoparticles are perhaps the most studied nanostructure. Metal nanoparticles are of particular interest due to the role of plasmonic effects in their optical interactions. As described in the theory section (Section 18.2.1), LSP resonances in gold and silver nanoparticles may be excited by incident light in the visible range resulting in increased field intensities at the surface of the nanoparticles, and wavelength-specific absorption. These properties make metal nanoparticles attractive in spectroscopy (see also Chapter 6). Nonspherical nanoparticles can exhibit additional phenomena such as enhanced field localization and distinct resonance frequencies corresponding to differing characteristic lengths. Examples include metallic nanorods and triangular metallic nanoparticles. Composite nanoparticles have also been developed, such as core-shell spherical nanoparticles, that exhibit optical properties dependent on the shell thickness. An illustrative example of the tunability of metallic nanoshell particles is given in Figure 18.8.

Analogous to other plasmonic substrates, the spectral location of a nanoparticle resonance red-shifts in response to an increased refractive index at the surface of the particle. The combination of their small size, highly confined and enhanced electromagnetic field, and spectral response to the near-surface refractive index motivate the application of metal nanoparticles to sensing (Stewart et al. 2008). At the heart of many nanoparticles-based sensing methods are plasmonic coupling between nanoparticles, and between nanoparticles and surfaces. Some of these sensing modes are discussed next.

Plasmonic coupling between nanoparticles increases as the distance between them decreases. At very small distances, typically less than the particle diameter, the region between the particles develops a

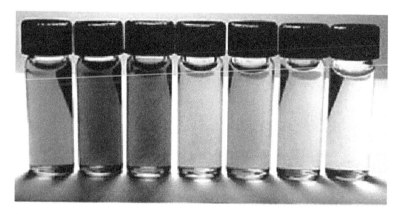

FIGURE 18.8 (See color insert following page 11-20.) Visual example of absorption characteristics of metallic nanoparticles as a function of nanoparticle properties. This case involves Au shell–silica core particles with differing shell thickness. (Reprinted from Loo, C. et al., *Technol. Cancer Res. Treat.*, 3, 33, 2004. With permission.)

strongly enhanced field as a function of frequency. Colorimetric detection is one sensing mode that exploits these properties. Specifically, suitably functionalized particles can aggregate when bonded to an analyte of interest, such as an oligonucleotide, and result in a visible color change (Mirkin et al. 1996). Alternatively, a color change may be observed when an analyte breaks linkages between previously linked nanoparticles. While many studies involving colorimetric detection using metal nanoparticles have focused on DNA detection, many other protocols have been developed, for instance to detect cocaine (Liu and Lu 2006) and lead (Liu and Lu 2003).

Less common, but potentially more powerful detection methods are possible when only the scattered light is detected as opposed to the extinction (combination of absorption and scattering) employed in colorimetric detection (Stewart et al. 2008). One approach is to evanescently excite the nanoparticles spotted onto a planar waveguide and detect the scattered light. The spectrum from each spot provides an indication of the degree of nanoparticle aggregation, which in turn indicates the presence or absence of a binding analyte (Storhoff et al. 2004a). The sensitivity and simplicity of nanoparticle-binding, light-scattering-based detection makes it a strong candidate for point-of-care diagnostics, particularly as the method may be applied without PCR amplification and associated complexities. The scattering properties of plasmonic nanoparticles also make them excellent candidates as labels where fluorescent probes have most commonly been employed. Plasmonic nanoparticles have been employed as labels in DNA microarrays (Storhoff et al. 2004b).

The methods discussed in this section have focused on exploiting plasmonic nanoparticle optical properties and interactions for sensing. Nanoparticles can also serve to enhance other plasmonic sensing modes such as propagating SP waves excited with a prism in Kretschmann configuration (Teramura and Iwata 2007). The addition of nanoparticles that are functionalized as labels, that is to bind specifically to target analyte, can improve sensitivity of traditional SPR significantly. Dielectric particles can increase sensitivity by increasing the local refractive index several folds more than the shift associated with the analyte alone. Metal nanoparticle labels can additionally electromagnetically couple to the SP on the base substrate and induce further red-shifting, and thus improved sensitivity.

Two additional noteworthy applications of plasmonic nanoparticles deserve mention here. First, with respect to imaging, functionalized plasmonic nanoparticles can serve as labels to bind and congregate near cells that are expressing a particular protein, such as a cancer marker, for instance (Kumar et al. 2007). Second, the large optical absorption cross section of plasmonic nanoparticles presents opportunities for *in vivo* therapy. Specifically, nanoparticles agglomerated in a specific region of the body, such as a cancerous tumor, may be excited by near-infrared light, and the resulting photothermal heating can cause local cell death. It is also possible to use suitably designed nanoparticles for both imaging and therapy. In one application, Loo et al. (2005) employed core-plasmonic nanoparticles to detect and destroy cells that over express HER-2, an established cancer marker. As shown in Figure 18.9, the plasmonic nanoparticles functionalized with the specific antibody label (right column) preferentially bind and aggregate to the cancerous cells resulting in increased scattering (observed by darkfield imaging). Exposure to near-infrared light results in photothermal cell death in the labeled cells (middle row, right). A silver stain confirms the increased concentration of specifically bound nanoparticles (bottom row).

18.2.4 Nanohole-Based Surface Plasmons

An interesting topic in plasmonics that has been receiving a lot of attention in the last few years is the phenomenon of "extraordinary optical transmission" (EOT) (Genet and Ebbesen 2007). EOT was first observed in 1998 by Ebbesen and coworkers (Ebbesen et al. 1998), and it is related to an unexpected increase in the amount of light transmitted by very small holes in a metal film. Intuitively, it is easy to imagine that the amount of light transmitted by a small circular aperture will decrease with the hole diameter. In fact, it has been established more than 60 years ago (Bethe 1944) that the amount of transmitted light decreases quickly as the diameter of the hole becomes smaller than its wavelength. In other words, a 200 nm hole should block almost all light incident at 600 nm, for instance. Ebbesen's

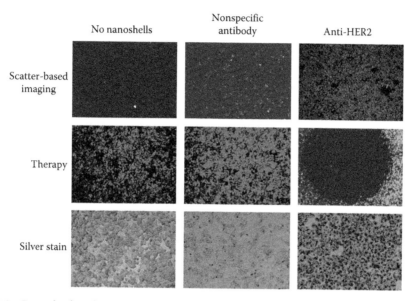

FIGURE 18.9 Example of combined cancer cell detection and therapy, via photothermal cell death, achieved with plasmonic nanoparticles (Loo et al. 2005). Detailed description provided in the text. (Reprinted from Loo, C. et al., *Nano Lett.*, 5, 709, 2005. With permission.)

group used periodic arrays of small holes (around 150 nm in diameter) in silver thin films (200 nm) and observed a surprisingly significant amount of transmitted light at long wavelengths (Ebbesen et al. 1998). They noticed that the wavelength positions of the peaks in light transmission were related to the periodicity (distance between the holes) of the square array of nanoholes. It was quickly recognized that SPs play an important role in EOT (Martin-Moreno et al. 2001). The general aspects of EOT are summarized in Figure 18.10a. In that scheme, white light is incident on the surface of the arrays of nanoholes generating

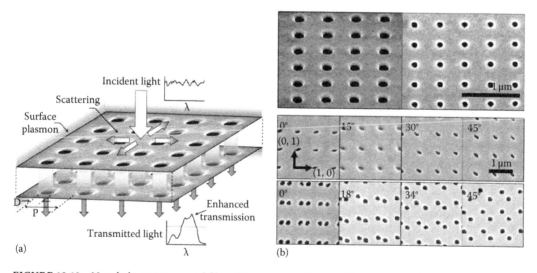

FIGURE 18.10 Nanohole arrays in metal films: (a) conceptual diagram illustrating incident light scattering into SP model and enhanced transmission at select wavelengths; and (b) scanning electron microscope images of several nanohole arrays of varying periodicity, and double hole arrays with varying lattice orientations. (Reprinted from Sinton, D. et al., *Microfluid. Nanofluid.*, 4, 107, 2008. With permission.)

SPs at certain wavelengths that propagate parallel to the surface. The SPs scatter into the apertures and couple to the other side of the film. The actual transfer of the surface waves (SP) from one side of the film to the other involve a Fabry–Perot resonance of the evanescent wave inside the hole, which couples with the structure on the other side and excites SP (Martin-Moreno et al. 2001). The SPs generated at the other side are converted back to transmitted light, generating enhanced transmission at certain frequencies characterized by peaks in the transmission spectrum. As discussed earlier in this chapter, a free photon cannot directly be converted to SPs due to a momentum mismatch. In the case of nanohole arrays, the requirements for photon—SP conversion is fulfilled by the periodic arrangement of the nanostructures. The position of the resonance peaks (λ_{SP}) will then follow the Bragg diffraction conditions of the grating and it is given by Equation 18.12. For comparison with other photonic-crystal-related phenomena see also Chapter 4.

The direct application of Equation 18.12 is not always straightforward, since the contributions from other types of surface waves, structural defects, and LSPs are not taken into account (Liu and Lalanne 2008). However, Equation 18.12 does provide an approximate prediction of the resonance position and explains the dependence of the resonance on the periodicity and on the refractive index of the medium in contact to the metal surface.

The arrays of nanoholes are fabricated using focused ion beam milling. Typical arrays are shown in Figure 18.10b. The diameters of circular nanoholes generally range between 100–250 nm, the thickness of the gold films are between 100–200 nm, and periodicities between 300–600 nm are appropriate to produce resonances in the visible range. The geometric parameters of the arrays have a profound effect on the properties of the transmitted light. For instance, the intensity of the transmitted light was shown to decrease exponentially with film thickness (Degiron et al. 2002); however, this dependence is only observed for films thicker than 200 nm. When the hole depth is less than 200 nm, the SPs from both sides of the film can couple, increasing the efficiency of the transmission. The properties of the transmitted light can also be tailored by the hole shape and orientation (Gordon et al. 2004, Koerkamp et al. 2004). Figure 18.10b shows a set of elliptical holes oriented at different angles relative to the lattice (Gordon et al. 2005). The shaped nanoholes shown in Figure 18.10b are polarizing structures, which means that they allow preferential transmission of light at a certain polarization direction. In both cases, for the elliptical holes and the double holes, light polarized parallel to the major axis will be blocked, and the structures work as nano-polarizers (Gordon et al. 2005). The direction of polarization of the transmitted light is also influenced by the orientation relative to the lattice for the case of the elliptical holes, but this effect is not observed for the double holes. This interesting behavior can be explained by considering the coupling between the surface waves (SPs) and the shaped nanoholes (Gordon et al. 2005).

Figure 18.11 shows zero-order (normal incidence) EOT spectra from nanohole arrays on gold films before and after surface modification (Brolo et al. 2004b). The amount of light transmitted at the peak is much larger than predicted by classical aperture theories. The minimum in the transmission curves, just before the transmission peaks (marked with asterisks in Figure 18.11), correspond to a diffraction phenomenon known as Wood's anomaly, which also depends on the geometry of the arrays, ε_d, and the angle of incidence (Kim et al. 1999). It can also be seen from Figure 18.11 that the resonance peaks present a Fano-type line shape resulting from the interference between the incoherent (direct transmission through the nanoholes) and the coherent (SP-mediated transmission) paths that contribute to EOT (Genet et al. 2003).

Figure 18.11 clearly shows a shift in the EOT resonance due to adsorption events. The red-shifts provoked by molecular adsorption are predicted by Equation 18.12, which shows a direct relationship between the resonance wavelength and the dielectric properties of the dielectric medium in contact to the metal. The bare array of nanoholes was cleaned in a plasma oven prior to the experiments and presented a distinct resonant peak at 645 nm. The transmission spectrum from the same array modified by a monolayer of mercaptoundecanoic acid (MUA) by a 24 h incubation shows a characteristic shift in the wavelength of maximum transmission (to 650 nm). Further modification of the surface by a protein (albumin) provoked an additional wavelength shift to 654 nm. The spectrum characteristic of a bare

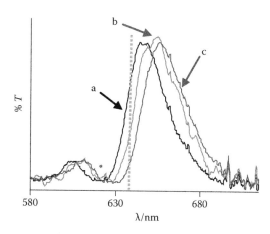

FIGURE 18.11 Detection of binding events on the surface of a nanohole array through red-shifts in the enhanced transmission peak. Normalized transmission spectra of white light at normal incidence are shown for (a) bare gold surface; (b) gold with monolayer of MUA; and (c) gold–MUA layer with BSA. The asterisk indicates transmission minimum resulting from Wood's anomaly, and the dashed line indicates a wavelength that is well suited to detection using single-line excitation. (Adapted from Brolo, A.G. et al., *Langmuir*, 20, 4813, 2004b.)

gold surface was again obtained after the surface species were removed by a plasma-cleaning treatment (Brolo et al. 2004b).

The results in Figure 18.11 marked the first demonstration of arrays of nanoholes being used to detect monolayer adsorption of organic and biological molecules. The shift in the resonance peak was similar to the response of commercial SPR sensors that operate in reflection geometry (Kretschmann configuration). However, the normal transmission geometry of the nanohole-based sensing scheme is much more appropriated for miniaturization and integration in lab-on-chip technologies than the Kretschmann geometry. This inherent advantage of nanohole arrays was recognized early, and several groups have since reported applications of nanohole arrays as sensing elements in microfluidic devices (Stark et al. 2005, Lesuffleur et al. 2007) (Sharpe et al. 2008). A significant amount of spectral information needs to be recorded to measure the wavelength shift precisely, as shown in Figure 18.11. This required specialized spectrometers and detectors. Alternatively, adsorption events can also be monitored using nanohole arrays by observing changes in light intensity at a fixed wavelength. This is illustrated by the dashed line in Figure 18.11 that emphasizes how the intensity of the transmitted light would change with the excitation at that fixed wavelength. This type of detection scheme can be easily implemented using a laser source and a simple photodiode detector. Alternatively, the transmitted laser light can be measured using an imaging CCD (Ji et al. 2008, Lesuffleur et al. 2008). In this case, the transmission from several arrays can be measured simultaneously, allowing the multiplexed detection of several biological markers at the same time. The small sensing areas of the arrays of nanoholes are conducive to highly integrated devices with potential for real-time monitoring of the dynamics of thousands of surface events in parallel.

A perceived disadvantage of nanohole arrays is their sensor output sensitivity, which is about an order of magnitude smaller than observed from commercial angle-resolved SPR systems (De Leebeeck et al. 2007). However, it is important to point out that the sensing area of the arrays of nanoholes is smaller and the SPR shift in this case is from a smaller number of molecules. In fact, nanohole-based detection systems with better sensitivity and resolution than commercial SPR devices have been reported (Stark et al. 2005, Tetz et al. 2006). In the case of Tetz et al. (2006) crossed polarizers were used to select between the coherent and the incoherent contributions to the EOT (these are the contributions that lead to the Fano lineshape observed in Figure 18.11). This scheme allows the recovery of a symmetrical line

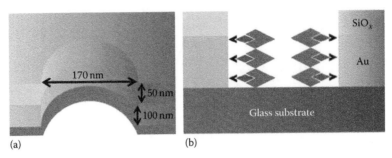

(a)   (b)

FIGURE 18.12 Schematic illustrating in-hole sensing. (a) Cross-section representation of a nanohole, illustrating the 100 nm gold layer sandwiched between the glass substrate and the SiO$_x$ nanolayer (50 nm thick). (b) Another cross section of a nanohole showing that the molecular binding only occurs inside the holes. The experiments were realized using 15 μm × 15 μm arrays contained 170 nm diameter nanoholes and a periodicity of 500 nm. (Adapted from Ferreira, J. et al., *J. Am. Chemical Society*, 131, 436, 2009.)

shape for the transmission which provides better sensitivity. The molecular confinement and the low detection limit of arrays of nanoholes was well illustrated in experiments that introduced the concept of "in-hole" sensing (Ferreira et al. 2009). In this case, the top surface of the gold was coated by a dielectric layer (SiO$_x$) using a sol-gel procedure. The nanoholes were milled through the SiO$_x$ layer and the gold film, exposing only the walls inside the holes. This concept is illustrated in Figure 18.12a below.

Figure 18.12b shows how the gold area inside the nanoholes was used to detect the binding of organic and biological molecules, while the outside gold surface was blocked. This new approach is more efficient than the previous surface-based sensitivity method, where the response was related to binding events taking place inside of the holes and on the top gold surface. Surprisingly, the sensitivity of this device was comparable to uncoated arrays of nanoholes. The detection of 3 amol of proteins was estimated from this scheme. Since the measurement of EOT through smaller arrays (with fewer holes) is straightforward, "in-hole" sensing can be easily extended to the detection of sub-attomolar amounts of proteins in the future.

Other nanohole configurations have been described that provide increased sensitivity to submonolayer amounts of adsorbed species. For instance, the introduction of a metal disk inside the nanoholes led to significant changes in the field distribution inside, as illustrated in Figure 18.13 (Stewart et al. 2006).

Nanoholes in metals can also be thought of as nanochannels for fluid transport. The combination of the transport characteristics through these channels and the plasmonic properties of the nanostructures

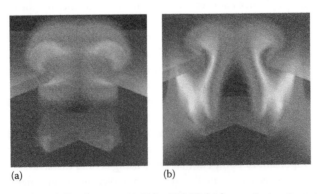

(a)   (b)

FIGURE 18.13 (See color insert following page 11-20.)   Hole/disk plasmonic structures. Computational results for the EM field distribution for 883 nm separation (a) and 1138 nm (b). The larger-wavelength excitation results in strong coupling between levels. (Reprinted from Stewart, M.E. et al., *Proc. Natl. Acad. Sci. USA*, 103, 17143, 2006. With permission.)

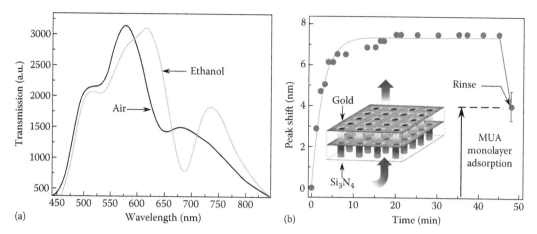

FIGURE 18.14 Nanoholes as nanochannels: flow-through nanohole-array-based sensing. (a) Transmission spectra for air and ethanol ($n = 1.359$) with a 500 nm periodicity array of 300 nm holes. (b) Measured peak shift (following 625 nm peak) as a function of time during flow through of an ethanol/MUA solution. As indicated in the inset, the sensor is operated with flow from the nonparticipating silicon nitride side to the active gold surface. Flow-through transport results in a red-shift in response to bulk refractive index change and surface absorption of MUA. For 45 min the array is flushed with pure ethanol resulting in a blue shift that indicates the sensor's surface response to the MUA monolayer alone. (Reprinted from Eftekhari et al., *Analytical Chemistry*, 81, 4308, 2009. With permission.)

opens the door for several opportunities (Eftekhari et al. 2009). Figure 18.14 shows the response of a flow-through nanohole array to changes in refractive index. Transmission spectra are provided for air and ethanol in Figure 18.14a. SPR binding curves obtained by flowing an ethanolic solution of mercaptoundecanoic acid (MUA) through the nanoholes (Eftekhari et al. 2009) are shown in Figure 18.14b. The MUA solution was initially in contact with a Si_3N_4 support, and the molecule just adsorbed at the gold surface after passing through the channels. In this case, the MUA adsorbs to the surface inside and at the top of the gold film. Future experiments should combine the "in-hole" sensing described in Figure 18.14 with the flow-through. Since the transport of analytes to the surface walls should be enhanced due to the nanometric size of the holes, that new method should accelerate analysis time and decrease the amount of species being detected. The flow-through implementation combines the advantages of analyte and light confinement with the enhanced properties of nanofluidics, and it is an exciting and promising new technology for optofluidic sensing.

A particular limitation to the routine application of nanohole arrays in integrated devices is the serial character of the FIB fabrication. However, several methods for large-area patterning have been described in the literature. For instance, UV photolithography can produce features down to 150 nm. Even smaller hole sizes are possible using phase-shifted lithography (Henzie et al. 2006, Henzie et al. 2007). Molding and printing techniques have also been used to produce plasmonic sensors (Gates et al. 2005). Self-assembled polystyrene spheres can also be used as templates for nanohole array fabrication (Ctistis et al. 2007). The increased development of these modern fabrication methods points to a possible scenario where integrated chips containing arrays of nanoholes as chemical detectors would be mass fabricated and consequently become economically viable for commercial applications.

18.2.5 Surface-Enhanced Raman Scattering

The excitation of SPs leads to a tightly focused electromagnetic field at the surface of the metal (Barnes et al. 2003). Adsorbed molecules experiencing this increased local field present enhanced spectroscopic response (Aroca 2006). SP excitation is thus the main contribution to surface-enhanced Raman scattering (SERS), which is a large increase in the Raman signal observed from molecules adsorbed on nanostructured surfaces of free-electron metals (mainly Ag, Au and Cu) (Aroca 2006). For additional information in this area, the reader is also referred to Chapters 17 and 6. Other resonance effects might also contribute to the SERS response of a particular species in certain conditions (Moskovits 1985). SERS effect has been observed from a variety of nanostructures of different sizes and shapes. Classical SERS substrates are roughened electrode surfaces (Brolo et al. 1997), aggregated metallic colloidal particles (Kneipp et al. 1999), and cold-deposited metal films (Otto et al. 1989). Recently, with the advent of new nanofabrication methods, SERS substrates with controlled geometric parameters are becoming more common (Abu-Hatab et al. 2008, Ru et al. 2008). For instance, Figure 18.15 shows the SERS spectra of a probe dye molecule adsorbed on arrays of nanoholes in a gold film (Brolo et al. 2004a). As discussed in Section 18.2.4, arrays of nanoholes support EOT, which is an SP-based phenomenon. The excitation of SPs in EOT also leads to the localization of the optical field that can be used for enhanced spectroscopy. The results from Figure 18.15 were obtained using a fixed laser excitation at 632.8 nm. The laser illuminated the back of the slide containing the nanoholes; that is, the laser radiation needed to the transmitted through the holes to excite the adsorbed species on the gold surface. It can be seen in Figure 18.15 that the efficiency of the SERS signal depends on the periodicities of the arrays. In fact, the best SERS was obtained for the array that allowed the most transmission at the laser frequency (Brolo et al. 2004a).

The magnitude of the SERS enhancement factor depends on several parameters, including the geometric characteristics of nanostructures and the nature of the adsorbed molecule. Sharp tips and valleys allow for strong-field localization and yield very efficient SERS. The interstices between nanoparticles are also highly localizing when excited at the proper polarization (Kelly et al. 2003). These regions of high-field localization in the plasmonic surface are known as the "hot spots." Only molecules adsorbed at the hot spots contribute significantly to the overall signal (Ru et al. 2007). Therefore, SERS actually probes a small fraction of the nanostructured surface. The average enhancement of the SERS signal from a classical SERS substrate is of the order of 10^6–10^7 relative to the molecule in solution (not in contact to the surface) (Aroca 2006). Individual hot spots can support very large enhancement in certain conditions, which allow the observation of SERS from single molecules, as discussed in Chapter 17. Considering that Raman is an extremely weak effect, the observation of single-molecule SERS is a very exciting result that has captivated the imagination of the analytical chemical community in the last

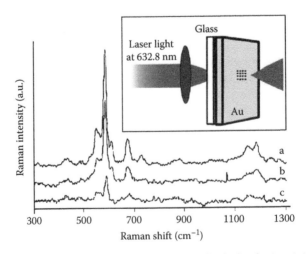

FIGURE 18.15 Surface enhanced Raman scattering with oxazine adsorbed at the Au–air interface obtained using a metal nanohole array. The spectra were obtained from arrays with different periodicities: (a) 560 nm, (b) 590 nm, and (c) 620 nm. The spectra are offset for clarity, and a schematic of the experimental setup is shown in the inset. (Reprinted from Brolo, A.G. et al., *Nano Lett.*, 4, 2015, 2004a. With permission.)

10 years. Single-molecule SERS combines the ultimate limit of detection with the unique fingerprint identification provided by vibrational Raman spectroscopy. Several advancements have been reported in that area, but the complexity of the SERS substrate and the random nature of the hot spots have precluded the application of this technique (Moskovits et al. 2002). Recently, designed hot spots for SERS have been prepared using modern nanolithographic methods, such as FIB and e-beam lithography. For instance, SERS from a few thousands of molecules have been reported from bow-tie structures (Jackel et al. 2007) and touching double holes surrounded by a concentric Bragg grating (Min et al. 2008).

Due to its high sensitivity, selectivity, and spatially localized nature, SERS is an excellent candidate for applications in integrated microfluidics and for multiplex analysis (Abu-Hatab et al. 2007, Ackermann et al. 2007). Several examples of SERS detection of trace amounts in microfluidics have been reported, and a recent review in the subject has been published (Chen and Choo 2008). A perennial problem in SERS is with reproducible quantification of the analyte. The problem arises from the random nature of the hot spot distribution in the most common SERS substrates. Some approaches for the solution of this problem in lab-on-chip conditions have been reported (Cialla et al. 2008), including an interesting scheme using an optical trap to selectively aggregate the nanoparticles for SERS in the microfluidic channel (Tong et al. 2009). A comprehensive review of Raman detection in microfluidics is provided in Chapter 17. The extremely high sensitivity of SERS certainly warrants more research in its on-chip integration as detection element. The combination of novel nanofabrication and imprinting technologies might provide the required tools for the implementation of reproducible SERS in future microfluidic technologies.

18.2.6 Plasmonic Integrated Circuits

The ability to guide and manipulate light at the surface of a metal provides the possibility of greater optical functionality, in the context of plasmonic integrated circuits. Many different approaches have been taken toward plasmonic waveguiding in two and three dimensions, such as guiding at the surface of strips, or in grooves, or from arrays of particles. On the general topic of waveguiding in optofluidic systems, see also Chapter 16.

As compared with conventional dielectrics, the main benefit of plasmonics is to confine the light to smaller volumes. This can allow for focusing light below the conventional diffraction limit in the dielectric, which also enables enhanced interaction with materials. From a sensing perspective, this provides a smaller sensing volume and a potentially greater response. At the same time, there is a trade-off between enhanced confinement and enhanced losses, so that there is a rapid reduction in the signal over length scales that are comparable to the optical wavelength when confinement goes significantly below the optical wavelength.

The ability to nanostructure metals has enabled plasmonic analogues of traditional optical devices, such as lenses, gratings, and interferometers. These components would enable greater functionality, for example, to provide spectral characterization, imaging or sensing, potentially in a multiplexed environment. Furthermore, plasmonics offers the potential for enhanced interaction with materials at the quantum level.

18.2.6.1 Planar Optics

Many plasmonic components have been demonstrated in the plane using conventional SPPs. These can be achieved by nanostructuring the metal or by dielectric loading (which can also be fluidic based). For example, as shown in Figure 18.16, a simple droplet on gold has been used to create a small imaging microscope (Smolyaninov et al. 2005). When operated close to the plasmon resonance, $\text{Re}\{\varepsilon_m\} = -\varepsilon_d\}$ has a divergent wave vector in the plane (see Section 18.2.1). The result is shorter wavelengths, and thus the conventional diffraction limit may be extended. In that demonstration, the wavelength was shortened to 70 nm, which allowed for the imaging of 40 nm gaps between nanostructures (Smolyaninov et al. 2005).

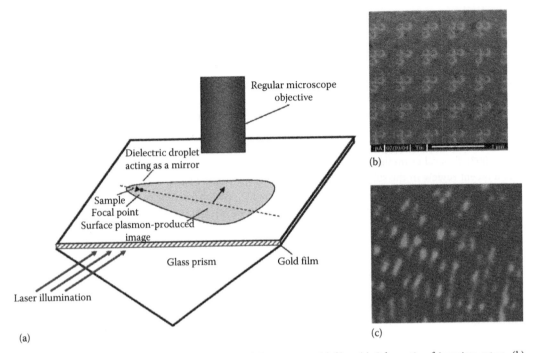

(a)

(b)

(c)

FIGURE 18.16   In-plane microscope from a liquid drop on a gold film. (a) Schematic of imaging setup. (b) Scanning electron microscope image of 100 nm triplet nanohole arrays. (c) Image of nanoholes produced by SPP imaging in-plane visible with regular microscope objective. Operating near the high-index region allows for imaging below the free-space usual diffraction limit. (Reprinted from Smolyaninov, I.I. et al., *Opt. Lett.*, 30, 382, 2005. With permission.)

Focusing elements and lenses have also been achieved within the plane by using curved scattering structures. In one demonstration, the scattering of an SPP off of a condenser arrangement of surface-relief dots allowed for focusing down to 400 nm and the excitation of a dot array guide (Nomura et al. 2005). In a similar demonstration shown in Figure 18.17, a focusing SPP was launched by a hole array and coupled to a stripe geometry that was only 250 nm (half the SPP wavelength) (Yin et al. 2005).

Bragg reflectors have also been demonstrated within the plane to redirect light. As mirrors, Bragg reflectors also offer the potential to feedback to resonant structures, such as hole arrays

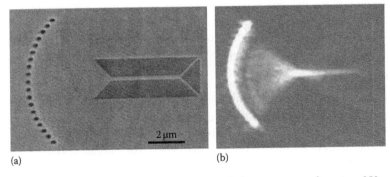

(a)   (b)

FIGURE 18.17   Focused SPP launching from a curved nanohole array to couple onto a 250 nm wide stripe. (a) Imaging of the focusing array with strip guide. (b) Image of SPP intensity with focusing and guiding. (Reprinted from Yin, L.L. et al., *Nano Lett.*, 5, 1399, 2005. With permission.)

(Marthandam and Gordon 2007), or to isolate large-scale arrays of devices (Lindquist et al. 2009) (See also Figure 18.24b and discussion in Section 18.2.8).

18.2.6.2 Waveguide Components and Resonators

As described in the introduction, SPP waveguides can involve a single metal interface, MIM, or IMI structures. As shown in Figure 18.18, there is a general trade-off in waveguiding between the length of propagation and the confinement due to material losses (Maier 2006). IMIs that allow for centimeter propagation lengths only localize light to the order of the optical wavelength. By contrast, MIM structures that localize light to 20 nm gaps can only propagate a few microns (Dionne et al. 2006).

There has been considerable work on laterally structured versions of each of these geometries to provide waveguiding in 3-D. These works include stripes of metals, gaps between two metal stripes, grooves in metal, and ridges in metal. In addition, there has been work on the coupling between plasmonic particles to transfer energy in a discrete version of the plasmonic waveguide. With waveguiding demonstrated, it becomes possible to build more functional waveguide components such as bends, multimode interferometers, waveguide Bragg reflectors, y-junctions, and Mach–Zender interferometer ring resonators (e.g., (Bozhevolnyi et al. 2006)). In some cases, these structures can promise greater functionality than their dielectric counterparts; for example, a 90° bend is possible with limited losses in a subwavelength groove structure (Pile and Gramotnev 2005). MIM or slot waveguides may also present interesting possibilities as nanoscale fluidic channels.

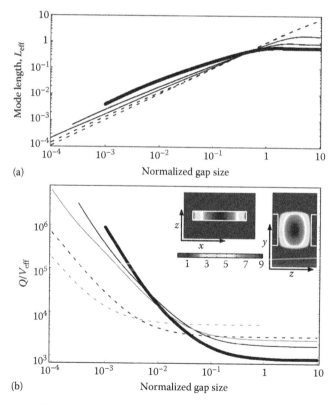

FIGURE 18.18 (a) Trade-off between effective length of propagation and confinement in an MIM gap. (b) Quality (*Q*) factor of a plasmonic cavity normalized to effective cavity volume (V_{eff}) for different MIM gap sizes. While confinement leads to increased losses, higher Q/V_{eff} ratios can be obtained, which is of interest to enhanced light–matter interactions. (Reprinted from Maier, S.A., *IEEE J. Sel. Top. Quantum Electron.*, 12, 1214, 2006. With permission.)

As discussed in other chapters of this book, interferometric configurations, such as the Mach–Zender can also be used as sensors in a microfluidic environment. For example, sensing is achieved by via interferometric changes between a reference arm and one exposed to an environment in which analyte alters its effective length. Loss is an important factor in such devices, which rely on greater propagation lengths for higher sensitivity, assuming a given solution concentration.

Resonators present a particular opportunity for interaction with materials using plasmonic structures. An important figure of merit in the interaction of photons in a cavity with matter is the quality factor of the cavity divided by its volume. While plasmonic structures have poor cavity quality due to their intrinsic losses, they do allow for volume reductions well below the optical wavelength in a conventional dielectric. As a result, Q/V has a favorable scaling as the cavity is reduced, and an enhanced photon–matter coupling at the quantum level is a promising area of research for plasmonics.

18.2.6.3 Gain

There have been efforts to introduce gain and thereby overcome losses that limit the scalability of plasmonic integrated circuits (Noginov et al. 2008). In this case, the dielectric material close to the metal should provide enough gain to overcome the losses of the metal. To quantify the gain required, it is necessary to ensure that the propagation constant of the plasmonic waveguides is purely real when including the imaginary dielectric response of the dielectric that represents the gain. For example, for an SPP on a metal–dielectric interface, the gain of the dielectric should be

$$\text{Im}\{\varepsilon_d\} = -\frac{\text{Im}\{\varepsilon_m\}\,\text{Re}\{\varepsilon_d\}^2}{\text{Re}\{\varepsilon_m\}^2} \tag{18.18}$$

to overcome the losses. It is clear from this formulation that the gain of the dielectric can be made smaller by using a dielectric with a smaller dielectric constant (as compared to the negative real part of the metal's relative permittivity). It is also advantageous to space the active dielectric material away from the metal by a few nanometers to avoid quenching of the gain (which actually enhances the loss). Opportunities exist within the context of nanofluidics to provide dye solutions with lower refractive index as active materials.

18.2.7 Plasmonics for Trapping

The controlled manipulation of particles and molecules using light was first demonstrated in 1970 (Ashkin 1970) (see also Chapter 14). External optical control provides a valuable handle with which to study a wide variety of physics at small scales, such as the physical properties of large biomolecules, and for the analysis of single cells. Traditional optical trapping makes use of a combination of optical scattering and gradient forces to manipulate particles exhibiting refractive indexes distinct from their surrounding environment, and has evolved to include multiple traps in structured light fields (Dholakia and Lee 2008). A strength of traditional optical trapping is the ease with which it may be incorporated into established microscope infrastructure and thus made accessible to a broad audience beyond that of the optics community. A weakness of traditional optical trapping is the severe dependency of trap stability with particle size, and the limitation on minimum particle size imposed by the diffraction limit.

Employing evanescent light for trapping particles close to surfaces has received much recent interest (Righini et al. 2008, Yang et al. 2009). Most notably, the evanescent field from subwavelength structures may be utilized for trapping, circumventing the minimum particle size imposed by the diffraction limit in traditional optical trapping. SPs are particularly attractive in the context of optical trapping for two reasons (Righini et al. 2008, Yang et al. 2009): The intensity enhancement of SP fields leads to decreased intensity requirements, and the subwavelength confinement of the SP field presents an opportunity to trapping on the nanometer scale.

The optical forces on a particle in an arbitrary geometry are rigorously derived by using the Maxwell stress tensor (Jackson 1998). For the special case of small (Rayleigh) scattering objects, the forces can be written in terms of the scattering and gradient forces (for detailed analysis of forces involved in trapping, see Chapter 14):

$$F_{opt} = F_{scat} + F_{grad} = \left[n_b \sigma \frac{I}{c} \right] + \left[\frac{n_b \alpha}{4} \nabla \left(|E|^2 \right) \right] \tag{18.19}$$

where

n_b (equivalent to $\sqrt{\varepsilon_d}$ as described in Section 18.2.1) is the refractive index of the surrounding medium, $\sigma = \sigma_{scat} + \sigma_{abs}$ (see Section 18.2.1)

I is the intensity

c is the speed of light in vacuum, $\alpha = p/E_0$

E is the electric field amplitude associated with the oscillating electromagnetic wave

The gradient force pulls the particle toward the region of highest intensity and so it is useful for optical trapping. About a stable equilibrium point, the optical forces acting on the particle can be linearized to resemble Hooke's law: $F_{opt} = -kx$. This spring force is damped by Stokes drag, and considering the low Reynolds number, the inertial mass of the object can be neglected. Thermal energy, $k_B T$, will cause the particle to oscillate, and the variance in the trapped particle's position is given by $k_B T/k$. Therefore, a stiffer trap from a more localized electric field intensity will result in a more localized particle position.

Plasmonics offers significant potential for trapping smaller objects. In particular, it allows for local field enhancements of SPPs and around nanostructures, with greater electric field intensity gradients, resulting in more localized trapping. An example of trapping and organization of particles in response to SPPs is given in Figure 18.19. In addition, plasmonic particles can have greater scattering cross sections

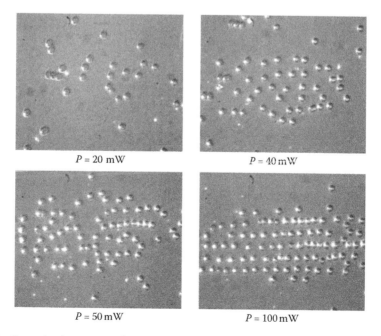

FIGURE 18.19 Example of trapping and organization of particles in response to SPPs. (Reprinted from Garces-Chavez, V. et al., *Phys. Rev. B*, 73, 085417, 2006. With permission.)

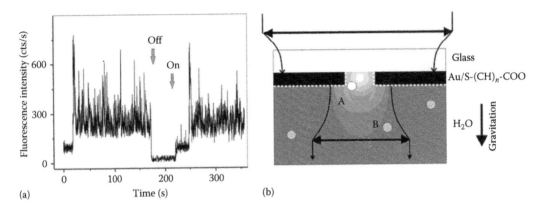

FIGURE 18.20 Trapping nanoparticles (200 nm polystyrene) with a nanohole. (a) Fluorescence emission with high intensities corresponding to time with the incident light applied indicating trapping of a fluorescent particle in the hole. (b) Schematic of the optofluidic setup. (Reprinted from Kwak, E.S. et al., *J. Phys. Chem. B*, 108, 13607, 2004. With permission.)

close to the plasmonic resonance, so they can experience greater optical forces, which can be useful in mechanical manipulation.

Many plasmonic nanostructures are well suited to trapping applications as the electromagnetic field is localized in multiple dimensions. The plasmonic nanohole structure (also discussed in Sections 18.2.1 and 18.2.4) may be employed for trapping. An example of particle trapping against gravity employing a nanohole structure is given in Figure 18.20 (Kwak et al. 2004). In that work, the time evolution of fluorescence from labeled 200 nm latex beads in a 500 nm diameter hole in a gold film was measured as the incident light intensity was varied. Figure 18.20b shows the fluorescence signal showing enhanced emission corresponding to trapped particles while the laser is on. The surface properties of the hole and the surrounding solution were conditioned to ensure a negative surface charge (and thus a repulsion of the particles from the hole) and to avoid agglomeration of particles (Kwak et al. 2004).

Figure 18.21 shows a recent work in which metallic nanoantennas patterned on a glass substrate were employed to trap polystyrene particles as well as *E. coli* bacteria (Righini et al. 2009). Arrays of optical gap antennas, spaced 10 μm, were illuminated at 800 nm by total internal reflection. As shown in Figure 18.21a through d, the localized plasmonic-trapping scheme provided stable trapping of the particles with the absence of 25% of particles attributed to missing antennas or structural defects. Figure 18.21e shows calculated forces on the particles near two different antenna types. An important advantage of this method over traditional optical trapping is the low incident light power employed, and thus it is expected to find application in the study of single-cell dynamics and related topics in life sciences. Specifically, the 10^7 W/m^2 at 800 nm was applied for particle and *E. coli* trapping in this work, as compared to 10^9 W/m^2 typical of 3D optical tweezers, and the 10^{10} W/m^2 threshold associated with cell damage (Righini et al. 2009).

18.2.8 Optofluidic Integration and Multiplexing with Plasmonics: Plasmofluidics?

While the combined term "plasmofluidics" has been resisted, at least until now, there are strong synergies between the fields of plasmonics and fluidics. All of the plasmonic effects, devices, and applications discussed in this chapter are relevant to optofluidics. Plasmonics provide confinement of the electromagnetic field, and are thus an excellent match with fluidics that can provide highly localized and controlled delivery/removal of analytes and drag forces. Integration efforts to date have developed in concert with

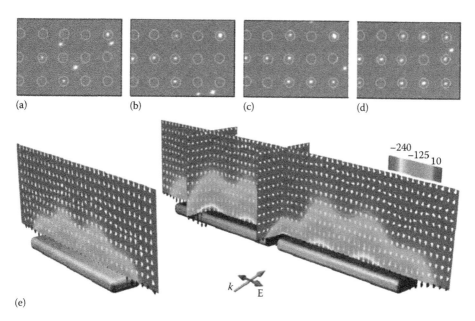

FIGURE 18.21 Trapping using plasmonic antennae near resonance. (a–d) Time sequence of fluorescence images showing the loading of the traps over time (order of minutes). (e) Simulation results showing the predicted force on a 200 nm polystyrene particle near the antenna: a bar antenna (left) and a gap antenna (right). This study also showed trapping of *E. coli* bacteria. (Reprinted from Righini, M. et al., *Nano Lett.*, 2009. DOI: 10.1021/nl803677X, With permission.)

plasmonic device development, such as sensors, and it is difficult to separate them entirely. Thus many of the studies already discussed in this chapter have exemplified optofluidic integration in the spirit of a lab-on-chip approach (see also Chapter 7). In this section, a short overview of optofluidic integration and multiplexing with plasmonics is provided with a few examples.

Traditional SPR sensing using the Kretschmann-style plasmonic excitation with a prism has been combined with fluidics in many cases. The most common configuration employs a chip to confine fluid flow near the surface of the active gold sensing surface. In order to cover a relatively large area, the fluidic layer thickness is typically in the microfluidic range, and thus much larger than the penetration depth of the SP into the solution. Such microfluidic service layers are particularly valuable when the gold surface is divided into discrete sensors to be addressed independently, as in the case of SPR imaging. A recent example of multiplexed analysis enabled by the combination of microfluidics and SPR imaging is shown in Figure 18.22 (Luo et al. 2008). An equally important role for fluidics in these cases is in the functionalization of the sensor surface prior to assembly of the device (Kanda et al. 2004). This is achieved with a specifically designed sacrificial chip that may be applied to facilitate functionalization of the surface with appropriate probes or capture antibodies, and subsequently removed. Polydimethylsiloxane (PDMS) chips are a favorite for such applications due to their low cost and elastomeric nature that facilitates reversible bonding. It is noteworthy, however, that the Kretschmann geometry is not, in general, as amicable to miniaturization as other SP-based detection methods due to the prism optics, geometry, and associated hardware.

Localized SPs and LSPRs provide confinement of the electromagnetic field in two or three dimensions, and thus open additional opportunities for multiplexing and optofluidic integration. With respect to plasmonic nanoparticles, their application has been so broad that they have been incorporated in all common fluidic system configurations. Two aspects are particularly noteworthy: First, selectively functionalized nanoparticles that may be uniquely identified, for instance, by their plasmonic resonance, may be mixed with otherwise functionalized and identifiable plasmonic nanoparticles directly

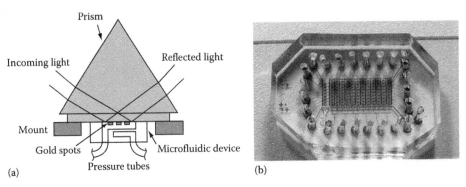

FIGURE 18.22 An example of multiplexed analysis enabled by the combination of microfluidics and SPR imaging: (a) Schematic of the optofluidic setup and (b) image of the chip that achieves multiplexed fluidic assignment to the active sensor surface. (Reprinted from Luo, Y.Q. et al., *Lab Chip*, 8, 694, 2008. With permission.)

into a sample of interest (Tong et al. 2009). Binding events for each particle type can be read through imaging or sequential readout by passing the solution through a microfluidic manifold. Microfluidics may also be employed in such cases to ensure exposure of the analyte to functionalized nanoparticles as shown schematically in Figure 18.23a (Chen and Choo 2008). Alternatively, analytes may be transported through SERS nanoparticles collected by means of a nanofluidic weir as shown in Figure 18.23b (Wang et al. 2007). As shown in Figure 18.23c, metal coated strip waveguides may also be integrated with fluidics in order to locally enhance fluorescence in the evanescent field (Ong et al. 2007). In that case a bimetallic silver–gold layer was used to ensure a balance between evanescent-field enhancement (silver) and stability of the sensing surface (gold).

Early in the development of nanohole arrays as plasmonic sensors (Brolo et al. 2004b) it was noted that the collinear optical mode offered by the enhanced optical transmission was very well suited to on-chip integration and multiplexing. Figure 18.24 shows some of the integration and multiplexing efforts to date, starting from the first on-chip implementation of nanohole arrays (De Leebeeck et al. 2007). As shown in Figure 18.24a, a gold-on-glass substrate with arrays of nanohole arrays was reversibly bonded to a microfluidic layer. The arrays were used to detect a concentration gradient, and in a proof-of-concept biosensing experiment. Figure 18.24b shows a recent work in which very small arrays of nanoholes, as small as 3 × 3, were packed to very high densities (Lindquist et al. 2009), by using the isolating and focusing effects of Bragg mirrors surrounding each array (Marthandam and Gordon 2007). In the context of sensors, nanohole arrays have been incorporated in a flow-over format. That is, the holes have been dead-ended, and thus failed to harness the rapid cross-stream transport associated with nanofluidics. Figure 18.24c shows the flow-through nanohole-array-based sensing scheme, as well as the streaming of a fluorescent dye through a nanohole array (Eftekhari et al. 2009). Proof-of-concept sensing results from that study were discussed in Section 18.2.4. Flow-through nanohole arrays are one example of highly integrated plasmonics and fluidics that enable enhanced functionality (in this case, solution sieving and rapid transport of reactants to active sites). The authors feel there are many more such couplings possible between plasmonics and fluidics that have promise.

18.3 Conclusion

In this chapter, an overview of plasmonics as relevant to the emerging field of optofluidics is presented. The central objective of this chapter is to introduce the field and highlight its relevance to optofluidic systems. An overview of the physics of different plasmonic phenomena relevant to optofluidic applications was presented first. Subsequent sections focused on typical plasmonic modes of operation and/or

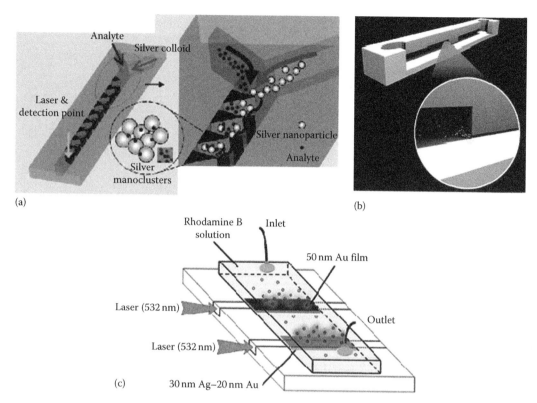

(a)

(b)

(c)

FIGURE 18.23 Schematics illustrating optofluidic integration: (a) Plasmonic nanoparticles mixed with an incubated analyte via a sawtooth channel structure. (Reprinted from Chen, L.X. and Choo, J.B., *Electrophoresis*, 29, 1815, 2008. With permission.) (b) Analytes flow through and react on the surface of SERS nanoparticles collected by means of a nanofluidic weir. (From Wang, M. et al., *Lab Chip*, 7, 630, 2007. With permission.) (c) Metal-coated strip waveguides integrated with fluidics to locally enhance fluorescence. (From Ong, B.H. et al., *Lab Chip*, 7, 506, 2007. With permission.)

applications that are enabled by the fundamental plasmonic phenomena, specifically the following: propagating SP modes, nanoparticle-based SPs, nanohole-based SPs, amplification of Raman scattering via plasmonics, plasmonic waveguiding, plasmonic trapping, and optofluidic integration and multiplexing with plasmonics. While the relatively broad scope of this chapter precludes a detailed review of individual contributions, many relevant studies were highlighted and references provided for the interested reader. Early studies that have involved both fluidics and plasmonics, some of which were highlighted in this chapter, have shown much promise in a variety of applications.

As noted through the text, there are many synergies between this chapter and the other chapters in this book. For instance, there are many linkages between plasmonics and spectroscopic methods (Chapter 6), particularly as many spectroscopic methods employ plasmonics on some scale. Likewise, the plasmonic nanohole arrays section in this chapter (Section 18.2.4) is also in the family of devices described in the chapter on photonic crystals (Chapter 4). The chapter dedicated to lab-on-a-chip (Chapter 7) is relevant to the section on optofluidic integration and multiplexing in this chapter (Section 18.2.8). As the majority of plasmonic applications to date focus on sensing there are many connections between this work and the optofluidic switches and sensors chapter (Chapter 10). Lastly, in Part III, the bioanalysis section of the book, there are many synergies, particularly with the Raman detection chapter (Chapter 17), the integrated optofluidic waveguides chapter (Chapter 16), and the optical trapping and manipulation chapter (Chapter 14).

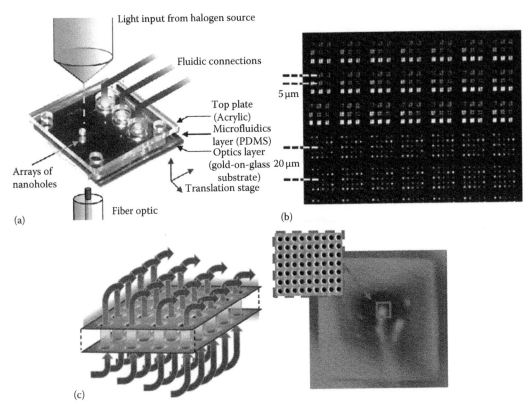

FIGURE 18.24 Examples of optofluidic integration and multiplexing of nanohole-array-based plasmonic sensor. (a) Simple on-chip integration of service microfluidics with a nanostructured gold-on-glass substrate containing arrays of nanohole arrays. (From De Leebeeck, A. et al., *Anal. Chem.*, 79, 4094, 2007. With permission.) (b) High packing density of nanohole array sensors achieved by surrounding each array (3 × 3 nanohole array) with plasmonic Bragg mirrors. (From Lindquist, N.C. et al., *Lab Chip*, 9, 382, 2009. With permission.) (c) Schematic and fluorescence image showing streaming in flow-through nanohole arrays. (From Eftekhari et al., *Analytical Chemistry*, 81, 4308, 2009. With permission.)

The authors are enthusiastic about the future of plasmonics, particularly in the context of optofluidics, and hope that the many reasons for that optimism are conveyed in this chapter. The electromagnetic field confinement offered by plasmonics is at the heart of all successful applications in this area. In many cases, it is the combination or overlap of effects that leads to the most exciting and useful phenomena, for instance, the collective effects that lead to SERS. Similarly, we see exciting future applications resulting from the combination of fluidic confinement and the associated control of molecules and particles of interest with plasmonic electromagnetic fields.

References

Abu-Hatab, N. A., J. F. John, J. M. Oran, and M. J. Sepaniak. (2007). Multiplexed microfluidic surface-enhanced Raman spectroscopy. *Applied Spectroscopy*, 61, 1116–1122.

Abu Hatab, N. A., J. M. Oran, and M. J. Sepaniak. (2008). Surface-enhanced Raman spectroscopy substrates created via electron beam lithography and nanotransfer printing. *ACS Nano*, 2, 377–385.

Ackermann, K. R., T. Henkel, and J. Popp. (2007). Quantitative online detection of low-concentrated drugs via a SERS microfluidic system. *Chemphyschem*, 8, 2665–2670.

Aroca, R. (2006). *Surface-Enhanced Vibrational Spectroscopy*. Hoboken, NJ: Wiley.

Ashkin, A. (1970). Acceleration and trapping of particles by radiation pressure. *Physical Review Letters*, 24, 156–159.

Barnes, W. L., A. Dereux, and T. W. Ebbesen. (2003). Surface plasmon subwavelength optics. *Nature*, 424, 824–830.

Bethe, H. A. (1944). Theory of diffraction by small holes. *Physical Review*, 163–182.

Bozhevolnyi, S. I., V. S. Volkov, E. Devaux, J. Y. Laluet, and T. W. Ebbesen. (2006). Channel plasmon subwavelength waveguide components including interferometers and ring resonators. *Nature*, 440, 508–511.

Brolo, A. G., D. E. Irish, and B. D. Smith. (1997). Applications of surface enhanced Raman scattering to the study of metal-adsorbate interactions. *Journal of Molecular Structure*, 405, 29–44.

Brolo, A. G., E. Arctander, R. Gordon, B. Leathem, and K. L. Kavanagh. (2004a). Nanohole-enhanced Raman scattering. *Nano Letters*, 4, 2015–2018.

Brolo, A. G., R. Gordon, B. Leathem, and K. L. Kavanagh. (2004b). Surface plasmon sensor based on the enhanced light transmission through arrays of nanoholes in gold films. *Langmuir*, 20, 4813–4815.

Chen, L. X. and J. B. Choo. (2008). Recent advances in surface-enhanced Raman scattering detection technology for microfluidic chips. *Electrophoresis*, 29, 1815–1828.

Cialla, D., U. Hubner, H. Schneidewind, R. Moller, and J. Popp. (2008). Probing innovative microfabricated substrates for their reproducible SERS activity. *Chemphyschem*, 9, 758–762.

Ctistis, G., P. Patoka, X. Wang, K. Kempa, and M. Giersig. (2007). Optical transmission through hexagonal arrays of subwavelength holes in thin metal films. *Nano Letters*, 7, 2926–2930.

De Leebeeck, A., L. K. S. Kumar, V. de Lange, D. Sinton, R. Gordon, and A. G. Brolo. (2007). On-chip surface-based detection with nanohole arrays. *Analytical Chemistry*, 79, 4094–4100.

Degiron, A., H. J. Lezec, W. L. Barnes, and T. W. Ebbesen. (2002). Effects of hole depth on enhanced light transmission through subwavelength hole arrays. *Applied Physics Letters*, 81, 4327–4329.

Dholakia, K. and W. M. Lee. (2008). Optical trapping takes shape: The use of structured light fields. *Advances in Atomic, Molecular, and Optical Physics*, 56, 261–337.

Dionne, J. A., L. A. Sweatlock, H. A. Atwater, and A. Polman. (2006). Plasmon slot waveguides: Towards chip-scale propagation with subwavelength-scale localization. *Physical Review B*, 73, 035407.

Drude, P. (1900). Zur elektronentheorie der metalle. *Annalen de Physik*, 306, 566.

Ebbesen, T. W., H. J. Lezec, H. F. Ghaemi, T. Thio, and P. A. Wolff. (1998). Extraordinary optical transmission through sub-wavelength hole arrays. *Nature*, 391, 667–669.

Ebbesen, T. W., C. Genet, and S. I. Bozhevolnyi. (2008). Surface-plasmon circuitry. *Physics Today*, 61, 44–50.

Eftekhari, F., C. Escobedo, J. Ferriera, X. Duan, E. M. Girotto, A. G. Brolo, R. Gordon, and D. Sinton. (2009). Nanoholes as nanochannels: Flow-through plasmonic sensing. *Analytical Chemistry*, 81, 4308–4311.

Fang, N., H. Lee, C. Sun, and X. Zhang. (2005). Sub-diffraction-limited optical imaging with a silver superlens. *Science*, 308, 534–537.

Farre, M., E. Martinez, J. Ramon, A. Navarro, J. Radjenovic, E. Mauriz, L. Lechuga, M. P. Marco, and D. Barcelo. (2007). Part per trillion determination of atrazine in natural water samples by a surface plasmon resonance immunosensor. *Analytical and Bioanalytical Chemistry*, 388, 207–214.

Ferreira, J., M. J. Santos, M. M. Rahman, A. G. Brolo, R. Gordon, D. Sinton, and E. M. Girotto. (2009). Attomolar protein detection using in-hole surface plasmon resonance. *Journal of American Chemical Society*, 131(2), 436–437.

Garces-Chavez, V., R. Quidant, P. J. Reece, G. Badenes, L. Torner, and K. Dholakia. (2006). Extended organization of colloidal microparticles by surface plasmon polariton excitation. *Physical Review B*, 73, 085417.

Garcia-Vidal, F. J., L. Martin-Moreno, E. Moreno, L. K. S. Kumar, and R. Gordon. (2006). Transmission of light through a single rectangular hole in a real metal. *Physical Review B*, 74, 153411.

Gates, B. D., Q. B. Xu, M. Stewart, D. Ryan, C. G. Willson, and G. M. Whitesides. (2005). New approaches to nanofabrication: Molding, printing, and other techniques. *Chemical Reviews*, 105, 1171–1196.

Genet, C. and T. W. Ebbesen. (2007). Light in tiny holes. *Nature*, 445, 39–46.

Genet, C., M. P. van Exter, and J. P. Woerdman. (2003). Fano-type interpretation of red shifts and red tails in hole array transmission spectra. *Optics Communications*, 225, 331–336.

Gordon, R. (2008). Surface plasmon nanophotonics: A tutorial. *IEEE Nanotechnology Magazine*, 2(3), 12–18.

Gordon, R. and A. G. Brolo. (2005). Increased cut-off wavelength for a subwavelength hole in a real metal. *Optics Express*, 13, 1933–1938.

Gordon, J. G. and S. Ernst. (1980). Surface-plasmons as a probe of the electrochemical interface. *Surface Science*, 101, 499–506.

Gordon, R., A. G. Brolo, A. McKinnon, A. Rajora, B. Leathem, and K. L. Kavanagh. (2004). Strong polarization in the optical transmission through elliptical nanohole arrays. *Physical Review Letters*, 92, 4.

Gordon, R., M. Hughes, B. Leathem, K. L. Kavanagh, and A. G. Brolo. (2005). Basis and lattice polarization mechanisms for light transmission through nanohole arrays in a metal film. *Nano Letters*, 5, 1243–1246.

Henzie, J., J. E. Barton, C. L. Stender, and T. W. Odom. (2006). Large-area nanoscale patterning: Chemistry meets fabrication. *Accounts of Chemical Research*, 39, 249–257.

Henzie, J., M. H. Lee, and T. W. Odom. (2007). Multiscale patterning of plasmonic metamaterials. *Nature Nanotechnology*, 2, 549–554.

Ho, C. K., A. Robinson, D. R. Miller, and M. J. Davis. (2005). Overview of sensors and needs for environmental monitoring. *Sensors*, 5, 4–37.

Homola, J. (2008). Surface plasmon resonance sensors for detection of chemical and biological species. *Chemical Reviews*, 108, 462–493.

Jackel, F., A. A. Kinkhabwala, and W. E. Moerner. (2007). Gold bowtie nanoantennas for surface-enhanced Raman scattering under controlled electrochemical potential. *Chemical Physics Letters*, 446, 339–343.

Jackson, J. D. (1998). *Classical Electrodynamics*. New York: Wiley.

Ji, J., J. G. O'Connell, D. J. D. Carter, and D. N. Larson. (2008). High-throughput nanohole array based system to monitor multiple binding events in real time. *Analytical Chemistry*, 80, 2491–2498.

Johnson, P. B. and R. W. Christy. (1972). Optical-constants of noble-metals. *Physical Review B*, 6, 4370–4379.

Kanda, V., J. K. Kariuki, D. J. Harrison, and M. T. McDermott. (2004). Label-free reading of microarray-based immunoassays with surface plasmon resonance imaging. *Analytical Chemistry*, 76, 7257–7262.

Kelly, K. L., E. Coronado, L. L. Zhao, and G. C. Schatz. (2003). The optical properties of metal nanoparticles: The influence of size, shape, and dielectric environment. *Journal of Physical Chemistry B*, 107, 668–677.

Kim, T. J., T. Thio, T. W. Ebbesen, D. E. Grupp, and H. J. Lezec. (1999). Control of optical transmission through metals perforated with subwavelength hole arrays. *Optics Letters*, 24, 256–258.

Kneipp, K., H. Kneipp, I. Itzkan, R. R. Dasari, and M. S. Feld. (1999). Ultrasensitive chemical analysis by Raman spectroscopy. *Chemical Reviews*, 99, 2957–2975.

Koerkamp, K. J. K., S. Enoch, F. B. Segerink, N. F. van Hulst, and L. Kuipers. (2004). Strong influence of hole shape on extraordinary transmission through periodic arrays of subwavelength holes. *Physical Review Letters*, 92, 4.

Kreibig, U. and M. Volmer. (1995). *Optical Properties of Metal Clusters*. Berlin, Germany: Springer-Verlag.

Kretschmann, E. (1971). Determination of optical constants of metals by excitation of surface plasmons. *Zeitschrift Fur Physik*, 241, 313324.

Kumar, S., N. Harrison, R. Richards-Kortum, and K. Sokolov. (2007). Plasmonic nanosensors for imaging intracellular biomarkers in live cells. *Nano Letters*, 7, 1338–1343.

Kwak, E. S., T. D. Onuta, D. Amarie, R. Potyrailo, B. Stein, S. C. Jacobson, W. L. Schaich, and B. Dragnea. (2004). Optical trapping with integrated near-field apertures. *Journal of Physical Chemistry B*, 108, 13607–13612.

Lesuffleur, A., H. Im, N. C. Lindquist, and S. H. Oh. (2007). Periodic nanohole arrays with shape-enhanced plasmon resonance as real-time biosensors. *Applied Physics Letters*, 90, 243110.

Lesuffleur, A., H. Im, N. C. Lindquist, K. S. Lim, and S. H. Oh. (2008). Laser-illuminated nanohole arrays for multiplex plasmonic microarray sensing. *Optics Express*, 16, 219–224.

Levene, M. J., J. Korlach, S. W. Turner, M. Foquet, H. G. Craighead, and W. W. Webb. (2003). Zero-mode waveguides for single-molecule analysis at high concentrations. *Science*, 299, 682–686.

Li, K. R., M. I. Stockman, and D. J. Bergman. (2003). Self-similar chain of metal nanospheres as an efficient nanolens. *Physical Review Letters*, 91, 227402.

Lindquist, N. C., A. Lesuffleur, H. Im, and S. H. Oh. (2009). Sub-micron resolution surface plasmon resonance imaging enabled by nanohole arrays with surrounding Bragg mirrors for enhanced sensitivity and isolation. *Lab on a Chip*, 9, 382–387.

Liu, H. T. and P. Lalanne. (2008). Microscopic theory of the extraordinary optical transmission. *Nature*, 452, 728–731.

Liu, J. W. and Y. Lu. (2003). A colorimetric lead biosensor using DNAzyme-directed assembly of gold nanoparticles. *Journal of the American Chemical Society*, 125, 6642–6643.

Liu, J. W. and Y. Lu. (2006). Fast colorimetric sensing of adenosine and cocaine based on a general sensor design involving aptamers and nanoparticles. *Angewandte Chemie-International Edition*, 45, 90–94.

Loo, C., A. Lin, L. Hirsch, M. H. Lee, J. Barton, N. Halas, J. West, and R. Drezek. (2004). Nanoshell-enabled photonics-based imaging and therapy of cancer. *Technology in Cancer Research & Treatment*, 3, 33–40.

Loo, C., A. Lowery, N. Halas, J. West, and R. Drezek. (2005). Immunotargeted nanoshells for integrated cancer imaging and therapy. *Nano Letters*, 5, 709–711.

Luo, Y. Q., F. Yu, and R. N. Zare. (2008). Microfluidic device for immunoassays based on surface plasmon resonance imaging. *Lab on a Chip*, 8, 694–700.

Maier, S. A. (2006). Plasmonics: Metal nanostructures for subwavelength photonic devices. *IEEE Journal of Selected Topics in Quantum Electronics*, 12, 1214–1220.

Marthandam, P. and R. Gordon. (2007). Plasmonic Bragg reflectors for enhanced extraordinary optical transmission through nano-hole arrays in a gold film. *Optics Express*, 15, 12995–13002.

Martin-Moreno, L., F. J. Garcia-Vidal, H. J. Lezec, K. M. Pellerin, T. Thio, J. B. Pendry, and T. W. Ebbesen. (2001). Theory of extraordinary optical transmission through subwavelength hole arrays. *Physical Review Letters*, 86, 1114–1117.

Min, Q., M. J. L. Santos, E. M. Girotto, A. G. Brolo, and R. Gordon. (2008). Localized Raman enhancement from a double-hole nanostructure in a metal film. *Journal of Physical Chemistry C*, 112, 15098–15101.

Mirkin, C. A., R. L. Letsinger, R. C. Mucic, and J. J. Storhoff. (1996). A DNA-based method for rationally assembling nanoparticles into macroscopic materials. *Nature*, 382, 607–609.

Moskovits, M. (1985). Surface-enhanced spectroscopy. *Reviews of Modern Physics*, 57, 783–826.

Moskovits, M., L. L. Tay, J. Yang, and T. Haslett. (2002). SERS and the single molecule. In Vladimir, M. Shalaev (ed.), *Optical Properties of Nanostructured Random Media*. Berlin, Germany: Springer-Verlag, pp. 215–226.

Noginov, M. A., G. Zhu, M. Mayy, B. A. Ritzo, N. Noginova, and V. A. Podolskiy. (2008). Stimulated emission of surface plasmon polaritons. *Physical Review Letters*, 101, 226806.

Nomura, W., M. Ohtsu, and T. Yatsui. (2005). Nanodot coupler with a surface plasmon polariton condenser for optical far/near-field conversion. *Applied Physics Letters*, 86, 181108.

Oh, B. K., W. Lee, W. H. Lee, and J. W. Choi. (2003). Nano-scale probe fabrication using self-assembly technique and application to detection of *Escherichia coli* O157: H7. *Biotechnology and Bioprocess Engineering*, 8, 227–232.

Okamoto, T. (2001). Near-field spectral analysis of metallic beads. In Kawata, S. (ed.), *Near-Field Optics and Surface Plasmon Polaritons*. Berlin, Germany: Springer-Verlag, pp. 97–122.

Ong, B. H., X. C. Yuan, Y. Y. Tan, R. Irawan, X. Q. Fang, L. S. Zhang, and S. C. Tjin. (2007). Two-layered metallic film-induced surface plasmon polariton for fluorescence emission enhancement in on-chip waveguide. *Lab on a Chip*, 7, 506–512.

Otto, A., T. Bornemann, U. Erturk, I. Mrozek, and C. Pettenkofer. (1989). Model of electronically enhanced Raman-scattering from adsorbates on cold-deposited silver. *Surface Science*, 210, 363–386.

Ozbay, E. (2006). Plasmonics: Merging photonics and electronics at nanoscale dimensions. *Science*, 311, 189–193.

Pendry, J. B. (2000). Negative refraction makes a perfect lens. *Physical Review Letters*, 85, 3966–3969.

Pile, D. E. P. and D. K. Gramotnev. (2005). Plasmonic subwavelength waveguides: Next to zero losses at sharp bends. *Optics Letters*, 30, 1186–1188.

Raether, H. (1988). *Surface Plasmons on Smooth and Rough Surfaces and on Gratings*. New York: Springer-Verlag.

Rakic, A. D., A. B. Djurisic, J. M. Elazar, and M. L. Majewski. (1998). Optical properties of metallic films for vertical-cavity optoelectronic devices. *Applied Optics*, 37, 5271–5283.

Righini, M., C. Girard, and R. Quidant. (2008). Light-induced manipulation with surface plasmons. *Journal of Optics A: Pure and Applied Optics*, 10, 093001.

Righini, M., P. Ghenuche, S. Cherukulappurath, V. Myroshnychenko, F. Garcia de Abajo, and R. Quidant. (2009). Nano-optical trapping of Rayleigh particles and *Escherichia coli* Bacteria with resonant optical antennas. *Nano Letters*. DOI: 10.1021/nl803677X.

Ritchie, R. H. (1957). Plasma losses by fast electrons in thin films. *Physical Review*, 106, 874–881.

Ru, E. C. L., E. Blackie, M. Meyer, and P. G. Etchegoin. (2007). Surface enhanced Raman scattering enhancement factors: A comprehensive study. *Journal of Physical Chemistry C*, 111, 13794–13803.

Ru, E. C., P. G. Etchegoin, J. Grand, N. Felidj, J. Aubard, G. Levi, A. Hohenau, and J. R. Krenn. (2008). Surface enhanced Raman spectroscopy on nanolithography-prepared substrates. *Current Applied Physics*, 8, 467–470.

Sharpe, J. C., J. S. Mitchell, L. Lin, H. Sedoglavich, and R. J. Blaikie. (2008). Gold nanohole array substrates as immunobiosensors. *Analytical Chemistry*, 80, 2244–2249.

Sinton, D., R. Gordon, and A. G. Brolo. (2008). Nanohole arrays in metal films as optofluidic elements: Progress and potential. *Microfluidics and Nanofluidics*, 4, 107–116.

Smolyaninov, I. I., C. C. Davis, J. Elliott, and A. V. Zayats. (2005). Resolution enhancement of a surface immersion microscope near the plasmon resonance. *Optics Letters*, 30, 382–384.

Sommerfeld, A. (1909). Propagation of waves in wireless telegraphy. *Annals of Physics*, 28, 665–736.

Stark, P. R. H., A. E. Halleck, and D. N. Larson. (2005). Short order nanohole arrays in metals for highly sensitive probing of local indices of refraction as the basis for a highly multiplexed biosensor technology. *Methods*, 37, 37–47.

Stewart, M. E., N. H. Mack, V. Malyarchuk, J. Soares, T. W. Lee, S. K. Gray, R. G. Nuzzo, and J. A. Rogers. (2006). Quantitative multispectral biosensing and 1D imaging using quasi-3D plasmonic crystals. *Proceedings of the National Academy of Sciences of the United States of America*, 103, 17143–17148.

Stewart, M. E., C. R. Anderton, L. B. Thompson, J. Maria, S. K. Gray, J. A. Rogers, and R. G. Nuzzo. (2008). Nanostructured plasmonic sensors. *Chemical Reviews*, 108, 494–521.

Stockman, M. I. (2004). Nanofocusing of optical energy in tapered plasmonic waveguides. *Physical Review Letters*, 93, 137404.

Storhoff, J. J., A. D. Lucas, V. Garimella, Y. P. Bao, and U. R. Muller. (2004a). Homogeneous detection of unamplified genomic DNA sequences based on colorimetric scatter of gold nanoparticle probes. *Nature Biotechnology*, 22, 883–887.

Storhoff, J. J., S. S. Marla, P. Bao, S. Hagenow, H. Mehta, A. Lucas, V. Garimella, T. Patno, W. Buckingham, W. Cork, and U. R. Muller. (2004b). Gold nanoparticle-based detection of genomic DNA targets on microarrays using a novel optical detection system. *Biosensors & Bioelectronics*, 19, 875–883.

Teramura, Y. and H. Iwata. (2007). Label-free immunosensing for alpha-fetoprotein in human plasma using surface plasmon resonance. *Analytical Biochemistry*, 365, 201–207.

Tetz, K. A., L. Pang, and Y. Fainman. (2006). High-resolution surface plasmon resonance sensor based on linewidth-optimized nanohole array transmittance. *Optics Letters*, 31, 1528–1530.

Tong, L. M., M. Righini, M. U. Gonzalez, R. Quidant, and M. Kall. (2009). Optical aggregation of metal nanoparticles in a microfluidic channel for surface-enhanced Raman scattering analysis. *Lab on a Chip*, 9, 193–195.

Wang, M., N. Jing, I. H. Chou, G. L. Cote, and J. Kameoka. (2007). An optofluidic device for surface enhanced Raman spectroscopy. *Lab on a Chip*, 7, 630–632.

Wu, L. P., Y. F. Li, C. Z. Huang, and Q. Zhang. (2006). Visual detection of Sudan dyes based on the plasmon resonance light scattering signals of silver nanoparticles. *Analytical Chemistry*, 78, 5570–5577.

Yang, A. H. J., S. D. Moore, B. S. Schmidt, M. Klug, M. Lipson, and D. Erickson. (2009). Optical manipulation of nanoparticles and biomolecules in sub-wavelength slot waveguides. *Nature*, 457, 71–75.

Yin, L. L., V. K. Vlasko-Vlasov, J. Pearson, J. M. Hiller, J. Hua, U. Welp, D. E. Brown, and C. W. Kimball. (2005). Subwavelength focusing and guiding of surface plasmons. *Nano Letters*, 5, 1399–1402.

19

Flow Cytometry and Fluorescence-Activated Cell Sorting

Chun-Hao Chen
Jessica Godin
Sung Hwan Cho
Frank Tsai
Wen Qiao
Yu-Hwa Lo

19.1 Introduction

Flow cytometry, a concept originated back in 1934 by Moldavan (Moldavan 1934), has evolved to become an indispensable bioanalysis tool, enabling researchers to study and characterize the physical (cell size, shape, and granularity) and biochemical (DNA content, cell cycle distribution, and viability) properties of cells in a highly quantitative manner. Besides its applications in basic research (e.g., immunology, and cell and molecular biology), this technology has allowed hematologists to detect and monitor the progression of diseases such as acute myeloid leukemia (AML) (Jennings and Foon 1997b) and AIDS (Shapiro 2003). A state-of-the-art flow cytometer (also known as a fluorescence-activated cell sorter, or FACS) can interrogate and sort cells with a throughput of tens of thousands of cells per second, making possible rare-event studies, such as the identification of bacterial cells (Casamayor et al. 2007) or the isolation of stem cells (Gratama et al. 1998). Currently, more than 30,000 flow cytometers have been employed in various research institutions and hospitals (Herzengberg et al. 2002), and this number has been growing as recent technological advances allow flow cytometers to become more sophisticated (multicolor detection and sorting capabilities), cost effective (<$50,000 for basic models), and less bulky. While the proliferation of these machines has been impressive, their basic operational principles have remained the same, and there is still room for paradigm-shifting improvements—especially with regard to the intrinsic limitations of the system (e.g., serial inspection and pressure–velocity relationship versus cell viability (Engh 2000)). In this chapter, we briefly discuss cytometric principles and system components of a flow cytometer as well as its applications in research, and clinical and biotechnological settings. Additionally, a survey regarding the performances of some of the state-of-the-art flow cytometers developed by different companies will be presented.

With a knowledge of the basic principles and the performances of today's state-of-the-art flow cytometers in mind, the remainder of this chapter will describe in greater detail the rationale and recent technological advances in developing a miniaturized version of FACS (i.e., µFACS), using the tools available from the fields of optofluidics and microfluidics. Even though state-of-the-art lab-on-a-chip (LOC) devices cannot compete with current flow cytometers in terms of performance, the gap is gradually being bridged. In addition, µFACS devices stand to gain from exploiting the unmatched benefits of miniaturization (ease of integration for different applications, cost reduction, and enhanced portability). To develop a microfluidic flow cytometer, three basic components, namely, fluidic, optical, and sorting components, need to be miniaturized and integrated on-chip. The ability to modularize allows each component to be built and characterized separately prior to integration. This adds the benefits of enhanced reliability and reproducibility of each component as well as offers significant reduction in development time. Besides looking at the major progress toward developing these three aforementioned components, this chapter will briefly discuss some recent developments in electronic control systems for data acquisition and noise reduction, and more importantly, developments in making sorting decisions, an integral, yet often overlooked, portion of a truly functional µFACS.

19.1.1 What Is Flow Cytometry?

Flow cytometry is a process that characterizes and analyzes cells' physical and biochemical properties based on their optical responses, as they are interrogated by external light sources in a serial manner (Givan 2001, Shapiro 2003). Emitted optical signals, such as forward scattering (FSC), side scattering (SSC), and fluorescence, are registered by a photomultiplier tube (PMT), processed by electronic circuitry, and eventually stored in computer memory for post-processing analysis. Figure 19.1 shows a simplified version of a four-parameter (two scattering signals plus two fluorescent signals) flow cytometer. The basic components of a benchtop flow cytometer are

- A fluidic system to introduce and confine samples to the center of a stream, typically by hydro-dynamic focusing
- An optical system for sample illumination and collection of emitted light
- A sorting system to deflect cells of interest into respective collection locations
- An electronics system for real-time analysis and decision making (if sorting is needed)

The fluidic system is responsible for the transport of samples through the flow chamber (typically ~50–250 µm in diameter). Pressure is applied to the reservoirs of both cell sample and sheath fluids. The higher pressure of the sheath fluid confines the sample stream to a thin coaxial flow, and then directs the cells through a laser beam. This confinement minimizes the problem of "cross talk" (simultaneous excitation of two or more particles that are too close to one another) during illumination. There are different interrogation mechanisms, some of which form a jet-in-air while others maintain flow in a flow cell. In the jet-in-air configuration, cells leave the orifice of the flow cell at speeds on the order of 10 m/s, forming the jet-in-air configuration before being excited by the laser beam. During excitation (generally performed by an elliptical beam spot on the order of 20 by 60 µm), the cells scatter light and emit fluorescence (Figure 19.2). To collect FSC (usually with an angle of bending >5°), a beam stop is required to protect photodetectors from direct illumination (which would result in measuring light extinction, a dip rather than a peak in intensity as a cell passes through the interrogation beam). Side scatter and fluorescence lines are conventionally placed perpendicular to the axis of illumination. Typically FSC and SSC carry implications for the physical characteristics of the cell, as they have some relation to characteristics such as size, refractive index, and internal granularity. Nearly all cytometers further characterize and identify cells by the collection of fluorescence signals emitted from various fluorescent tags. To collect fluorescence signals, the implementation of high-precision, well-aligned, lens-mirror-PMT systems is required. Lenses collect the emitted light from some region defined by the system's

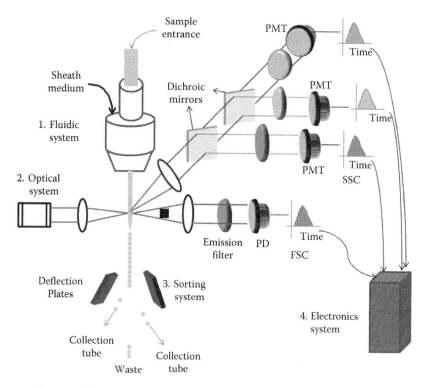

FIGURE 19.1   Schematic of a "four-parameter" benchtop flow cytometer.

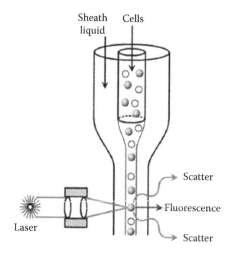

FIGURE 19.2   Schematic of a flow cell. Cells are hydrodynamically confined to a single-file stream by the adjacent sheath fluid. Cells are then serially excited by a laser source, resulting in the emission of signals (both scattering and fluorescence) that are captured by the PMTs and sent to the electronics for real-time analysis. (Reprinted from Huh et al., *Physiol. Meas.*, 26, R73, 2005. With permission from Institute of Physics publishing Ltd.)

numerical aperture, and then fluorescence signals are spectrally separated by dichroic mirrors before passing through optical filters (Figure 19.1). Each PMT detector will register and send the fluorescence signal from a single passband to the electronics for data analysis (namely, differentiating cell samples into different subpopulations).

Cell sorting (especially on the single-cell level) has become an essential feature in the field of flow cytometry as researchers and clinicians have become more interested in studying and purifying rare cells, such as stem cells (Shizuru et al. 2005) and circulating tumor cells (Gross et al. 1995). The principle of sorting generally relies on the breakage of jet flow into droplets. By applying vibrations to the jet flow from the nozzle end, droplets will be formed a fixed distance away from the nozzle after the cells have been interrogated. Cell-containing droplets will then be deflected by electrically charged plates into their respective collection tubes (droplets of no interest flow straight down to the waste tube without deflection). Theoretically, sort rates can be increased indefinitely by increasing vibration frequency, and jet flow velocity and pressure (Ibrahim and Engh 2003) (Figure 19.3); however, cell viability cannot be sustained when the jet pressure is too high due to the damage incurred during a cell's transit from jet to collection tubes. This pressure limitation puts a cap on the maximum sort rates at roughly 40,000 cells/s. A more detailed analysis is given in a book chapter by Van Den Engh (Engh 2000).

19.1.2 Flow Cytometry Applications

Flow cytometric analysis has emerged as one of the mainstream tools in the biological and biomedical fields, especially in the area of clinical studies, basic science research, and biotechnology (see Table 19.1), due to the high-quality quantitative data that the modern flow cytometer can provide. Among clinical applications, the monitoring of HIV progression (e.g., in developing countries) (Shapiro 2003), the diagnosis of hematologic disorders (e.g., different kinds of leukemia) (Jennings and Foon 1997a), and the enumeration and isolation of stem cells (Huang and Terstappen 1994, Gee and Lamb 2000, Lagasse et al. 2000) have made significant progress as a result of the high-throughput nature of the flow cytometer. Counting of CD4 (one of the main surface proteins on HIV virus-targeted lymphocytes) bearing T lymphocytes has become a widely used technique for determining the clinical stage of an HIV infection and for initiating antiretroviral therapy and evaluating its effectiveness. For instance, a normal individual usually has ~700 CD-positive lymphocytes in the peripheral blood, but in the case of an HIV infection, this number can gradually decrease to <100, signifying a severe risk of clinical

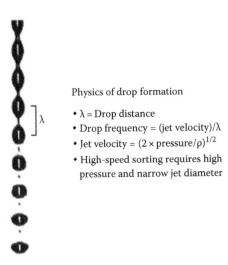

Physics of drop formation

- λ = Drop distance
- Drop frequency = (jet velocity)/λ
- Jet velocity = $(2 \times \text{pressure}/\rho)^{1/2}$
- High-speed sorting requires high pressure and narrow jet diameter

FIGURE 19.3 The frequency of droplet formation is linearly correlated with jet velocity, which depends on the operating pressure. The magnitude of the applied pressure is the limiting factor to the sorting throughput due to the pressure-induced cell damages. (Reprinted from Ibrahim, F.S. and Van den Engh, G., *Curr. Opin. Biotechnol.*, 14, 5, 2003. With permission from Elsevier.)

TABLE 19.1 List of Applications Using Flow Cytometry

Clinical Applications	Basic Science Research Applications	Biotechnology Applications	Others
HIV progression (Shapiro 2003)	Cell cycle analysis (Nunez 2001)	Optimizing antibody production (McKinney et al. 1995)	Oceanography (Legendre et al. 2001)
Hematologic malignancies (Jennings and Foon 1997a)	Apoptosis (Vermes et al. 2000)	Yeast cell cycle (Dien et al. 1994)	Environmental monitoring (Tay et al. 2002)
Gene therapy (Eliopoulos et al. 2002)	Intracellular cytokine measurement (Pala et al. 2000)	Cell growth and death (Wilson 1994)	
Stem cell enumeration (Gee and Lamb 2000)	Telomere length (Brummendorf et al. 2001)	Vaccine analysis (Roederer et al. 2004)	
Sperm quality (Graham 2001)	Phagocytosis (Lehmann et al. 2000)	Drug target discovery (Nolan et al. 1999)	

AIDS (Givans 2001). Thus, although the progression of HIV infection varies from patient to patient and cannot be predicted, proper treatment may be prescribed by measuring the CD4 count to monitor the clinical stage of infection.

As modern flow cytometric analysis has evolved to become highly multiparametric, its cell differentiation capabilities have been significantly enhanced, leading clinicians to diagnose various types of leukemic disorders (Maeda et al. 1993, Jennings and Foon 1997b). Figure 19.4 illustrates the separation of bone marrow mixtures in normal (Figure 19.4a) and leukemic (Figure 19.4b through d) blood. Distinct subpopulations can be visualized and extracted by plotting fluorescent signal intensities from cells stained with a CD45 surface marker against cells' SSC intensities. The collected signals enable the quantification of relative percentages of blood cell populations allowing researchers to perform diagnostic tests on leukemic diseases. Similarly, flow cytometry can be used to screen and purify hematopoietic stem cells (the stem cells that give rise to all blood cells) based on the existence (or nonexistence) of CD34, CD38, and CD45 antigens. In addition to extracellular labeling, intracellular DNA can also be stained (Hoechst 33258 and 33342, DAPI, chromomycin, and propidium iodide are among the nucleic acid probes that are commonly used) to determine cells' DNA content through cytometric analysis. Researchers have used this technique to study the growth of eukaryotic (Ohtsubo and Roberts 1993) and prokaryotic (Natarajan et al. 1999) cells, the change of cellular activities when subjected to physical or chemical perturbation (Fink and Meyer 2002, Berridge et al. 2003), and phases of the cell cycle (Nunez 2001). Using flow cytometry in investigating the phases of yeast cell cycles is an especially important application in industrial settings, since changes in the cell cycle can be correlated with other fermentation parameters (e.g., CO_2 production rate, ethanol production, and respiration coefficient) (Walker 1999) to optimize the cultivation process inside a bioreactor. Besides process optimization, flow cytometry has been used for quality control across various sectors within the food industry. For example, an obligatory quality control step is the application of flow cytometric methods to detect the presence of bacteria in milk (Gunasekera et al. 2000) and chicken (i.e., *Salmonella*) (Wang and Slavik 1999).

19.1.3 Main Features of a Modern Flow Cytometer: FACS

Due to demand from researchers in the biomedical fields, multiparametric analysis, high-speed screening, and sorting are three essential features for modern flow cytometers. As the result of advances in fluorescent probes, most modern flow cytometers can acquire and analyze >eight parameters simultaneously

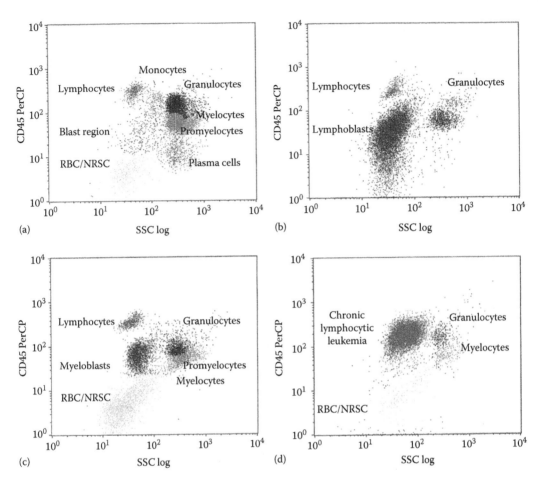

FIGURE 19.4 (See color insert following page 11-20.) Analysis of normal and leukemic bone marrow by CD45 side scatter (SSC) analysis. (a) Normal bone marrow showing normal populations of blood, (b) abnormal increase of lymphoblasts and myeloblasts as seen in acute lymphoblastic leukemia (ALL), (c) chronic myelogenous leukemia (CML), and (d) chronic lymphocytic leukemia (CLL). (From Jennings, C.D. and Foon, K.A., *Cancer Invest.*, 15, 384, 1997b. With permission. © American Society of Hematology.)

to enhance cell identification capabilities. This multiparametric analysis facilitates the identification process in two ways. First, individual surface markers often lack specificity. As a result, not only are cells of interest stained, but all other background cells could also become partially stained (due to nonspecific binding), causing an increase in overall signal noise. Thus, it is desirable to employ another fluorescent probe to act as the "negative control" to suppress noise (i.e., increase overall signal-to-noise ratio [SNR]), allowing users to better discriminate cells of interest from other cell types. Second, the ability to process multiple colors permits researchers to identify more cell types or stages, such as the maturing stages of a given cell type. For example, a CD34 probe is often used to identify human hematopoietic stem cells; however, additional probes, such as CD34 and CD45RA, can be used to identify cells at different differentiation stages (e.g., early/late progenitor cells) during hematopoiesis (Palsson and Bhatia 2004). In addition to its multiparametric capabilities, a flow cytometer typically exhibits high throughput, screening tens of thousands of cells each second. Clinicians or researchers often screen tens of millions of cells per run, and, thus, without a decent throughput, running experiments (e.g., screening for circulating tumor cells) involving a large sample size becomes impractical. The last essential feature, cell sorting, combined with a multiparametric and high-throughput analysis, has opened up a wide range of applications, including genetic studies of mutant bacteria (Link et al. 2007) and the purification and study of a primitive form of hematopoietic stem cells (Lagasse et al. 2000).

19.1.4 Comparison of State-of-the-Art FACS

Recent advances in laser technology (e.g., solid-state lasers) and signal processing electronics (e.g., in memory and processing speed) have revolutionized current FACS machines in terms of throughput (both screening and sorting), parameter number, sensitivity, and size. Table 19.2 shows some of the major players in the FACS-manufacturing business, with Becton Dickinson (BD) Bioscience being the current leader in sales. Dakocytomation typically manufactures high-end user-configurable flow cytometers, and, as of today, Dakocytomation's MoFlo offers the best-performance FACS machine, able to screen and process cells at a throughput of 100k cells/s for up to 16 parameters. Not surprisingly, such superior performance comes at the expense of high cost. In contrast, companies such as Partec, Guava Technologies, and Optoflow AS offer less powerful, cheaper, and more compact flow cytometers. A very compact (33.3 cm × 33 cm × 16 cm) microcyte flow cytometer developed by Optoflow AS employs a single-diode laser to perform a two-parameter analysis at a throughput of 5k cells/s. BD Bioscience developed the three-laser FACSAria (Figure 19.5) that offers a 15-parameter analysis and high-throughput screening (70k cells/s) and sorting (25k cells/s at a purity of 98%). In addition, relative to other high-end flow cytometers with sorting capabilities, the FACSAria gives higher sensitivity and is more compact and user friendly.

It is important to consider the specific needs of the application at hand when selecting flow cytometers because often the best-performing flow cytometer does not always translate to the best selection. For instance, if the application involves nonhazardous rare-event sorting (10^{-6} to 10^{-8} frequency), as in stem cell studies, it would be worthwhile to obtain a high-end FACS machine with very fast sorting capabilities (e.g., FACSAria), allowing users to obtain data in a reasonable amount of time. On the other hand, if the biological samples are hazardous (e.g., HIV-infected cells), it would be sensible to buy a FACS machine that employs a catcher tube mechanism for sorting (at a much lower rate of 300 cells/s), since this mechanism operates with a safer closed fluidic system. Taking another example, if the application requires counting of a specific cell type in a developing country (e.g. counting of CD4 cells for HIV monitoring), it would be appropriate to purchase a low-end, compact, lightweight, and vibration-proof flow cytometer, such as BD's FACS Count or Guava Technology's EasyCD4 (weighing just 35 lb).

19.2 Moving toward Miniaturization

As we move toward LOC flow cytometry, it is (somewhat counterintuitively) important to keep in mind the idea that smaller is not necessarily always better. With any device, it is important to consider the benefits of miniaturization and whether they fit with the desired application. Certainly there are applications that will not benefit from miniaturization, such as the use of flow for studying plankton and other large organisms (Thyssen et al. 2008). There are, however, many applications that would benefit from a miniaturized flow cytometer (Janossy et al. 2002, Chin et al. 2007).

The goals of miniaturization generally include any or all of the following:

1. Enabling mass fabrication/cost reduction
2. Enabling size reduction/portability
3. Automation/improving performance reliability
4. Creating devices that can meet the needs of new applications

TABLE 19.2 Comparison of Flow Cytometers across Manufacturers

Model	Price (in $1000s)	Sorting	Excitation Source	Detection Parameters	Speed	Sensitivity (MESF)	Sample Loading
BD Biosciences (bdbioscience.com)							
FACSAria	~$400	Yes	407 nm diode, 488 nm solid-state, 633 nm HeNe	15	70k events/s	125 fluorescein	Manual
FACSArray	—	No	532 nm green solid-state, 635 nm red diode	6	15k events/s	200 PE	Automated
FACSCalibur	—	Yes	488 nm Argon-ion, 635 nm HeNe	6	300 cells/s	750 fluorescein	Automated
Beckman Coulter (www.beckmancoulter.com)							
Epics Altra	$265–360	Yes	Customer-interchangeable air-/water-cooled laser	8	20k events/s	300 fluorescein	Automated
Cytomic FC 500	$150	No	488 nm argon-ion and 633 nm HeNe	7	3.3k events/s	600 fluorescein	Automated
DakoCytomation (www.dakocytomation.com)							
MoFlo	$350–750	Yes	Custom-configurable, choice of air-/water-cooled and solid-state lasers	Custom-configured (up to 16)	100k events/s	200 fluorecein	Automated
CyAn	$175–200	No	350 and 488 nm argon-ion, 635 and 405 nm diode laser	11	50k events/s	200 fluorescein	Automated
Partec (www.partec.com)							
Cyflow space	$55–156	Yes	488 nm solid-state, 405 and 375 nm diode laser	7	300 cells/s	100 fluorescein	Automated
CyFlow ML	$45–230	No	488 nm solid-state, 405 and 375 nm diode laser, and UV lamp	16	10k events/s	100 fluorescein	Manual

Sources: Data from Minerd, J., *Scientist*, 19, 28, 2005; Shapiro, H.M., *Practical Flow Cytometry*, John Wiley & Sons, Inc., Hoboken, NJ, 2003; the Companies' Websites.

We look to miniaturization as a way to make highly integrated (allowing necessary complexity), mass-fabricated (lower-cost) devices with reliably made elements fixed perfectly in position relative to one another. As discussed, flow cytometers consist of a number of components (lenses, channels, sorting components, etc.) that must operate in concert to form a functional device. The development of a process to reliably create the components of a cytometer inexpensively, in an integrated fashion, will mean a lower-cost device that is more accessible to researchers and clinicians. And at the same time, the development of ways to seamlessly integrate different components opens up possibilities for a range of devices with novel capabilities and more complex, perhaps parallel, devices.

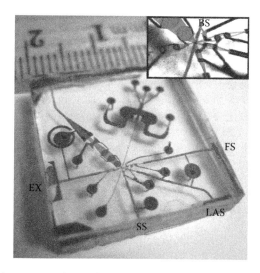

FIGURE 19.5 Prototype of an integrated LOC flow cytometer. The chip includes a microfluidic channel for fluid flow, waveguides for light transport, and lenses for beam shaping and light collection. The chip shown here has a port for bringing in excitation light (EX) and ports for measuring light scatter perpendicular to the excitation (SS), at a large non-perpendicular angle (LAS), and in the forward direction (FS) through the use of an on-chip beam block. A large numerical aperture light collection system has also been included for fluorescence detection. (Reproduced from Godin, J. et al., *J. Biophoton.*, 3, 1, 2008a; Godin, J. et al., *J. Biophoton.*, 1, 355, 2008b. With permission. Copyright Wiley-VCH Verlag GmbH & Co. KGaA.)

The benefits of miniaturization of the flow cytometer are many. A smaller device can reside in a clinic for quick, on-the-spot test results. Numerous studies indicate that blood transportation time and handling can affect test results (Dickover et al. 1998, Keeney et al. 1999); thus, shortening the route from the body to the cytometer can help producing more consistent results. A small cytometer can even become portable, allowing for testing in remote locations. Such an instrument could find field use, testing water contaminants or monitoring a soldier's health, or it could be deployed to remote villages where such medical testing is not widely available. A smaller device can also offer benefits such as lower sample volume use, more efficient reagent use, shorter light paths, and closer interaction distances for both light and cell-sorting forces. By employing microfluidics, conditions for laminar flow are almost always met, helping preserve cell viability during sorting. Microfluidics also offers a closed system, reducing biohazard risks. Eventually, a smaller device can offer benefits for truly single-cell handling and sorting, opening new avenues for biological and medical studies.

The lower cost of a mass-produced device would promote all of these applications, and would make the device more widely accessible to research labs, giving faster assay turnaround times and enabling real-time studies. An integrated device would have fewer opportunities for problems (misalignment, broken components, and gradual performance drift), reducing the need for costly service contracts. Furthermore, a mass-produced device would offer consistency from device to device, making it easier to repeat experiments or compare results from different labs.

An integrated device offers the potential for many novel advancements as well. With small size, low cost, and integration, parallel processing techniques may be exploited for higher throughput, even with lower flow speeds. The development of a "toolbox" of components or modules allows for custom-made chips to be created for unique experiments. This can offer the potential to automate more processes, removing the possibility of human error or contamination. For example, a chip could be made to incubate cells, changing media, and monitoring temperature until they have reached perhaps a certain optical density or a certain point in the cell cycle. The chip could then meter out and mix in an exact portion of reagents, with precision down to the nanoliter, and send the whole sample through the cytometer for readout. This type of an LOC device, requiring very little human interaction, could enable a researcher to run dozens or even hundreds of complex experiments simultaneously, and with virtually no differences in sample handling.

These are some of the main reasons for which researchers strive to miniaturize devices such as the flow cytometer. While today's state-of-the-art microfluidic cytometers may not be ready to replace traditional benchtop cytometers in the clinic, the necessary technologies are growing in leaps and bounds. In the remainder of this chapter, we will discuss the current state of microfluidics, micro-optics, and miniaturized cell-sorting devices, as they relate to microfluidic (and especially optofluidic) flow cytometers. We will also enumerate some of the key challenges that remain in this field, and look at the future of miniaturized flow cytometry.

19.3 Recent Development of Optofluidic μFACS

19.3.1 Introduction

In micro total analysis systems (μTAS), most components such as microfluidic channels and on-chip lenses are lithographically fabricated, and a high level of integration and flexibility can be achieved. Also, due to the micrometer-sized channel dimensions, most LOC devices operate under laminar flow conditions so that a cell-sorting event does not perturb cell viability. Importantly, μFACS systems may enable the isolation of a single cell from a large population of cells, allowing researchers to perform true single-cell analysis.

Several groups have developed microfabricated devices for performing basic flow cytometry (Schrum et al. 1999) or cytometry with cell sorting (Fu et al. 1999). Wolff et al. demonstrated a μFACS system integrated with various functional structures, including a microfluidic chamber for holding and culturing targeted cells and an optical system for detection. By confining samples to a microfluidic device, the risk of losing cells during cell handling processes can be eliminated. This approach may help resolve some of the ongoing challenges for developing a fully integrated micro cell-sorting and analyzing system. In the following sections, recent advances toward miniaturizing components of the flow cytometer (fluidic, optical, and sorting systems) as well as development toward electronic systems, which is a critical factor for the integration of on-chip optics with sorting modules (e.g., optofluidic μFACS system), will be discussed.

19.3.2 Fluidic System

In a commercial bench-top flow cytometer, the effect of cross-talking (e.g., two or more particles enter the optical detection region simultaneously) is minimized by flow focusing. This is achieved by confining the sample flow into a narrow stream using the pressure of the surrounding sheath-flow. Flow focusing further ensures uniform particle velocity by reducing the effect of sample velocity variation due to the parabolic flow behavior of liquid in a "pipe." Variations in particle velocities would not only undermine the reliability of the detected signals, they can also cause inaccurate cell sorting downstream (e.g., lower purity). Thus, the incorporation of effective focusing modules into a microfluidic flow cytometer is critical. For these reasons, significant efforts have been made toward developing focusing methods, which include electroosmotic flow focusing and pressure-driven flow focusing.

19.3.2.1 Dielectrophoretic and Electroosmotic Flow Focusing

Electroosmosis is a phenomenon in which fluid motion is caused by an applied DC voltage across a microchannel. In this approach, electrodes are inserted into fluidic channels, and as a DC bias is applied across the electrodes, electroosmotic flow is generated (Schrum et al. 1999). As sample flow enters the focusing region where the three streams cross, the electroosmotically-driven flow will confine the sample stream down to a width determined by the relative electric field strength between the sample and the sheath channels.

Schrum et al. have confined a sample stream containing 0.97 and 1.94 µm latex particles into an 8 µm stream inside of a 50 µm channel using a field strength ratio of ~0.15 (100 V/cm and 700 V/cm at the sample channel and side channels, respectively) at a screening throughput of 34 particles/s. Researchers have applied this electrokinetic focusing technique to a variety of applications, including detection and sorting of DNA molecules (Chou et al. 1999), controlled sample plug injections of various sizes (Fu et al. 2003), and flow switching (Lee et al. 2003).

Even though controlling the electrokinetically focused beam size is relatively straight forward, instantaneous, and accurate, the required input voltage is on the order of hundreds and even thousands of volts. The method might not be suitable in most flow cytometric applications since high electric field can cause severe damage to most biological agents.

19.3.2.2 Pressure-Driven Flow Focusing

Unlike electroosmotic flow focusing, pressure-driven flow focusing acts on the fluid rather than on the particles. Two-dimensional pressure-driven flow focusing, i.e., hydrodynamic flow focusing, has been the most widely used technique for particle confinement and has been applied to a number of applications, including multiple-outlet flow switching (Lee et al. 2001a), the enumeration of beads (Pamme et al. 2003) and cells (Lin and Lee 2003), and the deformation of DNA (Wong et al. 2003). By adjusting the relative pressures of the sample inlet and the two sheath-flow channels, which lie perpendicular to the sample channel, Knight et al. have focused the sample flow stream down to ~50 nm inside a 10 µm channel (Knight et al. 1998). They have also developed a mathematical model to relate the focused width to the flow resistances of the inlet, side, and outlet channels. Both numerically and experimentally, Lee et al. have investigated the effect of flow rates and channel dimensions on the width of the focused stream for both circular (Lee et al. 2001b) and rectangular (Lee et al. 2006) channel configurations. An example of 2D confinement is shown in Figure 19.6. Rhodamine dye is used for the visualization of flow confinement. Hydrodynamic focusing can also be achieved by applying suction at the device outlet (Huh et al. 2002, Stiles et al. 2005), allowing fluid focusing using only a single syringe pump (Stiles et al. 2005). In addition, rather than using sheath liquids, air can be used as the sheath fluid to confine sample flow (Huh et al. 2002). Under the same outlet suction pressure (~45 mm Hg), as the sample flow rate decreases down to a critical level (~5 mL/h), the flow stream becomes unstable and begins to break into individual droplets downstream. This feature may be attractive for microfluidic flow cytometers, as the formation of cell-containing droplets can facilitate the screening and sorting

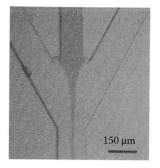
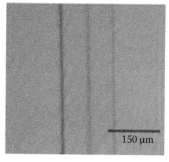

FIGURE 19.6   2D flow focusing. Sample sheath-flow rate ratio of 1:10 produces sample confinement to the center 20 µm of the fluidic channel.

of single particles or cells. Moreover, this method can eliminate the need for large sheath reservoirs, lowering the cost of operation and also eliminating the need for a clean fluid source for on-site operation in nonurban settings.

The aforementioned techniques have focused on flow confinement in x-y plane. However, without flow confinement in vertical (e.g., z axis) direction, there is still a wide distribution of sample flow velocities due to the parabolic flow profile induced by pressure-driven flow. Variation in flow velocities can lead to significant variations in detected signals from an otherwise homogenous population, which can severely undermine the reliability of the cytometric analysis. Also, to guarantee high-speed and high-purity sorting, it is essential that cells travel at same speed.

All these factors contribute to a strong interest in developing 3D focusing devices capable of confining small quantities of cells/particles in both horizontal and vertical directions.

Klank et al. simulated and fabricated a "chimney" structure in silicon by reactive ion etching (Klank et al. 2002, Wolff et al. 2003). The coaxial sample sheathing was obtained by injecting a sample into the sheath flow in a perpendicular direction. However, the disadvantage is that the fabrication process is very complicated. 3D focusing has also been achieved using multilayered 3D microfluidic devices (Sundararajan et al. 2004, Simonnet and Groisman 2005, Yang et al. 2005, Chang et al. 2007). Mao et al. have also demonstrated a "microfluidic drifting" technique that enables 3D hydrodynamic focusing with a simple single-layer planar microfluidic device fabricated via standard soft lithography (Figure 19.7) (Mao et al. 2007). This chip, is easy to fabricate and demonstrate superior 3D focusing capability, but the imposed flow rates (~337 μl min^{-1} sheath flow) could be too large for most microfluidic applications.

19.3.3 Methods of Optical Detection and Bioanalysis

Attempts at miniaturization of the flow cytometer (and indeed many other types of LOC devices) can be classified as either (1) the creation of a fully integrated device (entirely on-board pumps, lasers, detectors, etc.) (Balslev et al. 2006) or (2) the creation of a chip reader (with at least some permanent components, such as lasers, optics, detectors, and pumping systems) and a replaceable (often disposable) sample chip (Fu et al. 1999, McClain et al. 2001, Yang et al. 2006, Kostner and Vellekoop 2007). Of course, one must keep in mind that the device is only as small as its largest components, and, thus, for a portable device, it isn't enough to just miniaturize the optics—the size and weight of lasers, detectors, and pumping and perhaps sorting systems must be considered as well. The size of each of these components has been shrinking, however, and a very small chip reader is not beyond imagination.

A chip with on-board lasers has the benefits of being small and lightweight, but if these elements cannot be created at a low cost, then a disposable chip with on-board lasers is not an attractive approach. Similar arguments can be made against on-chip pumping systems, which, at this time, will generally require off-chip manifolds to control on-chip valves; thus, the benefits of size reduction for pumping elements are not truly achieved. For these reasons, we do not cover on-chip lasers and pumping systems in this chapter.

In terms of optics, arguments can be made both for and against on-chip integration, and ultimately the decision comes down to the application. Generally, optics will need to be either entirely off-chip (Pamme et al. 2003, Kostner and Vellekoop 2007), integrated into the chip (more or less necessitating 2D optics) (Seo and Lee 2004), or integrated onto the chip, similar to the lenses of microlens arrays (Chang and Yoon 2004, Chang and Yoon 2005). This last configuration can probably be eliminated at this time since it only includes a single-lens system, incapable of competing with the complex optical systems in today's cytometers. Completely off-chip optics is not of interest here due to the bulk nature of this scheme. More information regarding the optical system of a benchtop cytometer can be found in Shapiro's *Practical Flow Cytometry* (Shapiro 2003).

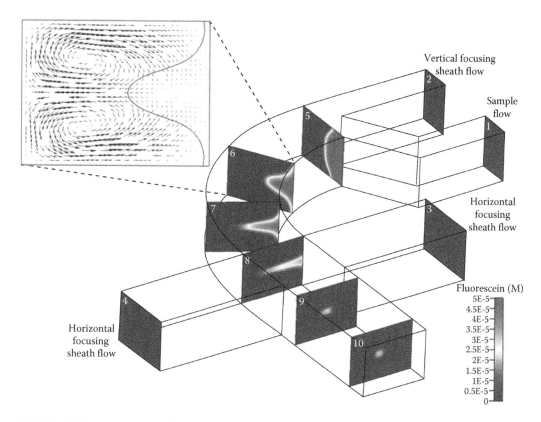

FIGURE 19.7 (See color insert following page 11-20.) Schematic of the 3D hydrodynamic focusing process by employing the microfluidic drifting technique. Slices are the cross-sectional profiles of the fluorescein dye concentration in the focusing device. Inset: the simulation of the secondary flow velocity field shows the formation of Dean vortices in the 90° curve. An iso-curve of fluorescein concentration=25 μM is arbitrarily chosen as the boundary of the sample flow. (Reproduced from Mao, X. et al., *Lab Chip*, 7, 1260, 2007. With permission from The Royal Society of Chemistry.)

Integrating the optical system chip (in-plane optics) offers a number of advantages. Such an optofluidic device is more compact and robust, without problems of misalignment by gradual drifts or by sudden jolting due to travel (in the case of a portable device). By employing an optical system that is fully machine-made and assembled, one expects that the results from two different chips should be considerably more similar than results from two bulk cytometers, since all positioning tolerances should be submicron. Laser and detector differences, of course, would still be present, but these are more easily accounted for by calibration. This optofluidic chip could certainly be expected to have a lower per-optic cost than individually created components, especially when utilizing highly customized lenses (parabolics and aspherics). On-chip optics also significantly shortens the optical path length, which may improve device sensitivity. In the following sections, we discuss methods to create and implement various integrated optical components (e.g., optical waveguides, beam blocks, and lenses), as well as a number of practical issues (e.g., cost, reliability, and performance) and challenges faced by the researchers creating optofluidic flow cytometry chips.

19.3.3.1 Integrated Waveguides

Similar to optical fibers, optical waveguides offer light confinement and routing by total internal reflection (see Chapter 3 on Passive integrated optics). Since waveguides and microfluidic channels can be lithographically defined through microfabrication, predefined local illumination of the targeted region can be achieved without the need for high-precision optical alignment (i.e., free-space optical alignment). Optical waveguides were first implemented and fabricated by oxide deposition (Mogensen et al. 2001), and by an ion exchange (McMullin et al. 2005) method under a silicon substrate for on-chip detection. Although optical coupling between the waveguides and the microchannels can be achieved, these methods require lengthy and complex fabrication processes, and the materials used are usually not suited for biological applications or for visible light. However, most of these shortcomings can be alleviated by using polymer-based methods. Monolithic integration of polymer waveguides made by PDMS (polydimethylsiloxane) alongside microfluidic channels is an attractive approach due to simple channel sealing, tunable optical properties (i.e., adjustable refractive indexes), and freedom from laborious optical alignment. Using a capillary filling technique, Lien et al. have integrated optical waveguides using "core" channels with a higher refractive index PDMS ($n \approx 1.42$) surrounded by cladding layers of a slightly lower refractive index PDMS ($n \approx 1.407$) (Lien et al. 2004). Figure 19.8 shows the pre-aligned waveguide structure and the fluorescent coupling capability of the waveguide. Based on a similar fabrication concept, Lien et al. later created an array of waveguides that allowed multiplexed signal processing for sensitivity enhancement (Lien et al. 2005). In this device, each analyte is optically interrogated eight times by eight equally spaced sets of optical waveguides. By employing a time-delay cross-correlation algorithm, they have achieved an 80 dB SNR enhancement. In Figure 19.9, extremely noisy signals can be readily extracted. This method could help enable researchers to cut down the cost and the size of the detector (e.g., from PMTs to photodiodes) without sacrificing sensitivity.

Instead of introducing/collecting light perpendicularly to the sample channel, researchers have also been working to deliver light along the microfluidic channel. In this way, biological samples can be directly illuminated inside the fluidic channel. Wolfe et al. suggested the idea of a liquid-core/liquid-cladding (L2) waveguide in 2004 (Wolfe et al. 2004). In the L2 waveguide, deionized (DI) water ($n=1.335$) and $CaCl_2$ aqueous solution ($n=1.445$) were used as the cladding layer and the core layer of the waveguide, respectively. Both layers are liquid, and thus they are reconfigurable through dynamic control. However, the L2 waveguide may suffer from the mixing of the core- and cladding-layer liquids, and the concentration of $CaCl_2$ may affect the cell viability. Lim et al. (2008) demonstrated a liquid-core/air-cladding (LA) waveguide. They used air as the cladding layer and DI water as the core layer. This

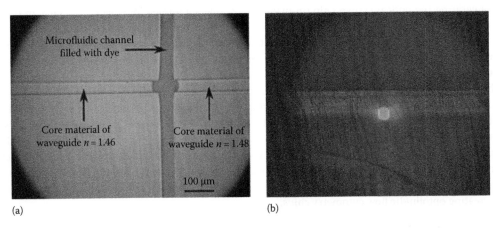

(a)   (b)

FIGURE 19.8 Images showing (a) a pre-aligned capillary-filled PDMS polymer waveguide oriented perpendicular to the microfluidic channel (top view) and (b) fluorescent light guided by the PDMS waveguide (side view). The method was later modified to completely separate the waveguide and the fluidic channels. (From Lien, V. et al., *IEEE Photon. Technol. Lett.*, 16, 1525, 2004. With permission. © 2004 IEEE.)

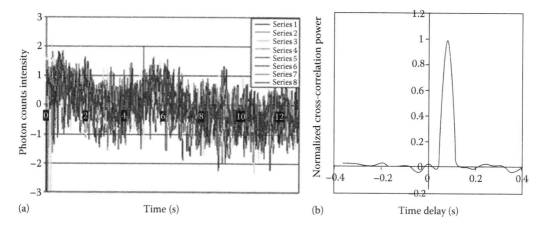

FIGURE 19.9 (a) Raw signals from eight waveguides. (b) After processing through a time-delay cross-correlation algorithm, a single pronounced peak showing the passage of a single 1 μm fluorescent bead. (From Lien, V. et al., *IEEE J. Sel. Top. Quantum Electron.*, 11, 827, 2005. With permission. © 2005 IEEE.)

method, however, requires an external pump, and the waveguide architecture may be limited. Both the L2 and LA waveguides demonstrated light confinement in only two dimensions. Unless the cladding layer is incorporated vertically, significant vertical light loss is unavoidable. Antiresonant reflecting optical waveguide (ARROW) structures with hollow cores were demonstrated by Yin et al. (2004). The ARROW waveguides are fabricated using SiO_2 and SiN layers with PI (polyimide layers). As they require multiple thin-layer coatings on the substrate, the complex fabrication process hinders the development toward low-cost and mass-producible LOC applications in PDMS (one of the most widely used polymers in optofluidic devices).

Another example of a liquid-filled waveguide can be achieved by using a low refractive index material as the cladding layer. Datta et al. (2003) demonstrated a 3D liquid-core waveguide by coating Teflon AF (Dupont Inc.) on either glass or silicon substrates. Teflon AF is an amorphous fluoropolymer that is chemically stable. Unlike other polymers, it is optically transparent from UV to IR, and more importantly, it has a refractive index ($n=1.31$) that is lower than the refractive index of water ($n=1.33$) (Dupont website 2008). A Teflon AF coating layer can therefore be exploited to clad a liquid core waveguide. When the Teflon AF–coated channels are filled with water (or almost any other aqueous solution), light will now be delivered along the same physical path as the fluid flow. The Teflon AF–coated channel confines light in three dimensions, and, as a result, minimizes vertical light loss.

The fabrication process for glass or silicon substrates is relatively expensive and time consuming, making these devices less attractive toward fabricating portable and disposable LOC devices. Recently, Cho et al. (2009a) have demonstrated a novel fabrication method of Teflon AF–coated optofluidic waveguides in PDMS microchannels. This process selectively coats Teflon AF onto PDMS channel walls by flowing the Teflon AF solution through the microchannel and removing it via the application of vacuum. The formation of a Teflon-coated layer again creates the cladding layer of a lower refractive index. Figure 19.10a shows the light output from the cross-sectional area of such an optofluidic waveguide. The light output from a channel without the waveguide is shown in Figure 19.10b. The dotted box shows the wall of the PDMS channel, and the solid line shows the boundary between the Teflon AF cladding layer and the liquid core. The comparison between these figures (widely diffused light versus narrowly confined light) verifies that the light is confined inside the liquid core of the Teflon AF–coated optofluidic waveguide. From the top view, Figure 19.11a shows the layout of a microfluidic channel, which includes a three-way splitting junction. Laser light ($\lambda=488$ nm) is fiber-coupled into the microfluidic channel, and then confined and guided by the fluid flow. At the three-way junction, as shown in the enlarged box, the

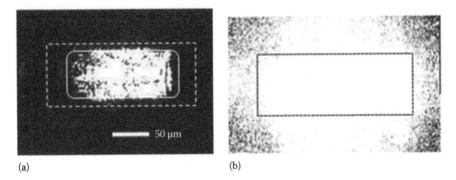

(a)   (b)

FIGURE 19.10 (a) Light output from a Teflon-coated optofluidic waveguide. The dotted box is the perimeter of the channel, and the solid line is the Teflon AF–coated core layer. (b) Light output from a channel without the Teflon AF layer.

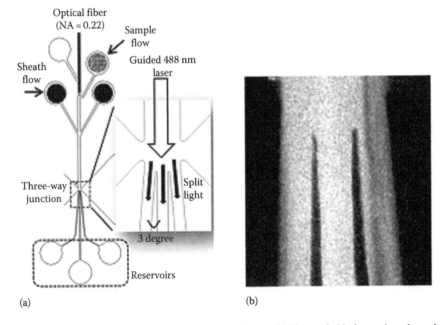

(a)   (b)

FIGURE 19.11 (a) The layout of a three-way split junction device. (b) The guided light can be split and guided at the three-way junction.

guided light is split in three ways, following the fluid flow toward the channel outlets. In Figure 19.11b, the emitted green fluorescence from a Rhodamine 6G solution is observed to demonstrate that light can be split and guided through all three channels.

The described method for waveguide fabrication has several advantages. First, the Teflon-coating process requires only a small amount of the Teflon AF solution (less than 10 μL), and the process can be completed rapidly, allowing savings in terms of both cost and time. Second, since water serves as both the sample carrier and the photon carrier through the Teflon AF–coated channel, the direct interaction between the optically confined photons and the samples (e.g., fluorophores) can enhance detection sensitivity. Third, the optical guiding capability further allows fluorescence detection at multiple locations using only a single light source, imparting a high degree of design flexibility to miniaturized optofluidic devices. Finally, in addition to delivering the light, the optofluidic waveguide can create a channel with an extremely low adsorption of biological samples due to the chemical inertness of Teflon AF. This can avoid the absorption problem common to many polymer-based LOC devices.

19.3.3.2 On-Chip Lenses

The first optofluidic cytometry chips consisted of fibers (Tung et al. 2004) or waveguides (Lien et al. 2005) integrated with a microfluidic channel. Such technologies are excellent stepping stones, but one must keep in mind the end application being worked toward. The optical system of a flow cytometer, as discussed above, is rather complex; it cannot be replaced simply by fibers or waveguides (Shapiro and Hercher 1986). If we are to create a truly integrated optofluidic cytometer, a system for creating on-chip lenses and other optics must be devised. Lenses are needed to shape interrogation light into a small, collimated beam. Lenses are also needed for high-numerical-aperture light collection from the cell of interest while simultaneously avoiding the collection of any background noise. These are tasks that simple waveguides or fibers, or even their lensed counterparts, will not be able to adequately accomplish. Some crafty signal processing might present some options to avoid the use of lenses, but the throughput would suffer tremendously. Thus, we find ourselves in need of a method of creating not simply a single spherical lens, but a small number of lenses whose shape and size can be chosen to suit our needs.

Miniaturized on-chip lenses also more readily enable multiparameter detection due to their small size, especially when one desires data from multiple scattering angles. The basic optical measurement parameters of a flow cytometer, as mentioned above, are forward light scatter, orthogonal light scatter, and fluorescence (of one or more colors). While a few specialized microfluidic devices may be able to get by with only one or two of these parameters, in general, we should aim for the microfluidic cytometer to offer all of the measurement parameters of a benchtop flow cytometer. This poses a problem in terms of chip real estate: simply including forward and orthogonal scatter (along with an excitation) requires 180° of working space around the microfluidic channel. The further inclusion of a separate fluorescence line complicates matters even further. In the standard flat-chip configuration, it can be difficult to work this many bulk lens systems around the chip, especially with refraction at low chip incidence/exit angles, the unworkable side faces of the chip, and the corners of the microfluidic channel. Here, again, integrated optics offers a potential advantage in exploiting chip real estate: nearly the full 360° panorama is available to on-chip lenses. Refer back to Figure 19.5 for an example of how on-chip lenses enable multi-angle, multiparameter detection. An earlier version of this device included an extinction detection line (0° from the excitation line with no beam stop) and a 15° scatter line, enabling multiparameter data to be recorded (Godin and Lo 2007). For a single-bead population, scatter coefficients of variation (CVs) were measured as 12%–15% (see Figure 19.12a), a significant improvement over those measured by other microfluidic devices with lensless light collection (Pamme et al. 2003, Wang et al. 2004). These results are also not far from those measured by a benchtop cytometer (Figure 19.12b). While this bead-based experiment is quite simple (and low throughput) relative to the study of blood in a clinical setting, it demonstrates that on-chip lenses can take us one promising step closer toward one day meeting this standard.

A few methods of fabricating on-chip lenses have already been demonstrated. In this section, we present some of the most promising techniques and examples. In moving forward in the pursuit of practical, reliable, and mass-producible on-chip lenses, we need to keep in mind the challenges and needs of an optofluidic flow cytometry system. Thus, in this section we will consider some of the unique challenges that researchers face, introducing techniques and examples in the context of the problems they help solve.

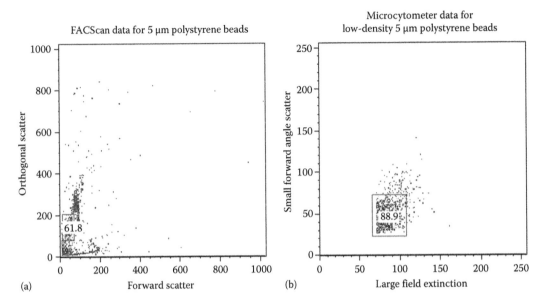

FIGURE 19.12　Comparing data for 5 μm polystyrene beads from a benchtop cytometer (a) with data for the same beads from a microfluidic flow cytometer with integrated on-chip optics (b). The gates (boxes enclosing relevant data) are chosen to encompass the population of singlets (individual beads). While the scatter angles on each axis are not the same and thus cannot be directly compared, the CV offers a means of performance comparison. For this relatively simple test, the microfluidic cytometer, with CVs of 11.6% and 24.7%, for extinction and small forward-angle scattering, performs quite well relative to the benchtop cytometer, with CVs of 12.5% for forward scatter and 14.8% for orthogonal scatter. (From Godin, J. and Lo, Y.H., Microfluidic flow cytometer with on-chip lens systems for improved signal resolution, *Sensors IEEE*, Atlanta, GA, pp. 466–469, 2007. With permission. © 2005 IEEE.)

19.3.3.3　Practical Issues toward Development of Integrated On-Chip Optics

What we look for in miniaturized optics:

1. Mass producibility
   - Minimal steps/high degree of integration
   - Few or no alignments
2. Reliable performance
   - Fixed lens positions (or high degree of tuning control)
   - Very similar chip-to-chip performance
3. Adequate performance with room for improvement
   - Ability to create multi-lens systems to begin to rival bulk cytometry systems
   - Ability to create optical quality elements
   - Possibility of high NA collection
   - Low reflection losses or reflection precautions
   - Potential to incorporate apertures, dichroics, filters, etc.

It is important to keep these needs in mind when assessing the value of any scheme of integration. Any system must be able to meet these needs or at least allow for further development to begin meet these needs. Any truly useful system created must have the potential to match the performance of at least the most basic flow cytometer, and ideally be expandable to one day rival state-of-the-art devices, or be capable of filling a yet unmet functionality (such as single-cell sorting, even if the performance of the optical detection system is rather limited).

19.3.3.4 Choosing an Optical Scheme and a Device Architecture

As mentioned above, it is important to develop a method that allows for the creation of a lens system: a multitude of lenses with the necessary shapes and positions to realize the needs of the designer. Lensed waveguides (Kou et al. 2004), fibers, fiber sleeves (Camou et al. 2003), and microfluidic channel walls (Wang et al. 2004) were all significant technical realizations, and these visions helped establish the early field. In looking forward, though, they lack the ability to form the necessarily sophisticated optical system. As mentioned above, the optical systems of a flow cytometer have a few specific needs: (a) collimation with a small beam size for interrogation and (b) collecting light from a very small area (a single cell) over a specified angular range (for scattering detection) or a very large numerical aperture (for fluorescence detection) while simultaneously rejecting as much light from the surroundings as possible. The optics to perform these tasks generally must be located tens of microns away from the cell to allow enough space in the fluidic channel for sheath flow, in order to ensure that cells flow past the interrogation point individually and at uniform velocities. In both cases (a) and (b), but especially in (b), which is the case of light collection, lenses will be needed to perform these tasks. The numerical aperture of an unlensed waveguide or fiber permits a great deal of light collection from the surroundings of the cell. A waveguide that is small enough and close enough to the cell to exclude much of this "noise" collection will generally suffer from a small collection cross section, and will also not possess the ability to restrict the angular range of collection, as is desired for light-scattering collection optics. Lastly, if such close proximity to the sample is required, the total number of light collection lines that can be included will be greatly reduced, likely prohibiting scattering parameters from being measured (in favor of fluorescence), resulting in a significant performance hit for the optofluidic cytometer, as light scatter is very often used in population gating.

Having determined that microfluidic flow cytometry chips (optics-free) will benefit from becoming optofluidic cytometry chips, we now must consider whether the lenses involved should be fixed (Seo and Lee 2004, Godin et al. 2006) or tunable (Dong and Jiang 2007). Tunable lenses offer adaptability and fine adjustment, seemingly giving more control to the user. At the same time, however, they largely undo some of the benefits of fabricating an optical system in fixed alignment by reintroducing some variability. The necessity of the inclusion of lens tunability (for a flow cytometer) should be carefully considered. For some applications, it may be necessary, for instance, to focus on different locations within the fluidic channel. Tunable lenses may also be needed to alleviate materials issues, perhaps to compensate for the performance drift due to heating. Of course, the fine control of the tuning system and the limitations on response time must be considered. Tunable lenses will likely find numerous uses, but it seems, for most general flow cytometry applications, that a fixed optical system is preferable for consistency of performance.

Another challenge for optofluidic flow cytometry chips is choosing a lens material. Multi-element fixed lens systems demonstrated to date have included air (Seo and Lee 2004), fluid (Godin et al. 2006, Tang et al. 2008), and, more recently, solid (Godin and Lo 2009) lenses. These lenses themselves are generally 2D in nature. Some clever 3D lenses have been created (Chang and Yoon 2005); however, the fabrication process does not currently exhibit the control and the scalability needed to be practical. 2D lenses leave one dimension for light to escape the system, requiring either (1) short optical path lengths to prevent significant losses or (2) waveguiding in the vertical dimension to retain light in the system. Air lenses more readily employ short track lengths, since the resulting index contrast is generally much larger than a filled lens, enabling a faster lens system. Air-filled lenses (or unfilled lenses) cannot be effectively fabricated inside of a slab waveguide, of course, due to their very low refractive indexes (resulting in leakage in the vertical direction due to a loss of waveguiding, since in this area the "core" of the waveguide, the lens, is now of a lower refractive index than that of the intended cladding). Air lenses also incur greater Fresnel losses, discussed later in this chapter.

Unlike air lenses, fluid and solid lenses generally have higher indexes than the device body, and material choices can often be made to allow for slab waveguiding. Several authors have demonstrated this concept (Wang et al. 2004 and Godin et al. 2006), including the authors' concept of an all-polymer device body, detailed in Figure 19.13 (Godin et al. 2006). Typical flow cytometry optics would employ lens systems for fluorescence collection with a light collection half angle of at least 30°. The slab-waveguided system mentioned above allows for light collection with a half angle of only about 4.8°. Improvements have been made through changes in the material choice, leading to slab waveguides with a light collection half angle of over 15°, and light collection is readily increased even further by increasing the index contrast between the core and cladding materials (Godin and Lo 2009). The core–cladding interfaces of the slab waveguide are created by molding against an optically smooth silicon wafer; thus, increasing the index contrast for the slab waveguide does not incur the significant Fresnel reflection loss penalties discussed later in this section.

When creating a fixed lens system, the benefits of using a fluidic lens are lessened. Typically, optofluidic lenses are intended to tune, and the fluid is employed to enable this tuning. There are, however, other benefits of fluid use to consider. It can be easier to locate optical grade fluids of varying properties instead of locating an appropriate solid material. Cost can also be a matter of consideration. In addition, the nature of in-plane optics, in particular, makes fluid-filling of a lens space very convenient; simple capillary force is often sufficient or low vacuum forces can be employed. Creating solid lenses, however, inside of a solid device body can prove quite challenging. A solid–solid architecture can pose problems such as air gap formation, if the materials don't bond perfectly or there are thermal mismatch issues due to laser heating. On the downside, fluids can be absorbed, spilled, or evaporated. In general, they will be more volatile than a solid, posing a problem for device reliability. Fluids, too, can be quite costly. Air bubble formation during filling is another potential issue.

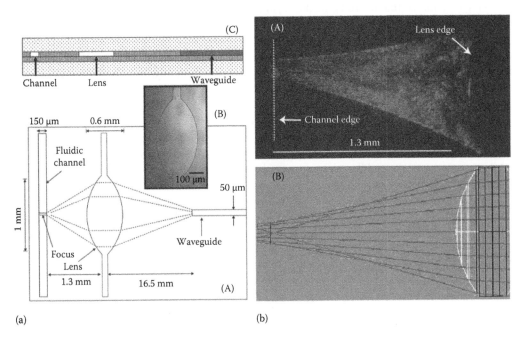

FIGURE 19.13   (a) (A–C) Top view of a schematic layout for a polymer-based optofluidic device employing 2D on-chip lenses integrated with microfluidic channels that are (B) slab-waveguided in the vertical dimension. The inset (C) is an image of an actual lens before fluid filling. The lens is fabricated as an empty channel with curved walls, and later filled with fluid to give the correct refractive index contrast. (b) A demonstration (A) of light focusing by such a 2D lens with a spherical profile, compared with ray tracing simulation results (B) used to design such lenses. (From Godin, J. et al., *Appl. Phys. Lett.*, 89, 061106, 2006. With permission. Copyright 2006, American Institute of Physics.)

Another useful solution to the liquid- or solid-lens debate is a compromise of sorts: the use of a fluid for ease of fabrication and filling, and then curing this fluid to a solid for long-term robustness and performance reliability. The main concerns are the bonding properties of the fluid-turned-solid to the body of the device, to ensure gap-free solid-lens formation. We have successfully employed an all-optical elastomer-based device using this concept (Cho et al. 2008). The body of the device uses Gelest OE41 as a lower-index ($n \sim 1.41$) cladding layer, Nusil LS-6946 as a higher-index ($n \sim 1.46$) slab waveguide to confine light in the plane of the device, and Nusil LS-6257 as a polymer for liquid phase filling of the lenses that can be cured to a higher-index solid ($n \sim 1.57$). The elastomers are fully bonded to one another either via cure bonding or UV/ozone-assisted bonding. Either method yields a strong, permanent bond.

Yet another consideration when choosing a materials system is the potential to include important elements, such as apertures, filters, dichroics, and AR coatings. A few in-plane apertures have been demonstrated (Tang et al. 2008), including a method from the authors' lab (Figure 19.14) that maintains an all-PDMS device architecture (Cho et al. 2008). Carbon-containing PDMS prepolymer is used to easily and completely fill aperture "channels," which are then heat-cured to seamlessly bond the blackened PDMS to the device body, preventing any air gaps and maintaining a smooth aperture face.

In-plane filters have also been demonstrated with dyed PDMS (Bliss et al. 2007). To be a practical solution, more filtering bands need to be demonstrated; however, this concept shows considerable potential. Antireflection coatings, to the best knowledge of the authors, do not exist at this time, and would create a significant engineering problem in terms of the need for finely controlled coating thicknesses on such small lenses. Solving or, at least, working around this problem will be a significant challenge for on-chip optics. Some improvement may be seen by minimizing index contrast, as discussed below. Dichroics, too, present similar issues; however, chromatic separation by other means may prove sufficient. We are working on the development of an on-chip light filtering and redirecting module based on the dispersive properties of PDMS and other elastomers. Such modules, composed of prism-like elements, may be used in place of dichroics and perhaps filters on-chip (Godin and Lo 2009).

Material choice is an issue not simply for lens fabrication; the materials system chosen has implications for device manufacturability and general practicality. Polymer or PDMS devices are often employed in the early development stages. Currently, no truly large-scale fabrication facilities exist for

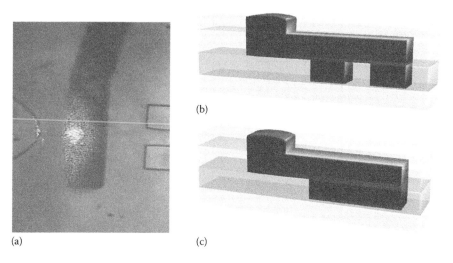

(a) (b) (c)

FIGURE 19.14 On-chip blocking elements are needed for stray light control. (a) A beam block created from carbon-containing PDMS demonstrates light-blocking abilities in an optofluidic chip. (b) Apertures or (c) beam blocks of this design can be easily manufactured alongside microfluidic channels and lenses. (Reprinted from Cho, S.H. et al., Microfluidic photonic integrated circuits, *Proceedings of the SPIE*, China, 7135, p. 71350M, 2008. With permission from SPIE.)

creating such devices in an efficient and a cost-effective manner. In addition, PDMS has well-known adsorption issues that can quickly contaminate a device. Recent work in this area shows promise for the development of coating processes to eliminate this issue (Cho et al. 2009b), but more work remains to be done. For optofluidic cytometers to push forward, either large-scale fabrication processes must be developed along with the development of a process for preventing adsorption by PDMS, or a new set of materials will need to be employed for commercialization. In the latter case, the transition from proto-type material choices to commercializable product material choices might prove to be a sticking point. This potential problem should be on the minds of researchers and developers. The problem is especially important when optics are involved due to the performance being dependent on refractive index, optical dispersion, optical absorption, and auto-fluorescence. In addition, sidewall roughness plays a major role in fabrication, posing significant constraints on available fabrication methods.

19.3.3.5   Creating Optical Quality Components

As mentioned above, a critical challenge for all on-chip optics is developing methods of creating optical quality lenses in a manner compatible with mass production. Roughness is an important matter to consider when fabricating optics, especially when antireflection coatings are not currently available for in-plane lenses. Fresnel reflection losses depend on the index contrast between the lens and its surroundings, and are given by

$$R = \left(\frac{\left(n_1 - n_0 \right)}{\left(n_1 + n_0 \right)} \right)^2 \tag{19.1}$$

Fresnel losses occur at each surface (i.e., twice per lens). Figure 19.15 shows the decrease in transmission as a result of Fresnel losses as a function of index contrast for 1–4 lenses (considering the possibility of a lens system for illumination as well as for detection). Note that these loss figures are only for Fresnel losses at the lens surface, and do not take into account Fresnel losses at the channel wall (approximately

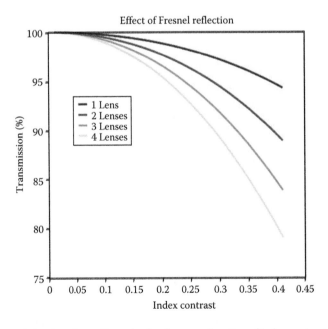

FIGURE 19.15 Transmission loss due to Fresnel reflections as a function of index contrast, for 1–4 lenses. Note that more powerful (higher-index contrast) lenses suffer much higher Fresnel losses. In designing on-chip lens systems, one must balance material properties, optical system power, and optical losses.

0.2% for each wall when filled with water), and absorption, radiation or coupling losses from waveguides, etc. An index contrast of 0.15 in a PDMS-based device results in Fresnel losses of roughly 1.2%, compared with losses of 11% for an air-based lens. Thus, while a higher-index contrast creates a faster lens system (i.e., more powerful lenses) and allows for higher NA light collection, Fresnel reflection is working in the opposite direction, resulting in higher losses as the index contrast increases. Reflection loss is problematic not only for the loss of signal but also for the potential of directing stray light into other light paths (when multiple collection paths are employed).

For flow cytometry, wavelengths employed are as low as 488 nm (for today's standard illumination) and may push toward 400 nm (for quantum dot excitation). For optical quality, we strive to achieve root mean square (RMS) roughness below the Marechal criterion of lambda/14. For 488 nm, this translates to ~35 nm RMS roughness in air, or ~25 nm RMS roughness in a material of refractive index 1.4 (such as PDMS). For 400 nm light, these figures become ~29 nm RMS and ~20 nm RMS, respectively.

Most optofluidic flow cytometry devices employ microfabrication to create their small features. Sidewall smoothness and verticality are not trivial to obtain for features that are fairly large (~20+μm) on the scales of microfabrication. Diffraction and heating issues make lithography thick enough to form microfluidic channels (typically ~50 μm) difficult; however, there are photoresists developed for these applications, such as MicroChem's SU8 line of resists. Etching materials, such as glass or silicon, provide a much more robust mold; however, etching techniques generally produce rough or scalloped sidewalls that are problematic for in-plane optics (Seo and Lee 2003). One notable exception is cryogenic plasma-assisted etching. For this method, fluorine-based etching is paired with sidewall passivation, such as an SF_6/O_2 chemistry, at cryogenic temperatures ($T \leq -100°C$). Since the etching and passivating gases exist simultaneously in the chamber, roughness from switching, such as in the Bosch process, is avoided. This process can produce highly vertical (>89.5°) and smooth (1–10 nm roughness) sidewalls (Pruessner et al. 2007). We have successfully used low-pressure cryogenic etching with the SF_6/O_2 chemistry to etch features 50 μm in height with surface roughness ~30 nm (see Figure 19.16) (Godin and Lo 2009), and expect to achieve significant improvements with continued process refinement.

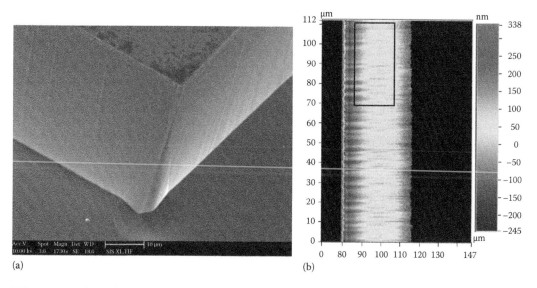

FIGURE 19.16 (See color insert following page 11-20.) (a) SEM image of a cryogenically etched mold feature. Features are 50 μm deep, etched into a silicon wafer for mold durability. Sidewall roughness is measured by optical profilometry (Veeco NT1100). (b) A sample of the profilometry measurement, along with a sample measurement area (black rectangle). Roughnesses on the order of 30 nm have been measured for cryogenically etched features. (From Godin, J. and Lo, Y.-H., Advances in on-chip polymer optics for optofluidics, *Conference on Lasers and Electro-Optics*, 2009. With permission from Optical Society of America.)

19.3.4 Sorting Methods

19.3.4.1 Introduction

Significant development of the optofluidic flow cytometer has been focused on bio-sensing and bio-analytical applications. However, *to develop a truly optofluidic μFACS*, in addition to the integrated optofluidic circuits, one needs to (2) incorporate a sorting module at the downstream of the system (e.g., detection occurs at the upstream) for cell manipulation on-chip and (2) develop a robust electronic control system for real-time signal processing and decision making. Cell sorting is essential to biological applications because analysis of cells in bulk (as in traditional biochemical assays) can often lead to conclusions very different from the conclusions of single-cell studies. For example, the cellular response (e.g., transient fluctuation of Ca^{2+} concentrations) due to certain chemical stimuli yields completely different results when examining cells in bulk versus single cells (Fink and Meyer 2002, Berridge et al. 2003, Lewis 2003). Therefore, cell sorting on a single-cell level with high throughput, purity, and cell viability has been the goal for most sorter technologies. In recent years, a variety of μFACS applying different sorting mechanisms have been developed and characterized. The major sorting mechanisms include electrokinetic switching (Fu et al. 1999, Fu et al. 2004), DEP (Voldman et al. 2002, Holmes et al. 2005), hydrodynamics flow switching (Fu et al. 2002, Krüger et al. 2002, Wolff et al. 2003), and piezoelectric-actuated flow switching (Chen et al. in press) (see Table 19.3). For a discussion of the development of various other non-fluorescence-activated sorting mechanisms (e.g., magnetic, acoustic, optical, and gravitational), which will not be discussed in this chapter, readers are encouraged to see a

TABLE 19.3   List of Most Common μFACS

Sorting Mechanisms	Methods	Pumping	Sort Rates	Materials	References
Electrokinetic flow switching	Flow redirection by forward/reverse electrokinetic flows	Electroosmotic flow	~20 samples/s	PDMS	Fu et al. (1999)
	Flow redirection by electrokinetic injection	Electroosmotic flow	N/A (manual flow switching)	PMMA, glass, and SU-8	Fu et al. (2004)
Dielectrophoresis	Dielectrophoretic deflection of particles	Syringe pump	~300 samples/s	Glass, Ti, gold, and polyimide	Holmes et al. (2005)
	Dielectrophoretic cell trapping and release	Syringe pump	N/A (manual cell sorting)	Glass, Ti, gold, and SU-8	Voldman et al. (2002)
Hydrodynamic flow switching	Fluid redirection by externally driven syringe pumps (response time: 0.26 s)	Syringe pump	N/A (manual flow switching)	Silicon, SU-8, and glass	Krüger et al. (2002)
	Fluid redirection by pneumatically controlled membrane valves (response time: 5 ms)	Microfabricated membrane valves controlled by external pneumatic devices	~26–44 samples/s	PDMS	Fu et al. (2002)
	Fluid redirection by high-speed external check valves (response time: 2.5 ms)	Syringe pump	~200 samples/s	Silicon, gold, and glass	Wolff et al. (2003)
Piezoelectric-actuated flow switching	Fluid redirection by high-speed on-chip piezoelectric actuation (response time: 0.1–1 ms)	Syringe pump	>1000 samples/s	PDMS and lead zirconate titanate (PZT)	Chen et al. (2009)

number of recently published review articles (Huh et al. 2005, Chung and Kim 2007, Pamme 2007, Godin et al. 2008b). The development of a robust electronic control system, which has been often overlooked by researchers, serves as a critical link between the optofluidic circuits and the sorting module. Based on the design of the optofluidic circuit, which dictates the nature of the collected signals, electronic algorithms can be implemented to drastically increase the SNR, facilitating detection of particles/cells in real time. For instance, the aforementioned optofluidic arrayed waveguide detection (Lien et al. 2005) allows a sensitivity enhancement of 80 dB after digital signal processing (DSP). Besides real-time signal enhancement, in recent developments, the electronic control systems can be further programmed to activate sorting events and characterize sorting performance (e.g., sorting efficiency and purity) in real time, making post analysis and verification (e.g., using the conventional FACS machine) unnecessary. Thus, in this section, in addition to discuss various versions of microfluidic μFACS (a precursor of optofluidic uFACS), recent developments in μFACS-specific electronic control systems and their implementations will be presented.

19.3.4.2 Recent Development of μFACS

The first generation of μFACS was developed using the electroosmotic effect to redirect fluid flow. Electroosmotic flow is induced based on the migration of charges, as a DC voltage is applied across two outlets (with inserted electrodes). In a T-shaped channel design, by controlling the states of the voltage applied between the two channel outlets and the channel inlet, Fu et al. demonstrated sorting of fluorescent *E. coli* at a throughput of ~20 cells/s (Fu et al. 1999). In a similar approach, another research group (Fu et al. 2004) shows manual sorting of red blood cells in a three-outlet-to-one-inlet channel design under 300–500 V operation. In general, electroosmotic flow manipulation is easy to control and the device fabrication is relatively simple (e.g., insertion of Pt electrodes into respective inlet/outlet reservoirs). However, this sorting scheme has many drawbacks including electrolysis-induced bubbles at the electrodes surface, frequent change of voltage settings due to ion depletion (Fu et al. 1999), low sort rates, high power consumption, and potentially low cell viability as a result of damages inflicted by a high electric field (i.e., high voltage setting).

Sorting by the dielectrophoretic principle, on the other hand, has recently gained momentum since the approach can alleviate some of the aforementioned limitations. DEP moves and manipulates fluid particles by exploiting the relative polarizability between the particle and the surrounding solvent. In contrast to electroosmosis, DEP uses AC voltage (typically at MHz regime) to deflect or move particles by the dielectrophoretic force (positive/negative DEP causes particles to migrate toward the high/low electric field region). Fielder et al. first used the concept to trap L929 cells under an eight-electrode quadrupole cage system (Fielder et al. 1998). Following this work, Voldman et al. have designed an array of dielectrophoretic traps that allow transient holding/releasing of cells against pressure-driven fluid flow in a noncontact manner (Voldman et al. 2002). By switching electrodes (on or off), users can isolate cells of interest (e.g., fluorescent cells) and wash away unwanted cells. One advantage of the hold/release sorting scheme is that a single-cell assay can be readily performed by exposing trapped cells to chemical stimuli. However, the sorting throughput is low and an extended exposure to DEP electric fields can potentially cause cell damage. In addition to trapping, sorting under dielectrophoretic deflection has been carried out by Holmes et al. in a T- and Y-junction channel design (Holmes et al. 2005). By having two embedded electrodes placed perpendicular to the fluid flow, the system can deflect 6 μm particle speeds of up to 300 particles/s under 10 MHz and 20 V_{p-p} operation. However, under automated sorting, the throughput is only 10 particles/s due to the lack of a high-speed electronic control system. The dielectrophoretic deflection-based sorting eventually faces a throughput limitation because the DEP force becomes small relative to the hydrodynamic drag force as flow rates increase. In summary, the DEP-based method can be low-powered, but the fabrication process is highly complex (e.g., deposition of Au and precise alignment for bonding) and the solvent used requires additional consideration (e.g., buffer with specific conductivity).

Sorting under hydrodynamic flow manipulation is typically achieved using external syringe pumps and either external or integrated valves. In an X-junction channel configuration, Kruger et al. control flow switching by manual activation of a syringe pump, which causes injection of nanoliter plugs of fluid from the side channel (Krüger et al. 2002). Consequently, the sample flow is temporarily redirected from the waste channel to the collection channel. Even though the fluid control system is simple, the slow mechanical response of the syringe pump (0.26 s) suggests that the sort rate (i.e., the number of targeted particles being sorted per second) cannot exceed 5 particles/s. Instead of controlling syringe pumps, Wolff et al. demonstrate fluorescence-activated sorting by actuating an external check valve (2.5 ms response time), which is attached to the outlet of the collection channel, as shown in Figure 19.17 (Wolff et al. 2003). Fluorescence signals upstream trigger the opening of the check valve, causing temporal flow redirection to the collection channel. Under high-speed valving, the potential sort rate can be as high as ~250 cells/s. In this scheme, the sorting of rare fluorescent beads from a chicken's red blood cells at a screening throughput (i.e., number of cells passing through the optical interrogation region per sec) of 12,000 cells/s has been demonstrated. However, without a robust electronic control system (such as feedback control with low jitter), the sorting of a highly concentrated sample can compromise the purity of the sorted sample. By integrating a valve system on-chip using multilayer soft lithography, Fu et al. have demonstrated sorting using PDMS-based membrane (~40 μm) valves (theoretical response time of ~5 ms) to manipulate flow direction (Fu et al. 2002). By pneumatically controlling membrane depression (thus blocking fluid flow), they demonstrated automated cell trapping and sorting at a throughput of ~26–44 cells/s. Although sorting under hydrodynamic flow redirection causes minimal cell damage and is not subject to buffer restrictions, the mechanical components (e.g., membrane, check valve, syringe pumps) have a limited response speed and are usually bulky, which can undermine progress toward miniaturization.

Recently, Chen et al. have achieved sorting based on piezoelectric-actuated flow switching. The piezoelectric (PZT) actuator is integrated on-chip by a UV-ozone bonding methodology in a configuration that maximizes power transmission by allowing direct contact between the stainless steel of the PZT surface and the fluid. Under a push-pull mechanism, a transverse displacement of fluid on the order of nanoliters can be introduced by the bending motion (upward or downward) of the actuator controlled by the input voltage waveform (Figure 19.18). As targeted cells/particles arrive at the sorting junction, they are being deflected (toward the left/right) by the drag force of the displaced fluid, eventually exiting toward the collection channels. When the PZT actuator is inactive, unwanted particles exit straight

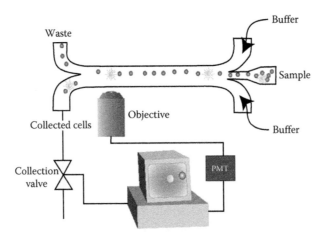

FIGURE 19.17 Schematic of a check-valve based μFACS. Sorting of cells is based on the triggering of external check valves, which in turn redirects cell-containing fluid into the collection channel. (Reproduced from Wolff, A. et al., *Lab Chip*, 3, 22, 2003. With permission from The Royal Society of Chemistry.)

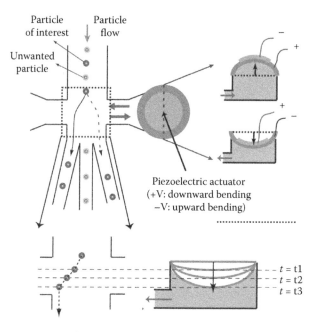

FIGURE 19.18 Operating principle of the PZT sorter. As particle enters the sorting junction, the bending motion of the PZT actuator will temporarily disturb fluid flow (either to the right or the left), causing particles to be deflected to the left/right channels. The bending orientation (e.g., upward or downward) and the amount of bending of the PZT actuator are controlled by the polarity and the magnitude of the input voltage, respectively. In the absence of PZT actuation, unwanted particles stay in the center streamlines, which travel straight down to the waste channel.

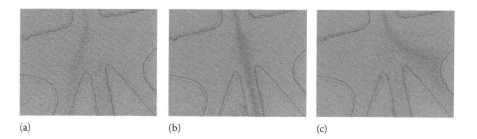

FIGURE 19.19 Images showing deflection of a rhodamine dye as a result of PZT actuation. (a) The rhodamine stream switches to the left as the PZT disk bends downward. (b) No stream deflection is exhibited when the PZT actuator is off. (c) The rhodamine stream is deflected to the right channel as the PZT disk bends upward.

down the center waste channel. This flow redirection capability can be seen in Figure 19.19 using a rhodamine dye for visualization. As the PZT actuator is bending upward/downward (i.e., due to upward/downward ramping of input voltage), the stream assumes a temporal flow switching toward the right/left collection channel. The magnitude of the stream deflection can be precisely controlled by adjusting the amplitude of the input voltage, showing the potential to extend the device architecture beyond the three-outlet system to sort cells of multiple types. Furthermore, such controlled flow-switching behaviors are observed for an operation frequency >1 kHz, implying the sort rate can be >1000 cells/s. Manual sorting of single *E. coli* cells at a rate of 330 cells/s under 200 Hz actuation frequency and $6V_{p-p}$ AC voltage has been demonstrated, showing the potential for high-purity sorting. In addition to a fast sort rate (response time of 0.1–1 ms), the actuator is low-cost and low-powered ($<10 V_{p-p}$ and <1 mW) and the fabrication process is simple. Moreover, the integration of an actuator on-chip eliminates bulky external actuators, taking the μFACS a step closer toward miniaturization.

To achieve automated sorting (i.e., true fluorescence-activated sorting) at high throughput and purity, the same group has recently developed a novel fluorescence detection strategy in combination with highly robust electronics using FPGA implementation, to not only amplify detected signals but also to provide a method to monitor fluorescence-activated sorting in real time. In addition to piezo-electrically–actuated sorter, the proposed detection strategy can also be applied to various other sorting modules and specially designed optofluidic circuits. Before describing its implementation, a brief background will be given in the following section.

19.3.4.3 Digital Signal Processing for Optofluidic μFACS

By integrating techniques from optics, modulation theory, and real-time DSP hardware, the performance of the μFACS can be improved. Furthermore, with the integration of real-time DSP algorithms and hardware, the overall cost of this optical system can be reduced.

To improve system performance, that is, to increase the chances of detecting the cells and reduce the chances of faulty detection, the noise characteristics of the system need to be studied. Armed with the knowledge of where noise originates, methods toward enhancing signals can be designed to reduce the effect of noise (Kay 1993, Kay 1998, Van Trees 2001, Ziemerm and Tranter 2001). Based on the analysis of the noise from the optical system, we have identified three major forms of noise: (1) nearly white Gaussian noise (WGN)—which mainly comes from thermal noise of the detection circuit; (2) impulse noise—which comes from the dark count of single-photon detectors; (3) low frequency near direct current (DC) noise—which comes from the intensity fluctuations of the laser source as well as any stray light from the surroundings. The noise spectra have the characteristics shown in Figure 19.20. In a typical flow cytometer system, the optical system is designed to produce a single pulse for each event being detected. There are several problems with this type of output signal in a μFACS. First of all, the bandwidth of an impulse is very wide. Therefore, all the WGN noise energy will be incorporated into the output response, reducing the SNR. Second, the impulsive dark count noise from the optical detector interferes with the impulse signal from the cells, causing the registration of false positive signals. Finally, the DC noise also degrades the output signal.

Understanding the characteristics of the noise spectrum, one can design a signal that has a narrower frequency bandwidth than a single pulse signal. The trick is to apply an optical spatial filter to the detection area in the microfluidic channel. When the fluorescence signal of a traveling cell is detected, the spatial optical filter converts the fluorescence into a temporarily modulated signal of a desired frequency

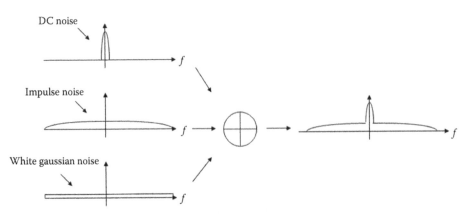

FIGURE 19.20 Forms of noise from a flow cytometric system.

spectrum for maximum S/N ratio. The concept is shown in Figure 19.21. The pulse signal has a very wide frequency component; therefore, it overlaps with a lot of noise. On the other hand, the narrow-band signal only shares a small bandwidth with the noise. Under the assumption that both signals have the same amount of energy, the narrow-band signal will have a significantly higher SNR under the condition that the bandwidth supports the desired throughput.

The most straightforward method of creating the spatial filter is to place a spatial mask with the designed slot patterns near or attached to the microfluidic channel. Alternatively, the spatial mask may be placed on the image plane of the microfluidic channel in a microscope setup. When multiple fluorescence detection locations exist on the μFACS device for parallel operation, the above scheme suffers from low laser excitation intensity and subsequently low signal intensity. A more attractive approach is to use a computer-generated phase mask to diffract the excitation laser to the locations of fluorescence detection (O'Shea et al. 2003) As an example, a four-phase diffractive optical element is shown in Figure 19.22a, and the corresponding output pattern is shown in Figure 19.22b. Furthermore, the added benefit of using diffractive optics is that more advanced signal patterns can be generated.

According to Kay, the optimal detector under WGN is the match filter (Kay 1998), a filter having a response that is the inverse reciprocal of the original signal. The match filter is essentially a filter having exactly the same frequency component as the signal. Under the constraint that the noise is white, the filter amplifies the frequencies of a high SNR and attenuates the frequencies of a low SNR, providing improvement in the overall SNR. Another observation is that fluorescence signals going through a different spatial filter at another location along the microfluidic channel can also be detected by the same optical detector. In other words, using a single PMT detector that has nanosecond response time, one can detect and distinguish multiple fluorescence signals. One signal waveform may be used to classify the samples to make sorting decisions and the other signal waveform may be used to verify the result

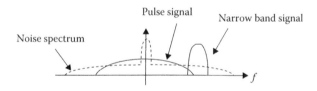

FIGURE 19.21 Comparison of pulse signal and narrow-band signal buried in noise spectrum.

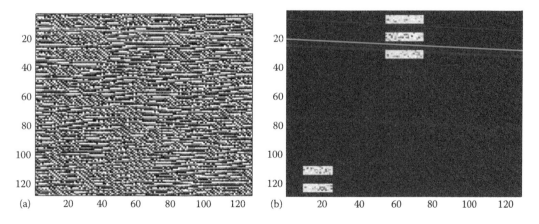

FIGURE 19.22 Images showing (a) 4-level phase diffractive optical element and (b) the resulting output beam pattern generated.

of sorting, thus producing information in real time on sorting efficiency and sorting error ratio. The latter is another attractive feature possessed only by μFACS systems, since integration makes it easy to introduce extra components and add redundancy to the system.

19.3.4.4   Real-Time Detection and Feedback Control with Hardware Implementation

The experimental setup for automated sorting is shown in Figure 19.23. For electronic control and signal processing, LabView cRIO from National Instruments (NI) is employed. cRIO is an embedded system—microcontroller with a real-time operating system (RTOS) to provide device drivers and a TCP–IP interface for the connection to a PC. The control algorithm is implemented in an FPGA (field programmable gate array) for low-jitter (<10 μs) real-time control.

As mentioned previously, various forms of noise can be suppressed by the concept of a match filter. For implementation, this translates into the use of a spatial filter (e.g., photolithographic transparency mask) that allows fluorescence only from certain areas in the channel to reach the detector. To achieve both particle detection and verification of a sorting event, the mask is encoded with triple slits and double slits. Hence, the passing of the fluorescent particle into the sorting channel would cause the PMT detector to register a three-lobed signal followed by a two-lobed signal (Figure 19.24a). The flow of the signal processing algorithm, which is implemented in the embedded FPGA chip, is described in Figure 19.24b. Random high pulse noises from the PMT (e.g., caused by sporadic discharge of the device) are removed before running the signal amplification algorithm based on finite impulse response (FIR) filtering. With an FIR-matched filter, the SNR can be increased by 18 dB (Figure 19.25). After SNR enhancement, a threshold and a search of maximum signal criteria are applied to determine the presence of the detected particle. A signal above the threshold indicates that a particle/cell to be sorted is found, triggering the following actions: (a) a delay counter delays the firing of the pulse generator, (b) a preprogrammed output voltage signal is fired to drive the on-chip PZT actuator, (c) after a certain time delay the system should detect the "verification" signal from the sorted sample traveling through the "verification zone," and (d) the records of the sorting efficiency and sorting error are updated. The amount of time delay equals the travel time of the particle from the optical detection zone to the sorting junction. Until the sorted particle is verified, the PZT actuator will not be fired again. This avoids

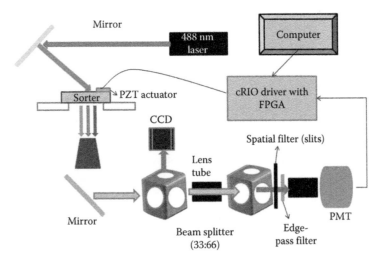

FIGURE 19.23 Experimental schematic for fluorescence-activated cell sorting using a PZT actuator. A spatial filter is placed at the image plane of the light-collecting microscope objective to allow region-specific transmission of fluorescence emitted by the passing-by particles. A cRIO external driver with embedded FPGA processes the input signals and, upon detection of a particle of interest, the driver sends a designed output waveform to drive the PZT actuation.

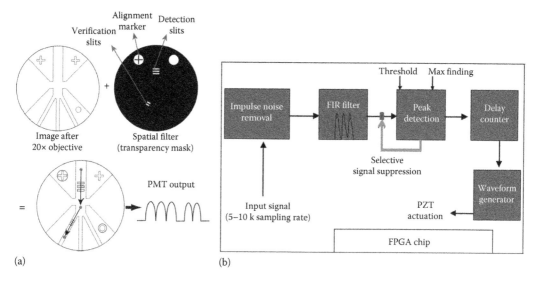

FIGURE 19.24 Design of a spatial filter and the electronics algorithm within the FPGA chip. (a) The spatial filter is designed to purposefully coincide with the image plane of the objective lens. (b) The control-loop algorithm first amplifies the detected signal and, after the peak detection algorithm, the output waveform is delayed and sent to the PZT actuator for sorting.

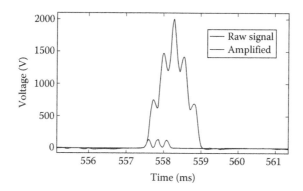

FIGURE 19.25 Comparison of the three-lobed raw signal with amplified signal after running through the FIR-matched filter algorithm. The resulted amplified signal shows an 18 dB SNR enhancement.

the problem of confusing the verification signal with the signal from particles traveling too close to the particle being sorted.

By implementing the proposed signal processing algorithm and electronic control, fluorescence-activated sorting of 10 μm fluorescent beads from a sample containing both fluorescent 10 and 5 μm beads (ratio of ~10⁻²) has been performed. Since 5 μm beads emit significantly less fluorescent light than 10 μm beads (~10–20 times less), setting a threshold for sorting 10 μm beads (but not 5 μm) becomes possible. The preliminary result shows a 75% sorting efficiency (i.e., percentage of targeted particles being sorted), ~50-fold enrichment, and a screening throughput of 110 particles/s (Figure 19.26). A high processing speed (50–100k sampling rate) and low electronic jitter, along with system optimization (e.g., flow condition, particle concentration, input waveform, and time delay), allows the sorter to achieve >80% sorting efficiency, ~1000-fold enrichment, and a screening throughput of >1000 particles/s.

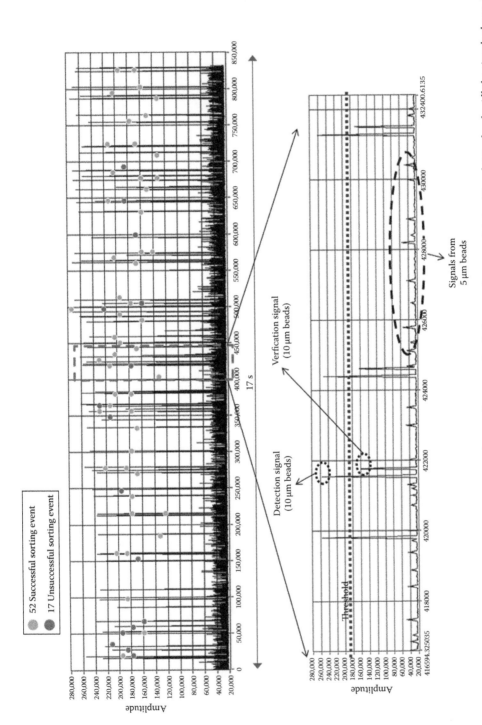

FIGURE 19.26 (See color insert following page 11-20.) Sorting of 10 μm fluorescent beads from a mixture containing 10 and 5 μm beads. All the signals shown here have been amplified by the match filter. Firing of the PZT actuator occurs only when the detected signal intensity goes above threshold. For every single sorted particle, the detected signal (upstream) is always followed by a verification signal (downstream). Within the time period (~17 s), 52 out of 69 (~75% sorting efficiency) 10 μm beads have been successfully sorted. Most importantly, no beads were mistakenly sorted, yielding a high enrichment (~1000-fold) ratio for rare events.

In summary, the developed signal processing and electronic control system has three main advantages. First, the developed methodology can easily be applied to many other micro-sorting systems, such as the dielectrophoretic and check-valve based sorters mentioned above. Second, signals can be amplified in real time, resulting in more accurate decision making. Third, the system is able to provide real-time information on sorting efficiency and sorting error ratio, allowing operators to monitor the health and operation condition of the device and to make necessary adjustments and interventions during the experiment.

Even though the DSP algorithms and the electronic control system are designed for conventional μFACS (a precursor of optofluidic μFACS), they can be readily implemented to integrate the two systems—on-chip optical components and sorting modules. Device structures such as Lien's arrayed waveguides (Lien et al. 2005) (signal amplification based on the concept of oversampling) and Cho's Teflon-coated liquid-core waveguides (Cho et al. 2009a) (enable fluorescence/scattering detection at multiple locations) can be readily integrated with the piezoelectric-actuated sorting module with slight modifications in the proposed DSP algorithms (match filter designs) and spatial filter designs for real-time signal enhancement and sorting performance characterization. The development, thus, can lead to a self-contained optofluidic μFACS.

19.4 Conclusion

Due to tremendous technical advancements, the state-of-the-art FACS can now perform bio-analysis upto 16-parameters at a throughput of 100,000 events/s and sorting at high purity. As a result, there has been an explosion of flow cytometric applications ranging from assessing the brewing process quality in the brewery industry (Muller and Hutter 1999) to the isolation of adult stem cells for single-cell studies. However, in general, most benchtop flow cytometers are still bulky, expensive, mechanically complex, intolerant to vibrations (e.g., causing misalignment of optics), and not very user friendly. On the other hand, μFACS technologies have yet to be widely adopted in the biology community, partially due to researchers' unawareness of the technologies and a lack of practical commercial products. Nonetheless, μFACS offers the benefits of significant size and cost reduction as well as reduction in required sample usage and improvement toward portability (resistance to vibrations, etc.). In addition, due to the ease of integration, functional components such as sample pretreatment (e.g., cell lysing or staining) and post-sort cell culturing could be incorporated into a μFACS, providing further time and cost savings for researchers and clinicians. Finally, the disposable nature of LOC devices eliminates the chances for cross contamination and may reduce the need for routine maintenance.

Even though a performance gap still exists between the conventional μFACS and high-end benchtop flow cytometers, the optofluidic μFACS devices, particularly those with integrated photonics and PZT on-chip sorters, show great promise in real applications and offer capabilities such as single-cell sorting that are unavailable on commercial systems. To be truly useful, chip-based flow cytometers with optical systems that can conceivably compete with today's basic cytometers need to be developed. This generally requires the ability to form a lens system with optical quality (low roughness) components. In addition, LOC cytometers need to be capable of performing multiparameter analysis. Also, for any high-throughput applications such as HIV/AIDS monitoring, the screening throughput of a μFACS should reach a level of at least 10,000 cells/s (i.e., ~15 min to run a sample containing ~10^7 cells). One common strategy to increase the throughput in microfluidic devices is to use multiplexing (i.e., parallel processing). In other words, instead of analyzing cells in just one fluidic channel, the analysis could be performed simultaneously in multiple channels. Finally, a μFACS requires a sorter with a sort rate of >1000 particles/s while maintaining high reliability (purity >90% and running time of >1 h). Again, this could be achieved using a fast-response deflector (e.g., PZT actuator) combined with a high-speed, robust, and closed-loop electronic control system with low-timing jitter. Upon satisfying these stringent yet achievable requirements, in a foreseeable future, the optofluidics field holds promise for the development of a low-cost, handheld optofluidic μFACS, affordable to individual clinics and hospitals, research labs, and military units (for the detection of biowarfare agents, etc.). More importantly, such an emerging technology could provide point-of-care analysis or diagnosis in the remote areas of Africa and Asia that continue to struggle with widespread epidemics, such as Malaria and HIV.

Acknowledgment

We would like to thank the Nano3 staff for technical support and NIH (Grants 1R01HG004876-01 and 1R21RR024453-01) for financial support.

References

Arvind Natarajan, A., D. Boxrud, G. Dunny, and F. Srienc. 1999. Flow cytometric analysis of growth of two *Streptococcus gordonii* derivatives. *Journal of Microbiological Methods* 34: 223–233.

Balslev, S., A. M. Jorgensen, B. Bilenberg et al. 2006. Lab-on-a-chip with integrated optical transducers. *Lab on a Chip* 6: 213–217.

Berridge, M. J., M. D. Bootman, and H. L. Roderick. 2003. Calcium signaling: Dynamics, homeostasis and remodeling. *Nature Reviews Molecular Cell Biology* 4: 517–529.

Bliss, C. L., J. N. McMullin, and C. J. Backhouse. 2007. Integrated wavelength-selective optical waveguides for microfluidic-based laser-induced fluorescence detection. *Lab on a Chip* 8: 143–151.

Brummendorf, T. H., N. Rufer, T. L. Holyoake et al. 2001. Telomere length dynamics in normal individuals and in patients with hematopoietic stem-cell-associated disorders. *Annals of the New York Academy of Sciences* 938: 293–303.

Camou, S., H. Fujita, and T. Fujii. 2003. PDMS 2D optical lens integrated with microfluidic channels: Principle and characterization. *Lab on a Chip* 3: 40–45.

Casamayor, E. O., I. Ferrera, X. Cristina, C. M. Borrego, and J. M. Gasol. 2007. Flow cytometric identification and enumeration of photosynthetic sulfur bacteria and potential for ecophysiological studies at the single-cell level. *Environmental Microbiology* 9: 1969–1985.

Chang, S. I. and J. B. Yoon. 2004. Shape-controlled, high fill-factor microlens arrays fabricated by a 3D diffuser lithography and plastic replication method. *Optics Express* 12(25): 6366–6371.

Chang, S. I. and J. B. Yoon. 2005. A high efficiency 3D planar microlens for monolithic optical interconnection system. *The 18th Annual Meeting of the IEEE Lasers and Electro-Optics Society*, Sydney, Australia.

Chang, C.-C., Z.-X. Huang, and R.-J. Yang. 2007. Three-dimensional hydrodynamic focusing in two-layer polydimethylsiloxane (PDMS) microchannels. *Journal of Micromechanics and Microengineering* 17: 1479–1486.

Chen, C. H., S. H. Cho, F. Tsai, A. Erten, and Y.-H. Lo. 2009. Microfluidic cell sorter with integrated piezoelectric actuator. *Biomedical Microdevices*, DOI: 10.1007/s10544-009-9341-5.

Chin, C. D., V. Linder, and S. K. Sia. 2007. Lab-on-a-chip devices for global health: Past studies and future opportunities. *Lab on a Chip* 7: 41–57.

Cho, S. H., J. Godin, C. H. Chen, F. S. Tsai, and Y.-H. Lo. 2008. Microfluidic photonic integrated circuits. *Proceedings of the SPIE*, China, 7135, p. 71350M.

Cho, S. H., J. Godin, and Y.-H. Lo. 2009a. Optofluidic waveguides in teflon-coated PDMS microfluidic channels. *IEEE Photonics Technology Letters* 21(15): 1057–1059.

Cho, S. H., J. Godin, and Y.-H. Lo. 2009b. Optofluidic waveguide using teflon-coated microfluidic channels. *CLEO/QELS*, Baltimore, MD, May 31–June 4 (in press).

Chou, H.-P., C. Spence, A. Scherer, and S. Quake. 1999. A microfabricated device for sizing and sorting DNA molecules. *Proceedings of the National Academy Science United States of America* 96: 11–13.

Chung, T. D. and H. C. Kim. 2007. Recent advances in miniaturized microfluidic flow cytometry for clinical use. *Electrophoresis* 28: 4511–4520.

Datta. A., I.-Y. Eom, A. Dhar et al. 2003. Microfabrication and characterization of Teflon-coated liquid core waveguide channels in silicon. *IEEE Sensors Journal* 3: 788–795.

Dickover, R. E., S. A. Herman, K. Saddiq et al. 1998. Optimization of specimen-handling procedures for accurate quantitation of levels of human immunodeficiency virus RNA in plasma by reverse transcriptase PCR. *Journal of Clinical Microbiology* 36: 1070–1073.

Dien, B. S., M. S. Peterson, and F. Srienc. 1994. Cell-cycle analysis of *Saccharomyces cerevisiae*. *Methods Cell Biology* 42: 457–475.

Dong, L. and H. Jiang. 2007. Tunable and movable liquid microlens in situ fabricated within microfluidic channels. *Applied Physics Letters* 91: 041109.

Dupont website. http://www2.dupont.com/Teflon_Industrial/en_US/products/product_by_name/teflon_af/

Eliopoulos, N., A. Al-Khaldi, C. M. Beausejour, R. L. Momparter, L. F. Momparler, and J. Galipeau. 2002. Human cytidine deaminase as an ex vivo drog selectable marker in gene-modified primary bone narrow stromal cells. *Gene Therapy* 9: 452–462.

Engh, Van den. 2000. High-speed cell sorting. In *Emerging Tools for Single-Cell Analysis: Advances in Optical Measurement Technologies*, ed. G. Durack, and J. P. Robinson. New York: John Wiley & Sons, Inc., pp. 21–48.

Fiedler, S., S. G. Shirley, T. Schnelle, and G. Fuhr. 1998. Dielectrophoretic sorting of particles and cells in a microsystem. *Analytical Chemistry* 70: 1909–1915.

Fink, C. C. and T. Meyer. 2002. Molecular mechanisms of CaMKII activation in neuronal plasticity. *Current Opinion in Neurobiology* 12: 293–299.

Fu, A. Y., C. Spence, A. Scherer, F. H. Arnold, and S. R. Quake. 1999. A microfabricated fluorescence-activated cell sorter. *Nature Biotechnology* 17: 1109–1111.

Fu, A.-Y., H.-P. Chou, C. Spence, F. M. Arnold, and S. R. Quake. 2002. An integrated microfabricated cells sorter. *Analytical Chemistry* 74: 2451–2457.

Fu, L.-M., R.-J. Yang, and G.-B. Lee. 2003. Electrokinetic focusing Injection methods on microfluidics devices. *Analytical Chemistry* 75: 1905–1910.

Fu, L.-M., R.-J. Yang, C.-H. Lin, Y.-J. Pan, and G.-B. Lee. 2004. Electrokinetically driven micro flow cytometers with integrated fiber optics for on-line cell/particle detection. *Analytica Chimica Acta* 507: 163–169.

Gee, Adrian P. and Lawrence S. Lamb. 2000. Enumeration of CD34-positive hematopoietic progenitor cells. In *Immunophenotyping*, ed. Carleton C. Stewart and Janet K.A. Nicholson. New York: John Wiley & Sons, Inc.

Givan, A. L. 2001. *Flow Cytometry First Principles*. New York: John Wiley & Sons, Inc.

Godin, J. and Y. H. Lo. 2007. Microfluidic flow cytometer with on-chip lens systems for improved signal resolution. *Sensors IEEE*, Atlanta, GA, pp. 466–469.

Godin, J. and Y.-H. Lo. 2009. Advances in on-chip polymer optics for optofluidics. *Conference on Lasers and Electro-Optics*, 2009.

Godin, J., V. Lien, and Y.-H. Lo. 2006. Demonstration of two-dimensional fluidic lens for integration into microfluidic flow cytometers. *Applied Physics Letters* 89: 061106.

Godin, J., C.-H. Chen, S. H. Cho, W. Qiao, F. Tsai, and Y.-H. Lo. 2008a. Microfluidics and photonics for bio-system-on-chip: A review of advancements in technology towards a microfluidic flow cytometry chip. *Journal Biophoton* 3: 1–22.

Godin, J., C.-H. Chen, S. H. Cho et al. 2008b. Microfluidics and photonics for bio-system-on-a-chip: A review of advancements in technology towards a microfluidic flow cytometry chip. *Journal of Biophotonics* 1: 355–376.

Graham, J. K. 2001. Assessment of sperm quality: A flow cytometric approach. *Animal Reproduction Science* 68: 239–247.

Gratama, J. W., A. Orfao, D. Barnett et al. 1998. Flow cytometric enumeration of CD34+hematopoietic stem and progenitor cells. *Cytometry (Communications in Clinical Cytometry)* 34: 128–142.

Gross, H. J., B. Viewer, D. Houck, R. A. Hoffman, and D. Recktenwald. 1995. Model study detecting breast cancer cells in peripheral blood mononuclear cells at frequencies as low as 10^{-7}. *Proceedings of the National Academy of Sciences USA* 92: 537–541.

Gunasekera, T. S., P. V. Attfield, and D. A. Veal. 2000. A flow cytometry method for rapid detection and enumeration of total bacteria in milk. *Applied and Environmental Microbiology* 66: 1228–1232.

Herzengberg, L. A., D. Parks, B. Sahaf, O. Perez, and M. Roederer. 2002. The history and future of the fluorescence activated cell sorter and flow cytometry: A view from Stanford. *Clinical Chemistry* 48: 1819–1827.

Holmes, D., M. E. Sandison, N. G. Green, and H. Morgan. 2005. On-chip high-speed sorting of micron-sized particles for high-throughput analysis. *IEE Proceedings-Nanobiotechnology* 152: 129–135.

Huang, S. and L. W. Terstappen. 1994. Lymphoid and myeloid differentiation of single human CD34+, HLA-DR+, CD38- hematopoietic stem cells. *Blood* 83: 1515–1526.

Huh, D., Y.-C. Tung, and H.-H. Wei. 2002. Use of air-liquid two-phase flow in hydrophobic microfluidic channels for disposable flow cytometers. *Biomedical Microdevices* 4: 141–149.

Huh, D., W. Gu, Y. Kamotani, J. B. Grotberg, and S. Takayama. 2005. Microfluidics for flow cytometric analysis of cells and particles. *Physiological Measurement* 26: R73–R98.

Ibrahim, F. S. and G. v. d. Engh. 2003. High-speed cell sorting: Fundamentals and recent advances. *Current Opinion in Biotechnology* 14: 5–12.

Janossy, G., I. V. Jani, M. Kahan et al. 2002. Precise CD4 T-cell counting using red diode laser excitation: For richer, for poorer. *Cytometry* 50: 78–85.

Jennings, C. D. and K. A. Foon. 1997a. Recent advances in flow cytometry: Application to the diagnosis of hematologic malignancy. *Blood* 90: 2863–2892.

Jennings, C. D. and K. A. Foon. 1997b. Recent advances in diagnosis and monitoring of leukemia. *Cancer Invest* 15: 384–399.

Kay, S. M. 1993. *Fundamentals of Statistical Signal Processing, Volume 1: Estimation Theory*. Upper Saddle River, NJ: Prentice Hall, Inc.

Kay, S. M. 1998. *Fundamentals of Statistical Signal Processing, Volume 2: Detection Theory*. Upper Saddle River, NJ: Prentice Hall, Inc.

Keeney, M., I. Chin-Yee, R. Nayar et al. 1999. Effect of fixatives on CD34+cell enumeration. *Journal of Hematotherapy* 8: 327–329.

Klank, H., G. Goranovic, J. P. Kutter, H. Gjelstrup, J. Michelsen, and C. H. Westergaard. 2002. PIV measurements in a microfluidic 3D-sheathing structure with three-dimensional flow behavior. *Journal of Micromechanics and Microengineering* 12: 862–869.

Knight, J. B., A. Vishwanath, J. P. Brody, and R. H. Austin. 1998. Hydrodynamic focusing on a silicon chip. *Physical Review Letters* 80: 3863–3866.

Kostner, S. and M. J. Vellekoop. 2007. Cell analysis in a microfluidic cytometer applying a DVD pickup head. *Sensors and Actuators B: Chemical* 132: 512–517.

Kou, Q., I. Yesilyurt, V. Studer, M. Belotti, E. Cambril, and Y. Chen. 2004. On-chip optical components and microfluidic systems. *Microelectronic Engineering* 73: 876–880.

Krüger, J., K. Singh, A. O'Neill, C. Jackson, A. Morrison, and P. O'Brien. 2002. Development of a microfluidic device for fluorescence activated cell sorting. *Journal of Micromechanics and Microengineering* 12: 486–494.

Lagasse, E., H. Connors, M. Al-Dhalimy et al. 2000. Purified hematopoietic stem cells can differentiate into hepatocytes in vivo. *Nature Medicine* 6: 1229–1234.

Lee, G.-B., B.-H. Hwei, and G.-R. Huang. 2001a. Micromachined pre-focused M X N flow switches for continuous multi-sample injection. *Journal of Micromechanics and Microengineering* 11: 654–661.

Lee, G.-B., C.-I. Hung, B.-J. Ke, G.-R. Huang, B.-H. Hwei, and H.-F. Lai. 2001b. Hydrodynamic focusing for a micromachined flow cytometer. *Transactions of the ASME* 123: 672–679.

Lee. G.-B., L. M. Fu, R. J. Yang, Y. J. Pan, and C. H. Lin. 2003. MxN micro flow switches using electrokinetic forces. *Solid State Sensors and Actuators* 2: 1895–1898.

Lee, G.-B., C.-C. Chang, S.-B. Huang, and R.-J. Yang. 2006. The hydrodynamic focusing effect inside rectangular micro-channels. *Journal of Micromechanics and Microengineering* 16: 1024–1032.

Legendre, L., C. Courties, and M. Troussellier. 2001. Flow cytometry in oceanography 1989–1999: Environmental challenges and research trends. *Cytometry* 44: 164–172.

Lehmann, A. K., S. Somes, and A. Halstensen. 2000. Phagocytosis: Measurement by flow cytometry. *Journal of Immunological Methods* 243: 229–242.

Lewis, R. S. 2003. Calcium oscillations in T-cells: Mechanisms and consequences for gene expression. *Biochemical Society Transactions* 31: 925–999.

Lien, V., Y. Berdichevsky, and Y.-H. Lo. 2004. A prealigned process of integrating optical waveguides with microfluidic devices. *IEEE Photonics Technology Letters* 16: 1525–1527.

Lien, V., K. Zhao, Y. Berdichevsky, and Y.-H. Lo. 2005. High-sensitivity cytometric detection using fluidic-photonic integrated circuits with array waveguides. *IEEE Journal of Selected Topics in Quantum Electronics* 11: 827–834.

Lim, J.-M., S.-H. Kim, J.-H. Choi et al. 2008. Fluorescent liquid-core/air-cladding waveguides towards integrated optofluidic light sources. *Lab Chip* 8: 1580–1585.

Lin, C. H. and G. B. Lee. 2003. Micromachined flow cytometer with embedded etched optic fibers for optical detection. *Journal of Micromechanics and Microengineering* 13: 447–453.

Link, A. J., K. J. Jeong, and G. Georgiou. 2007. Beyond toothpicks: New methods for isolating mutant bacteria. *Nature* 5: 680–688.

Maeda, K., R. C. Alessio, and R. C. Hawley. 1993. Recent advances in diagnosis of leukemia. *Japanese Journal of Clinical Oncology* 23: 79–84.

Mao, X., J. R. Waldeisen, and T. J. Huang. 2007. "Microfluidic drifting"—Implementing three-dimensional hydrodynamic focusing with a single-layer planar microfluidic device. *Lab on a Chip* 7: 1260–1262.

McClain, M. A., C. T. Culbertson, S. C. Jacobson, and J. M. Ramsey. 2001. Flow cytometry of *Escherichia coli* on microfluidic devices. *Analytical Chemistry* 73: 5334–5338.

McKinney, K. L., R. Dilwith, and G. Belfort. 1995. Optimizing antibody production in batch hybridoma cell culture. *Journal of Biotechnology* 40: 31–48.

McMullin, J. N., H. Qiao, S. Goel, C. L. Ren, and D. Li. 2005. Integrated optical measurement of microfluid velocity. *Journal of Micromechanics and Microengineering* 15: 1810–1816.

Minerd, J. 2005. Buy's guide to flow cytometers. *Scientist* 19: 28–29.

Mogensen, K. B., N. J. Petersen, J. Hubner, and J. P. Kutter. 2001. Monolithic integration of optical waveguides for absorbance detection in microfabricated electrophoresis devices. *Electrophoresis* 22: 3930–3938.

Moldavan, A. 1934. Photo-electric technique for the counting of microscopical cells. *Science* 80: 188.

Müller, S. and K.-J. Hutter. 1999. Prozessoptimierung von Reinzucht- und Anstellverfahren mittels Flusscytometrie in sächsischen Brauereien. Monatsschr Brauwiss 3:40–48.

Nolan, J. P., S. Lauer, E. R. Prossnitz, and L. A. Sklar. 1999. Flow cytometry: A versatile tool for all phases of drug discovery. *DDT* 4: 173–180.

Nunez, R. 2001. DNA measurement and cell cycle analysis by flow cytometry. *Current Issues in Molecular Biology* 3: 67–70.

Ohtsubo, M. and J. M. Roberts. 1993. Cyclin-dependent regulation of G1 in mammalian fibroblasts. *Science* 259: 1908–1912.

O'Shea, D. C., T. J. Suleski, A. D. Kathman, and D. W. Prathe. 2003. *Diffractive Optics: Design, Fabrication, and Test*. Bellingham, WA: SPIE Publications.

Pala, P., T. Hussell, and P. J. Openshaw. 2000. Flow cytometric measurement of intracellular cytokines. *Journal of Immunological Methods* 243: 107–124.

Palsson, B. O. and S. N. Bhatia. 2004. *In Tissue Engineering*. Upper Sadel River, NJ: Pearson Education, Inc.

Pamme, N. 2007. Continuous flow separations in microfluidic devices. *Lab on a Chip* 7: 1644–1659.

Pamme, N., R. Koyama, and A. Manz. 2003. Counting and sizing of particles and particle agglomerates in a microfluidic device using laser light scattering: Application to a particle-enhanced immunoassay. *Lab on a Chip* 3: 187–192.

Pruessner, M. W., W. S. Rabinovich, T. H. Stievater, D. Park, and J. W. Baldwin. 2007. Cryogenic etch process development for profile control of high aspect-ratio submicron silicon trenches. *Journal of Vacuum Science & Technology B: Microelectronics and Nanometer Structures* 25: 21–28.

Roederer, M., J. M. Brenchley, M. R. Betts, and S. C. De Rosa. 2004. Flow cytometric analysis of vaccine responses: How many colors are enough? *Clinical Immunology* 110: 199–205.

Schrum, D. P., C. T. Culbertson, S. C. Jacobson, and J. M. Ramsey. 1999. Microchip flow cytometry using electrokinetic focusing. *Analytical Chemistry* 71: 4173–4177.

Seo, J. and L. P. Lee. 2003. Fluorescence amplification by self-aligned integrated microfluidic optical systems. *TRANSDUCERS, Solid-State Sensors, Actuators and Microsystems, 12th International Conference*, Berkeley, CA, 2, 1136–1139.

Seo, J. and L. P. Lee. 2004. Disposable integrated microfluidics with self-aligned planar microlenses. *Sensors and Actuators B: Chemical* 99: 615–622.

Shapiro, H. M. 2003. *Practical Flow Cytometry*. Hoboken, NJ: John Wiley & Sons, Inc.

Shapiro, H. M. and M. Hercher. 1986. Flow cytometers using optical waveguides in place of lenses for specimen illumination and light collection. *Cytometry* 7: 221–223.

Shizuru, J. A., R. S. Negrin, and I. L. Weissman. 2005. Hematopoietic stem and progenitor cells: Clinical and preclinical regeneration of the hematolymphoid system. *Annual Review of Medicine* 56: 509–538.

Simonnet, C. and A. Groisman. 2005. Two-dimensional hydrodynamic focusing in a simple microfluidic device. *Applied Physics Letters* 87: 114104-1–3.

Stiles, T., R. Fallon, T. Vestad et al. 2005. Hydrodynamic focusing for vacuum-pumped microfluidics. *Microfluidics and Nanofluidics* 1: 280–283.

Sundararajan, N., M. S. Pio, L. P. Lee, and A. A. Berlin. 2004. Three-dimensional hydrodynamic focusing in polydimethylsiloxane (PDMS) microchannels. *Journal of Microelectromechanical Systems* 13: 559–567.

Tang, S. K. Y., C. A. Stan, and G. M. Whitesides. 2008. Dynamically reconfigurable liquid-core liquid-cladding lens in a microfluidic channel. *Lab on a Chip* 8(3): 395–401.

Tay, S. T., V. Ivanov, S. Yi, W. Q. Zhuang, and J. H. Tay. 2002. Presence of anaerobic bacteroides in aerobically grown microbial granules. *Microbial Ecology* 44: 278–285.

Thyssen, M., G. A. Tarran, M. V. Zubkov et al. 2008. The emergence of automated high-frequency flow cytometry: Revealing temporal and spatial phytoplankton variability. *Journal of Plankton Research* 30: 333–343.

Tung, Y.-C., M. Zhang, C.-T. Lin, K. Kurabayashi, and S. J. Skerlos. 2004. PDMS-based opto-fluidic micro flow cytometer with two-color, multi-angle fluorescence detection capability using PIN photodiodes. *Sensors and Actuators B: Chemical* 98: 356–367.

Van Trees, H. L. 2001. Representation of random processes. In *Detection, Estimation, and Modulation Theory, Part I.* New York: John Wiley & Sons, Inc.

Vermes, I., C. Haanen, and C. Reutellingsperger. 2000. Flow cytometry of apoptotic cell death. *Journal of Immunological Methods* 243: 167–190.

Voldman, J., M. L. Gray, M. Toner, and M. A. Schmidt. 2002. A microfabrication-based dynamic array cytometer. *Analytical Chemistry* 74: 3984–3990.

Walker, G. M. 1999. Synchronization of yeast cell populations. *Methods in Cell Science* 21: 87–93.

Wang, X. and M. F. Slavik. 1999. Rapid detection of Salmonella in chicken washes by immunomagnetic separation and flow cytometry. *Journal of Food Protection* 62: 717–723.

Wang, Z., J. El-Ali, I. R. Perch-Nielsen et al. 2004. Measurements of scattered light on a microchip flow cytometer with integrated polymer based optical elements. *Lab on a Chip* 4: 372–377.

Wilson, G. D. 1994. Analysis of DNA-measurement of cell kinetics by the bromodeoxyuridine/anti-bromodeoxyuridine method. In *Flow Cytometry: A Practical Approach*, ed. M. G. Ormerod. Oxford, U.K.: IRL Press, pp. 137–156.

Wolfe, D. P., R. S. Conroy, P. Garstecki et al. 2004. Dynamic control of liquid-core/liquid-cladding optical waveguides. *Proceedings of the National Academy of Sciences of the United States of America* 101: 12434–12438.

Wolff, A., I. R. Perch-Nielsen, U. D. Larsen et al. 2003. Integrating advanced functionality in a microfabricated high-throughput fluorescent-activated cell sorter. *Lab on a Chip* 3: 22–27.

Wong, P. K., Y.-K. Lee, and C.-M. Ho. 2003. Deformation of DNA molecules by hydrodynamic focusing. *Journal of Fluid Mechanics* 497: 55–65.

Yang, R., D. L. Feeback, and W. Wang. 2005. Microfabrication and test of a three-dimensional polymer hydro-focusing unit for flow cytometry applications. *Sensors and Actuators A* 118: 259–267.

Yang, S.-Y., S.-K. Hsiung, Y.-C. Hung et al. 2006. A cell counting/sorting system incorporated with a microfabricated flow cytometer chip. *Measurement Science and Technology* 17: 2001–2009.

Yin, D., H. Schmidt, J. Barber, and A. Hawkins. 2004. Integrated ARROW waveguides with hollow cores. *Optic Express* 17: 2710–2715.

Ziemerm R. E. and W. H. Tranter. 2001. *Principles of Communication: Systems, Modulation and Noise*. New York: John Wiley & Sons, Inc.

Appendix A: Optical Properties of Water

Mikhail I. Rudenko

A.1 Absorption of Water

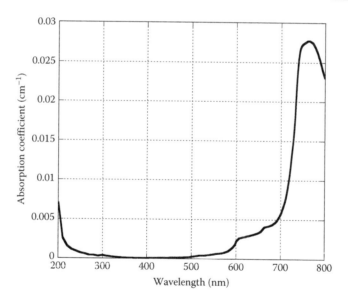

FIGURE A.1 Absorption of pure water in the UV, visible, and near-infrared spectrum. (Data from Hale, G.M. and Querry, M.R., *Appl. Opt.*, 12, 555, 1973.)

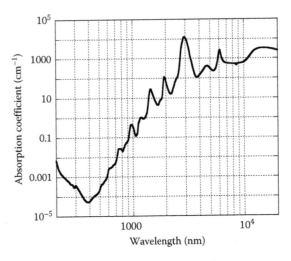

FIGURE A.2 Absorption of pure water from 200 nm to 20 μm. (Data from Hale, G.M. and Querry, M.R., *Appl. Opt.*, 12, 555, 1973.)

A.2 Index of Refraction of Water

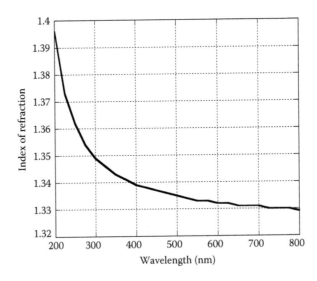

FIGURE A.3 Index of refraction of pure water in the UV, visible, and near-infrared spectrum. (Data from Hale, G.M. and Querry, M.R., *Appl. Opt.*, 12, 555, 1973.)

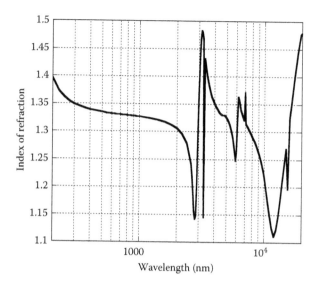

FIGURE A.4   Index of refraction of pure water from 200 nm to 20 μm. (Data from Hale, G.M. and Querry, M.R., *Appl. Opt.*, 12, 555, 1973.)

A.3 Raman Spectrum of Water

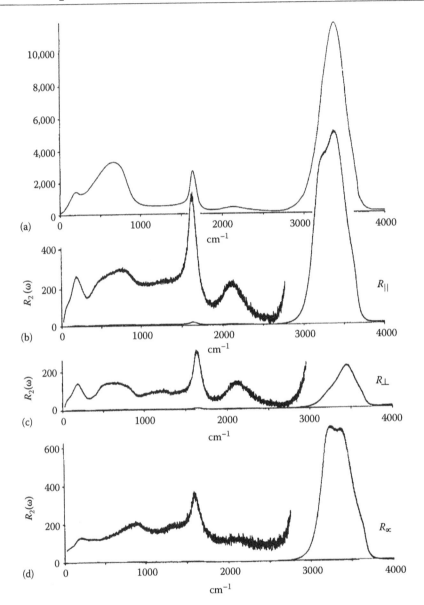

FIGURE A.5 Vibrational spectra of $H_2^{16}O$ at 298 K. (a) Infrared spectrum; (b) Raman spectrum plotted as normalized Raman intensity, $R_2(\omega)_{\parallel}$; (c) Raman spectrum plotted as normalized Raman intensity, $R_2(\omega)_{\perp}$; and (d) Raman spectrum plotted as normalized Raman intensity, $R_2(\omega)_{\propto}$. (Reproduced from Brooker, M.H. et al., *J. Raman Spectrosc.*, 20, 683, 1989. With permission.)

TABLE A.1   Peak Centers of H_2O at 256 bar and 295 K

Peak Center (cm⁻¹)	Assignment
65	Translational
162	
430	Librational
650	
795	
1581	OH bending
1641	
3051	OH stretching
3233	
3393	
3511	
3628	

Source: Data obtained from Carey, D.M. and Korenowski, G.M., *J. Chem. Phys.*, 108, 2669, 1998.

TABLE A.2   Assignments for Raman Overtones and Combinations from Water

Observed Position (cm⁻¹)	Assignment
4000 ± 10	$\upsilon_s + \upsilon_{L(3)}$
5065 ± 10	$\upsilon_s + \delta$
5065 ± 10	$\upsilon_{AS} + \delta$
5065 ± 10	$\upsilon_s + \delta + \upsilon(O{-}O)$
5700 ± 20	$\upsilon_s + \delta + \upsilon_{L(3)}$
6740 ± 20	2υ

Source: Data obtained from Walrafen, G.E. and Pugh, E., *J. Solution Chem.*, 33, 81, 2004.

References

Brooker, M.H. et al., 1989. Raman frequency and intensity studies of liquid H_2O, $H_2^{18}O$ and D_2O. *Journal of Raman Spectroscopy*, 20(10), 683–694.

Carey, D.M. and Korenowski, G.M., 1998. Measurement of the Raman spectrum of liquid water. *The Journal of Chemical Physics*, 108(7), 2669–2675.

Hale, G.M. and Querry, M.R., 1973. Optical constants of water in the 200-nm to 200-μm wavelength region. *Applied Optics*, 12(3), 555–563.

Walrafen, G.E. and Pugh, E., 2004. Raman combinations and stretching overtones from water, heavy water, and NaCl in water at shifts to ca. 7000 cm⁻¹. *Journal of Solution Chemistry*, 33(1), 81–97.

Appendix B: Refractive Index of Liquids and Solids

TABLE B.1 Refractive Index of Selected Liquids (at 589 nm, 293 K)

Name	Formula	n_D
Acetic acid [1]	CH_3COOH	1.3720
Acetone [1]	$(CH_3)_2CO$	1.3588
Benzene [1]	C_6H_6	1.5011
Bromine [1]	Br_2	1.6590
Butanoic acid [2]	$C_4H_8O_2$	1.3960
1-Butanol [2]	$C_4H_{10}O$	1.3990
Calcium chloride (aqueous, 1 mol kg^{-1}) [1]	$CaCl_2 \cdot H_2O$	1.3575
N,N-Dimethylformamide (DMF) [1]	C_3H_7NO	1.4306
Dimethyl sulfoxide (DMSO) [1]	$(CH_3)_2SO$	1.4170
1,2-Ethanediol (ethylene glycol) [1]	$(CH_2OH)_2$	1.4318
Ethanol [1]	C_2H_5OH	1.3611
Glycerol [1]	$CH_2OHCHOHCH_2OH$	1.4746
Hydrogen peroxide [1]	H_2O_2	1.4061
Methanol [1]	CH_3OH	1.3288
2-Propanol (isopropyl alcohol, ISO) [1]	$CH_3CHOHCH$	1.3776
Styrene [2]	C_8H_8	1.5440
Toluene [1]	C_7H_8	1.4961
Trichloroethylene [3]	C_2HCl_3	1.4760
Water [1]	H_2O	1.3330

TABLE B.2 Refractive Index of Selected Aqueous Solutions (at 589 nm, 293 K)

Name	Formula	n_D
Acetic acid	CH_3COOH	$n_0 = 1.3327$; $n_1 = 0.00083732$; $n_2 = -7.627e-06$; $n_3 = 1.2519e-07$; $n_4 = -9.2919e-10$; (0.5–100)
Acetone	$(CH_3)_2CO$	$n_0 = 1.333$; $n_1 = 0.00072202$; (0.5–10)
Ammonia	NH_3	$n_0 = 1.3329$; $n_1 = 0.00050103$; $n_2 = 2.6592e-06$; (0.5–30)
Calcium chloride	$CaCl_2$	$n_0 = 1.333$; $n_1 = 0.0023642$; $n_2 = 9.2496e-06$; (0.5–40)
Ethanol	C_2H_5OH	$n_0 = 1.333$; $n_1 = 0.00052044$; $n_2 = 1.9089e-05$; $n_3 = -6.4691e-07$; $n_4 = 7.1086e-09$; $n_5 = -2.7805e-11$; (0.5–98)
Ethylene glycol	$(CH_2OH)_2$	$n_0 = 1.3325$; $n_1 = 0.0010072$; (0.5–60)
D-Fructose	$C_6H_{12}O_6$	$n_0 = 1.3331$; $n_1 = 0.0013927$; $n_2 = 6.1224e-06$; (0.5–48)
D-Glucose	$C_6H_{12}O_6$	$n_0 = 1.3331$; $n_1 = 0.0013906$; $n_2 = 6.2353e-06$; (0.5–60)
Glycerol	$CH_2OHCHOHCH_2OH$	$n_0 = 1.3326$; $n_1 = 0.0012017$; $n_2 = 2.1907e-06$; (0.5–100)
Lactose	$C_{12}H_{22}O_{11}$	$n_0 = 1.3327$; $n_1 = 0.0015925$; (0.5–18)
Magnesium chloride	$MgCl_2$	$n_0 = 1.3331$; $n_1 = 0.0024818$; $n_2 = 8.0122e-06$; (0.5–30)
Methanol	CH_3OH	$n_0 = 1.3331$; $n_1 = 0.00015328$; $n_2 = 9.8027e-06$; $n_3 = -2.849e-07$; $n_4 = 2.6204e-09$; $n_5 = -9.4532e-12$; (0.5–100)
Nitric acid	HNO_3	$n_0 = 1.3326$; $n_1 = 0.0013577$; (0.5–40)
Phosphoric acid	H_3PO_4	$n_0 = 1.3331$; $n_1 = 0.00087844$; $n_2 = 3.3513e-06$; (0.5–40)
Potassium chloride	KCl	$n_0 = 1.3328$; $n_1 = 0.0013886$; (0.5–24)
2-Propanol (isopropyl alcohol)	$CH_3CHOHCH$	$n_0 = 1.3321$; $n_1 = 0.0010803$; $n_2 = -6.9459e-06$; (2–100)
Sodium chloride	$NaCl$	$n_0 = 1.3328$; $n_1 = 0.0017818$; (0.5–26)
Sucrose	$C_{12}H_{22}O_{11}$	$n_0 = 1.3333$; $n_1 = 0.0013732$; $n_2 = 7.3562e-06$; (0.5–80)

Source: Original values of refractive indices data obtained from Lide, D.R., *CRC Handbook of Chemistry and Physics*, 90th ed, CRC Press, Boca Raton, FL, 2009.

Note: The refractive index is represented by the coefficients of a polynomial of the form $n_D = n_0 + n_1 \times mp + n_2 \times mp^2 + \cdots + n_m mp^m$, where mp is the mass percentage of a substance in the solution. Validity region is given in parentheses.

TABLE B.3 Refractive Index of Selected Solids

Name	Formula	n (488 nm)	n (589 nm)	n (633 nm)	n (785 nm)
Silicon [4]	Si	4.367	3.969	3.882	3.705
Gallium arsenide [4]	GaAs	4.392	3.940	3.856	3.693
Silicon nitride [1] (noncrystalline)	Si_3N_4	2.041 (496 nm)	2.027 (estimated)	2.022 (620 nm)	2.008 (827 nm)
Silicon dioxide [4] (glass)	SiO_2	1.463	1.458	1.457 (643 nm)	1.454 (852 nm)
Teflon AF 2400 [5]		1.296 (510 nm)	1.295	1.294 (670 nm)	1.293 (750 nm)
Polydimethylsiloxane (PDMS) [6]			1.370–1.399 (0.65–10 cs) 1.4030–1.4036 (100–60,000 cs)		
Polymethyl methacrylate (PMMA) [7]		1.497 (486 nm)	1.491 (588 nm)		1.484 (833 nm)
SU8 (3000, calculated) [8]		1.586	1.574	1.571	1.564
Fused silica [9]	SiO_2	1.463 (486 nm)	1.458	1.457 (644 nm)	1.452 (852 nm)
Quartz (crystal) [9]	SiO_2		1.544 n_o, 1.553 n_e		1.539 n_o, 1.548 n_e (768 nm)

TABLE B.3 (continued)   Refractive Index of Selected Solids

Name	Formula	n (488 nm)	n (589 nm)	n (633 nm)	n (785 nm)
Borosilicate glass (13.5 mol% B_2O_3) [10]	SiO_2-B_2O_3	1.4632 (480 nm)	1.4579	1.4560 (644 nm)	1.4523 (808 nm)
Borofloat glass [11]			1.4713	1.4695 (644 nm)	
Polytetrafluoroethylene (PTFE) [6]			1.376		
Paraformaldehyde (PFA) [12]			1.35		
Fluorinated ethylene propylene (FEP) [12]			1.344		
Polysterene [13]		1.604	1.590	1.587 (630 nm)	
Tellurium [4]	Te	3.46 $n_{\parallel}$, 3.76 $n_{\perp}$ (489 nm)	5.45 $n_{\parallel}$, 5.49 $n_{\perp}$ (586 nm)	6.21 $n_{\parallel}$, 5.97 $n_{\perp}$ (627 nm)	6.75 $n_{\parallel}$, 5.88 $n_{\perp}$ (775 nm)
Arsenic selenide [4] (amorphous)	As_2Se_3				3.05
Titanium dioxide [4] (rutile)	TiO_2	3.08 $n_{\parallel}$, 2.75 $n_{\perp}$ (480 nm)	2.92 $n_{\parallel}$, 2.62 $n_{\perp}$ (580 nm)	2.88 $n_{\parallel}$, 2.59 $n_{\perp}$ (620 nm)	2.80 $n_{\parallel}$, 2.52 $n_{\perp}$ (780 nm)
Indium antimonide [4]	InSb	3.512	4.136	4.249	4.534
Indium phosphide [4]	InP	3.851	3.585	3.530 (639 nm)	3.467 (775 nm)
Gallium phosphide [4]	GaP	3.638	3.3675	3.3132 (630 nm)	3.209 (775 nm)
Aluminum oxide [4]	Al_2O_3	1.77547 (486 nm)	1.76808	1.76547 (643 nm)	
Sapphire [9]	Al_2O_3		1.769 (579 nm)	1.765 (643 nm)	1.759 (852 nm)
Phosphate glass (17.4% P_2O_5) [10]	SiO_2-P_2O_5		1.471		

References

1. Lide, D.R. *CRC Handbook of Chemistry and Physics*, 90th edition. CRC Press, Boca Raton, FL (2009).
2. Wohlfarth, C. and Wohlfarth, B. *Refractive Indices of Inorganic, Organometallic, and Organonon-metallic Liquids, and Binary Liquid Mixtures. Landolt-Börnstein—Group III Condensed Matter*, Vol. 38A. Springer-Verlag Berlin/Heidelberg (1996).
3. Pavoni, B. et al. Assessment of organic chlorinated compound removal from aqueous matrices by adsorption on activated carbon. *Water Research* **40**, 3571–3579 (2006).
4. Palik, E.D. *Handbook of Optical Constants of Solids*, Five volume set. Academic Press, New York (1998).
5. Lowry, J.H., Mendlowitz, J.S., and Subramanian, N.S. Optical characteristics of Teflon AF fluoroplastic materials. *Optical Engineering* **31**, 1982–1985 (1992).
6. Mark, J.E. *Polymer Data Handbook*. Oxford University Press, New York (1999).
7. Kasarova, S.N. et al. Analysis of the dispersion of optical plastic materials. *Optical Materials* **29**, 1481–1490 (2007).
8. Michrochem, SU-8 3000 Data Sheet, viewed March 14, 2009. <http://www.microchem.com/products/pdf/SU-8%203000%20Data%20Sheet.pdf>
9. McClatchey, R.A. et al. *Handbook of Optics*, eds. W.G. Driscoll and W. Vaughan. McGraw-Hill, New York (1978).

10. Bansal, N.P. and Doremus, R.H. *Handbook of Glass Properties*. Academic Press, New York (1986).

11. SCHOTT, BOROFLOAT®—Product Properties—Optical Properties. <http://www.us.schott.com/borofloat/english/attribute/optical/>

12. Bloch, D.R. *Polymer Handbook*. John Wiley & Sons, Inc, New York (1999).

13. Grigoriev, I.S., Meilikhov, E.Z., and Radzig, A.A. *Handbook of Physical Quantities*. CRC Press, Boca Raton, FL (1997).

Appendix C: Viscosity and Surface Tension of Typical Liquids

TABLE C.1 Dynamic Viscosities and Surface Tension of Selected Newtonian Liquids (298 K)

Name	Formula	η (mPa s)	γ (mN m^{-1})
Acetic acid [1]	CH_3COOH	1.056	27.10
Acetone [1]	$(CH_3)_2CO$	0.306	22.72
Aniline [1]	$C_6H_5NH_2$	3.85	42.12
Benzene [1]	C_6H_6	0.604	28.22
Butanoic acid [1]	$C_4H_8O_2$	1.426	26.05
1-Butanol [1]	$C_4H_{10}O$	2.54	24.93
Calcium chloride	$CaCl_2$	1.2189 [2] (aqueous, 1.09 mol kg^{-1})	74.70 [3] (aqueous, 0.992 mol kg^{-1})
Diethyl ether [1]	$(C_2H_5)_2O$	0.224	16.65
N,N-Dimethylformamide (DMF) [1]	C_3H_7NO	0.794	35.74
Dimethyl sulfoxide (DMSO) [1]	$(CH_3)_2SO$	1.987	42.92
1,2-Ethanediol (ethylene glycol) [1]	$(CH_2OH)_2$	16.06	47.99
Ethanol [1]	C_2H_5OH	1.074	21.97
Glycerol [1]	$CH_2OHCHOHCH_2OH$	934	63 (293 K)
1-Hexanol [1]	$C_6H_{14}O$	4.58	25.81
Methanol [1]	CH_3OH	0.544	22.07
Mercury	Hg	1.526 [1]	487 (288 K) [4]
Olive oil		86.62 [5]	
Pitch		2.3×10^{11} [6]	
1-Propanol [1]	$CH_3CH_2CH_2OH$	1.945	23.32
2-Propanol (isopropyl alcohol, ISO) [1]	$CH_3CHOHCH$	2.04	20.93
Styrene [1]	C_8H_8	0.695	34
Toluene [1]	C_7H_8	0.560	27.73
Water [1]	H_2O	0.890	71.99

TABLE C.2 Dynamic Viscosities and Surface Tension of Selected Non-Newtonian Liquids

Name	Temperature (K)	η (mPa s)	γ (mN m^{-1})
Blood	310	3–6 [7]	52 [8]
Castor oil	298	985 [1]	39 (293 K) [9]
Corn syrup	298	1380.6 [1]	75 (312 K) [10]

References

1. Lide, D.R. *CRC Handbook of Chemistry and Physics*, 90th edn. CRC Press, Boca Raton, FL (2009).
2. Zhang, H., Chen, G., and Han, S. Viscosity and Density of $H_2O + NaCl + CaCl_2$ and $H_2O + KCl + CaCl_2$ at 298.15 K. *Journal of Chemical & Engineering Data* **42**, 526–530 (1997).
3. Cupples, H.L. The surface tensions of calcium chloride solutions at 25° measured by their maximum bubble pressures. *Journal of the American Chemical Society* **67**, 987–990 (1945).
4. Amiji, M. and Sandmann, B. *Applied Physical Pharmacy*. McGraw-Hill Medical, New York (2002).
5. Fasina, O. et al. Predicting temperature-dependence viscosity of vegetable oils from fatty acid composition. *Journal of the American Oil Chemists' Society* **83**, 899–903 (2006).
6. Edgeworth, R., Dalton, B.J., and Parnell, T. The pitch drop experiment. *European Journal of Physics* **5**, 198–200 (1984).
7. Rosenson, R., McCormick, A., and Uretz, E. Distribution of blood viscosity values and biochemical correlates in healthy adults. *Clinical Chemistry* **42**, 1189–1195 (1996).
8. Rosina, J. et al. Temperature dependence of blood surface tension. *Physiological Research* **56 Suppl. 1**, S93–S98 (2007).
9. Johnson, W. Final report on the safety assessment of *Ricinus communis* (castor) seed oil, hydrogenated castor oil, glyceryl ricinoleate, glyceryl ricinoleate SE, ricinoleic acid, potassium ricinoleate, sodium ricinoleate, zinc ricinoleate, cetyl ricinoleate, ethyl ricinoleate, glycol ricinoleate, isopropyl ricinoleate, methyl ricinoleate, and octyldodecyl ricinoleate. *International Journal of Toxicology* **26 Suppl. 3**, 31–77 (2007).
10. Williams, J.G., Morris, C.E.M., and Ennis, B.C. Liquid flow through aligned fiber beds. *Polymer Engineering & Science* **14**, 413–419 (1974).

Appendix D: Common Fluorescent Dyes

TABLE D.1 Common Fluorescent Dyes Sorted by the Absorption Maximum

Name	Absorption, Max (nm)	Emission, Max (nm)	Extinction $(cm^{-1} M^{-1})$	Molecular Weight $(g\ mol^{-1})$	Supplier	Note
Alexa Fluor 350 [1]	346	442	19,000	410	Invitrogen	IgG conjugated
Hoechst 33258 [2]	352	455		624	Invitrogen	DNA bound, AT selective
Pacific Orange [3]	400	551	24,500		Invitrogen	
Alexa Fluor 405 [1]	401	421	34,000	1028	Invitrogen	IgG conjugated
Pacific Blue [3]	404	455	46,000	406	Invitrogen	
Alexa Fluor 430 [1]	434	541	16,000	702	Invitrogen	IgG conjugated
SYTOX Blue [4]	444	480		400	Invitrogen	
Cy2 [5]	489	506	150,000	714	GE Healthcare	
YOYO-1 [3]	491	509	98,900	1271	Invitrogen	
Alexa Fluor 488 [1]	495	519	71,000	643	Invitrogen	IgG conjugated
Fluorescein	495	519	68,000	389		pH dependent
Alexa Fluor 500 [1]	502	525	71,000	700	Invitrogen	IgG conjugated
Acridine Orange [3]	502	526/650			Invitrogen	DNA/RNA
Rhodamine Green [3]	502	527	68,000		Invitrogen	
SYTOX Green [3]	504	523		600	Invitrogen	DNA
TOTO-1 [3]	514	533	117,000		Invitrogen	
Alexa Fluor 514 [1]	517	542	80,000	714	Invitrogen	IgG conjugated
Rhodamine 6G [3]	530	566	116,000	479		
Alexa Fluor 532 [1]	532	554	81,000	721	Invitrogen	IgG conjugated
SYTOX Orange [3]	547	570		500	Invitrogen	DNA
Cy3 [5]	550	570	150,000	767	GE Healthcare	
Alexa Fluor 555 [1]	555	565	150,000	1250	Invitrogen	IgG conjugated
Alexa Fluor 546 [1]	556	573	104,000	1079	Invitrogen	IgG conjugated
Cy3B [5]	558	572	130,000	1102	GE Healthcare	
Lissamine Rhodamine B [3]	570	590			Invitrogen	

(*continued*)

TABLE D.1 (continued)   Common Fluorescent Dyes Sorted by the Absorption Maximum

Name	Absorption, Max (nm)	Emission, Max (nm)	Extinction ($cm^{-1}\,M^{-1}$)	Molecular Weight ($g\,mol^{-1}$)	Supplier	Note
Rhodamine Red [3]	570	590	120,000		Invitrogen	
Alexa Fluor 568 [1]	578	603	91,300	792	Invitrogen	IgG conjugated
Cy3.5 [5]	581	596	150,000	658	GE Healthcare	
Alexa Fluor 594 [1]	590	617	90,000	820	Invitrogen	IgG conjugated
Texas Red [3]	595	615	80,000	625	Invitrogen	
Alexa Fluor 610 [1]	602	624	132,000	1285	Invitrogen	IgG conjugated
Alexa Fluor 633 [1]	632	647	100,000	1200	Invitrogen	IgG conjugated
TOTO-3, TO-PRO-3 [3]	642	661	154,100	1355	Invitrogen	
Cy5 [5]	649	670	250,000	792	GE Healthcare	
Alexa Fluor 647 [1]	650	665	239,000	1300	Invitrogen	IgG conjugated
Alexa Fluor 660 [1]	663	690	132,000	1100	Invitrogen	IgG conjugated
Cy5.5 [5]	675	694	190,000	1128	GE Healthcare	
Alexa Fluor 680 [1]	679	702	184,000	1150	Invitrogen	IgG conjugated
Alexa Fluor 700 [1]	702	723	192,000	1400	Invitrogen	IgG conjugated
Cy7 [5]	743	767	200,000	818	GE Healthcare	
Alexa Fluor 750 [1]	749	775	240,000	1300	Invitrogen	IgG conjugated
Alexa Fluor 790 [1]	785	810	260,000	1750	Invitrogen	IgG conjugated
Alexa Fluor 350 [1]	346	442	19,000	410	Invitrogen	IgG conjugated

References

1. Invitrogen 2007, Alexa Fluor® Succinimidyl Esters, viewed February 12, 2009, <http://probes.invitrogen.com/media/pis/mp10168.pdf>

2. Invitrogen, Product Spectra—Hoechst 33258/DNA, viewed February 12, 2009, <http://probes.invitrogen.com/media/spectra/data/1398dna.txt>

3. Invitrogen, Spectral characteristics and recommended bandpass filter sets for Molecular Probes' dyes, viewed February 12, 2009, <http://www.invitrogen.com/site/us/en/home/References/Molecular-Probes-The-Handbook/tables/Spectral-characteristics-and-recommended-bandpass-filter-sets-for-Molecular-Probes-dyes>.

4. Invitrogen, SYTOX® Blue Dead Cell Stain, viewed February 12, 2009, <http://probes.invitrogen.com/media/pis/mp34857.pdf>

5. GE Healthcare Life Sciences, CyDye Fluors Properties, viewed February 12, 2009, http://www5.gelifesciences.com/aptrix/upp00919.nsf/Content/drugscr_applic_cydye~DrugScr+CyDye+Fluors+Modalities

Appendix E: Common Physical Constants

TABLE E.1 Common Physical Constants

Quantity	Symbol	Value	Unit	Relative Std. Uncertainty
Speed of light in vacuum	c, c_0	299,792,458	m s^{-1}	Exact
Magnetic constant (vacuum permeability)	μ_0	$4\pi \times 10^{-7} = 12.566,370,614... \times 10^{-7}$	N A^{-2}	Exact
Electric constant (vacuum permittivity)	ε_0	$8.854,187,817... \times 10^{-12}$	F m^{-1}	Exact
Newtonian constant of gravitation	G	$6.674,28(67) \times 10^{-11}$	m^3 kg^{-1} s^{-2}	1.0×10^{-4}
Planck's constant	H	$6.626,068,96(33) \times 10^{-34}$	J s	5.0×10^{-8}
$h/2\pi$	$\hbar$	$1.054,571,628(53) \times 10^{-34}$	J s	5.0×10^{-8}
Elementary charge	e	$1.602,176,487(40) \times 10^{-19}$	C	2.5×10^{-8}
Electron mass	m_e	$9.109,382,15(45) \times 10^{-31}$	kg	5.0×10^{-8}
Proton mass	m_p	$1.672,621,637(83) \times 10^{-27}$	kg	5.0×10^{-8}
Avogadro constant	N_A	$6.022,141,79(30) \times 10^{23}$	mol^{-1}	5.0×10^{-8}
Faraday constant $N_A e$	F	$96,485.339,9(24)$	C mol^{-1}	2.5×10^{-8}
Molar gas constant	R	$8.314,472(15)$	J mol^{-1} K^{-1}	1.7×10^{-6}
Boltzmann constant R/N_A	k_B	$1.380,650,4(24) \times 10^{-23}$	J K^{-1}	1.7×10^{-6}
Stefan–Boltzmann constant $(\pi^2/60)k^4/\hbar^3 c^2$	σ	$5.670,400(40) \times 10^{-8}$	W m^{-2} K^{-4}	7.0×10^{-6}
Molar volume of ideal gas RT/p; $T=273.15$ K, $p=101.325$ kPa	V_m	$22.413,996(39) \times 10^{-3}$	m^3 mol^{-1}	1.7×10^{-6}
Characteristic impedance of vacuum	$Z_0 = \mu_0 c$	$376.730,313,461...$		Exact
Non-SI units				
Electron volt (e/C) J	eV	$1.602,176,487(40) \times 10^{-19}$	J	2.5×10^{-8}
(Unified) atomic mass unit $1\,u = m_u = 1/12\,m(^{12}C) = 10^{-3}$ kg mol$^{-1}/N_A$	u	$1.660,538,782(83) \times 10^{-27}$	kg	5.0×10^{-8}

Reference

National Institute of Standards and Technology 2006, Fundamental Physical Constants—Complete Listing (2006), viewed February 15, 2009, <http://physics.nist.gov/cuu/Constants/Table/allascii.txt>.

Appendix F: Common Biological Buffers

TABLE F.1 Effective pH Range and pK_a of Selected Biological Buffers Sorted by the Effective pH Range

Effective pH Range	pK_a (298 K)	Name
1.2–2.6	1.97	Maleate (pK_1)
1.7–2.9	2.15	Phosphate (pK_1)
2.2–3.6	2.35	Glycine (pK_1)
2.2–6.5	3.13	Citrate (pK_1)
2.5–3.8	3.14	Glycylglycine (pK_1)
2.7–4.2	3.4	Malate (pK_1)
3.0–4.5	3.75	Formate
3.0–6.2	4.76	Citrate (pK_2)
3.2–5.2	4.21	Succinate (pK_1)
3.6–5.6	4.76	Acetate
3.8–5.6	4.87	Propionate
4.0–6.0	5.13	Malate (pK_2)
4.9–5.9	5.23	Pyridine
5.0–6.0	5.33	Piperazine (pK_1)
5.0–7.4	6.27	Cacodylate
5.5–6.5	5.64	Succinate (pK_2)
5.5–6.7	6.1	MES
5.5–7.2	6.4	Citrate (pK_3)
5.5–7.2	6.24	Maleate (pK_2)
5.5–7.4	1.70, 6.04, 9.09	Histidine
5.8–7.2	6.46	Bis-tris
5.8–8.0	7.2	Phosphate (pK_2)
6.0–12.0	9.5	Ethanolamine
6.0–7.2	6.59	ADA
6.0–8.0	6.35	Carbonate (pK_1)
6.1–7.5	6.78	ACES
6.1–7.5	6.76	PIPES
6.2–7.6	6.87	MOPSO
6.2–7.8	6.95	Imidazole
6.3–9.5	6.80, 9.00	BIS-TRIS propane
6.4–7.8	7.09	BES

(*continued*)

TABLE F.1 (continued) Effective pH Range and pK_a of
Selected Biological Buffers Sorted by the Effective pH Range

Effective pH Range	pK_a (298 K)	Name
6.5–7.9	7.14	MOPS
6.8–8.2	7.48	HEPES
6.8–8.2	7.4	TES
6.9–8.3	7.6	MOBS
7.0–8.2	7.52	DIPSO
7.0–8.2	7.61	TAPSO
7.0–8.3	7.76	Triethanolamine (TEA)
7.0–9.0	0.91, 2.10, 6.70, 9.32	Pyrophosphate
7.1–8.5	7.85	HEPPSO
7.2–8.5	7.78	POPSO
7.4–8.8	8.05	Tricine
7.5–10.0	8.1	Hydrazine
7.5–8.9	8.25	Glycylglycine (pK_2)
7.5–9.0	8.06	Trizma (tris)
7.6–8.6	8	EPPS, HEPPS
7.6–9.0	8.26	BICINE
7.6–9.0	8.3	HEPBS
7.7–9.1	8.4	TAPS
7.8–9.7	8.8	2-Amino-2-methyl-1,3-propanediol (AMPD)
8.2–9.6	8.9	TABS
8.3–9.7	9	AMPSO
8.4–9.6	9.06	Taurine (AES)
8.5–10.2	9.23, 12.74, 13.80	Borate
8.6–10.0	9.5	CHES
8.7–10.4	9.69	2-Amino-2-methyl-1-propanol (AMP)
8.8–10.6	9.78	Glycine (pK_2)
8.8–9.9	9.25	Ammonium hydroxide
8.9–10.3	9.6	CAPSO
9.5–11.1	10.33	Carbonate (pK_2)
9.5–11.5	10.66	Methylamine
9.5–9.8	9.73	Piperazine (pK_2)
9.7–11.1	10.4	CAPS
10.0–11.4	10.7	CABS
10.5–12.0	11.12	Piperidine

Reference

Sigma-Aldrich 2009, Biological Buffers, viewed March, 05, 2009, <http://www.sigmaaldrich.com/life-science/metabolomics/bioultra-reagents/biological-buffers.html>.

Index

Printed and bound by CPI Group (UK) Ltd, Croydon, CR0 4YY

18/10/2024

01776249-0017